SILVICULTURE

CONCEPTS AND APPLICATIONS

SILVICULTURE

CONCEPTS AND APPLICATIONS

Ralph D. Nyland
Professor of Silviculture
State University of New York
College of Environmental Science and Forestry

Boston, Massachusetts Burr Ridge, Illinois Dubuque, Iowa
Madison, Wisconsin New York, New York San Francisco, California St. Louis, Missouri

McGraw-Hill

A Division of The ***McGraw-Hill*** *Companies*

This book was set in Times Roman by Ruttle, Shaw & Wetherill, Inc.
The editors were Anne Duffy and John M. Morriss;
the production supervisor was Margaret Boylan;
Project supervision was done by Tage Publishing Service, Inc.
R. R. Donnelley & Sons was printer and binder.

Frontispiece

Under appropriate silviculture, forests of all kinds provide many values that help to sustain our lives.

Jacket

Designed and illustrated by Christopher Brady.

SILVICULTURE

Concepts and Applications

This book is printed on acid-free paper.

3 4 5 6 7 8 9 0 DOC DOC 9 0 9 8 7

ISBN 0-07-056999-1

Library of Congress Catalog Card Number: 96-75292

To students everywhere,
young and old.

McGraw-Hill Series in Forest Resources

Paul V. Ellefson, *University of Minnesota, Consulting Editor*

Avery and Burkhart: *Forest Measurements*
Dana and Fairfax: *Forest and Range Policy*
Daniel, Helms, and Baker: *Principles of Silviculture*
Davis and Johnson: *Forest Management*
Duerr: *Introduction to Forest Resource Economics*
Ellefson: *Forest Resources Policy: Process, Participants, and Programs*
Harlow, Harrar, Hardin, and White: *Textbook of Dendrology*
Klemperer: *Forest Resource Economics and Finance*
Laarman and Sedjo: *Global Forests: Issues for Six Billion People*
Nyland: *Silviculture: Concepts and Applications*
Piirto: *Study Guide to Accompany* Sharpe, Hendee, Sharpe, and Hendee: *Introduction to Forest and Renewable Resources*
Sharpe, Hendee, Sharpe, and Hendee: *Introduction to Forest and Renewable Resources*
Sinclair: *Forest Products Marketing*
Stoddart, Smith, and Box: *Range Management*

ABOUT THE AUTHOR

RALPH D. NYLAND is Professor of Silviculture at State University of New York, College of Environmental Science and Forestry. After earning his BS and MS degrees in forestry and silviculture from State University of New York, College of Environmental Science and Forestry at Syracuse, he worked for 5 years as a forester with the NYS Department of Environmental Conservation. He enrolled at Michigan State University, and earned his Ph.D. in forest management and silviculture. In 1967, he joined the faculty at SUNY-ESF, and helped to form the Applied Forestry Research Institute. After serving 10 years as Senior Research Associate, Professor Nyland was appointed to the teaching faculty, and currently serves as Professor of Silviculture within the Faculty of Forestry. He offers both undergraduate and graduate courses in silviculture, participates as an instructor in continuing education projects, and maintains an active program of silviculture research. In addition, he serves as coordinator of the Forest Ecosystem Management and Productivity Program Area of the New York Center for Forestry Research and Development.

Professor Nyland's studies of different silvicultural systems have focused primarily on northern hardwoods and transition oak communities. Past and ongoing research includes studies of natural regeneration in the context of both even- and uneven-aged silviculture, and tending of intermediate-aged stands and age classes. He has written extensively about uneven-aged silviculture, as well as early even-aged stand development following clearcutting and shelterwood methods. The work with oak-dominated communities has employed prescribed burning as a method of site preparation.

Professor Nyland has authored or coauthored more than 80 technical papers and one book, and served as editor of another. His current research includes the long-term monitoring of natural stands following alternative silvicultural treatments, and explores ways to enhance wildlife habitat and other values of importance in the emerging area of ecosystem management.

CONTENTS

1 Silviculture As an Orderly Discipline 1

ITS PLACE IN FORESTRY 1
A PHILOSOPHY OF SILVICULTURE 2
THE CHANGING CONTEXT OF FORESTRY 3
NO CUTTING AS AN APPROPRIATE RECOMMENDATION 6
SILVICULTURE DEFINED—AN ART AND A SCIENCE 7
SETTING THE OBJECTIVES AND PAYING THE COSTS 8
OWNERSHIPS, OBJECTIVES, AND ECOSYSTEMS 10
TAKING AN ECOLOGIC PERSPECTIVE 13
WHAT SILVICULTURISTS DO 16
CONTROLLING ESTABLISHMENT, COMPOSITION, AND GROWTH 17
SILVICULTURE AS A DELIBERATE STRATEGY 18

2 The Silvicultural System 19

THE SILVICULTURAL SYSTEM AS A PLAN FOR MANAGEMENT 19
CHARACTER AND OBJECTIVES OF SILVICULTURAL SYSTEMS 21
SILVICULTURE FOR SUSTAINED YIELD OF MULTIPLE VALUES 23
CONSERVATION, SILVICULTURAL SYSTEMS, AND THE ETHICS OF RESOURCE USE 25
COMPARISON OF EVEN- AND UNEVEN-AGED SYSTEMS 28
ADJUSTING THE INTENSITY OF MANAGEMENT 32
FINANCIAL MATURITY AND OTHER ECONOMIC CONSIDERATIONS 36
JUDGING NONMARKET VALUES 41

3 Harvesting As a Tool of Silviculture 43

TIMBER HARVESTING AND SILVICULTURE 43
THE NATURE OF TIMBER HARVESTING 44
SILVICULTURAL REQUIREMENTS OF LOGGING 45

EFFECTS OF TIMBER HARVESTING 46
BEST MANAGEMENT PRACTICES TO PREVENT ENVIRONMENTAL POLLUTION 49
PLANNING TO CONTROL EROSION POTENTIAL AFTER HARVESTING 54
PROTECTING VEGETATION AND SPECIAL HABITATS 56

4 Concepts of Regeneration 58

OBJECTIVES OF FOREST REGENERATION PRACTICES 58
INSURING TIMELY AND EFFECTIVE REGENERATION 58
REGENERATION AS A PROCESS 60
IMPEDIMENTS TO SUCCESS FOLLOWING SEED DISPERSAL 62
KINDS AND SOURCES OF REGENERATION (NATURAL AND ARTIFICIAL) 63
REGENERATION BY VEGETATIVE METHODS 64
REGENERATION FROM SEED 65
GERMINATION AND DEVELOPMENT 68
ADVANTAGES AND DISADVANTAGES OF DIFFERENT REGENERATION METHODS 70
CHOOSING BETWEEN NATURAL AND ARTIFICIAL REGENERATION 71
THE IMPLICATIONS OF SILVICULTURAL PRACTICE TO FOREST TREE IMPROVEMENT 72
USE OF EXOTIC SPECIES 75
IMPORTANCE OF SEED SOURCE SELECTION 78

5 Site Preparation 79

ROLE AND SCOPE OF SITE PREPARATION 79
PASSIVE SITE PREPARATION 82
ACTIVE SITE PREPARATION 83
MECHANICAL SITE PREPARATION 84
USING HERBICIDES IN SITE PREPARATION 90
CONDITIONS FOR SUCCESS WITH HERBICIDES 95
FIRE AND ITS EFFECTS IN FORESTS 97
PRESCRIBED BURNING AS A TOOL OF SILVICULTURE 99
METHODS FOR CONTROLLING THE EFFECTS OF PRESCRIBED FIRE 100
FIRING METHODS FOR PRESCRIBED BURNING 102
SELECTING A SITE PREPARATION METHOD 104

6 Planning for Artificial Regeneration 106

TREE PLANTING IN SILVICULTURE 106
PLANNING A TREE-PLANTING PROJECT 108
PURPOSES AND ADVANTAGES OVER NATURAL METHODS 109
REQUIREMENTS FOR SUCCESSFUL ARTIFICIAL REGENERATION 112
SPECIES SELECTION 113
USE OF EXOTIC SPECIES 115
EXAMPLES OF SUCCESSFUL EXOTIC SPECIES 116

WINDBREAKS AND AGROFORESTRY IN TREE-POOR REGIONS 118
FIELD ASSESSMENT OF SITE CONDITIONS 120
IMPORTANCE OF SOIL CONDITIONS 121
KINDS OF PLANTING STOCK 122
COMPARING BARE-ROOT AND CONTAINER STOCK 124
OTHER FACTORS RELATED TO SEEDLING QUALITY 125
SELECTING A SPACING 126
ARRANGEMENT AND OTHER OPERATIONAL CONSIDERATIONS 129
GETTING READY FOR THE PLANTING 130

7 Nursery and Tree Planting Operations 133

SEED SELECTION AND HANDLING 133
SEED CERTIFICATION 136
SEED PROCESSING AND STORAGE 136
SEEDLING PRODUCTION 138
NURSERY SOIL MANAGEMENT 138
SOWING 140
PROTECTION AND HEALTH MANAGEMENT 141
CULTURAL PRACTICES 142
NURSERY LIFTING AND HANDLING OF BARE-ROOT STOCK 143
TIME OF PLANTING 145
FIELD STORAGE AND HANDLING OF BARE-ROOT STOCK 147
CONTAINER STOCK 148
METHODS OF HAND PLANTING 152
MACHINE PLANTING 156
SOME SPECIAL CONSIDERATIONS IN PLANTING 157

8 Regeneration from Seed and Direct Seeding 159

UNDERLYING PREMISE 159
ECOLOGIC CONCEPTS OF IMPORTANCE 160
PLANT SUCCESSION AND FOREST REGENERATION 161
DIRECT SEEDING FOR FOREST REGENERATION 163
PREPARATION FOR DIRECT SEEDING 166
SOWING THE SEED 169
APPLICATION OF DIRECT SEEDING 173

9 Reproduction Methods and Their Implications 175

REPRODUCTION METHODS AND THEIR ROLE IN SILVICULTURE 176
DIFFERENCES BETWEEN EVEN- AND UNEVEN-AGED METHODS 178
CHARACTERISTICS OF DIFFERENT REPRODUCTION METHODS 179
SOME REGULATORY CONSIDERATIONS IN SILVICULTURE 181
DETERMINING THE GROWTH 185

DEPICTING DEVELOPMENT PATTERNS IN FOREST STANDS 186
PHASES OF EVEN-AGED STAND DEVELOPMENT 190
SOME PATTERNS OF UNEVEN-AGED STAND DEVELOPMENT 192
FINANCIAL MATURITY AND DIFFERENT PRODUCT OPTIONS 193
RELATIONSHIP TO A REGENERATION STRATEGY 196

10 Selection System and Its Application 198

THE CHARACTER OF SELECTION SYSTEM 198
CHARACTERIZING CONDITIONS AMONG SELECTION SYSTEM STANDS 201
DEFINING A RESIDUAL STRUCTURE 205
BALANCED STRUCTURES AND OTHER GUIDES FOR SELECTION SYSTEM 209
APPLYING SELECTION SYSTEM 213
PREPARING A MARKING GUIDE 215
ENHANCING QUALITY AND VALUE 217

11 Uneven-aged Reproduction Methods 221

THE ROLE OF A REPRODUCTION METHOD IN SELECTION SYSTEM 221
SINGLE-TREE SELECTION METHOD 222
ALTERNATIVE APPROACHES TO SINGLE-TREE SELECTION METHOD 224
GROUP SELECTION METHOD 226
HYBRID UNEVEN-AGED METHODS 229
PATCH AND STRIP CUTTING METHODS 231
EXAMPLES OF UNEVEN-AGED SILVICULTURE IN A SYSTEMS CONTEXT 234

12 Growth and Development in Selection System Stands 241

ADVANTAGES AND DISADVANTAGES OF SELECTION SILVICULTURE 241
REQUIREMENTS FOR SUCCESS IN SELECTION SYSTEM 243
EFFECTS OF STAND STRUCTURE CONTROL ON TREE GROWTH RATES 245
CONTRASTS BETWEEN SELECTIVE AND SELECTION CUTTING 246
REGULATING STAND STABILITY AND GROWTH 248
ASSESSING STAND STRUCTURAL CHANGE UNDER SELECTION SYSTEM 249
AN EXAMPLE OF SELECTION SYSTEM 250
INTEGRATING CUTTING CYCLE LENGTH, RESIDUAL DENSITY, AND STRUCTURE 252
APPLICATIONS FOR NONCOMMODITY OBJECTIVES 256
SETTING THE MAXIMUM DIAMETER FOR NONCOMMODITY OBJECTIVES 260
SOME OTHER CONSIDERATIONS FOR SERVING NONCOMMODITY GOALS 262

13 Clearcutting 263

CLEARCUTTING AS A REPRODUCTION METHOD 263
ATTRIBUTES OF CLEARCUTTING AS A REPRODUCTION METHOD 266
IMPORTANCE OF SEED SOURCE AND ADVANCED REGENERATION 267
EFFECT ON ENVIRONMENTAL CONDITIONS NEAR THE GROUND 271
ACCOMMODATING SPECIES TOLERANCES 275
ALTERING THE CONFIGURATION OF CLEARCUTTING 276
EFFECTS ON SOME OTHER ECOSYSTEM ATTRIBUTES 282
SITE PREPARATION AND OTHER ANCILLARY TREATMENTS 285
CLEARCUTTING WITH ARTIFICIAL REGENERATION 286
OPERATIONAL CONSIDERATIONS IN CLEARCUTTING 288
VALUES AND LIMITATIONS OF CLEARCUTTING AS A REPRODUCTION METHOD 289

14 Shelterwood and Seed-tree Methods 292

A RATIONALE FOR ALTERNATIVE EVEN-AGED REPRODUCTION METHODS 292
CHARACTERISTICS OF SHELTERWOOD METHOD 292
CHARACTERISTICS OF SEED-TREE METHOD 295
SELECTING SEED TREES FOR SHELTERWOOD AND SEED-TREE CUTTING 299
DETERMINING AN APPROPRIATE LEVEL OF RESIDUAL STOCKING 301
RESERVE, GROUP, AND STRIP SHELTERWOOD METHODS 304
REMOVAL CUTTING AND SEEDLING DAMAGE 308
SITE PREPARATION WITH SHELTERWOOD AND SEED-TREE METHODS 310
SOME NONCOMMODITY ASPECTS OF SHELTERWOOD AND SEED-TREE METHODS 312

15 Early Stand Development 314

EVEN-AGED COMMUNITY ESTABLISHMENT AND FORMATION 314
EARLY DEVELOPMENT OF EVEN-AGED COMMUNITIES AT MESIC SITES 316
INTERMEDIATE TREATMENTS FOR EVEN-AGED STANDS 321
JUDGING SUCCESSFUL ESTABLISHMENT 323
VEGETATION AND REGENERATION SURVEYS 325
SOME SAMPLE GUIDELINES FOR JUDGING REGENERATION SUCCESS 327

16 Release Cuttings 328

EARLY RELEASE TREATMENTS 328
WEEDING 330
CLEANING 332
RESPONSE OF CROP TREES TO RELEASE 339

LIBERATION CUTTING 340
SOME BROAD ECONOMIC CONSIDERATIONS 344

17 Thinning and Its Effects upon Stand Development 349

SOME DYNAMICS AMONG INTERMEDIATE-AGED EVEN-AGED STANDS 349
CROWN CLASSES AND THEIR SILVICULTURAL IMPORTANCE 352
SOME IMPLICATIONS OF TREE AGING 358
THE PRODUCTION FUNCTION AS A MODEL OF EVEN-AGED STAND DEVELOPMENT 361
INTERMEDIATE TREATMENTS TO TEMPER STAND DEVELOPMENT 362
A PRODUCTION FUNCTION FOR THINNED STANDS 364
STOCKING AND RELATIVE DENSITY AMONG UNMANAGED STANDS 365
CONTROLLING RELATIVE DENSITY THROUGH THINNING 372

18 Methods of Thinning 376

COMPARING METHOD AND INTENSITY OF THINNING 376
LOW THINNING 378
CROWN THINNING 380
SELECTION THINNING 384
MECHANICAL THINNING 387
FREE THINNING 391
SELECTING A METHOD 393
CONTROLLING SPACING AND SERVING DIVERSE LANDOWNER VALUES 394

19 Thinning Regimes in the Even-aged Silvicultural System 395

CHARACTERISTIC RESPONSES AMONG THINNED STANDS 395
INFLUENCING YIELD BY THINNING 397
EFFECTS ON INDIVIDUAL TREE GROWTH 400
EFFECTS OF THINNING ON STAND VOLUME PRODUCTION 410
MANAGEMENT SCHEMES FOR FIBER PRODUCTION 414
MANAGEMENT SCHEMES FOR SAWTIMBER 414
SOME CONSIDERATIONS FOR TIMING A FIRST THINNING 417
SETTING A THINNING INTERVAL 420

20 Managing Quality in Forest Stands 423

MARKETING AND QUALITY FACTORS IN SILVICULTURAL PLANNING AND OPERATIONS 423
DETERMINING QUALITY IN TREES AND STANDS 426
QUALITY REQUIREMENTS FOR DIFFERENT WOOD PRODUCTS 428
RECOGNIZING PRODUCT QUALITY IN STANDING TREES 429

FACTORS THAT PROMOTE QUALITY AMONG TREES AND STANDS 434
PRUNING 437
PRUNING TOOLS AND EQUIPMENT 440
MANAGING PRUNING AS AN INVESTMENT 442
JUDGING THE PROFITABILITY OF PRUNING 444
PRUNING LANDSCAPE TREES 446
PRUNING AND NONMARKET VALUES 447

21 Stand Protection and Health Management 450

SILVICULTURE AND STAND HEALTH 450
INTEGRATED HEALTH MANAGEMENT AND SILVICULTURE 454
HEALTH MANAGEMENT THROUGH SILVICULTURE 459
FACTORS THAT INFLUENCE FOREST TREES AND STANDS 460
HOW TREES BECOME UNHEALTHY 461
EFFECTS OF NONCATASTROPHIC AGENTS 462
SPECIES AND GENETIC SELECTION AS A CRITICAL PREVENTIVE MEASURE 463
DAMAGE, DISCOLORATION, AND DECAY IN LIVING TREES 467
DIFFERENCES BETWEEN SELECTION SYSTEM AND THINNING 470
MEASURES FOR CONTAINING LOGGING DAMAGE 471

22 Improvement, Salvage, and Sanitation Cuttings 473

SILVICULTURAL RESPONSES TO EFFECTS OF INJURIOUS AGENTS 473
THE ROLE OF IMPROVEMENT CUTTING 474
ASSESSING OPPORTUNITIES FOR IMPROVEMENT CUTTING 475
INTEGRATING IMPROVEMENT INTO OTHER TREATMENTS 480
CAVITIES, CULLS, AND DEAD TREES AS HABITAT ESSENTIALS 482
SALVAGE AND SANITATION CUTTING 484
REGENERATION AFTER SALVAGE AND SANITATION CUTTING 486
SOME ECONOMIC CONSIDERATIONS IN IMPROVEMENT AND SALVAGE CUTTING 487
OTHER ASPECTS OF IMPROVEMENT, SALVAGE, AND SANITATION CUTTING 490

23 Other Partial Cuttings 491

DELIBERATE SILVICULTURE AND CONSERVATION 491
MAINTAINING AN ECOLOGIC STANDARD 492
TWO-AGED AND TWO-STORIED SILVICULTURE 494
ADVANTAGES AND DISADVANTAGES OF TWO-AGED SILVICULTURE 500
TWO-AGED TREATMENTS AS A PATHWAY TO SELECTION SILVICULTURE 501
NONSYSTEMS HARVEST CUTTINGS 502
DIAMETER-LIMIT CUTTING IN EVEN- AND UNEVEN-AGED COMMUNITIES 504

SELECTIVE CUTS RESEMBLING DELIBERATE SILVICULTURE 504
COMPARISON OF STANDS UNDER SILVICULTURE AND CASUAL CUTTING 508
EXPLOITATION VERSUS SILVICULTURE 511

24 Coppice Silviculture 513

ROLE IN FOREST STAND MANAGEMENT 513
COPPICE METHODS BASED UPON STUMP SPROUTING 515
SHORT-ROTATION COPPICE SYSTEMS 518
COPPICE METHODS BASED UPON ROOT SUCKERS 522
SETTING ROTATION LENGTH IN COPPICE SYSTEMS 524
ADVANTAGES AND DISADVANTAGES OF SIMPLE-COPPICE SYSTEMS 525
COPPICE-WITH-STANDARDS SYSTEMS 527
CONVERSION FROM COPPICE TO HIGH-FOREST SYSTEMS 531

25 Adjusting to Administrative Demands 533

SILVICULTURE IN PERSPECTIVE 533
A PROBLEM-SOLVING PROCEDURE FOR SILVICULTURE 533
SILVICULTURE FROM AN ADMINISTRATIVE PERSPECTIVE 536
ADMINISTRATIVE CONSIDERATIONS FOR INITIAL PLANNING 538
FACTORS THAT FORCE UNSCHEDULED CHANGES IN SILVICULTURAL PRACTICE 540
EXAMPLES OF CHANGING ADMINISTRATIVE DEMANDS 541
TAKING AN ECOSYSTEM APPROACH 544
SILVICULTURE IN RETROSPECT 548

REFERENCES 550

INDEX 610

PREFACE

A PHILOSOPHY OF SILVICULTURE

This book defines and describes modern silviculture. It characterizes the conditions and responses that silviculturists create by applying different methods and treatments. It includes analytic methods and management guidelines that foresters use in making judgments about what silviculture to try. It illustrates the responses from applying different silviculture, including some obvious and subtle differences between the methods. Further, it outlines the advantages and shortcomings of each practice to help foresters evaluate the alternatives and decide how to treat a forest to serve a landowner's objectives.

Silviculture: Concepts and Applications does not recommend a specific set of practices for treating forests at different locations. It presents no standard recipes that guarantee successful practice. Nor does it promote any single silvicultural approach in preference to others. Rather, the book reflects a belief that all of the tools of silviculture have a useful role in modern forestry. The choice depends upon the circumstances and conditions, and a landowner's needs. It espouses a philosophy that each forester must decide how to best integrate the biologic-ecologic, economic-financial, and managerial-administrative requirements and opportunities at hand. It promotes the notion that by creative planning, foresters can address a wide array of commodity and nonmarket interests and opportunities while maintaining dynamic and resilient forests. The book reflects my belief that well conceived and deliberate management must prove ecologically sound, and economically viable.

Silviculture: Concepts and Applications espouses a simple truth that students of silviculture must adjust to: people owning similar kinds of forest may have divergent interests, and follow a different course of management. They cannot learn a simple set of generic prescriptions, and then routinely apply them from one situation to another. Rather, they must learn to deliberately evaluate each case for its own merits and requirements, and then decide what to do and when to do it. Particularly where landowners want to integrate both commodity and nonmarket values, silviculturists must transcend common practice. That becomes possible only if they understand how different forests grow and develop following an array of treatments, and how the resulting vegetal conditions might serve some particular management objective.

Silviculturists must also adapt to changing interests of those landowners, new developments from science, improvements in technology, and evolving attitudes of society toward natural resources. The new priorities may make some old practices less useful, and give some new ones a high priority. This occurred in the 1940s and early 1950s, when public concerns about world-wide timber supplies became manifest in public policy to intensify timber management on both public and private lands. In response, forestry focused predominantly on commodity production. By the 1960s, mushrooming demands to use forests for leisure activities triggered a new focus on noncommodity benefits and uses, and fostered a new philosophy called *multiple use management.* Foresters again adjusted their silviculture and other management activities to balance potentially competing demands, and put a new emphasis on improving the habitat for popular wildlife, the opportunities for recreation, and the scenic values of actively managed forests.

By the early 1990s, concerns for landscape-scale ecological balances and biological diversity had come into prominence, bringing still another new focus to silviculture. Foresters began looking at techniques to maintain some prescribed set of vegetation conditions of ecological importance, and taking a new look at ways that treatments in one area might affect the balance, diversity, and connectedness of important ecosystem elements across a broader landscape. Timber production and recreation became anticipated side effects of having established and perpetuated some desirable set of conditions in the ecosystem, and sustained them for long periods of time and across large expanses of space. Eventually, this philosophy will also yield to other new ideas about natural resource conservation, and silviculturists will again adapt by bringing emerging concepts from science to the art of managing forests to serve the changing interests of people.

THE APPROACH TAKEN

Silviculture: Concepts and Applications borrows ideas and concepts from a wide array of researchers, teachers, and practitioners. Even so, the literature cited does not provide a totally comprehensive listing of all silviculture resources. Rather, it credits authors whose ideas I incorporated in the book, or whose writings provided useful illustrations for the practices or concepts under review. The citations include several old books and papers to illustrate the roots of silviculture, and to acknowledge the early silviculturists who articulated many of the basic ideas that modern foresters still rely on. I also extensively used papers published in the proceedings of recent symposia, and many management guides appropriate to the forests of North America. These represent a special kind of expert system, where knowledgeable people summarized their study and experience, and shared their views for all foresters to consider. The proceedings and management guides also provide extensive lists of publications dealing more thoroughly with silviculture and stand dynamics.

ACKNOWLEDGMENTS

Clearly, a book of this kind reflects the shared wisdom of many people who aided my study of silviculture, and the diverse experiences that helped to shape my understanding of the discipline. I first learned about forestry as a youth by going to the forest with the late David B. Cook (see Frontispiece), a forester who lived in Albany, New York. He specialized in growing larches and willingly shared his excitement for and ideas about plantation culture and silvicultural practice. David helped to lay an intellectual foundation for me to develop into a personal concept of silviculture. Later at the New York State College of Forestry in Syracuse, I studied with Svend O. Heiberg, Eugene C. Farnsworth, and John W. Barrett. All encouraged my interests and shared ideas and opportunities to help nurture my understanding and study. During that time, and for many years since, I read from early editions of *The Practice of Silviculture* by Ralph C. Hawley and David M. Smith (1954), and progressive editions by David M. Smith (1962, 1986). Those writings and my periodic contact with David M. Smith over a period of years certainly influenced my thinking, and I am grateful for the way he helped to shape my view of silviculture.

My perceptions of ways that concepts of forest regulation and finance interface with silviculture grew out of graduate studies with Professor Victor J. Rudolph at Michigan State University. He helped me to see a dimension of silviculture that had previously eluded me, and I acknowledge his mentoring. Later following my appointment to the faculty of the State University of New York College of Environmental Science and Forestry, I had many opportunities to share ideas with many colleagues who work in a wide array of forestry disciplines and associated technical fields. Other forestry practitioners whom I visited from time to time, those I met during technical meetings, and foresters who hosted field trips with my classes, all influenced my sense of silviculture. I gratefully acknowledge the encouragement and contributions of all those colleagues and professional associates.

Probably no single group forced me to ponder ideas and struggle to articulate them more than the members of my silviculture classes, and the graduate students who came to work at Syracuse over the years. They brought sharp and inquisitive minds, and raised demanding questions that required studied responses. They challenged old interpretations with their own fresh ideas, and never accepted well-worn answers in response to their inquiries. Their reports, papers, and other writings continually introduced new notions about silviculture, and offered challenging ideas for study and deliberation. Their inquisitiveness and creativity pressed me to revise old ways of thinking as research and innovative practice pushed back the boundaries of knowledge and experience. In fact, they inspired me to write this book. In many respects it reflects our collective experiences, and our mutual interest to encourage others in the study and excitement of silviculture.

I acknowledge the way that all of these people and all of these experiences contributed to my interests in silviculture, and my preparation for this book. Others gave advice and offered comments about different chapters, and made suggestions about the content and coverage. In particular, I acknowledge the reviews and other contributions of Douglas C. Allen, Lawrence P. Abrahamson, Craig J. Davis, Paul D. Manion, Charles A. Maynard, Norman A. Richards, Daniel J. Robison, Edwin H. White, James J. Worrall, James L. Williamson, Robin Hoffman, and Flora May W. Nyland of SUNY College of Environmental Science and Forestry; Laura M. Alban of Maryland Department of Natural Resources; Laura W. Kenefic and Linda Thomasma of the U.S. Forest Service, Northeastern Forest Experiment Station; Robert S. Seymour of the University of Maine; and Margaret Shannon, Syracuse University.

McGraw-Hill and I would like to thank all those who acted as reviewers, namely: Todd Bowersox, Penn State; Philip Burton, The University of British Columbia; Clive David, University of Washington–Stevens Point; Don Dickman, Michigan State University; Paul Ellefson, University of Minnesota; Douglas Frederick, North Carolina State; Andrew Gillespie, Purdue University; Ronald Hay, University of Tennessee; George Parker, Purdue University; Douglas Piirto, California Polytechnic State University; Klaus Puettmann, University of Minnesota; David Wm. Smith, Virginia Tech; Frederick Smith, Colorado State University; John Tappeiner, Oregon State University.

THE CHALLENGE AHEAD

Ideally, silviculture has developed as a dynamic discipline that will continually evolve in response to new knowledge and new experience. Similarly, *Silviculture: Concepts and Applications* will become just one more step along a pathway of personal life-long study. It presents my current interpretation of silviculture based upon information available to me, and tempered by my own practice and research. It reflects my present judgment about what ideas and interpretations to incorporate from the rich array of published materials and shared experiences that have characterized the history of silviculture. Hopefully, the book will encourage others to also evolve in their appreciation of silviculture. In time, new understanding will swell the body of theory and the storehouses of experience. Then new books will offer additional interpretations of silviculture. Ideally, this one will motivate you to participate in that process, and encourage you to continue the exciting adventure of creatively practicing silviculture among the forests where you work.

Ralph D. Nyland

1

SILVICULTURE AS AN ORDERLY DISCIPLINE

ITS PLACE IN FORESTRY

Forestry encompasses the science, business, art, and practice of purposefully organizing and managing forest resources to provide continuing benefits for people. *Silviculture* involves the methods for establishing and maintaining communities of trees and other vegetation that have value to people. These include benefits derived either directly or indirectly from the trees themselves, other plants, water, wildlife, and minerals found in forested areas—and also a host of intangible benefits that people realize through recreation and other noncommodity uses (Nyland et al. 1983; Ford-Robertson 1971). Silviculture also ensures the long-term continuity of essential ecologic functions, and the health and productivity of forested ecosystems.

As organized disciplines, forestry and silviculture differ from operations that simply extract timber and other commodities from forested areas whenever opportunities permit. Like agriculture, forestry developed as a practical art that foresters continually improved through experience, research, and experimentation (Boerker 1916). Also, both agriculture and forestry work with the inherent conditions of climate and soil to sustain a continuous production of desired crops and other values over the long run. They both deal with the complex ecologic web of plants and other associated biologic resources that landowners can sustain in perpetuity through appropriate methods for regular tending and periodic regeneration—and only if they practice good stewardship over the soil and other physical resources of the land. Also like agriculture, forestry provides people with opportunities for the timely and controlled use of the goods and other values that management provides. As a consequence, landowners realize continuing benefits (tangible or otherwise) to repay their investment in long-term ownership and management.

A PHILOSOPHY OF SILVICULTURE

Silviculture has its foundations in biologic and ecologic sciences, yet it also responds to economic and administrative concerns of a landowner when determining the best way to manage the forested lands under consideration (Kostler 1956; Assmann 1970; Smith 1986). The biologic and ecologic factors determine what alternatives a landowner can reasonably pursue on a particular parcel of land, and how different management practices will likely affect the growth and reproduction of both indigenous and introduced plants and animals (Lutz 1959). The economic (financial and institutional) and administrative requirements determine which of the ecologically appropriate options to follow, if any. They also indicate the level of investment and the intensity of management that the land will support, or that a landowner will accept. To succeed over the long run, silviculturists must address a complex of questions related to ecologic, economic, and administrative matters of concern simultaneously. They must resolve any potential conflicts among these before undertaking active management.

In their role as silviculturists, foresters deal primarily with the subdivisions or parts of forests called *stands* (after Ford-Robertson 1971; Soc. Am. For. 1989):

- communities or groups of trees that grow together at a particular place, and that foresters can effectively manage as a unit
- communities or groups of trees with some unique vegetal characteristic that landowners can maintain by a particular series of treatments

Within these individual stands, silviculturists can often trigger more than a single kind of ecologically acceptable response by simply using a variety of alternative treatments, applying them with different intensities, and arranging them in a different sequence over time. From one stand to another, the effects of a treatment or series of treatments may also vary with differences in the mixture of species present and the inherent characteristics of the physical environment (the *site*).

The different methods and practices used to manage the vegetation in forest stands comprise the *technology of silviculture.* In drawing upon this technology, foresters must often decide whether to (1) increase the quantity and quality of goods and values already available; or (2) alter the character of a stand to provide an entirely different set of benefits. This may mean choosing between a management scheme that promises a broad array of outcomes, and trying to enhance the yields of one specific product or use opportunity. In some cases, the best management might even mean doing nothing at all because the values derived from unmanaged stands would best satisfy a landowner's interests. Under other circumstances, silviculturists might prescribe a complex of treatments to apply in a particular sequence, and with a specific degree of intensity over a long period of time. The final choice depends upon the interests and requirements of a landowner, tempered by the biologic potentials of the forest, and the ecologic limitations of a site.

Since many different kinds of owners control forest land, and since ecologic and economic conditions differ so widely from place to place, this can prove challenging. In all cases, the management must respond to the economic and administrative circumstances, consistent with the ecologic limitations and potentials. In fact, laws in many countries or other legal jurisdictions require managers, at least of public lands, to accommodate a multiplicity of public concerns. This may mean integrating management for nonmarket values with traditional timber production. Such an integration may help to sustain a variety of social interests for people in local communities, and also generate revenues to cover costs of active management and contribute to the treasury. Public forests also usually serve an important role in maintaining some special set of ecological conditions across larger landscapes, and for sustaining unique communities of plants and animals. In developing countries, public forests must often provide commodities for local use as construction materials and fuel, and to sell in export markets as foreign exchange.

By comparison with public lands, industrial private owners want a continuous and steady supply of trees to merchandize or use in manufacturing the products that they sell. They must sustain suitable ecosystem conditions to insure long-term access to the requisite goods, and many ancillary values as well. These include steady yields of high-quality water, good habitats for an appropriate array of wild creatures, and recreational opportunities to make available through leases and other formal arrangements. Alternatively, while some nonindustrial private landowners also want to produce commodities, many use their forests primarily for recreation. They often harvest products only occasionally, or if these operations will enhance (or not interfere with) the nonmarket values that they prize. Some manage their forests quite actively for recreation, and use silviculture to create better conditions for their leisure pursuits. Still others want to sustain a particular set of ecosystem conditions without thought of commodities or in-place uses of the land. At the same time, many of these same owners, and others as well, would not preclude tree cutting if it enhanced the nonmarket or ecosystem values that they deemed important. Consequently, foresters must learn to deal with all kinds of landowner interests by tailoring their silviculture to fit the circumstances at hand, and using a variety of methods to create and sustain different kinds of vegetation communities appropriate to the conditions and needs (Broun 1912; Osmaston 1968; Soc. Am. For. 1981; Smith 1986).

THE CHANGING CONTEXT OF FORESTRY

Historically, forestry throughout the world has focused primarily upon organizing and managing lands to grow and utilize wood products and other commodities that forested ecosystems can provide in perpetuity (Osmaston 1968; Davis and Johnson 1987). Foresters working with many different kinds of landowners also manipulated stands to

1 improve wildlife habitat, particularly for a target species, or group of them;
2 manage watersheds to enhance water quality and yields, or to protect soils against erosion;
3 provide a better setting for different kinds of recreation opportunities;
4 manage visual qualities of a forest at both the stand and landscape scales; and
5 maintain a particular set of ecological conditions, like the habitat for threatened and endangered plants and animals, or particular community types that interest a landowner.

Through a process called *multiple-use management,* they learned to integrate commodity production with demands on forests to provide other values as well, and even to preclude cutting where an unmanaged stand would best serve the objectives. The approaches taken and intensity of management depended upon the interests of each landowner, and on landowners' priorities for emphasizing one group of potential opportunities over others.

In this context, some landowners may decide to emphasize commodities. Others want to integrate them with other uses, or have no policy against cutting and selling trees. Then foresters can schedule regular timber sales to provide a sustained and steady yield of revenues. These repay costs of the management, and provide a profit on the long-term investment in a property (Matthews 1935, Knuchel 1953, Davis and Johnson 1987). In these cases, foresters can use the timber sales program to manage the vegetation for nonmarket values as well as timber. They use the levels and consistency of volume and value increment, and the income derived from harvesting and selling that growth, to judge success of their silvicultural and forestry operations. In fact, the long history of research and practice in managing forests for commodities has helped foresters to identify the kinds of stand conditions that will yield the greatest and most valuable outputs of salable volume. They can use silviculture to create and maintain this ideal set of conditions, and can draw on past experience to forecast fairly accurately the kinds and frequency of yields that landowners might expect from the management.

In contrast to historic timber management programs, people who use their properties primarily for recreation-related benefits must judge the worth of ownership by the pleasures gained from seeing birds and other animals, or the spiritual value of walking in the woods. Many can derive these benefits from unmanaged forests, and some consider tree cutting inappropriate, even to enhance use opportunities. Where they encourage management, landowners might judge the benefits subjectively, using criteria like the availability of habitat for the wild creatures of interest or the frequency of songbird sightings per visit to their woodlot. Others might evaluate the management in terms of their success in hunting game species, the hours spent hiking and skiing, or the less tangible measures of pleasure derived from using forested land for different recreation pursuits. The possibilities depend upon each individual's interests.

On the surface, the difficulty in quantifying the benefits from noncommodity uses can make forestry more complex than concentrating on commodity production only. Even so, silviculturists continue to follow an important basic premise that has served them well from earliest times: that forestry has the primary purpose of producing something of value to landowners, and of doing it in a way that these individuals consider profitable. Foresters work to satisfy the particular interests of each different landowner, and to let each one define the criteria for success. This objective pertains to nonindustrial and corporate private lands, those controlled by institutions and foundations, and to public forests under the stewardship of government agencies at all levels.

With public and corporate holdings, officials who set the objectives, and the ways they make choices between alternatives, will differ. Yet in each case, silviculturists must first compare the management objectives with existing characteristics of a vegetation community, or the forest as a whole. They must then decide how to bring it to some optimal condition in a specified time, at an acceptable cost, and without jeopardizing environmental conditions and other values critical to long-term site productivity (Tappeiner and Wagner 1987). In this way they work to create and maintain a combination of plant community conditions that furthers a well-defined set of management objectives, and enhances opportunities for the uses of interest. This includes management for traditional commodities, and to promote other uses of less easily quantified worth.

Where landowners focus on nonmarket options as a primary goal for ownership, they often cannot find ways to generate revenues to pay costs of the management. In fact, most of the benefits that they cherish have no marketplace value, so income must come from other sources. Some can and will open their lands to public use, and charge fees or offer leases for people to hunt, fish, camp, hike, ski cross-country, or pursue some other recreational activity on the property. Some industrial firms have also marketed opportunities to build cabins as a term of a lease, perhaps along a lake or stream. At the same time, features like the visual qualities of a property, the yields and purity of water from it, the role it plays as habitat for indigenous wildlife, or the potential it offers for low-intensity dispersed outdoor recreation may have no salable value, per se. These may primarily create an environment that people find attractive, and thereby enhance the opportunities for recreational leases of different kinds. Even so, most landowners must absorb the costs of managing for these benefits—or they use other sources of revenue, like timber sales, to compensate for their investments.

The increasing interest in forests for their nonmarket values has challenged foresters to identify the plant community attributes that will sustain different nonmarket values and uses in perpetuity. These often depend upon personal preferences of a landowner or fee user, and the mix of values sought. Conceptually, foresters can identify the stands that seem to provide the benefits of interest, and then quantify the structural attributes of those communities. To do this, they might include the following measures:

1 the best levels of residual stand density, and even the thresholds of minimum stocking of some key species;

2 the optimal sizes (heights and diameters) and numbers of trees per unit of area;

3 the diversity of age classes present, and if a community contains primarily young or old trees; and

4 the spatial distribution of stems within a community—either uniform or dispersed in some kind of irregular pattern.

They can create communities with a high degree of uniformity or diversity in the features of interest, or strive for something in between. A community could have trees of many different sizes, or mostly trees of similar diameters and heights. This might mean having trees of similar or widely different ages, or simply different growth characteristics.

All these features determine the structural attributes of a community, how it looks, and the degree to which it contributes to the nonmarket values of interest. A silvicultural program that created and maintained the desirable conditions would allegedly satisfy a landowner's objectives, and provide sufficient benefits to compensate for the costs of management, yet economists have just begun to devise the means for quantifying the worth of different nonmarket benefits—particularly in tangible terms that landowners recognize as a payoff from investments in owning and managing a property (see Chapter 2). Forestry still relies primarily upon measures of volume and value to evaluate investments from a business perspective (Duerr 1960; Brodie et al. 1987), and in comparing different alternatives for managing a forest and repaying the investments at some requisite compound rate of return.

NO CUTTING AS AN APPROPRIATE RECOMMENDATION

Among both public and private forests, untreated stands will sometimes best provide the values of interest; then foresters should recommend no manipulation. They will identify ways to protect the trees from harm by insects, disease, fire, and other agents. They may suggest ways to capitalize most effectively on the nonmarket values a landowner can derive through various kinds of in-place uses—often related to some kind of recreational activity. They will help landowners to find some intangible compensation for the investments in ownership and property maintenance, and they may help landowners to improve the access through a property by developing a network of roads, trails, and pathways. These will usually facilitate some kind of personal use, and make a forest more valuable for recreation. However, where the naturally occurring vegetation adequately serves the interests of a landowner, or provides an appropriate setting for some noncommodity benefit of concern, *no* silviculture will often become the *best* silviculture.

Though appropriate in some cases, this no-management option carries a cost. It forces landowners to pay for the protection, administration, and maintenance of a property without benefit of revenues derived from sale of commodities. In some cases, landowners can rent or lease different use opportunities instead of cutting trees. The leases might give exclusive rights for recreational activities like hunting and fishing. In some circumstances public agencies and citizen groups might offer conservation easements to preserve a social value they associate with uncut forests. These would provide annual payments or tax credits to a landowner who agrees to forego tree cutting or certain other management activities. Otherwise, a landowner must pay the costs of protection and administration out of revenues derived from other parts of the forest, or by using other specially designated internal funds to cover expenditures or compensate for the lack of revenues.

SILVICULTURE DEFINED—AN ART AND A SCIENCE

As its ultimate goal, practical silviculture aims to create and maintain tree communities that serve specific objectives. In most cases, silviculturists also hope to provide those values at the fullest level that a site can sustain in perpetuity (Troup 1921). In fact, most foresters would agree to the general premise that silviculture

1 enhances the sought-after values, or improves upon the ways a landowner can benefit from them;
2 provides an effective and appropriate means for managing a stand to provide the sought-after benefits, and to sustain them in perpetuity;
3 keeps the management appropriate to the biologic and physical limitations of a site; and
4 maintains or enhances the basic productivity of the land in the process.

In practice, this translates into knowing (Broun 1912)

1 how different factors related to ecosystem structure and function influence individual tree and forest community establishment and development (e.g., soils, climate, locally adapted vegetation, indigenous fauna);
2 the kinds and magnitudes of biologic and physical environment responses that result from different kinds of treatments;
3 the treatments that will prove ecologically appropriate over the long run; and
4 the responses that will best serve a landowner's economic interests.

Silviculture brings economic, administrative, and ecologic dimensions into the process of planning for the management and use of forest resources for the benefit of a landowner.

In both theory and practice, silviculture draws heavily upon forest ecology, or the specialized branch of it known as *silvics.* By definition, silvics deals with the biologic characteristics of individual trees and communities of them. It studies how trees grow and reproduce, as well as the ways that the physical environment influences their physiology (Ford-Robertson 1971; Daniel et al. 1979; Smith 1986). Silvics also includes the study of ways that the physical environment tempers the makeup and character of a forest community, and the interactions of biologic components in those communities. Further, it investigates how communities of trees in turn modify conditions of the physical environment that supports them, and the never-ending interaction between vegetation and physical environment as forests mature and change over time (Berglund 1975; Smith 1986). In general, silvics translates this scientific knowledge into practical information about the habitat or site requirements for the successful reproduction and growth of forest stands, and of the different tree species that comprise them.

Silviculture is the science and art of growing and tending forest crops. More particularly, the term *silviculture* means the theory and practice of controlling the establishment, composition, character, and growth of forest stands to satisfy specific objectives (Broun 1912; Kostler 1956; Ford-Robertson 1971; Daniel et al. 1979; Smith 1986). To serve these purposes, silviculturists draw upon the principles and theory amassed through scientific inquiry in botany, zoology, soil science, physical sciences, ecology, silvics, managerial science, economics, and quantitative methods. All contribute techniques and information that foresters can use in identifying and evaluating the options for practical management, and in forecasting the likely outcomes from a given set of silvicultural treatments (Fig. 1-1, from Assmann 1970 and Berglund 1975). In turn, silviculturists contribute to decision making in the broader discipline of *forest management* (Ford-Robertson 1971): the practical application of scientific, economic, and social principles to the administration and working of a forest for specific objectives. It integrates the biological and ecological concerns of silviculture with considerations for policy, law, administration, business, economics, and similar aspects of the general profession called forestry.

In this sense, silviculture incorporates both biophysical and economic (institutional, managerial, and social) concerns in devising ways to manage forest communities to better serve the interests of a landowner. It involves the art of devising and prescribing ways to change the character and condition of a forest stand to make it more useful for whatever purpose a landowner deems important, or to sustain a set of values already available. Stated another way, silviculture includes the methods for handling forest stands in view of silvics, and modified by economic factors (Baker 1950). In most cases, a practitioner's judgment and experience play a critical role in identifying when to manipulate the character and condition of a forest community, and how to do it to best serve the economic interests of importance.

SETTING THE OBJECTIVES AND PAYING THE COSTS

Economic interests involve more than just financial matters; they also mean institutional and social interests. These include personal preferences of a nonfinancial nature, operational requirements of an enterprise, and any other opportunity or constraint that will influence the choices an owner makes about

1 what values to seek;
2 how to do it;
3 with what intensity; and
4 in what quantity.

Where the benefits come as revenues via commercial sales or rental fees, the income must often repay the costs compounded at a prevailing rate of return determined by the enterprise or landowner.

With management objectives of this kind, foresters have historically used silviculture to ensure the perpetual supply, timely use, and optimum yield of whatever marketable values a forest can provide, and a landowner demands. Yet in modern parlance, *marketable* has taken on a broader meaning that may include almost any

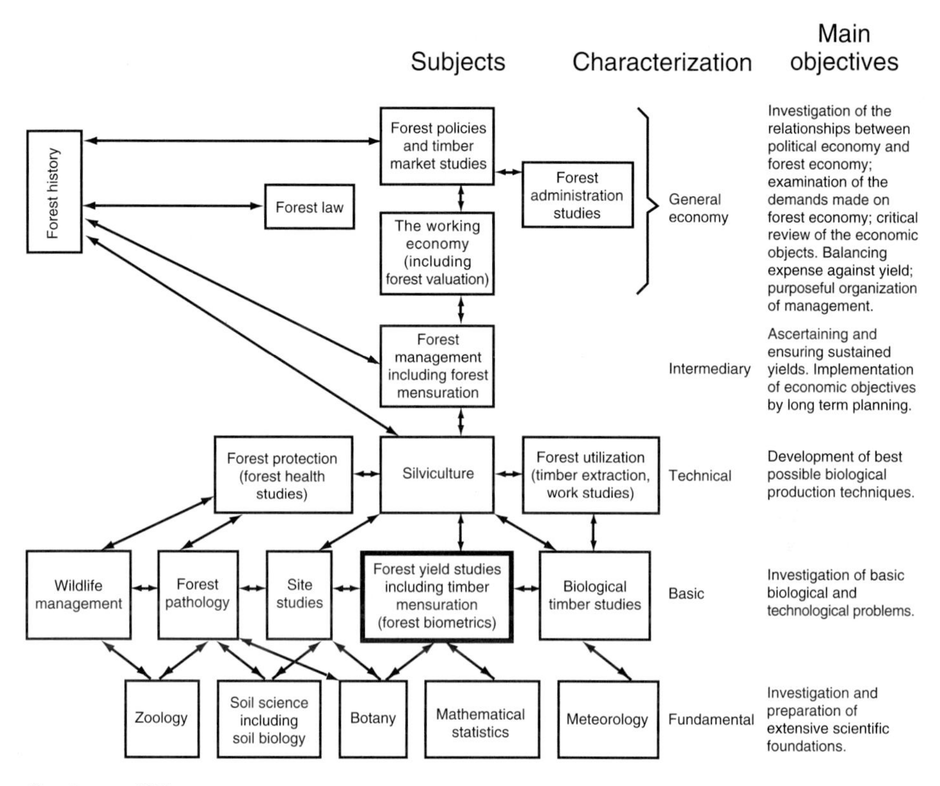

(from Assmann 1970)

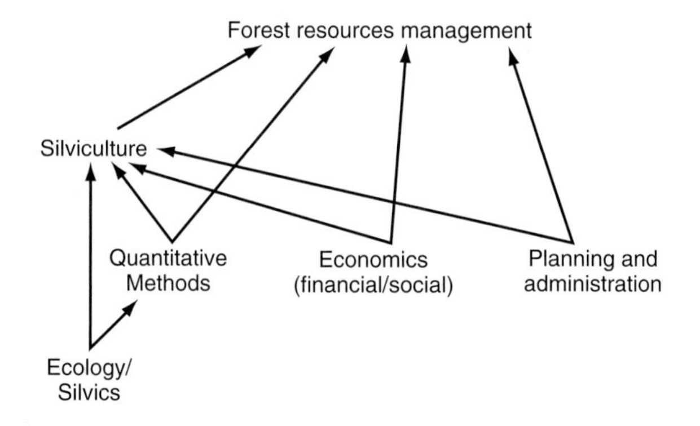

(from Berglund 1975)

FIGURE 1-1
Practical silviculture rests on a foundation of basic science, and contributes to the managing of forest resources in the broadest sense.

set of outcomes of interest, as long as a landowner does not violate basic requirements for long-term ecosystem sustainability in realizing them. This generally includes two broad classes of benefits (after Osmaston 1968):

1 intangible values—encompassing a multiplicity of in-place uses and benefits that have no direct marketplace value, including the protection and regulation of ecosystem structure and function

2 tangible values—including a broad array of harvestable commodities like logs and pulpwood to sell or to utilize as a raw material for some manufacturing process, and any in-place use opportunity or service that a landowner can merchandise or rent.

In concentrating on nonmarket values, landowners themselves must ultimately decide if the intangible benefits that they realize adequately compensate for their investments in management. In these circumstances, foresters often have great difficulty in quantifying the benefits by using common measures of output like those historically accepted under a timber management philosophy. This often hinders attempts to communicate with landowners about the success of silviculture, and to determine what alternatives would better serve their management objectives. Even so, the management must provide some package of financial or intangible values that adequately compensate a landowner for the investments made. As a general rule of thumb, silviculture must provide sufficient benefits to balance the costs of getting them, and even pay some form of interest on the investment in a forest.

OWNERSHIPS, OBJECTIVES, AND ECOSYSTEMS

The ownership of forested lands varies widely throughout the world, between and across continents, and even within nations. In some areas, governments control most or virtually all of the forest. They may maintain staffs of foresters to plan and arrange for management programs to serve their objectives, and administrators or representatives of the people set the goals in response to the demands of society at large. In some cases, they offer concessions for private corporations to oversee the management, and receive payments to the treasury for products or use opportunities sold as an outcome. In still other countries, government agencies at several levels may control large acreages in all or parts of the country, and private corporations and individuals own the remainder. Countries with the latter arrangement include the United States, Sweden, and Canada. There different units of government have jurisdictional responsibility for most of the land in some regions, but private ownership prevails elsewhere. Within North America, private forest land ownership is most common along coastal and accessible inland areas of the West, and throughout the East.

In countries and regions with a preponderance of governmental forests or many large corporate ownerships, managers can plan programs of silviculture for vast contiguous areas. As a consequence, they can affect the character and use of resources at landscape and ecosystem scales in an organized and coordinated way. They can integrate the timing and kind of silvicultural treatments across sufficient area to substantially influence population characteristics and dynamics for many woodland creatures and groups of them, and to establish a mixture of vegetation communities that will provide an array of ecological conditions and use opportunities across entire watersheds or larger expanses of space.

By contrast, foresters face a different set of tasks in regions dominated by private lands, and with a preponderance of limited-acreage ownerships. There the interspersion of relatively small private farms or ranches, forests, and unmanaged open areas gives the working landscape a unique character. Landowners decide what type and intensity of management to pursue in satisfying their individual purposes. As a result, the exact characteristics of a watershed or landscape depend on what the many different landowners decide independently to do with their land, and the kinds of vegetation conditions that they create by those activities. This complicates the potential for devising and implementing any ecosystem-scale management strategy, unless groups of private owners join in cooperative ventures.

In areas close to urban centers across North America, many people and firms who control large blocks of forest continue to subdivide and resell their lands to capitalize on the demand for sites where people can build vacation cottages or second homes, and for recreational developments. This parcelization exaggerates the historic discontinuity of management and use in regions dominated by private ownership, and makes future ecological conditions within the landscape difficult to forecast. In addition, ecological fragmentation has also become a concern in some regions of expansive and unbroken tree cover. In this sense, *fragmentation* means interspersion of vegetation communities with distinctly different composition or structures, and often with different resultant conditions of the physical environment as well. In this context, fragmentation creates extensive *edges,* or boundaries where communities of distinctly different structural attributes come together (Hunter 1990; Patton 1992). Edges can occur as a result of natural phenomena (*inherent edges*) or as an aftermath of some management activity or human action (*induced edges*). Further, either one can create an abrupt (sharp) or transitional (gradual) interface between the adjacent communities (Thomas et al. 1979).

Fragmentation can occur either on a broad landscape scale, or in smaller areas. On the larger scale (landscape fragmentation), it commonly results from

1 differences in land use practices across a landscape;

2 natural features of the physical environment that limit the kinds of vegetation communities that can develop on different areas of land; or

3 natural phenomena (like fire, losses to insects and diseases, and blowdown) that kill or extensively damage some communities, trigger successional changes, or otherwise moderate conditions across a landscape (after Thomas et al. 1979; Hunter 1990; Patton 1992).

Fragmentation can also take place within a forest (internal fragmentation) as the result of management practices that create an interspersion of stands having different structural characteristics (like tall forest adjacent to young stands), or that remove some of the tree cover to create permanent open spaces (Temple and Wilcox 1986). This changes the composition or structural characteristics of the vegetation on a broad horizontal plane, and on a multistand or landscape scale. In this context, silviculture serves to alter the structural attributes of an existing community, or to establish ecological conditions where a replacement community will regenerate. On an ecological time scale, silviculture causes only a temporary interruption in the continuity of conditions on a broad spatial scale, or only temporarily alters the balance of habitats that had previously existed across a landscape of some appreciable size (Haila 1986).

The cutting of previously unmanaged ancient or old-growth forests in northwestern North America exemplifies one commonly cited case of forest fragmentation on both the forest and landscape scales (Maser 1994; Noss and Cooperrider 1994). In at least some cases, commodity objectives encouraged silvicultural practices that removed all the standing trees from fairly large areas to insure efficient harvesting, and to simultaneously create conditions for regeneration of fairly shade-intolerant species of high commercial value. Landowners (private and government alike) often coupled these techniques with forestwide management systems that programmed a sequence of cuttings to provide regular and steady supplies of timber crops from each property over time (Osmaston 1968; Davis and Johnson 1987). They also often dispersed the treatments in a checkerboardlike pattern to create a matrix of contrasting vegetation conditions (induced edges) across a large expanse of area (Fig. 1-2).

Though successful in establishing replacement crops of economically desirable trees, these age-honored schemes have promoted two disparate outcomes. First, they created an interspersion of stands with sharply contrasting sizes and ages of trees (extensive edge), and with differences in several other vegetal attributes as well. These schemes generally enhance the habitat for wildlife adapted to a combination of open areas, dense communities of young trees, and adjacent mature forest (Roach 1974a; Gullion 1972; Hunter 1990; Patton 1992). At the same time, the patchwork of cut areas reduces the extent and connectedness of remaining old-growth or mature (tall) vegetation, and the habitat for species better adapted to expansive forest cover of old and large trees.

For some specialized fauna and flora, keeping limited areas of old-growth timber interspersed among the second-growth stands will not maintain adequate habitat for self-perpetuating communities or populations. Some other ecologic values associated with the old-growth or mature forest conditions may diminish as well (Franklin 1989; Brooks and Grant 1992; Maser 1994). By contrast, some recreationists like the improved access, the complex interspersion of different vegetation conditions, and the structural and biological diversity that active management created across large areas. Others find large, contiguous areas of inaccessible old stands preferable. This has triggered considerable public debate about how to appropriately determine the desired outcomes, and how to set priorities for management. It has often polarized interest groups having divergent perspectives about appropriate objectives for managing and using a forest (Brooks and Grant 1992).

In some of these cases, a fairly obvious and straightforward change in the management strategy may address both the commodity and nonmarket interests of a

FIGURE 1-2
Past clearcutting in many regions created a checkerboard pattern of cut and uncut areas across a property, creating an artificial landscape pattern as with these conifer forests in western North America *(courtesy of Thomas J. Temple).*

landowner or group of people—over both the short and long run. Yet frequently an approach that may seem best to one group will conflict with important institutional or personal preferences of others—or it may not optimize the package of benefits that an individual, agency, or firm would find desirable or necessary. Then silviculturists must devise a unique approach that maintains the essential character of existing forest conditions across an appropriate expanse of space, or they must change it to provide other kinds of sustainable ecosystem conditions that have long-term value to a landowner, or to society. Modern silviculture faces this major challenge for the future: to determine what combination of treatments would prove ecologically and institutionally best over the long run, and to find ways to implement them in an economically attractive manner.

TAKING AN ECOLOGIC PERSPECTIVE

In an ecologic sense, the term *forest* means an extensive area, landscape, or ecosystem dominated by trees and other woody vegetation growing closely together (Ford-Robertson 1971). Where ownership patterns have divided the countryside into a complex matrix of different condition classes and land uses, *forest* often means something akin to an *ownership* or estate. So from a management or silvicultural perspective, the term *forest* also means an area managed for timber and other resources, or maintained in woody vegetation for other indirect benefits (Ford-Robertson 1971).

In a broader landscape, open lands may divide the forested parcels in irregular patterns, creating a patchwork of disparate cover conditions (with inherent or induced edges) that add a special richness to plant and animal species diversity across a landscape. Silviculture can help to maintain these conditions by appropriately managing forest stands in an ecologically and economically sound manner. In areas with highly profitable agriculture (e.g., the central United States, central California, and parts of south-central Canada), some owners still continue to clear forested lands for cultivation (Sharpe et al. 1981; Burgess and Sharpe 1981). Then forest fragmentation continues as a result of woodland destruction (permanent clearing for different land uses). Under these circumstances, silviculture may make only a limited contribution in tempering or maintaining a landscape. In yet other regions, agriculture has declined (e.g., the eastern United States and Canada), cropping and grazing have decreased, and many old farms and ranches have naturally regenerated to a new tree cover. This has created a renewing landscape of greatly expanded and more consolidated forest cover, and less diversity of condition classes on a large scale (Nyland et al. 1986; Zipperer et al. 1990). In these circumstances, silviculture provides the principal means to create a diversity of other conditions, and to help landowners capitalize on the opportunities that those emerging forests provide, both at the stand and at the landscape levels.

Such variations in the degree of landscape diversity and the extent of ownership parcelization present a challenge to foresters trying to decide what comprises an *ecosystem.* In one sense, it should encompass a fairly complex, self-sustaining, and interdependent collection of interdependent plant and animal communities, with the requisite physical environment to support them (Spurr and Barnes 1980). As such, ecosystems have no standard minimum size or easily delineated physical boundaries. Rather, they constitute functional entities wherein changes of one component will often alter the condition of others as well (Kimmins 1987). In a practical sense, ecosystems do cover a minimum amount of space. At the least, they must encompass a critical mass of biotic elements and sufficient amounts of the requisite physical environment to sustain the component organisms and communities in perpetuity.

Fully functional ecosystems may have fairly easily recognized edges where distinctly different kinds of systems interface. Yet the ambiguity between ecosystems as functional entities, and forests as places, complicates human efforts at forest ecosystem maintenance and renewal. The greatest difficulties arise in settings where society has conveniently divided the landscape into relatively small and well-defined properties with distinct physical boundaries that ignore ecologic conditions, and allowed the owners independence in managing the resources on those lands. Under such circumstances, most landowners do not control sufficient area to effectively influence

1 ecosystem-scale communities and populations of many mammals, insects, other animals, fungi, and other plants;
2 the quality and yields of water from whole watersheds; or
3 general visual characteristics across the countryside (Nyland 1991).

Then any effort to maintain a specific set of sustainable forest conditions at the landscape scale will depend upon some kind of institutional arrangement that encourages collaboration among landowners.

On both a regional and national scale, forest scientists and laypeople alike have become more aware of the importance of this multi-ownership collaboration. More particularly, they have come to recognize the need for planning an appropriate interspersion of silvicultural treatments (including cuttings) across broad expanses of space. Only then can they truly insure the stability of many landscape-scale values derived from forests. In addition, many foresters and landowners in highly industrialized countries have begun to view silviculture as a process for *influencing forest ecosystem maintenance and renewal,* rather than as a procedure for simply producing commodities and nonmarket benefits of interest. Under this new philosophy, values like enhanced visual qualities and recreational use, continuity of diverse populations of vertebrates, stability of habitats for specialized plants, and consistent yields of wood products all represent desired outcomes from maintaining and renewing sustainable forested ecosystems having specific attributes (Kessler 1992). Yet neither a multi-owner approach to management planning nor a change of philosophy about the purpose of silviculture will succeed if landowners think only about what they will accomplish one stand at a time. Rather, they must determine how each treatment might help to create some desired set of vegetation conditions across a broader landscape, and to sustain a desirable pattern of forest condition classes over time on the landscape scale. That applies in both highly fragmented and parcelized landscapes in the more urbanized regions of the world, and in planning a management system for large ownerships within heavily forested regions having a low population density.

To silviculturists, this philosophy of *ecosystem management* means finding ways to use stand-level treatments to help sustain the integrity, diversity, and resiliency of ecosystem conditions on a landscape scale, and over the long run. It challenges individual landowners to view different kinds of forest uses and products as the reward of sound ecosystem management (Kessler 1992). It means moderating some historic schemes that have greatly simplified the complexity of ecosystem conditions so landowners could maximize production of a single value at the greatest financial efficiency. It will require landowners to ensure that all of their management and use maintains some minimal pool of nutrients to cycle through the system, an adequate stockpile of carbon stored in the living and dead organic material on a site, and a continuity in the critical balance between surface and subsurface water storage and moisture cycling. Maintaining these and many other essential conditions will sustain the diverse communities of plants and animals that give forested ecosystems their essential character. This, in turn, will keep open a wide array of future options for managing and using those systems for human benefit.

WHAT SILVICULTURISTS DO

While silviculture historically focused on producing commodities at sustainable levels, its foundation in ecology and economics readily accommodates concerns for intangible values and the integrity of ecosystem conditions as well. In fact, the Society of American Foresters incorporated this breadth of thinking in the following definition of silviculture (after Ford-Robertson 1971; Soc. Am. For. 1989):

- the art of producing and tending forest stands by applying scientifically acquired knowledge to control forest stand establishment, composition, and growth
- applying different treatments to make forests more productive and more useful to a landowner
- integrating biologic and economic concepts to devise and carry out treatments most appropriate in *satisfying the objectives of an owner*

In fulfilling these roles, silviculturists draw upon their knowledge of ecology, plant physiology, economics, logging practice, pathology, soil science, forest influences, silvics, management science, and many other disciplines to devise and subsequently implement a rational plan for manipulating individual forest stands, and controlling their long-term character and development (Florence 1977).

From an ecologic perspective, a rational plan takes cognizance of the interaction and interdependence of a full range of ecosystem components, rather than just seeking to maximize one value. It means evaluating the effects of different alternatives at a range of spatial scales from microsites, through stands, to watersheds, and across landscapes. The assessment should also include effects from alternative treatments over time, and even beyond periods that most landowners might consider long-term from a traditional economic perspective. Additionally, the plan must discriminate against practices that lead to irreversible changes in the character of an ecosystem, and the loss of potentially important options for future use (Brooks and Grant 1992). The management and use must cause no long-term decline in the capacity of a site to support the growth and reproduction of trees, and the associated plants and animals that inhabit forested ecosystems. In modern parlance, this translates into providing stability and sustainability of desired ecosystem values, ensuring the timely maintenance and renewal of appropriate biologic communities, and protecting physical site components needed to support those organisms over the long run.

From an economic perspective, silviculturists must concurrently create and maintain tree communities that will provide landowners with the goods or values of interest. Further, they will commonly strive to

1. shorten the time it takes to realize the benefits of management;
2. increase the package of options or values compared to what an owner could get from an unmanaged forest;
3. improve the quality of benefits or uses that interest a landowner; and
4. provide the full measure of sought-after values at the least cost.

They must integrate ecologic, economic, and administrative concerns into a unified approach to the management and use of resources to insure both an appropriate level of current use, and a renewal and replacement of those resources to satisfy future needs.

In working toward these goals, silviculturists perform four primary functions (after Cheyney 1942; Smith 1964):

Control—establishment, composition, and growth
Facilitate—harvesting, management, and use
Protect—site
—maintain continuous vegetative cover
—insure stable soils
—protect natural land forms
—trees
—from insects, disease, fire, and other agents
—from deterioration
Salvage—dead and diseased trees, and stands
—potential mortality before it occurs

These functions serve as the means for maintaining ecologic continuity, sustaining the benefits, insuring full production, realizing maximum usefulness, and providing an adequate return of whatever outcome a landowner demands. Silviculturists accomplish this by deciding how to treat various stands, when to do the work, and with what intensity to apply a combination of treatments.

CONTROLLING ESTABLISHMENT, COMPOSITION, AND GROWTH

Silviculturists control forest stand and age class establishment by periodically altering conditions of the physical environment in a particular stand, primarily by cutting trees, or controlling their density and character in other ways. At times when it seems inappropriate to regenerate a new cohort of trees, they maintain a high stand density to restrict the light and other site resources needed for germination, or to sustain the development of new seedlings. To trigger regeneration at other times, they will cut some or all of the trees present, or undertake other treatments that alter the character of an existing vegetation community and the environment near the ground surface. This will favor seed germination and seedling establishment. Silviculturists may also decide to create new communities of trees or add to others by artificially sowing tree seed or planting young trees in an area, including sites that may not support forest cover at the time. They may use these artificial methods to enrich the species composition of a naturally occurring community, to supplement the density of seedlings that develop naturally, or to substitute a particular mixture of tree species for some other kind of community that would normally regenerate at a site by natural means. The artificial regeneration methods may include woody shrubs as well as trees, if those species would enhance the values to a landowner.

Besides intervening at the time of stand establishment, silviculturists can also affect community composition later on through a variety of intermediate treatments—usually, by cutting unwanted trees to limit the numbers of species present.

They may also reduce stand density at times to improve the growth of the most desirable trees, independent of species. These intermediate treatments may favor trees of the best form and growth relative to a landowner's objectives, as well as promote a mixture of species that will best serve the needs.

In large measure, silviculturists control stand and tree growth by regulating stand density. They recognize that in communities of high tree density, the foliage of most trees receives less than full direct solar energy due to shading by neighboring trees. The low light levels limit the photosynthetic output, and may cause death of heavily shaded trees that photosynthesize too poorly to sustain life. To promote better diameter growth and limit mortality, silviculturists reduce stand density to lessen intertree crowding and brighten the environment around crowns of the residual trees. Yet to maintain high levels of standwide production and control different kinds of vegetation responses, they keep sufficient numbers of well-distributed trees for complete site utilization. Further, by controlling stand density, they also concentrate the growth potential on trees of desired species or character.

SILVICULTURE AS A DELIBERATE STRATEGY

To fulfill all of these expectations, silviculturists must apply different kinds of treatments at varying stages of stand or age class development. They must also intervene at the stage of development when a treatment will have the sought-after effect. So silviculturists must understand the following (after Troup 1921; Kostler 1956; Smith 1986):

1 the silvical characteristics of the target species;
2 the objectives of a landowner;
3 the natural patterns of community regeneration and development;
4 the ways conditions of the physical environment affect tree growth and reproduction;
5 the economic forces that influence a landowner's choices; and
6 the potential for realizing sufficient payoff to balance the costs of ownership and management.

Silviculturists will devise deliberate measures for intervening into the natural scheme of ecosystem dynamics to control the nature and composition of a community, and the way it grows and develops.

2

THE SILVICULTURAL SYSTEM

THE SILVICULTURAL SYSTEM AS A PLAN FOR MANAGEMENT

Any review of silviculture must recognize at least four basic concepts or premises that underlie its practice:

1 Foresters can change the character of a tree community by manipulating its composition and density, and often to better serve the special interests of a landowner.

2 They can affect the results by applying different kinds and intensities of treatments at varying stages of stand or age class development, and arranging them in a unique sequence.

3 The treatments must fit characteristics of the species of interest and the physical site conditions within the stand under management, and prove ecologically acceptable.

4 To adequately control stand or age class establishment, composition, and development (growth), silviculturists must plan both an appropriate intensity and an optimal time for applying each treatment to insure the sought-after effects.

Conceptually, both the type of manipulation and the time of its use influence how a stand or age class will develop afterward. For that reason, timing and sequence may prove as important as the actual kind of treatment applied at any juncture. Time in this sense means the stage of development, rather than season of year or even the chronological age of the trees. In some cases, a particular method may have time-of-year constraints (e.g., when using a pesticide to control a damaging insect). Silviculturists may limit the season of application to realize a special effect, to avoid conflicts between competing uses, or to protect a site. Sometimes they may want to cut trees during summer months to insure adequate seedbed disturbance. To protect a site at high latitudes, they may restrict logging to winter months when snow covers the site and the soils freeze. Winter logging may also reduce competition between harvesting activities and some recreational uses or essential faunal activity. However, time of year has only secondary importance in the planning of most silvicultural events.

To integrate these concerns and guarantee the appropriate timing, sequence, and kind of treatments that will produce the desired outcomes, foresters need a conceptual framework for their management. The *silvicultural system* serves this role. It describes the long-term plan for managing an individual stand to sustain a particular set of values of interest (Champion and Trevor 1938; Ford-Robertson 1971; Van Lear 1981; Smith 1986). It links the needed treatments in sequence, and programs them at appropriate times to insure a continuous forest cover of trees and associated plants having value to a landowner. It reflects the silviculturist's concept of how to control, facilitate, protect, and salvage within a stand. These plans may change from time to time to better reflect how the resources have grown and developed, what new reasons a landowner may have for managing them, and how to better capitalize on the benefits from management (Champion and Trevor 1938; Smith 1986). All serve as genuine reasons to alter the approach, as long as the methods fit the species (plants and animals of interest) in a community and the physical and biotic environment that supports growth.

Conceptually, foresters develop a unique silvicultural system for each forest stand. Yet all silvicultural systems (Fig. 2-1) include three basic component treatments or functions (Champion and Trevor 1938; Cheyney 1942; Ford-Robertson 1971; Soc. Am. For. 1981; Smith 1986): (1) regeneration; (2) tending; and (3) harvest. In some cases, tending may simply take the form of protection, or landowners may use methods other than timber harvesting to implement the treatments. Also, the timing, sequence, kinds, and intensity of actual treatments often differ from one stand to another, based upon a landowner's objectives and the ecologic conditions. Social attitudes may also influence the choices, and these often become a part of the management objectives. But all systems for all stands will give due attention to all components in some form or another, and arrange them to fit the needs of a particular stand and a particular ownership.

In both practice and concept, silviculture applies to forest *stands,* and the trees in those stands. In contrast, the discipline of *forest management* works with the collection of stands that comprise a forest or property, and integrates the silvicultural plans for individual stands into an overall program of management for an entire enterprise, an ownership, or an administrative unit. The standard definition of a forest *stand* conveys this notion (Ford-Robertson 1971; Soc. Am. For. 1989): a contiguous community of trees sufficiently uniform in composition, structure, age and size class distribution, spatial arrangement, site quality, condition, or location to distinguish it from adjacent communities—so forming a silvicultural or management en-

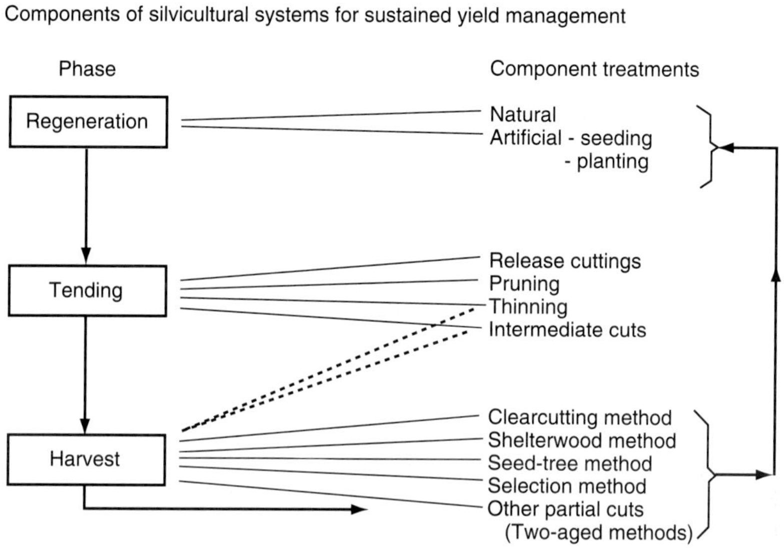

Components of silvicultural systems for sustained yield management

Phase

Component treatments

Regeneration

Natural
Artificial - seeding
- planting

Tending

Release cuttings
Pruning
Thinning
Intermediate cuts

Harvest

Clearcutting method
Shelterwood method
Seed-tree method
Selection method
Other partial cuts
(Two-aged methods)

FIGURE 2-1
Components and character of silvicultural systems (after Nyland et al. 1983).

tity. A silvicultural system describes the means for effectively regenerating, tending, and harvesting these ecologic and management units in a timely and economically viable manner.

CHARACTER AND OBJECTIVES OF SILVICULTURAL SYSTEMS

Fig. 2-1 indicates that silvicultural systems commonly incorporate harvesting techniques to implement both the intermediate treatments and the methods for regenerating a new age class at appropriate times. This applies to all kinds of stands, and with different age arrangements. Fig. 2-2a highlights the cyclic nature of silvicultural systems. Normally, events move from regeneration of an age class, to the tending of it during intermediate ages, and finally to its regeneration again at maturity. Fig. 2-2b suggests the interdependence of these components, and the necessary linkage of different parts to ensure a systems approach to management. Skipping one part makes the program incomplete, just as taking away one leg of a tripod lets it collapse.

As suggested by Fig. 2-1, regeneration may arise from natural or artificially introduced sources. Once established, these trees might receive a series of tending treatments to nurture their growth and development, particularly in light of a landowner's objectives. Release cuttings, pruning, thinning, and other intermediate treatments serve this purpose. Later, when an age class reaches a predetermined condition or stage of maturity, silviculturists again use harvesting techniques to initiate regeneration to replace the mature age class. These treatments include a family

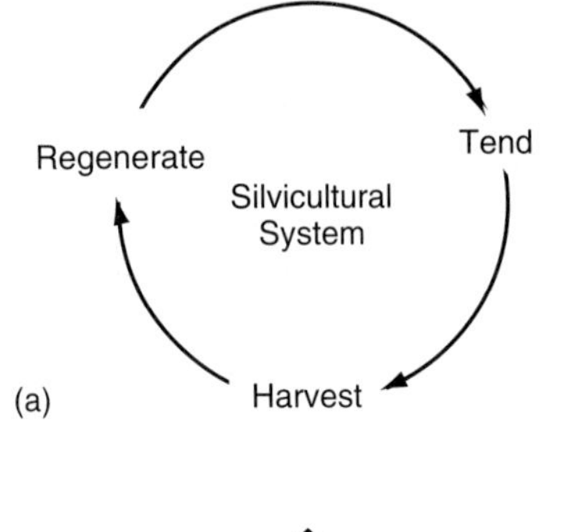

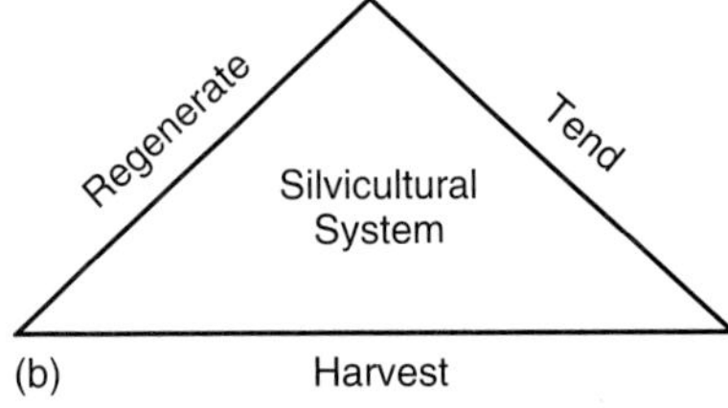

FIGURE 2-2
Alternate views of the silvicultural system, emphasizing its continuous nature and the interdependence of its component treatments.

of *reproduction methods* called clearcutting, shelterwood, seed tree, selection, and two-aged methods. In combination, the sequence and nature of possible treatments that comprise it delineate the nature of a silvicultural system, and in turn shape the character and development of the resultant forest stand. They determine the distinctive form of a tree community, and the degree to which it will serve a landowner's interests.

Aside from simply conveying information about the intended long-term program of management for a stand, the silvicultural system also gives a silviculturist the means for insuring some long-term continuity in the program of treatments for it (Wenger 1984). This should translate in practical terms into realizing the sought-after biologic and economic benefits (requisite levels of goods and values) in a timely fashion and in appropriate amounts. In fact, an appropriate silvicultural system should

1 optimize the yields—capitalize upon the full productive potential of a site to serve a landowner's interests;
2 improve the quality—provide the kind of stand and trees best suited to a landowner's needs, and to the fullest extent possible;
3 shorten the investment period—bring a crop or tree community to the desired condition or stage of usefulness without needless delay;
4 contain the costs—minimize the investments to optimize those sought-after values; and
5 sustain ecosystem health and productivity—limit practices to those that appear ecologically and biologically appropriate.

In general, silviculture provides trees and stands that landowners can put to some kind of desired use, and at the levels of utilization that will repay costs of ownership and management at a required rate of return. Silvicultural systems articulate the means for attaining these goals.

SILVICULTURE FOR SUSTAINED YIELD OF MULTIPLE VALUES

Historically, foresters have looked at silvicultural systems to provide predictable crops of commodities at regular intervals, and to sustain them over the long run. Appropriately designed systems have also sustained nonmarket values as well. These might include nontimber benefits like the following (Cheyney 1942; Osmaston 1968; Florence 1977; Daniel et al. 1979; Soc. Am. For. 1981; Randall and Peterson 1984):

1 protecting sites and ecosystems;
2 stabilizing soils and inhibiting erosion;
3 maintaining indigenous populations of insects, fungi, and essential microorganisms;
4 improving habitat for wild animals and many kinds of native plants;
5 enhancing water quality and yield from large areas, and insuring suitable habitat for indigenous fishes;
6 producing forage and other habitat essentials for domesticated and wild animals; and
7 enhancing visual qualities of landscapes, and creating better opportunities for recreational use of forested areas.

Foresters still lack extensive experience with silvicultural systems to specifically enhance the more intangible values, or to maintain some essential ecosystem conditions. In many cases, they must still plan nonmarket silvicultural systems one treatment at a time, while learning from each experience how to sustain the desired conditions that landowners wish to maintain.

For the present, foresters can judge the ecological appropriateness of alternative treatments on a conceptual scale, and identify some practical effects of the intended management. Basic tenets for selecting both ecologically and economically acceptable treatments might include the following (Brooks and Grant 1992):

1 Account for the objectives of ownership, and the values of interest to the person, firm, or institution that has the ownership rights.
2 Look for ways to balance various ecosystem components and conditions to ensure a minimum necessary level of integrity, resiliency, and diversity.
3 Take into account the effects that a treatment in one stand has upon conditions of the surrounding landscape, and how it might alter critical balances.
4 Consider effects and outcomes of a treatment on an ecological time scale (the permanency and potential effect on resiliency for the long term and for the present).
5 Favor actions that insure a variety of options for future management and use, and cause no irreversible ecologic change.
6 Assess the financial viability of both the treatment and any requisite practices

to ameliorate potentially negative side effects, and insure compliance with the economic requirements of an owner.

Essential ecosystem conditions might include the stability, health, and continued productivity of

1 vegetation and faunal communities over the long run;
2 the soil and nutrient capital;
3 conditions that insure normal water cycling, yield, and quality;
4 suitable habitats for indigenous plants and animals, or acceptably altered communities of them; and
5 acceptable visual qualities for intended noncommodity uses in a stand and the surrounding landscape (after Florence 1977).

In practice, foresters must guard against manipulations that might upset the long-term stability of an ecosystem in unacceptable ways, and use only those that appear ecologically appropriate.

Because silvicultural systems engineer stand and age class development along a deliberately chosen ecological pathway, they provide continual access to the kinds of conditions that a landowner desires over the long run. This has historically happened with deliberate management for timber crops, and will become increasingly feasible with nonmarket values as well. Foresters call this *sustained yield*—providing perpetual and continuing access to the goods, services, and values sought by a landowner under a given intensity of management (after Ford-Robertson 1971; Nyland et al. 1983). In that silvicultural systems provide for a continuance of trees after trees, they give landowners the assurance that later management can bring an age class to some specified character (distinctive form) and desired stage of utility at a reasonably predictable time. Further, silvicultural systems keep use and harvest in balance with the capacity of a site to sustain a dynamic forest and the desired values over the long run.

To provide continuing access to the goods and values of interest, and at fairly consistent levels over time, foresters deal with ideas and techniques from two related but distinct disciplines of forestry. Specifically, they manage (1) the kind, quantity, and quality of products and other values or use opportunities available in individual stands (through silviculture); and (2) the time, intensity, and spatial distribution of different uses across an entire forest and over the long run (through regulation activities of forest management). To the degree that foresters succeed in their silviculture, they can enhance potentials for realizing a continuing and steady level of forestwide uses through their forest management plans and programs. The marriage of appropriate silviculture with an effective program for integrating use opportunities over both time and space provides a potential for *sustained yield* (Fig. 2-3). Further, if the management provides fairly consistent levels of production and use over time, landowners realize what foresters call *even-flow sustained yield* (Fig. 2-3). The term implies a never-ending access to the desired values in fairly consistent amounts over the long run. This should apply to nonmarket values and opportunities as well as timber crops. However, foresters must first determine what set of

Use Options by Degree of
Even-Flow Recovery of Products, or Sustained
Availability of Non-Market Values

Types of possible management policies

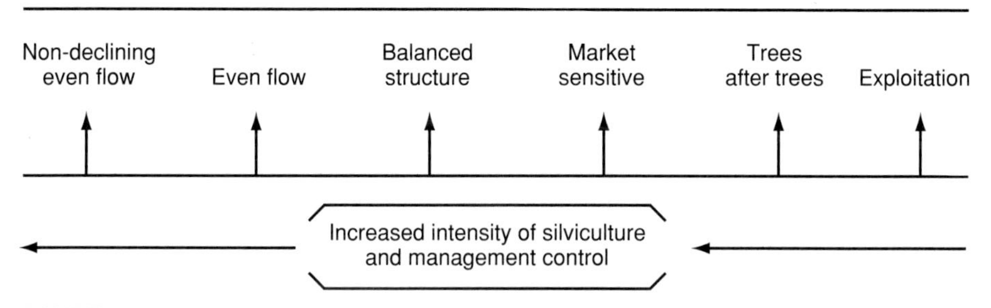

FIGURE 2-3
The marriage of appropriate silviculture with an effective program for integrating use opportunities over both time and space provides a potential for sustained yield, and if the management provides fairly consistent levels of production and use over time, landowners realize what foresters call even-flow sustained yield (from D. Koten, SUNY ESF, 1994).

tree community attributes will provide the desired noncommodity benefits before they can develop a system to create and maintain those conditions in perpetuity (Nyland 1991).

Above all, both the economic goals (commodity and nonmarket) and the intended scheme of management reflected in the silvicultural system must prove compatible with the biological or silvical attributes of the species in a stand, and the physical capacity of the site to sustain their growth and development (Broun 1912; Troup 1921; Champion and Trevor 1938; Soc. Am. For. 1981; Smith 1986). Foresters must feel confident that the soil, moisture, incoming solar energy, and other ecosystem components will not limit the kind and level of growth and plant community structure that will produce the conditions of interest, and also have sufficient value to repay the costs of ownership and management. Also, effects of indigenous insects, diseases, and pests must remain at levels consistent with the economic goals. Within this broad constraint, a silvicultural system must assure harmony between a landowner's objectives and the capacity of the land to produce the sought-after values in perpetuity (Smith 1986).

CONSERVATION, SILVICULTURAL SYSTEMS, AND THE ETHICS OF RESOURCE USE

An appropriate silvicultural system should reflect a sense of *conservation,* or a determined effort to provide future yields of goods and other values even while harvesting or using those available at the moment. Foresters often link this notion of conservation with the idea of wise use, suggesting that landowners should feel free to utilize available forest resources now, but in ways that also insure adequate access for the future. They also imply that over the long run, benefits should serve the needs of society as a whole, rather than only provide personal gain for a few. Most of the time, conservation includes a decision about how much and in what way to utilize resources at present, and how to regulate current use to guarantee adequate and continuous supplies for the future (Pinchot 1947; Barlowe 1958; Duerr 1960; Nyland et al. 1983). The period of interest may cover only a relatively short time in some cases, then decisions may follow fairly traditional concepts of economics. Alternatively, the planning horizon may seem almost infinite and uncertainty may complicate decisions, particularly since choices about how to allocate resources at one point of time potentially influence opportunities for access later on. Under these circumstances, conservation may mean deciding to restrain current use to reduce the risk of an irreversible future (Randall 1981). At times, landowners must make an ethical choice, between maximizing short-term economic gains irrespective of future needs and moderating current use to insure adequate opportunities over the long run.

When conservation focuses primarily on the time of use, it pertains mainly to fixed-quantity *fund* or *exhaustible* resources like coal and oil that occur in limited amounts at specific places, and have no renewable nature (Barlowe 1958; Randall 1981). Once producers exhaust available supplies, they must look elsewhere for new reserves to exploit. If that search uncovers no additional stockpiles on a global scale, society must turn to science and technology to develop substitutes and make them available at an affordable price. The people and corporations who control fund resources—and society as a whole—must decide how long existing quantities should last, and what amounts they can wisely withdraw each day without exhausting available supplies too rapidly. They must choose between the desire to maximize current economic interests, and their responsibility to future generations.

The wise use philosophy of conservation also covers *flow* resources (e.g., natural supplies of water, sunlight, or air). With these, natural ecosystem processes provide a renewing long-term supply, with continual inputs to replace quantities that naturally and regularly leave the system (Barlowe 1958). Flow resources are naturally replenished without human intervention, and independent of particular economic interests. Water arrives at the earth's surface via precipitation. Some accumulates for a period of time in soil and surface sources, and the remainder naturally flows out of a site and down a stream or river. Eventually all the water arriving at one point of time evaporates from the earth, yet periodic precipitation repeatedly recharges the groundwater reserves. It also continues the long-term flow through soil into various kinds of surface sources, though not always at a consistent rate or in the volumes that people want.

Flow resources qualify as renewable. Yet at any one time, people can extract only as much of the flow resource as arrives at their point of use. Further, they have ethical or legal obligations

1 to leave a reasonable portion of accessible supplies for others to utilize downstream;

2 to safeguard the quality of a flow resource; and

3 to moderate use so that withdrawals do not exceed the rates of natural long-term renewal.

Conservation of water means leaving a fair and high-quality share for other people to use, as well as sufficient reserves to sustain viable biotic communities in the streams and ponds. In areas where supply tends to fluctuate over short periods of time (e.g., rainfall varies appreciably between seasons, or from one year to the next), society or individuals may choose to divert a portion of the passing resource into artificial reservoirs. In such cases they treat the impounded pool of water like a fund or exhaustible resource. However, the planning horizon covers only the relatively short period between periodic natural events that replenish supplies, and even excessive short-term use will not necessarily preclude renewal over time.

Though economically and philosophically applicable to some biologic resources, these concepts of conservation do not suffice as a basis for silviculture and forestry. They fail because forested ecosystems do not remain static in condition or character. They depend upon fund resources like soil as a substratum for plants and fauna, and flow resources like water and sunlight to sustain their growth and development (Randall 1981). They continually change in density and in many other attributes, even if left alone, so what people find in the future may not resemble present conditions. In fact, simply refraining from use now will not necessarily insure adequate opportunities in the years ahead.

As biologic systems, forests also regenerate naturally following most kinds of disturbance (or withdrawal or use). Yet with forests, unlike flow resources, the rate of replenishment, the species that become established, and the community structure may change. In fact, in the absence of management to control this renewal, future forests might not adequately serve the long-term needs of a landowner, or of society. Careless exploitation can damage the physical environment in ways that prevent or delay renewal, or reduce the reserves and diversity of plants and animals that remain and regenerate. Both would interrupt the potential for steady and sustained use (Barlowe 1958; Randall 1981). On the other hand, landowners who employ reasonable methods that do not kill the indigenous plants and animals, destroy essential life-sustaining elements of the environment, or prevent regeneration from occurring can expect their forests to remain resilient and renewable. Further, with the aid of silviculture they can channel that renewal and development to assure the kinds of values and commodities that people will need in the future, while still making reasonable use of those same resources at the present.

Overall, silviculture (or rather the silvicultural system) provides a means to assure perpetual supplies of forest resources that landowners find economically attractive. It promises the deliberate manipulation and use of tree resources to control their growth, development, and timely regeneration. Thus to silviculturists, applying an appropriate silvicultural system equates to the conservation of forests, and the ethical use of resources that people need. Further, it guarantees an appropriate blending of the biologic-ecologic opportunities and constraints with a landowner's economic-institutional needs. In the process, the silvicultural system

1 provides for timely and recurring availability of suitable resources;
2 insures predictable kinds of use opportunities at regular intervals, and well into the future;
3 addresses both biologic and economic concerns, and balances them to assure continued renewability of essential resources;
4 adds a long-term perspective to management planning, stand manipulation, and resource use; and
5 satisfies the minimum requirements for properly sequenced regeneration, tending, and harvest.

Providing a sustained and consistent access for the future is the silvicultural imperative, and the ethical responsibility of foresters and landowners alike.

COMPARISON OF EVEN- AND UNEVEN-AGED SYSTEMS

All silvicultural systems provide for timely regeneration, tending, and harvest in a stand. Yet a silvicultural system appropriate to any one community may not adequately serve the needs in another. Rather, a landowner's interests will substantially influence the long-term scheme of management. So will the species composition, the nature of the physical environment, and the potential pest and disease problems. Along with these and many other factors, the age arrangement in a stand most strongly determines the nature of the system, the treatments needed to bring a community to some desired condition in the most efficient manner, and the means for replenishing it at an appropriate future time (Muelder 1966; Roach 1974b).

Both cutting and natural disturbance may change the age arrangement within a tree community. They can also create some unique combination or interspersion of age classes (Fig. 2-4), and the effects may persist for long periods. Yet most forest stands have one of two basic age arrangements: (1) *even-aged,* where the trees growing within the stand have only small differences in their ages; and (2) *uneven-aged,* where the trees in a stand differ markedly in their ages. By convention, the spread of ages in an even-aged community does not differ by more than 20% of the intended rotation. *Rotation* means the planned number of years between the time a stand regenerates, and its final cutting at a specified stage of maturity (Ford-Robertson 1971). It includes sufficient time to establish a community of trees (regenerate it) and grow it to a desired condition of maturity under the intended program of management. So, if a forester planned to grow an even-aged community for a 100-year rotation, the maximum spread of ages in it would not exceed 20 years.

By contrast, uneven-aged stands contain trees with a wide range of ages, and have more than one age class. The term *age class* means a collection of trees that grow in a single stand, and that have about the same age—a cohort (after Ford-Robertson 1971). The trees in a single age class all became established following some management activity or natural disturbance. So an uneven-aged community has two or more cohorts interspersed within the stand. And by convention, the spread of ages will exceed 25% of the intended life span for any one of them. So, if a forester planned to grow individual trees or age classes to 100 years of age, the spread of ages in an uneven-aged stand would necessarily exceed 25 years.

Since an even-aged stand contains a single age class, foresters will tend it as a unit and apply a single treatment throughout (Fig. 2-5). During early stages when even-aged stands contain no mature trees, silviculturists do not want to establish a new age class. Instead, they use various kinds of tending treatments to nurture the development of the existing community until it reaches maturity. They harvest excess trees as part of the tending operation, and sell merchantable ones to generate revenues for a landowner. Eventually, they apply another reproduction method to replace the even-aged community with a new one. Since no more immature trees remain, they will not schedule another tending operation until the new age class develops to at least some minimal degree. Hence, in even-aged systems foresters may harvest trees to help tend an immature community, or harvest mature trees to regenerate a new one. However, they regenerate and tend an even-aged stand at separate times in its development, rather than concurrently.

Uneven-aged stands contain two or more age classes at different stages of maturity. For the most practical management, uneven-aged stands should have at least three age classes (Smith 1986). Among these communities, foresters periodically tend the trees of immature ages to foster their future growth and development, perhaps using different techniques for each separate age class (Fig. 2-4). They concurrently remove the mature age class to regenerate a new one across part of the stand area. In these combined operations, they harvest excess immature trees as well as the mature ones, and merchandise the logs and pulpwood to generate revenues. In all cases, regeneration, tending, and harvest take place concurrently, and with each scheduled management activity. Further, the time interval between subsequent treatments and the number of years it takes to grow a tree to maturity determine the number of age classes in a stand, and the difference in ages between them. Ideally, foresters return to treat each uneven-aged stand at regular time intervals (called a *cutting cycle*), thereby developing and maintaining a uniform spread of age classes within the community.

In some cases, foresters may create communities having only two ages of trees. These special uneven-aged stands require a hybrid approach that borrows from both even- and uneven-aged silviculture (see Chapter 23). Normally, silviculturists create a two-aged arrangement by removing at least one-half or more of the stocking from a middle-aged, even-aged community, and ideally halfway through a rotation. This leaves a low-density stand of widely spaced tall trees, and uses the cutting as a means to trigger the regeneration of a second age class underneath the older residuals. Foresters may return to the stand periodically to tend the youngest age class, without necessarily cutting in the low-density overstory. When the oldest age class reaches maturity, they remove all of it, and also reduce the younger age class (now middle-aged) to a low-density residual stand of widely spaced trees. Once more this procedure helps to regenerate a new age class underneath the residual overstory, reestablishing the two-aged structure. So, a two-aged silvicultural system has special treatments that regenerate and tend each age class separately, using methods similar to those for even-aged silviculture. However, the community will include a complement of older trees, with a much younger age class underneath it—and some aspects of regeneration and tending may converge in time as with uneven-aged silviculture.

(a)

(b)

(c)

FIGURE 2-4
Both cutting and natural disturbance may change the age arrangement within a tree community, creating some unique combination of age classes, as in these (a) even-aged, (b) two-aged, and (c) uneven-aged communities.

Historically, foresters have used the term *silvicultural system* to separate the scheme of management into only two broad classes based upon the age structure of the trees in a stand: namely even- and uneven-aged systems (Muelder 1966; Roach 1974b). Some consider two-aged silviculture a third broad system, because of its hybrid nature. Also, they have traditionally named a silvicultural system after the reproduction method used within it. This stresses the importance of timely regeneration in perpetuating a community, and conveys information about its character and the methods used to manage it. Thus, throughout most parts of the world foresters subdivide the even-aged systems into the clearcutting, shelterwood, or seed-tree systems. They call the systems for uneven-aged stands the single-tree or group se-

FIGURE 2-5
Since an even-aged stand contains a single age class, foresters will tend it as a unit and apply a single treatment throughout.

lection systems. The two-aged systems have become known as reserve shelterwood system and leave-tree system (Smith 1995).

ADJUSTING THE INTENSITY OF MANAGEMENT

Conceptually, not all silvicultural systems include identical kinds of tending and regeneration operations, even among stands with a similar age arrangement. Rather, differences may reflect site and species characteristics, or the financial and institutional constraints of a landowner. Generally, these factors limit the *intensity* of management, as reflected in the degree of effort expended (and money invested) in controlling the stand or age class establishment and development (Duerr 1960; Florence 1977; Smith 1986). In practical terms, the intensity of management reflects the landowner's financial objectives. It depends upon

1 the level of investments a landowner will make for noncommercial cultural treatments among immature age classes;
2 the frequency and character of treatments implemented via commercial timber harvesting operations;

3 the amounts of money a forester can spend to control the composition and character of regeneration; and

4 the relative importance assigned to taking immediate profits from a stand, compared with realizing long-term benefits at some sustained level.

For private enterprises that must balance financial costs and returns over the long run, the relative profitability of alternate practices governs the choice of silvicultural options and the intensity of these treatments. With public agencies, choices often reflect a judgment about how best to use available funds, and how to get the most social benefit from those operations.

Several factors—both ecologic-biologic and financial-institutional—affect the economic prospects for stand management. Important elements include

1 accessibility—whether a landowner can economically get to the stand to harvest or otherwise realize the yield or benefit;

2 markets—the kinds of products demanded, the quantities that consumers want, and the prices they will pay;

3 site quality—the amounts of different products and benefits the land can produce;

4 tree character and condition—the potential to supply available markets and command a reasonable price for the products, or support various kinds of noncommodity uses; and

5 the nature and requirements of an ownership—the financial requirements (the required alternate rate of return) and the kinds and minimum levels of values sought as a result of the management.

To this end, a silvicultural system must delineate the most efficient series of treatments to capitalize on the productive capacity of a site to the fullest extent, and for whatever purpose a landowner espouses (either commodity or nonmarket values). Also, it must insure the required minimum level of payoff to a landowner—sustained indefinitely—and consistent with the desired outcomes.

While most sites will generally support extensive silviculture, only the best promise adequate yields to pay off high-level investments. Further, to the degree that a landowner matches the silvicultural inputs with the capacity of a soil to produce, the more favorable the outcome in an economic sense. While this should hold true for nonmarket values as well as commodities, it has great importance to commercial forestry ventures. To illustrate, in some cases the productive capacity of a site limits uses to fairly low-level activities (Fig. 2-6, after Stone 1975), and management may primarily involve protection operations. These wild forests most readily provide for low-level recreational uses, watershed protection, and habitats for naturally occurring communities of plants and animals. On somewhat better sites, conditions may support timber management and higher levels of nonmarket activities, but repay only limited investments to control regeneration or early stages of age-class development and character. Landowners will take steps to insure adequate regeneration within these *exploited* or *opportunistic* forests, but accept whatever species naturally reproduce. That applies to the indigenous plants, and the

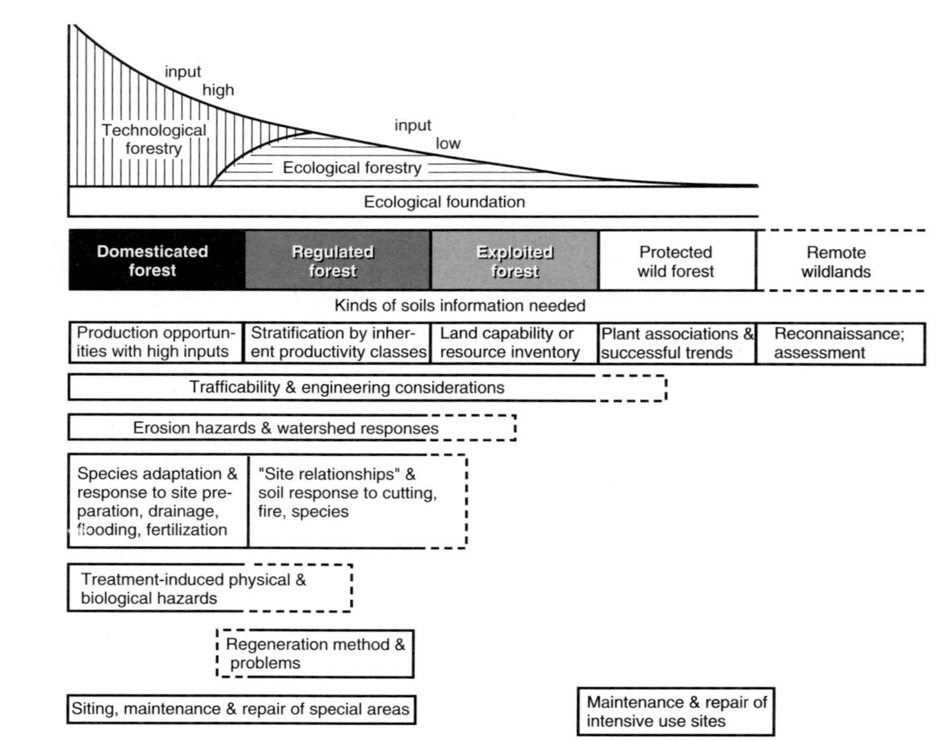

FIGURE 2-6
Conditions of the soil and other environmental factors that regulate productivity determine the intensity of forestry and many use opportunities that a site will support at a sustained level (after Stone 1975).

community of wild animals that lives in the forest. Among even better sites, silvicultural systems will include tending operations, controls over regeneration and stand stocking levels, and other moderately intensive practices that promise a reasonable payoff to a landowner. These *regulated* forests typify most cases where the economic objectives allow at least moderate investments, provided that the long-term outcome repays the costs at an acceptable rate of return. Here landowners might also include some deliberate measures to enhance the habitat for wildlife or special plants, and even try to increase the population of some target species. For fewer sites with the best growing conditions, landowners may make fairly high levels of investments and practice highly intensive silviculture to establish and maintain a *domesticated* forest. For these cases, the management activities will often have considerable technical complexity, include steps to introduce species not common to a stand, and make provisions to establish the trees in nonnatural arrangements (Stone 1975). Operations on these domesticated forests may resemble agricultural enterprises, and usually interest forest products companies that require particular kinds of raw materials to support their manufacturing facilities. They might also include intensively used recreational sites, and special measures to support plants and animals not indigenous to an area.

The concept portrayed in Fig. 2-6 has value in managing for nontimber values, including those that landowners cannot merchandize via leases and use fees. Depending upon the actual array of potential uses, soil factors may or may not limit the intensity of a noncommodity activity, or the approach to management. In other cases, it will. To illustrate, intensive recreational use would not prove ecologically acceptable at sites with severe soil or other important biophysical limitations. The quality and capacity for recreational use of these areas will often decline over time, forcing landowners to divert the activities to other sites and to carefully regulate the kinds and frequency of uses, and the numbers of people using an area. In such a case, they could not justify investments for intensive on-site developments. By contrast, places having better soils with few physical limitations will support reasonably high levels and more diverse kinds of sustained recreational activity. They will often justify investments for intensive site development. In some cases, landowners may even domesticate appropriate areas by constructing campsites or other intensive-use facilities, similar to creating artificial forests to maximize wood production on highly productive soils.

Several ecologic conditions other than edaphic ones might also limit activities—such as needing to protect special habitats for threatened or endangered plants or animals. Then only low-level uses of any kind might prove appropriate, despite the overall productive capacity of a soil. The horizontal axis on Fig. 2-6 could reflect a fuller array of ecologic factors (biological or physical) that logically influence the intensity and kinds of activities a site will support in perpetuity. The model would then serve as a useful planning tool for evaluating an array of alternate commodity and nonmarket uses, and in helping landowners to balance decisions about investments in management with the long-term capacity of a site to sustain the values of interest at an acceptable rate of return.

FINANCIAL MATURITY AND OTHER ECONOMIC CONSIDERATIONS

Across the full range of wild to domesticated forests, most forestry investments involve a long lapse of time before landowners can realize any appreciable tangible payoff (Osmaston 1968; Gregory 1987). These delays exceed periods common to other investments available in industrialized societies. Further, even when they want to generate revenue, landowners cannot always quickly liquidate the assets represented by their standing timber, due to factors like the following:

1 lack of easy access;
2 weather conditions that prevent logging;
3 disinterest by local buyers;
4 time required to prepare a sale and secure a contractor; and
5 similar operational, financial, or environmental factors that cause delays.

Landowners also have no certainty that existing markets will not change with reductions (or increases) in demand for the products they have available, or changes (up or down) in their market price. Further, value increases also depend upon the rates of tree and stand-level production, and many factors affecting growth remain outside the control of silviculture (e.g., soil quality, and effects of climate). Standing timber serves both as a product to sell and the biological machinery to grow more of it. Landowners who wish to continue producing at steady rates over long periods must keep some minimal portion of their trees as a base for future production. This limits the amount that they can merchandise at times of peak prices. To maintain a cash flow to pay costs of operations, landowners must cut timber in times of low prices as well. Further, uncontrollable natural factors like windstorms and wild fire could badly damage or kill a crop before it becomes merchantable or reaches its best market price. Therefore, silvicultural investments carry considerable uncertainty, including the risk of premature loss or a decline in value.

Despite these many factors that add to uncertainty, research shows that historic price data provide one valid index to the potential of forestry investments. Analysts look both at the trends and the periodic fluctuations (Baumgartner and Hyldahl 1991). In fact, the history of prices for red oak and sugar maple sawlogs illustrates how investors might balance those two conditions in making decisions (Kingsley and DeBald 1987; Nolley 1991). In the case of red oak, prices for high-quality logs remained low to modest from the 1950s into the 1970s in northeastern North America (Fig. 2-7). Then after the mid-1970s, red oak showed cyclic short-term rises and declines in price, but with a substantial long-term increase in overall value. In retrospect, this made early investments in management quite profitable, and would encourage continued interest in red oak. During the same period, sugar maple prices started above those for red oak in many regions, and increased steadily through about 1980. Prices then leveled off for several years, dropped between 1984 and 1992, and finally began to increase again. These data give no clue that the late improvements in price will persist. Rather, they suggest considerable long-term uncertainty about the payoff from sugar maple management.

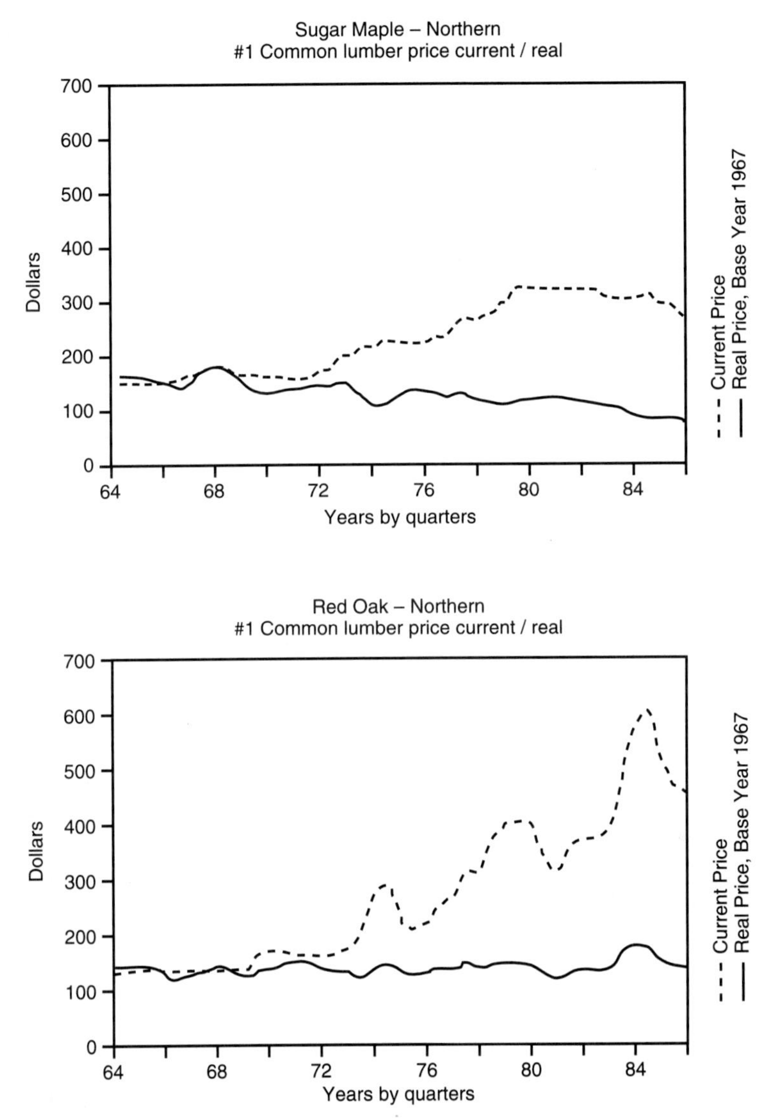

FIGURE 2-7
Change in prices for red oak and sugar maple lumber from the 1960s through the mid-1980s (from Kingsley and DeBald 1987).

Nonindustrial landowners with young even-aged stands deal with an additional degree of uncertainty in these regards. They must wait extra long times for the trees to grow to useful sizes for common commodity markets, and they must make guesses about forest health and log prices much further into the future. Yet during this interim, young even-aged stands often have considerable value as

1 places for recreation;
2 areas of great beauty;
3 sources of high-quality water;
4 critical habitats for an array of plants and animals; and
5 protective cover for soil and other physical site resources.

Even so, such intangible values prove difficult or impossible to assess in traditional monetary terms. That confounds attempts to articulate their real benefits to a landowner, and to demonstrate their contributions in repaying investments in stand maintenance and regeneration (Osmaston 1968; Randall and Peterson 1984).

In addition to dealing with uncertainty, individuals or firms who enter forestry as a part- or full-time commercial venture usually want to know the best way to invest their capital among the alternatives afforded by their forests. That holds true for public agencies as well, even though they have no profit motive. In a totally ruthless financial environment, most investors would most likely opt for whatever opportunity promised the highest possible payoff in the shortest time. In other cases, different kinds of institutional demands dictate a more moderate philosophy, and landowners will simply require some minimum payoff from the management. In either case, foresters usually have a requisite rate of return to realize, and any investments must repay the costs compounded at that interest rate. Under these circumstances, nonmarket benefits have only an indirect bearing on decision making. They would most likely influence choices between opportunities promising equal rates of return, or at least offering some minimum required payback.

Primarily, business decisions focus on how to generate revenues to recoup investments and provide the required interest for profit. For a forestry enterprise, investors often assume that they will

1 find forestry profitable;
2 want to maximize the present net worth of their woodlands;
3 use the most financially rational management alternative from among the ecologically reasonable options; and
4 apply a single guiding interest rate when evaluating investments and reinvestments to maintain the enterprise (after Davis and Johnson 1987).

Thus, when deciding upon the silvicultural system for any stand, foresters will consider only those treatments that appear immediately profitable (generate more revenues than they cost), and that promise at least the minimum required rate of return on any investment (costs in excess of immediate income). In this context, *alternate rate of return* means the interest rate a forest owner expects to earn from the capital invested, and the rate of value increase an individual tree or stand must produce by growth (Duerr et al. 1956). It should reflect the likely payback from alternative investments actually available to a firm, that have an acceptable risk, and that have a payoff period comparable to the forestry activity under consideration (Duerr 1960, 1988).

Landowners who enter forestry as a business need a steady flow of income to pay costs of purchasing equipment and forest land, paying salaries and wages, maintaining facilities, paying taxes, and securing other resources essential to the enterprise. An entrepreneur will also require a profit as compensation for the risk of investing. Foresters call this balance of costs and returns the *financial yield*—the compound interest realized over a period, taking into account all items of expenditure (direct and capital) and income (Ford-Robertson 1971).

Generally, the term *capital* means the accumulated value of money and property owned or used by an individual or corporation. In forestry, this includes the *forest capital*—the value of land, the trees and other resources, and any improvements made by an enterprise, or in an estate. Ideally, the program of management for each stand will enhance the overall worth of the forest capital, and the financial yield from investments made.

The decision about when to consider an age class *mature* (or profitable) ranks among the most important economic choices that silviculturists must make. It determines when a landowner can no longer grow the trees profitably. Foresters call this condition *financial maturity* (after Duerr et al. 1956; Duerr 1960; Mills et al. 1981):

- the time when an age class or stand no longer increases in value at the required rate of interest
- the point when current value growth would drop below the alternate rate of return
- the stage when costs of keeping a tree or even-aged stand exceed the expected gain from future value growth

When using the concept of financial maturity in this way, foresters assume that they can access the best-paying local markets, and that landowners will sell the trees at an appropriate time.

As part of their financial assessments, landowners often forecast future income by periodically determining the amount of salable timber on the land, and estimating its probable worth at the end of an investment period (e.g., a cutting cycle). They recognize that trees change in value in three principal ways (Mills et al. 1981): (1) growth in size and volume; (2) changes in quality, and the effect on market value; and (3) inflation or decline in market prices. As a practical measure, foresters usually assume that prices will increase or at least remain stable, and that quality will not decline. They use simulation techniques, data derived from permanent sample plots, or information amassed by periodic reinventory to project the growth potential for specific stands, and to determine the time of financial maturity.

These financial maturity determinations differ for even- and uneven-aged stands in at least one important respect. In the former case, it assesses the composite increase in value for an entire stand, and not for individual trees in the community. With uneven-aged stands, it evaluates the rate at which individual trees of the oldest age class increase in value, and does not consider standwide changes in value per se. Otherwise all financial maturity assessments generally assume that

1 as individual trees grow larger, they will contain more useable volume, as does an entire stand that fully utilizes the site resources;

2 trees continue to add volume as long as they remain alive, and become more valuable unless something damages the wood;

3 the tallest trees of the largest diameters and best-quality trunks have the highest commercial value, as do stands with the greatest numbers of high-value trees;

4 the larger and more valuable a tree or stand becomes, the more rapidly it must increase in absolute value to continue yielding some minimum required compound rate of return; and

5 the rate at which value increases depends upon how rapidly the trees increase in size, quality, and volume.

In general, landowners want to know if the volume and value will accumulate at an acceptable rate, or if they would earn more by harvesting the trees and investing the cash value in another way. A financial maturity analysis provides this information.

Figure 2-8 illustrates the concept of financial maturity, using information for Douglas-fir growing on a good site in western Oregon (after Duerr 1960, based upon McArdle et al. 1961). The value growth curve portrays changes in the compound rate of return over a period of time up to 120 years. The horizontal line depicts the guiding or alternate rate of return (ARR), set at 6% for this example. An even-aged stand reaches financial maturity when the composite value growth drops below this line. For this case, a firm would terminate the rotation at no sooner than

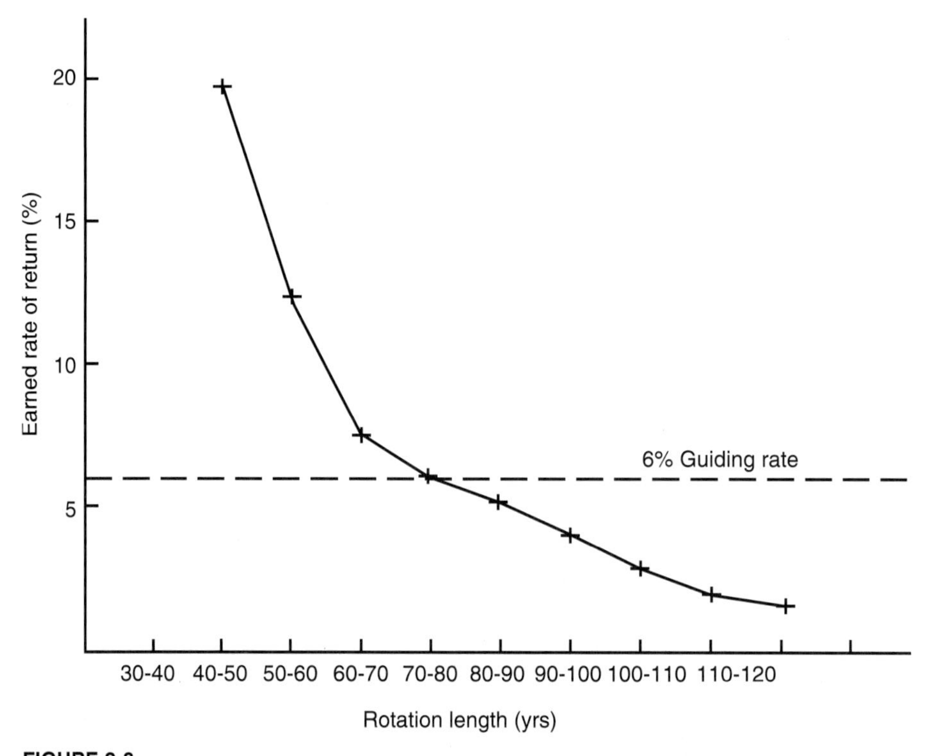

FIGURE 2-8
An example of financial maturity in Douglas-fir growing on a good site in western Oregon (after Duerr 1960, based upon McArdle et al. 1961).

70 years of age, and not after 80 years. Lowering the guiding rate to 5% would extend the rotation to 80–90 years. Increasing it to 7% would shorten the rotation to 60–70 years. Graphs representing the value growth for individual trees in uneven-aged stands would look much like that shown in Fig. 2-8, but based entirely on expected changes of individual trees in the mature age class.

Sometimes, institutional demands and other kinds of financial considerations force landowners to harvest trees before financial maturity, or later. Factors like unusual market conditions, needs to supplement cash flow, and wood supply requirements of a mill change the outlook. In essence, these push the alternate rate of return up or down, at least in the short run. Yet under normal circumstances, foresters use financial maturity assessments for long-term forecasting, and as a fairly straightforward means to determine when to regenerate an age class (as in Fig. 2-8).

JUDGING NONMARKET VALUES

Foresters assign *values* to things in order to make decisions about how to allocate funds and other resources to optimize the outcomes of interest. In a broad sense, putting values on different options helps in (1) deciding whether to alter the quality, quantity, and kinds of opportunities available; and (2) measuring the payoff from opportunities used, taken, invested in, or relinquished. In the absence of some assigned or marketable worth, landowners and foresters alike have no reasonable way to judge the effectiveness of a management scheme, or to determine what future investments might yield different benefits (Canham 1986; Kaiser et al. 1988).

For traditional commodities, foresters use measures like *stumpage value* (the price offered or paid for standing or uncut timber and other roundwood products) and *user fees* (the payment for recreational and other on-site uses). They use local prices, and only include products that existing or anticipated markets will absorb. Common examples of nontraditional sales include lease arrangements for exclusive rights to hunt and fish on designated parts of a commercial forest property, or entrance fees for campsites and picnic areas. This allows foresters to assess the worth of some nontraditional benefits that landowners might merchandise in the open market, and in the amounts that people actually demand.

Many amenity values defy simple qualification. Their worth depends upon subjective factors that many people can only describe in terms of feelings and emotional benefit. Foresters cannot often translate this worth into tangible measures that have a marketable value. Instead, they can only rank the alternatives by user preference, without defining how much people would pay for different kinds of experiences, or gain by having them. This has encouraged some landowners to use a surrogate like the numbers of people visiting a site, and to presume some correlation between the frequency of use and the intangible worth. Yet these approaches only rank alternatives on a relative scale, unless landowners set some monetary value on each person's visit (Kaiser et al. 1988). On the other hand, people can often quite readily tell when some management activity destroyed or lessened an intangible value, or what areas do not provide quality experiences of some inexplicable kind. Then foresters can attempt to quantify the structural differences and visual attributes of stands before and after a silvicultural treatment, or between ones offering or lacking some desired nonmarket value. Over time, accumulated empirical evidence of this kind will indicate how different silvicultural systems might enhance amenity values, or reduce them under some circumstances.

By contrast to even these approaches, foresters and forest scientists also cannot readily assign a monetary value to the importance of maintaining healthy and sustainable forest ecosystems across a property, landscape, or region. They cannot articulate the ecological value of having those forests, or the costs of losing them. Yet most people would agree to the importance of maintaining sustainable balances among the living and abiotic components of a forest system, and insuring noncommodity benefits that people derive from forested lands. Also, they would oppose uses that irreversibly alter the physical environment, the plant community, or the capacity for timely replacement of existing biota at appropriate times.

In reality, resource managers can use silvicultural techniques only to create a setting wherein people derive some special kind of in-place value garnered from the indigenous physical environment, the existing vegetation community, and the associated fauna. In this context, forests and the stands in them become *greenspace*—lands and associated vegetation communities that provide benefits of no direct commodity value (Richards 1975). Their worth comes from the way people use a forest or stand for different kinds of in-place activities, and the monies they spend (called *acquisition expenditures,* or *consumer costs*) for services to support their recreation (Canham 1986; Kaiser et al 1988; Randall and Peterson 1984). In this respect, the worth of nonmarket benefits may parallel the concept of *value added* in manufacturing, where a firm transforms raw material to a consumer good of higher value. At best, approaches like these may fix the worth only in *relative* terms, and indicate only a priority for allocating funds among the different alternatives for management and use.

3

HARVESTING AS A TOOL OF SILVICULTURE

TIMBER HARVESTING AND SILVICULTURE

People within the forestry community often view timber harvesting from two different perspectives. For those who own or manage forested lands, it serves primarily as a mechanism for implementing the requisite silvicultural treatments, and for recovering the yield (and revenues) from successful silviculture. By contrast, wood products firms see timber harvesting as the beginning of a process for converting trees to consumer goods. It must bring in a steady and adequate supply of desirable logs and other roundwood pieces, and deliver them to a mill at competitive prices. To serve both interests, timber harvesting must

1 remove excess and mature trees as a way to tend immature age classes and regenerate mature ones, and to sustain the vegetation conditions of interest over the long run;
2 protect the residual trees, advance regeneration, and other vegetation, as appropriate to the long-term silvicultural plan;
3 protect against damage to the soil, drainage courses, and other features of the physical environment;
4 provide the contractor with sufficient volume of desirable material to make the operation profitable;
5 allow crews to operate efficiently; and
6 leave conditions that facilitate future uses of a stand and its environs—including those for an array of nonmarket values.

Timber harvesting links the silvicultural objective of sustaining a forest's commodity and nonmarket values, and the industrial purpose of harvesting trees to produce goods for human consumption. Under the best of circumstances, it serves both sets of goals, and both landowners and contractors profit in the process (Conway 1982).

From a silviculturist's perspective, timber harvesting provides a no-cost option for implementing a variety of treatments that change the character of a vegetation community, enhance some alternative uses, create some special conditions, or insure the continuous production of desirable commodities. Each time they schedule a stand for treatment, foresters must also directly address the needs of a harvesting contractor in several ways. These may include

1 delaying reentry until a stand regrows sufficient volume to sustain another commercial harvest;
2 cutting enough volume per unit area for a minimum operable cut throughout the stand;
3 treating sufficient area to insure the minimum total volume for cost-effective setup and operation;
4 providing efficient entry into, through, and out of a stand; and
5 requiring no unnecessary work nor unreasonable expenditures.

Foresters normally allow some flexibility in the way contractors organize the logging, and when they start the work. Yet they reserve the right to shut down an operation that does not comply with terms of a contract. They may also limit the sizes and kinds of equipment used by a contractor, and not permit logging during certain seasons to protect the residual trees or the site. They normally compensate a contractor for any extra (unprofitable) silvicultural work required by the contract (e.g., felling unmerchantable trees), and require payment for the timber before any cutting begins, or before any logs leave a site. Further, the plan of operations must include measures to protect soil stability and the integrity of all natural drainages, and to leave the roads and trails in a safe condition and ready for future use. In fact, the timber sales contract should list all of the necessary conditions and requirements to guarantee a successful operation. Then both the contractor and landowner know in advance what to expect, and how each will comply with terms of the agreement.

THE NATURE OF TIMBER HARVESTING

Timber harvesting involves a fairly complex series of work steps to insure the efficient handling and transport of heavy and bulky products, and to bring those materials to a mill or some other primary processing plant at a low cost. To succeed, contractors must use techniques that prove (1) efficient to implement with the machinery and personnel available; (2) flexible to conditions associated with the time of year and site attributes; and (3) economical to apply, with adequate payback for the costs incurred. They hope to accomplish these tasks safely and with the least possible wear and tear on their equipment. They also want to maintain a good working relationship with the landowner, and with the forester who helps to manage the lands scheduled for harvest (Conway 1982).

Foresters often use the terms *harvesting* and *logging* interchangeably. Both mean removing some kind of product from the forest (Ford-Robertson 1971). In a strict silvicultural sense, a *harvest cutting* means felling the *final* crop, either in a single operation or a series of regeneration cuttings by removing financially or physically mature trees (Ford-Robertson 1971). By common usage, it may include any intermediate cutting that extracts salable trees from an area (after Wackerman et al. 1966; Ford-Robertson 1971; Conway 1982; Stenzel et al. 1985). Standard definitions also cover several specific components of logging operations, as follows:

Extraction—removing forest produce from its place of growth to some delivery point

Skidding—hauling logs by sliding them on the ground from stump to roadside

Yarding—hauling logs to a landing or collection point, traditionally by pulling them using cable systems rather than behind tractors

Forwarding—moving logs from stump to a landing for further transport, typically by lifting the pieces off the ground and carrying them to a point of deposit

Felling—cutting the trees

Delimbing—cutting off branches and limbs

Topping—cutting off the upper stem and side branches that comprise the crown from the merchantable portion (main trunk) of a tree

Bucking—cutting a tree trunk or stem into shorter logs or bolts

Contractors must complete all these tasks efficiently. They have little direct relation to the silvicultural objectives for a stand, but foresters must take measures to insure that the logging fulfills the short- and long-term management objectives as intended.

SILVICULTURAL REQUIREMENTS OF LOGGING

Silviculture strives to capture the full productive capacity of a site, to safeguard its integrity, and to enhance growing conditions where economically feasible. Foresters do this by creating and maintaining stands of an appropriate character and density, and replacing each age class at the time of its maturity. They work to maintain the inherent site productivity by protecting the *physical environment,* and primarily the soil. To forward these goals, timber harvesting and all other uses must insure that

1 trees of desirable species continue to occupy a site over the long run;

2 soils remain free of accelerated erosion, and suitable as rooting media and a source of nutrients and moisture; and

3 natural drainage systems and other natural landforms remain intact and free of mass soil movement (Hough, Stansbury & Assoc. Ltd. 1973).

Responsible use will preclude an irreversible ecologic change at a site, and insure its long-term stability and resiliency as part of the broader ecosystem. Treatments may change the species composition and stand character for as long as silviculture controls the vegetation community via appropriate treatments. However, if silviculturists relax that control, natural processes eventually move a community back to the ecologic norm for the physical environment and area.

EFFECTS OF TIMBER HARVESTING

Harvesting timber from forested areas has both direct and indirect environmental effects. Those directly associated with logging include

1 cutting or destroying all or a portion of the trees and other vegetation on an area, and adding large amounts of logging debris or slash (branches and unused portions of the trunks) to the surface;
2 removing the organic surface layer from portions of the treatment area, exposing the underlying mineral soil; and
3 disturbing (churning, mixing, compacting, and displacing) the mineral soil during skidding, forwarding, and yarding—and when building roads and landings.

Research has shown that tree cutting, in itself, will not increase the potential for surface erosion off a site (Lull 1965; Lull and Reinhart 1972; Armson 1977; McClurkin et al. 1987; Kochenderfer and Edwards 1991). Complete tree removal does have some indirect effects. It will lessen the site protection afforded by foliage and roots to some degree, and significantly reduce interception by the crowns and subsequent transpirational use of soil moisture. All the incoming precipitation then falls to the surface after complete tree removal, and more infiltrates the surface. The soils also remain saturated for a longer period following heavy storms. As a result, they might more readily slump and move on steep and unstable slopes, particularly during earthquakes (Chamberlin 1982) and as the roots die and decompose. At less severely affected sites, tree cutting primarily increases soil moisture and subsurface flow, but does not heighten the danger of mass soil movement.

If the cutting removes most of the crown cover from a fairly large area, subsurface flow will increase following timber harvesting. The degree and duration of this indirect effect depends upon many factors, including

1 the size of a cutting area and amount of tree cover removed;
2 the rates of revegetation, and subsequent increases in interception and evapotranspiration;
3 the amount and distribution of yearly precipitation, the periodicity of heavy storms, and rates of early season snowmelt at high latitudes;
4 the soil texture, porosity, depth, and storage capacity; and
5 the topography, and rates of internal soil drainage from a site.

Research in eastern North America showed that cutting all the trees from small forested watersheds increased downstream water yields substantially, though the effect declined within five years as a result of natural revegetation of the area (Lull and Reinhart 1967, 1972). In the northern Rocky Mountains of North America (Montana), removing all the trees from 20% to 33% of several small conifer-covered watersheds increased annual stream flow by an average of 69% over a four-year period (King 1989). Following similar treatments in Colorado (Fig. 3-1), stream flow rose by 40% over a thirty-year period (Troendle and King 1985).

If timber removal appreciably increases the volume of water leaving a site and flowing through downstream channels, it may have another indirect effect of accelerating undercutting of the banks, putting more suspended particles into the water,

FIGURE 3-1
Research in the northern Rocky Mountains of North America has shown that removing all the trees from 20% to 33% of small conifer-covered watersheds will increase annual stream flow appreciably for an extended period of years (from Troendle and King 1985; *courtesy of U.S. Forest Service*).

and increasing siltation into pools and at slow-moving stretches along the channel. The degree depends upon the topography and the velocity and volume of water flow, with greatest effects in rapidly running and high-volume streams. In fact, for one steep-gradient stream at a high-elevation site in Arizona, the cross-sectional area of active streamflow increased by 10% over nine years following timber harvesting, compared with 2.5% along another stream downslope from an uncut area. Natural stream bed structures (logs and gravel bars) also decreased by 45% below the harvested area, but increased by 23% downstream from the uncut forest. Both effects gradually stabilized over time after cutting (Heede 1991).

Other effects of logging accrue primarily to the site as a result of activities that disturb the surface and the soil. Tree cutting, by itself, will not have these effects. Nor will the logging slash left during delimbing and topping, except if dropped into a stream channel. Then it may dam or divert the water flow, often causing unnatural digging of the bottom or undercutting of the banks. This changes the course of a stream channel, and may increase siltation downstream. Within a logged area, site disturbance occurs largely in two ways: (1) skidding, yarding, and forwarding removes the protective organic layer along a trail, may cut deep ruts into the surface, and may displace the soil by mechanically plowing or pushing it into adjacent areas; and (2) the machines compact soil along the trail system, increasing bulk density and decreasing total pore space underneath. These disturbances potentially expose the mineral soil and reduce the rates of infiltration and percolation, increasing the chance for erosion. Also, severe disturbance cuts the roots of residual vegetation, and leaves ruts and small (relatively shallow and narrow) gullies in the surface. This could damage the health of any affected trees, channelize the water by creating new drainage paths, and make the soil more susceptible to slumping on the steeper slopes.

Several studies of logging illustrate some of these effects. They also suggest ways that careful job layout and different work practices may contain the extent and severity of disturbance. For example, among ten steep sites in the Appalachian Mountains of eastern North America, conventional ground skidding left only 3% to 10% of the surface in skid trails, but disturbed up to 20% to 30% of the total area. Altogether, logging severely disturbed 6% to 14% of the surface, primarily by constructing the trails, roads, and landings. Even so, most of the risk occurred at a relatively few vulnerable spots, and contractors could mitigate postlogging soil movement by simple measures to control surface water flow across the exposed soil (Stuart and Carr 1991). On granitic soils in the Rocky Mountains of central Idaho, skidding with crawler and rubber-tired tractors disturbed 29% and 34%, respectively, of the total area within a harvested site. Bulk density increased by 1.3 times on main skid trails, and 1.2 times on secondary trails with both methods of skidding (Clayton 1990). On medium-texture soils at upland sites in New York, tree-length skidding with rubber-tired tractors removed or greatly disturbed the surface litter across about 8% of the area in partially cut stands, and 15% to 20% in strip and patch clearcuts. Multiple trips increased bulk density by 7% to 15%. Greatest compaction occurred at the most severely disturbed areas closest to the landings. On average, macroporosity under the trails decreased by 46% (Nyland et al. 1976).

On saturated bottomland soils in Alabama, bulk density actually decreased by 15% where the skidders had high-flotation tires, but increased by 1.3 times along trails used by machines with more conventional tires. On average, bulk density increased only slightly with one skidding pass, by 1.1 times with three passes, and by 1.2 times after seven passes. Total pore space decreased in proportion to amount of traffic over a trail section (Aust et al. 1993). Along the southeast coastal plain of North America, most of the compaction also occurred during the first three passes. There bulk density increased by 1.2 times with three trips of a rubber-tired tractor, and 1.3 times after nine passes (McKee et al. 1985). In Mississippi, the bulk density of loam soil also increased with each skidder pass, and with most change occurring in the first three trips. It reached a maximum with six passes, increasing by 22% and 32%, respectively, under conditions of approximately 3% and 14% to 16% soil moisture content. Total porosity also decreased with increased trafficking, mostly because of reduced macroporosity (Guo and Karr 1989). In loessial soil, bulk density increased by about 23% as a consequence of skidding during wet conditions, but by only 18% in dry periods. Rutting also increased during wet periods, and the greatest surface displacement (rutting and soil displacement) occurred along primary trails at middle positions of steep slopes (Rachal and Karr 1989). Similarly, tests within the coastal plain of southeastern Georgia showed that while skidding did little measurable rutting or churning in soils at 18% moisture content, the effects increased significantly following more than a single pass when moisture content reached 30%. Skidder performance (efficiency) also decreased under these wetter conditions (Burger et al. 1989).

Such observations highlight the importance of limiting operations to periods with relatively low soil moisture, and concentrating the skidding into a few well-placed trails. They suggest that single passes with a load of logs will have only nominal effects, and that operators reduce the total degree of compaction across a site by making multiple passes over the main trails only. Findings also indicate that removing the surface litter and compacting the soil reduces infiltration rates and increases the surface flow. This could increase the potential for erosion along the trail surface. Other studies also indicate that certain kinds of postlogging site treatments to enhance chances for regeneration or reduce fire hazards may also expose the mineral soil over an extensive contiguous area, increasing the erosion potential even in the space between the skid trails (Wood et al. 1987; Blackburn et al. 1987).

The importance of all these effects depends upon (1) the extent of disturbance and compaction; (2) the topographic conditions across a site; (3) whether preventive measures keep the loss within environmentally acceptable standards; and (4) how rapidly new vegetation occupies the exposed sites and stabilizes the surface. On the positive side, exposed mineral soil and mixed humus may serve as an excellent seedbed for many herbs, given the greater understory brightness after cutting all or some of the overstory trees. In fact, on one wetland site in southeastern North America, timber harvesting, coupled with ground skidding, increased the herbaceous biomass from 2 to 196 g/m^2 over a one-year period, and herb species diversity increased by 1.2 times (Pavel and Kellison 1993). These plants help to stabilize the exposed surfaces, and may also serve as an important food source for wildlife using the area. The extent of colonization varies with local conditions.

BEST MANAGEMENT PRACTICES TO PREVENT ENVIRONMENTAL POLLUTION

Clearly, foresters and harvesting contractors must exercise particular care in planning timber harvesting operations for slopes, or where the machines might cut into or displace the soil along trails on a hillside. The ruts channel water as it flows downhill. This increases the velocity and force of flow across an exposed surface, or against unstable road banks and fills. Skidding, forwarding, and yarding on wet soils may also cut roots of the trailside residual vegetation. This reduces the stability of residual trees, and increases the chances of blowdown that turns over and exposes additional soil. The mounds gradually become flattened by erosion during rainstorms. Extreme (deep and extensive) soil disturbance over large areas, including improperly designed cut banks along roads, might even trigger localized (onsite) landslides in some cases (Chamberlin 1982). The degree varies widely with (1) topography, (2) soil conditions, (3) proximity to flowing and ponded waters, and (4) the intensity and methods of logging (Froehlich and McNabb 1984; Tesch and Mann 1991). To mitigate these potential effects, contractors can grade severely disturbed places, smooth rutted surfaces, or divert surface water off the trails and away from exposed banks along cuts and fills.

While logging may remove the protective organic layer and compact the soil, it will not necessarily accelerate the rates of natural erosion substantially. The risk actually changes little on flat or gentle slopes, or even on gentle-to-moderate slopes with well-drained soils and little surface disturbance. Some erosion will occur continually within all terrestrial ecosystems, even when people do not venture into them. In fact, any time that water flows across an exposed soil it will detach particles, eroding the surface. Erosion also happens continually as a natural ecologic phenomenon as water flows through streams and rivers, undercutting the banks (particularly at turns in the course), and loosening materials from the sides and bottoms. The potential for this natural base level of soil movement increases with the steepness of a slope, and the rate at which water flows down or across a pitched surface. However, on sites with a continuous surface organic layer, most of the water infiltrates the soil and moves from a site as subsurface flow (Kittredge 1948; Cleary 1967; Nyland et al. 1983).

In addition to its potential effect on surface erosion, careless logging and road building might artificially impound water in the depressions and water channels. They could also inadvertently drain a soil by creating deep ditches and other artificial channels that speed water movement from a site, and might disturb the banks and bottoms of streams and channels (Brown 1973; Chamberlin 1982). In fact, state and federal laws throughout the world regulate many kinds of land use practices to prevent these kinds of damages, and to protect the purity of ponded and flowing waters from other human disturbance (Siegel and Cubbage 1990; Siegel and Haines 1990). Generally, the laws aim to prevent pollution, and to maintain water at its naturally occurring quality.

In the United States, government regulations classify the origin of a pollutant as either a *point* or a *nonpoint* source. Point sources have a specific and easily identified place of origin like a pipe or ditch, and the pollution emanates from that distinct source. Nonpoint pollution has no easily identified single point of origin. It comes from large areas as a result of activities like silvicultural treatments, agriculture practices, mining, and construction (Siegel and Cubbage 1990, Siegel and Haines 1990). These can trigger pollution of different kinds. Important potential problems (U.S. Environmental Protection Agency 1973) include

- mineral sedimentation—due to nutrient runoff from fertilizers or decomposition byproducts from sites
- addition of solid materials—as would occur with erosion of soil and organic matter from the logged area
- debris—in the form of logging slash and trash from trails and landings
- foreign chemicals—from dumping and leakage of petroleum byproducts and other chemical materials
- changes in water temperature—due to exposure of streams or the adding of heated or cooled substances

All might follow forestry and silvicultural practices, including timber harvesting and road construction and maintenance. Federal and state laws also regulate the application of herbicides and pesticides in preparing sites for regeneration, or as part of forest protection operations.

In the United States, landowners and timber harvesting contractors must prevent unnatural changes to water quality by utilizing procedures known as *best management practices,* or *BMPs.* Section 208 of the Federal Water Pollution Control Act (1972) defines these as "optimal operating methods and practices for preventing or reducing water pollution, and protecting wetlands." They include a variety of voluntary measures for controlling potential sources of nonpoint pollution. Section 404 of the law also controls dredge-and-fill operations in navigable waters and adjacent wetlands, and activities that might drain them, even by rather limited grading of the surface soils during road building and other timber harvesting activities. In addition, Section 319 of the Federal Clean Water Act requires all states to develop water quality management plans and pollution control mechanisms appropriate to their local conditions. The states may establish more stringent rules and compliance standards, but not weaken the federal regulations (Siegel and Cubbage 1990; Siegel and Haines 1990). BMPs resulting from these programs mostly include fairly simple measures like those illustrated in Table 3-1.

In many respects, foresters circumvent most potential erosion by (1) carefully

TABLE 3-1
COMMON BEST MANAGEMENT PRACTICES FOR MINIMIZING SOIL EROSION AND PROTECTING WATER QUALITY DURING TIMBER HARVESTING

	When and where appropriate				
Best management practice	Landings	Skid trails	Truck roads	Stream and wetland crossings	When harvesting by water
Install water bars and culverts		x	x	x	
Leave tree-covered buffer strip	x	x	x		x
Plan the design and location in advance	x	x	x	x	x
Seed with stabilizing plants after use	x	x	(x)		
Control season of use	x	x		x	
Control slash and debris	x			x	x
Undertake periodic maintenance	x	x	x		
Retire after use	x	x	(x)	x	

(after Kochenderfer 1970; Lantz 1971; Brown 1973; Hartung and Kress 1977; Megahan 1977; Haussman and Pruett 1978; NYS Dep. Environ. Conserv. 1981, 1993; Chamberlin 1982) locating and using the roads, skid trails, and landings; (2) properly maintaining them during use; (3) minimizing soil disturbance, especially with multiple passes; and (4) applying appropriate retirement practices following logging (Kochenderfer 1970; Haussman and Pruett 1973; NY Soc. Am. For. 1975; Stuart and Carr 1991; Lynch and Corbett 1991; Jones 1993).

Relevant BMPs include measures like the following:

1 locating skid trails and roads to fit the terrain (minimize cut and fill operations, and insure good drainage), and bypassing areas with thin and poorly drained soil;

2 locating landings on a gentle slope to insure good surface drainage, but isolating them from streams and ponds by at least 200 ft (61 m) to prevent direct runoff into the water;

3 avoiding long, steep grades with a high potential for rapid surface runoff and erosion;

4 avoiding long level sections with no grade to drain the trail or road surface naturally, and that might have poorly drained soil underneath;

5 filling unavoidable perennial wet spots along the trails with a suitable natural material, or surfacing the trails with fibrous mats or a temporary covering of cull logs and logging slash;

6 installing ditches and cross drainages (water bars and low-based dips) at strategic locations along slopes greater than 5%;

7 allowing filled areas to settle before use, or compacting them artificially to insure a firm and stable base;

8 minimizing the numbers of stream crossings, and using temporary culverts or bridging devices to protect the banks and bottoms;

9 prohibiting logging machinery from streamside buffer zones, except at approved crossing points;

10 maintaining shading along streams and ponds by permitting no cutting or only light partial cutting within a distance equivalent to at least one tree height;

11 requiring directional felling away from streams and ponds, or pulling logging slash back into the buffer zone and away from the banks and water;

12 shutting down skidding and truck traffic when soils become saturated, and will not support the machines;

13 grading all disturbed trails and roads after use, and repairing water bars and dips as needed; and

14 sowing grasses and other perennial herbs on trails and landings where natural colonization will not occur rapidly.

Most states and provinces have regulations to minimize disturbance of the banks and bottoms, and also require permits for stream crossings. Under some conditions, contractors must use temporary bridging devices or culverts, then remove these and stabilize the crossing area following use. For surface structures on trails, the spacing depends upon slope steepness, with water bars, low dips, and culverts placed at the following intervals (after Haussman and Pruett 1973):

Percent slope	Spacing
2–5%	300–500 ft (91–152 m)
6–10%	200–300 ft (61–90 m)
11–15%	100–200 ft (30–60 m)
16–20%	<100 ft (<30 m)

Generally, this will adequately control surface flow and prevent needless disturbance. In regions with freezing temperatures, foresters can schedule the logging of fragile or erosion-prone sites for winter season when the soil becomes hard and snow blankets the surface. This reduces the chance for surface disturbance, rutting, or compaction. In general, these and other simple BMPs leave nothing to chance. However, foresters must insure timely use of the requisite control measures, and completion of the postlogging stabilization work before a contractor leaves a site.

Uncut or lightly cut buffer strips along streams, ponds, and wetlands help measurably in protecting water resources (Brown 1973; Chamberlin 1982; Kochenderfer and Edwards 1991). Common buffers cover 50–100 ft (15–30 m), depending on the slope. Withholding even this narrow strip of timber from sale (e.g., one times mature tree height) minimizes stream bed and bank disturbance, and circumvents potentials for accelerated surface erosion close to a stream channel or other water body. It also reduces the potential for compaction and removal of the organic surface layer, insuring continued high rates of infiltration. The irregular texture of surface litter in the streamside buffer zones also traps sediments that might wash from nearby roads and trails. In addition, trees in the buffer strips shade the water and prevent thermal pollution by direct sunlight, and deposit insects and detritus necessary for much aquatic life. Landowners would consider the value of unsold timber within these streamside management zones as a routine cost for the necessary environmental protection.

Carelessly constructed truck roads can become point sources for erosion and sedimentation. The general design, kind of surfacing, and treatment of cut banks and fill slopes all affect the potential. To illustrate, in the southern Appalachians of eastern North America, road surfaces of bare soil surface can lose 44 tons/ac of material/ac (93.6 mg/ha) annually, compared with 5 tons/ac (11.2 mg/ha) if surfaced with 3 inches (7–8 cm) of clear gravel or 1 inch (2–3 cm) of crusher-run gravel (Kochenderfer et al. 1984). In addition, features like surface mounds and depressions, logs, rocks, trees and stumps, slash and brush, the litter layer, and herbaceous vegetation within downslope filter strips all trap sediments that might wash from a road (Packer 1967). The closer the spacing of these surface features, the greater the effect (Fig. 3-2). Engineered coverings and mulches also can help to stabilize cut banks and fill slopes (Fig. 3-3, after Burroughs and King 1989). So will a variety of plant materials established by direct seeding (e.g., grasses and forbs) and transplanting (e.g., shrubs) of species that persist for long periods. These plantings also beautify the roadsides, and may serve as food sources for wildlife (Blauer et al. 1993).

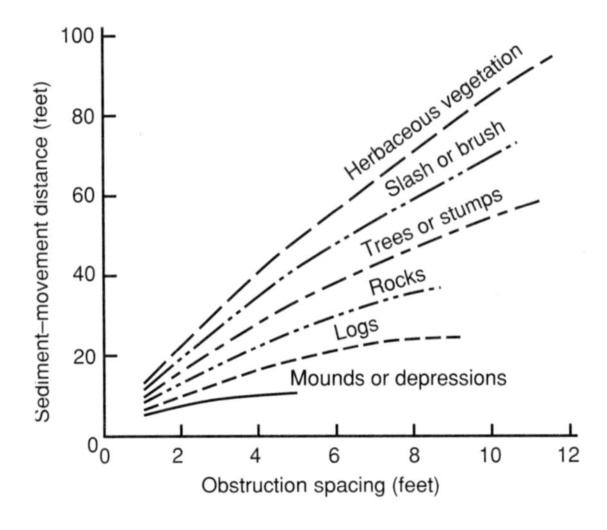

FIGURE 3-2
Effect of different kinds of surface obstructions in trapping sediments downslope from the shoulders of forest roads on soils derived from basalt (from Packer 1967).

PLANNING TO CONTROL EROSION POTENTIAL AFTER HARVESTING

Forests at a regional or larger scale include a wide diversity of physical and biologic conditions, as well as many different kinds of ownerships. Their management requires many unique combinations of treatments and logging practices. The natural biophysical diversity also complicates attempts to effectively control soil stability and water quality by any generalized set of detailed rules and regulations that stipulate exactly how contractors must operate, what BMPs they must use, or how to operate under any given set of circumstances. Instead, many government agencies have set minimum standards for soil stability and water quality, and stipulate that contractors and landowners must comply with those requirements. They usually permit flexibility in deciding how to perform the work, as long as contractors maintain environmentally acceptable conditions in the process. In fact, most governments in the eastern United States permit timber harvesting even in wetlands, with only a few restrictions over methods or procedures. They largely rely on voluntary compliance with the legally defined standards (Cubbage et al. 1987; Siegel 1989).

In general, logging practices that safeguard soil stability and protect the banks and bottoms of streams and other flowing water will satisfy most government requirements for water quality. Actual practices will necessarily differ with each operation, and foresters and contractors must customize these to match the silvicultural treatment and site conditions. Yet common rules of thumb for planning (Kochenderfer 1970; Lantz 1971; NY Soc. Am. For. 1975; Hartung and Kress 1977; Megahan 1977; Haussman and Pruett 1978; Nyland 1994) often include

1 scheduling operations to avoid critical times
—limiting work to seasons with dry or frozen soil to control rutting and surface disturbance

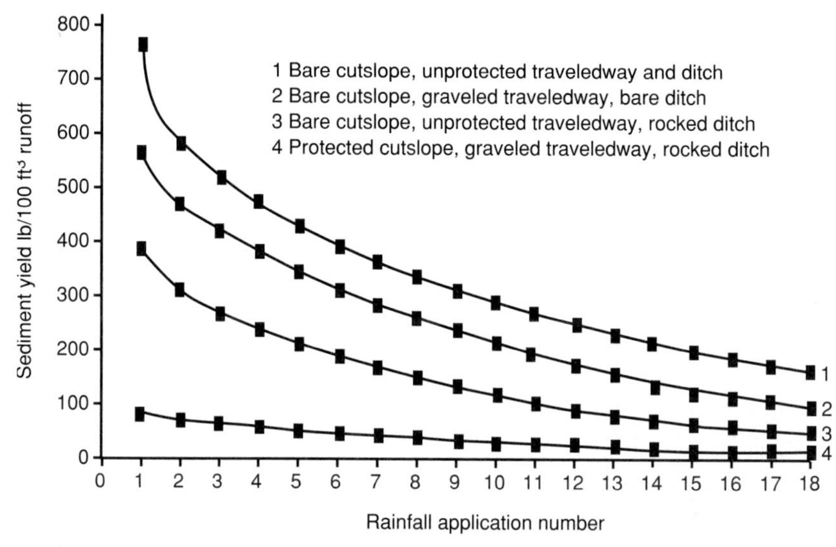

FIGURE 3-3
Effect of different surface treatments on erosion potential from cut banks and fill slopes along forest roads (from Burroughs and King 1989).

—relying on snow cover and frozen ground where possible to protect the surface organic layers and support the machinery
—terminating operations when conditions deteriorate

2 using skidding, forwarding, and yarding strategies that minimize the extent of litter layer disturbance
—limiting major skidding trails to 5%–10% of the surface area
—bypassing areas with impeded drainage or erodible soil
—circumnavigating areas with surface features that complicate efficient machine operation

3 designing trail and road features to keep the surface dry
—locating roads and trails on sites with good internal soil drainage
—designing trails and roads to accommodate the topographic conditions within operational limitations of the available machinery
—keeping reasonable setbacks between trails or roads and flowing or ponded water, and using appropriate methods for stream crossings
—including drainage ditches and other devices to keep water off the roads and trails during use, and from directly flowing into streams and ponds
—closing temporary trails and roads after logging, and stabilizing the surface of all access corridors

4 siting and maintaining landings to prevent erosion and protect flowing and ponded water
—locating landings at places with good internal soil drainage

—installing drainage ditches and other devices to keep water off the landings, or from flowing from them directly into streams and ponds
—requiring reasonable setbacks from surface water, with vegetated buffer strips along the shores
—keeping landings off steep slopes whenever possible, and providing appropriate drainage devices to safely control surface runoff
—periodically grading the surface to reduce rutting and water impoundment
—collecting and properly disposing of trash and waste materials to avoid contamination of the soil or nearby water

5 matching equipment and practices to the needs and conditions, and using the machines in appropriate ways
—utilizing machinery sufficient to the task, but not unnecessarily oversized for it
—selecting equipment fitted to the topographic and soil conditions, and that will cause minimal disturbance when used appropriately
—overseeing work practices to insure compliance with requisite standards for environmental protection

In many regions, forestry schools, government agencies, and many timber harvesting associations offer programs for logger education and training. These include formal educational programs in schools, with both classroom training and supervised practical experience. Harvesting contractors and foresters can also participate in workshops and demonstrations that deal with specific practices. They can access videotapes and a wide variety of published materials about environmentally sensitive effective logging (Jones 1993). When taken seriously, the procedures minimize logging disruption and the chance of an irreversible ecological change.

PROTECTING VEGETATION AND SPECIAL HABITATS

In addition to protecting water quality, many federal, state, and provincial laws protect rare and endangered species of plants and animals, and the habitat that they depend upon. Companion laws may also cover special local needs, and even require unique silviculture if a landowner decides to manage the habitat for a special purpose. In some cases, a specific treatment may actually enhance the physical environment and make conditions more suitable for a threatened species. Generally, the management must accommodate

1 the kinds and balance of habitat elements a threatened species requires for survival, growth, and reproduction;

2 the kinds and intensities of disturbance that might endanger a target species;

3 the times of the year and stages of development when a threatened organism or habitat might suffer the greatest danger if disturbed in different ways and with various intensities;

4 how different management techniques might help to expand or improve a critical habitat and reduce the danger of losing a special plant or animal; and

5 where such special needs occur within a property and in the surrounding landscape, and how a specific set of treatments might make one of the critical habitats better or more stable for the species of interest.

Silviculturists deal with these matters as a professional ethic to insure ecosystem resiliency and stability, maintain diversity of species and habitats for indigenous plants and animals, and insure their existence at different stages and conditions of development in perpetuity. In the long run, their decisions contribute substantially to the goal of managing forest health, and enhancing the diversity of opportunities for landowners to benefit from their forests and indigenous resources.

4

CONCEPTS OF REGENERATION

OBJECTIVES OF FOREST REGENERATION PRACTICES

To many foresters, the methods for regenerating mature communities and age classes comprise the most critical part of a silvicultural system, and the central focus of silviculture. They argue that landowners should never harvest mature timber without simultaneously establishing a new age class of appropriate density and species composition. Otherwise, cutting might permanently alter ecosystem conditions or deplete the reserves of standing timber, with no assurance of timely renewal. It will also disrupt the continuity of benefits landowners can realize over the long run (Brodie and Todder 1982). However, they can preclude a problem by using timber harvesting as a tool of silviculture to replace mature trees with new ones, or to implement cultural treatments that eventually follow.

INSURING TIMELY AND EFFECTIVE REGENERATION

From a silvicultural perspective, foresters regenerate a stand or age class when it matures, when some natural or human event abruptly kills or badly damages it, or when a landowner wants to establish forest cover in areas lacking trees. Each of these situations demands deliberate analysis and planning to

1 clarify the objectives for a new stand or age class;
2 determine if the physical environment will support germination of seed or growth of young trees planted at a site;
3 identify the requisite methods for successfully establishing a new age class under the prevailing conditions; and

4 decide at what time, in what sequence, and with what intensity to implement those techniques for the greatest effectiveness and efficiency.

In each case, they must create an age class suited to the intended needs, and these often differ appreciably between owners.

Conceptually, foresters may need to establish many different kinds of communities to serve the objectives for even a single ownership. In the process, they would use a variety of techniques, depending upon the species, the environmental conditions, and the nature of existing vegetation at each site. Yet in most circumstances, a regeneration program must generally

1 make the results predictable
—control natural succession during the regeneration period by regulating environmental conditions near the forest floor
—insure prompt replacement of the mature trees at an appropriate time
2 secure utilitarian species
—influence natural succession to favor the species that interest a landowner
—insure a species mix suited to the site conditions, and the long-term management objectives
3 realize full site utilization by sought-after species
—regenerate an appropriate number of trees for the management objectives
—insure a uniform distribution of new seedlings across the regeneration area
4 set the stage for future management
—insure that later investments will prove constructive, rather than remedial
—preclude needless effort to reinforce poor stocking, or reduce overstocking
5 minimize chances of failure
—insure an adequate supply of seed or seedlings at an appropriate time
—maintain an environment conducive to germination and early survival of the new trees
—control harmful natural agents
—take necessary follow-up measures should the prescribed treatment fail or prove inadequate to the needs.

In general, silviculture attempts to reduce the risk of failure by insuring an adequate source of new trees, and suitable conditions for their growth and development. It also controls losses to natural agents that might interrupt the regeneration process, or harm the seedlings.

Forest regeneration programs geared to these objectives generally fall into three broad groups (after Ford-Robertson 1971):

Forestation—establishment of forest growth on areas that either had supported a forest, or previously lacked it

Reforestation—reestablishment of forest cover on areas where it once occurred, even if missing for many years

Afforestation—introduction of trees to sites that never supported forests, or had no forest cover for a long period of ecologic time

In this context, *forestation* includes both *afforestation* and *reforestation.* Afforestation specifically applies to areas lacking a source of seed, sprouts, or suckers. Moreover, these conditions exist over a large area, precluding any natural influx of seed from nearby trees. This forces foresters to use artificial methods for afforestation (e.g., in prairie or grassland savannah). Reforestation may also require artificial methods if a site currently lacks both forest cover and an appropriate supply of seed from adjacent sources, as with programs to plant conifers on unused agricultural sites. *Natural reforestation* has also restocked vast acreages of open lands throughout the world, independent of any silvicultural treatment (Zipperer et al. 1988, 1990). So the term reforestation applies to any process for replacing former or existing tree communities with new ones.

When foresters want to reforest or regenerate an existing forest community, they must first remove the tall trees, or reduce their density by timber harvesting (Daniel et al. 1979; Smith 1986). This changes conditions at the forest floor, and makes the environment more favorable for seed germination and new seedling growth. Since the cutting also removes mature trees from a stand, landowners can sell them to help repay costs of ownership and management. However, the timber harvesting serves as a mechanism for encouraging regeneration, rather than as an end unto itself.

REGENERATION AS A PROCESS

All plants regenerate naturally, and most of them from seed. In fact, (Cheyney 1942; Toumey and Korstian 1947) new forest communities form whenever (1) the trees and associated plants produce abundant amounts of viable seed or vegetative propagules; (2) soil and seedbed conditions enable germination of seed, or help to induce shoot development off parent trees; and (3) environmental conditions foster the survival and growth of the new trees. Generally, mature trees in or immediately adjacent to a stand serve as the source of seed for natural regeneration. Yet these must grow to an appropriate stage of biologic maturity before producing adequate quantities to insure regeneration success. With artificial reproduction methods, foresters substitute planted seedlings or artificially sown seed for natural sources. This makes the supply (seed or seedlings) more certain, insures a good stocking by trees that can develop well in the local environment, and promises a composition and arrangement matched to the interests of a landowner.

Once biologically mature, some species and forest community types regenerate readily by natural methods, as long as local conditions do not prevent germination (or sprouting) and early survival. For these species, regeneration seems to occur as a simple event, and new trees develop quickly following overstory cutting. Examples in North America include even-aged aspen stands that regenerate by sprouting and suckering (Perala 1977). Community types that regenerate readily from seed include northern hardwoods treated by either an even- or uneven-aged reproduction method (Tubbs 1977), and even-aged lodgepole pine stands killed by fire (Baumgartner et al. 1985). Oak communities that cover vast areas of eastern North America exemplify the difficult-to-regenerate forest community types that require special conditions for success (Clark and Watt 1971; Sander and Clark 1971).

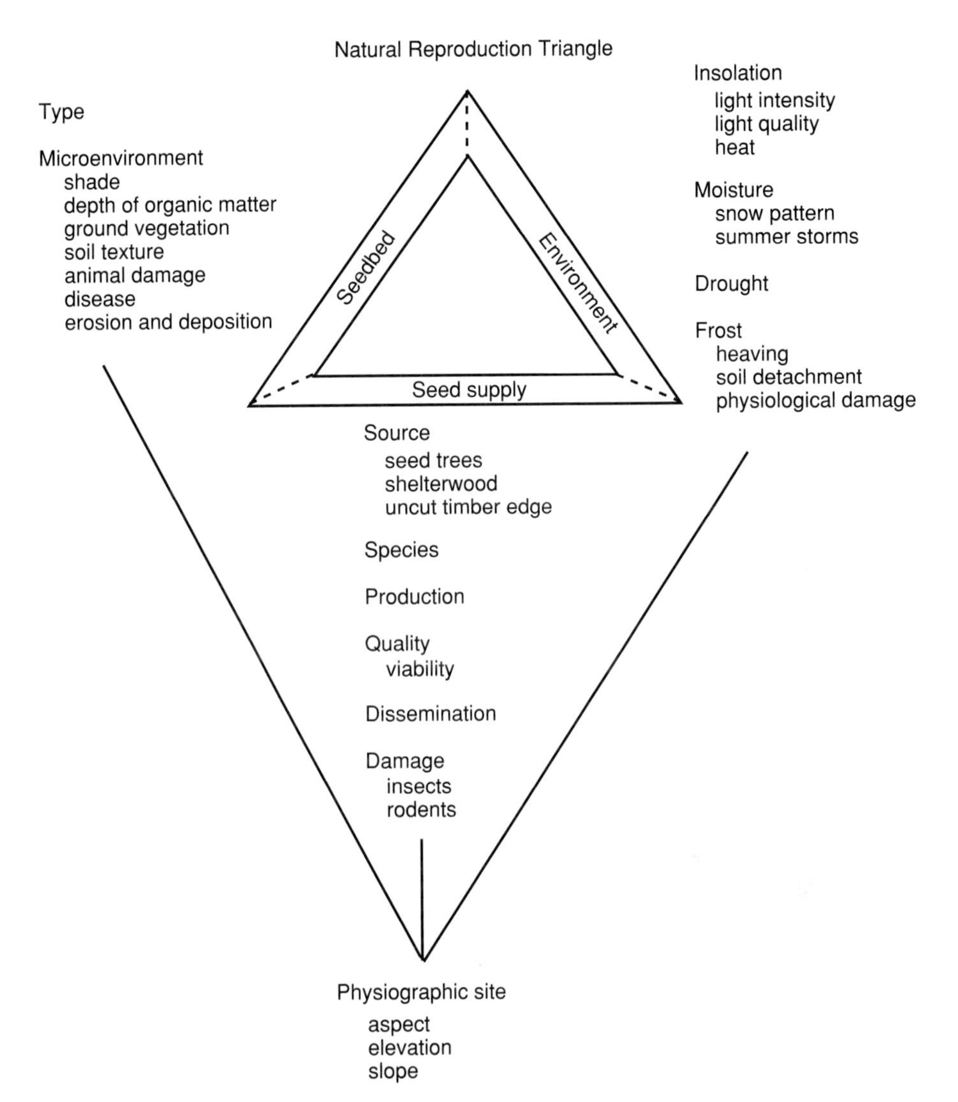

FIGURE 4-1
Both the easily regenerated and the more exacting species actually depend upon a fairly common chain of events that must remain unbroken, and follow in a particular sequence. When legs of the triangle come together in a biophysical sense, conditions favor regeneration.

Both the easily regenerated and the more exacting species actually depend upon a fairly common sequence of biophysical events that must remain unbroken, and follow in a particular sequence (Smith 1986). Figure 4-1 emphasizes this interconnectedness. When legs of the triangle come together in a biophysical sense, conditions favor regeneration. If disrupted by some natural phenomenon or the activities of people, the regeneration process fails, or too few seedlings become established for full site occupancy. Then the cycle must start again, usually in another year, and remain unbroken in order for regeneration of adequate composition and density to develop (Cheyney 1942; Toumey and Korstian 1947; Roe et al. 1970; Alexander 1987).

In some cases, the necessary conditions converge almost annually, and abundant regeneraton occurs rather readily (e.g., northern hardwoods). In other cases, many years may pass before all conditions became favorable, and regeneration succeeds (e.g., the oaks). In both cases, success from seed generally requires

1 convergence of suitable conditions for flowering, pollination, and seed maturation by desirable species;

2 production of abundant viable seed, and assurance of a means for uniform dispersal across a site;

3 favorable conditions at the forest floor, including suitable seedbeds and an environment that promotes germination;

4 limited losses to pathogens, insects, birds, and mammals, both before and after seed dispersal;

5 adequate light, moisture, temperature, and nutrients for seedling establishment and survival; and

6 appropriate conditions for the growth and long-term survival of preferred species.

As long as nothing interrupts the process, a new age class will develop. When even one critical element fails, so will the regeneration.

IMPEDIMENTS TO SUCCESS FOLLOWING SEED DISPERSAL

Both external agents and intrinsic site factors could block germination, threaten new seedlings, and cause a failure. *External agents* include animal browsing and feeding, disease, insect attack, and fire. In some cases, foresters use countermeasures to protect the trees (e.g., fencing), or to mitigate potentially deleterious conditions (e.g., fuel reduction, or pest control). They also try to regenerate species with a known tolerance to the problem, or from parents that seem least susceptible to the danger. Other external agents like flooding, drought, and a variety of other harmful weather conditions affect forests at a landscape scale, and defy human control. Foresters can take cognizance of how commonly they occur and with what intensity, determine if the desired species have a natural resistance or tolerance, and judge the probability for success. They can sometimes revert to artificial regeneration methods (especially tree planting) where natural regeneration seems likely to fail. Otherwise, landowners can only retain the existing community, rather than trying to regenerate it at all.

Several *intrinsic site factors* also affect germination, survival, and growth. These include effects of temperature, moisture, nutrients, and light near the forest floor (Fig. 4-1). Above all, foresters must insure that the species can grow well in the soil and physical environment common to a site. If not, they manipulate the amount of overstory cover or the conditions of soil to mitigate against a threatening condition (e.g., by fertilization, draining, or cultivation). In some cases, they will introduce species and seed from trees of known tolerance, or use unusual measures to protect the trees during critical stages of early development (e.g., irrigation).

The complexity of these factors has encouraged silviculturists to take a systems approach in assessing the nature of conditions essential to success, and in controlling the variables over time and space (Daniel et al. 1979). It makes the regeneration program a process, rather than a simple event. In either case, foresters will take deliberate steps to

1 insure an adequate and suitable natural supply of seed, or supplement it by artificial means;
2 manipulate the overstory, improve the seedbed, and reduce the impediments to germination and early survival by appropriate cultural treatments;
3 protect the newly established seedlings and control the hazards to long-term age class development by timely tending.

In many regards, these measures make the regeneration process to a *deliberated* program of *risk management*, allowing foresters to influence

- what species make up the new age class, and the condition of those seedlings;
- where the new trees become established;
- when a regeneration process commences and culminates; and
- how many trees develop across the regeneration area.

Silviculturists work to influence the timing, kind, type, and distribution of both natural and artificially established regeneration, giving a landowner the preferred species, in preferred places, and at preferred times.

KINDS AND SOURCES OF REGENERATION (NATURAL AND ARTIFICIAL)

Forest regeneration originates or develops from several different sources that foresters can use with either natural or artificial methods. These include *sexual reproduction* (seed, and advance seedlings from germinated seed), and *vegetative or asexual reproduction* (sprouts, shoots, suckers, cuttings, grafts, and clonally propagated plantlets).

Generally, foresters classify the silvicultural programs that include regeneration from seed or seedlings as *high forest systems*, and those based upon vegetative sources as *low forest or coppice systems* (Hawley 1921; Daniel et al. 1979; Hocker 1979; Spurr and Barnes 1980; Smith 1986). In high forest systems, genetic diversity results from natural or artificial crossing or mating different combinations of parent trees. The cross-pollinating passes genetic traits from both parents, and the dominant traits become manifest in the *phenotype* (outward appearance) of their progeny. By contrast to the sexual methods, the low forest methods use trees that originate vegetatively via *sprouts* from a stump, or *suckers* off the root system (Ford-Robertson 1971). These have a genetic makeup identical to that of the parent.

REGENERATION BY VEGETATIVE METHODS

Most broadleafed species will regenerate naturally from vegetative sources, as will a few conifers like redwood. In some cases like aspen, communities may regenerate more readily by vegetative means than from seed. Further, since most broadleafed species sprout more profusely when young, vegetative propagation provides a convenient and dependable means for regenerationg short-rotation crops for fuelwood and fiber from fast-growing species like *Eucalyptus, Gmelina arborea*, the hybrid poplars, and the willows (Chapter 24). With the majority of species, stump sprouting usually declines as a tree ages, and foresters must vegetatively regenerate them before a stand becomes too old (Spurr and Barnes 1980; Smith 1986).

Some sprouts and suckers may develop if disease, an injury, or some other factor reduces the vigor of a tree. With species like American beech, injuries to exposed or superficial roots may also induce suckering (Jones and Raynal 1986; Jones et al. 1989), and these often persist in the understory for long periods. Even so, large numbers of vegetative shoots normally arise only after something (e.g., wind or ice) breaks off the top of a tree, or someone cuts it. That gives foresters a fairly simple way to initiate regeneration by vegetative means. They just fell the trees within a stand when it reaches an appropriate stage of maturity (see Chapter 24).

In some cases, foresters use vegetative sources for artificial regeneration. For this, they normally take a *cutting*—a short length of young stem, branch, or root cut from a living tree and used to propagate it vegetatively (Ford-Robertston 1971). These develop roots and shoots when put into an appropriate rooting medium. In some cases, foresters plant the unrooted cuttings directly at a forest site following an overstory removal, or in open areas. In other cases, they first root the cuttings at a nursery, and then transplant the new trees to the forest.

Where an owner or agency can access a forest tree improvement program, foresters may use cuttings (rooted or unrooted) from genetically selected trees, plantlets developed by clonal propagation, or rootstocks with branches grafted on them from older parent trees of desirable characteristics (Wright 1976; Zobel and Talbert 1984). In some cases, forest tree improvement specialists may take tissue samples from parent trees having desirable genetic or phenotypic characteristics, and propagate the cells into new trees using tissue culture techniques (Fig. 4-2). Once the plantlets reach some threshold size, they transplant them to regular soil or a rooting medium, and grow the new trees to a size suitable for outplanting (Tricoli et al. 1985; Maynard 1994). Yet because of the expense in producing clonally propagated trees by these tissue culture techniques, foresters mostly use them for enrichment planting or to establish special-use plantations.

Overall, clonal forestry has proven beneficial for many needs (e.g., establishing intensively managed wood production areas at low cost). At the same time, it fails to maintain genetic diversity within a stand if foresters do not mix cuttings from several parent trees growing at widely dispersed locations. Also, they would run the risk that a single genotype may not grow well in the local physical environment, or may lack adequate resistance to important insects and diseases. Foresters have learned to reduce these risks by simply using multiclone mixtures, and selecting

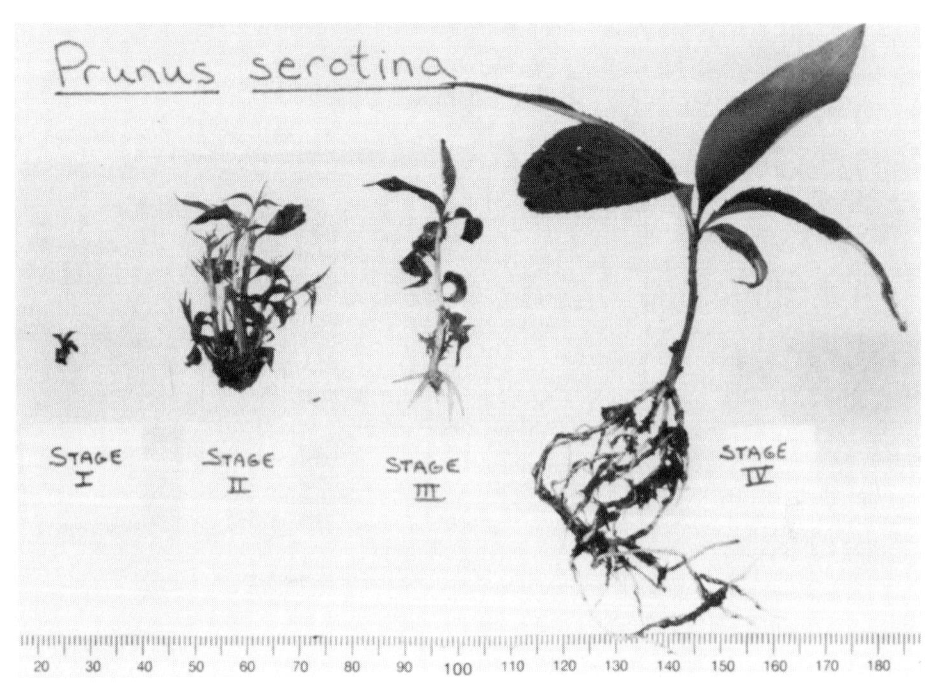

FIGURE 4-2
In some species, forest tree improvement specialists may take tissue samples from parent trees having desirable phenotypic characteristics, and propagate the cells into new trees using tissue culture techniques (from Tricoli et al. 1985; *courtesy of Charles A. Maynard*).

sources with a proven capacity to grow well under the conditions that prevail at a site (see Chapter 24).

REGENERATION FROM SEED

For methods using sexual reproduction, new communities arise by germination from *seed banks* of viable seed stored in the litter. These represent dynamic sources with regular inputs of new seed, and regular losses to predation and a decline in viability (Fig. 4-3). Additions come from seed dispersed from parent trees within or immediately adjacent to the regeneration area. Then germination occurs when some environmental stimulus (e.g., temperature, moisture, or some chemical factor) triggers physiological activity that breaks dormancy within the seed. This may occur in the absence of any kind of overstory disturbance, and contribute *advance regeneration* to the understory. Other trees originate after a cutting by germination from seeds stored in the seed bank, or new seeds that fall onto the forest floor afterward. Sources may include residual trees, trees in adjacent stands, or viable seed held on the logging slash (Spurr and Barnes 1980). Among broadleafed species, some other new trees may emanate from vegetative sources, but these only supplement the

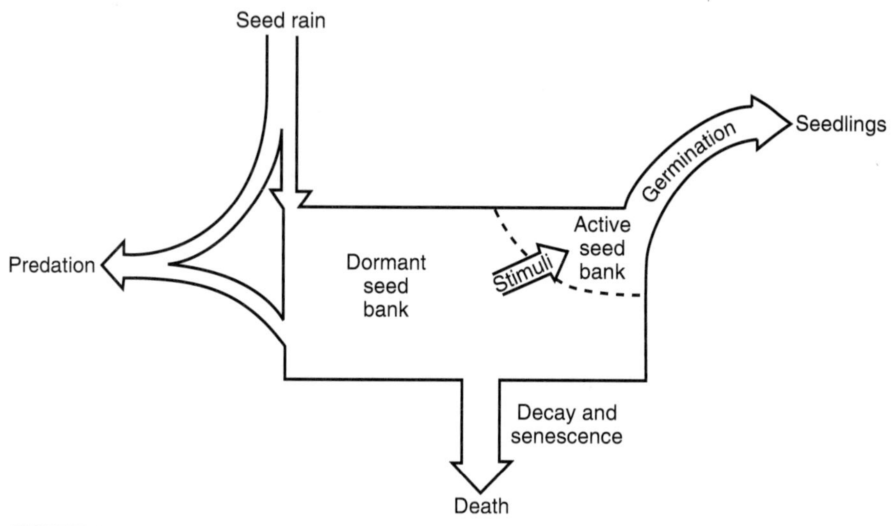

FIGURE 4-3
The character and dynamics of seed banks in the forest floor (from Harper 1977).

larger numbers that arise from seed. When necessary, foresters can also utilize both artificially established and natural regeneration.

Following germination, woody plants must develop through a juvenile or vegetative period before they too begin to produce abundant flowers and seed. This initial period may pass in relatively few years with some species (e.g., lodgepole pine, jack pine, white spruce, pin cherry, and striped maple), but lasts longer for most (Burns and Honkala 1990a, 1990b). The time of first flowering and seed production depends upon a combination of tree age, physiological condition, and environmental influences. By the time an age class reaches at least timber size, the trees will normally produce sufficient seed to serve the regeneration needs. Seed production usually continues until senescence (Krugman et al. 1974).

Dissemination may occur by more than a single means in most species, and depends to a large degree on the architecture of the seed itself. In North America (Baker 1950), the most common kinds of seeds and mechanisms of transport include

1 airborne and floating on water for small seeds with cottonlike tufts (e.g., willows, aspens, poplars, and cottonwoods);

2 airborne for seeds with relatively large terminal wings, and that whirl in flight and glide relatively long distances (e.g., larch, spruce, Douglas-fir, hemlock, maple, ash, and most pines);

3 airborne for seeds with relatively small terminal wings, or that do not whirl in flight and glide for only relatively short distances (e.g., incense cedar, hornbeam, and yellow-poplar);

4 airborne for seeds with specialized bracts of varying aerodynamic effectiveness (e.g., basswood and hophornbeam);

5 airborne for small seeds with large marginal wings, and that glide for relatively long distances (e.g., elm, birch);
6 airborne for small seeds with small marginal wings, and that glide for relatively long distances (e.g., redwood, red gum, and alder);
7 airborne for angled seed with an aerodynamic shape or some winglike structure that provides lift sufficient for only limited dispersal (e.g., bald cypress, true cypress, and beech);
8 animal-carried for seeds ingested as food (e.g., juniper, yew, cherry, gum, and dogwood);
9 animal-carried for seeds cached or buried by animals as reserves for later eating (e.g., pine with large seeds and small wings, oak, hickory, walnut, butternut beech, hackberry, or any species used for food); and
10 no particular means for seeds with no specialized structures (e.g., black locust, other *Leguminosae*, magnolia, buckeye).

Foresters can most readily predict the disperal patterns and distances for species with wind-disseminated seed.

In general, the amount of seed raining (falling) into a partricular area depends upon several important factors (after Harper 1977), including

1 the height of seed release;
2 the distance from a seed source;
3 the abundance or concentration of seed at the source;
4 how readily the seed disperses (its weight and structure), or its adaptation for dispersal; and
5 the activity of its dispersing agent (wind speed and direction, or the numbers and movement patterns of animals that distribute it).

For wind-disseminated seed, topographic features (e.g., slope, aspect, and position), the direction and strength of prevailing winds, height of parent trees at edges of an opening, and the weight and aerodynamic characteristics of the seed generally determine the dissemination patterns and distances by wind (Harlow et al. 1978; McDonald 1980; Burns and Honkala 1990a, 1990b). Glide distances usually differ for seeds falling within closed stands, and from border trees along openings. Greater wind speeds and fewer obstructions in the openings normally result in greater dispersal distances. Most seeds actually fall quite close to the parent source, and seed density declines sharply beyond one to two times the height of the seed trees for most species. On the other hand, small seeds with a large marginal wing (e.g., birch and elm) or other structure that facilitates floating (e.g., the cotton-like tufts of the poplars) usually disperse for much greater distances (Baker 1950; Harper 1977).

Foresters also need to think about how frequently a tree community produces a good crop of seed. These fluctuate with year-to-year variations in the amounts of blooming and the flower structure, the rates of pollination and fertilization, and the success of seed maturation. Light and heavy flowering commonly occur in alternate years, or with longer and somewhat irregular cycles after a year of heavy seed production. Since some trees flower profusely in years when others do not, and some species seem to flower by an inherent rhythm, good production years do not necessarily correlate with site or climatic conditions. Even so, communities of many species produce and mature abundant seed only in years with favorable climatic conditions (Krugman et al. 1974; Daniel et al. 1979; Spurr and Barnes 1980).

Foresters have recognized some general patterns that they can take into account when planning their regeneration programs. In fact, (Baker 1950) experience has suggested that

1 levels of annual production may vary widely in some species, but little in others;
2 good seed years may occur on short cycles in some species and long cycles in others, but not necessarily at regular intervals in either case;
3 some individual trees may produce little seed during years when the species supports a bumper crop, and some trees will still bear seed in poor years;
4 most species produce at least some seed annually, and total failures rarely occur;
5 good seed years do not normally coincide across all species, nor do complete failures;
6 during good seed years, a species may produce abundant seed over a rather large geographic area; but
7 due to variations in weather patterns and other critical factors, seed production may largely fail in some stands or localized areas, even when bumper crops develop across a larger region.

For many tree species and communities, an appropriate combination of conditions for peak seed production occurs in cycles that often vary from three to ten years, depending upon the species and location. However, foresters cannot predict in advance when good seed years will occur, and when trees will produce only limited amounts (Baker 1950; Krugman et al. 1974; Daniel et al. 1979; Spurr and Barnes 1980).

Species that have historically produced abundant seed only at long or irregular intervals present a special challenge in planning a reproduction method. In these cases, foresters must often delay cutting until a good seed year, or retain a seed source on the site or within a dependable dissemination distance until the new cohort forms. They must also take measures to insure a suitable seedbed and freedom from competition by an interfering understory when a good seed crop finally develops. Alternatively, they can sow the seed artificially or plant seedlings to circumvent the possibility of a failure.

GERMINATION AND DEVELOPMENT

Once seed dispersal occurs, other critical environmental conditions control the rates of germination, and of subsequent survival. These factors vary at standwide and even regional scales (the *macroenvironment*), and in close proximity to individual seedlings (the *microenvironment*) (Radosevich and Osteryoung 1987; Tesch and Mann 1991). Macroenvironment features include general conditions of climate and soil, and the influence of topography and aspect across a site. The microenvironment reflects localized edaphic conditions, and effects of features like debris, existing plants, and microtopography on environmental resources available to individual tree seedlings. Sites appropriate for high levels of germination and subsequent seedling development have a favorable balance of standwide and microenvironmental conditions to promote germination, sustain germination and early plant development, and keep the seed and seedlings free from damage.

Different silvicultural methods can modify many of these features (both standwide macroenvironment, and more localized microenvironment). In fact, in designing a reproduction method, foresters will combine an appropriate combination of tree cutting and site surface treatments to minimize or alter the limiting physical environment (at both standwide and local scales), and optimize conditions for success of the target species. They will also use a variety of precutting and postgermination measures to protect the seedlings, and to contain the agents that might damage the seed or the seedlings. In addition, foresters must account for periodic extremes of common macroscale climatological events, or nonmanageable limitations of soil conditions. These affect the array of species that will germinate and survive at a site, as well as their rates of long-term development.

Important macroenvironmental features like slope, aspect, elevation, soil, and climatic regimes give a site its characteristic balance of light, temperature, moisture, atmospheric gases, wind, and nutrients (Radosevich and Osteryoung 1987; Tesch and Mann 1991). These vary spatially across a landscape and within a stand, and influence the composition and density of indigenous plants. Different species have either a wide or narrow range of tolerance to variations of the critical environmental conditions, and to interference and competition with other plants that grow near them. As suggested by the *Law of Minimum* (also known as *Liebig's Law*, as modified by Mitscherlich 1921):

1 the best adapted species become a dominant component of the regeneration, given an adequate source of seed; and
2 the environmental factor near or below the minimum level for successful growth and development determines what species regenerate at a site.

If foresters modify the microenvironment through various kinds of treatments, they can temper the patterns and kinds of regeneration (given an adequate seed source).

Physical site features critical to regeneration success often reflect important differences in growing-season radiation levels, and the effect on related temperature and moisture regimes. To the degree that tree cutting and other vegetation control practices affect these environmental factors, as well as the rates of evapotranspiration, they also potentially influence regeneration success (Waring and Schlesinger 1985, Radosevich and Osteryoung 1987; Tesch and Mann 1991). Often, these conditions reflect topographic features, as well as the abundance and distribution of growing-season precipitation. Generally, high-elevation sites have more favorable moisture regimes and lower temperatures. In many regions, steeper slopes also have coarser-textured soil, with a lower moisture-holding capacity and poorer water balances on the more southern aspects in regions like the Pacific Northwest of North America (Graham et al. 1982). In fact, availability of water during a hot and dry growing season determines the potential for seedling survival and plant growth at many locations within the region (Radosevich and Osteryoung 1987; Tesch and Mann 1991). Further, even a difference in the aspect may result in more favorable conditions of moisture, so that south and west slopes have higher temperatures and greater losses to evapotranspiration than north and east aspects. Slope, aspect, and soil texture often prove far less consequential in other regions with more favorable growing-season climates, and foresters often have greater flexibility in selecting reproduction methods that will succeed under the conditions. Where physical site conditions might deter germination or lead to poor initial survival (but not necessarily long-term growth), foresters must often transplant well-developed seedlings from nurseries, rather than rely on natural regeneration.

ADVANTAGES AND DISADVANTAGES OF DIFFERENT REGENERATION METHODS

Each natural and artificial technique has distinct advantages and disadvantages that may make it preferable in some cases and poorly suited to others. At times, foresters can compensate for the shortcomings by adding different ancillary treatments, or mixing natural and artificial methods. These mixed-mode approaches may help a forester to enrich the naturally occurring species or genotypes, or to reinforce the distribution or density of seedlings that naturally regenerate in a stand. With *enrichment plantings*, foresters may add a limited complement of species or genotypes that would not otherwise regenerate in a community. They apply *reinforcement plantings* to sites lacking sufficient regeneration for full site utilization.

These different approaches to regenerating a forest stand have the following advantages:

Natural

- least expensive to apply
- usually results in species and trees well adapted to the site

Artificial

- direct control over the genotypes, species, and placement of trees in the new stand
- complete control over time of establishment, and duration of the establishment period
- applicable on repeated occasions to insure success at difficult sites and circumstances

Mixed

- intermediate costs, if carefully planned
- allows some influence over genotypes, species, and placement of trees in the new stand
- excellent for salvaging otherwise hopeless or inadequate situations

The different methods have these relative disadvantages:

Natural

- often produces stands with some variation in species composition, stocking, and time of establishment
- limited to indigenous genotypes and species
- subject to natural forces of disaster
- dependent upon coincidence with a good seed crop

Artificial

- expensive to implement, with a long wait until revenues repay the investment
- impractical to use in remote areas, except at a high cost
- requires a major logistical support effort, and especially for growing, shipping, and planting seedlings

Mixed

- often requires investments to control naturally occurring plants or unwanted trees
- if needed to reinforce natural regeneration, may indicate unsuitable environmental conditions that could impede success

In some respects, natural regeneration methods serve as the most opportunistic means for replacing mature forests stands. But they often require ancillary measures to control conditions of the site, and to insure the prompt seedling establishment. This will make the outcome more certain, but also increases the costs.

CHOOSING BETWEEN NATURAL AND ARTIFICIAL REGENERATION

Most deliberate regeneration programs have distinct *aims* that guide a forester's decisions. These include

- promptly establishing a new community or age class following the removal of an old one, or whenever a landowner needs to create a new forest stand;
- achieving full site utilization by sought-after species, and trees of a quality suited to the intended use; and
- replenishing the stand with reasonable economy.

Within this broad framework, foresters begin their planning by determining the purpose and objectives, and then deciding what combination of species and techniques might satisfy those interests.

In general, foresters tend to use natural regeneration methods, except where naturally occurring species will not serve the purpose, or when an adequate density and arrangement of new trees would not naturally develop across a site in a timely fashion. Nevertheless, to insure a high probability of success they must assess

• *availability* of adequate seed for suitable species and genotypes, or of abundant and desirable advance regeneration;
• *assurance* of a suitable and reliable means of seed dispersal, or spread of vegetative parts;
• *potential* for a suitable seedbed or ground conditions at the appropriate time; and
• *likelihood* of available conditions essential for germination, initial survival, growth, and development.

Where these appear unlikely, foresters must take measures to correct the deficiency—perhaps substituting artificial regeneration methods.

THE IMPLICATIONS OF SILVICULTURAL PRACTICE TO FOREST TREE IMPROVEMENT

In many regions of the world, agencies and corporations maintain active forest tree improvement programs to select and breed trees of ideal characteristics, and use them to replace or supplement natural regeneration. They may use fairly intensive practices like controlled breeding, production of genetically improved seed in special orchards, and a variety of clonal propagation methods (Fig. 4-4). Forest regeneration programs to complement them necessarily involve artificial measures, if only by enrichment plantings. Where past practices have exploited a forest by simply cutting timber without regard to the quality or condition of trees left as a seed source, such artificial methods may also prove necessary. Yet where landowners cannot access these special programs, foresters must necessarily look to silviculture for at least a base level of improvement in the genetic character of a community, and the generations of trees that follow from it (Daniel et al. 1979; Johnson 1984; Zobel and Talbert 1984; Maynard et al. 1987).

All forest tree improvement programs capitalize upon inherent genetic differences between trees in a population, and the natural tendency for inherited traits to control the character, growth and development of the progeny (Larsen 1956; Wright 1976; Zobel and Talbert 1984). This becomes manifest in the *phenotype*. It represents the interaction of a tree's *genotype* (genetic makeup) with other factors (environmental) that influence its growth and development. Even though the phenotype does not actually reveal anything definite about the genotype, foresters and tree improvement specialists generally assume that (1) trees with genetic attributes matched to the environment of a site will grow the best (good phenotypes); and (2) these traits will pass on to the progeny through pollination, fertilization, and seed dispersal at the time of biological maturity. Even so, trees that grow well in one environment may do poorly in another, or vice versa. As a result, tree improvement specialists must test the suitability of different parent sources (trees or stands) before moving them to new locations through artificial forestation methods, and particularly over long geographic distances and to notably different environments.

For some needs, tree improvement specialists and silviculturists look for entire communities having better-than-average characteristics, and use these as a source

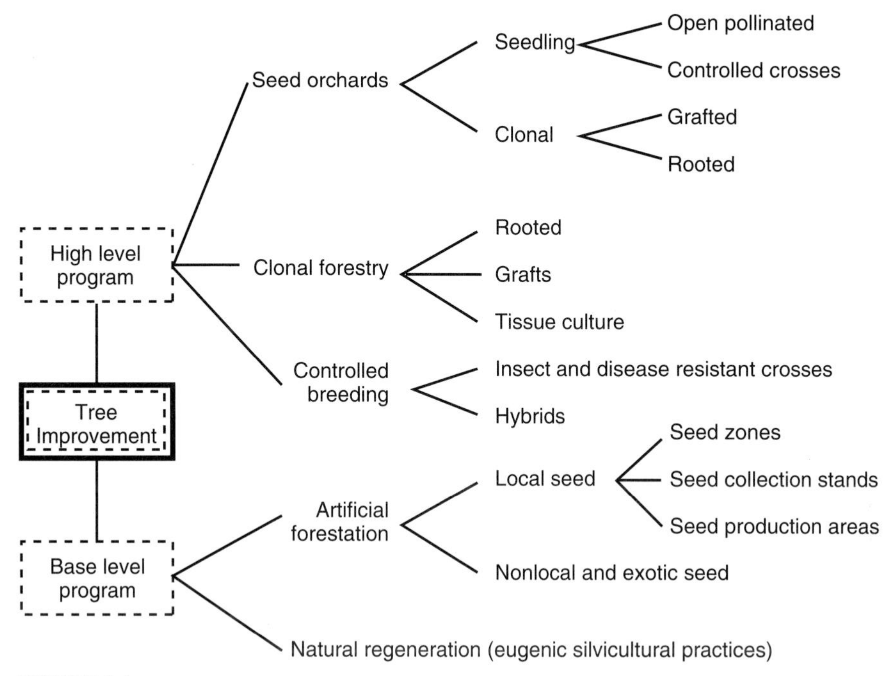

FIGURE 4-4
Effective forest tree improvement involves both high-level programs of fairly intensive selection and propagation, and appropriate silviculture to capitalize upon inherent genetic potential of tree populations in a forest (from Maynard et al. 1987).

of seed (*seed production areas*). They may remove the poorest individual trees from the stand, with the thought of limiting cross-pollination to trees of good phenotypes. In other cases, they look for especially good individual trees in many different stands, collect cuttings from the select trees (often called *plus* trees), and artificially propagate them in special seed-producing areas called *seed orchards*. By bringing the choicest trees from many locations into a single stand, they hope to improve chances for producing seed that carries the best genetic traits of a species (Wright 1976; Zobel and Talbert 1984). These orchards should include sources from environmental conditions similar to the planting sites and managed similar to the intended silvicultural program.

In selecting for phenotypic characteristics, foresters must use stands with a high proportion of good trees and not degraded by past cutting. Characteristics like wood specific gravity and adaptability to environmental conditions generally have a high heritability. Stem form and disease resistance have intermediate heritability, and growth characteristics a lower heritability in individual trees (Zobel and Talbert 1984). Even so, adaptability to conditions of a site becomes manifest in the rate of growth, and the size of a tree in relation to its neighbors.

Many choices for establishing a new age class or stand directly influence its genetic makeup (Daniel et al. 1979; Zobel and Talbert 1984; Maynard et al. 1987). With natural reproduction methods, foresters use trees of the best phenotypic characteristics as a seed source, and expect the new community to have genetic attributes at least as good as those of the parents (Fig. 4-4). Conversely, when they cut the good trees and leave poor trees as a sole source of seed, the genetic quality and the general character of the new stand will suffer. In selecting seed sources for tree planting or direct seeding, they can insist upon seed from high-quality sources in zones compatible with local conditions. In accepting nonlocal seed or exotic species, they can look for proven sources already growing in the area, and under the intended management regime. These measures increase the chances that a new community will fully capitalize upon the productive potential of a site, and also produce high-quality trees fitted to the needs of an ownership.

Many intermediate treatments also directly influence the genetic character of a tree community or age class (Daniel et al. 1979; Zobel and Talbert 1984; Maynard et al. 1987). Where silviculture concentrates the growth potential of a site on trees of excellent phenotypic characteristics, it will likely enhance the overall genetic quality of a stand (Fig. 4-4). Eventually, this becomes manifest in the makeup of regeneration derived from those parent trees. Likewise, in protecting the health and vigor of the residual trees, foresters insure that the best trees will live through the intended period of management, and serve as a seed source for future generations. In these ways, they directly influence the long-term character of a forest, and the future generations of trees that will eventually occupy each site.

Foresters face many obstacles in actually realizing such genetic gains across any forest ownership. Common deterrents include both biologic and economic factors. To illustrate:

1 They cannot control the cross-pollination that occurs in natural settings, and the actual genetic composition of a new stand.

2 They can leave good trees as a seed source, and presume that these will produce good-quality trees.

3 They have no certainty that the seed trees will actually pass desired traits to their progeny.

4 Because of the slow growth of trees, foresters will have no immediate verification that the progeny actually have the sought-after attributes.

5 To realize the most gain for artificial regeneration programs, they must select superior trees by screening many individuals growing across a wide geographic area.

6 They must collect and propagate each select tree or seed source, and test the offspring in long-time growth trials under a variety of site conditions.

7 They must eventually plant the selected trees and sources in seed production stands or seed orchards, and wait for them to mature sufficiently to produce abundant amounts of seed.

Altogether, improvement through routine silviculture requires long periods to realize, and considerable investments to implement. It also takes administrative determination to provide continuing fiscal resources to bring such a program to fruition (after Daniel et al. 1979).

In addition to their encouragement of these silvicultural methods to improve a forest, many forestry agencies have organized more intensive tree improvement programs for the express purpose of enhancing the productive potential of the stands that they manage, and particularly for commodity production purposes. In general, these intensive tree improvement projects (after Larsen 1956; Evans 1982; Zobel and Talbert 1984) will

1 improve growth, form, quality, and other desirable tree characteristics;
2 increase resistance to diseases, pests, and drought;
3 identify new sources and varieties of trees better adapted to growing conditions in an area of interest, and particularly sites degraded by past land uses;
4 breed new strains of trees for specialized products or uses, or better fitted to a desired program of silviculture; and
5 provide genetically improved planting stock for use in forests degraded by past exploitive cutting.

To realize the full benefit, landowners must also use appropriate cultural practices to ameliorate the limiting environmental factors, and limit their planting to strains of trees fitted to the local environment (Zobel and Talbert 1984). They should concentrate on the most productive sites that have the fewest inherent limitations, and that promise the best potential for long-term growth and development.

Despite their advantages, intensive selection and breeding programs do carry some risks. Primarily, they screen large numbers of communities and select a relatively small proportion of the total population for propagation and breeding. This reduces the genetic base or diversity (Zobel and Talbert 1984), and may affect the tolerance to some environmental factors—particularly if the parent trees or stands grow under only a narrow range of site conditions. That might prove especially important in programs with vegetatively propagated species, and no opportunity for crossing between the selected clones. It also poses a potential risk when foresters move a species from its native environment to new regions or continents. Including a large number of selections in the program, selecting parent trees and stands from a wide range of site conditions, and adding sources from across a large geographic area within a species' natural range will help to minimize these risks.

USE OF EXOTIC SPECIES

With artificial reproduction programs, foresters may introduce exotic species to supplement or replace those common to an area. *Exotic* broadly means any species grown outside the limit of its natural range (Wright 1962, 1976; Zobel and Talbert 1984; Zobel et al. 1987). These include trees imported from other regions of a country or continent, or trees from a foreign land. Some already occur in a region as a result of past importation, giving foresters much insight into adaptability and growth. In other cases, foresters continue to introduce and test new exotics because local species seem inadequate for a particular need, or because of their adaptability to plantation forestry (Zobel et al. 1987; Zobel and Talbert 1984).

Introduction and testing of new exotics continues in tropical and subtropical countries, and particularly in areas that lack conifer timber (Zobel et al. 1987). Exotic tree planting also has great importance in areas with only a sparse tree cover due to harshness of the environment (e.g., areas of grassland and scrub forest), in regions deforested by shifting agriculture and other causes, and where people have a shortage of accessible fuelwood. Difficulty of working with indigenous tropical hardwood forests, and the lack of information about suitable methods for their management, has often encouraged replacement with plantations of species that foresters understand better and have learned to successfully domesticate and manage (Evans 1982; Zobel and Talbert 1984; Longman and Jenik 1987; Zobel et al. 1987). Continued deforestation and population growth, particularly in developing countries, will probably make plantation forestry and exotic species increasingly important in addressing basic human needs in those regions.

In temperate countries, foresters have mostly favored native species for large-scale reforestation planting (Little 1949; Zobel and Talbert 1984). They have largely used trees with a proven capacity to grow well in the region, and with known and manageable pest and disease problems. Even so, some exotic species have become a valuable part of many successful regeneration programs. Examples include Norway spruce, Scotch pine, and several species of larch for reforestation of former agricultural sites in northeastern North America. Foresters also imported a variety of trees and shrubs for shelterbelt planting in the Great Plains. Throughout much of the east, landowners have extensively planted Douglas-fir from the Pacific Northwest for use as Christmas trees. European foresters imported species like Douglas-fir and Sitka spruce to Great Britain and the continent, or lodgepole pine to Sweden. They have used Monterey pine from California extensively in New Zealand and Australia. In fact, this introduction has particularly interested the forestry community, since the species occurs in such a limited geographic area and has grown so rapidly as an exotic plantation tree outside its natural range (Little 1949; Wright 1962, 1976; Zobel and Talbert 1984; Zobel et al. 1987).

By carefully selecting a species suited to the local environment, foresters can often realize some important benefits with exotics (Wilde 1958; Wright 1964; Evens 1982; Zobel and Talbert 1984; Smith 1986; Zobel et al. 1987). Important gains include the following:

1 The wide variety of species growing throughout the world increases chances of finding an exotic to substitute for poorly suited native species, particularly on degraded sites or ones with a harsh environment.

2 Tree improvement programs have identified some exotic species with preferable genetic and phenotypic characteristics.

3 Exotic species may lend themselves better to mass propagation as needed with extensive plantation programs.

4 Many exotic trees grow more rapidly, provide higher yields of wood with better technical attributes, and provide fruits and other products of higher economic value.

5 Foresters may not have access to native seed nor know how to germinate many native tropical species, but can often more easily establish uniform plantations of exotics.

6 Uniform stands of planted exotics often better lend themselves to intensive culture over short rotations, and may not naturally regenerate in adequate numbers.

7 Experience in other regions or parts of the world may already provide dependable silvicultural methods for the culture and use of many exotic species.

8 Exotics may have greater resistance to local diseases and other damaging agents in some new environments, making protection easier.

9 Foresters may have few other options for extensive afforestation in nonforested regions or reforestation in partly forested areas, and often find exotic species well suited to the needs.

10 Movement of a species to new locations within a continent may extend its natural range, or reintroduce it to sites where it may have once grown as part of the original forest.

11 Establishment of exotic plantations in areas lacking tree cover allows landowners to realize many nonmarket benefits that forests offer.

Benefits may also accrue from using exotic plantations to establish new kinds of wildlife habitat, expand watershed management and erosion control programs, and develop recreation opportunities and aesthetic values missing from tree-poor or partially forested landscapes. Past afforestation for shelterbelts in the Great Plains of North America and the savannahs of Africa, and tree or shrub planting of riparian zones in semi-arid regions throughout the world, illustrate the use of exotic species to provide a variety of noncommodity benefits.

Despite these clear benefits from using exotic trees, indigenous species may also offer important biological advantages (after Evans 1982) like the following:

1 The growth of indigenous species in natural stands gives some indication of their potential, reducing the risk of a real failure when grown in plantations.

2 Indigenous species have a natural adaptation to the environment, and a resistance to known diseases and pests.

3 Indigenous species fill an ecological niche, and may prove important to the habitat for local fauna and flora.

4 Plantations of indigenous species conserve the native flora, particularly in forest-poor regions or areas of extensive deforestation.

Also, local markets and industries already use the native trees, and know how to process and utilize the wood and fruits. Even so, indigenous species may lack important wood characteristics. In addition, foresters may have difficulty securing sufficient seed from native trees, and may lack knowledge about nursery and planting techniques. Native trees may not grow well where past land uses rendered the soil unsuited to many native species, but not to some exotics.

Experience clearly shows the importance of careful seed source and species selection and testing before landowners extensively plant an exotic tree. Under the best circumstances, these trials should span an entire rotation and include test sites having a range of environmental conditions. Common problems revealed by past test plantations (Zobel and Talbert 1984; Zobel et al. 1987) include

1 inferior results from planting of exotic species on sites unsuited to their long-term development, and where native species actually would grow better under the intended silvicultural regime;

2 failure of extensive single-species stands due to slowly developing effects of site, delayed attack by pests and harm from other destructive agents, lack of conditions essential to good long-term growth (e.g., necessary mycorrhizal fungi), or eventual occurrence of adverse weather and climatic conditions not uncommon to the regions;

3 unacceptable stem form, undesirable wood qualities, and poor adaptability to environmental conditions due to careless selection of a species or seed source suited to the needs and conditions; and

4 development of wood characteristics unsuitable for local markets, despite high rates of survival and rapid growth of individual trees and entire plantations.

In regions with established plantations and arboreta of exotic species, foresters can draw upon the experience of others in making selections for their own forestation programs. In fact, where landowners have used a species extensively for long periods, successful plantations also provide local sources of seed, and serve as a proving ground for silviculture. More often than not, the transfer of a species out of its natural range has proven successful when foresters took cognizance of natural growing conditions, and used species compatible with the local environment. Successful examples include the introduction of red pine and eastern white pine from North America to suitable parts of Europe and Japan, and the planting of American red oak in Europe (Wright 1962, 1976).

IMPORTANCE OF SEED SOURCE SELECTION

Artificial regeneration methods, including use of exotic species, commit a landowner to long-term investments of considerable magnitude, so landowners must have assurance that the trees will grow well. This means knowing what the parent trees looked like, and some key attributes of the seed source (the location, elevation, quality, and site conditions of the parent stand). Common desirable attributes include (1) high survival, vigorous growth, and high yields; (2) desired form, suitable wood characteristics, and other physical attributes suited to the intended uses; (3) adaptability to the climate and soil; and (4) resistance to common diseases and pests (after Nienstaedt and Snyder 1974; Einsphar 1982; Smith 1986). For non-market purposes, tree form and wood qualities may have little importance. However, in all cases foresters must insist on seed sources with a proven capacity to grow in *their* area, suited to *their* use and markets, appropriate to *their* expected management, and resistant to *their* local pest and disease problems.

5

SITE PREPARATION

ROLE AND SCOPE OF SITE PREPARATION

The most critical time in the life of a tree seedling comes immediately after the radicle emerges from underneath the seed coat and begins to take root in the soil (Fig. 5-1). This period lasts until (1) a root system develops and begins functioning; (2) the primary leaves photosynthesize to sustain growth and development; and (3) the new shoots harden sufficiently to withstand environmental conditions. Before this process commences, a favorable balance of moisture, temperature, oxygen, and light must trigger the chemical processes that spark life into a seed's dormant embryo.

In some cases, foresters may decide that a particular kind of overstory treatment, in itself, will create the proper environment for seedling establishment. Otherwise, they must actively mitigate conditions in favor of the target species. Silviculturists call this family of practices *site preparation* (Ford-Robertson 1971; Soc. Am. For. 1989):

- removing unwanted vegetation, slash and stumps, roots and stones from a site before the regeneration method commences
- any treatment that modifies existing vegetative or physical site conditions to improve germination, survival, and subsequent growth of desired seedlings

In this context, site preparation includes any measure that makes the physical environment suitable for germination, and later survival and growth of seedlings. The terms *reproduction method* and *regeneration method* mean procedures for actively establishing a stand or age class by natural or artificial means (Soc. Am. For. 1989). This includes cutting the mature trees to make ecologic space for new seedlings, and to promote the growth of advance regeneration or seedlings planted at a site

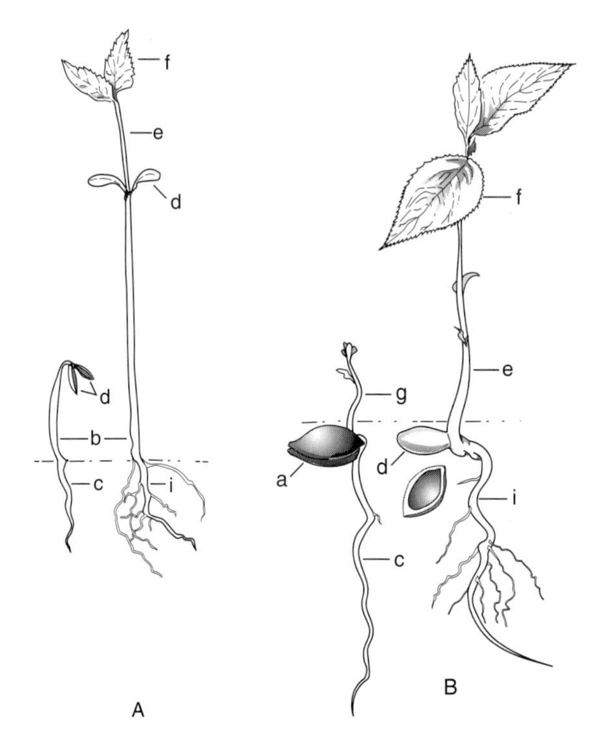

Seed germination

A, Epigeal germination of pin cherry (*Prunus pensylvanica* L.f.) seedlings at 1 and 10 days; and B, hypogeal germination of Allegheny plum (*P. alleghaniensis* Porter) seedlings at 1 and 9 days. *a,* seed; *b,* hypocotyl; *c.,* radicle; *d,* cotyledons; *e,* epicotyl; *f,* leaves; *g,* plumule; *i,* primary root. The plumule on both types of seedlings consists of the epicotyl and the emerging leaves.

FIGURE 5-1
Young seedlings with limited root systems and succulent tissues remain highly susceptible to fluctuations of physical environment conditions, and particularly to extremes of temperature and moisture (from Krugman et al. 1974).

following some kind of overstory treatment. A reproduction method may include site preparation in some cases, but not in others. At times, site preparation comes prior to the reproduction method, and in other cases afterward, depending on the needs and the method used (Fig. 5-2).

Conceptually, site preparation interferes with natural succession by making environmental conditions suitable for the species that a landowner desires, rather than ones that might otherwise regenerate. More broadly, foresters use site preparation (after Chavasse 1974; Gutzwiler 1976; Balmer and Little 1978; Stewart 1978; Sajdak 1982; Smith 1986; Williams 1989; Lowery and Gjerstad 1991) to

1 control competing ground-level vegetation that would interfere with the new tree community;

2 add new kinds of vegetation to control erosion or inhibit potentially interfering plants;

3 remove or mix the litter and other upper soil horizons to improve the seedbed, and the texture, temperature, and permeability of the soil;

4 promote decomposition of the surface litter, expose mineral soil, and modify the structure and porosity of even fairly deep subsoil layers;

5 alter nutrient balances within the rooting zone by adding elements in short supply;

6 alter the habitat for potentially damaging agents, including harmful insects, diseases, and animals;

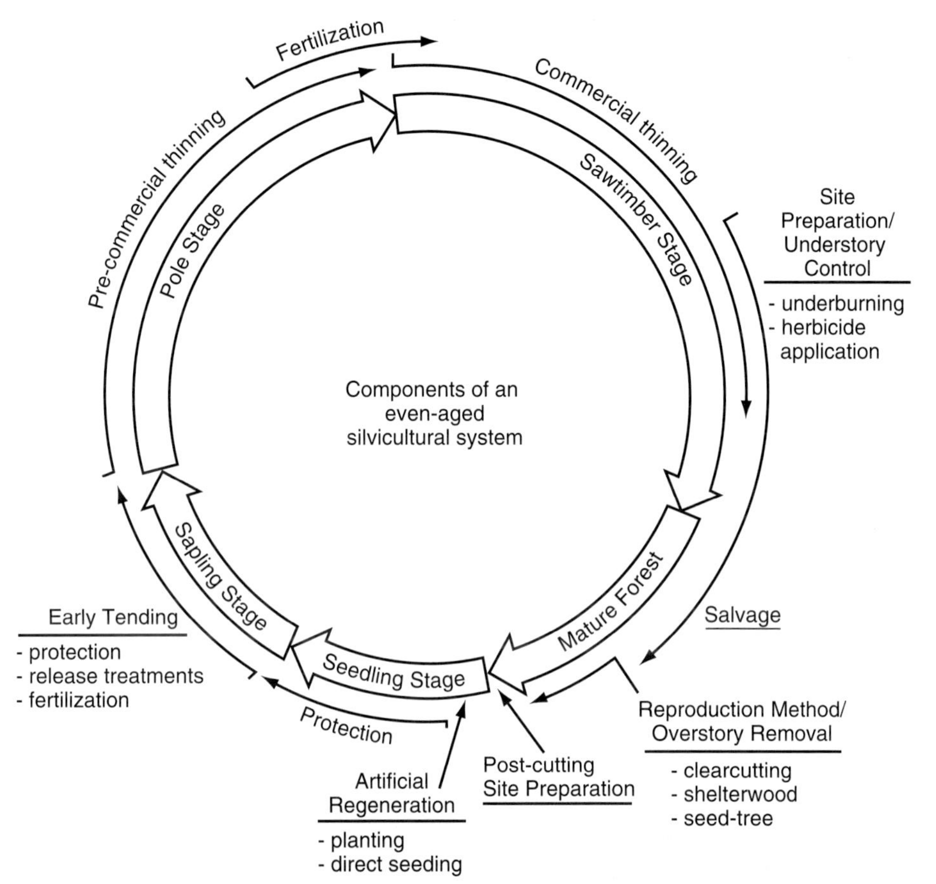

FIGURE 5-2
The timing and juxtaposition of site preparation and of the silvicultural practices in the rotation of an even-aged forest community (after Tappeiner and Wagner 1987; Walstad and Selder 1990).

7 enhance conditions for wildlife, and improve forage and browse for domesticated animals;

8 reduce fuels that potentially increase the risks of damage from future wildfire; or

9 remove, reduce, or rearrange impediments to human activities, and improve access for subsequent operations.

Foresters seek all of these effects in some cases, or only some or none of them in others.

Many site preparation practices affect the soil surface layers, either to modify seedbed conditions or alter physical attributes of the rooting zone. These include *scarification* (Ford-Robertson 1971):

- loosening of the upper soil, or breaking up of the organic layer;
- removing the undecomposed litter and humus to expose mineral soil; or
- mixing surface organic materials with mineral layers beneath them.

Scarification improves seedbed conditions, and facilitates root penetration and infiltration. Further, mixing the organic materials with mineral soil hastens their decomposition and the concurrent release of nutrients. Incorporating organic matter into the mineral soil increases the moisture-holding capacity, and should enhance nutrient transport into the seedlings.

Since several damaging agents threaten seedlings, many reproduction methods incorporate site preparation to reduce the risks. These generally make the habitat unsuitable for harmful insects, animals, diseases, and parasites. By reducing the volume of logging slash or breaking it apart, a treatment may also lessen the danger of fire. Many methods also facilitate movement of workers and equipment through a stand, or improve other operations by rearranging or reducing the slash and surface debris or altering the surface microrelief. Foresters may apply the treatments across an entire stand, or just to selected parts of it (spots, rows, or bands).

PASSIVE SITE PREPARATION

Successful regeneration of some species and forest community types may not require active site preparation. Conditions may naturally support germination and survival of the target species, or the new community may emanate from advance regeneration that already exists in adequate amounts and distribution. Foresters may elect to safeguard ground conditions to inhibit some unwanted species, or to prevent surface erosion. Then they would deliberately decide against site preparation. However, the choice should come as a deliberated decision, not because a forester forgot to consider it.

Logging will disturb the surface and rearrange the logging debris to some degree, apart from any active site preparation. If this proves beneficial, foresters rely on logging as a *passive site preparation measure*—one that accrues as a side effect of another activity or phenomenon. For example, yellow and paper birch of northeastern and transcontinental boreal North America (Erdmann 1990; Safford et al. 1990), or Engelman spruce and western redcedar of western United States and Canada (Alexander and Sheppard 1990; Minore 1990), regenerate best on scarified seedbeds like the edges or surfaces of skidding trails, or a mixture of humus and mineral soil. Then foresters may design the trail system to increase or disperse disturbance or mixing. They may also select a skidding or yarding method that causes a greater extent of ground disturbance, or a different intensity of it. Some species such as sugar maple and black cherry of eastern North America (Godman et al. 1990; Marquis 1990), or Sitka spruce and noble fir in western areas of the continent (Harris 1990; Franklin 1990), do not need a scarified seedbed. Then foresters will minimize logging disturbance by limiting the numbers of trails and using methods that have the least effect on the surface litter. They would first make a careful assessment of the needs, and decide if such passive measures would produce the desired effects with a high degree of certainty.

Past research has shown that both traditional ground skidding and high-lead yarding will often disturb less than 20% of the stand area with partial cuts, and not more than 30% with complete stand removal (Marquis and Bjorkbom 1960; Dryness 1965; Miller et al. 1974; Nyland et al. 1977; Balmer and Little 1978). That may not prove adequate. Also, the effects of logging differ from one season to another, such as when the soil freezes and snow blankets the ground. Also, foresters cannot control effects from natural fires, which may burn irregularly across a site, partly reducing the litter in some places and completely consuming it in others (Miller et al. 1974; Agee 1990). As a result, passive measures may prove only partially effective, or even harmful.

Some kinds of disturbance may actually interfere with the regeneration process. In these cases, foresters must take measures to prevent the disturbance. For example, among forest communities that regenerate best from well-established advance regeneration, they must protect the seedlings already growing in the understory. Likewise, if a target species regenerates most effectively in natural litter or surface conditions, foresters must minimize disturbances prior to or in conjunction with the harvesting. Also, clearing or reducing the logging slash will reduce obstacles to animals like deer or elk, and browsing may increase. A plan should minimize such undesirable effects, and increase the beneficial ones, even where foresters decide to take a passive approach.

ACTIVE SITE PREPARATION

In some circumstances foresters must take active measures where

1 conditions favor less-preferable species;
2 choice species regenerate under a narrow range of environmental conditions;
3 interfering vegetation or soil conditions might limit subsequent growth and development of the new tree community.

Active site preparation implies some on-site activity to lessen impediments to operations, or protect the trees. It includes (1) *mechanical methods* to alter the soil or litter, or reduce unwanted vegetation and debris; (2) *chemical applications* to kill or deter interfering organisms, or supplement soil nutrients; and (3) *prescribed fire* to kill interfering vegetation or reduce organic debris. Foresters must always make detailed on-site analyses before deciding which of these treatments to apply.

Besides using different methods, and with different intensities, foresters can also treat only narrow strips or bands, or relatively small spots. Normally, this limits the site preparation to mechanical or chemical methods. Important determinants in prescribing broadcast or areal treatments, vs. spots or bands, include

1 the kind and size of machinery or equipment available, its mobility, and experience with its use;
2 characteristics of the reproduction method, including the choice of either natural and artificial regeneration;
3 whether the site has some residuals from the original tree community, and if they will remain as short- or long-term components of the stand;
4 unique features of the physical environment or biota that preclude use of a particular site preparation method, or a specific combination of them; and
5 the required density of suitable microsites.

With natural regeneration or artificial sowing, site preparation must normally cover most of the surface area in a stand. Also, where interfering plants dominate a site, a fairly complete treatment may prove essential.

Whether a method lends itself well to spot or band treatments largely depends upon the equipment used, the proportion and amount of area treated, the topography and surface conditions, conditions of soil drainage, and how much a landowner will invest in site preparation. Examples include

1 using pairs of tractor-drawn toothed disks to mechanically scalp (clear litter and vegetation) parallel narrow strips (Botti 1982);

2 using tractor-mounted sprayers to treat narrow parallel strips along the planting rows (von Althen 1979);

3 using hand tools and portable sprayers to scalp or spray herbicides within 2 to 3 feet (0.6–0.9 m) of each seedling (Chavasse 1974; von Althen 1979; Alexander 1987);

4 row mulching or piling (via plowing) logging slash prior to planting (Koch and McKenzie 1977; Stewart 1978); and

5 contour stripping and terracing prior to planting in mountainous terrain (Gutzwiler 1976; Stewart 1978).

Leaving untreated areas between the seedlings helps to reduce costs, preserve a protective organic layer or vegetation cover over the intervening space, maintain herbs and shrubs that might have value as food plants for wild creatures, or for other purposes. On the other hand, spot or band treatments leave sufficient habitat to maintain some harmful pests like the pine webworm or tip moth (Hertel and Benjamin 1977). Also, too few seedlings may succeed on partially prepared sites, as with Douglas-fir in the coastal region of western North America. There site preparation should expose mineral soil across at least 70% of the area (Buck 1959, reported by Strothmann and Roy 1984). Also, vegetation left between the prepared strips or spots may readily colonize prepared soil around the new trees, reducing survival and impairing development (Stewart 1978).

MECHANICAL SITE PREPARATION

Mechanical site preparation serves primarily to push over or cut off unwanted trees and other vegetation, to move or break down logging slash and debris, or to cultivate the upper or subsurface soil layers. By combining different kinds of commonly available machines and implements, foresters can often accomplish the first two objectives in a single operation. In some cases, mechanical treatments may precede other active site preparation measures, such as piling logging slash in rows to facilitate subsequent burning once the fuels dry. Also, one kind of mechanical preparation may set the stage for another, such as piling slash in rows to clear an area for disking or other means of cultivation. More specifically (Balmer and Little 1978; Stewart 1978; Sajdak 1982; Smith 1986; Lowery and Gjerstad 1991), the techniques allow foresters to

1 reduce or rearrange surface debris and litter by chopping it, piling it, and extracting stumps;

2 dispose of vegetation by felling and girdling it, chopping it, pulling it from the soil, raking it into piles, or scalping (scraping) it from the ground surface;

3 alter soil structure by disking, plowing, auguring, and subsoiling;

4 change soil drainage by ditching and breaking or puncturing an impervious surface or subsoil layer;

5 scarify or mix the surface litter by disking, plowing, raking, and chaining (dragging heavy chains or other objects across the surface);

6 modify the microrelief by plowing, ditching, grading, mounding, and bedding; and

7 alter the general surface topography by contouring, terracing, and leveling.

Some of these methods could increase soil compaction and erosion, or siltation into nearby water (Gutzwiler 1976; Pritchett and Wells 1978; Stewart 1978; Williams 1989). Foresters must carefully weigh the benefits and costs (financial and otherwise), and select a method appropriate to the needs at hand. They must make site preparation a deliberated process.

Historically, foresters have used mechanical site preparation with artificial reforestation methods to reduce slash and debris and facilitate the planting operations. Many combination techniques also alter soil structure to promote drainage or improve potential for root penetration. For example, along the southeastern coastal plain of North America, foresters may use bulldozer-mounted blades to uproot unmerchantable standing trees and shrubs (called *blading*), or to push the debris into rows (called *windrows*). After the debris dries, they may burn it. Then they disk the site to incorporate any remaining organic material into the mineral soil (Fig. 5-3), and form cultivated beds with a slightly elevated center (called *bedding*). This technique has at least short-term advantages in survival and tree growth of pines common to southeastern North America (Derr and Mann 1977; Haywood et al. 1990). In the northern Rocky Mountain region of North America, western larch grew better over a 25-year period following site preparation by slash piling and burning, while Douglas-fir did somewhat less well (Cole and Schmidt 1968). Such differences emphasize the importance of carefully assessing species characteristics before selecting a site preparation method.

Despite the apparent early gains, removing and burning the slash and other organic materials prior to bedding mobilizes some nutrients that may leach from a site. This reduces the available pool of some elements essential to good tree nutrition, and lessens the long-term productive capacity of a site. At least with southern pines, tree growth and development diminish during subsequent rotations (Pritchett and Wells 1978; Wilhite and Jones 1981; Wells 1983; Tiarks 1987). Similar effects have become apparent in other regions as well, even when not linked to any tilling of the soil. Examples include litter removal treatments in ponderosa pine stands of California (McDonald and Fiddler 1989), and prior to planting or direct-seeding Norway spruce after cutting 35-year-old red pine plantations on former agricultural sites in New York (Nyland et al. 1979).

Some studies suggest that windrowing and burning may affect soil nutrient status more than any other possible site preparation treatment (Wells 1983), yet simply leaving the litter layer intact or incorporating it into the mineral soil would preserve

FIGURE 5-3
Along the southeastern coastal plain of North America, foresters use heavy-duty disks to incorporate and mix any remaining organic material with the mineral soil and form cultivated beds having a slightly elevated center (called *bedding*), (*courtesy of John Bartlam, Savannah Forestry Equipment, Inc.*).

much of the nutrient capital at a site. Alternatively, foresters can add fertilizers at the time of planting, or later as a stand develops (Pritchett and Wells 1978; Jokela et al. 1991). This will prevent nutrient deficiencies from developing, but not prevent effects of organic material loss on soil physical properties and water-holding capacity. Fertilization does increase the cost of management.

In many areas of North America, foresters use heavy rolling choppers to crush the debris or break it apart with a procedure called *chopping* (Fig. 5-4) (Gutzwiler 1976; Stewart 1978; Smith 1986; Lowery and Gjerstad 1991). They may also shear off (*shearing*) unmerchantable residual trees using a sharpened swept-back blade horizontally mounted on a crawler tractor, or push over and break off these stems using some other kind of machine. The shearing and chopping flattens and masticates the residual trees and branches, and leaves the materials lying flat on the ground. This generally facilitates later movement through a stand, and reduces costs of tree planting or other operations. Flattening the debris also alters fuel characteristics of the logging slash, and reduces the fire hazard. It promotes decomposition of the woody debris, and destroys the habitat for many important pests. Additionally, bars on the rollers cut into the ground, loosening the organic material and upper mineral soil. This reduces surface compaction, facilitates infiltration, and may enhance seedling root penetration under some circumstances. However, it does not prevent stumps of broken and cut deciduous shrubs and trees from sprouting.

On flat sites with a high water table, landowners may construct ditches (*ditching*) to improve soil drainage within the rooting zone. They may also rip or plow the

FIGURE 5-4
As a common mechanical practice in many areas of North America, foresters use heavy rolling choppers to simply crush the debris or break it apart through a procedure called *chopping* (*courtesy of John Bartlam, Savannah Forestry Equipment, Inc.*).

FIGURE 5-5
For *chaining,* workers attach a heavy anchor chain between two large bulldozers, and drive them parallel to each other across a site, uprooting the vegetation and scarifying the surface to some degree (*courtesy of U.S. Forest Service*).

subsoil (called *subsoiling*) to break an impervious layer that impedes percolation or root penetration (Stewart 1978; Smith 1986; Lowery and Gjerstad 1991). Across the Northern Hemisphere, ditching has converted many poorly drained sites (e.g., peat lands and muskeg in boreal zones) into productive tree-growing areas (Trettin 1986; Olszewski 1989). At the same time, it alters many important hydrologic characteristics of a site, and this changes the vegetation community to species that grow in better-drained soil. Consequently, landowners must weigh carefully the overall ecologic consequences before draining large areas, and adhere to government rules and regulations that pertain to areas classified as wetlands.

Foresters working in shrub communities of southwestern North America sometimes use a method called *chaining*. They attach a heavy anchor chain between two large bulldozers (Fig. 5-5), and the operators drive parallel to each other across a site. The chains uproot the vegetation and scarify the surface to some degree. They may follow with chopping to reduce the woody debris prior to planting. Within boreal forests of North America, foresters have also scarified areas by dragging lengths of anchor chain behind crawler or wheeled tractors. The chains bounce along the ground, disturbing the litter, exposing mineral soil in places, and mixing the organic and mineral horizons. Since this method rearranges and mixes the litter and humus, it does not affect the nutrient status, except to promote decomposition of the incorporated organic material. To increase the degree of scarification, foresters may drag an area with heavy barrellike devices that have stubby metal fins or blades welded along the outside of each drum (Fig. 5-6). Foresters use these techniques primarily in preparing stands for natural regeneration.

Usually, the large machinery needed to push over and chop small unmerchantable

FIGURE 5-6
To increase the degree of scarification, foresters may drag an area with heavy barrellike devices that have stubby metal fins or blades welded along the outside of each drum.

residual trees and logging slash dictates broadcast or complete area treatments. So does chaining and barrel scarification. All require complete removal of the mature tree community, since most of the methods would damage or destroy any residual trees and their root systems. Even scarification has proven impractical with partial cuttings.

To successfully plant many broadleafed species (e.g., eastern cottonwood, hybrid poplars, and other eastern hardwoods) on former agricultural sites, foresters must usually till the entire area much as farmers fit a field for agricultural crops (Erdmann 1967; von Althen 1981; Hansen and Netzer 1985). They clear any old vegetation, and cultivate the surface using plows or disks. They may also continue to cultivate annually until the tree crowns form a closed leafy canopy across the site. Many times, a landowner reduces these follow-up operations by applying a herbicide to the soil surface just prior to planting. This slows the colonization by herbaceous species. However, they must also often return periodically with foliar sprays to control the invasion of weeds until crown canopy closure.

Some techniques work particularly well for enhancing natural regeneration where a forester wants to scarify the surface or to reduce the density of interfering understory vegetation. Yellow and paper birch illustrate this opportunity. They regenerate in greater numbers at scarified sites, although the practical long-term importance of having such large numbers of seedlings diminishes over time (Bjorkbom 1967; Roberge 1977). Other reports indicate that abundant birch will also become established in undisturbed leaf litter following removal of interfering understory plants by chemical methods (Kelty and Nyland 1981). This suggests that the disking may enhance regeneration success primarily by removing competing plants. Surface scarification has aided natural regeneration of several conifers, such as red and white pine in the upper Lake States and Ontario (Eyre and Zehngraff 1948; Horton 1962); red, white, and Engelmann spruce across northern United States and in Canada (Lees 1963; and DeJarnette 1965; Hughes 1967; Zasada and Grigal 1978); lodgepole pine in the Rocky Mountain region of North America (Johnson 1968; Perry and Lotan 1977); and Douglas-fir in western North America (Buck 1959, reported by Strothmann and Roy 1984). In these cases, site preparation improves seedbed characteristics, and probably also reduces competition from shrubs and small trees already growing at a site.

Mechanical methods have some undesirable side effects. Examples (after Erdmann 1967; Stewart 1976, 1978; Gratkowski 1975, 1977; Balmer and Little 1978; von Althen 1981; Strothmann and Roy 1984; Hungerford and Babbitt 1987; McDonald and Fiddler 1989) include the following:

1 They remove or disrupt the protective organic covering and loosen upper soil layers, thereby increasing and erosion potential on slopes.

2 They require large investments in machinery and equipment, and take considerable time to apply across large areas.

3 Obstacles like rocks, stumps, and features of the topography hinder movement of the equipment, and some may reduce the effectiveness of a treatment.

4 Heavy machines compact surface layers in some soils and during some seasons, thereby reducing infiltration, percolation, and soil aeration.

5 The hazard of operating machines on slopes exceeding about 30% makes many mechanical methods impractical.

6 Surface disturbance creates a good seedbed for many herb and shrub species, and these may interfere with the new trees.

7 Mechanical methods will not prevent sprouting of broadleaved shrubs and tree species, and may only temporarily reduce competition by interfering plants.

8 Removing the organic layers may reduce nutrients in the soil and soil solution, necessitating replacement by fertilization.

9 Reducing logging slash removes its shade, increasing the maximum and minimum surface temperatures, and may reduce seedling survival.

While mechanical site preparation may increase chances for erosion, some treatments do improve soil permeability and speed infiltration and percolation. Others reduce the coarse woody debris and herbs, and reduce the habitat of some fauna and flora that a landowner may wish to encourage or maintain.

USING HERBICIDES IN SITE PREPARATION

Use of herbicides in conjunction with regeneration programs in forestry differs markedly from their application in agriculture. First, farmers usually apply one or more herbicides each year, and sometimes more than once per season. Generally they apply sprays over the soil, or to the foliage of actively growing weed plants. By contrast, the long rotations for most forest crops lengthen the interval to once in five years for short-rotation biomass crops, or even in as few as one year in one hundred for large-diameter sawlogs. The long intervals and low dosages applied in most forestry uses minimize the amounts of foreign substances introduced to an ecosystem as part of silviculture. This does not reduce the importance of careful planning and control to prevent unwanted side effects, or even contamination.

Foresters use chemical site preparation for an array of purposes (After Chavasse 1974; Stewart 1976, 1978; Jokela et al. 1991; Lowery and Gjerstad 1991), including to

1 kill dense or unwanted vegetation that might impede workers, or interfere with the survival and development of desired tree species;
2 control harmful insects and animals by poisoning or repelling them during the regeneration period;
3 alter the habitat for a variety of insects and animals by enhancing the carrying capacity of a site or reducing it;
4 kill or desiccate ground-level vegetation to improve fuel conditions for prescribed burning;
5 prevent the development of understory vegetation that would serve as fuels for wild fires;
6 inhibit germination of seed for various herbaceous weed species;
7 defoliate a tree crop prior to harvesting; and
8 supplement available nutrients by fertilization.

In these ways, foresters can increase light levels at the forest floor, and increase other site resources available to the regeneration. Other chemical measures help to safeguard the seedlings from damaging agents, or otherwise facilitate survival and development of a desired tree community.

Foresters most commonly use the chemical methods to kill undesired plants and inhibit their establishment (Chavasse 1974; Stewart 1976, 1978; Balmer and Little 1978; Daniel et al. 1979; Smith 1986; Walstad et al. 1987). This includes use of a herbicide by (1) spraying or mistblowing it onto the foliage; (2) injecting it into or spraying it onto the bark of target woody plants; (3) spraying it over the stumps of cut trees and shrubs that might resprout; or (4) spreading the substance over or incorporating it into the soil. Foresters can apply foliar or basal sprays and soil-active compounds to individual plants, to strips or spots within a stand, or as broadcast applications. They normally use mistblowing as a broadcast treatment to kill understory plants prior to a reproduction method cutting. Individual stem treatments substitute for tree cutting with noncommercial tending operations (e.g., in immature stands), and as an alternative to sharing and bulldozing scattered unmerchantable trees. Spot methods often prove necessary where safety, or habits of a pest, dictate a more selective approach. On the other hand, foresters might fertilize an entire stand using aircraft or tractor-drawn spreaders, or treat the soil immediately surrounding selected individual seedlings only. The appropriate method will depend upon the scale of the operation, and the effect desired.

Compared with other methods for controlling vegetation, herbicides offer some distinct advantages. The more important ones (Walstad et al. 1987) include the following:

1 They effectively kill a broad array of weeds and other interfering plants.
2 They prevent sprouting from stumps and root systems.
3 They do not disturb the surface, or affect the inherent site productive potential.
4 They work equally well in treating large or small areas in a number of ways, and at many times of the year.
5 They have proven cost effective, particularly for broadcast and other mechanized applications over large areas.

In addition, foresters can use aerial applications in steep terrain that would preclude many other vegetation control methods.

Individual stem herbicide treatments normally cost more than broadcast applications if a landowner must remove more than about 400–500 stems per acre (990–1235/ha) (Sage 1987). Commonly used compounds include picloram, tricopyr, and glyphosate (Walstad et al. 1987; Sage 1987). Along with spot and band applications, individual stem treatments generally have greater efficacy than the broadcast methods (Walstad et al. 1987). Common techniques include injecting a chemical into frills or cuts that encircle the main stem (Fig. 5-7), applying it to the bark as a basal spray, or spraying the top and sides of a stump (Daniel et al. 1979; Smith 1986). The frilling technique delivers a herbicide directly to the inner bark, cambium, and xylem through a series of spaced or overlapping cuts. Then the compound translocates throughout the plant, taking its effect. In most cases, herbicides used for stem injection have a water-based carrier (used as a carrying solution for the active ingredient). By contrast, a basal spray must translocate into the tissues through the bark before taking effect, and this works best in an oil solution. Further, basal sprays usually prove effective only for thin-bark plants, like small-diameter trees and shrubs.

Foresters mostly apply broadcast spraying or mistblowing for foliar treatments across large areas, yet backpack mistblowers or low-volume sprayers make effective treatment of small areas feasible as well. They use solutions or emulsions of compounds like glyphosate, phenoxy, picloram, triclopyr, dicamba, and fosamine in water, and can apply these materials at relatively low volumes per unit of area (Sage 1987; McCormack 1991). For example, effective control with glyphosate requires only 0.8–1.6 gal (3–6 l) of active ingredient in a total spray volume of as little as 5.3 gal/ac (20 l/ha). Experienced crews can treat 1,000–1,800 ac (400–730 ha) per day by aircraft operations (McCormack 1991). For small- to moderate-sized stands, or for understory treatments prior to a reproduction method cutting, they must apply the chemicals using ground equipment, often mounted on large four-wheel-drive tractors or with backpack mistblowers when treating small areas or for band or spot treatments. These ground applications do have some disadvantages (McCormack 1991):

1 Workers can treat only smaller areas per day and costs run higher than for aerial operations, especially in treating large areas.
2 The rugged and irregular terrain at most forest sites makes exact calibration of

FIGURE 5-7
By frilling a tree with spaced axe cuts and injecting a chemical into the cuts, workers can effectively kill trees without danger of spilling the herbicide or contaminating surrounding vegetation (*courtesy of U.S. Forest Service*).

spray equipment impractical, and uniform application of the spray solution nearly impossible.

3 It takes more herbicide and greater volumes of spray solution to realize efficacy equivalent to that of aerial applications.

4 Tractors used to transport the sprayers or mistblowers may damage existing regeneration.

5 Terrain, debris, standing trees, and other obstacles may impede access.

6 The treatment may not effectively kill interfering vegetation disturbed by the spray equipment.

7 Workers run a higher risk of exposure in applying the herbicide, compared with aircraft application.

With broadcast application to large areas, foresters mostly use helicopters or fixed-wing aircraft to complete the work while they have favorable atmospheric and weather conditions, and when a target species has its greatest susceptibility to the chemical (Chavasse 1974; Strothmann and Roy 1984). Ground applications work better for smaller areas with less dense and short (less than 15–20 ft or 5–6 m tall) vegetation (Wenger 1984; Walstad et al. 1987). These techniques cost more than aerial applications (Chavasse 1974; Balmer and Little 1978), but have proven especially effective for precutting treatments of interfering understory vegetation (Kelty and Nyland 1981; Sage 1987).

Under most circumstances, landowners will delay broadcast spraying via aircraft until after they have harvested the mature trees from an area. Since most herbicides delivered by aerial application enter a plant through its foliage, foresters may actually postpone a treatment for up to three years to realize maximum benefits at high-quality sites, and five years on poorer ones (McCormack 1991). Granular herbicides absorbed through root systems may provide some added flexibility in the timing of an aerial treatment, since they must dissolve into the soil solution prior to uptake. Foresters may also broadcast granular herbicides using spreaders mounted on wheeled and tracked vehicles, or use hand spreading to treat individual trees and shrubs for small-scale operations.

Soil-active chemicals also serve some other site preparation needs. They generally work in two different ways. Some (e.g., granular hexazinone) enter the root systems, translocate into tissues of target plants, and then take effect. Foresters can broadcast the granules across a site, apply them in bands or strips, or spread them around the base of an individual shrub or interfering tree. This method will effectively kill deciduous hardwood species. Other soil-active herbicides prevent germination of seeds, primarily of herbaceous species. Examples include atrazine, hexazinone, oxyfluorofen, simazine, and sulfometuron. All have proven effective as preemergence treatments for herbaceous weed control after intensive mechanical cultivation, particularly in the southern and Lake States regions of North America. Foresters often use these herbicides as capping compounds following site preparation for short-rotation biomass plantations with poplars, cottonwood, willows, and other broadleaved species.

In some cases foresters will not want to control the vegetation across an entire site. Instead, band or spot applications will suffice, such as prior to planting conifers in old fields, or after bedding on cutover sites. For these cases, foresters can sometimes attach a sprayer behind a tree planting machine, and treat a narrow band of vegetation immediately adjacent to the planted dormant seedlings. Alternatively, they could apply the herbicide during the growing season prior to planting, and then set out the trees in the treated strips or spots. For small-scale operations, landowners could use backpack sprayers to treat 2–3 feet (0.6–0.9 m) on either side of the planted trees.

Exact purposes for a herbicide treatment depend upon conditions at a site, the kinds of vegetation present, and the degree that interfering plants might impair the development of desirable seedlings. Examples of common uses include

1 control of perennial grasses and other persistent herbaceous plants that form old-field or grassland communities at a planting site (Erdmann 1967; von Althen 1981; Michael 1985);

2 inhibiting regrowth of unwanted woody plants and herbs following overstory removal or mechanical site preparation treatments on forest lands (Boyd 1986; Creighton et al. 1987; Wendel and Kochenderfer 1984; McCormack 1991); or

3 deterring woody species establishment by natural forestation in grasslands and fields and utility rights-of-way (Sturges 1983; Vallentine 1983; McDaniel and WhiteTrifaro 1987).

Factors like the scale of operations, cost of the herbicide, conditions of the topography and obtrusive vegetation, purposes for the work, the species involved, and the amount a landowner will spend for the treatment influence the choices.

CONDITIONS FOR SUCCESS WITH HERBICIDES

In general, successful control of the target vegetation depends on a combination of factors that foresters control in their planning, and when eventually applying the chemicals. One formula for effective control includes (after Stewart 1976):

$$\text{Success} = \text{Proper } H,F,D + \text{Proper } T + \text{Proper } C,V + \text{Proper } A,S$$

where H = herbicide selection
F = formulation
D = dosage
T = timing or application
C = carrier type
V = volume
A = application technique
S = supervision

Most foliar treatments have highest kill rates when applied early in a growing season, but following full leaf development and during active growth. They also do not work effectively during droughty periods or if washed off too quickly following application. Soil-active compounds depend upon adequate levels of soil moisture and good rates of uptake by the target species after herbicide application. If foresters apply foliar or granular herbicides after a harvesting or some mechanical site preparation treatment, they should wait until the vegetation resprouts, or regenerates from seed (after Stewart 1976; Daniel et al. 1979; Lowery and Gjerstad 1991).

Generally, foresters consider the individual stem treatments environmentally safe, and relatively easy to use. Workers will apply the compounds directly into or onto the tree, with little chance for drift or dripping onto nontarget plants. The compounds quickly translocate into the stem after application and degrade within the treated trees and shrubs as they die and decompose (Daniel et al. 1979). With foliar sprays, weather and atmospheric conditions dictate the days and parts of days when an application can proceed safely. Contractors must use especial caution to prevent drifting, or inadvertent application outside the targeted site. Careful navigation and ground control, limiting flying or ground spraying to calm periods, and keeping droplet size greater than 100 μm normally prevents drift off the intended site. Also, many chemicals work effectively only if applied to dry foliage, and if not washed or vaporized from the plant during some minimum period thereafter. Using appropriate carriers and applying sprays and mist treatments during calm and cooler periods of the day (early mornings) usually controls these factors (Daniels et al. 1979). With preemergence sprays or capping compounds applied to cultivated areas, the soil surface must remain undisturbed during the germination period, or the effectiveness of control diminishes.

Such constraints demand special knowledge about how different compounds work, and the conditions for success. This includes controlling the carrier, concentration, rate and method of application, and timing. In addition, users must know and strictly follow all safety measures to prevent injury to people, wild and domesticated animals, or plants not intended for treatment. They must also safeguard against contamination of soil in the mixing and storage areas, and drift into nearby wetlands and water bodies. Further, since people in neighboring areas may worry about potential negative side effects of chemical applications, foresters should take account of the social concerns in deliberating the appropriate site preparation to employ, and in preparing for and controlling the application of any chemical treatments (after Stewart 1976, 1978; Lowery and Gjerstad 1991).

Foresters can access many different guidelines to appropriate use of herbicides for silvicultural purposes. Responsible agencies update these frequently as technology changes and new information becomes available. Companies that manufacture herbicides also provide help to users in determining formulation and dosage for different methods of application. Product labels identify potential hazards to people and the environment, and describe methods for safely handling the chemicals. At least throughout North America, federal, state, and provincial governments regulate herbicide use. Both the manufacturers and the users must comply with the regulations, even to the point of providing specially trained staff to oversee all herbicide applications.

In North America, chemicals currently approved for foliar treatments break down into harmless compounds within a few days following application, either by the effects of sunlight or by reacting with common organic compounds. As a result, they do not persist in the ecosystem. This provides an important environmental safeguard, as well as limiting the duration of toxicity to the vegetation. Nevertheless, users must still carefully plan all uses to protect the nontarget plants and animals. They must also attend to the safety of people who do the work, and others who live in the vicinity.

Despite these potential concerns, chemical site preparation will have few long-term deleterious effects on the soil or its nutrition when applied in approved ways (McCormack 1991; McDonald and Fiddler 1989). Desirable features of the methods (e.g., relatively low cost for treating large areas and sites with difficult terrain, and the lack of site disturbance that could substantially increase the soil erosion potential) also give the chemical methods real advantages for site preparation. Yet chemical treatments also have some decided shortcomings that foresters must consider (after Chavasse 1974; Balmer and Little 1978; Strothmann and Roy 1984; Stewart 1976, 1978; Walstad et al. 1987; McDonald and Fiddler 1989; Lowery and Gjerstad 1991; McCormack 1991):

1 Available chemicals will not necessarily kill all target species, especially if not applied at an appropriate time of year or in proper dosages.

2 Differences in sensitivity of various species, plus operational problems in delivering the herbicide uniformly to all the undesirable plants on large areas, may result in uneven effects across a site.

3 Foresters must often wait for one or more years for the woody plants to sprout before a herbicide application has maximum effectiveness.

4 Individual stem treatment or application of herbicides to narrow bands or spots generally proves expensive, and operationally impractical for large areas.

5 The chemical methods serve only the narrow purpose of killing unwanted vegetation, poisoning or repelling pests, and supplementing nutrient deficiencies.

6 By itself, a chemical treatment will not reduce accumulated slash or the thickness of a surface litter layer, nor expose the soil or scarify the surface to improve seedbed conditions.

7 Problems with application techniques may cause skips and double dosing, reducing the efficacy and selectivity of a herbicide.

8 Herbicide methods leave the dead standing woody materials killed by the treatment and do not reduce other obstacles to subsequent operations.

9 Dead standing shrubs and other brush serve as protective cover for small animals that may damage the new tree community.

10 Chemical methods leave combustible material that may serve as a fuel and increase the fire hazard at a site.

11 Foresters cannot use chemical treatments in or adjacent to sensitive areas, including surface waters, or if the treatment would kill both desired and unwanted plants.

12 Removal of one vegetation community by chemical treatment may only improve conditions for a new plant association that may interfere even more with the regeneration.

13 Effective applications by spraying and mistblowing depend upon favorable weather conditions before, during, and after the treatment.

14 Chemical applications often stir social controversy.

These limitations often recommend in favor of mechanical methods, or perhaps prescribed burning.

FIRE AND ITS EFFECTS IN FORESTS

Fires have influenced the character of forested ecosystems throughout the world. In fact, early peoples deliberately set fires to enhance farming and hunting, and to increase visibility around their settlements. They let natural barriers or fuel conditions contain the fires, and periodically reburned many areas to prolong the benefits. Natural fires also burned freely in many other ecosystems, at variable recurring intervals. In fact, many forest communities around the world persist only under the influence of periodic wildfires. They changed considerably when people excluded fire by active prevention programs (Kauffman 1990; Agee 1993).

Three kinds of fires burn in forested areas when conditions of fuel and weather permit ignition and sustained combustion (Davis et al. 1959; Kimmins 1987). *Surface fires* burn in the upper litter layer and small branches that lie on or near the ground. These usually move rapidly through an area, and do not consume all the organic layer. Moisture in the organic horizons often prevents ignition of the humus layer, and protects the soil and soil-inhabiting organisms from the heat. The heat pulse (high temperatures) generated at the burning front of these fast-moving fires does not normally persist long enough to damage tissues underneath the thick protective bark of large trees. It will girdle the root collar of small trees and shrubs, and reduce small-diameter branches and other fine surface fuels. Even so, many shrubs and hardwood trees resprout after a single surface fire, and lasting effects follow only after two or more successive burns (Martin 1976; McDonald and Fiddler 1989; Wade and Lundsford 1990; Van Lear and Waldrop 1991).

During periods of protracted drought, the entire organic layer under a forest may dry sufficiently to support a *ground fire*. These normally smolder or creep slowly through the litter and humus layers, consuming all or most of the organic cover, and exposing mineral soil or underlying rock (Davis et al. 1959; Kimmins 1987). Ground fires may burn for weeks and months until precipitation and low temperatures extinguish the fire, or they run out of fuel. Because of the long and high temperature heat pulse generated by these slowly moving fires, they generally kill the large and small trees and other vegetation.

Ground fires release considerable amounts of nutrients from the burned fuels, destroy many small organisms and fungi that live in the humus and organic layers, consume seed stored in the litter, and kill roots in all but deep soil layers. They increase the chance of surface flow and erosion on slopes, and leave a baked and hardened seedbed that may prevent rapid revegetation by all but especially fire-adapted species. Increased surface runoff across the exposed surface may carry away ash and dissolved nutrients, making conditions even less favorable for plant growth. As a consequence, foresters normally consider ground fires unsuited for site preparation, and restrict prescribed burning to periods when the humus remains too moist for complete combustion (Martin 1976; Stewart 1978; McDonald and Fiddler 1989; Harrington and Sackett 1990; Wade and Lundsford 1990; Van Lear and Waldrop 1991).

A third type of fire may occur in conifer forests during periods of drought and low relative humidity, particularly in areas with heavy accumulations of understory material called *ladder fuels* (e.g., fallen trees, logging slash, and combustible understory vegetation). Under these circumstances a surface or ground fire may ignite slash piles and dead or living lower branches of the standing trees. Then the entire tree crown becomes engulfed in flames, and the fire spreads to nearby trees (Davis et al. 1959; Kimmins 1989). These *crown fires* generate tremendous heat that rises in a strong convection column, drawing in brisk surface winds that fan the flames even more. Heated air blowing across the flames also warms and drys the fuels ahead of the fire, and releases volatile gases from vegetation ahead of the flaming front. Fast-moving crown fires also commonly ignite ground-level fuels below them. They kill all the trees and shrubs in their path, and usually consume the surface organic layers as well. As a consequence, crown fires have environmental effects similar to those of ground fires. They, too, often stop only by running out of fuel, or if precipitation cools the fire and wets the fuels sufficiently to inhibit burning. Overall, crown fires have little value for site preparation within forested ecosystems, and have not become a common part of silviculture.

All of these fires favor the regeneration of fire-adapted and early-succession species. Historically (after Kauffman 1990; Norris 1990; Agee 1993; Brennan and Hermann 1994; McRae 1994; Mutch 1994), they have

1 prevented woody vegetation from succeeding in several herbaceous ecosystems;
2 maintained fire-dependent community types that would have changed in composition and structure in the absence of periodic burning;
3 prevented dense and tall understory vegetation from developing;
4 periodically reduced the load of dead fuels, and recycled nutrients locked in organic materials on the forest floor; and
5 controlled the populations of several insects and diseases, thereby contributing substantially to forest health.

Because organized fire suppression programs reduced the extent of area burned by wildfires, the character and structure of many community types has changed, and some important pest and disease problems have intensified (Brennan and Hermann 1994; McRae 1994; Mutch 1994).

PRESCRIBED BURNING AS A TOOL OF SILVICULTURE

During the last few decades, foresters have promoted prescribed burning to re-create the more beneficial effects of the burns. In fact, fire has become a regular part of silviculture in many regions (Heinselman 1971; Stewart 1978; Crow and Shilling 1980; Johnson 1984; Smith 1986; McDonald and Fiddler 1989; Clark and Starkey 1990; Folliott 1990; Harrington and Sackett 1990; Martin 1990; Pieper and Wittie 1990; Wade and Lundsford 1990; Walstad and Seidel 1990; Van Lear and Waldrop 1991), and serves to

1 reduce slash, debris, and undecomposed litter, and release nutrients stored in accumulated organic materials;
2 kill back interfering vegetation, or reduce understory plants to alter visual qualities in a stand (maintaining a parklike appearance);
3 kill colonizing trees to prevent natural reforestation of grassland and shrub ecosystems;
4 open cones and other fruiting structures to release the seed (e.g., lodgepole and jack pine);
5 increase water yields by altering the kind and size of vegetation, and reducing transpiration;
6 destroy pests and harmful disease organisms, and the habitats that sustain them;
7 induce sprouting of surviving vegetation to improve cover, browse, or forage production;
8 influence plant succession or increase ecological diversity by perpetuating fire-dependent plant communities;
9 reduce accumulated hazardous fuels; and
10 reduce the thickness of the forest floor or expose mineral soil.

In contrast to mechanical and chemical methods, foresters can use prescribed burning only for broadcast treatments. They must wait until fuel and atmospheric conditions will support a burn of adequate intensity, and take cognizance of effects on air quality. This limits their flexibility in using fire as a silvicultural tool. In this context, the terms *prescribed burning* and *controlled burning* mean the same thing—

controlled use of fire under conditions that permit its containment to a predetermined area, and still produce a specified intensity of heat and rate of spread to satisfy certain planned objectives, and deliberate use of fire to produce the desired benefits with minimum damage, and at an acceptable cost (Ford-Robertson 1971; Mobley et al. 1978; Soc. Am. For. 1989).

To maintain the control and insure the positive effects, foresters must plan why and how to burn. This includes

1 writing a detailed prescription based upon a complete analysis of the site and needs, a clearly described set of objectives, and a list of cautions to insure safety;

2 preparing fire lines to contain the burn, and assembling sufficient personnel and equipment to extinguish it if necessary;

3 implementing the prescription according to the plan, but making on-the-spot adjustments as conditions change during a burning period; and

4 returning after the burn to evaluate its effectiveness.

Where foresters and landowners have followed these steps, prescribed burning has often proven effective as a site preparation tool in modern silviculture (Mobley et al. 1978; Harrington and Sackett 1990; Martin 1976, 1990; Wade and Lundsford 1990; Van Lear and Waldrop 1991).

METHODS FOR CONTROLLING THE EFFECTS OF PRESCRIBED FIRE

Foresters light prescribed fires at times of their preference, under conditions they select, and to achieve a particular kind of result. They have frequently used fire as a site preparation tool to kill back small trees and shrubs in the understory of maturing stands. Yet since hardwood species and most shrubs resprout from the root collar or produce root suckers when killed back to ground level, a single prescribed burn may not eliminate the understory altogether. In fact, a single burn may increase the numbers of small understory stems. Consequently, foresters may use a series of successive burns at intervals of three to five years to periodically kill back the understory, and more permanently control its size and density. By contrast, most conifers will not sprout or sucker, and a single prescribed burn permanently removes them as a living part of the community. In both cases the fires increase light levels near the ground by reducing low shading of understory plants, and killing small trees with crowns in lower portions of the main canopy. In fact, because it will kill the small trees and not the larger ones with thicker bark, some foresters use prescribed surface fire to kill the smaller trees of poor crown canopy positions, and eliminate low-vigor hosts for damaging insects (Martin 1976; Crow and Shilling 1980; McDonald and Fiddler 1989; Harrington and Sackett 1990; Wade and Lundsford 1990; Van Lear and Waldrop 1991).

Foresters know that environmental factors—like relative humidity, air temperature, and wind speed—influence the way a fire burns, and ignite prescribed fires only when environmental conditions limit the intensity. In fact, at air temperatures of 50°–75° F (10°–24° C) and relative humidity of 35%–50%, fine fuels with a moisture content of 7%–20% will burn at acceptable rates. Fires in fuels with a moisture content below 5% burn rapidly when relative humidity drops below 20%, making conditions uncontrollable and dangerous. A fire will also burn slowly and irregularly in fuels with a moisture above 30% and at a relative humidity over 50%, producing inconsistent results. By taking account of these and other factors, foresters can decide whether they can sustain a fire of acceptable intensity during the period of interest (Mobley et al. 1978; Harrington and Sackett 1990; Martin 1990; Wade and Lundsford 1990; Van Lear and Waldrop 1991).

To kill vegetation, a fire must heat the cambium or leaves to 120°–140° F (49°–60° C), with the actual thermal death point varying somewhat among species (Mobley et al. 1978; Wright and Bailey 1982; Wenger 1984). Even fast-moving surface fires generate heat far in excess of these lethal levels, if the heat pulse persists long enough to increase the cambial temperature to a lethal level. During prolonged warm periods above 80°F (27°C), heat exchange will already have increased the bark and wood to fairly high temperatures. Then even the relatively short-lasting heat pulse from a fast-moving surface fire will often kill both small- and large-sized trees. Experience indicates that setting a fire during cooler periods with air temperatures below 50° F (10° C) will limit the killing effect, and protect the larger trees from damage (or death) (Martin 1976; Mobley et al. 1978; Harrington and Sackett 1990; Wade and Lundsford 1990; Van Lear and Waldrop 1991).

Foresters often rely on prescribed burning to reduce the abundance of logging slash and other down woody material: (1) as a fire prevention measure; (2) to facilitate subsequent activities (e.g., planting); or (3) to control pests and diseases that proliferate rapidly in logging slash and fallen trees. Slash fires usually consume most of the woody fuels less than about 3 in (7–8 cm) diameter, but only char the larger pieces. They generate sufficient heat to kill back shrubs and trees near the slash, or may leave an inverted V-shaped basal wound on the heated face of even large trees.

Since logging contractors do not spread the slash uniformly, foresters often rely upon a surface fire to carry the burn across a stand. In fact, the discontinuity of fuels may make a slash-reduction fire effective on less than one-half of the area. For that reason, foresters often use mechanical means to push the slash into piles or rows before burning. This improves the slash disposal. On the other hand, the burning piles generate a long-lasting heat pulse that dries and ignites the surface organic material underneath, and may bake the top 2–3 in (5–8 cm) of mineral soil on up to 5% of the area.

The variability of fuels and physical environment across a site prevents foresters from precisely regulating a prescribed fire. Still, they can manipulate some factors that influence the intensity. For example, they might ignite a fire during periods of low air temperatures and with fuel moisture somewhat above 20%. This will dampen the effect. To intensify a burn, they wait for warmer days with drier fuels, and when a light breeze will help fan the flames. To limit burning intensity on windy days, they burn into the wind or downhill. To increase the intensity they let the fires burn with the wind, or uphill. Even these simple techniques affect the outcome to an important degree, and enhance the versatility of prescribed fire as a site preparation tool (Mobley et al. 1978; Harrington and Sackett 1990; Martin 1976, 1990; Wade and Lundsford 1990; Van Lear and Waldrop 1991).

FIRING METHODS FOR PRESCRIBED BURNING

Foresters can temper fire intensity and the effect on fuel and vegetation in an area by altering the patterns of ignition relative to topographic gradients, the direction of the wind, and the size and shape of the burning area (Mobley et al. 1978; Wade and Lundsford 1990; Van Lear and Waldrop 1991). They might use a *ring fire, center fire, or perimeter fire* for slash reduction fires, establishing a firebreak around the area, and igniting the fuel all along its perimeter (Fig. 5-8). This creates a strong convection column that draws in air from outside the burning area, and produces a hot fire. The convection column may carry embers downwind for considerable distances, so foresters must usually refrain from using ring fires during periods of high fire danger and with unstable atmospheric conditions (Mobley et al. 1978; Harrington and Sackett 1990; Martin 1976, 1990; Wade and Lundsford 1990; Van Lear and Waldrop 1991).

For uncut stands with lighter fuels, and where they want to control only the understory vegetation with a lower intensity burn, foresters might employ various kinds of *strip burning* techniques. During days with wind in excess of 5 mph (8km/hr), they may elect a *strip backfire technique* so the flames burn into the wind at only a slow rate of spread. In addition, by igniting *progressive strip backfires*, they can limit the distance any single flaming front travels before running into an already burned area and dying out. This, too, helps to regulate fire intensity. So will backfire methods that move a fire progressively down a slope, thereby regulating the rate of spread and making the fire more manageable (Mobley et al. 1978; Harrington and Sackett 1990; Martin 1976, 1990; Wade and Lundsford 1990; Van Lear and Waldrop 1991).

During times with only slight winds and in fairly flat topography, foresters will often employ a *progressive strip headfire method* (Fig. 5-8). In this case, they will first light a backfire to create a protection buffer at the downwind end of a stand. Then they ignite a progressive series of strip head fires across the site, but upwind from the backing fire. The head fires burn with the wind until they meet a backfire, and run out of fuel. By regulating the distance between successively burned strips, foresters can control the strength of a resultant convection column, and the burning intensity along each fire front. A *spot fire technique* will produce similar effects if a crew carefully regulates the distance between individual fire spots along progressive lines through a stand (Fig. 5-8). To use this and several other techniques (e.g., *flank fire* and *chevron fire*) safely and effectively, foresters must often make on-the-spot adjustments to control the intensity and effects (Mobley et al. 1978; Harrington and Sackett 1990; Martin 1976, 1990; Wade and Lundsford 1990; Van Lear and Waldrop 1991).

Despite its usefulness, prescribed burning has some shortcomings compared with mechanical and chemical methods. (Gratkowski 1977; Stewart 1978; Strothmann and Roy 1984; McDonald and Fiddler 1989; Daniel 1990; Ffolliott 1990; Walstad and Seidel 1990; Van Lear and Waldrop 1991), including the following:

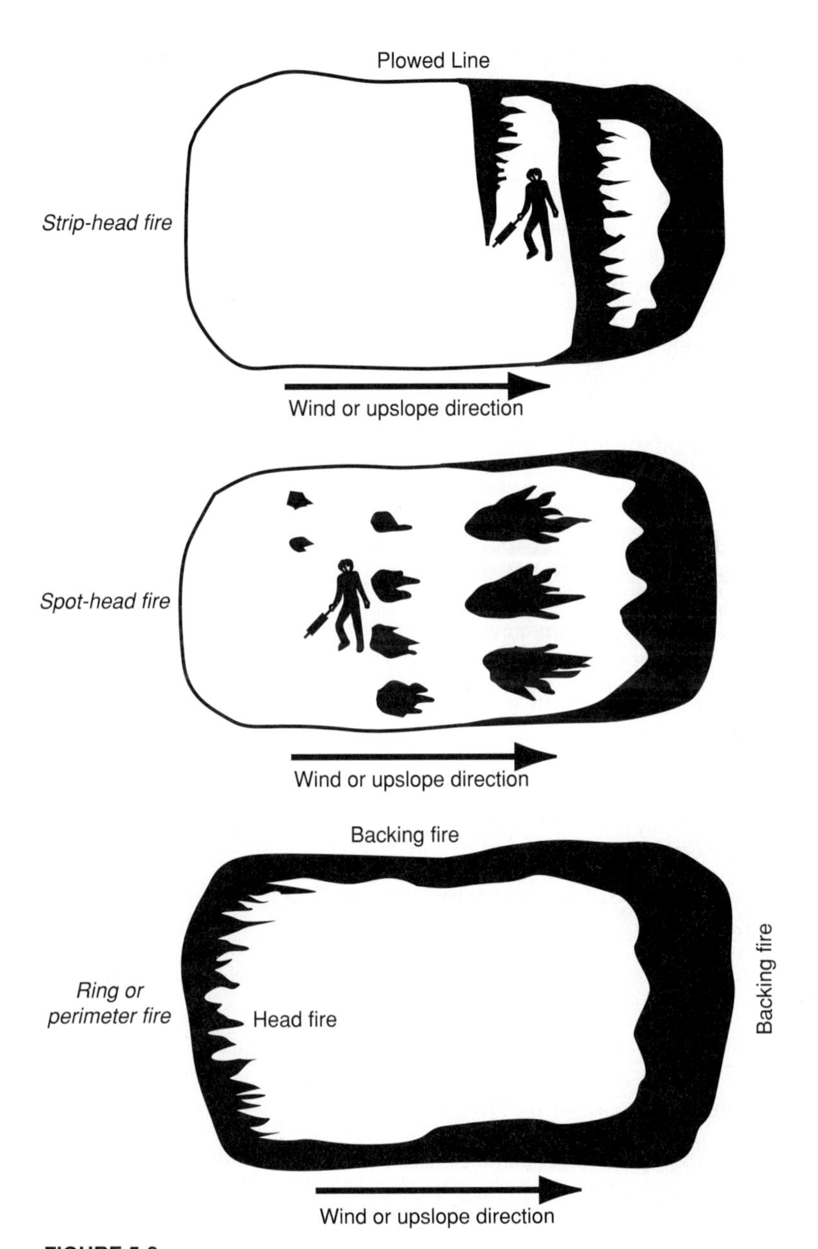

FIGURE 5-8
Foresters can use different approaches to influence the intensity and nature of a prescribed burn, and to affect the outcome in addressing various kinds of site preparation needs (after Mobley et al. 1978).

1 Foresters can start fires only under a rather narrow range of weather and fuel conditions.

2 Fires can escape into adjacent stands if not properly contained, or damage a site if ignited under inappropriate conditions.

3 Prescribed burning requires skill in planning and implementation, and considerable logistical support to prevent escape and unwanted side effects.

4 Laws and government regulations may prevent burning under all but carefully controlled circumstances, or require considerable time in preparing permit applications.

5 Many people find burned areas unsightly until new vegetation masks the blackened surface and burnt debris.

6 Fire will seldom have lasting effects on interfering understory vegetation.

7 Burning may promote colonization by fire-adapted species that interfere with the regeneration.

8 Severe fires will alter soil physical properties, creating a crusty and compacted surface with poor infiltration capacity or seedbed characteristics.

9 Removal of the protective organic horizon may increase the soil erosion potential on some sites.

10 Combustion releases nutrients from organic material, and increases the potential for loss from a site.

11 Burning the logging slash will reduce ground-level shade and increase surface temperatures, sometimes to harmful levels.

12 The fire may damage or kill residual trees.

13 Smoke reduces visibility and air quality.

14 Fires may burn too intensely for effective control on steep slopes, except during periods of low to moderate risk.

15 With all logistical costs considered, fire may cost more than other methods that produce similar effects.

One or more of these shortcomings may recommend against the use of prescribed burning in all but selected circumstances.

SELECTING A SITE PREPARATION METHOD

Under most circumstances, foresters have considerable latitude in selecting a site preparation method. None works well for all cases. Further, not all regeneration programs require the same preparatory treatments, and some species become established in abundance with only minimal manipulation of the site. The choice should follow an assessment of factors like the following: (1) the type of plant community present and wanted at a site; (2) the type, abundance, and distribution of unwanted vegetation and debris; (3) the conditions of terrain, surface relief, and soil physical features; (4) the intensity and kinds of subsequent management activities planned; and (5) the costs relative to any expected benefits. In many cases mechanical, chemical, or prescribed burning methods have similar effects (Table 5-1). In fact, the preferred method will often reflect local preferences (customs), or prevailing

TABLE 5-1
USES OF MECHANICAL, CHEMICAL, FIRE, AND BIOLOGICAL METHODS FOR SITE PREPARATION (AFTER DINGLE 1976; CLEARY 1978; STEWART 1978).

	Purpose[a]				
Method	Reduce woody materials	Control interfering plants	Prepare seedbeds	Modify soil conditions	Alter micro-topography
Mechanical	***X***	*x*	***X***	***X***	***X***
Chemical		***X***			
Fire	***X***	*x*	***X***	*x*	
Biological		*x*			
Combinations	***X***	***X***	***X***	***X***	***X***

[a]A lower-case *x* indicates an effective result, and a capital ***X*** even more effect of using the treatments for the designated purpose. Methods not marked generally prove unsatisfactory for the need.

conditions that make some approaches impractical or too costly to apply. Many mechanical methods requiring large machinery often prove impractical on slopes exceeding 25%–35% (Chavasse 1974; Gutzwiler 1976; Stewart 1978; Miyata et al. 1982), and may not produce the desired effects on rocky surfaces or frozen and snow-covered soils (Paddock 1882). Chemical treatments may prove ecologically unacceptable in sensitive areas or adjacent to ponds and streams (Stewart 1976; Williams 1989). Prescribed burning has risks of escape, and produces smoke that may prove unacceptable in some environments. The heat may also kill residual trees a landowner wants to preserve, and a burn may have only temporary effects on the vegetation community (Smith 1986; Wade and Lundsford 1990; Walstad and Seidel 1990; Van Lear and Waldrop 1991). Ultimately, the choice depends upon the long-term financial prospect of doing or not doing the work, and the importance of site preparation in regenerating a plant community of desirable composition.

6

PLANNING FOR ARTIFICIAL REGENERATION

TREE PLANTING IN SILVICULTURE

Tree planting has served as a common means to replace, supplement, or extend tree cover throughout the world—adding trees for landscaping in urban settings and throughout the urban-wildland interface, afforesting or enriching the species composition of tree-poor landscapes, reforesting open areas in forest-rich regions, and replacing mature forest communities to address a multiplicity of values and objectives. In forestry operations, landowners primarily use tree planting (after Limstrom 1962; Pritchett 1979) to (1) reforest or afforest open areas that lack tree or other woody vegetation cover; (2) reestablish tree cover on sites following removal of a mature crop; (3) improve the density of tree cover within poorly stocked stands; and (4) restore a tree community where some natural phenomena (e.g., fire or insect devastation) destroyed the original trees. Foresters call an artificially created community of trees a *plantation*—a forest community or stand established artificially by sowing or planting (after Ford-Robertson 1971). In many cases, foresters must first reduce the density of an existing stand (apply a reproduction method), and control an understory community (site preparation) that might interfere with the new trees. The site preparation may control herbs and shrubs that already occupy a site, alter physical soil characteristics that might impede root development (e.g., drain an area, or break apart an impervious compacted layer in the soil), and reduce slash and debris (see Chapter 5).

All factors considered, tree planting makes biologic sense when natural means would likely fail. Common circumstances (after Boyd 1976; Pritchett 1979; Barnett et al. 1990) include

1 stands with an inadequate seed source of desirable species or genetic constitution (e.g., on extensive cutover or naturally deforested sites);
2 sites with conditions that make natural regeneration in a reasonable period of time unlikely;
3 community types or sites where natural regeneration commonly produces too many seedlings, and intense intertree competition will slow growth and make a favorable economic outcome unlikely;
4 community types where conditions for establishment remain favorable for only a limited time following a reproduction method cutting; and
5 stands with poorly distributed natural regeneration and unstocked areas that warrant filling through artificial means.

Even when natural regeneration might provide a cohort of adequate density, landowners may rely on tree planting for some special reasons. These include (1) introducing new species and genotypes having greater economic-financial worth, and more uniform rates of growth; (2) creating a tree community of specific attributes, such as a simplified structure and arrangement suited to mechanized harvesting and other cultural techniques; and (3) regulating stand density and other environment conditions to foster better individual tree growth rates, or promote greater uniformity in tree sizes and lower bole characteristics. Where business interests dominate, expected long-term revenues and other values must repay the costs, compounded at an acceptable rate of return. This includes firms who need specific roundwood products that they could not otherwise secure via natural reproduction (Schweitzer and Shuster 1976; Pritchett 1979). In other cases, natural regeneration may not satisfy important nonmarket objectives like enhancing the habitat for wildlife, introducing visual diversity to a landscape, or providing special settings for some kinds of in-place use of a forest. Then foresters must simply demonstrate that natural regeneration would not provide the desired values or nonmarket benefits.

Before proceeding with artificial regeneration, foresters must first evaluate site conditions to determine if the desired species would grow well after establishment, or if an anticipated delay in the establishment of natural seedlings might signal some environmental condition detrimental to a new plantation as well. Particularly in treeless or tree-poor regions, a deficit between growing season precipitation and evapotranspiration often does not favor survival and long-term development of closed forest stands. Then afforestation would likely fail, unless it included unusual measures to mitigate or accommodate the limiting environmental factor, such as (1) using especially drought-resistant species; (2) planting the trees at an unusually wide spacing; (3) reducing and subsequently controlling competing vegetative cover by site preparation and other measures; (4) irrigating the trees during times of moisture stress; and (5) deep planting of large seedlings grown in containers. Though helpful, these techniques may only partially or temporarily compensate for the limited moisture supply, and they increase the costs. Further, the trees may not develop a good form or grow at acceptable rates to repay the costs at an acceptable rate of return. Such factors usually force foresters and landowners alike to question

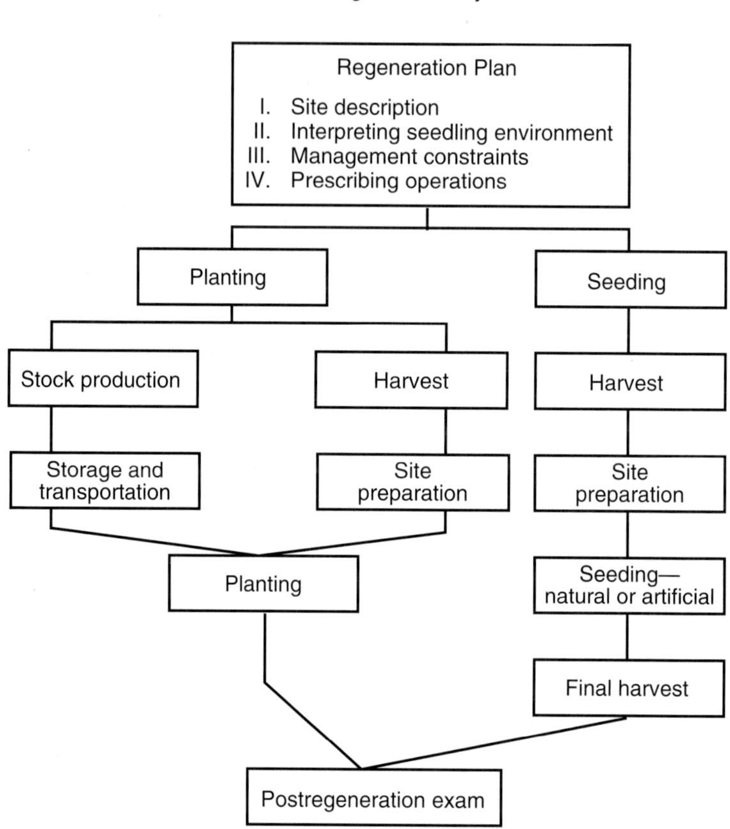

FIGURE 6-1
Some important steps in a natural or artificial regeneration program (from Clearly and Greaves 1976).

the wisdom of large-scale tree planting in treeless regions, except to address special social or environmental needs that justify the costs independent of any return-on-investment criteria.

PLANNING A TREE-PLANTING PROJECT

To make tree planting a truly deliberated process, foresters must evaluate a series of interdependent factors that could influence the outcome. These include choices and operations that nursery managers oversee (Fig. 6-1). The more critical ones (after Soc. Am. For. 1961; Sandvik 1974; Cleary and Graves 1976; Daniel et al. 1979; Morrow et al. 1981; Ek 1982; Sutton 1982; Tuttle 1982; Wenger 1984; Smith 1986; U.S. For. Serv. 1989) include

1 evaluating the site and selecting an appropriate species;
2 harvesting any mature trees that already grow in the planting area;

3 determining if site preparation must precede the planting, and completing the work;

4 finding a seed source of suitable tolerance to environmental conditions, tree form, growth, and other important attributes;

5 specifying the kind of planting stock (seedlings and cuttings) needed, including its size, condition, and other qualities essential to planting success;

6 selecting a reliable nursery to produce the trees in the numbers needed;

7 determining the best time to lift (dig) the trees from a nursery, and replant them at the site;

8 planning a means for transporting the stock in a healthy and dormant condition to the planting site;

9 arranging for an appropriate way to safely store the seedlings until planting;

10 selecting a method for planting;

11 planning an appropriate spacing, arrangement and row orientation to best satisfy the objectives;

12 selecting a well-trained crew to do the work;

13 arranging for appropriate safety equipment, as well as other measures to protect worker health;

14 supervising the operation to insure careful work and consistent use of an appropriate planting technique; and

15 determining how to protect the plantation from damaging pests and other harmful environmental conditions after establishment, and monitoring its condition.

Wise choices minimize the chances of failure. They make the regeneration effort more predictable, provide landowners with utilitarian species and trees suited to the needs, insure timely development and full site utilization by the new community, and set the stage for efficient future management and use throughout the life of a plantation.

Tree planting commences a long-term program of silviculture, including a progressive series of tending operations to insure the health of a plantation, and to nurture its development. In many cases, planting implies a commitment to fairly intensive silviculture, and long-term investments in tree growing (Smith 1986). In fact, landowners must often follow with additional expenditures for early cultural operations, and wait for even these investments to pay back the compounded costs. In many respects, a forester's choices during early stages of planning determine how soon a plantation will provide some desired value (return), when it will support commercial operations to supplant the noncommercial tending, and what kinds and quantities of values it will eventually provide to a landowner.

PURPOSES AND ADVANTAGES OVER NATURAL METHODS

Tree planting and other modes of artificial regeneration may serve a variety of purposes, and many do not involve timber production as a primary or even secondary goal. Instead, landowners may justify the investment primarily for noncommodity reasons, and even plant trees in a setting not traditionally associated with forestry. Common purposes (after Toumey and Korstian 1931; Minckler 1948; Limstrom 1962; Morrow et al. 1981; Laarman and Sedjo 1992) include

1 products—to produce timber, pulpwood, Christmas trees, nuts or fruits, fodder, and other harvestable commodities or materials;
2 protective cover—to stabilize open sites against erosion and other means of deterioration, improve soil conditions through addition of organic matter and nitrogen fixation, or provide protective cover for watershed and other values;
3 landscaping—to create new images on or adjacent to a site, or increase visual diversity across a landscape;
4 animal food and cover—to enhance or create special habitats for a target animal species or guilds of them, or increase plant community diversity of a habitat;
5 windbreaks—to interrupt and redirect wind to a landowner's advantage, and inhibit surface evaporation and erosion from exposed soil; and
6 alter ecosystems—to substitute one kind of plant association (including exotic species) for another, reintroduce tree species that formerly occurred in an area, reestablish forest cover where past land uses removed it, or alter the genetic character of a forest community.

In many circumstances, tree planting serves more than a single purpose (e.g., providing both protective cover and products, or concurrently serving commodity and wildlife needs). In fact, most plantations generally have multiple values, even though a landowner may view many as secondary or incidental to the primary reason for making the investment.

On the surface, these uses of tree planting seem most relevant to forested environments, and to working forest landscapes. Yet foresters' knowledge about the ways trees grow, and the conditions for survival and successful development, apply to intensively used and highly developed areas as well. These include recreation sites and parks, greenbelts along transportation corridors and water courses, and many other kinds of greenspaces found in an urban setting. Such plantings (Richards 1992) may serve to create

1 urban groves, consisting of small closed tree communities in underdeveloped parts of an urban area;
2 urban savannahs, comprising strips of streetside trees, widely dispersed individual yard trees, isolated small clusters of trees and shrubs, or open-grown groves in parks and buffer areas; or
3 paving trees, including widely dispersed individual trees and shrubs (often potted) in parking lots and other artificially surfaced areas.

The order suggests a gradient of increasing harshness in environmental conditions (e.g., of soil physical properties, nutrient status, and moisture stress), and in the difficulty of insuring long-term success. Generally, the importance of tree-to-tree interactions and the effects on the urban-open microclimate and watershed decrease as the space between the trees increases. Conversely, the amenity value (e.g., visual effect) of each individual tree increases as the distance between them increases, and their canopy cover decreases. So does the importance of individual tree survival, appropriate growth and form, and good health in satisfying the long-term objectives (Richards 1992).

Conceptually and functionally, similar principles apply to the transition from closed woods, to fairly open-canopy tree communities, to isolated trees in pots or open spaces in any environment, and at a range of scales. At a broad geographic level, this includes the change in vegetation conditions from an urban center, through the suburbs, and into the more rural hinterland. On a smaller scale, it may represent features within or around a park, a recreation site, or a building complex nestled in the woods. On both scales, as well as in an urban setting, the amenity value of isolated or widely spaced trees and shrubs will usually require foresters to

1 choose a species for both its appearance, and its tolerance to environmental conditions at the exact place for planting;

2 determine how each tree will affect the surrounding space and facilities as it grows older and larger, including the implications of its form (habit) and potential size;

3 design the placement (juxtaposition) of each tree within the available space, and in relation to the surroundings;

4 take cognizance of ways to minimize long-term hazards to the trees, to the structures around them, and to the people who use the space; and

5 incorporate appropriate measures to minimize the costs of postplanting maintenance for both the trees and the grounds and facilities surrounding them.

Landowners will usually spend more for landscape plantings consisting of isolated or widely spaced trees, for maintenance to keep them in good health and condition, and for protection from damage. This will generally hold true for tree plantings around individual structures, for tree groves and clusters of them in parks and recreation areas, and for scenic buffers along roadside corridors and other greenbelts within and around populated areas.

In more traditional forestry, landowners may also appreciate the amenity value of planted trees, yet they generally look more for benefits that accrue from an entire plantation or group of them, rather than from individual trees. This affects the kinds of plantations that foresters commonly deal with, the conditions that they work to create through a forestation effort, and the amounts of time and money they invest in establishing and nurturing any single tree. In fact, for general reforestation purposes, foresters generally justify the costs as follows:

1 Planting (or direct seeding) provides the only means to establish tree cover in drastically altered environments (e.g., open fields), and for suitable sites among prairie, savannah, and other tree-poor ecosystems.

2 It allows complete control over stand attributes like species composition, spacing, arrangement, genetic composition, and time of establishment.

3 Planting lets landowners regulate tree community characteristics where natural regeneration would likely prove suboptimal to the needs, or from an operational or financial perspective.

4 It gives landowners a certain means to successfully establish a new forest community *on time*, and to realize the desired degree of site occupancy *without delay*.

5 Tree planting gives the new stand a *head start* by setting out well-developed, healthy, and vigorous plants, thereby insuring their rapid growth and development.

Primarily, tree planting increases the probability of establishing a new community of trees *when* desired, *where* wanted, and *in a way* that guarantees a sought-after spacing and arrangement. It also often helps to *shorten* the time until a community provides the sought-after values (amenity or commodity), including an earlier repayment on the investment of plantation establishment.

REQUIREMENTS FOR SUCCESSFUL ARTIFICIAL REGENERATION

An artificial regeneration program that includes tree planting will unfold in three stages: (1) developing a plan; (2) readying for the operation; and (3) implementing the prescription. It goes most smoothly when foresters first review the landowners' objectives, and identify the economic and operational constraints that govern the choices. Foresters translate these decisions into a prescription called the *planting plan*—a plan designating the general layout, access, species, methods of planting, site preparation, and time of establishment for a specific area scheduled for planting (after Ford-Robertson 1971). The prescription sets the financial bounds for an operation, designates the areas scheduled for planting at specific times, and helps foresters select the most suitable of the ecologically acceptable options (after Toumey and Korstian 1931).

Preparations commence months or years in advance, with a careful assessment of many biologic and operational factors that might affect success of the program. Specific aspects of the evaluation will differ from site to site, and for different sets of landowner objectives. Even so, foresters must account for four major factors (Toumey and Korstian 1931; Linstrom 1962; Wenger 1984; Smith 1986; Dougherty and Duryea 1991):

1 the conditions of physical environment and existing biota that influence the methods they must use for successful stand establishment, and for sustaining long-term development of the plantation;

2 the species and kind of planting stock best suited to a site and its growing conditions, and that serve the intended purposes;

3 the methods, timing, and design for planting operations that will prove biologically reasonable and economically efficient to apply, and that appear geared to long-term success; and

4 the arrangement of trees (distribution and spacing) to insure consistency with future uses, and to facilitate future management.

Foresters must also compare the potential benefits (financial and nonmarket) from each alternative against the probable costs. This will insure an acceptable economic outcome, as well as biologic success.

Consistent with the institutional requirements of a landowner, foresters must determine what kind of plant materials to use, and how to introduce them to a site. The options include (1) planting—seedlings (bare-rooted or containerized), cuttings or suckers (rooted or unrooted), or grafted seedlings (containerized or bare-rooted); and (2) seeding—broadcast across the site or sown or drilled in spots or rows. Foresters may combine different techniques and sources of materials for any single operation to address a special need. Yet to gain operational efficiency, they usually employ only a single method for each plantation.

SPECIES SELECTION

Each species has a given ecological amplitude, reflected by a range of site and environmental conditions where it will survive and grow at acceptable rates, and remain healthy. Conditions at a site ultimately limit the array of species that a forester can appropriately select (Limstrom 1962; Pfister 1976; Einspahr 1982; Smith 1986). While the makeup of nearby native vegetation communities often reflects a general regional environment suitable to those species, conditions at a particular site may preclude optimal growth and long-term development of some. Also, foresters often find that species growing in adjacent natural stands may not best suit the landowner objectives. Further, past land use and cutting practices may have eliminated some desirable native species from an area, or left them only on a particular class of sites. In other cases, foresters might determine that they can better serve the needs by introducing exotic species that grow in similar environments. Whatever the purpose, the species must fit conditions of the site and the needs of a landowner. No combination of silvicultural methods will mitigate mistakes in planting an inappropriate species, or setting them out in an improper arrangement across a site (Toumey and Korstian 1931).

Several features of a site influence the likely success of either planted trees or seedlings from naturally distributed seed. Important ones (after Limstrom 1962; Stone et al. 1970; Sandvick 1974; Pfister 1976; Morrow et al. 1981; Coffman 1982; Einspahr 1982; Smith 1986; Gholz and Boring 1991; Morris and Campbell 1991) include

1 cover, such as the presence or absence of overstory trees that would shade new seedlings, and the density of herbaceous plants and shrubs that might interfere with seedling survival and growth;

2 surface litter layer, and particularly its completeness and the degree it mulches the mineral soil and protects it against desiccation and erosion;

3 past and current land use, to the extent that these may have altered soil physical and chemical conditions, modified surface relief, and introduced hazards to postplanting survival and development;

4 climatic conditions, such as temperature extremes and regime, the length of a growing season, and distribution of precipitation during critical periods of a year;

5 aspect and slope, as these might influence moisture and temperature levels and regimes, soil depth and physical characteristics, and potentials for efficient and safe machine operation during planting and subsequent tending or harvest;

6 edaphic factors, like the abundance and consistency of available soil moisture, texture and structure (consistence and tilth) of the soil, depth to impervious layers, drainage potential and aeration, and soil nutrient availability and exchange capacity; and

7 biotic agents, with primary emphasis on those that live in, on, or near the site and might affect the growth, development, health, and vigor of trees in both positive and negative ways.

Species matched with the conditions will grow best over the long run, and promise the greatest economic potential by virtue of higher-quality products, more volume per unit of area, and shorter rotations. Those sites also tend to support the best responses to site preparation and subsequent tending (Carmean 1982; Coffman 1982). The stands also have fewer health problems, and may better provide many of the nonmarket benefits.

In some regions, landowners react to predominant market conditions by growing a single species, or small number of them, across large areas of their forest. Under these circumstances, foresters must take special care in judging if all the sites will support those species, and whether planting only one or two over large areas has important health risks. They must also consider the ecologic consequences of establishing extensive areas of fairly simple communities with little diversity of vegetation composition or structural characteristics. This can affect the potential habitat for some other plants and many animal organisms, and may prove consequential for long-term ecosystem stability and diversity. Establishment of single- or limited-species mixtures (*monocultures*) over extensive areas may intensify some local pest problems, primarily if the homogeneous conditions would encourage a rapid buildup of some pest populations to dangerously high levels.

In localities where markets take a wide variety of species, foresters will usually look at a greater array of ecologically suitable species to capitalize upon more diverse market opportunities. Yet in all cases, foresters must take a holistic view (after Toumey and Korstian 1931; Limstrom 1962; Pfister 1976; Coffman 1982; Einspahr 1982; Smith 1986) and look for

1 a proven capacity to grow in the area and under current site conditions, especially for species moved long distances from their natural biologic range;

2 a general suitability to the intended use, including a promise to enhance sought-after noncommodity values, or to offer a good financial return in the available commodity markets;

3 an adaptation to the intended silvicultural system, including a responsiveness to planned cultural operations and methods of harvesting;

4 a compatibility with the growth characteristics of other species in the plantation;

5 a potential for establishment at an acceptable cost by using reasonable methods of planting and site preparation; and

6 a natural resistance to known injurious agents, including the effects of people who would use the plantation for some nonmarket purposes.

For many of these, foresters must use caution in selecting exotic species, and avoid large plantings of those not previously tested in the region. With windbreaks and shelterbelts, foresters must also consider features like crown density, branch and leaf retention, tree form (habit), and pest or disease resistance (Baughman 1991).

USE OF EXOTIC SPECIES

Generally, foresters have best results with exotic species when they carefully match them to conditions of the local environment, particularly with reference to climate and soils. They also benefit from using species with a proven record of growth and development in a new environment (Toumey and Korstian 1931; Wilde 1958; Evans 1982). They should recognize that exotics do not normally withstand extremes of local conditions as well as local species, and that the new environment must prove favorable in all respects to insure high rates of survival and good rates of development (Toumey and Korstian 1931; Wilde 1958; Smith 1986). Prior testing should screen for a range of common edaphic and climatic conditions, and include seed sources from diverse habitats within the exotic species' natural range. It should evaluate resistance to important local pests and diseases, the degree that wood and fruits serve the needs of local markets, and the suitability as habitat for indigenous flora and fauna.

Exotics may include genetically engineered trees, and artificial crosses between species that do not commonly hybridize in their natural environments. The parent species may grow on the same continent, or even at nearby regions or sites (e.g., loblolly x shortleaf pines, or various crosses within *Populus* or *Eucalyptus*). Other crosses involve species from distant parts of the world (e.g., eastern x Himalayan white pines, or European x Japanese larches). In addition, some species brought together through exotic introductions may eventually hybridize naturally, forming a new species with many valuable attributes and an excellent adaptation to the local environment. Foresters might use any of these as potential sources in their reforestation programs.

With both natural crosses and artificial ones made by tree breeders, the resultant gene combinations may give a tree unusual vigor or growth (called *hybrid vigor*), special fitness to difficult environmental conditions within a locality or region, or unique resistance to some pests and diseases. If foresters can propagate those hybrids by vegetative means (as with *Eucalyptus* and *Populus*), they can readily mass-produce the new trees for widespread planting on suitable sites. However, with genera that reproduce primarily from seed, tree improvement specialists normally rely on controlled pollination to insure appropriate crossings, and plant special seed orchards and seed production areas to support nursery operations for large-scale outplanting (Zobel et al. 1987).

In many respects, exotics also include trees (or their seeds) taken from one geographic area or altitudinal zone, and moved considerable distances to other locations within the species' natural range. Benefits from making these transfers might accrue (1) when past cutting practices genetically degraded a local population, and exotic sources would grow better; (2) when trees from one area appear particularly resistant to a disease or pest that also occurs at another locality; or (3) when an exotic source tolerates a particular localized environmental condition better than the native trees (e.g., moving northern sources southward for planting on north-facing slopes, or to accommodate hazards from freezing temperatures early in the growing season). In some cases, foresters face special risks in moving a source long distances along an environmental gradient (e.g., moving northern sources to southern sites, or those from low altitudes to high elevations), and should test the nonnative sources before attempting any large-scale planting. Some forestry agencies have established seed transfer guidelines to minimize these risks.

EXAMPLES OF SUCCESSFUL EXOTIC SPECIES

Historically, silviculturists have established plantations of many different exotic species throughout the world. General purposes (Stoeckeler and Williams 1949; Evans 1982; Zobel et al. 1987; Laarman and Sedjo 1992) included

1 wood, fuel, Christmas trees, fruits, nuts, and other products;
2 protective cover to stabilize eroded and nutrient-depleted sites;
3 landscape plantings within urban areas, and in the rural hinterland;
4 shade trees around homes and settlements, and in grazing areas;
5 shelterbelts and windbreaks to protect crops and stop erosion;
6 fodder production for wild and domesticated animals; and
7 living fences to delineate field boundaries and control the movement of wild and domesticated animals.

Many plantations have additional value as nesting sites for wildlife, and offer food and cover for both wild and domesticated animals. On the other hand, custom may deter local industries from using an exotic for traditional products, even though the wood or fruits have properties similar to a native species. In other cases, the exotic will have poorer qualities and may not serve local needs as well, even if it survives and grows adequately in the new environment (Laarman and Sedjo 1992). Exotic plantations may also not serve some important noncommodity uses, particularly if a species has poor form or other attributes that hinder recreational and other social uses.

Exotic plantation forestry for wood and other commodities in temperate regions has dominantly used imported species of pines, spruces, larches, poplars, and eucalyptus (Zobel et al. 1987). This includes some special hybrids as well. For example, foresters have successfully planted Norway spruce from northern Europe, and larches from Europe and Japan, for reforestation of former agricultural sites in northeastern North America. Many landowners in the region also extensively planted Scotch pine and Douglas-fir for Christmas trees. A loblolly x shortleaf pine hybrid has shown a high degree of resistance to fusiform rust in the southeastern United States, and a pitch x loblolly pine hybrid seems to grow well on dry sites in the same region. Foresters in New Zealand, Australia, and Chile have successfully planted Monterey pine from California over extensive areas. Others have grown Caribbean pine and *Gmelina* in both Africa and South America, and Douglas-fir and Sitka spruce from the northwestern United States in Great Britain (Johnson 1976; Pritchett 1979; Evans 1982; Zobel et al. 1987; Laarman and Sedjo 1992).

Foresters have also used exotics extensively for landscaping in cities, villages, and around home sites. These included species moved from one region of a continent to another, and across oceans. Successful introductions in northeastern North America include Norway maple, Chinese elm, Norway spruce, Shumard oak, ginkgo, European beech, Douglas-fir, and Colorado blue spruce (Waterman et al. 1949). In the southeast, European and American beech, camphor-tree, mimosa, and West Indies mahogany have grown well (Lindgren et al. 1949). Siberian elm, ailanthus, European white birch, Chinese elm, Russian-olive, eastern black walnut, ponderosa pine, white fir, Douglas-fir, and Austrian and Scotch pine have succeeded as landscaping species among the plains states (Wright and Bretz 1949). In areas of the Rocky Mountains and Pacific Coast, species like ailanthus, Siberian elm, Norway maple, Russian mulberry, Lombardy poplar, Scotch pine, Camphor-tree, red ironbark, ginkgo, sweetgum, southern red oak, tulip-poplar, northern red oak, European linden, Deodar cedar, and Himalayan pine have succeeded (Childs 1949; Wagener 1949; Gill 1949).

Not all experiments with exotic species proved satisfactory. In fact, foresters have often seen new problems developing with exotic plantations (Zobel et al. 1987) as a result of

1 unfavorable soil conditions related to acidity, nutrient deficiencies (including trace elements) or excesses (toxicity), structure and depth, and the absence of essential mycorrhizae;

2 limitations of climate due to drought, extremes of heat and cold, and violent weather (blowdown and breakage); and

3 pests and diseases not common to the species in its natural range, or imported into the new environment.

In some cases, the plantations fail soon after establishment. Other problems may not become apparent for many years, and then only as substandard growth and development. In some cases, a pest that caused only minor damage in the exotic's natural habitats may also appear in the new environment, for reasons not clearly understood. Then because of the lack of natural controls over its development, the pest or disease may cause severe damage. In some cases, environmental conditions at a new location may affect tree vigor to an extent that a species becomes particularly susceptible to devastation by pests that seemed only minor in the natural habitat, or by local pests and diseases that normally do little damage to healthy indigenous trees (Wilde 1958; Smith 1986; Zobel et al. 1987). Foresters deal with these potential problems by selecting an appropriate species or seed source, by timely cultural operations throughout a rotation (Zobel et al. 1987), and by

1 using only resistant species or seed sources, and species with other attributes of local interest;

2 including seed or cuttings from several sources or clones to insure a high degree of genetic diversity;

3 taking precautions with sanitation measures when moving new plant materials into and through an area, and when growing the nursery stock and later transporting it for outplanting;

4 limiting the plantings to particularly suitable sites that will sustain adequate long-term growth and development of the species; and
5 programming special protection measures to safeguard stands of exotic trees after their establishment.

These measures can only supplement the advantages of appropriately matching the species and site, and choosing sources with a known tolerance of resistance to local pest and disease problems.

WINDBREAKS AND AGROFORESTRY IN TREE-POOR REGIONS

Forestation programs in treeless and tree-poor regions have generally required extensive use of exotic species, including both trees and shrubs. These often served as shelterbelts and windbreaks (Stoeckeler and Williams 1949; Frank and Netboy 1950; Davis 1976; Cassel and Wiehe 1980; Nyland et al. 1983; Brandle and Hintz 1987; Baughman 1991; Laarman and Sedjo 1992) that

1 interrupt or redirect air movement to reduce wind erosion, or slow surface evaporation;
2 shelter or protect crops, domesticated animals, home sites, roads, and recreation areas;
3 provide unique habitats for wild creatures;
4 create visual diversity across a landscape;
5 provide recreation sites for landowners and people who live nearby;
6 yield nuts, fruits, and other foods for people, cattle, and wildlife; and
7 provide limited amounts of wood for fuel and other purposes.

At high latitudes, windbreaks also trap snow, so it accumulates in drifts on the downwind site (Dronogen 1984). This indirectly adds moisture to the soil, and may improve the overall water balance to the benefit of crops in adjacent fields. People have even planted windbreaks to trap drifting soil, hoping to slow the migration of desert into grassland ecosystems (Evans 1982; Laarman and Sedjo 1992).

Generally, a windbreak forces the air upward (Fig. 6-2), reducing wind speeds to the leeward side for up to twenty times the average height of trees in a belt (Brandle and Hintz 1987). The actual distance depends upon the heights of trees at the center of the break, the density of the canopy, and the cross-section shape of the planting (Stoeckeler and Williams 1949; Drongen 1984; Brandle and Hintz 1987). For best effects, foresters must also orient the plantings at right angles to the direction of prevailing winds. They put the fastest- and tallest-growing trees at the center, with parallel rows of shorter tree species toward the upwind and leeward sides, and outer rows of shrubs. This creates a vegetation slope to divert the wind upward as it passes over a windbreak.

People throughout the world have also integrated tree planting with the growing of agricultural crops to serve many different needs. These agroforestry systems (Evans 1982; Matthews 1989; Nair 1991; Laarman and Sedjo 1992) include

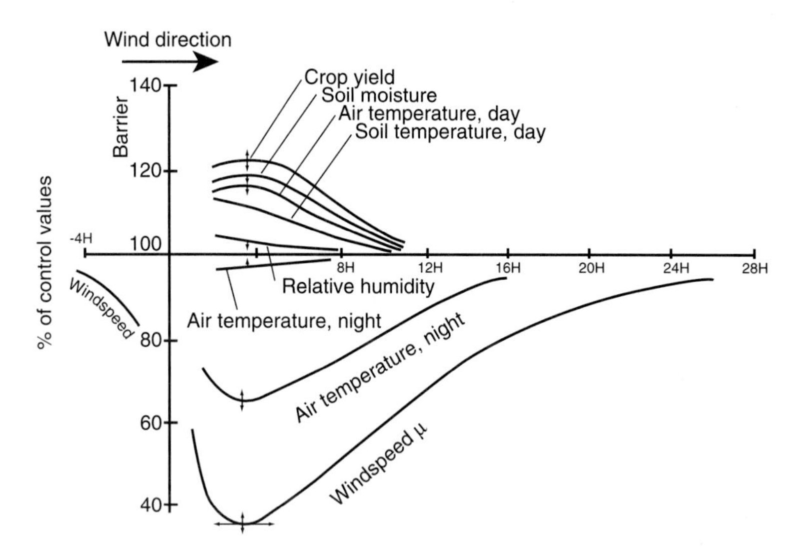

The arrows indicate the directions that values of different factors vary relative to the central values measured in unsheltered areas. H = height of barrier.

FIGURE 6-2
Windbreaks and shelterbelts reduce wind speed and temper environmental conditions in adjacent fields and grasslands (from Brandle and Hintz 1987).

1 agri-silvicultural systems—involving concurrent production of agricultural and tree crops;
2 sylvo-pastoral systems—integrating wood production and rearing of domesticated animals in a forested setting;
3 agro-sylvo-pastoral systems—combining production of agricultural and forest crops with raising of domesticated animals; and
4 multiple-purpose systems—incorporating methods of cultivating trees for wood, and for leaves and fruits to use as food and fodder.

Most agroforestry systems have served as a low-input means for subsistence agriculture and wood production in developing countries of the tropics and subtropics. They often involve relatively small and dispersed areas that serve local needs, upgrade land use, and promote local development. Plantings include timber trees, bamboo, palms, rubber trees, shrubs, and other woody plants. These generally have both economic and ecologic objectives, and often require intensive culture to insure success. The plantings have additional value as windbreaks, as greenbelts along roads and other transportation corridors, as plantings in village commons, and as protective cover on waste areas.

Conceptually, agroforestry incorporates the simultaneous cropping of two or more species on the same land. This may involve multiple cropping in two forms (Evans 1982): (1) *sequential cropping* of two or more species on the same land in a single year; and (2) *intercropping* of two or more species on the same land at the same time. Generally, the trees outlast the food crop, and people can often grow the food plants only during early stages of plantation establishment. This has proven beneficial to the forestry objectives in many cases. In fact, sequential cropping has served as an effective means for site preparation and early weed control until the tree community dominates a site.

FIELD ASSESSMENT OF SITE CONDITIONS

Many species survive in a broad range of physical environments (climatic, edaphic, and topographic features). Yet each grows best under a more limited range of soil pH, texture, depth, nutrition status, and moisture supply. Trees also grow best when the soil contains *mycorrhizal organisms*. These fungi form a symbiotic association with the tree to enhance absorption of moisture and nutrients through the roots. Most trees also have little tolerance to climatic conditions greatly different from those in their natural range. In addition, the indigenous biotic community may either interfere with their development or enhance their growth and health.

On a regional or landscape scale, an ecosystem or habitat classification serves to integrate effects of many site factors, and to reveal broad geographic areas where a species regularly grows well (Pfister 1976; Barnes 1982; Kotar and Coffman 1982; Gholz and Boring 1991). Particularly in many mountainous regions, conditions vary considerably with altitude and slope steepness, often causing a natural stratification of species by life zones related to elevation (Holridge et al. 1971). On a smaller landscape scale in these zones, and in regions of more moderate topography, soil texture, depth, and drainage may also change along topographic gradients. These usually affect the water and nutrient holding and exchange capacities of a soil, the total volume of moisture and nutrients available to the trees, the potential for root penetration, and the aeration. In less hilly areas of moderate topography, past land often altered these features, particularly where careless agricultural practices led to erosion of the upper soil horizons.

Frequently, local experience has identified broad landscapes having growth-limiting conditions for some species, including effects of important pests and diseases. Then foresters must exclude a species from their reforestation projects. For example, imperfectly drained flatwood sites along the Atlantic and Gulf Coastal Plain of North America have longleaf and slash pines as a predominant tree species, while evergreen hardwoods naturally occur on the well-drained sandy hammocks within the same region (Wilde 1958). Compared with slash pine, longleaf has less tolerance to flooding during juvenile stages, and this can limit plantation success at low sites (Shoulders 1990). In the Rocky Mountains, blister rust kills western white pine and causes important economic losses in pure stands (Adams 1995). In these cases, local foresters have both empirical evidence and past experience to identify a general array of unacceptable species for different physiographic zones and sites, or find ways to minimize the potential problems through appropriate management practices.

IMPORTANCE OF SOIL CONDITIONS

Edaphic features often prove most consequential to both initial survival and satisfactory long-term growth and development (Toumey and Korstian 1931; Wilde 1958; Stone et al. 1970; Morris and Campbell 1991). These include water supply and nutrients, and some soil organisms. The supply of available growing season moisture often proves a major determinant. Depth to the water table, and effects of soil texture on the extent of capillary fringe, influence this condition. Where a soil includes a compacted or impervious layer, water may inundate the rooting zone for part or all of the year, and impair the growth of all but shallow-rooted species. For example, in medium- to fine-textured soils of south-central New York, sites with gentle slopes and a shallow impervious layer have high water tables early in the growing season, and following prolonged precipitation. This makes the site unsuitable for red pine, but not Norway spruce (Stone et al. 1970). By contrast, an impervious layer in a coarse-textured soil would slow percolation and improve growing conditions during dry periods.

Soil texture and rooting depth also influence aeration, the ease of root penetration, the degree of anchoring, and the volume of soil that tree roots spread through (Wilde 1958; Pritchett 1979). This latter feature determines both the adequacy and consistency of moisture supply, and the quantity of nutrients and moisture available to the vegetation community. Amounts of organic matter incorporated into the upper horizons or lying on the surface temper a soil's moisture-holding capacity, nutrient supply, texture and structure, pH and cation-exchange capacity, and permeability to roots and water. The organic layer also mulches the soil, reduces surface evaporation, holds free water from running off, enhances infiltration, and tempers daily and seasonal changes of soil temperature. However, slope steepness and position influence many hydrologic characteristics of soil, and particularly soil drainage via subsurface flow.

Foresters often dig pits or use an auger to examine general physical attributes of the different soil horizons, determine the rooting depth, and measure depth to any limiting feature. Trained observers can identify areas of poor drainage by looking for mottling within the soil profile, and comparing the depth to mottles with the requisite rooting zone for a species of interest. They study the topographic features for signs of changes in soil depth and drainage (Wilde 1958; Stone et al. 1970; Pritchett 1979; Morris and Campbell 1991). These vary with the shape and steepness of a slope, and the position of a planting site along the topographic gradient. To illustrate, convex slopes generally have better internal soil drainage than concave areas, but may have eroded more over geologic time. The detached fine particles will have accumulated in the concave areas farther down the hill, increasing the soil depth, moisture supply, nutrient status, and texture of coarse-textured soils along the lower slope. Other concave areas may have poor drainage because the fine soil materials formed into an impervious layer that substantially limits rooting depth. Exact conditions vary with locality and soil texture, and require on-site inspection to interpret. So will other features of the habitat and its suitability to the species under consideration (Barnes 1982; Kotar and Coffman 1982). Foresters must judge the extent that environmental conditions might preclude success of a species, and map the boundaries of areas suitable for each species under consideration.

KINDS OF PLANTING STOCK

The term *planting stock* means seedlings, transplants, and cuttings suitable for planting (Ford-Robertson 1971). Tree nurseries supply these as bare-rooted or container-grown seedlings, or unrooted cuttings. They may offer trees of different ages and sizes. The choice depends upon environmental conditions at a plantation site, and the degree and kind of mechanization used for the planting operation. Generally, foresters prefer to plant the largest and best-developed seedlings that will survive under site conditions of a given environment (Daniel et al. 1979; Wenger 1984; Duddles and Owston 1990; Owston 1990), and that they can handle efficiently and economically. They generally look for seedlings with a good balance between tops and roots.

In current parlance, foresters base their planting stock selection on a target seedling concept. *Target seedling* means a tree seedling with structural and physiological traits that insure high survival and appropriate development following outplanting (Rose et al. 1990). This normally translates into features (Barnett et al. 1990; Rose et al. 1990) like the following:

1 suitable height and stem diameter;
2 good balance of roots and foliage;
3 adequate functioning foliage and viable buds;
4 a root system with numerous tips capable of elongating;
5 an adequate internal supply of nutrients and moisture;
6 appropriate branching of tops and roots; and
7 a tolerance to adverse conditions of the field environment.

While inherent genetic factors affect many of these attributes, seedling conditions mostly reflect the regimes used to grow them, and effects of handling after lifting from nursery beds (Wilde 1958; Johnson and Cline 1991).

While ongoing research continues to reveal effective criteria for evaluating seedling condition, available information suggests that foresters can use readily measured morphological features as an index of seedling quality (Forward 1982, Johnson and Cline 1991). These include (1) seedling color; (2) shoot length; (3) root mass and architecture; (4) root collar diameter; and (5) the shoot-root ratio (the balance of biomass between roots and top). Exact criteria depend upon conditions at the planting site (particularly during the first year), and the species. Where any combination of environmental factors increases the chances of poor seedling growth and greater mortality, foresters may prefer to use seedling *transplants* (Owston 1990). Nursery managers grow these seedlings for one or two years in regular seedbeds.

FIGURE 6-3
By growing seedlings in small individual containers in climate-controlled greenhouses, nurseries can shorten the production time to just several months, and produce healthy and vigorous trees that tend to survive and grow well following outplanting, even at fairly harsh sites.

They then lift and replant the seedlings at a wider spacing, and grow the transplanted stock for another one or two years before final lifting and shipping (Ford-Robertson 1971). Alternately, foresters have had good success using container-grown seedlings for planting harsh or difficult sites (Sloan et al. 1987; Brissett et al. 1991). Nurseries produce container stock by growing individual seedlings in small containers, and normally in climate-controlled greenhouses (Fig. 6-3). They may either ship the seedlings in their containers, or remove the stock before shipping. In the former case, the containers protect the roots from damage during handling, and insure that the nutritious rooting medium remains intact around the root system (see Chapter 7).

To insure success with all kinds of seedlings, foresters must have healthy and vigorous planting stock, with full and fibrous root systems, and a general balance in the size of tops and roots (Benson 1976; Forward 1982; Mexal and South 1991). Nursery managers will regulate these attributes by

1 selecting appropriate seed sources;
2 controlling seedbed density to reduce intertree competition;
3 controlling harmful insects and diseases;
4 eliminating interfering plants from the seedbeds;
5 irrigating and fertilizing the stock to enhance seedling growth;

6 inoculating the soil with beneficial mycorrhizae fungi;
7 pruning the tops and roots to limit seedling height and insure compactness of the root systems;
8 controlling shoot growth, budset, and dormancy by regulating the photoperiod; and
9 growing the seedlings over longer rotations to produce larger plants.

They may choose special seed sources to accommodate unique site conditions, particularly where shortages of soil moisture might affect seedling survival and development, or where freezing temperatures might occur early in the growing season and kill or damage the young trees.

COMPARING BARE-ROOT AND CONTAINER STOCK

Overall, bare-rooted or container-grown planting stock each has certain distinct biologic and operational characteristics that weigh in its favor or prove a disadvantage, depending upon the circumstances. Foresters must consider these in selecting the kind of stock to use for a particular operation. For example, bare-rooted seedlings normally have these advantages:

1 lowest production costs, if grown over short rotations and using mechanized nursery operations;
2 reasonable cost for bulk storage, transport, and handling;
3 readily tailored for size and condition by altering seedling age and some cultural methods (e.g., root pruning and fertilization); and
4 easily handled during field operations.

Normally, nurseries require one to two years to grow bare-root seedlings to suitable sizes. For many species, they can produce container seedlings in one-third to one-half the time. Container stock also has certain biological and operational advantages (Stein 1976; Daniel et al. 1979; Wenger 1984; Barnett et al. 1990; Brissette et al. 1991), such as the following:

1 The seedlings remain suitable for planting over an extended season, with less need to strictly control dormancy.
2 The containers protect the roots from mechanical injury and desiccation until planted.
3 Leaving the rooting medium around the roots at planting provides an enhanced microsite that favors good root-soil contact.
4 The standardization of size (due to a common container diameter and length) makes container seedlings readily adapted to mechanized handling, and to certain kinds of machine planting.
5 Workers can plant the smaller individual container-grown trees more quickly and without bunching and twisting of the roots.

On the other hand, container stock takes more logistical support to produce, and to transport the seedlings to and around a planting site. As a result, container stock may cost up to two times more than bare-root trees.

OTHER FACTORS RELATED TO SEEDLING QUALITY

Experience throughout the world has clearly shown that seedlings and cuttings of poor quality and condition will not survive or grow well, even for a species suited to the prevailing site conditions. Local planting guides normally identify thresholds for critical seedling attributes, with special reference to the common species and general growing conditions of a region (Cleary and Graves 1976; Pfister 1976; Morrow et al. 1981; U.S. For. Serv. 1989; Morris and Campbell 1991; Pait et al. 1991). Most list minimum values for shoot length, root collar diameter, shoot-root balance, and other easily recognized seedling characteristics. Foresters can assess nutrient status and health by looking at foliage color. Healthy trees have a deep and uniform leaf color, and firm buds and foliage. Also, most low-vigor seedlings have small buds, and a short and slender stem. Those with nutrient deficiencies exhibit unusual coloration, and possibly dead tips to the foliage (Fig. 6-4). Among conifers, the needles may have become yellowed, sparse and short, or somewhat distorted. Knowledgeable observers can readily see these symptoms by inspecting the seedlings in a nursery during the growing season (Wilde 1958; Bigg and Schalau 1990; Johnson and Cline 1991).

FIGURE 6-4
Hardwood seedlings in poor health will have smaller leaves with unusual color patterns, like this paper birch with a potassium deficiency (*courtesy of U.S. Forest Service*).

Even seedlings of apparently good health and vigor will not readily survive outplanting once they break dormancy and the new root tips and shoots emerge (Hocking and Nyland 1971; Burr 1990; Johnson and Cline 1991). They do not withstand a variety of environmental stresses. Further, tender tissues of the newly emergent shoots often break off during routine seedling handling, reducing the proportion of branches and roots that can elongate and produce new functioning parts. Exposure to high air temperatures and the limited capacity for moisture uptake often also lead to desiccation, and eventual mortality. To guard against these problems, foresters should insist on fully dormant seedlings with firm and closed buds, no signs of active shoot elongation within the root system, and no sign of desiccation.

SELECTING A SPACING

In planning a planting operation, foresters must decide how to accommodate several long- and short-term operational needs. These include (1) spacing between the planted seedlings; (2) orientation of the rows; (3) locations of access lanes and no-plant areas; and (4) the method of planting. These depend upon the species as well as the kinds of products that the landowner wants, the intended noncommodity uses, the program of tending that will eventually follow, and the length of time to grow the trees to some desired condition or stage of maturity (e.g., minimum size). To some degree, the arrangement will also reflect limitations related to slope steepness and aspect, soil drainage and texture, and known obstacles to planting or subsequent use of a stand. Row orientation also depends upon the juxtaposition of a plantation to existing access roads, or features of the landscape that would block ingress or egress for management, use, or eventual harvest.

In practice, foresters plant trees at a greater spacing than common in naturally regenerated stands. This reduces the cost of seedlings and labor for stand establishment, and in some cases the investment or site preparation as well (e.g., for band, strip, and spot methods). Spacing also determines

1 when the crown canopy closes within a community;
2 when canopy shading suppresses understory vegetation;
3 when intertree competition intensifies;
4 when lower branches begin to die, and the live crown length begins to decline;
5 when rates of diameter growth begin to slow; and
6 when the weakest and most suppressed trees begin to die due to intertree competition.

To illustrate, 25-year-old European larch trees in plantations at a 14-foot (4.3m) spacing had 1.5 times deeper crowns than those planted at a 6.6-foot (2m) interval, yet lower branches had grown twice as large in diameter (Morrow 1978). Also, 12-year-old ponderosa pine trees at an 18-foot (5.5m) spacing had 1.3 times wider crowns than those at a 6-foot (1.8m) interval, and 1.8 times greater branch diameters (Oliver 1979). Figure 6-5 illustrates similar effects on loblolly and slash pine (from Smith and Strub 1991), ponderosa pine (from Oliver 1979), and Douglas-fir (from Reukema 1979). Generally, the closer the spacing, the shorter the time until the crown canopy closes, and the more rapidly the number of trees declines. With particularly close spacing, at least some species will also not grow as tall (Reukema 1979; Smith and Strub 1991).

Close spacing has some advantages when growing trees of high quality. The lower branches die sooner in closely spaced stands, and never become large in diameter. This increases the chances of exfoliation at an earlier age. In addition, closely spaced plantations produce more total wood volume during early stages of stand development, but primarily on large numbers of small and slow-growing trees. Early crowding also leads to high rates of mortality, and particularly among the shortest trees with the smallest crowns. With wider spacing the trees initially grow bigger faster, and reach minimum merchantable sizes sooner. This provides landowners with earlier revenues to help repay at least part of the initial investment for plantation establishment (Toumey and Korstian 1931; Daniel et al. 1979; Smith 1986).

Foresters must find a spacing that best satisfies a landowner's purposes, and fits the silvical characteristics of the component trees. In plantations intended for commodity production, the interval should (1) insure a maximum volume and highest quality of sought-after products in the shortest time; (2) minimize the cost for establishment, consistent with the goals; and (3) facilitate eventual harvesting during intermediate tending, and for the reproduction method at the end of a rotation. For nonmarket objectives, foresters must select a spacing that optimizes the hoped-for noncommodity values, even if it makes no sense from the perspective of traditional plantation forestry. This might include some kind of irregular, clumped, or random spacing that more closely mimics natural stands, or curved or crooked rows to make the plantation appear less artificial and more complex.

Foresters generally set out the minimum number of trees that will insure full site occupancy in due course, but not sacrifice stem and stand quality relative to the management goals. This depends on a variety of factors (after Toumey and Korstian 1931; Limstrom 1962; Morris et al. 1981; Smith 1986):

1 objectives for the planting;
2 numbers and kinds of intermediate tending operations planned;
3 probable rates of individual tree mortality, and the effect on plantation density and tree vigor;
4 risks from pests and diseases, and how spacing might influence the success of various harmful agents;
5 requisite tree characteristics and wood quality to satisfy the long-term objectives;
6 likelihood that some natural regeneration will supplement the planted trees;
7 obstacles to planting related to site features, or caused by prior operations at the site;
8 branching characteristics of the species, its shade tolerance, and the ease with which dead branches fall from the trees;
9 methods of planting, and the equipment used; and
10 degree of mechanization likely for future tending and harvesting operations.

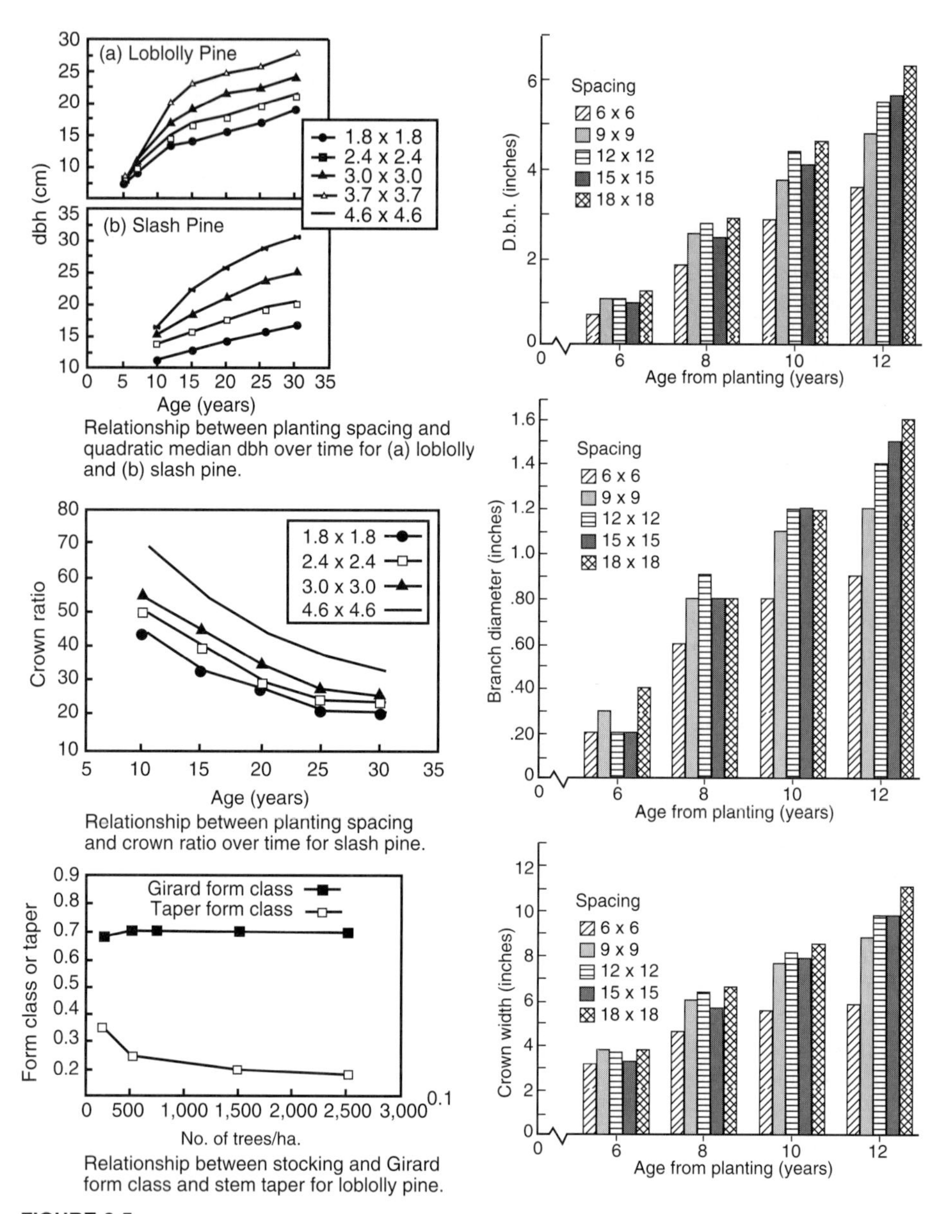

FIGURE 6-5
Intertree spacing will influence several individual tree attributes, like these for loblolly and slash pine (from Smith and Strub 1991), ponderosa pine (from Oliver 1979), and Douglas-fir (from Reukema 1979).

Foresters would also consider any other factors that influence the planting costs, and the potential for a satisfactory return on the investment.

Within these constraints, foresters can select a balanced or square spacing, or some other arrangement. Examples of the former might include a 6-ft x 6-ft (or

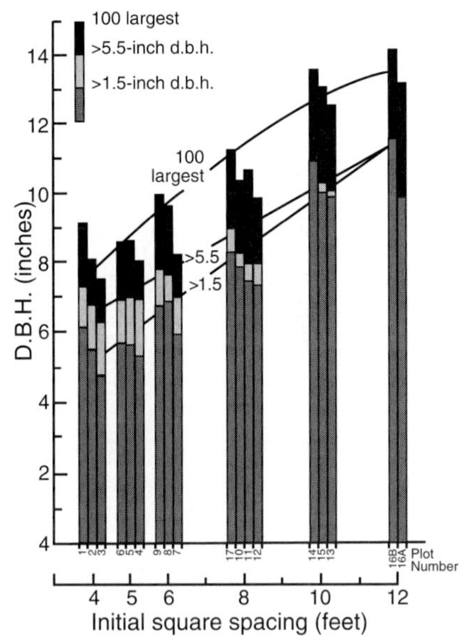

Average d.b.h. at age 53, by stand component.

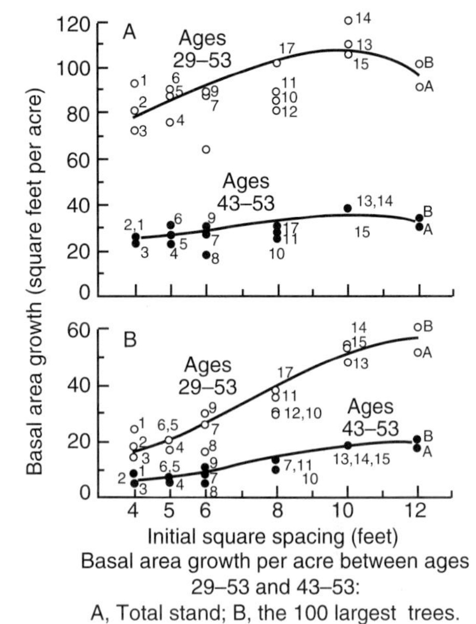

Basal area growth per acre between ages 29–53 and 43–53:
A, Total stand; B, the 100 largest trees.

FIGURE 6-5 (*continued*)

1.8-m x 1.8-m) or 10-ft x 10-ft (or 3-m x 3-m) interval. A rectangular spacing of 7 ft x 9 ft (2.1 m x 2.7 m) or 6 ft x 10 ft (1.8 m x 3 m) would give more space between the rows. This might facilitate future harvesting, or accommodate other special needs. It offers a practical compromise for insuring adequate space for different machines, plus timely crown closure. Foresters could also use a triangular spacing, with a regular distance between trees in all directions. Conceptually, this better accommodates the circular shape of a tree crown along a horizontal plane, and leads to more uniform competition all around a tree. It might also enhance the visual qualities by making a plantation appear less regular and more complex. Yet, in a practical sense, planting crews can control the spacing between and within rows reasonably well, but would have great difficulty aligning trees perfectly across rows. As a consequence, plantations rarely have an exact geometric spacing between the trees.

ARRANGEMENT AND OTHER OPERATIONAL CONSIDERATIONS

Foresters must also select an orientation for any rows, and decide how best to access a plantation for future tending or other purposes (e.g., for protection). In most cases, they can simply orient the rows at a 45 to 90° angle to an existing road and access lane. For moderate to steep sites, the rows must run up- and downhill to insure safe machine operation. This may preclude direct access to an existing road, and require an additional trail from the roadside to the beginning of each row. Obstacles like streams, swampy areas, ponded water, rock ledges, and other landscape features may also hinder machine operations. Then foresters must change the row orientation to facilitate movement around these features. They might also bend or curve some of the rows to bypass an obstacle, or leave special unplanted trails for that purpose. Most of the time, the topography and the existing road or trail system give foresters some clue to the best row orientation, and indicate where to reserve special unplanted access lanes to make entry easier in the future. Often, they simply leave unplanted rows at fixed intervals to facilitate access.

Foresters will commonly designate some areas as no-plant zones. These have the purpose of (1) precluding useless investments, by excluding microsites that will not sustain good tree growth; (2) facilitating future management and use, by providing efficient and unobstructed access through a site; (3) accommodating special needs and interests of a landowner, by maintaining some kind of vegetation or surface condition other than trees; or (4) permitting natural vegetation to dominate part of a planting site. Experience has also shown that young trees often grow poorly within 1–2 times the height of an older adjacent community, apparently due to shading or withdrawals of soil moisture and nutrients by the taller trees. This includes hedgerows along fields and roads. Leaving these areas unplanted provides access around a plantation, and space to turn machinery and stockpile logs. Unplanted strips may also serve as firebreaks, and buffers around buildings. Exact reasons differ with the circumstances and the objectives of ownership.

GETTING READY FOR THE PLANTING

After carefully evaluating the opportunities and assessing the site conditions, foresters must articulate their decisions in a planting plan that a landowner or supervisor can study and eventually approve. This plan will normally include maps to show

1 the location, boundaries, and position of a plantation relative to important features of the landscape and the forest as a whole;
2 adjacent roads and existing access corridors;
3 the proposed orientation of rows;
4 the location and orientation of access lanes and other no-plant areas within the plantation;
5 subdivisions of a site where crews will plant different species, use a different spacing, or change the orientation of rows; and
6 cultural features within and adjacent to the planting area.

For reforestation operations involving multiple sites, a larger areawide map should also show boundaries of the different plantations, and highlight the species composition and general characteristics of each one. Such information reduces the written materials included with a planting plan, and gives field supervisors a practical means for controlling operations once the planting begins.

Table 6-1 lists an array of factors that foresters must deal with in formulating a planting plan. Each requires a deliberate decision. In some cases, a preliminary field

TABLE 6-1
CHECKLIST OF FACTORS TO CONSIDER IN PLANNING A TREE PLANTING OPERATION.

Species Selection
- owner purposes
- anticipated markets
- soil conditions
- climatic factors
- pest problems
- on-site competition
- intended rotation
- intended management

Spacing
- species habit and shade tolerance
- time to first treatment
- intended products
- facilitate operations
- enable general access
- matched to planned silviculture

Arrangement
- fitted to stand area
- recognize cultural features
- facilitate access
- match slope constraints
- suited to intended uses
- facilitate machine operations

No-plant Areas
- due to on-site competition
- edge effects
- physical site constraints
- cultural features
- needed access
- alternate uses
- trafficability
- relevant to species

Site Preparation
- soil amelioration
- vegetation control
- pest management needs
- cost factors

Operational Considerations
- slope conditions
- obstacles
- wet spots
- potential site harm
- machine requirements
- supplemental activities
- planting method
- intended harvest system

Special Considerations
- habitat needs and effects
- alternate uses of the site
- site protection needs
- postplanting protection
- associated with obstacles

Economic Factors
- eventual payoff and costs
- least-cost alternative explored
- cost-saving measures considered
- markets potentially certain
- purposes clearly understood

Seedling Handling
- efficient and prompt delivery
- storage in cool place
- damage during planting prevented

Forgotten Items
- related to objectives
- about the site
- involved in the management
- directed toward use
- specific to pests
- related to operations
- anything else of importance

All factors deserve deliberate attention in planning each tree-planting project.

assessment will indicate what factors require no further attention. Others take a detailed evaluation, including assessments of the soils, existing vegetation, climate, indigenous pests, markets, and landowner preferences. At times, the assessment uncovers a problem that limits the choices or dictates extra measures to insure success. Foresters eventually summarize their choices in a separate written plan. These written documents will usually include

1 a summary of the species and number of seedlings required;
2 a description of the requisite site preparation, the method of planting, and other ancillary measures to insure success;
3 a proposal for transporting and storing the seedlings, and protecting them during planting;
4 a list of labor, machinery, and other logistical support to accomplish the work;
5 an enumeration of costs for all work and materials; and
6 any other information not readily conveyed by maps or worksheets.

Well in advance, foresters must order the seedlings, arrange for the necessary equipment, hire the work crews, and complete any site preparation. For large planting programs, they may need to elicit support from payroll and personnel staff, and arrange for other logistical services. They must have a commitment for adequate numbers of well-trained and experienced staff to serve as crew chiefs, and to fulfill other technical requirements.

7

NURSERY AND TREE PLANTING OPERATIONS

SEED SELECTION AND HANDLING

The health and physical condition of tree seedlings at the time of planting determine the initial rates of survival and development. The long-term growth potential depends in large measure upon their genetic constitution, and their adaptability to environmental conditions at a planting site (Nienstaedt and Snyder 1974; Rehfeldt and Hoff 1976; Zobel and Talbert 1984). This potential may reflect genetic differences related to the *provenance*—the broad geographic origin or location of a seed source. Provenance implies some effect of the environment on the phenotype, and in determining the genotypes that succeed or fail under the prevailing conditions. Within a provenance, additional genetic differences may result from variations of site conditions, among different communities of trees growing in similar physical environments, and between individual trees in a single stand (Zobel and Talbert 1984). Poorly adapted trees may suffer damage from adverse climatic conditions or indigenous diseases and pests, or they may just grow poorly or die (Rehfeldt and Hoff 1976).

Generally, foresters depend upon nursery managers and forest tree improvement specialists to identify the seed sources appropriate to a region. In addition, they require provenances that satisfy a wide array of landowner objectives. For example, they might want a different source for Christmas trees than for growing sawlogs over long rotations. Also, some provenances may have a resistance to important pests and diseases. In fact, nurseries may provide different sources of a single species. Then foresters may recommend the most appropriate one for a particular plantation. When a firm or agency has its own nursery, its foresters will often oversee the seed collection, and can select from the sources that they consider best suited to the need.

For field expedience, foresters assume that the outward appearance mirrors a tree's genetic capacity for growth and development in the physical environment at a planting site. Tree improvement specialists can judge the real potential only after planting trials (called *provenance* or *progeny tests*) across a range of site conditions common to their region (Nienstaedt and Snyder 1974). Commonly, these plantings screen for features (after Nienstaedt and Snyder 1974; Bonner 1991; Pait et al. 1991), like (1) individual tree characteristics of silvicultural importance; (2) desirable attributes for specific products, or other end uses; and (3) resistance to locally important insects, diseases, and extremes of climate. For the most part, foresters do not collect from isolated trees. Instead they normally select fully stocked communities with tall and straight trees having small-diameter branches, and rapid growth rates.

Foresters generally assume that seed from stands in or near their region have a natural adaptation to local growing conditions. Typically (Daniel et al. 1979; Isaacson 1984; Zobel and Talbert 1984), they look for sites (1) within 100 mi (161 km) north and south of the planting area; (2) at elevations within 500 ft–1000 ft (150 m–305 m) of the planting site, varying somewhat by region; and (3) with climate, soil, topography, and vegetation communities (e.g., habitat type) similar to those of the planting site. Guidelines for western North America also suggest not using seed from northern slopes for plantations on southern slopes, from river bottoms for use on side slopes, and from low elevations to regenerate stands in high areas (Isaacson 1984; Hibbs and Ager 1989). Also, a shift of 1° in latitude has the same effect as moving a plant 330 ft (100 m) of elevation (Zobel and Talbert 1984). Exact limits vary somewhat among species. However, foresters can usually move seed within these ranges without expecting an appreciable difference in population performance (Morgenstern and Roche 1969).

Appropriately adapted seed sources seem particularly critical for species that have an extensive natural range. Though their broad geographic distribution suggests great adaptability to environmental conditions, trees from different locations may actually vary widely in the genetic basis for growth rates, resistance to some pests and diseases, and tolerance of environmental extremes. To illustrate, loblolly pine grows in the United States from Delaware to Florida, and west to the eastern part of Texas. Tests indicate that across this range, eight distinct seed sources show differences in growth rates, resistance to fusiform rust, susceptibility to drought and subfreezing temperatures, and likelihood of suffering ice damage (Pait et al. 1991). Thus, foresters must avoid the temptation of taking seed from a particularly promising stand at one part of a species' natural range, and using it at a site far removed from the parent stand, particularly where the physical and biotic environment may differ considerably.

Foresters can generally feel more comfortable about transferring seed for even wide distances toward the center of a species' natural range, and across areas with narrow natural environmental gradients (Zobel and Talbert 1984). For many species, adaptive variation occurs with substantial differences in elevation and latitude (e.g., with ponderosa and Scotch pines, and among spruces), or even elevation and aspect at a more localized scale (e.g., Douglas-fir in southern Oregon), (Rehfeldt and Hoff 1976). To aid foresters and landowners in making wise selections,

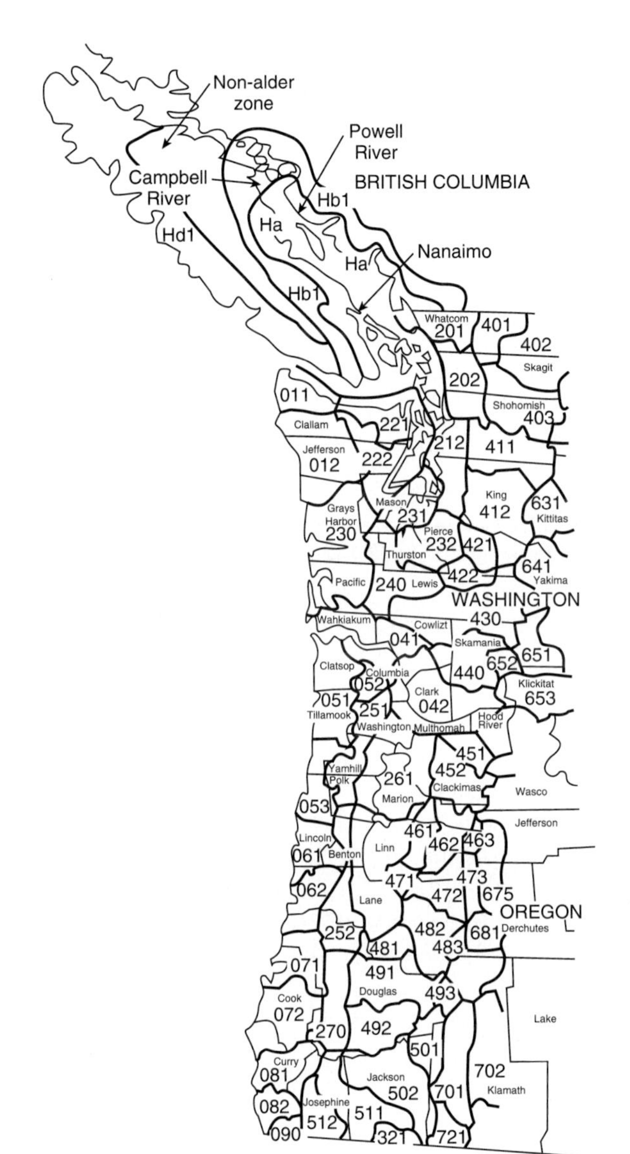

FIGURE 7-1
To aid foresters and landowners in making wise selections, forest tree improvement specialists have drawn on field tests to establish *seed zones* for many commonly planted species like these for red alder in western Oregon and Washington, and the recommended biogeoclimatic zones of coastal British Columbia (from Hibbs and Ager 1989).

forest tree improvement specialists have drawn on field tests to establish *seed zones* for many commonly planted species (Fig. 7-1). These delineate geographic areas where nursery managers can confidently move seed from one source to another planting site, and the boundaries not to cross without further testing. The zone boundaries often differ between species, and may even reflect special limitations for gathering seed to use at sites having severe environmental conditions (Rehfeldt and Hoff 1976; Johnson 1984).

SEED CERTIFICATION

Nursery managers who purchase seed from outside sources commonly use vendors who can certify the source. In fact, many states and federal governments have tree seed certification laws that local nurseries use to verify the seed source, and assure landowners that the seedlings will have acceptable traits. The Organization of Economic Cooperation and Development also has a certification program that many countries use, but no one program has yet received worldwide acceptance (Evans 1982).

Generally, certification requires vendors to supervise the collection and handling of forest tree seeds, and use consistent and accepted methods for seed processing and labeling. The suppliers must identify the species, seed zone (location) and elevation, quality of tree seed, and other characteristics of importance (Rudolf 1974; Evans 1982; Zobel and Talbert 1984). This (after Matthews 1964) identifies the seed as

1 *unclassified seed*—with little known about origin and quality;
2 *source-identified seed*—taken from good natural stands and plantations, and registered as to location;
3 *selected seed*—gathered from superior phenotypes, and likely as above-average in genetic quality; and
4 *certified orchard seed*—taken from clonal trees in seed orchards, or from elite trees with genetic qualities proven by progeny testing.

Most vendors handle the first two categories of seed, but most nurseries would generally use source-identified and selected seed for general forest planting. If they have access to a special tree improvement program, they would substitute selected and certified orchard seed to the extent possible. Nursery managers maintain precise records to keep track of each *seed lot*—a quantity of seed gathered at a single source, and in one collecting year. These records identify the location and time of collection, and who gathered or supplied the seed. They also show the seed lot history from its initial processing and testing, into storage, through sowing and tending, and to eventual harvesting. Managers will label each seedling bundle with its seed lot number so landowners can trace the source and screen different provenances for adaptability to the environmental conditions on their lands.

SEED PROCESSING AND STORAGE

If nursery managers collect or purchase cones or fruits, they must extract the seed or pits, or remove the wings or fleshy covering, and screen out impurities. Then they test the potential for germination, package the seed, and place it into storage until sowing. The tests indicate the numbers per pound (kg), the amount of impurities and hollow seed per pound (kg), and the probable rate of germination. The germination potential may change somewhat over time, necessitating periodic tests to monitor for signs of deterioration during long-term storage, and again prior to sowing. Then nursery managers can determine an appropriate *sowing rate* (density of seed to sow per unit of area) for a specific postgermination density within the seedbeds (Bonner 1974).

If they use proper methods, managers can keep seed viable for up to ten times longer than under uncontrolled field conditions. This allows them to secure large quantities of seed during years with bumper crops, and hold it for later use in poor seed years. Exact requirements differ between species, and with the quality of seed (Schopmeyer 1974; Young and Young 1992). *Quality* in this sense includes freedom from mechanical injury during processing and handling, and from uncontrolled environmental factors that might reduce the germination potential below an acceptable level.

For successful long-term storage, managers must carefully control seed moisture content and the storage temperature (Stein et al. 1974; Tanaka 1984; Bonner 1991), and protect the seed from fungi, insects, rodents, and mechanical damage. Generally, seed from most conifers will remain viable for long periods if reduced to a 5%–10% moisture content, and kept under refrigeration at less than 20° F (−7° C). For storage at temperatures as low as 0° F (−18° C), the moisture content must not exceed 5% or 6%. Managers must first dry the seed, put it into a moisture-proof container, seal the top, and place the container into a refrigerated chamber. Leaving a small air space inside the container will insure adequate aeration for seed respiration, as long as someone turns the container periodically to mix the contents (Stein et al. 1974; Belcher 1982; Evans 1982; Tanaka 1984; Bonner 1991).

In a natural environment, seed matures and falls from a tree, and lies on or in the litter layer for several weeks or months before germination. This naturally conditions the seed so that germination occurs promptly at the beginning of a growing season. By contrast, nursery managers must collect seed before it falls from a tree naturally, and then store it in an environment that maintains the dormancy. As a result, the seed of many species will not germinate promptly after managers remove it from storage. Instead, they must often *stratify* the seed—condition it to maintain or overcome dormancy (Ford-Robertson 1971). For species with hard shells or coverings, workers may also need to soak the seeds in water or acid, or mechanically scarify them to insure high rates of germination. In this context, *scarification* means wearing down or removing the outer coats of hard seed by abrasion or chemical means to improve their germination (Ford-Robertson 1971). When needed, scarification precedes stratification. In some cases, workers must scarify the seed to remove its hard shell, but need not stratify it prior to sowing.

As many as 60% of temperate tree species benefit from stratification—in a moist environment and at a particular temperature for some minimum period. Exact requirements differ by genera and species. For many, storage at a temperature of 34°–40° F (1.1°–4.4° C) for twenty to sixty days will suffice (Bonner et al. 1974). This helps to trigger the after-ripening biochemical changes that would occur naturally after the seeds fall from a tree. It breaks seed dormancy, rehydrates the seed following long-term cold storage, and keeps it ready for germination until sowing. In most cases, the bulk of seeds in a stratified lot will germinate shortly after sowing, insuring a greater degree of uniformity in seedling size and rate of development (Bonner et al. 1974; Tanaka 1984; Bonner 1991; Mexal and South 1991).

Prior to sowing, nursery managers make a final series of tests for germination and quality, and to measure (1) *physical characteristics*—proportion of sound (full) seed, percent of weight represented by impurities (debris like resin, parts of cones, and weed seeds), weight, number of seeds per pound (kg), and moisture content; and (2) *biological potential* (viability)—as reflected by the germination percent and rate of germination. From these data managers can calculate sowing rates (density of seed to spread) (Bonner 1974, 1991; Tanaka 1984).

SEEDLING PRODUCTION

To insure success in growing high quality bare-root stock, nurseries must have suitable soils, and managers must use fairly specialized and intensive cultural practices. They must produce fairly large numbers of seedlings to realize economy of scale, and require sufficient space for some fields or beds to lie fallow between successive seedling crops. Further, they must have facilities to store the seedlings and keep them fresh until shipping. Managers must also deal with a variety of natural conditions that they cannot control, like vagaries of the weather and the effect on nursery operations (Tinus and McDonald 1979; Guldin 1981).

The quality, condition, and uniformity of bare-root seedlings depends on a combination of factors that include the methods for storing and processing the seed prior to sowing. Important elements (Duryea 1984; Barnett 1989; Cordell et al. 1990, Mexal and South 1991) include

1 proper preparation, testing, and storage of the seed;
2 sorting the seed by size, keeping only the medium- and large-sized seed and discarding the small seed;
3 preconditioning the seed for prompt germination once seedbed conditions become favorable; and
4 precision sowing at an optimal time into a properly prepared and controlled seedbed.

Good practices should insure at least 90% germination, and postemergence losses of less than 10%. They also give a manager the means to reduce postgermination remedial measures to compensate for inadequate seedbed densities, or to relieve crowding as the seedlings mature.

NURSERY SOIL MANAGEMENT

In large measure, inherent conditions of site (e.g., soil, climate, and weather) determine the postgermination rates of seedling growth and development, and the kinds and intensity of practices needed to produce a crop of suitable seedlings. Managers can modify the growing conditions to an important degree, but cannot change many inherent characteristics of a nursery site. Largely, they can modify physical and chemical attributes to upper soil layers, and irrigate the beds to reduce moisture stress during dry periods.

Experience with forest tree nurseries throughout the world has shown that no site offers a perfect combination of conditions. The best locations (Wilde 1958; Prichett 1979; Evans 1982; Morby 1984; Liegel and Venator 1987; Davidson et al. 1994) have

1 a favorable climate, with no extremes of regular occurrence;
2 suitable and manageable soils (good nutrients and texture), free of rock, and with good drainage;
3 abundant supplies of good water;
4 clean air, free of harmful pollutants;
5 fairly level topography (<3% slope), or conditions suited to terracing;
6 no deleterious residual effects of past land use;
7 a readily accessible source of affordable skilled labor and support services;
8 easy access from a good transportation network; and
9 close proximity to customers.

Of these, conditions of the soil (workability, drainage, depth, fertility, and texture), a safe and abundant water supply, and a favorable growing season climate have proven most critical from a biologic perspective. Best sites have medium- to coarse-texture soil, a pH of 5.0–7.0, at least 2%–5% organic matter, no deficits in critical nutrients, and a good cation-exchange capacity. In reality, few sites have optimal levels of all these edaphic attributes, so a preferred location would have the fewest suboptimal features (Stoeckeler and Jones 1957; Evans 1982; Morby 1984; Davidson et al. 1994).

To insure rapid growth and healthy seedlings, nursery managers normally apply many different soil management practices throughout a seedling rotation. They also schedule some after harvesting one crop of seedlings, or when preparing the site for the next sowing. For example, they might grow a short-term crop of oats or buckwheat to protect the soil after harvesting. These plants die during early winter, leaving a light mulch of dead vegetation. In other cases, they may grow grasses, alfalfa, millet, soybeans, sorghum, legumes, or brassicas as a green manure crop for one to two seasons, then plow the plants into the soil. As an alternative, nursery managers might spread various kinds of dead organic material over the seedbeds (e.g., sawdust, peat, manures, sludges, or compost), and incorporate them into the soil prior to the next sowing. Many nurseries also add a thin organic mulch following sowing to protect the soil from drying and wind erosion. At high latitudes, they also often mulch the seedbeds again after one year to protect the trees and insulate the soil against periodic wintertime freezing and thawing (Armson and Sadreika 1974; Pritchett 1979; Davey and Kraus 1980; McGuire and Hannaway 1984; Davidson et al. 1994).

Adding organic materials improves the soil physical, chemical, and biologic properties (Davey and Kraus 1980). It

1 serves as building blocks for humus;
2 contains nutrients and holds water;
3 releases some nutrients upon decomposition;
4 enhances soil structure and tilth;

5 contributes to the habitat of many soil-inhabiting organisms (both beneficial and harmful ones); and
6 enhances mycorrhizal development.

Organic matter decomposes in the soil, and rather rapidly in tropical and semitropical environments. Nursery managers must repeat the treatments each rotation (Mexal and Fisher 1987).

The decomposition of organic matter, the activity of various soil organisms, the uptake of nutrients by the trees, and the leaching by water as it percolates through a soil collectively remove more nutrients than cover crops and other organic supplements replenish. To compensate for the losses, managers must add chemical fertilizers at the time of seedbed preparation, and also as a top dressing throughout a rotation. Commonly used fertilizers contain nitrogen, phosphorus, potassium, calcium, magnesium, and several critical microelements (e.g., iron, manganese, boron, and copper). Managers may also apply acids or ground limestone to adjust soil pH to suitable levels. The kinds and rates of fertilization, and the need for other soil amendments, become apparent from continuous soil testing and tissue analysis that managers also use as a routine part of their nursery operations. They commonly work with soil scientists to interpret the tests and develop prescriptions for the kinds of fertilizers to use, and the time for their application to optimize the benefits (Wilde 1958; Armson and Sadreika 1974; Pritchett 1979; Davey and Kraus 1980; van den Driessche 1984; Youngberg 1984; Davidson et al. 1994).

As part of the seedbed preparation, workers will also harrow and smooth the surface, and form the soil into raised beds. This promotes better drainage within the rooting zone and enhances seedling survival, health, and growth. They will often fumigate the seedbeds or apply preemergence herbicides and other biocides to kill weeds, pathogens, insects, and other pests. In some cases, nursery managers may reduce the amounts and frequency of toxic chemicals (such as methyl bromide) by relying on steam or heat for seedbed sterilization. The high temperatures kill weed seeds, pathogenic fungi, insect pupae and larvae, and other harmful organisms. All these treatments also kill many beneficial microorganisms like mycorrhizal fungi, and may upset the biological equilibrium of a soil. Workers may later inoculate a seedbed with spores of beneficial fungi, or add forest litter as a protective mulch and a source of mycorrhizae (Stoeckeler and Jones 1957; Wilde 1958; Armson and Sedreika 1974; Pritchett 1979; Thompson 1984; Cordell et al. 1990; Aim et al. 1991; Mexal and South 1991).

SOWING

Sowing follows in spring or fall after seedbed preparation. The actual time depends upon the species, climate, soil conditions, and nursery practices (Toumey and Korstian 1931). In northeastern North America, nurseries largely use fall sowing for most species, because the seedlings become larger and better developed by the end of the first growing season. In addition, the seed becomes naturally stratified as it lies in the seedbed over winter, and germination occurs as soon as environmental conditions become suitable in springtime (Stoeckeler and Jones 1957; Williams and Hanks 1976). Most nurseries in northwestern North America sow in springtime, just as the soil warms to an average of 30° F (−1.1° C) at a 4-in. (10 cm) depth. This minimizes over-winter losses to birds and small mammals, damage from frost heaving, and losses of early germinants to freezing (Toumey and Korstian 1931; Thompson 1984). For southern North America, mid- to late-April sowing usually insures early growing season germination and establishment, and larger seedlings by the end of the first year (Barnett 1989). In fact, available information indicates that first-year seedling biomass may decrease by 1% for each day of delay in springtime sowing past the optimal time (Mexal and South 1991).

Managers must carefully regulate sowing depth, in that seed sown too shallow or too deep may not germinate or grow uniformly. The optimal depth varies with species, and somewhat with soil texture and season of sowing as well. Generally, managers get the best results with a depth not exceeding 1 1/2 to 2 times seed diameter, or just deep enough to cover it with a thin layer of soil (Williams and Hanks 1976; Thompson 1984; Liegel and Venator 1987). In most cases, workers should bury the seed slightly more for fall sowing (Stoeckeler and Jones 1957; Williams and Hanks 1976), and put it about one-third deeper in sandy soils than in fine-textured soils (Williams and Hanks 1976). At sites prone to wind erosion, and in most cases with fall sowing, nursery managers must also mulch the surface or cover the beds to protect the seed until soil moisture and temperature become ideal for germination at the next growing season (Stoeckeler and Jones 1957; Williams and Hanks 1976). This mulch should cover the soil in a layer about equal to the diameter of the seed (Barnett 1989).

Tests of germination percent, soundness, and purity tell managers what proportion of a seed lot will germinate, and how many seeds they have per pound (kg). Then experience indicates what proportion of germinated seed will likely grow into suitable seedlings. Based upon this information, managers can determine how closely to sow the individual seed within a row to realize an appropriate spacing (Stoeckeler and Jones 1957; Williams and Hanks 1976; Thompson 1984).

PROTECTION AND HEALTH MANAGEMENT

Nursery managers must protect the seedlings from pests, diseases, physical harm, and desiccation as part of a general integrated health management program. It will include (1) active measures for prevention and early detection; and (2) control by chemical, cultural, genetic, and biologic measures (Nicholls 1989). Exact components depend upon the circumstances. They may differ between years at a single nursery, from one species to another, and with the kind of nursery crop produced. Managers must also continually monitor the seedbeds for early signs of harmful insects and diseases, indications of feeding by animals, environmental stress, mechanical damage, and poor nutrition. When problems first appear, managers must act promptly and decisively.

Important losses can occur immediately after germination from harmful fungi or desiccation. Critical diseases include a major group of organisms called *damping-*

off fungi. These proliferate during warm and damp weather, and can damage seedlings until the succulent stem tissues begin to harden. Common species belong to the genera *Rhizoctonia* and *Pythium*. These attack new seedlings at or slightly below the *root collar* (where the seedlings emerge from the soil), killing the stem tissues. The seedlings fall flat on the surface and soon die (Daniel et al. 1979). Managers control damping-off fungi by applying fungicides before sowing or after germination, by regulating rates of irrigation to control moisture in upper soil layers, and by soil management to improve drainage (Williams and Hanks 1976; Wall 1978; Southerland 1984). Wide spacing and good air flow through the seedbeds also help to reduce the risks.

Common diseases of older seedlings include stem cankers, needle casts, root diseases, and rusts. Harmful insects include aphids, insects that girdle the stem, shoot moths, and defoliators. The particular diseases and pests of importance vary with the location of a nursery, and the species grown. Most nurseries must rely on herbicides, fungicides, and insecticides to prevent or contain major problems of these kinds. In addition, weed species frequently germinate in the seedbeds, interfere with the tree seedlings, and inhibit their development. For these, managers rely on appropriate programs of soil management (cultivation and mulching), herbicide applications, and proper seedling culture to control the density and kinds of weeds. They must carefully integrate the strategy with other aspects of nursery management to insure success, to minimize the potential for environmental harm, and to preclude inadvertent damage to the tree seedlings (Wall 1978; Owston and Abrahamson 1984; Southerland 1984; Davidson et al. 1994).

High temperatures and dry winds can also damage or kill seedlings of any age, usually by desiccation. Managers prevent these problems by timely irrigation. Watering also protects seedlings and succulent shoots against freezing temperatures early in a growing season, or heat injury during later periods. Some nurseries also dissolve fertilizers in the water, and use the irrigation system as one means for managing soil fertility and seedling nutrition (Williams and Hanks 1976; Day 1984; McDonald 1984; Mexal and South 1991). In some cases, managers have also used irrigation to leach potentially toxic substances from the rooting zone, and to reduce surface erosion by wind.

CULTURAL PRACTICES

Long-term tree seedling health and development depends upon an abundant and steady supply of nutrients, water, temperature, and light. These interact with plant physiological factors (e.g., carbohydrate reserves, hormone levels, frost hardiness, and dormancy) in a complex way to control the rates of plant growth (Cleary et al. 1978; Daniel et al. 1979; Barnett 1989). Each species has a different set of optimal conditions, and simply increasing one element without dealing with other potentially limiting ones will not necessarily improve survival and development (Cleary et al. 1978). Instead, nursery managers must take a holistic approach, using a seedling production system that balances all the critical elements within practical limits. Such a program would include periodic addition of organic materials to the soil, appropriate site preparation, control of seedbed density via carefully controlled sowing rates and methods, timely fertilization, periodic irrigation, well-programmed cultural practices, and continual surveillance and protection (after Lavander 1984; Barnett 1989).

Seedbed density profoundly affects the vigor, growth, size, form, and uniformity of bare-root seedlings (both tops and roots). Optimum levels vary with species, but most conifers grow best at densities of 10–20 trees/ft^2 (110–215/m^2). Some may tolerate densities as great as 30/ft^2 (325/m^2) (Cleary et al. 1978; Duryea 1984; Stein 1988; Cordell et al. 1990; Aim et al. 1991; Brissette 1991; Mexal and South 1991). Regulating sowing rates and methods (e.g., drilling vs. broadcast sowing) with reference to anticipated germination rates allows managers to avoid later remedial measures to reduce crowding. Then timely fertilization, regular irrigation, and a series of root culturing techniques will maintain a good shoot-root ratio and enhance seedling development and quality. This improves seedling survival and growth following outplanting (Duryea 1984; Randall 1984; Simpson 1988; Cordell et al. 1990; Mexal and South 1991).

Common cultural practices in conifer nurseries include root pruning and wrenching. *Root pruning* means undercutting the seedling roots by drawing a thin blade horizontally along the seedbed at a fixed depth below the surface (e.g., mostly 4–8 in. or 10–20 cm deep). Nursery managers may also draw coulter blades lengthwise through the beds to vertically sever the roots between rows. In both cases, pruning keeps the root systems compact, and makes them easier to handle. Managers normally do root wrenching during the second growing season, and at a greater depth than undercutting. *Root wrenching* means drawing a thick inclined blade under the seedbed to lift and loosen the soil, and break off the ends of new roots. Both practices cut or break off the deeper roots. During subsequent growth they branch into a more compact and denser root system and develop more fine absorbing roots (Cleary et al. 1978; Duryea 1984; Simpson 1988; Cordell et al. 1990; Aim et al. 1991; Mexal and South 1991). As a consequence, workers can better fit the tree into a normal-sized planting hole. The roots have more actively growing tips and a higher absorption capacity. This translates into better survival and development following outplanting. On the other hand, shallow or late-season root pruning and wrenching may somewhat reduce first-year height and diameter growth in some species. That may help to adjust the shoot-root ratio (Cleary et al. 1978; Duryea 1984; Simpson 1988; Cordell et al. 1990; Aim et al. 1991; Brissette 1991; Mexal and South 1991).

For warm climates and fast-growing species, a rotation may last only one growing season. Managers may need two or three seasons to produce suitable stock at higher latitudes. In all cases, they must program the cultural practices to insure uniform growth and form throughout the entire rotation. Then the seedlings will reach the requisite minimum size before harvesting, but not become too large for efficient handling.

NURSERY LIFTING AND HANDLING OF BARE-ROOT STOCK

Success in planting requires good coordination between the manager of a nursery and each landowner who receives the stock. Within any region, sites at different elevations and latitudes, and with opposing or contrasting aspects, may become ready for planting at different times. This commonly occurs in planting zones with distinctly cold winters, but not in regions of milder climate. To complicate the situation even more, the first opportunity for lifting (loosening and removing seedlings from the seedbeds) and the time when dormancy breaks within seedbeds depend upon environmental conditions at the nursery. In some cases, the ideal time for lifting may not match conditions at some planting sites within a region.

This discontinuity can cause a problem. First, landowners who can begin their field operations before a nursery starts spring lifting will miss the optimal time for planting. The temperatures may become too hot for the new shoots, and the soils may dry before the seedlings develop a well-functioning root system. This could lead to heat damage or desiccation, and lower survival. Conversely, nursery soils often thaw earlier than sites at higher elevations and colder climates, necessitating special storage to keep the stock dormant until landowners can use it. On the other hand, many shrubs used for wildlife and shelterbelt planting, and some tree species like the larches, tend to break dormancy even before a nursery staff can complete spring lifting, and the seedlings arrive at a planting site after active growth commences. The tender root tips and newly elongated branch shoots readily wilt if exposed to hot and dry air or planted with poor root-soil contact. The tender shoots also easily break off during handling. By contrast, the root tips and buds of fully dormant seedlings normally withstand proper lifting, processing, and transport without suffering extensive damage. Also, they do not desiccate rapidly if exposed to warm and dry air for short periods. With species that commonly break dormancy early, and to supply stock for early shipment to warmer geographic areas, nursery managers often lift the seedlings after the onset of dormancy in fall, and put the seedlings into over-winter cold storage (Hocking and Nyland 1971; Williams and Hanks 1976; Cleary et al. 1978). This gives them the option of staggering the shipping so that the trees and shrubs arrive at different sites as conditions become optimal for planting.

Once lifted from the soil, tree seedling quality can deteriorate rapidly due to desiccation, molding, metabolic activity (e.g., respiration), and growth. Since the latter three depend substantially on temperature (Burdett and Simpson 1984), managers in temperate zones can schedule their lifting after temperatures drop in autumn, or before hot weather arrives in the spring. They then put the seedlings into bulk storage containers or shipping packages to guard against desiccation and damage, and move them into cold storage to keep the trees dormant. With spring lifting, the storage facility must have a high relative humidity to inhibit moisture loss (particularly conifers), and temperatures below the threshold for physiological activity. The low temperature also inhibits the spread of harmful mold fungi. A range of 35°–40° F (1.7°–4.4° C) will suffice for short-term storage of up to two to four weeks (Hocking and Nyland 1971; Cleary et al. 1978; Burdett and Simpson 1984). For longer periods, molding may become a problem unless managers keep the temperatures close to or at 32° F (0° C), but managers should not refreeze seedlings once they thaw in springtime.

Over-winter cold storage of bare-root stock has succeeded when managers delayed fall lifting until after the full onset of dormancy, placed the seedlings into protective containers or packages to reduce moisture losses, and then put the containers into a dark storage room with near-freezing temperatures. Optimal storage temperatures for over-winter or other long-term storage of dormant seedlings (Nyland 1972; Hocking and Nyland 1971; Cleary et al. 1978; Jenkinson and Nelson 1985; Carlson 1991) include 32°–34° F (0–1.1° C) for Douglas-fir and southern pines; or below freezing—28°–30° F (−2.2 to −1.1° C)—for most boreal or northeastern conifers. Managers must maintain these levels throughout the storage period, and minimize the cyclic warming and cooling of the storage chamber. Keeping a fairly consistent temperature prevents desiccation from the foliage and roots, even throughout a fairly long storage period (Nyland 1974b; Jenkinson and Nelson 1985; Dunsworth 1988). In fact, nurseries have successfully stored a variety of conifer and deciduous species for up to four to six months, as long as they used fully dormant stock and carefully controlled the storage temperature within acceptable limits (Hocking and Nyland 1971). Seedlings stored over winter must go through a physiological conditioning period after removal in the spring, and this may delay bud break after outplanting (Nyland 1974a; Burdett and Simpson 1984; Dunsworth 1988).

Nurseries use a variety of packaging techniques to protect bare-root stock during handling, storage, and shipping (Duryea and Landis 1984; Edgren 1984; Carlson 1991; Wenger 1984) (Fig. 7–2). One traditional method, *bales* or *jelly-roll bundles*, consists of rolling the seedlings inside a waterproof wrapping and a layer of water-absorbing packing material (e.g., sphagnum moss, cellulose batting, shingle toe, or peat moss) to protect the roots. The tops remain exposed. Some nurseries package stock in rigid boxes made of waxed cardboard or having a plastic liner. Boxes lacking a liner offer little protection, and the seedlings may dry after just a short storage period. During the past decade or so, most nurseries in the United States and Canada began using polyethylene-lined kraft bags. These do not cost much. Workers can also quickly fill them with seedlings, and easily seal the tops using stapling or sewing machines. If properly sealed, the bags (and lined boxes) keep the seedlings moist, even if a nursery staff adds no supplemental packing material around the tree roots (Cleary et al. 1978; Duryea and Landis 1984; Edgen 1984; Carlson 1991). If a planting crew keeps the closed boxes and bags cool and shaded, the seedlings will remain fresh and dormant for a long period.

TIME OF PLANTING

The initial survival and postplanting growth of bare-root seedlings depend largely upon their condition at the time of lifting (Armson and Sadreika 1974; Duryea 1984; Ritchie 1984; Barnett et al. 1990; Rose et al. 1990; Mexal and South 1991). Seedling physiological condition and particularly dormancy critically influence reforestation success (Cleary et al. 1978). Once stimulated by warm temperatures and ample soil moisture, the seedlings break dormancy and begin to grow. First, the roots elongate into the surrounding soil, improving the root-soil contact, insuring adequate moisture uptake before new foliage forms on the elongating shoots, and

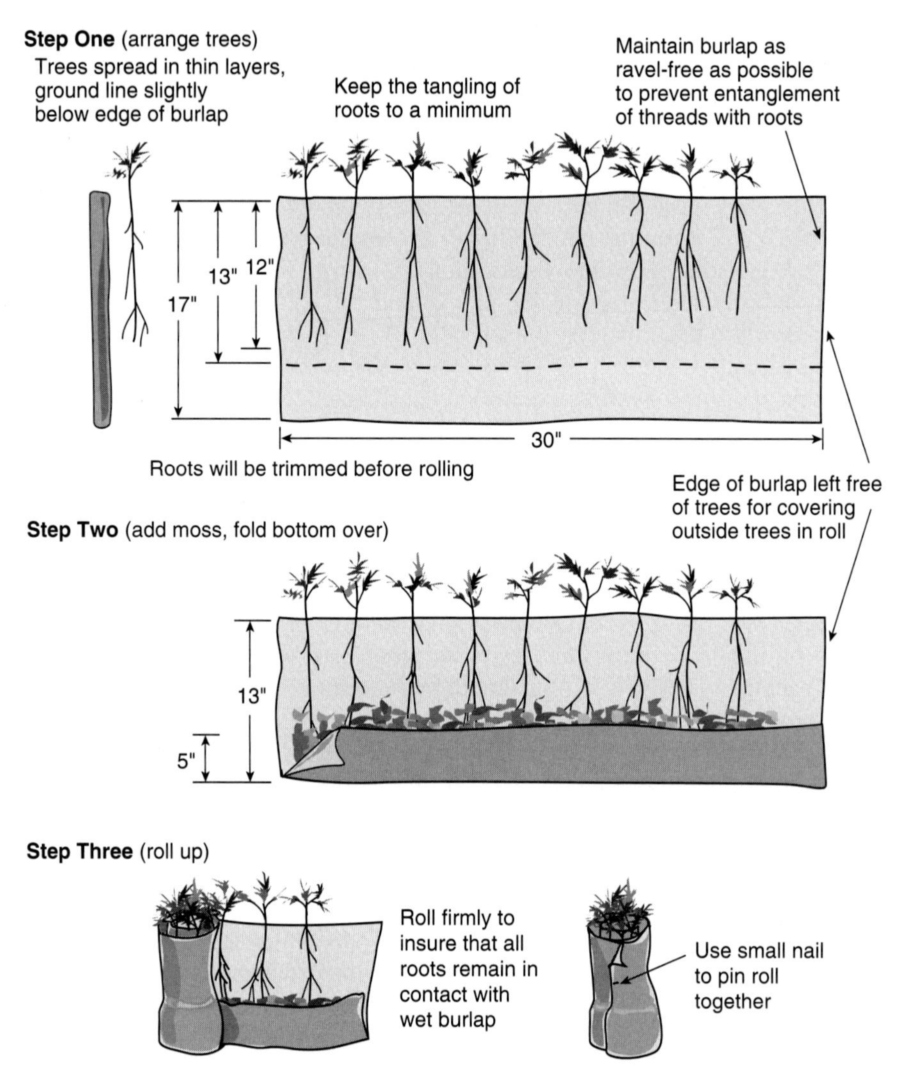

FIGURE 7-2
Procedures for packaging tree seedlings in traditional jelly-roll bundles (from Wenger 1984).

providing firm anchoring as the seedlings develop. Old needles on conifer seedlings begin to transpire moisture, and photosynthesis adds new carbohydrates to help sustain growth and other physiological functions. After a few days the foliage buds break, and the branches and terminal shoots begin to elongate. As long as freezing or overheating does not harm the seedlings, new needles or leaves eventually form. Once at this critical stage of development, a young tree has a good chance of surviving, as long as environmental conditions remain favorable.

Planting season, planting weather, and soil conditions have historically proven decisive for planting success (Long 1991). The ideal time to plant bare-root stock comes just as environmental conditions (e.g., air and soil temperature, soil moisture, and day length) would trigger physiological activity within the trees, and root elongation would commence (Toumey and Korstian 1931; Smith 1986; Long 1991). In temperate zones, the soil normally holds good quantities of free water during this time, insuring adequate supplies to balance transpirational losses. Favorable conditions may persist for a relatively short period. Scheduling the planting for just before the onset of growth by the native vegetation will normally insure a suitable environment for good survival (Toumey and Korstian 1931; Smith 1986; Long 1991).

In tropical and subtropical areas, and particularly those with distinct wet and dry periods, the ideal time for planting comes with the onset of a rainy season. If foresters wait until the beginning of a dormant period caused by high temperatures and dry weather, the seedlings often die due to desiccation, and the plantation fails. In temperate zones, the best conditions come in spring. Then foresters may have an optimal period that lasts for about two to four weeks during most years, but for as little as a few days during years with unusual weather. By contrast, they may plant successfully for as long as three months in areas with mild climates, cool growing seasons, or late snowmelt. In all of these diverse environments, most foresters begin planting operations as soon as site conditions allow, and complete the work in a relatively short period of time.

Under some circumstances, landowners in temperate zones may plant in autumn after daily temperatures drop below the threshold for active plant growth, or even throughout winter in areas with a mild climate. To succeed, they must delay fall lifting and planting until the stock becomes fully dormant. This window of opportunity lasts until active growth commences the following spring. Dormant-season planting has succeeded within regions and elevation zones where wintertime temperatures do not drop to extreme levels (e.g., southern United States, and the Pacific Northwest), causing the soil and seedlings to freeze (Greaves 1978; Long 1991). By contrast, fall planting of coniferous seedlings at upland and mountainous sites in northern North America often leads to winter damage and considerable mortality. This mostly occurs during years with too little persisting snow cover to protect the trees from repeated freezing and thawing at the beginning and end of winter, or from exposure to drying winter winds. In addition, frost heaving in fine-textured soil may force open the planting holes, and push the seedlings partly or wholly out of the ground (Schubert 1976).

FIELD STORAGE AND HANDLING OF BARE-ROOT STOCK

Even the best-quality planting stock will not withstand careless lifting, sorting, packaging, storage, shipping, and planting. In fact, damage to bare-root seedlings appears cumulative, and each instance of poor treatment further reduces the potential for good survival and growth (Duryea and Landis 1984). Damage can result primarily from mechanical injury to the roots and buds during lifting and processing, and from desiccation and the breaking of dormancy. Seedlings will remain dormant and fresh (adequate internal moisture) for long periods if kept moist, and stored at a temperature below 35°–40° F (1.7°–4.4° C) (Nyland 1974b; Duryea and Landis 1984). Foresters can use refrigerated trucks or railroad cars to move the seedlings to a planting site, and to temporarily store them prior to use (Edgren 1984; Carlson 1991). In some cases they may transfer the seedlings to a cool and dark storage place (e.g., a root cellar), or unwrap and heel-in the stock for temporary storage. *Heeling-in* means temporarily putting (planting) the seedlings in a trench, and covering the roots or rooting portions with soil (Ford-Robertson 1971). Shaded areas with moist soil adjacent to streams serve as good places to heel-in planting stock.

In cool climates, cellars of barns and unheated buildings normally suffice for temporary and short-term springtime storage of properly packaged trees and shrubs. In warmer climates, foresters must often use some kind of refrigerated facilities. In areas with cold winters, workers may also find residual snowdrifts on shaded slopes, and can dig small caves into these drifts. Placing reflective tarps over the bundles also helps to prevent heating, and will suffice as a field-expedient measure for up to a few hours (Toumey and Korstian 1931; Duryea and Landis 1984; Edgren 1984; Carlson 1991).

CONTAINER STOCK

Beginning in the 1950s and 1960s, foresters began to try some new methods for forestation. These included (after Stein 1976; Hallett and Murray 1980; Guldin 1981; Hahn 1982; Reese 1982; Brissette et al. 1991)

1 shortening the nursery rotation to gain greater flexibility in seedling production;
2 mechanizing nursery and planting operations to facilitate large-scale artificial forestation and reduce planting costs;
3 gaining greater economy in the use of the high-value seed from tree improvement programs;
4 extending the planting season at locations with a short window of opportunity for bare-root planting;
5 improving survival and growth on difficult sites by protecting the seedling from damage and desiccation, and insuring better root-soil contact; and
6 enhancing potentials for raising some species that proved difficult to grow under traditional bare-root nursery systems.

While planting of container-grown trees has not necessarily reduced bare-root planting, yet it has become operational in many parts of the world. Generally cited advantages (after Cayford 1972; Randall 1984; Brissette et al. 1991) include

1 *better control of germination and growing conditions*, insuring more efficient use of seed, and better results with some species;
2 *shorter rotations*, allowing production adjustments with relatively short notice;
3 *intensive cultural techniques*, including control of the rooting media and seedling tending to speed growth;

4 *more uniform size and condition of the stock*, leading to a higher proportion of plantable seedlings, and better consistency in plant size;
5 *compact root systems*, insuring a good shoot-root balance and facilitating planting;
6 *added protection of roots*, resulting from having the seedlings in containers during handling, and reducing damage prior to planting;
7 *faster early growth*, enhancing survival and promoting better development following outplanting; and
8 *planting of an undisturbed root system*, keeping the rooting material intact and enhancing survival over a longer planting season.

In this context, *container* or *containerized planting* means the use of seedlings grown in containers, and later outplanted with the rooting medium left undisturbed around the roots (Carlson 1979; Tinus and McDonald 1979; Ek 1982).

Containers can vary in design and size, and nursery managers can alter the rooting medium to provide better conditions for growth (Fig. 7-3). Generally, the container and rooting material (Tinus and McDonald 1979; Carlson 1979; Barnett and McGilvary 1981) must

1 prove suitable for use within the facilities at a nursery;
2 promote timely germination and insure high seedling survival;
3 provide good nutrition for rapid growth;
4 protect roots from damage and drying during production, storage, and shipping;
5 allow efficient handling and transport;
6 match the planting method and equipment; and
7 insure good survival and rapid growth following outplanting.

Within these constraints, nursery managers must select a combination of materials and methods that prove both biologically acceptable and cost-efficient.

Many types of containers work adequately. In tropical areas, nurseries have historically used biodegradable paper pots or sections of bamboo, rigid polyethylene tubes or pots, tarpaper pots, plastic bags, or Styrofoam blocks (Liegel and Venator 1987). Those in temperate regions tend to use rigid-wall *tubes* that workers plant with the seedling, reusable molds for growing *plug* seedlings that users remove from the container prior to storage or planting, and molded *blocks* of a preformed rooting medium commonly impregnated with nutrients, and having firm or compressed side walls. The *tubes* have break-away or mesh sides that roots penetrate or push open, or biodegradable walls (e.g., paper) that rapidly deteriorate following planting. Containers for *plugs* have rigid walls manufactured from plastic and Styrofoam. Workers may carry these around a planting site to protect the seedlings, and remove each tree just at the time of planting. Containers for *blocks* consist of low-density polyurethane, wood pulp, or mixtures of peat with vermiculite and cellulose fibers. Seedling roots often penetrate the walls, so workers plant the seedling and block as a unit (Carlson 1979; Barnett and McGilvary 1981; Brissette et al. 1990, 1991).

In mild and humid regions of the tropics and subtropics, nursery managers can often produce container-grown seedlings out of doors under trees, in shade houses, or

FIGURE 7-3
Containers for growing production seedlings can vary in design and size to fit growth characteristics of different species, and nursery managers can also alter the rooting medium to fit local needs (*courtesy of Georgia Pacific*).

beneath screens that reduce light intensity to some degree. They must regularly irrigate and fertilize the plants, and otherwise care for the trees to promote good growth and thrifty development (Liegel and Venator 1987). By contrast, nurseries in temperate environments commonly use heated greenhouses (Stein 1976; Carlson 1979; Tinus and McDonald 1979; Brissette et al. 1990, 1991). In all of these settings, they can sow the seed during a dormant season, and grow the seedlings under a controlled environment consisting of (1) optimum light intensity and photoperiod; (2) temperature ideal for maximum growth; and (3) intensive irrigation and fertilization regimes that facilitate maximum tree development. Attendants can also protect against insects and diseases by standard pest management techniques (e.g., fungicide and insecticide applications). They can alter the light and temperature to slow or accelerate growth following germination. As a result, managers can produce seedlings of fairly uniform characteristics, and well fitted to the planting requirements.

In the tropics, container-grown trees will normally reach suitable sizes in as few as one to two months. There managers must carefully time a sowing to keep the trees from becoming too large for efficient planting (Liegel and Venator 1987). In temperate zones, greenhouse managers can usually have the seedlings ready for outplanting within three to six months, but could grow them longer to increase the size. They, too, must carefully select a start-up date and control light level and quality, day length, moisture availability, and temperature regime to insure that the seedlings reach the requisite size in phase with conditions at a planting site. Also, prior to outplanting they will usually physiologically condition the stock for up to six weeks under a regime of light and temperature comparable to the natural environment. They may even move the stock to a partially shaded outdoor storage area for further acclimatization (called *hardening off*) before outplanting, or to temporarily hold it for shipment to the field sites (Tinus and McDonald 1979; Carlson 1979; Brissette et al. 1990, 1991).

Though container-grown seedlings grown under these carefully controlled conditions offer many advantages, they also have some important shortcomings (Stein 1976; Tinus and McDonald 1979; Randall 1984; Brissette et al. 1990). These include

1 Container seedlings usually cost more to produce due to high overhead charges, and the costs of energy for heating and cooling.

2 Greenhouse managers must have greater technical knowledge about plant propagation techniques and ways to control environmental conditions, and must use a more intensive production system than with bare-root stock.

3 Containers hold a limited amount of rooting medium, and seedlings may readily deplete the nutrients and moisture unless frequently fertilized and irrigated throughout the entire production period.

4 The medium will dry quickly if workers expose the stock to high temperatures and drying winds at a field site, and if crews fail to water and ventilate the containers periodically during temporary storage.

5 Containers of seedlings take far more space in storage, weigh more, and require a greater logistical effort for handling and transport.

6 The bulk and weight of containers make them more awkward to transport and distribute during planting.

Overall, the small size and well-contained root systems of container seedlings make them easier to plant. Further, putting an intact root system with the attached rooting material into the planting hole enhances survival, and promotes good root development after planting.

Production estimates for container-grown seedlings vary from 900 to 2,500 per person/day for planting, depending upon the type of container and the planting method. These rates compare favorably with 800 to 2,000/day for conventional planting of bare-root stock on favorable sites (Minckler 1948; U.S. For. Serv. 1989). In fact, field experience indicates that container planting may actually reduce planting time by as much as 25% to 65%, compared with bare-root seedlings (Stein 1976; Cayford 1972; Abbott 1982; Guldin 1982). The planting process does require some added care (Randall 1984; Brissette et al. 1991), in that

1 the small size of container seedlings makes them more subject to burial, and to overtopping by competing vegetation;

2 leaving the rooting medium exposed, or planting during periods of dry soil conditions, will deplete rooting medium moisture rapidly, and desiccate the seedlings; and

3 if grown too long in a container before outplanting, the roots may become bound and spiraled, making the seedling more prone to frost heaving following fall or winter planting, and also inhibiting root development later on.

Better planting success often outweighs the added logistical problems and attendant costs, especially at difficult sites. Also, special value has accrued from container planting at sites with rocky and shallow soils, in high elevation forests, and at high latitudes (Stein 1976; Reese 1982; Altsuler 1985). It has allowed foresters to lengthen the planting season (Guldin 1981), and this helps landowners who must plant a large area each year. Also, in many arid tropical areas with harsh site conditions, foresters get better survival than with bare-root seedlings (Liegel and Venator 1987).

METHODS OF HAND PLANTING

To insure good survival after planting of bare-root trees and shrubs (Linstrom 1962; Schubert 1976; Morrow et al. 1981; U.S. For. Serv. 1989; Long 1991), workers must

1 open a hole perpendicular to the surface and large enough for the root system in its natural configuration;
2 insert the roots into the hole without bunching or twisting them;
3 hold the tree upright, and with the root collar even with the ground surface;
4 fill around the roots with soil so no air holes remain; and
5 pack the soil firmly to insure good contact with the roots.

Workers must dig holes large enough to accommodate the root systems, and not bend them into a tight ball or J shape (Fig. 7-4). Otherwise the roots often become deformed, and this may affect the growth of some species (Daniel et al. 1979; Morrow et al. 1981; Long 1991). They may plant each seedling slightly deeper than it grew in the nursery bed, but shallow planting reduces the chances for survival (Linstrom 1962; Schubert 1976; Long 1991). A well-planted seedling should remain in the ground if gently tugged upward by the terminal shoot. Further, it should stand freely in a vertical position when not held upright.

For container planting, workers will put the rooting medium as well as the root system into the soil, and will usually make the hole with a specially designed tool called a *dibble* (Fig. 7-5). It has a metal end identical in shape and size to that of the container. The workers push the dibble into the soil to form a hole, and then insert the root mass of a tree. To plant unrooted cuttings, they will normally use a dibble constructed of a steel rod having a diameter comparable to that of the cutting. They push the dibble into the soil to a predetermined depth, and insert the cutting into the hole.

Foresters have devised several different methods for hand planting bare-root tree seedlings, and used numerous kinds of tools for digging the holes. The preferred technique often varies with region and local custom. Crew members may work as individuals to both dig the holes and plant the trees, or in teams of two with one per-

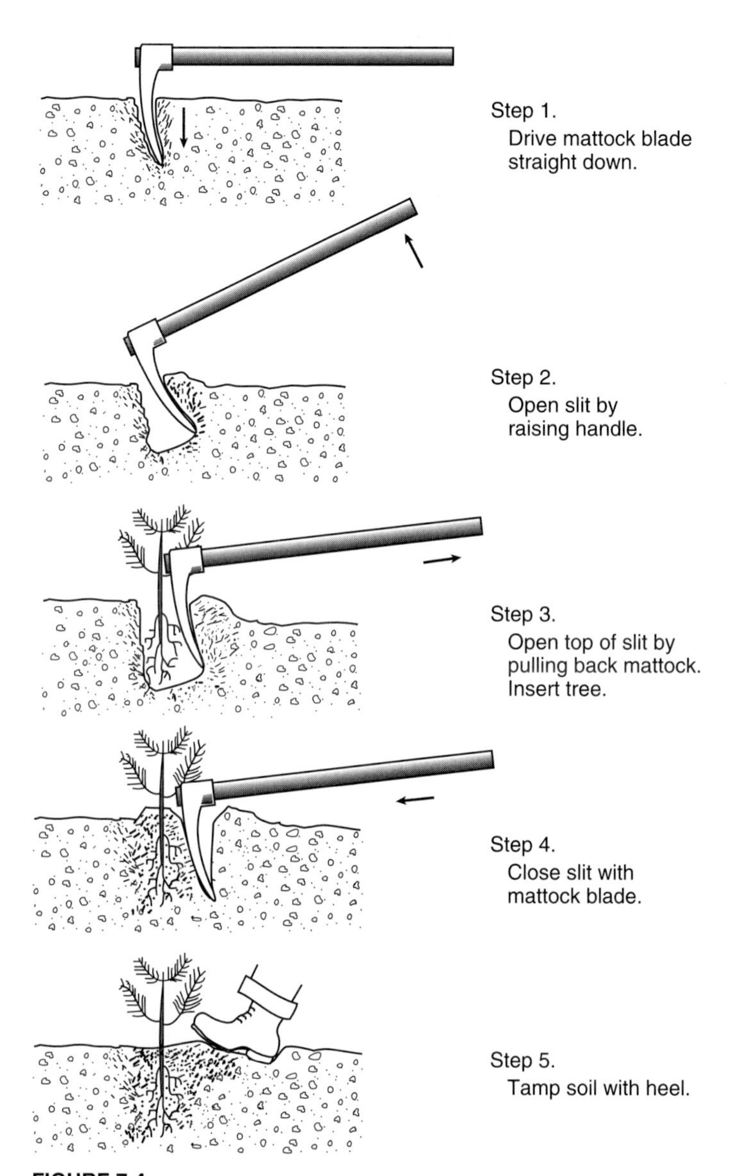

FIGURE 7-4
In hand planting, workers must dig a hole large enough to accommodate the root system without bunching or bending it into a tight ball, and insert the tree up to the root collar before firmly packing the soil back into the hole (from Wenger 1984).

FIGURE 7-5
In planting container stock and unrooted cuttings, workers use a *dibble* that makes a hole the size and shape of the container hole or the unrooted cutting.

son digging and the other planting. Common tools include shovels, mattocks or grub hoes, and planting bars (Linstrom 1962; Long 1991). In some areas of the western United States planting crews have used power augers to more quickly open a hole of adequate depth and width (Schubert 1976) (Fig. 7-6). In the Northeast, foresters often mount lugs onto wheeled or crawler tractors to dig the holes mechanically, and then have workers insert the trees by hand (Morrow et al. 1981). This speeds the planting by reducing labor costs.

To insure high rates of survival, workers must not break the shoots, damage the buds, or plant the seedlings too shallow or too deep. They must also protect bare-root stock by carrying the seedlings in a bucket or planting bag with some water to

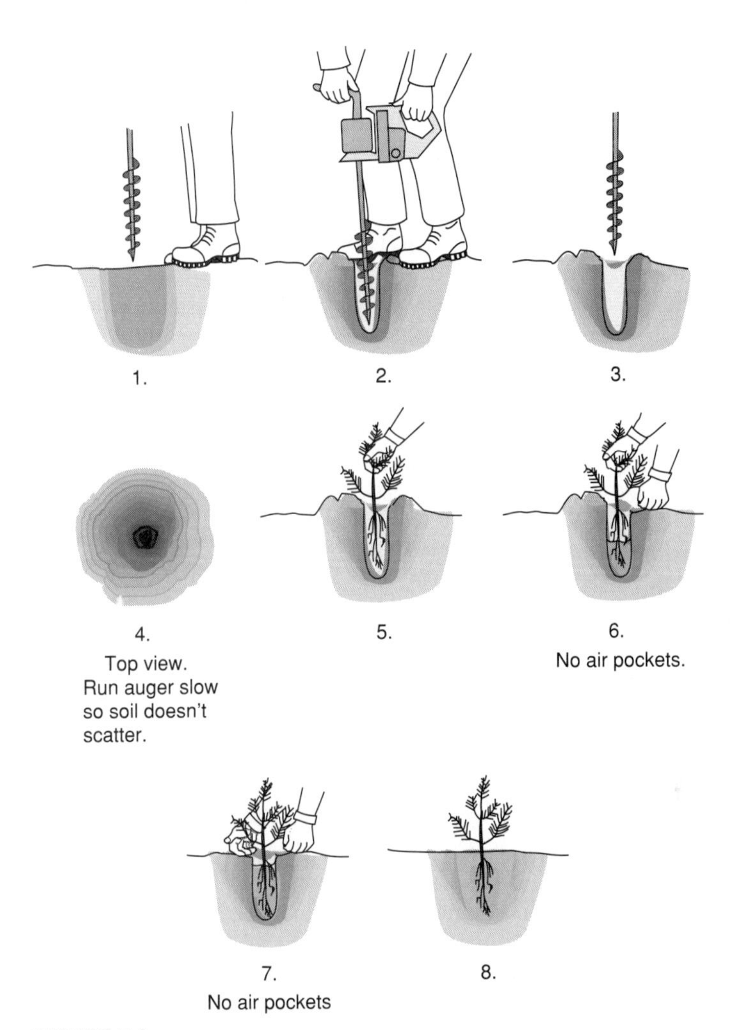

FIGURE 7-6
Planting technique using a power auger to dig holes for the seedlings (from Wenger 1984).

keep the roots moist, or in the nursery packaging. In fact, most planted seedlings die from damage or exposure during lifting or handling, or because of inadequate moisture uptake as a result of poor soil-root contact. Exceptions include losses from damage by an injurious agent not controlled through site preparation, or extreme environmental conditions that develop during a growing season (drought, freezing, and excess heat). In northeastern North America, foresters normally expect three-year survival of at least 85% following well-done hand planting of dormant bare-root stock (Armitage 1974). A survival rate of 80% has been considered suitable as a first-year standard for the Pacific Northwest and Southwest regions of the United States (Jenkinson and Nelson 1985).

MACHINE PLANTING

Except to mechanise the opening and closing of the planting hole, machine planting involves the same basic steps as hand planting. Most planting machines (Fig. 7-7a) have a coulter disk that cuts the surface and a single narrow plow that opens a deep slit in the soil. Continuous-furrow planting machines of this kind usually have a cage to protect workers against injury from debris, and covered trays or storage boxes for carrying the seedlings. For forest planting, foresters use a crawler tractor or skidding machine to pull the planter, but farm tractors will often suffice in reforesting open agricultural sites. A worker sits on a seat mounted over the plow. Wings on the plow hold the slit open while the worker reaches down to insert a tree, letting the roots hang freely in the open slit. At the rear of the plow, packing wheels compress the soil back into place around the roots as the machine passes. The worker holds the tree upright until the packing wheels pass by the seedling, then lets go of the newly planted tree, and readies another one for planting at the next spacing interval. More modern versions may include a semiautomatic tree handler that holds the seedling, and moves it into position between the packing wheels (Fig. 7-7b). Workers simply lay the seedlings into holders mounted onto a rotating planting chain, and the mechanical device regulates spacing within a row by virtue of the speed with which the planting chain turns relative to the speed of the tractor (Fig. 7-7c). As with hand planting, the crew must protect the stock before and during the operation, keep the packing wheels properly aligned, and maintain sufficient pressure on the packing wheels to insure good contact between the soil and roots of the planted tree. Also, even with careful operation some skips will occur, so a worker must walk behind the machines to fill the occasional gaps by hand planting (Linstrom 1962; Schubert et al. 1970; Schubert 1976; Benson 1982; U.S. For. Serv. 1989; Long 1991).

Despite its apparent advantages, machine planting will not work well at sites with steep and gullied slopes, numerous soil and surface rocks and heavy logging debris, areas already vegetated with brush and small trees, and other conditions that impede trafficability (Linstrom 1962; Schubert 1976; Benson 1982; Long 1991). Also, in poorly drained fine-texture soils the packing wheels may not exert sufficient pressure to insure good root-soil contact, and the soil may crack and open along the planting slit as it dries. At cutover sites with heavy logging debris, foresters must often clear paths for the tree-planting machine. In some cases, they might mount a V-shaped blade on the front of a tractor pulling the planting machine. Then the machine pushes the debris aside to clear a path for the planter, accomplishing site preparation and planting in a single operation. In some regions they will use prescribed burning to reduce the logging slash.

Due to its higher cost, foresters will generally limit hand planting to sites unsuitable for mechanized operations, to small areas that do not justify the use of planting

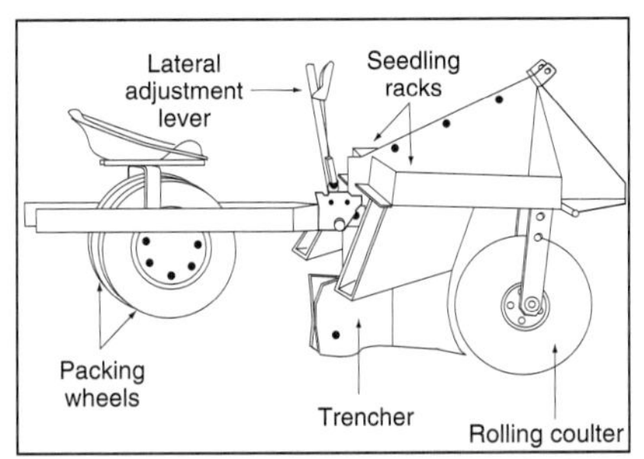

a. Simple continuous-furrow, mechanical tree planter, attached to a farm tractor through a 3-point hitch. Seedlings carried in the racks are manually placed in the planting slit by the operator (adapted by Slusher)

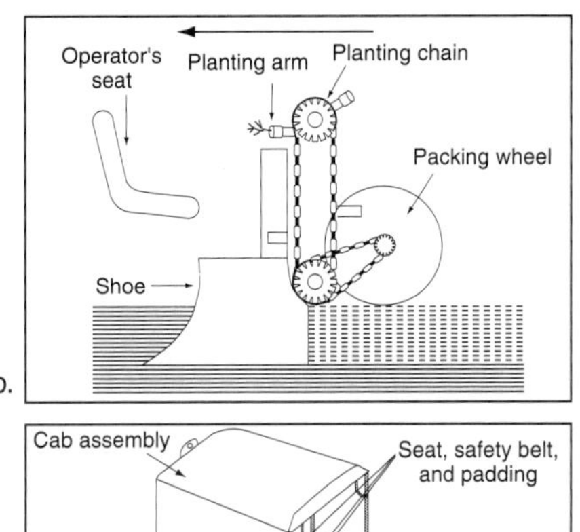

b.

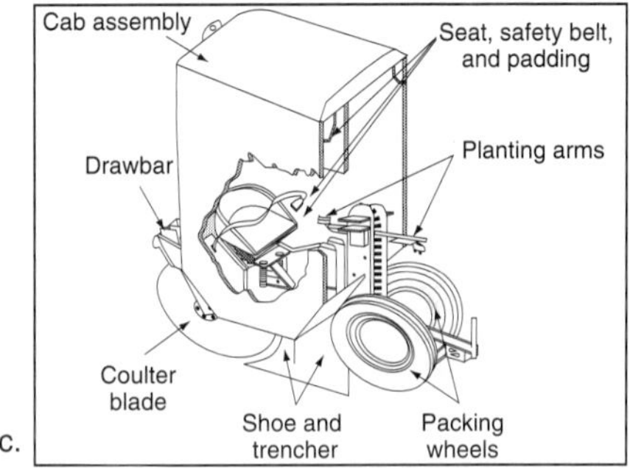

c.

Continuous-furrow mechanical tree planter, with (b.) a semiautomatic device which sets seedlings in the planting slit after the operator has placed them in the fingers of the device, and (c.) a cab assembly for safety. (Design by Whitfield Manufacturing Co.)

FIGURE 7-7
Machine planting usually proves effective in soils that can be plowed to a depth to accommodate the tree root systems (from Long 1991).

machines, and to interplanting in stands with a partial tall tree or seedling stocking (Linstrom 1962; Schubert et al. 1970; Schubert 1976; Benson 1982; Long 1991). Production rates for tree planting machines vary widely, with crews planting

1 2,000–6,000 seedlings per day per planting machine on clear areas in the Inland Northwest of North America (Schubert 1976);
2 7,000–20,000 per day per machine in the Upper Lake States, depending on the degree of site preparation and soil conditions (Benson 1982); and
3 8,000–12,000 for forest land planting following site preparation in the Southeast (Long 1991).

The actual production varies with site conditions, with greatest efficiency at open sites (old fields) with coarse-texture soils and gentle topography.

SOME SPECIAL CONSIDERATIONS IN PLANTING

An optimum day for planting (U.S. For. Serv. 1989; Long 1991) will have temperatures between 35° and 60° F (1.7–15.5° C), more than 40% relative humidity, and a windspeed of less than 10 mph (16 /kmh). Seedlings often suffer heat damage or desiccation during periods with stronger wind, a lower relative humidity, or temperatures above 70° F (21° C). The soil should have a moisture content of at least 50%

of field capacity (Long 1991). This insures good packing and sufficient cohesion to keep the planting hole closed.

Even with these conditions some sites prove difficult to reforest, especially where

1 accelerated surface erosion or frost heaving would expose the roots of newly planted seedlings;
2 slash and other cover enhance the habitat for harmful pests and herbivores;
3 people and domesticated animals may damage the young trees;
4 wildfire, drought, unfavorable weather, and any other uncontrolled environmental condition damage the trees; and
5 the site has rocky, shallow, and excessively drained soil prone to mid-summer moisture deficits.

Foresters must take special preventive measures at sites with coarse-texture soils of dark color, low soil moisture and no shade, or important competition from residual vegetation (Hite 1976). Especially in dry and well-drained soils, the most favorable microsites have (1) protection from direct insolation (as on the shady side of stumps, downed logs, and rocks); (2) good soil drainage (but off the tops of dry hummocks or mounds); and (3) good soil moisture (as in shallow depressions, but not at the risk of inundation or siltation) (Schubert et al. 1970). To enhance survival, workers can move pieces of logging debris to the south side of each newly planted seedling (to cast shade), or place natural or synthetic mulches on the ground around it (Wenger 1984; McDonald and Helgerson 1990).

Even for favorable sites, foresters must take preventive measures to protect a plantation. Important threats (after Schubert et al. 1970; Stoszek 1976; Minogue et al. 1991) include

1 diseases and insects;
2 competition from preexisting and newly developing vegetation;
3 animal feeding and other activities;
4 weight of snow and ice, and breakage by people and animals;
5 drying winds and extremes of temperatures and moisture supply;
6 frost heaving;
7 nutrient deficiencies and absence of mycorrhizal organisms; and
8 fire.

With periodic monitoring they can detect emergent problems during incipient stages, and take prompt preventive and control measures.

8

REGENERATION FROM SEED AND DIRECT SEEDING

UNDERLYING PREMISE

Direct seeding simply means artficially spreading seed over an area where a landowner wishes to regenerate a new forest stand artificially (after Ford-Robertson 1971). While silviculturists classify direct seeding as an artificial regeneration method, to use it successfully they must follow all of the steps necessary for natural regeneration. In addition they must

1 account for most factors important to a tree planting operation, like selecting a species and arrangement for the new stand;
2 identify what might cause poor germination or seedling failure, and plan preventive measures to minimize the dangers;
3 determine the seeding rates for a desired density and spacing;
4 select a method for efficiently dispersing the seed; and
5 take cognizance of possible follow-up actions if the sowing provides too many or too few seedlings.

As with other means for artificial regeneration, foresters must take a systems approach to (1) apply appropriate site preparation; (2) carefully handle the seeds; (3) use proper sowing techniques; and (4) monitor the needs for postgermination care (Barnett et al. 1990). They must also account for some additional factors that affect the success of natural regeneration, and this makes the planning and implementing fairly complex.

From a biologic or ecologic perspective, natural and artificial regeneration from seed differ primarily with respect to the source of seed, the time of sowing or dispersal, the mode of delivery, and the dispersion (spatial arrangement) of seeds. In natural regeneration, the seed must come from trees growing within or immediately adjacent to the stand, or transported to a site by birds and mammals. By contrast, foresters can gather seed for direct sowing from carefully selected trees and stands, including those at widely dispersed locations and from species not represented in a regeneration area. They can also control the spacing and arrangement, and when to make the sowing.

Generally, direct seeding enables foresters to address a broader range of management options than they can sometimes accommodate through natural regeneration. At the same time, it forces them to deal with a more complex array of ecological factors than common to tree-planting. In essence, they must manage two kinds of ecologic space (Smith 1986): (1) *above-ground*, to insure adequate levels of light and control some other important determinants to success; and (2) *below-ground*, to insure adequate moisture and nutrients. This normally means reducing the overstory and near-ground shading of already-forested sites, and enhancing soil conditions for timely germination and optimal growth of the new cohort. In addition, foresters must secure adequate amounts of viable seeds, and plan a reliable means to disperse the seed uniformly across the regeneration site at an appropriate time.

ECOLOGIC CONCEPTS OF IMPORTANCE

Foresters must insure adequate physical site resources to bring about timely germination, and to sustain vigorous growth and development of the seedlings. Critical factors (Krugman et al. 1974; Cleary et al. 1974; Daniel et al. 1979; Smith 1986) include

1 *light* in an appropriate intensity and quality (from direct and indirect solar radiation);
2 *temperature* of an appropriate level at and near the surface and within the upper soil horizons;
3 *oxygen, CO_2, and other essential gases* in proper supply, balance, and exchange rates;
4 *moisture* within an acceptable range of supply at the surface and in the rooting zone of the upper soil; and
5 *nutrients* in adequate supply and at the critical times.

To succeed, foresters must insure at least a minimum requisite level of each critical environmental factor, an appropriate balance among mutually dependent ones, and no excesses that might inhibit germination or discourage seedling growth and development.

Among forested sites, the density of any already-established vegetative community often determines the levels of light (including surface temperature) and available moisture through shading and transpirational use, respectively. In addition, seedbed conditions profoundly affect the amount and stability of moisture and nutrient supplies, the ease of root penetration, and the surface temperatures. As a consequence, foresters must often apply both a well-designed cutting method to reduce shading from tall trees, and site preparation techniques to eliminate understory shrubs and herbs and modify the soil surface. They also must protect the seeds and new seedlings from pests, diseases, and fire in order to insure a favorable long-term outcome.

In many cases, foresters will elect silvicultural methods that insure an adequate stocking and also favor a particular group of species. With both objectives they must insure an adequate supply of viable seed, and create an appropriate environment for germination and development of the chosen species. To do so, they draw upon well-established principles of ecology in judging the needs and forecasting likely outcomes. Site elements at intolerably high or low levels will limit survival and development. A site will support only those plants that tolerate the range of conditions commonly present, even though extremes of critical factor may persist for relatively short periods (annually, or only at some single critical juncture in the life of a tree).

Foresters must also consider the ways levels of one critical element may compensate for otherwise dangerously high or low levels of another (Hocker 1979; Spurr and Barnes 1980). Plants will survive and grow robustly only if the environment has an appropriate balance of critical factors, with an abundance of one compensating for low levels of others. To illustrate, low temperatures at high elevations might regulate transpiration sufficiently to compensate for limited soil moisture. Or transpirational use of abundant soil moisture at low elevation sites may cool plants sufficiently to compensate for high temperatures during some periods. In fact, the concept of *site* embodies the notion that atmospheric, edaphic, and biotic components of the environment create a prevailing set of conditions that determines the potential for a species at a particular location.

Shortages or imbalances of critical site factors may mainly limit population densities, or the rates of growth and development. Sites with limited growing season soil moisture may support only widely spaced trees (e.g., low-density stands of ponderosa pine in southwestern United States). Also, the trees may grow slowly and become stunted. Likewise, some species may survive but grow poorly in soils having limited internal drainage capacity (e.g., red pine on imperfectly drained soils of northeastern North America). Foresters must remember that these principles apply throughout the life of a stand or age class, and not just during the regeneration period. They must take deliberate steps to control the critical factors within tolerances of the selected species, or look for a species that will otherwise grow successfully at the site. Much depends upon a landowner's objectives and the financial limitations.

PLANT SUCCESSION AND FOREST REGENERATION

The *Law of Minimum* (Chapter 4) helps to explain at least one way that plant succession commonly affects the composition and timing of natural regeneration in forested environments. Though complex in scope (West et al. 1981), classical use of the term *succession* (after Daniel et al. 1979; Hocking 1979; Spurr and Barnes 1980; Kimmins 1987) means the process whereby one plant community replaces another, with conditions of the physical environment, growth characteristics of the different plants, effects of herbivory and other biologic factors, and the available sources of regeneration determining the species that become established and eventually dominate a site.

These changes (Spurr and Barnes 1980; Kimmins 1987) occur as the consequence of two basically different processes:

1 new trees develop underneath mortality-caused openings (canopy gaps) between trees in the overstory, leading to a gradual long-term replacement (or maintenance) of the existing community; or

2 a major disturbance (human or natural) abruptly alters an existing community, or greatly reduces its density, triggering widespread regeneration, and often of different species than previously occupied the site.

In the absence of a major disturbance, unmanaged stands continually and slowly modify the site conditions as they grow and develop. Eventually some trees mature physiologically, weaken, and die (Fig. 8-1). This leaves a void in the upper canopy, or thins the density of its leafy cover. Then light and other conditions change at ground level. Given a source of viable seed, regeneration will usually fill the opening. In the absence of a more drastic external disturbance, a fairly limited array of shade-tolerant species will normally succeed, and become increasingly more abundant over time. Gradually, individuals from within these groups of trees grow tall and reach overstory positions. Eventually they, too, mature and die to form new gaps that become the centers for other clusters of seedlings. These, too, eventually mature and die, in a long-term cycle that creates and maintains an uneven-aged community dominated by shade-tolerant species. Such dynamics typify many old-growth communities of long-lived species.

In some dense stands with inadequate understory light and poor seedbed conditions, many species fail to germinate at all, or new seedlings survive for only short periods. These communities have only a low-density understory, or advance tree regeneration limited to the most environmentally tolerant species. In fact, when regeneration forms underneath only small canopy gaps, light often remains limiting. Then the shade tolerance, differential growth rates of competing species, and abundance and distribution of viable seed primarily influence the outcome. Tree species diversity within such communities remains fairly stable over protracted periods, unless some external force substantially alters environmental conditions near the ground. By contrast, if groups of adjacent or nearby trees succumb at about the same time, understory brightness and some edaphic conditions may change to a greater degree. Then a wider array of species will succeed, given an adequate distribution of seed across the regenerating area.

Replacement of one seral stage by another occurs slowly and gradually, usually when the community reaches an advanced age and the trees begin to decline physiologically (at an old-growth stage). At any time, natural disturbances may disrupt the continuity of a crown canopy and change conditions sufficiently for new plants to regenerate in a fairly continuous layer across a site. Insect-caused mortality and blowdown during storms illustrate this kind of change in unmanaged stands. Since these disturbances occur at unpredictable times, silviculturists do not depend upon them to trigger natural regeneration. Instead they attempt to create controlled ecologic disturbances (e.g., overstory cutting and site preparation) at selected times

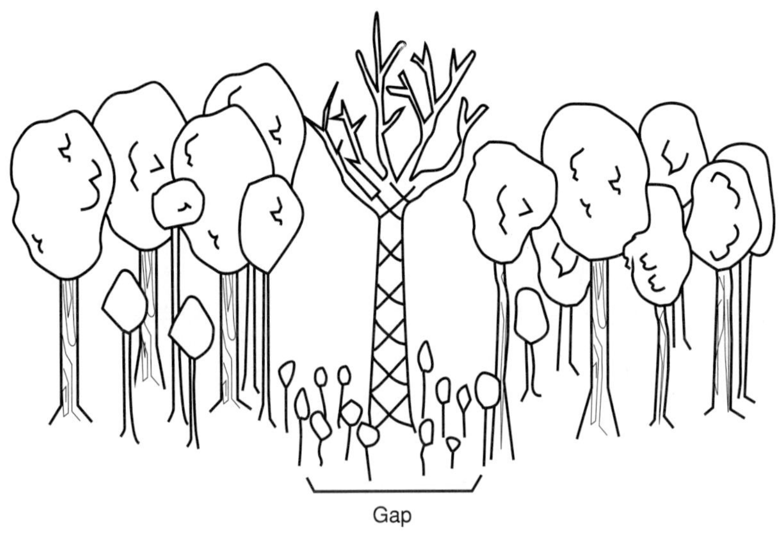

FIGURE 8-1
Death of a large tree in unmanaged stands opens a small gap in the upper canopy, brightening a small spot at the ground surface, and enhancing conditions for seed germination and seedling survival underneath the opening (after Runkle 1992).

and different intensities to influence what species replace the old community of trees. In this, they temper the kind of magnitude of environmental change (after Starr 1965; Natl. Res. Counc. 1981; Johnson 1982) by

1 varying the types and intensities of overstory cutting;
2 eliminating unwanted species and stimulating the production of the preferred seed;
3 using direct seeding or other artificial regeneration methods to supplement natural seed distribution, or substitute for it;
4 reducing the density of understory vegetation that might inhibit tree seedling establishment and growth;
5 scarifying the forest floor to make seedbeds more favorable to the target species;
6 indirectly controlling pests and predators, or disrupting the habitat that supports them;
7 preventing uncontrolled fire by reducing the slash and other fuels; and
8 tending the new community to influence the species that succeed.

Foresters use these treatments to overcome some natural limitation to regeneration, and to alter environmental conditions to favor preferred species.

DIRECT SEEDING FOR FOREST REGENERATION

Direct seeding has not replaced tree planting as a common artificial forestation method (Fig. 8-2). It primarily complements or substitutes for it when the needs,

FIGURE 8-2
By direct seeding, foresters can establish a new cohort of trees with a more uniform density and more optimal spacing than often occur with natural regeneration.

sites, and species make tree planting less advantageous (Natl. Res. Counc. 1981). Compared with natural methods, however, direct seeding gives foresters added control over (1) the genetic constitution and species of the seed source; (2) the losses to natural agents; and (3) the time of seed dispersal. Generally, they prefer direct seeding to treat large areas at a reduced cost. It also works well for sites with poor access, difficult terrain, and limited trafficability. In addition, direct seeding often proves advantageous when some natural disaster destroys a forest over an expansive area, reducing the seed source and making timely natural reforestation unlikely (Mann 1965a; Daniel et al. 1979; Natl. Res. Counc. 1981; Benzie 1982; Smith 1986; Barnett et al. 1990).

Direct seeding will not succeed when ecologic conditions prove unsuitable to prompt germination and long-term survival of a desired species (Mann 1965a, 1965b; Natl. Res. Counc. 1981; Benzie 1982; Smith 1986; Barnett et al. 1990; Barnett and Baker 1991). Neither will natural regeneration. In fact, direct seeding will fail

1 at unsuitable seedbeds, or those that dry rapidly during spring and early summer;

2 at droughty sites;

3 on highly erodible surfaces where the seed would wash off or the seedlings become uprooted;

4 on soils that become inundated during critical times in the growing season; and

5 at sites where seed-eating vertebrates consume large amounts of the dispersed seed.

Generally, it will fail any place where the light, moisture, nutrients, and essential gases makes germination, survival, and long-term growth improbable.

Direct seeding has certain distinguishing characteristics. They suggest its greatest advantages *and* shortcomings (after Mann 1965a, 1965b; Rudolph 1973; Natl. Res. Counc. 1981; Benzie 1982), such as when it

1 increases flexibility, with no long lead time for growing seedlings at the nursery, and more opportunity to cancel the regeneration operation due to unforeseen events;

2 allows sowing during periods with high demand on labor;

3 requires less labor to produce the stock, and this helps to contain the costs;

4 produces seedlings with good root systems, since they develop directly on site and remain free of damage from handling and transplant;

5 requires site preparation to improve a seedbed for germination and early survival, increasing the costs;

6 provides flexibility in regenerating species that produce bumper seed crops only infrequently, have really small seeds, or prove difficult to handle and store;

7 carries a higher risk of failure due to uncontrolled vagaries of the weather, losses of seeds to pests, limitations in placing the seed in favorable microsites, and problems with implanting the seeds in the seedbed medium;

8 makes inefficient use of high-value or genetically improved seed;

9 results in an irregular spacing and arrangement within the new community, unless a landowner invests in special measures to control these factors; and

10 depends on germination at the site, which may lengthen the establishment period compared with planting.

The potential to lower the cost offers the most compelling advantage over tree planting, but the higher risk of failure generally deters the widespread use of direct seeding. Effective site preparation will reduce the chances of failure, but foresters cannot insure against unfavorable weather just before or after germination. The type, intensity, and extent of requisite site preparation also vary greatly from one location to another, and this affects the costs. Yet for appropriate sites, landowners can efficiently direct seed in any kind of terrain and across any amount of land area at a reasonable cost.

Foresters commonly object to direct seeding because broadcast methods do not control seedling spacing, arrangement, and density. They can change the spacing and arrangement by altering the seeding method, but weather conditions largely determine the density. Too many seedlings often survive during seasons with extremely favorable weather, and too few develop during unfavorable years. As with natural regeneration, failures result from unseasonable drought, too much moisture, high temperatures, and freezing weather. Other critical factors (after Mann 1965a; Millar 1965; Schubert et al. 1970; Campbell and Mann 1973; Rudolph 1973; Barnett and Baker 1991) include

1 inadequate control over the time and rate of sowing, the delivery (seeding) method, rodents and other animals that eat the seed or destroy the seedlings, and seedbed conditions;
2 using immature, improperly stratified, damaged, or poor-quality seed; and
3 attempting direct seeding or using inappropriate species at sites where environmental conditions make regeneration from seed difficult, including interference by a vegetation community already established at the site.

Failures often result from human error.

PREPARATION FOR DIRECT SEEDING

Direct seeding will succeed only when foresters adequately prepare the site and seed, and control the time and method of sowing. In fact, with any method of regeneration from seed, foresters must often use site preparation to expose mineral soil, or mix the mineral soil and humus. Further, by balancing the degree of seedbed preparation with an adjusted sowing rate, foresters can often control the spacing within a new cohort (Natl. Res. Counc. 1981).

Past experience underscores the importance of appropriate site preparation, and the likelihood of failure at sites lacking appropriate surface disturbance (Mann 1965a; Horton and Wang 1969; Brown 1974; Griffin and Carr 1973; Fraser 1981; Barnett et al. 1990). The methods of preparation and means for sowing may vary by region and site conditions. Yet in the upper Lake States of the United States, direct seeding succeeded in about one-half the cases where site preparation preceded the seeding, but at only 10% of the sites receiving no site preparation (Roe 1963). Failure at prepared sites resulted from unfavorable weather or other uncontrolled factors. In the western and southern United States, most successful seeding followed prescribed burning that reduced the litter and organic covering over the soil, and killed back competing vegetation (Natl. Res. Counc. 1981).

These treatments may require considerable investments, thereby increasing the costs. In fact, available guidelines suggest exposing mineral soil over at least one-quarter to one-half of the surface, using some combination of deliberate and passive measures (after Brown 1974; Horton and Wang 1981). This insures a fairly uniform distribution of favorable seedbeds. Foresters must also prepare the seed to insure adequate germination and to reduce the losses to fungi, birds, mammals, and insects (Derr and Mann 1971; Fraser 1974; Schmidt et al. 1976; Natl. Res. Counc. 1981; Benzie 1982; Barnett et al. 1990). Foresters often get help from nursery personnel for the seed preparation work (Schopmeyer 1974), and for information about germination, soundness, and purity to use in calculating sowing rates.

Direct seeding generally leaves the seed exposed on the surface, or just pressed into the seedbed. This makes the seed more vulnerable to birds, mammals, and insects. In many cases, site preparation treatments can make the habitat less favorable, and indirectly reduce the losses. In other cases, foresters treat the seed with repellents. In extreme cases they may also set out poisoned baits to reduce the population of animals like mice and voles. Government regulations control the use of these poisons and repellents, including the methods to protect nontarget animals and workers, and prevent unexpected environmental side effects.

Some species germinate best if first subjected to freezing and thawing, soaking, or some treatment that decomposes the seed coat (Schopmeyer 1974; Young and Young 1992). Many times, nursery managers can stratify the seed to produce the desired results artificially, permitting foresters to delay sowing and shorten the exposure to seed eaters, disease organisms, and other hazards. Generally, they must have the seed in place when conditions first become favorable for germination at the beginning of a growing season (Schmidt et al. 1976; Natl. Res. Counc. 1981; Barnett and Baker 1991), or to insure adequate in-place stratification. This usually means sowing during late winter or early spring. In fact, experience within the upper Lake States and eastern Canada of North America has shown that winter seeding reduces seed losses and yields favorable rates of germination (Abbott 1974; Hellum 1974; Natl. Res. Counc. 1981; Benzie 1982). In southwestern North America, sowing from mid-February through early March proves best, except at the most southern sites where dry spring weather recommends fall sowing (Barnett and Baker 1991). Spring sowing also works best for western larch, Douglas-fir, and Englemann spruce in northwestern North America (Schmidt et al. 1976).

Foresters can determine the rate of seeding by procedures commonly used at forest nurseries, but tempered by judgment to account for differences between field conditions and nursery seedbeds, and the method of seeding. These (Millar 1965; Burns 1965; Wenger 1984; Smith 1986; Barnett et al. 1990; Barnett and Baker 1991) can include

1 *broadcast seeding*, where they distribute the seed uniformly across a site;
2 *spot seeding*, where they sow the seeds in groups (e.g., 2–6 per spot) at preplanned intervals, as with tree planting; and
3 *drilling*, where they line out the seed in appropriately spaced rows, as with the sowing of many agricultural crops.

The latter two methods give foresters more control over arrangement and spacing, but at a higher cost. On the other hand, they can then utilize band or spot site preparation treatments, and that will cost less than other methods.

Rates for broadcast seeding depend upon several factors (after Stoeckeler and Jones 1957) such as the following:

$$W = \frac{A * D}{G * S * P * Y}$$

where W = pounds (kg) of seed cast per unit of area

A = area covered (ft^2 or m^2)

D = density of seedings desired per unit of area (e.g., #/ft^2 or #/m^2)

G = germination capacity in percent, based upon standard tests

S = number of seeds per pound (kg)

P = purity, as percent of weight in actual seed

Y = the estimated survival factor

For drilled or spot seeding, the formula reduces (after Stoeckeler and Jones 1957) to

$$N = \frac{D}{G * Y}$$

where N = number of seeds per lineal foot (m), or per spot

D = density sought per lineal foot (m), or per spot

G and Y = the same as above

Foresters can determine all except *Y* and *D* by fairly objective methods using measurements or tests of different kinds.

The survival factor (*Y*) represents a *best judgment* about what proportion of germinated seed will survive for some preset period of time, and usually one year. It will vary with site conditions, species, and weather patterns following germination. The desired density (*D*) indicates the number sought per unit area, and usually at the end of one growing season. Both *Y* and *D* reflect local experience. Poor judgments about these factors result in stands of inadequate stocking or too many trees. Foresters can adjust that later by early spacing treatments, but can only supplement poor stocking by additional seeding or reinforcement planting.

Actual seeding rates may differ from one stand to another within a region. Much depends upon the objectives, type of site preparation, and the species. For example, in the West Gulf region of the southern United States, foresters frequently use about 1 pound of seed per acre (1.1 kg/ha) for broadcast sowing, but only 0.75 pound per acre (0.8 kg/ha) for row sowing with slash pine on disked seedbeds. In the Southeast, they can reduce the rates by about one-third because of more favorable summer moisture, and because management objectives usually allow lower stand densities (Lohrey and Jones 1983). A lower seeding rate may also suffice for well-prepared, moist soils rather than on coarse dry sands, but foresters cannot merely increase the seeding rate to compensate for poor or inadequate site preparation at those sites (Campbell and Mann 1973).

Among many seeding trials in Canada between 1900 and 1972, sowing rates averaged around 50,000 seeds per acre (123,500/ha), but varied somewhat between species. With jack pine, sowing as few as 20,000–30,000 seeds per acre (49,400–74,100/ha) gave satisfactory results. Factors like favorable weather following seeding, the application of repellents and fungicides, the completeness of site preparation, and the timing of seeding seemed more critical than variations in the sowing rates (Waldron 1974). In Alberta, foresters have commonly used between 20,000 and 50,000 seeds per acre (49,400–123,500/ha), for jack pine, lodgepole pine, white spruce, and Douglas-fir. These conform closely to rates of natural seed dispersal in the region. Further, scarification must expose mineral soil across at least 25% of the surface to insure success. Best results follow ground seeding in winter or early spring on top of the snow (Hellum 1974).

SOWING THE SEED

For any seeding, foresters must have a means to transport the seed across a site, a mechanism for dispensing the seed, and a way to regulate the rate of dispersal. While they must mechanize these procedures for large areas, even simple hand-powered devices suffice in small stands. The degree and kind of control over spacing and arrangement will also temper the means used, and the costs.

Aerial seeding has historically proven most efficient in treating large areas in a relatively short period of time, areas with inaccessible terrain, and debris-covered sites (Lohrey and Jones 1983). Pilots fly progressive side-by-side paths at a low altitude (e.g., 75–200 ft, or 23–61 m) and speed (e.g., 50–80 mph, or 80–129 km/hr), depending upon the type of aircraft and the desired sowing rates. Ground crews use flags or radar guidance systems to mark the flight paths (Griffin and Carr 1974; Worgan 1974; Natl. Res. Counc. 1981; Lohrey and Jones 1983). Given proper logistical support, a fixed-wing airplane can broadcast seed about 1,500 ac (607 ha) per day. A helicopter can treat about 50% more area due to its operating capabilities (Derr and Mann 1971; Barnett et al. 1990; Barnett and Baker 1991). Neither method insures exact control over spacing or stand density (Natl. Res. Counc. 1981; Barnett and Baker 1991). However, by outfitting the aircraft with special seed dispensers and flying at 25–50 ft (8–15 m) in altitude, pilots can place the seed in rows and regulate seed spacing within rows (Mann et al. 1974).

Ground methods for direct seeding include broadcast, row, and spot seeding. For small areas a crew may simply use hand-held cyclone seeders, treating as much as 12 ac (5 ha) per day per seeder (Barnett et al. 1989; Barnett and Baker 1991). Workers control seeding rates by how fast they walk, and how quickly they turn the crank that operates a seed spreader. For larger areas, they can mount a cyclone spreader onto a tractor or a modified skidder, and motorize the spreader mechanism. These large seeders can cover about 1–2 ac/hr (0.4–0.8 ha/hr) (Brown 1974). For winter sowing, they might use snowmobiles or other light snow machines and broadcast seed at up to 6 ac/hr (2.4 ha/hr) (Abbott 1974; Brown 1974; Hellum 1974; Waldron 1974). For mechanized row seeding, they might combine site preparation and seeding in a continuous operation by (1) pushing aside the litter layer in a narrow band using a fireline plow; (2) dispersing seed onto the exposed soil with an agricultural row seeder; and (3) lightly pressing the seed into the seedbed with a packing wheel attached to the back of the seeder. Depending upon conditions, such a furrow-seeder will treat about 1–4 ac/hr (0.4–1.6 ha/hr), and about 15 ac (6 ha) per day, at a cost of about one-third of that for bare-root planting (Campbell 1965; Graber and Thompson 1969; Graber 1974; Barnett and Baker 1991).

Foresters do not normally do row seeding by hand, but an individual could drop the seed onto a prepared seedbed and get similar results (though with less efficiency). With longleaf pine in southeastern North America, a common row seeding will create bands of seedlings about 8–10 ft (2.4–3.0 m) apart, with individual seeds about 1–2 ft (0.3–0.6 m) apart in the rows (Campbell 1965; Barnett et al. 1990; Barnett and Baker 1991). With ponderosa pine in the southwest, foresters would sow the rows at 4- or 8-ft (1.2- or 2.4-m) intervals, but at one-half the spacing between seeds within a row (Schubert et al. 1970). As with broadcast sowing, foresters must adjust the sowing rate to compensate for losses of seeds (Schubert et al. 1970; Brown 1974; Jones 1974; Barnett and Baker 1991) that

1 fall onto unfavorable seedbeds;
2 fail to germinate due to excessively deep sowing;
3 rot due to deep sowing and other factors;
4 become damaged during sowing;
5 die from smothering by inundation or a deep cover of silt or debris;
6 die from desiccation after germination;
7 succumb to freezing and frost-heaving during the first winter; and
8 vertebrates or insects damage or eat.

The method works best on fine- to medium-textured soils with good internal drainage. For the less well-drained sites, foresters can add rotating disks that mound the seedbed 3–6 in. (7.6–15 cm) above the general ground surface (bedding), and then sow bands of seed on these slightly elevated seedbeds (Barnett and Baker 1991).

As another alternative, foresters may drop a small number of seeds (usually three to seven) on well-prepared spots ranging from 1–2 ft (0.3–0.6 m) in diameter, and spaced at intervals comparable to those for tree planting. Of all the direct seeding methods, this seed-spot method takes the greatest investment of labor, and one person can treat only about 2–6 ac (0.8–2.4 ha) per day (Jones 1965; Barnett et al. 1990). Foresters use spot sowing mostly in small areas, or to fill in or finish off broadcast or drill seeding operations (Jones 1965; Barnett and Baker 1991). To improve efficiency for spot seeding, foresters in Ontario use a mechanized scalper to prepare about 2–3 ac/hr (0.8–1.2 ha/hr), making about 1,000 scalps/ac (2,470/ha) (Brown 1974), and deposit the seed on these prepared spots. Guides like that shown in Fig. 8-3 (from Schubert and Fowels 1964) allow foresters to adjust the sowing rates based upon expected survivorship (the *Y* factor). Commonly, seedling losses (Jones 1974) result from

1 frost heaving during the first winter;
2 trampling by cattle and wild animals;
3 damage through human activity; and
4 washing out of new seedlings through erosion.

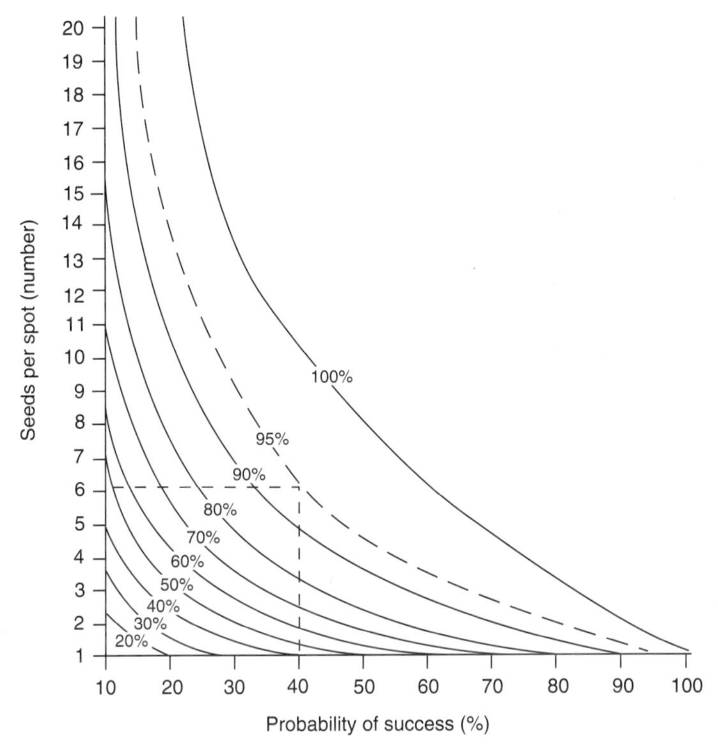

FIGURE 8-3
Sowing guides allow foresters to adjust the number of seeds deposited per spot based upon expected survivorship and the desired long-term stand density (from Schubert and Fowels 1964).

They must also consider the minimum number of successful spots per unit of area for acceptable future stocking. For example, if foresters require 95% successful spots with ponderosa pine in southwestern North America, they must sow six seeds per spot. For an 80% success rate, they could sow only three seeds per spot (Schubert et al. 1970).

Foresters can apply spot seeding by hand or mechanized methods. Workers scarify an area for each spot, deposit a small number of seeds onto the prepared spot, and press them into the soil. Seeders like the one shown in Fig. 8-4 will suffice for small acreages. The tool has

1 a seed bottle, to protect the seed from damage and also safeguard workers against contact with repellents and fungicides on the seed;
2 a dispensing mechanism, with changeable bushings to regulate the number of seeds that fall from the bottle;
3 a tube that the seeds fall through to the scarified spot; and
4 a metal claw to aid with scarification.

FIGURE 8-4
With spot seeders, workers can regulate the spacing of new trees in a way similar to tree planting, or they can fill gaps in areas regenerated by natural methods.

By using such devices, foresters can also do reinforcement seeding in areas with an irregular distribution of natural regeneration, or enrich the composition with species that would not naturally occur in a stand.

In spot seeding, workers scrape away the litter layer and mix the humus and mineral soil using the claw. Then they pull the trigger attached to the seed bottle to release a predetermined number of seeds down the tube onto the seedbed. Finally, they step on the seeds to lightly press them into the soil, or they cover them with less than 0.5 in. (1.3 cm) of loose soil or humus. Depositing too many seeds can result in multiple seedlings per spot, but the long-term consequences remain unclear. One study of loblolly and slash pines indicated that crowding reduced both height and diameter growth of seedlings in a spot (Campbell 1983). Other field assessments of loblolly and Virginia pine in Tennessee indicated that seedlings in these clusters grew as well as those more widely dispersed by broadcast seeding (Russell 1975).

APPLICATION OF DIRECT SEEDING

Direct seeding has not gained favor in most of North America. In fact, its fairly widespread use in the western United States prior to the mid-1970s (Abbott 1974) declined because

1 there are restrictions on use of chemicals for controlling losses to rodents, birds, insects, and fungi;
2 the need for extensive site preparation made the cost of direct seeding comparable to that of bare-root planting;
3 alternate methods permitted more efficient use of genetically improved seed; and
4 past results proved erratic.

Foresters have used direct seeding extensively in the U.S. Gulf States. Within the upper Great Lakes region of North America, foresters must use great care in selecting sites for direct seeding because of the sensitivity of newly germinated seedlings to excess heat, freezing, drought, inundation, and competition from surrounding plants (after Rudolph 1973; Benzie 1982). Appropriate site preparation must accompany direct seeding to ensure favorable seedbeds (Brown 1974; Johnston 1982). Also, foresters must protect the seed from insects, birds, and mammals to insure sufficient supplies for germination (Schubert et al. 1970).

Evaluations of areas regenerated to loblolly pine in Louisiana and Texas indicated that direct seeding, planting, and natural methods work successfully if applied properly. Planting proved most reliable, but less feasible for the prompt reforestation of large areas. Direct seeding worked better than natural regeneration methods when properly applied (Campbell and Mann 1973). Compared with 20-year-old planted stands in the central Gulf Costal Plain, those from direct seeding yielded less but not significantly different volume, and had smaller and less regularly spaced trees. Volumes with slash pines did not differ (Campbell 1983). For 132 test seedings in the Lake States, 40% succeeded. Foresters had best luck with jack pine sown on disked or burned sites. Red pine, white pine, black spruce, and balsam fir succeeded only with spot seeding on a suitable seedbed. Most failures resulted from interference and competition from other vegetation, seed losses to rodents and birds, or unfavorable weather (Roe 1963). In Canada, direct seeding has largely failed, except with jack pine, lodgepole pine, and white spruce on prepared sites using seed treated with repellents (Hellum 1974; Brown 1974; Waldron 1974). In the tropics, it has proven reliable only with large-seeded species like *Gmelina arborea, Swietenia* (mahogany), *Mora excelsa, Acadia arabica, Prosopis cineraria, Pinus caribaea, Pinus oocarpa* and *Leucaena leucociphala* on adequately prepared sites (Rimando and Dalmacio 1978; Edwards 1982; Evans 1982).

Compared with tree planting, direct seeding demands greater attention to microsite quality, and provides less control over intertree spacing and arrangement, except when foresters invest in row or spot sowing. Difficulties (after McQuilkin 1965; Rudolph 1973, Evans 1982; Barnett et al. 1990) include

1 the relatively slow growth of new seedlings, compared with interfering plants or planted seedlings;

2 problems in applying the requisite site preparation on sites with heavy logging slash, rough and stony surfaces, or in low and poorly drained areas;

3 the need to include site preparation methods that break and mix fairly deeply accumulated organic humus or litter on many sites;

4 losses of the seed on highly erodible sites and steep slopes;

5 competition from interfering vegetation following overstory removal;

6 the need to control pests that consume or damage the seed prior to germination; and

7 uncontrollable weather conditions prior to and after germination.

Where conditions permit and foresters can exercise adequate control, direct seeding can play an important role in meeting special regeneration needs at a low cost, and within a short period of time.

9

REPRODUCTION METHODS AND THEIR IMPLICATIONS

In planning a reforestation strategy by either artificial or natural means, foresters must deliberately deal with a variety of ecologically and economically important factors (after Daniel et al. 1979) including

1 the management objectives, and the stand structural attributes that will satisfy these purposes;
2 the managerial, institutional, financial, social, and ecologic perspectives;
3 the existing plant community and its characteristics, and conditions of the physical environment that support it;
4 the range of alternative species and age arrangements that might serve the needs; and
5 the treatments that could transform an existing community into one of desirable conditions.

Existing and desired tree community attributes of particular interest include the following:

1 the species composition
2 silvical characteristics of each species and their mutual compatibility
3 range of sizes
4 differences in ages
5 variations and levels of density or spacing among individual trees
6 health and vigor of different elements in the total plant community
7 agents that might harm individual trees or jeopardize the health and life of an entire community

Physical site conditions of importance include those essential to the growth and reproduction of individual trees and groups of them, their suitability to support management and use without danger of site degradation or limitation to operability, and any conditions that might heighten the danger of loss or damage by indigenous natural agents.

Under a philosophy of management by objectives, landowners set the criteria for success from an economic or institutional perspective. Foresters helping them must identify the best ways to bring the goals to fruition. Ethically, they must insure an ecologic soundness to the approach taken, and look for ways to (1) balance the costs and benefits, (2) optimize the values derived at some sustainable rate, (3) fully utilize the capacity of a site to provide the values of interest, and (4) promptly replace consumed values with new use opportunities for the future. In more traditional terms of commodity production, these equate to providing regular and sustainable yields at an affordable cost (Fig. 9-1), insuring full site occupancy by valuable trees, and promptly regenerating new crops when the existing ones reach financial maturity.

REPRODUCTION METHODS AND THEIR ROLE IN SILVICULTURE

Silviculturists can serve varying kinds of landowner objectives by (1) employing differing sets of manipulations; (2) rearranging the sequence over time; (3) altering the timing; and (4) changing the intensity to cause different kinds and magnitudes of biological responses. Each combination alters the tree species and sizes, stand density, age arrangement, and tree forms in some unique way. This gives silviculturists the opportunity to tailor the treatments to fit the objectives for management and to provide different kinds of economically important outcomes.

Many of the differences between alternative silvicultural strategies became particularly apparent when regenerating a forest stand, since the common reproduction methods substantially change the character of a community. Definitions of a *regeneration cutting*, or *reproduction method* (Ford-Robertson 1971; Smith 1986; Soc. Am. For. 1989) imply this:

- a procedure for establishing or renewing a forest community
- any removal of trees to assist regeneration already present, or to make regeneration possible
- a procedure for regenerating a forest community

Normally these treatments involve some kind of disturbance such as cutting and timber harvest. The methods may incorporate either natural or artificial means to secure the regeneration. All allow foresters to establish a new age class (*cohort*) to replace the trees deemed mature by a landowner.

In silviculture, foresters use the term *regeneration* (Ford-Robertson 1971; Soc. Am. For. 1989) interchangeably with *reproduction*. Either may denote the process of replacing one community of trees with a new one. Foresters may substitute the term *recruitment* to describe this process. *Regeneration* and *reproduction* may also

FIGURE 9-1
When landowners have commodity objectives, foresters must provide a steady and sustained yield of sawlogs, pulpwood, and other roundwood products that landowners can sell to cover costs and provide profits.

apply to the actual seedlings that result from applying a reproduction method. A related term, *regeneration period*, means the time required to renew or regenerate (or recruit) a new age class by either natural or artificial means. *Regeneration area* means the place or geographic space selected for regeneration or renewal. In silviculture, it normally includes an entire stand wherein the trees, or a portion of them, have reached a specified state of maturity. The reproduction method will remove the mature trees to recruit or regenerate a new age class. In this context, foresters may define *maturity* using financial criteria (as with commodity production goals), or other characteristics related to the purpose of management (e.g., in wildlife management when a stand no longer provides suitable habitat for a target species or community of animals).

In making choices about the approach to take, foresters usually opt for artificial methods (after Boyd 1976) when

1 they do not have an adequate or suitable seed supply within a stand;
2 they want a species or genotype not present among nearby natural sources;
3 they wish to control stocking or limit the time and duration of establishment; or
4 they anticipate that natural regeneration methods would not succeed for some other reason.

By contrast, natural methods usually prove best (after Boyd 1976)

1 in stands with abundant and well-distributed advance regeneration;
2 where foresters expect to regenerate a desirable new community by natural means within the time desired;
3 when economic or use limitations preclude site preparation methods deemed critical to an artificial regeneration system;
4 where costs of planting or direct seeding would prohibit their use; and
5 where foresters want to encourage natural patterns and processes of stand development.

The actual choice depends upon how well one method or another, or some combination of them, would satisfy the objectives, and whether some ecologic or economic factor makes any one risky or preferable. Most important, with natural regeneration under the high forest methods, foresters must have the assurance of adequate dispersal and germination of seed to insure a well-distributed community of new seedlings of an appropriate density (Hawley 1921; Toumey and Korstian 1931; Cheyney 1942; Smith 1986).

DIFFERENCES BETWEEN EVEN- AND UNEVEN-AGED METHODS

To convey information about the methods used to regenerate repeated cohorts of trees in a stand, and to stress the importance of securing adequate regeneration for full site utilization, foresters often name the silvicultural system after the reproduction method used in it (Wenger 1984; Smith 1986). Thus the terms *clearcutting system, shelterwood system,* and *selection system* each employ different reproduction methods. The names describe the approach taken to regenerate new trees at appropriate times in the long-term management of a stand. In all these cases, foresters remove all or part of the mature overstory to improve environmental conditions at ground level. This normally involves commercial timber harvesting as a mechanism for initiating the regeneration process. With clearcutting, they often substitute tree planting or direct seeding rather than rely upon natural vegetation.

Foresters classify the high-forest reproduction methods into broad groups (after Muelder 1966; Smith 1986; Matthews 1991; Smith 1995):

- *even-aged methods*, where they remove the entire community of mature trees in one or more cuttings within a short time interval, allocating all the growing space to the new age class, and creating a new community of trees all having comparable ages
- *uneven-aged methods*, where only some of the trees in a community have reached maturity, and where foresters remove only those mature trees to allocate part of the growing space to a new age class
- *two-aged method* (often called the *leave-tree method*), where foresters convert an even-aged community by removing most of the stocking, but leaving at least 20% of the basal area in vigorous trees of upper canopy positions

Foresters have devised many variants of these basic approaches to address special needs and interests. In fact, two-aged silviculture developed as a means to address concerns over visual qualities of even-aged reproduction methods, particularly for communities with an important component of shade-intolerant species (Beck 1987, Smith 1988, Smith et al. 1989).

These methods produce a community of trees having either similar or disparate ages and reflect the density and character of any residual overstory left after a treatment. In all cases, the associated timber harvesting will

1 change the quantity and quality of light near the ground by reducing shading from tall trees, either partly or completely;

2 increase surface temperatures at the litter layer to varying degrees, based upon the sizes, numbers, and distribution of spaces (gaps) the harvesting opens within an overstory canopy;

3 reduce withdrawals of soil moisture from upper horizons by decreasing the numbers of large trees (if any) that remain; and

4 stimulate the release of nutrients into the soil solution by warming the newly exposed surface organic layer, and increasing the activity of decomposers within the litter and humus.

The reproduction method may also incorporate supplemental site preparation to reduce or control undesirable competing vegetation or improve the seedbed.

CHARACTERISTICS OF DIFFERENT REPRODUCTION METHODS

Foresters must coordinate the long-term plans for tending (to influence age-class development) with needs to regenerate replacement crops at appropriate times. For even-aged systems, they never apply all components of a system at the same time. Rather they tend a community during immature stages of development, using harvesting techniques whenever possible to generate revenues. Then when a stand matures, they harvest the entire standing crop to allocate total space to the new age class. Once recruited, it will grow on the site over another rotation, and foresters will continue to tend the community as a single-age crop. With uneven-aged stands, they periodically harvest the mature trees to regenerate replacements, and also concurrently remove excess numbers of younger trees to nurture each residual age class. This dual-purpose cutting provides recurring revenues for a landowner at regular intervals, and periodically re-creates a predetermined set of stand structural attributes across a site. However, it never removes all the trees at any time, thereby maintaining the stand's uneven-aged character.

Each high-forest reproduction method has unique features (Fig. 9-2), and most produce characteristic results. In essence, the attributes depend upon the proportion of trees removed at any single time. For example, *selection methods* (uneven-aged) remove only some of the standing trees, leaving a plentiful reserve to supply seed, and only somewhat reducing withdrawals of moisture and soluble nutrients by tran-

FIGURE 9-2
Even- and uneven-aged reproduction methods differ in the proportion of area occupied by mature trees, and the amount of ecologic space allocated to the new age class.

spiration. Soil moisture and light intensity near the ground will increase following a treatment, but only to a modest degree. These conditions limit the numbers of trees in a new cohort, and often discriminate against shade-intolerant species. At the other end of the spectrum, *clearcutting* removes an entire standing tree community in one operation. This eliminates any subsequent on-site seed dispersal, placing complete dependence upon seed from trees growing adjacent to the regeneration area or seed already stored in the litter layer. The supplies of soil moisture and dissolved nutrients increase appreciably, because clearcutting dramatically reduces transpirational use. Light intensity, surface temperatures, and light quality increase similar to those of open areas. This creates environmental conditions suitable to a broader array of plant species, including shade-intolerant shrubs, herbs, and trees. Other even-aged reproduction methods (e.g., *shelterwood* and *seed-tree methods*) create intermediate conditions for many of these environmental characteristics, depending upon the amounts, extent, and duration of residual tree cover maintained during the regeneration period. So would the *leave-tree method*.

In principle, foresters use the reproduction methods to deliberately reduce site occupancy to less than full capacity. This creates a temporary void in the ecosystem, and changes conditions at ground level and within the soil in proportion to the extent and size of openings created in the upper canopy. Though largely applied to promote tree regeneration, foresters can use the different reproduction methods to recruit communities of shrub or herbaceous plants as well. In making choices, foresters recognize that each method influences the immediate and long-term use of a community for many nonmarket values, and the kinds of conditions or commodities a landowner can produce. Foresters can vary the level of residual density (degree of removal) with all but clearcutting, either as an intermediate step in securing the new age class, or in maintaining the uneven-aged character of a stand. In managing for nonmarket benefits they may select different kinds of trees to keep or cut, remove different amounts of the initial stocking at any one time, and concentrate on establishing a particular set of species that best fit the long-term interests of a landowner. In some cases, limitations of the physical environment bode against the use of some methods, or financial considerations may make one approach more favorable than others. Foresters will evaluate each alternative for compatibility with the ecologic and economic conditions at hand, identify the one most appropriate to the needs, and then decide when and how to best employ it.

SOME REGULATORY CONSIDERATIONS IN SILVICULTURE

Biologic-ecologic factors limit what any particular site will produce, both in quantity and character (e.g., in the stature of the standing trees, and their species composition). In assessing these, foresters must identify any method that might lead to an irreversible ecologic change, or bring a long-term degradation of the overall productive capacity, and dismiss these from further consideration. In addition, they must assess an array of economic criteria that deal largely with *ownership wants*. These include financial constraints and any requirements for realizing particular kinds of output from the management. Local markets may dictate how much and what kinds of products or fee uses an owner can merchandise at any time, and the prices consumers will pay. This may make some options unattractive to implement. These and other economic factors (both institutional and financial) usually dictate the method to use from among the ecologically viable options.

In the process of planning a program of management, foresters must deal with a variety of matters that effect the silvicultural choices, if only indirectly. Important ones (after Matthews 1991) include:

1 dividing the forest into operating units suited to the silvicultural and management needs;
2 programming timely harvests and other uses to insure regular and continuing yields and benefits over the long run;
3 identifying the methods of harvesting and use suited to the planned silviculture and other ongoing operations;
4 planning an access system and other cultural features critical to success; and
5 organizing the fiscal and human resources necessary for effective management.

These link silvicultural planning to the business aspects of an ownership, and influence the allocation of treatments over time and across the geographic area under management.

Historically, foresters have given a great deal of attention to the consistency and flow or availability of products and uses over the long run. This involves the *regulatory aspects of silviculture*—procedures for controlling the flow of timber, goods, products, and services off the land to realize regular returns and continually sustain the values of interest to a landowner. These concerns parallel the concepts of forest regulation, for scheduling of timber harvests (harvest scheduling), and timber management activities (timber management scheduling), as well as programming the timely development and use of the nonmarket values of interest (after Osmaston 1968; Davis and Johnson 1987). However, silviculture focuses on the planning and decision making for individual stands rather than across an entire forest.

Historically, forest regulation has served as one primary mechanism for realizing and sustaining the yields of goods and services from forested lands. In this sense, *sustained yield* (Knuchel 1953; Osmaston 1968; Ford-Robertson 1971; Matthews 1991) means

- the yield that a forest can produce regularly and continuously at a given intensity of management
- the continuous production that reflects a balance between increment and cutting, or availability and use

Increment has traditionally meant the growth or increase of some measurable attribute of individual trees or a community of them (e.g., the change in volume, basal area, or diameter) during a specified period of time (after Ford-Robertson 1971). It could just as well include any measurable nonmarket value that also increases in abundance or worth as a stand or age class develops. These might include some measure of important visual qualities in an area, the potential for outdoor recreation, and the stability of critical habitats for particular plants and animals.

From a silvicultural perspective, sustained yield requires a continuation of trees after trees in an unbroken sequence over time. This insures long-term biologic or ecologic continuity, and depends largely upon the reproduction method employed. In principle, adequate and timely recruitment insures sustained yield by insuring future benefits of the kind sought. As such, *sustained yield* does not deal explicitly with the periodicity of use, or the levels of value derived at any specified time. With uneven-aged systems a landowner can usually realize a recurring level of use at relatively short intervals (e.g., every twelve to twenty years). With even-aged systems, landowners may wait relatively longer periods after a reproduction method (e.g., fifty to one hundred years) before a new stand provides the values of interest.

The discipline of forest management puts a time constraint on when future supplies must become available through growth and change (Thompson 1966; Davis and Johnson 1987; Leuschner 1990). In this sense, regulation controls how a landowner recovers the yields and benefits over time and space to insure consistent levels at regular intervals, and to sustain them indefinitely. In this, foresters must mesh the types, frequencies, and timing of treatments in any one community with the overall plan of management for others on the property. They must also deal with these at the landscape level in managing for some particular benefits that depend upon critical vegetal conditions at a broader spatial scale (e.g., habitat for large mammals). Further they must insure that the actions prescribed never compromise ecosystem sustainability, or bring an irreversible ecologic change.

In practice, silviculture deals with the treatments to create and maintain forest *stands* of a specified character, and to bring them to a desired stage of maturity at a predetermined time. Forest management organizes the management (silvicultural systems) of all stands across an *ownership* or *landscape* into a well-orchestrated program that balances increment and loss (cutting, disturbance, consumption) to provide a continuity in the goods and services. Even so, silviculture must link the reproduction method and subsequent intermediate-aged tending to one of four basic regulatory approaches used in forest management (after Matthews 1935; Knuchel 1953; Davis and Johnson 1987; Leuschner 1990):

Area control—the schedule identifies an amount of area to cut (or otherwise use) per year, and foresters treat that area annually. This only indirectly controls the volume (quantities of output) harvested in any year, and *presumes* that foresters can treat different combinations of stands to provide nearly comparable amounts of the desired goods or values annually. Foresters can take this approach as one method for regulating forests consisting of even-aged stands, and for scheduling the total area treated by even-aged reproduction methods each year.

Volume control—foresters identify the volume that a forest produces (grows) annually, and this dictates how much to cut (or otherwise use) per year. Foresters then treat (or otherwise utilize) sufficient area to recover the predetermined volume of goods (or amount of services) annually. This does not directly control the area treated from one year to the next. Where an entire forest will support fairly uniform quantities of the desired output per unit area, foresters will coincidentally treat similar areas each year. To use this approach they must have reliable growth information pertinent to the forest. Foresters use this approach as another way to regulate forests consisting of even-aged stands, and as an indirect way to program the total area treated by even-aged reproduction methods each year.

Volume-volume control—foresters schedule the volume to cut (or otherwise use) per year, and treat sufficient area to recover that volume (or other services) annually. In addition, they use a volume or basal area control to regulate the intensity of cutting and residual stocking in a stand. Foresters use the approach in scheduling thinnings for even-aged stands, or for controlling operations in uneven-aged communities and with two-aged silviculture.

Area-volume control—foresters treat a fixed proportion of the area annually, and control the intensity of cutting in each stand using a basal area or volume constraint. This directly controls both the area treated and the volume removed from a stand at any one time. Foresters normally use a marking guide to regulate the intensity of cutting in each community. They use the approach in thinning even-aged stands to a specified residual basal area or relative density, and in regulating the basal area or volume and the diameter (age class) distribution within uneven-aged communities. It could serve as a basis for controlling two-aged silviculture as well.

With current computer techniques, foresters can use a variety of mathematical programming techniques (e.g., linear and dynamic programming) to integrate the reproduction method cuttings and intermediate treatments across a property (Leuschner 1990). However, the long-term program of annual treatments must sustain the yields of goods and other benefits, and facilitate the full range of uses demanded of a property.

From a silvicultural perspective, the first two approaches to forest regulation have an extensive character. Foresters apply them where an owner strives to (1) methodically control annual work loads; (2) control annual ecologic or hydrologic impacts of timber harvesting; (3) program a regular recovery of products already standing on the forest; or (4) access goods that become available in an opportunistic way through unregulated future growth and development across an ownership. Most commonly, foresters use these fairly simplistic regulatory schemes where landowners will not invest in tending the immature age classes. Instead, they opportunistically take the regeneration that results from a harvest cut, and the volumes and values that an unregulated stand provides. They often thin stands that would yield sufficient volume to pay the costs, plus some intermediate profit. The latter two regulating schemes imply more intensive management involving a deliberate precutting stand analysis, careful prescription writing, and skilled marking to bring each stand into some specified condition.

Under many kinds of opportunistic schemes, landowners may recover substantial yields of many goods and services over long periods, but not at regular intervals nor in consistent amounts from one time to another. As long as new trees regenerate to replace harvested ones, a landowner can eventually recover some unpredictable amount and kind of goods and benefits. This may suffice in some cases. Where landowners demand consistent yields at regular intervals, the management must regulate the growth and yield, and balance cutting with increment based upon the capacity of the land to produce the kinds of goods and services sought.

DETERMINING THE GROWTH

Most aspects of forest regulation (silviculturally linked, or not) require some measure of growth and yield to forecast future cuts and times for treatment. Particularly with volume control, forest managers must know the rate that volume accumulates and structural conditions change following a treatment. They basically strive to cut the growth annually across a management unit, but by treating one stand at a time. This includes the net accumulation on trees that survive (*accretion*), plus the volume added by new trees that grow up into minimum measurable sizes (*ingrowth*). Having only poor growth data often leads to excess cutting or undercutting at the forest level. Overcutting draws down the growing stock and may reduce future yields, while undercutting fails to fully capitalize upon the production potential of a property. It will often lead to uncontrolled death of many trees from overcrowding and other natural causes (*mortality*). In fact, foresters cannot realistically practice volume control unless they have reliable growth values relevant to the kinds of sites and stands that make up an ownership.

Area-volume regulation also requires growth data, though foresters can usually formulate at least a simplistic area-based cutting budget without having precise growth information for an ownership. In this sense, area control methods have particular relevance in programming treatments for many noncommodity objectives. They must delineate the residual stand conditions that best capture the growth potential of a site, bring a stand into some particular structural condition, and regenerate a new age class where appropriate. The growth data help foresters to judge stand responses, the likely structural changes, and the rates of diameter and volume growth (Fig. 9-3). Growth data also indicate how long a landowner must wait between subsequent treatments in any single stand (the *cutting cycle*). Growth data also indicate the amounts of products available at the end of each cutting cycle.

Depending upon the levels of financial and personnel resources available, foresters have three basic ways to assess growth. These include (after Husch 1963; Davis 1966; Avery 1975; Husch et al. 1982)

1 periodic reinventory of stands and larger areas, using temporary plots or points to delineate standing volume or basal area, and computing the difference between successive inventories as a measure of net change;

2 periodic remeasurement of permanent sample plots, and using growth information from successive remeasurements to generate predictor equations or adjustment factors describing growth per unit area, or per tree; and

3 projecting residual stand inventory values into the future by applying growth estimates generated from permanent sample plots or other relevant sources, and using the differences between measured and projected conditions as an indicator of probable growth.

Permanent sample plots yield the most precise information, but require large investments to establish and periodically remeasure. Also, foresters must use reasonably sophisticated analytical methods (often computerized) to convert the remeasurement data efficiently into useful management information. As a consequence many nonindustrial owners and smaller enterprises must limit the intensity of sam-

FIGURE 9-3
Regulation methods based upon growth require data that describe how volume changes over time, using periodic reinventory, or through assessment of permanent sample plots subjected to periodic remeasurement. (Courtesy U.S. Forest Service)

pling or range of stand conditions targeted by a permanent sample plot program, or use other methods to estimate growth.

Periodic reinventory using temporary plots often provides adequate information for area or area-volume regulation. In these cases, foresters use the growth data to judge the periodicity of treatment for a stand, and to forecast the general level of future yields. Foresters could also combine remeasurement data and projection techniques, or use growth information from outside sources, to judge how managed stands will develop over time under a specific management regime. Then prior to a scheduled treatment, they would reinventory a stand and develop specific marking guides for bringing it to a desired condition. In these cases, the growth information indicates when to treat a stand, and the precut inventory identifies the kind and intensity of silviculture to apply.

DEPICTING DEVELOPMENT PATTERNS IN FOREST STANDS

To regulate the use of forest stands adequately, foresters must understand how they develop over time. Further, they must appreciate how management can alter those developmental patterns, or the rates at which certain goods and values become available. Also, by knowing how stands change, foresters can often better judge the most critical time to implement some new phase of a silvicultural system, or know when not to manipulate a stand to realize some particular advantage to a landowner.

Most readily available information portrays age class development within unmanaged even-aged stands following a reproduction method cutting, and suggests a pattern reminiscent of the classic production function model for many input-output relationships (Fig. 9-4) (from Barlowe 1958). The basic graph portrays output on the vertical axis, and inputs on the horizontal one. It suggests that with each added unit of input the output will increase, but in changing amounts. At the beginning of production, small amounts of additional input result in rapidly increasing rates of output, until some point of diminishing returns when the rate of yield begins to increase less and less with each added level of input. With the increase of inputs, the total physical product reaches a peak, levels off, and begins to decline. Secondary curves (Fig. 9-4) portray the *average physical product* (average output at a given level of input), and the *marginal physical product* (added output from each additional amount of input).

Figure 9-5 seems to portray even-aged forest stand development adequately. Time

FIGURE 9-4
The classic function used in portraying expectations from increasing inputs to various levels in a production system (from Barlowe 1958).

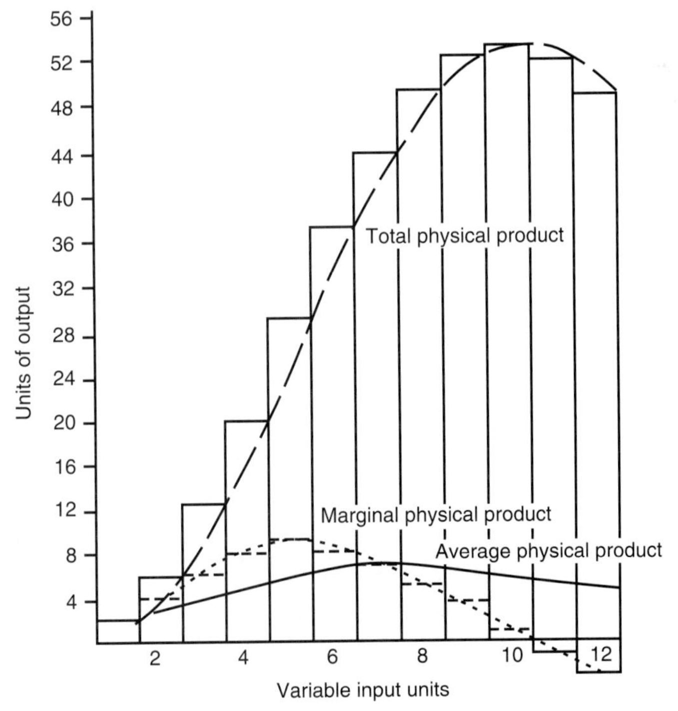

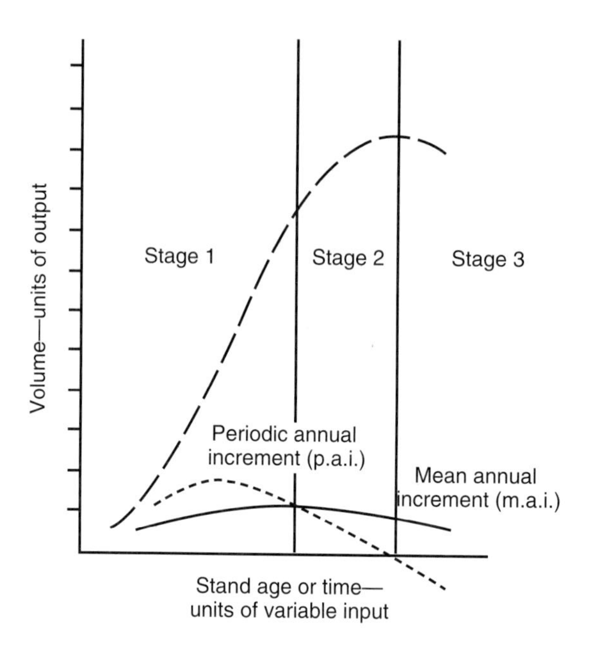

The three stages of rational action include:

1. the rotation to maximize fiber production, ending when m.a.i. equals p.a.i.;
2. the range of ages for a sawtimber rotation, with the exact length determined by economic criteria; and
3. the transition phase where mortality exceeds production, and standing volume declines progressively

FIGURE 9-5
Application of the production function to depict even-aged stand development, using stand age as a measure of inputs, and volume as the units of output over time.

(or stand age) serves as the measure of input, and stand volume or basal area can depict the output. The curve portrays how volume or basal area will increase as a stand matures (Matthews 1935; Davis and Johnson 1987; Leuschner 1990). The marginal change (called the *periodic annual increment*, or *p.a.i.*) represents the standwide balance between (1) *accretion*, (2) *mortality*, and (3) *ingrowth* as represented by the basic model (after Gilbert 1954; Beers 1962; Marquis and Beers 1969):

$$p = A + I - M$$

Foresters call this net change the *production*. They compute the p.a.i. (net change during a specific period) (after Leuschner 1990) by

$$\text{p.a.i.} = (Y_a + Y_{a+n})/\ n$$

where Y = production per year during the period of interest

a = the initial measurement date, or age

n = the number of years between successive measurements

Or foresters can divide the standing crop volume at any time of interest by stand age to determine the *mean annual increment* (*m.a.i.*). This represents the average amount produced by stands held to some specified age, as such:

$$\text{m.a.i.} = Y_a / a$$

Note that m.a.i. reaches a peak level at the point where it also equals p.a.i., or where the long-term average production equals that realized during a shorter-term measurement period.

For the general case depicted in Fig. 9-5, the total volume increases to a peak level at an age called *physiological maturity*. Then mortality (M) equals growth ($M = A + I$), and for a short period of years p.a.i. equals zero. Thereafter, mortality exceeds increment ($M > I + A$), and the total standing volume declines. Note also that during early stages of stand development, p.a.i. rises rapidly, and then begins to decrease steadily. After a period of years, p.a.i. equals m.a.i. This delineates the maximum level of m.a.i., often called the *culmination of mean annual increment.*

Throughout the time depicted by Fig. 9-5, surviving trees increase in size. Some develop slowly and end up in the shaded understory positions. Many of the poorest

FIGURE 9-6
Even-aged stands develop through four stages where the number of trees reaches a peak during the reorganization phase and declines thereafter, and the biomass builds continually until the time of physiological maturity (from Bormann and Likens 1979).

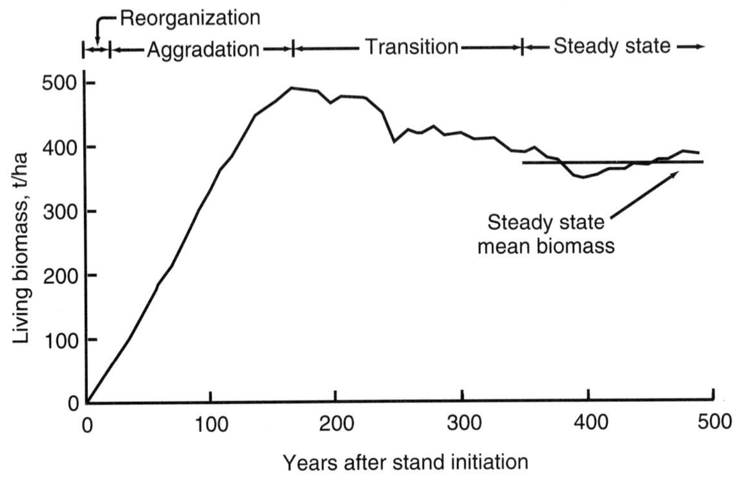

die, reducing the increment. Since small trees contain little volume, their death has little effect on volume growth during early stages of stand development. As a stand matures, mortality takes out larger and larger trees, and the loss of each one more profoundly reduces the change in volume. Height growth also slows, contributing proportionately less and less to volume increment. Thus as a stand develops, reduced individual tree diameter growth, reduced individual tree height increment, and losses of some trees causes p.a.i. to peak, and then decrease continually until physiological maturity for the stand. Thereafter, mortality exceeds the growth of surviving trees, and p.a.i. becomes negative.

To a large degree, the units of measure applied in an analysis of stand development determine when p.a.i. peaks, and when it reaches zero. Generally, the smaller the threshold diameter used in the assessment, the sooner that p.a.i. will peak. For example, it peaks earlier with basal area than volume growth, simply because volume estimates exclude trees smaller than the minimum merchantable size accepted in local markets. In like manner, the p.a.i. for cubic volume will peak sooner than that for sawtimber volume, because the latter includes only fairly large trees. In general, p.a.i. peaks first for basal area, then biomass, then cubic volume, and then sawtimber volume.

PHASES OF EVEN-AGED STAND DEVELOPMENT

Conceptual and computer models for even-aged stand development show patterns similar to that of the classic production function. They identify four different phases of long-term ecosystem development following an even-aged reproduction method, and express stand development in ecologic rather than economic terms. The phases (after Bormann and Likens 1979; Oliver 1981) include

1 a *reorganization phase* or *period of stand initiation* lasting up to two decades, wherein the ecosystem loses total biomass despite the rapid accumulation of living vegetation;

2 an *aggradation phase* or *period of stem exclusion* covering up to a century, with intertree competition and other factors causing the death of many stems, and with the total biomass accumulating to a peak level by the end of the phase;

3 a *transition phase* of variable length, wherein a permanent understory forms in gaps created by the death of individual old trees; and

4 a *steady-state phase* when the total biomass in an old-growth stand fluctuates about some mean level, but remains fairly stable over long periods.

Some major natural disturbance (e.g., fire, blowdown, or insect infestation) may destroy a stand at any stage, initiating another period of regeneration and development like that depicted in Fig. 9-5.

Fig. 9-6 represents the development of even-aged stands after the reorganization or stand initiation phase. During those early years, the trees have little volume by traditional standards. Living biomass increases rapidly, with continual height differentiation between the species present. Numbers of individual herbaceous and woody plants peak early, but living biomass continues to increase geometrically

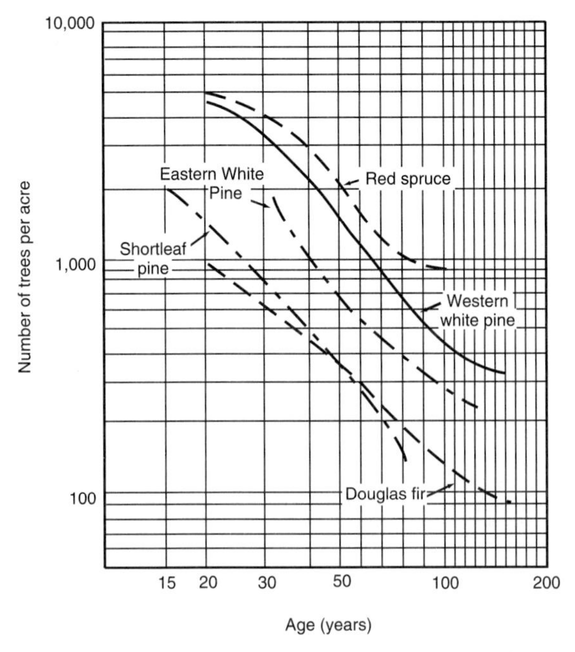

FIGURE 9-7
Change in numbers of trees over time in an unmanaged even-aged stand as it matures (from Baker 1950).

due to an increase in the size of those that remain alive (Bormann and Likens 1979). Thereafter, many light-sensitive herbs and shrubs (e.g., *Rubus*) decline as the rapidly growing shade-intolerant and mid-tolerant trees (e.g., yellow birch, pin cherry, and white ash) emerge into free-to-grow positions and shade the shorter plants (after six to eight years according to Kelty and Nyland 1981; Walters and Nyland 1989). Throughout this early ten- to fifteen-year period, a stand may have fairly stable numbers of trees (Kelty and Nyland 1981). Shade-intolerant species grow most rapidly in height, with many shade-tolerant trees remaining alive in shaded understory positions. After a new tree canopy forms, the number of trees declines continuously over several decades while living biomass increases annually, but not at a fixed rate (Bormann and Likens 1981).

During the late reorganization and early aggradation stages, individual trees differentiate by height due to unequal growth rates, and many in poor crown position eventually die (Oliver 1981; Oliver and Larson 1990). The number of trees per unit of area decreases as shown in Fig. 9-7. Actual rates depend upon how rapidly individual trees grow in height, as influenced largely by species characteristics and site quality. The process continues until the transition phase, when mortality largely results from old age and effects of natural agents (e.g., environmental factors, insects, and disease), rather than intertree competition.

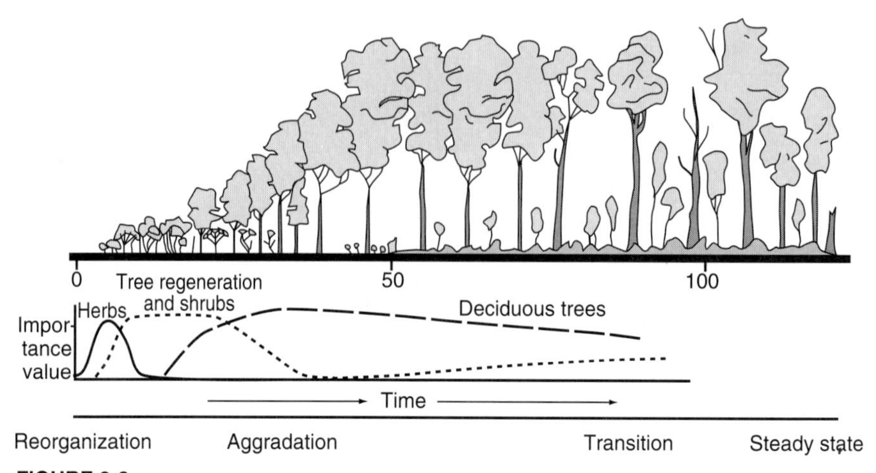

FIGURE 9-8
Patterns of development among unmanaged natural hardwood stands over long periods from stand initiation through steady state (after Kimmins 1987).

Figure 9-8 portrays changes in unmanaged stands based upon published data for hardwood stands of northeastern North America (Gilbert and Jensen 1958; Marquis 1965, 1967; Wilson and Jensen 1954; and Jensen 1943). Managed stands follow a different developmental pattern. Foresters would apply an even-aged reproduction method to replace a mature stand, often adding appropriate site preparation to alter conditions in favor of some particular mix of species. They would protect a stand from harmful agents, and might do some early cultural work during the ten- to twenty-year reorganization or stand initiation phase. They would later implement various silvicultural treatments to control intertree competition and influence species composition through about a 60- to one 100-year period of the aggradation or stem exclusion phase. Then once a stand reached some specified stage of maturity, they would regenerate it once more, usually many years before physiological maturity and the transition phase of stand develpment.

As a result of silvicultural intervention, managed stands do not pass out of the aggradation phase. Further, the different cultural treatments alter within-stand patterns of inter-tree tree competition. This improves individual tree vigor, increases their rates of growth, and controls mortality by cutting and removing trees that would otherwise die during a rotation. In effect, by reducing the numbers of trees in a stand and increasing the rates of individual tree diameter increment, foresters substantially change the development patterns. As a consequence, managed stands have attributes similar to those of naturally developing stands of much greater ages.

SOME PATTERNS OF UNEVEN-AGED STAND DEVELOPMENT

During the transition phase in unmanaged even-aged stands, total biomass declines steadily. The death of each large tree leaves a lasting opening within the crown canopy (a canopy gap), and a void in the ecosystem (Fig. 8-1). Contrary to earlier stages of development, when lateral branch extension quickly fills these spaces, the canopy gaps persist in stands past physiological maturity. Space underneath the gaps becomes brighter, though with subdued light. A new cohort forms in the brightened space (Rinkle 1981, 1982, 1990). The exact composition of the cohort depends largely on the available source of seed and the sizes of openings that develop within the overstory canopy.

As individual old-growth trees continue to die from time to time throughout the transition period in some spatially discontinuous pattern throughout a community, a continuing series of small clusters of new trees develops. In time the periodic ingrowth of these new age classes, and their subsequent development to larger and larger sizes, eventually gives the stand an uneven-aged character. Over the long run, volume added by the eventual upgrowth (increase in size) of each new cohort will balance losses by the periodic death of individual old trees. Once that happens, a stand reaches the steady-state phase that characterizes many old-growth forests. Thereafter, the production function curve levels, with only minor fluctuations over time unless some natural agent kills all or most of the trees in a community.

The basic model in Figs. 9-5 and 9-6 does not represent stands having trees of several different ages. This includes old-growth stands in a steady-state status, and other uneven-aged communities created and maintained by periodic cutting. In these communities, the separate cohort develops in the way portrayed in Fig. 9-9 with a unique curve for each age class. A total stand inventory could not readily separate the different production functions for each age class, but would describe a composite production function showing a fairly steady level of mean and periodic increment over long periods. The total standing crop will undulate up and down over time, reflecting short-term shifts in the balance of accretion, ingrowth, and mortality (or the removals by cutting). Conceptually it should resemble the steady-state condition often assumed for old-growth stands as depicted in Fig. 9-6.

FINANCIAL MATURITY AND DIFFERENT PRODUCT OPTIONS

Given a production function to represent the development of a single stand or age class, foresters can readily transform it into some economic measure by applying values to the units of physical product (Barlowe 1958; Duerr et al. 1956). Then the total physical product curve becomes a measure of cumulative value over time. Correspondingly, p.a.i. and m.a.i., respectively, transform to yearly and long-term average annual value increases. These measures give foresters a way to set rotation length for even-aged stands using financial rather than physical measures of production. Thus, if landowners elect to maximize total financial worth of an unmanaged stand (total standing value), they will hold it until the total value curve peaks and then apply an appropriate reproduction method. To optimize average annual value growth of otherwise unmanaged stands, they culminate a rotation at the peak of the average annual value growth curve. If they apply intermediate treatments, foresters normally manage the stand until the compounded rate of return from value growth drops below some required alternate level specified by the landowner. In contrast with uneven-aged stands, they should see a fairly constant long-term level

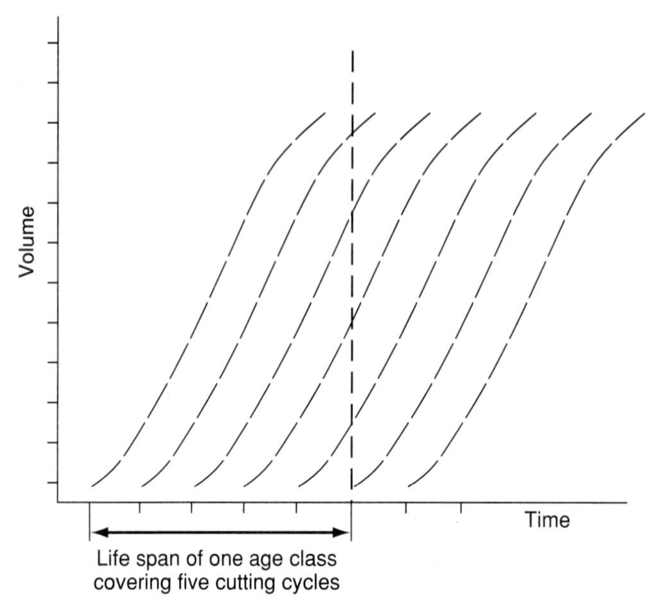

At any time, an uneven-aged community has several age classes, each initiating with a reproduction method cutting and developing in a predictable pattern as depicted by a separate production function.

FIGURE 9-9
Production function for multi-aged communities under selection system.

of value production that may rise or decline somewhat over short periods (the *cutting cycle*), depending upon the sizes of trees present and their value in current markets. In this case, financial maturity for the oldest age class depends upon the anticipated rates of value growth for individual trees in it.

When landowners want to grow wood fiber, foresters regenerate a community at some point of maturity, and grow repeated crops of wood without any tending, except to protect the stand from different kinds of natural agents (e.g., fire or harmful insects). To maximize the amounts of volume produced over long periods of time, they set rotation length at the time when m.a.i. for total wood volume reaches a peak level (the *culmination of m.a.i.*). They then initiate another rotation of comparable length (again matched to the culmination of m.a.i.). If they hold a stand longer, the average volume growth declines to some lower long-term average level, and a landowner does not realize maximum sustainable yields from the site. This assumes that markets place no major constraint on tree size, other than requiring a minimum merchantable diameter for each log or bolt delivered to a mill. Also, with fiber crops of this kind, buyers do not normally offer premium prices for large-diameter pieces or those with special wood quality. Hence, the increment of mean annual value closely parallels that for m.a.i., and volume and value increment peak at about the same time.

When available markets offer a premium price for large-diameter logs, and when landowners elect to grow trees to serve those markets, then the classic production function no longer serves as a useful guide to management. In these cases, landowners will often choose to manipulate stand density to increase individual tree diameter growth. Then the trees reach marketable sizes sooner, and they become larger for any given rotation period. Intermediate treatments remove excess trees to sell as intermediate products, and foresters add the yields into the inventories of volume in the residual stand. This transforms m.a.i. into a new function that represents the *mean annual yield (m.a.y.)*. It includes the volume in standing trees, plus the cumulative yield already realized from all past cuttings in the stand, or (after Davis and Johnson 1987):

$$m.a.y. = [V + (v_1 + v_2 + v_3 + \ldots v_i] / A$$

where V = standing volume

v_i = the volume removed in previous intermediate cuttings

A = stand age

Foresters can plot *m.a.y.* over time to determine the culmination of *m.a.y.* This represents the best rotation for maximizing total physical output from managed even-aged stands. In general, *m.a.y.* will peak before the stand reaches physiological maturity, but after the culmination of *m.a.i.*

As a counterpart to *m.a.y.*, foresters can also calculate *periodic annual yield (p.a.y.)*. It represents the change in cumulative volume between successive inventories (change in standing crop volume, plus that removed during past intermediate treatments). This has importance in assessing the rotation length where landowners want to grow trees to large diameters (sawtimber), and where buyers pay more per unit of volume (e.g., per board foot, or per m^3) for large-diameter logs of high quality. In those cases, landowners who want to maximize the production of sawtimber volume will hold a stand until *m.a.y.* equals *p.a.y.*, and then regenerate the community.

These measures do not depict patterns of value growth, and offer landowners little help in setting a rotation to maximize revenues from a stand. They could transform *m.a.y.* and *p.a.y.* to *mean annual value (m.a.v.)* and *periodic annual value (p.a.v.)*, and set rotation at the time when these match. The analysis would account for changing product values related to tree size, and tell a landowner the age when the increasing average value of a stand begins to decline to some lower long-term mean level. However, such measures provide little information about the profitability of investments.

Foresters who must evaluate choices based upon return-on-investment criteria set the rotation to match the time of *financial maturity* (see Chapter 2, and Duerr et al. 1956). For this, they determine the change in worth between successive periods (*p.a.i.* multiplied by the value per unit of volume), and often through the time when *m.a.y.* reaches a peak level. For each period, they calculate the compound interest represented by the change in standwide value, as such:

$$V_n / V_o = (1 + p)^n$$

where V_n = value at the end of the period
V_o = initial value at the beginning of the period
p = rate of return
n = the number of years in the period

They can look up the value of $(1+p)^n$ in a compound interest table to determine the value of p. When it falls below the alternate required rate of return, the stand becomes *financially mature*. Then foresters end the rotation with a reproduction method cutting.

Normally, foresters make these calculations for a time interval equivalent to the cutting cycle. Its length may change over time, or remain fixed, but by the end of each cutting cycle foresters must decide how best to treat the stand to further a landowner's management objectives, and with what intensity. They also must decide if a landowner will gain financially by making another intermediate treatment, and letting the stand grow longer. Alternatively, if they anticipate that the financial return would fall below the required rate, they should regenerate a new community and grow it for another rotation matched to the age of financial maturity. Generally, this comes after the peak of *p.a.i.* and *m.a.i.*, but before the culmination of *m.a.y.* or physiological maturity.

For uneven-aged stands, foresters must decide when the oldest age class reaches its financial maturity. To simplify the process, they traditionally evaluate the growth for individual trees of different diameters (as a surrogate for growth through a progression of ages), and convert individual tree volume increment to estimates of value growth. They compute the compound rate of return from holding an average tree over a long series of successive cutting cycles, and look for the size when value growth would likely drop below the required rate of return (see Chapter 2, Duerr et al. 1956). This tree size represents the diameter for financial maturity.

RELATIONSHIP TO A REGENERATION STRATEGY

In planning any silvicultural system, including the timing and nature of a reproduction method used in it, foresters must take cognizance of many factors. These include biologic-ecologic, managerial-social, and financial-institutional concerns. Important biologic factors include:

1 the age for onset of adequate seed production;
2 the time to secure adequate numbers of seedlings and grow them to a minimum size where they no longer require special care;
3 the patterns of seed dispersal and germination within a target species, and the periodicity of good seed crops; and

4 the silvical requirements of seedlings for protection by overhead shelter of larger trees.

These have particular relevance in even-aged systems, where a reproduction method removes the entire standing community (seed-bearing trees) once each rotation, and where a replacement stand will not usually reach seed-production age for a fairly long period of time. With uneven-aged stands, the large numbers of reproductively mature trees that remain on site over successive cutting cycles regularly produce abundant seed.

The timing of a reproduction method cutting must fit within the context of a forestwide harvesting schedule, or a landowner's scheme for marketing products to insure a regular and steady source of income to the enterprise. In some cases, it may need to occur at some critical time to provide the requisite habitat for wildlife, or to facilitate some other nonmarket value. At times, foresters may schedule a reproduction method to optimize some overall forestwide benefit or need, or to temper the spatial distribution of different condition classes across a landscape. Also, landowners may occasionally find just cause to alter the sequence or time of stand treatment simply to capitalize upon some unexpected market opportunity, or to minimize the losses if some species or product suddenly decreases in value. Again, these short-term financial and institutional needs will often require greater adjustments in the timing of a reproduction method cutting for even-aged stands. With uneven-aged communities, economic factors of the kind enumerated often force a change in the diameter for financial maturity, or in the age class arrangement by altering the frequency of cutting within a stand.

10

SELECTION SYSTEM AND ITS APPLICATION

THE CHARACTER OF SELECTION SYSTEM

The silvicultural systems for managing uneven-aged stands differ markedly from those for even-aged communities in one primary characteristic. With uneven-aged methods foresters never remove all the trees at any one point in time, or even over a relatively short period of years. Instead, with each entry they harvest the mature trees to *regenerate* a replacement age class across a portion of the stand area (Fig. 10-1). Additionally, they *tend* the intermediate (immature) age classes to nurture their continued growth and development. In the process, they recover (*harvest*) the felled trees to cover costs and repay investments in ownership and management (Knuchel 1953; Kostler 1956; U.S. For. Serv. 1978; Nyland 1987).

This linkage of regeneration, tending, and harvest at each entry adds a degree of complexity to planning for uneven-aged silviculture, and demands a thorough approach to stand analysis and management. Yet when foresters link an appropriate reproduction method, any requisite site preparation, and deliberate tending to control the density and spacing of each residual age class, selection system will provide

1 sustained and regular yields of products and values from *a stand*;

2 a stability of characteristic forest conditions and structure, both within each stand and across a broader area under selection silviculture;

3 full site utilization to insure optimum levels of wood production and other use opportunities over the long run;

4 replacement of mature with new trees across a fixed portion of the stand area at regular intervals; and

5 timely release of individual residual trees from competition to insure good levels of vigor and diameter increment.

FIGURE 10-1
Selection system uses timber harvesting to regenerate a new age class by removing the mature trees, and to nurture the continued growth and development of immature trees by cutting excess numbers of trees from each intermediate age class.

Further, selection system stands have a predictable structure and density that varies only within well-defined limits from one cutting cycle to the next, and over the long run.

The importance of simultaneous regeneration, tending, and harvest becomes apparent in several technical terms used to describe uneven-aged silviculture (after Ford-Robertson 1971):

Selection system—an uneven-aged silvicultural system involving the removal of individual trees here and there from a large area, and ideally across the annual cutting series (stand)

Cutting series—an area of sustained-yield management . . . to (1) distribute the felling and regeneration operations to suit local administration and markets, and (2) maintain or create a suitable distribution of age classes

Selection cutting—the periodic removal of trees individually or in small groups from an uneven-aged forest or working circle (stands) . . . to realize the yield, establish a new age class, and improve the forest

Working circle—a forest area organized for a particular objective under one set of working plans, and managed by one silvicultural system or specified combination of systems

These fit neatly into a simplified definition that describes selection system as any silvicultural program for *creating and maintaining* an uneven-aged community (after Smith 1986). In this context, selection system has the unique purposes and requirements of

1 creating a new age class with each entry;
2 maintaining a predetermined number of trees (diameter distribution) among the immature age classes;
3 recovering the volume in mature and excess trees once each cutting cycle; and
4 providing a consistent and sustained yield of desirable values at regular intervals.

Yields may include either nonmarket benefits or timber products.

Since each entry for selection cutting recruits a new cohort, foresters can represent the numbers of age classes in selection system stands by:

$$NAC = r / cc$$

where NAC = number of age classes

r = the planned age of a class at maturity

cc = the *cutting cycle*, or period of years between successive entries to a stand

With *selection method* (the reproduction method), foresters remove trees of the oldest age class to regenerate a new one. To make the treatment a *selection system*, they also concurrently tend (thin) the immature age classes to enhance their growth and concentrate the productive capacity of a site on trees that best satisfy a landowner's interests. By selling the cut trees they can generate revenues for a landowner.

Foresters separate selection system or selection cutting from more casual kinds of partial cuts that do not serve a specific purpose of regeneration or tending. They generally group these under the broad term *selective cutting* (after Ford-Robertson 1971):

- creaming, culling, or high-grading
- an exploitive cutting that removes only certain species, large trees, or ones of high value
- a cutting that largely disregards known silvical requirements and sustained yield principles

Selective cutting may maximize some short-term benefit (usually revenues), but without insuring regular long-term yields and other values that an uneven-aged community could provide. New trees may regenerate, but selective cutting does not deliberately control the species, the density, or the dispersal of seedlings across a regeneration area (the stand). It simply serves as a means for extracting available products, rather than for managing forest stands in a planned and sustained manner.

Ideally, in a *selection system* stand each age class occupies an equivalent amount of space or horizontal crown area. Such a condition makes the stand *balanced*—having an intermixing of age classes, with each occupying comparable amounts of area (Hawley 1921; Smith 1986; Nyland 1987). The nature, intensity, and frequency of selection cutting may differ between ownerships and stands, depending upon (1) objectives and interests of a landowner (including demands for nonmarket benefits); (2) ecologic and silvical characteristics of the tree community, its associated shrubs and herbaceous vegetation, and site (physical environment); and (3) economic factors (institutional and financial) that control the cutting and marketing of mature and excess trees. Whatever the plan, selection cuttings must create and maintain a balance of age classes within a community, insuring a consistency of its uneven-aged character in perpetuity. This includes making provisions for the periodic tending of the immature age classes. If landowners elect only to remove the marketable trees, they do a selective cutting (or high-grading) rather than selection system.

CHARACTERIZING CONDITIONS AMONG SELECTION SYSTEM STANDS

To facilitate management applications and offer regular and consistent yields (or other values) over time, each selection system stand should have at least three age classes (Smith 1986). Each one would develop much like a single even-aged stand (see Fig. 9-5), with a composite of overlapping age class development functions representing conditions in each selection system stand (Fig. 9-9). Yet, foresters do not usually try to separate the trees into distinct individual age classes during their inventory and management. Rather, they use a simplified version of the selection system production function as depicted in Fig. 10-2. Each dip in this saw-toothed line represents an entry to regenerate the mature age class and tend the immature ones. Each cutting reduces the standing volume or basal area to a specified level. Thereafter, the immature residual trees grow larger and new trees replace saplings that grew to bigger sizes. These processes rebuild the standing volume (and maintain the structure), and the production function curve rises once more. After a preset period (the cutting cycle) a landowner once more draws down the stocking to a level represented by the recurring bottoms along the saw-toothed line. This provides a sustained and consistent yield of commodities over the long run. In addition, structural attributes of the community remain stable over time, lending a consistency to the habitat for indigenous plants and animals, and the nonmarket values a landowner may derive from any single stand.

Foresters have some flexibility in planning the length of a cutting cycle. Much depends on the rates of volume growth, and the financial requirements of the logging contractor and landowner. Common cycles run as short as 8 to 10 years, to as long as 20 to 25 years. With the shorter cycles, foresters can remove only relatively

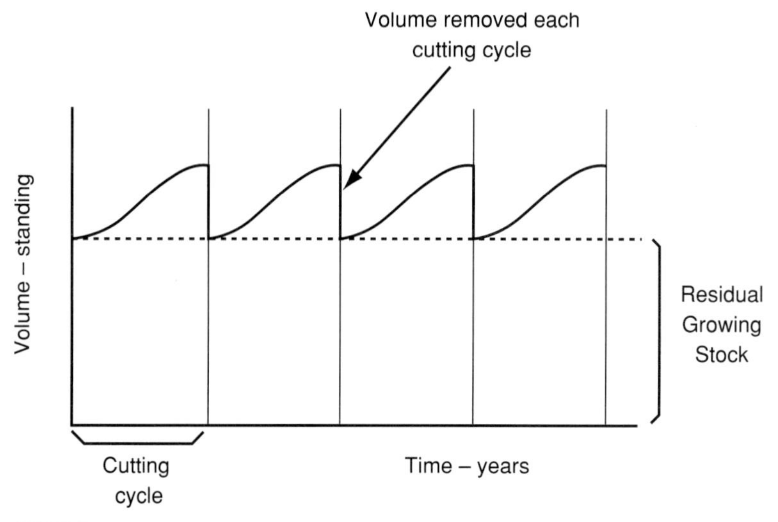

FIGURE 10-2
Simplified production function for selection system stands, representing the composite of multiple age class development resulting from concurrent regeneration and tending treatments at regular intervals.

small amounts of volume per cutting. With a longer cutting cycle, they remove more volume at each entry, but less frequently. In fact, the harvest would equal the amount of volume added since the last cutting (the cut equals the growth). At the same time, each landowner and harvesting contractor must remove some minimum volume per unit of area to make the operation profitable from their perspectives. An appropriate cutting cycle balances these economic requirements with the potential for growth based upon the productive capacity of a site and the growth characteristics of the component species.

Figure 10-3 depicts common structural features of uneven-aged stands containing four distinct age classes, under single-tree selection system. The stand includes

1 a mature age class of widely spaced sawtimber trees having the largest diameters and the greatest heights;

2 two intermediate (younger) age classes (pole and small-sawtimber stages) of shorter and smaller-diameter trees interspersed between the mature ones;

3 a young age class of small trees (saplings) much shorter than the others; and

4 advance regeneration primarily of shade-tolerant species.

Note that within selection system stands, tree size reflects its age. In fact, trained workers can judge tree age in many species fairly well based upon outward characteristics alone (McGee 1989), and need not document the actual age arrangement before proceeding with uneven-aged silviculture. Further, besides having distinctly different heights and diameters, trees of all sizes have a large live-crown-to-total-height ratio (called the *live crown ratio*, or *crown length ratio*), and good crown di-

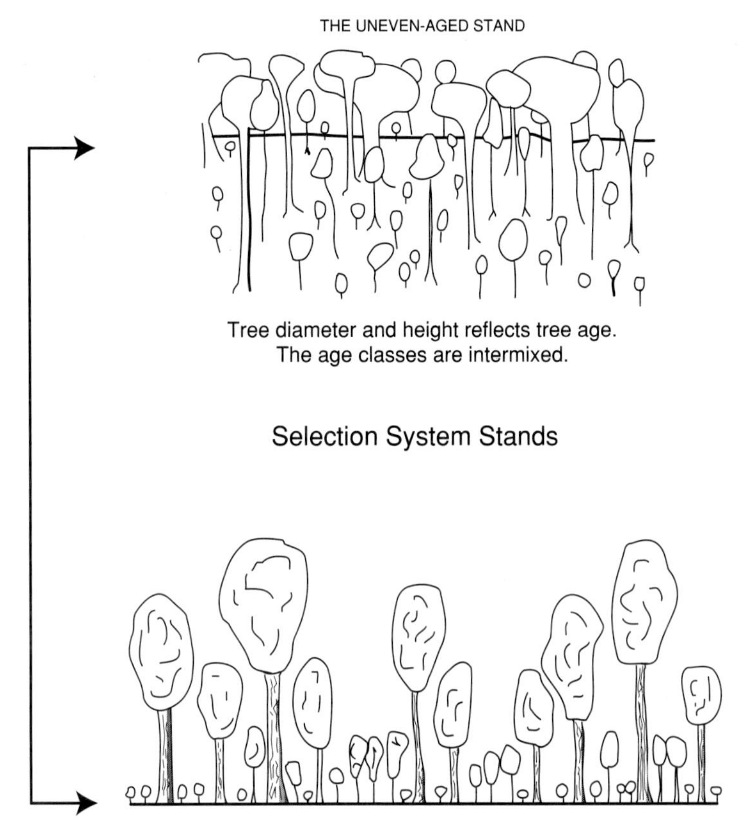

FIGURE 10-3
Structural attributes of an uneven-aged community having three or more distinct age classes interspersed throughout the stand area, and with tree diameter and height reflecting its age.

ameters. As a consequence, foresters can use the *diameter class* distribution of selection system stands as an index to the *age class* distribution. Further, they can balance the age classes by controlling the diameter distribution, working to ensure that each size (age) class occupies a fixed proportion of space in the stand.

Historically, foresters have represented an uneven-aged diameter distribution by using various kinds of reverse-J curves (Knuckel 1953; Kostler 1956; Meyer 1952). These depict communities with large numbers of small trees, and relatively few large-diameter trees (Fig. 10-4a). In reality, each separate age class has a distinct diameter distribution (Fig. 10-4b). So the traditionally used reverse-J curve represents a composite of curves for several age classes. Further, in a balanced stand, each age class occupies an equivalent amount of horizontal crown space, as shown in Fig. 10-4c. Thus, in a stand containing four age classes, the proportional space per age class equals 25%. Generally, young (small) trees each take up a fairly limited amount of horizontal space, while older trees occupy a large horizontal area (Fig.

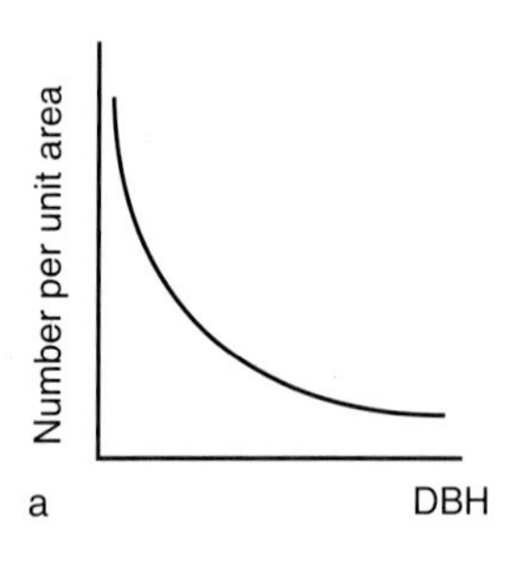

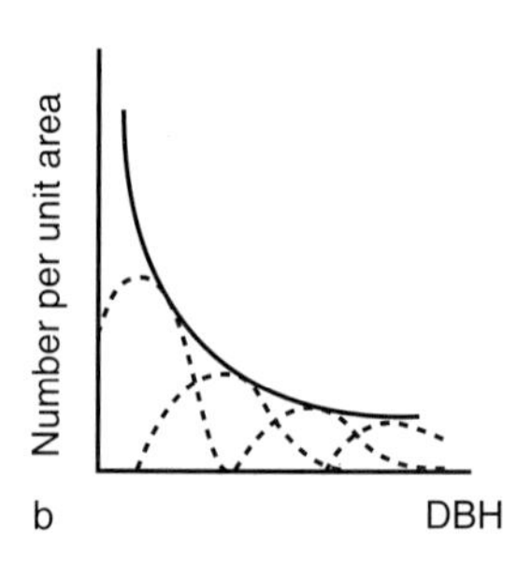

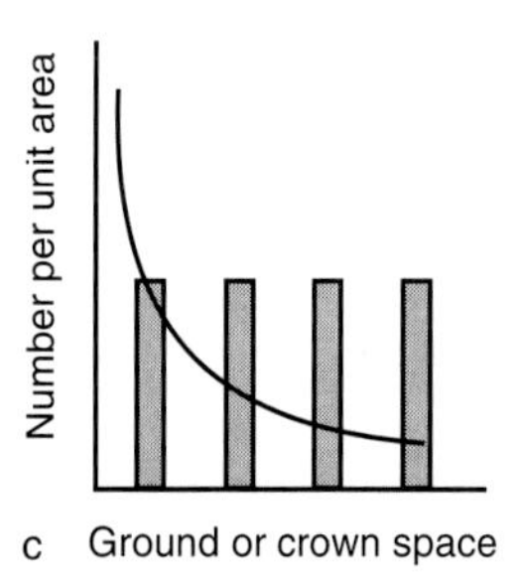

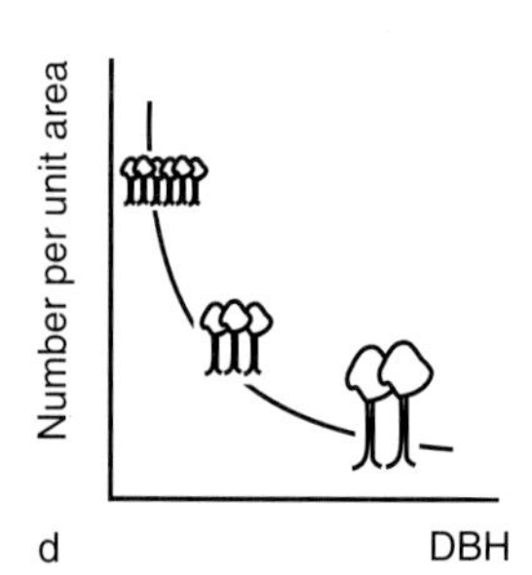

FIGURE 10-4
The reverse-J curve (a) that depicts the diameter distribution of a balanced uneven-aged stand represents a composite of the component age class distributions (b), where each age class occupies an equivalent amount of space (c), with the numbers of trees needed to fill that space dependent on the average crown spread for trees in each different age class (d) (from Nyland 1987).

10-4d). Once foresters know the amount required for vigorous trees of different ages, they can determine the numbers to balance the space (Fig. 10-4a).

This need to balance multiple age classes in a stand, and to manage each through progressive stages of development, makes uneven-aged silviculture somewhat complex. In fact, each selection system cutting (after Hawley 1921; Troup 1928; Champion and Trevor 1938; Knuchel 1953; Kostler 1956; Smith 1986; Nyland 1987; Matthews 1991) must

1 harvest an appropriate number of large (financially mature) trees to open just the right amount of space for a new age class, and to recover the yield;

2 trigger the regeneration of a new age class to replace the mature one; and

3 tend the immature age classes to reduce crowding, maintain a particular balance among the different residual age classes present, focus the growth onto the best of the immature trees, and recover the yield in excess trees.

Once *balanced* through these cuttings, stands will have long-term structural stability, and landowners can return at regular time intervals to re-create the diameter distribution of interest. Further, each cut should yield similar amounts of volume.

Stands under selection silviculture for two or more cutting cycles usually contain trees with a good correlation between tree diameter and age, and tree diameter and crown size. By contrast, trees in unmanaged old-growth stands (and other unmanaged stands as well) often show weaker correlations of diameter and age (Gates and Nichols 1930; Blum 1961; Betters and Woods 1981; Leak 1985). That complicates the management in making a transition to selection system. The differences appear to reflect a greater degree of variation in tree vigor and growth rates, as related to crown condition and size. In fact, to start a program of deliberate management in such stands, foresters may need an adjustment period of two-to-three cutting cycles before individual trees develop vigorous crowns and higher rates of growth. During this interim they will still cut back to the diameter distribution of a balanced structure in order to create sufficient space for a new age class, and to foster crown development and improved radial increment among the residual trees.

DEFINING A RESIDUAL STRUCTURE

Selection system had its origins in the *check method* devised for European forests by the mid-nineteenth century (Knuchel 1953; Kostler 1956; Assmann 1970). There foresters grew trees in fairly high-density stands to rather old ages and large diameters, using short cutting cycles of no more than six to ten years (Knuchel 1953; Osmaston 1968; Matthews 1991). Further, most managed forests had historically consisted of even-aged stands. Within these forests, foresters used fairly sophisticated regulation systems to equalize the amount of area in stands of each age class, and to control the stem density within the stands. They called these forests *balanced* (equal area per age class), and regularly tended the individual stands to keep the stocking consistent with a preferred set of *normal yield tables*. These describe appropriate numbers of trees to retain in stands of each different age class in order to optimize wood production, and to keep individual trees growing at acceptable rates. The tables also recommend the way to adjust stocking levels and treatment schedules to account for variations in site quality and species composition.

As foresters looked for appropriate diameter distributions to use with early applications of the *check method* (selection system), they apparently examined differences in numbers of trees listed in their yield tables for consecutive age classes within intensively regulated even-aged forests (see descriptions by Osmaston 1968 and Assmann 1970). From this evidence, they reasoned that uneven-aged stands having a similar spread of ages should also have a comparable proportion of residual trees per age class (Knuchel 1953; Osmaston 1986; Assmann 1970; Matthews 1991). Eventual studies showed that the numbers of trees among progressive age classes in stands managed under this philosophy followed a geometric pattern (after de Liocourt 1898; Meyer 1952; Knuchel 1953; Kostler 1956; Sammi 1961; Assmann 1970; Davis and Johnson 1987) of this form:

$$m, mq, mq^2, \ldots mq^n$$

where m indicates the numbers of trees in a diameter class. Foresters have referred to such distributions as Q structures, after the q value that defines the regular change of numbers across consecutive diameter classes. A logarithmic equation to describe such a structure has this form (after Meyer et al. 1961; Davis and Johnson 1987):

$$\log N = \log k - aD \log e$$

where N = the number of trees in a diameter class

D = the diameter class, or dbh

e = base of natural logarithm

k = the number of trees in the smallest diameter class accounted for in the inventory

a = the slope of the line, or rate that numbers decline across progressive diameter classes

Plotted on semilog paper, such a structure describes a straight line (Fig. 10-5), with the change between adjacent classes (the slope) equivalent to q, as described.

Available evidence indicates that q-based structures adequately represented conditions within intensively managed European selection forests. Individual stands had residual trees of large maximum diameters (the oldest age class), and a high residual density. Foresters maintained acceptable rates of diameter increment and controlled mortality by using short treatment cycles, and making fairly low-intensity cuttings each time (Knuchel 1953). The q-based diameter distributions also proved relevant to unmanaged old-growth mixed hardwood forests in eastern North America (Meyer and Stevenson 1943; Meyer 1952). Those stands exhibited at least seven different structural types. All had a q-like diameter distribution, a high stand density, and old trees with large diameters (e.g., up to 35–40 in., or 89–102 cm).

Advances in biometrics and computer modeling fostered several recent studies of uneven-aged silviculture using q-based structures. These explored effects of altering maximum diameter, residual density, cutting cycle length, and structural type (e.g., the value of q) to address different economic objectives (Moser 1976; Adams and Ek 1974; Buongiorno and Michie 1980; Haight et al. 1985; Hansen 1987; Hansen and Nyland 1987). Findings emphasize that different stand structures result in different patterns of stand development, and optimize different sets of production goals. These might include management to maximize total volume growth, total large sawtimber production, composite stand value increment, or compound rate of return. Also, the cutting cycle must reflect the rates of residual stand regrowth, and the time needed to rebuild stocking sufficiently to sustain another commercially viable treatment. Selection system has the flexibility to permit these adjustments, as long as foresters maintain a balance among the age classes.

Defining a residual structure using a q-type structure has the primary advantage of simplicity. Foresters need only select a q value, a maximum diameter for the structure, and the number of trees in the largest diameter class. Then they can quickly calculate the trees for the next smaller diameter by multiplying the numbers

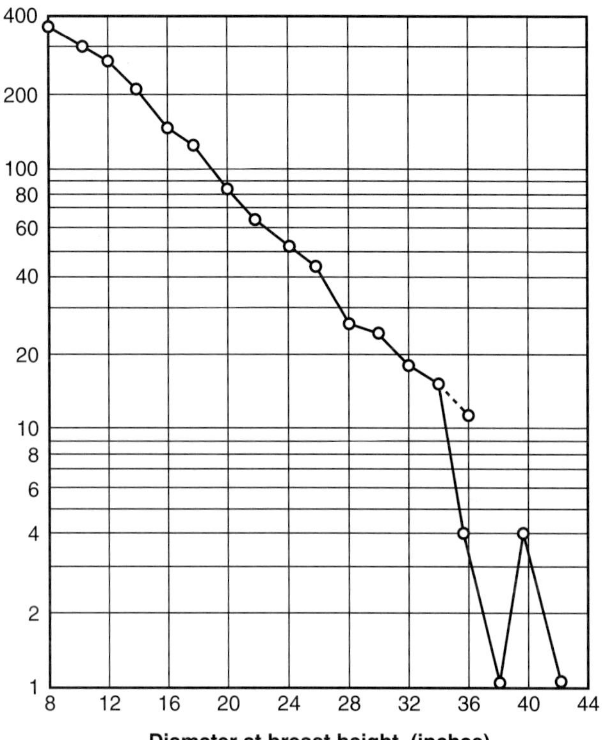

FIGURE 10-5
Unmanaged old-growth hardwood forests of eastern North America, as well as high density European stands managed under the check method, had diameter distributions that plotted a straight line on semi-logarithmic paper, with the slope of the line described by q value representing the change in numbers of trees in progressively smaller size classes (from Meyer *et al.* 1961).

in the largest class by the value q. In turn they multiply the result by q to determine numbers for the next smaller diameter, and repeat the process through each progressively smaller size class (commonly by two-inch or 5 cm classes). By adjusting the maximum diameter or the number of trees in the maximum size class, they can keep total basal area at an acceptable level.

In most cases, using q values to generate structures by one-inch (2-3 cm) classes gives unrealistically high values for total numbers and basal area, except with really low values of q (e.g., $q = 1.2$). Even so, such small q values underestimate the numbers that regenerate and grow into the sapling and small-pole classes in managed stands. With larger q values (e.g., $q = 1.5$) users must make the calculation by at least two-inch (5 cm) diameter classes. Otherwise, the numbers for saplings and small poles become unrealistically large. Because of these complications, many foresters just refer to published tables showing precalculated structures by two-inch (5 cm) classes. These normally give distributions for several separate q values, levels of residual density, and maximum diameter (Smith and Lamson 1982; Alexander 1986a, 1986b, 1986c, 1987).

Despite the popularity of this simple method, neither field research nor computer-based experimentation has demonstrated how to select a q value having both biological and economic relevance (Hann and Bare 1979). Some observations suggest that sustainable structures do not have a q-type distribution at all, and that managed stands have higher q values for different parts of the diameter distribution (Leak 1978). Other computer simulations with northern hardwoods indicate that single-q structures do not remain stable, even through a single cutting cycle. They may have too few small trees for normal upgrowth, so that deficiencies develop by the end of a cutting cycle. Conversely, a single q value may underestimate the actual number of small trees that enter the structure through recruitment following cutting. Foresters would then need to remove large numbers to reduce small-tree stocking to the levels of a q-type distribution. These studies suggest that for structural stability over repeated cycles, managed stands may need a reverse-J distribution based upon unique q values for the sawtimber (q=1.2), pole (q=1.5), and sapling classes (q=1.8) (Hansen 1987; Hansen and Nyland 1987). Some research even suggests that the q should increase incrementally with each diameter class from large to small (Hett and Loucks 1976). Proposed structures for selection stands of Norway spruce in Britain (developed from normal yield tables for forests consisting of even-aged stands) also have differences in q values across diameter classes, ranging from about 1.12 for large sawtimber to 1.54 for saplings (Osmaston 1968). However, earlier structures proposed for high-density spruce selection forests in Europe have single-q structures (Assmann 1970).

Early European assessments of selection cutting under the check method suggested that the residual structure should reflect actual rates of diameter increment within different age classes and species, and the influence of residual stand density and physical site attributes on radial growth (Biolley 1934; Prodan 1948, after Knuchel 1953). Later empirical studies among hardwood forests in the Upper Lake States of the United States demonstrated that structures of stands subjected to some kind of disturbances did not follow a typical q-type pattern. Instead, they had more trees in the smaller diameter classes, fewer at the midrange of sizes, and a dropoff among those of the largest diameters (after Adams and Ek 1974; Goff and West 1975; Lorimer and Frelich 1984). As shown in Fig. 10-6a, this gives the diameter distribution a rotated-S structure, with large numbers of small trees reflecting high levels of recruitment following cutting, and a reduction in midsize classes due to intertree competition. Other evidence (Eyre and Zillgitt 1953; Arbogast 1957) suggests that setting the maximum diameter below that common to old-growth areas simplifies the rotated-S curve to a reverse-J shape (Fig. 10-6b), and that northern hardwood stands under selection system commonly have reverse-J structures rather than single-q distributions (Berry 1981; Johnson 1984; Mader and Nyland 1984). This suggests that q-type distributions may have little biological relevance in depicting real patterns of recruitment or upgrowth among uneven-aged selection system stands (Davis 1966; Leak and Filip 1977).

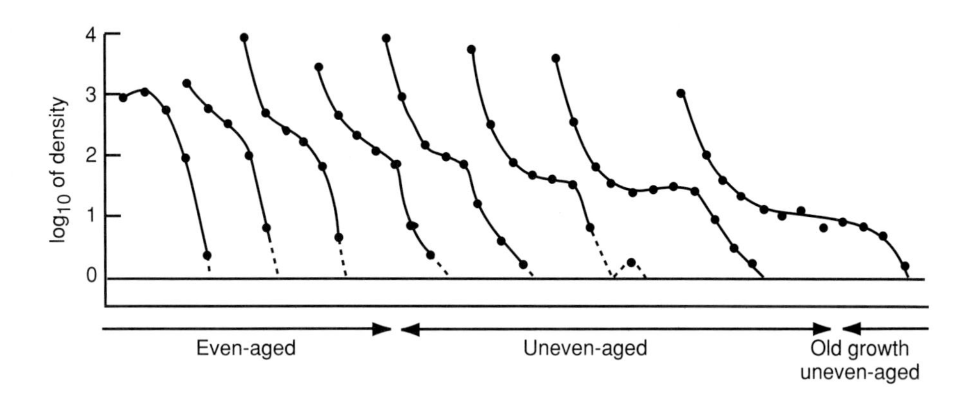

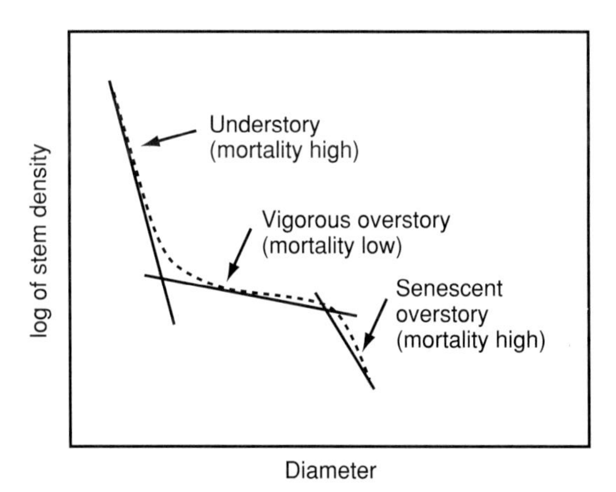

FIGURE 10-6
The diameter distribution of uneven-aged upland hardwood stands resembles a rotated-S or reverse-J structure with a changing rate in the difference of trees between progressive diameter classes (after Goff and West 1975).

BALANCED STRUCTURES AND OTHER GUIDES FOR SELECTION SYSTEM

Early research of selection systems in North America focused upon two sets of deliberate cutting trials among old-growth northern hardwoods in the upper Lake States and New England regions of North America (Eyre and Zillgitt 1953; Arbogast 1957; Gilbert and Jensen 1958; Leak et al. 1969). These cuttings removed financially mature trees to recruit a new age class, and simultaneously reduced crowding across the residual diameter classes. Yield data for as long as twenty-five years helped the researchers to target residual conditions that remained stable and maximized sawtimber production (Table 10-1). Later experience indicated that northern hardwood stands cut to the recommended structure and density remained

TABLE 10-1
RESIDUAL DIAMETER DISTRIBUTIONS FOR UNEVEN-AGED NORTHERN HARDWOODS UNDER SELECTION SYSTEM (AFTER ARBOGAST 1957; LEAK ET AL. 1969).

Tree	Residual stocking	
	Arbogast (1957)[a]	Leak et al. (1969)
(In)	(ft²/ac)	
2–5	11	–
6–11	20	20
12–17	30	28
18+	32	32
Total	90 (2 in. +)	80 (5 in. +)
(cm)	(m²/ha)	
5–13	2.5	–
14–28	4.5	4.5
29–43	6.8	6.5
44+	6.8	7.3
Total	20.6 (5 cm +)	18.2 (13 cm +)

stable and regrew sufficiently for repeated selection cutting at regular intervals (Crow et al. 1981; Nyland 1987). These diameter distributions have a reverse-J structure that resembles a rotated-S structure.

The similarity of these two independently developed northern hardwood guidelines suggests that foresters can apply them over a broad geographic area. Also, data show that users can reduce the residual stocking as much as 15 ft²/ac (3–4 m²/ha) to lengthen the cutting cycle, if they also maintain the same relative structure among the size classes (Eyre and Zillgitt 1953). Computer modeling has confirmed this notion (Hansen and Nyland 1987). It also identified alternative structures appropriate to cutting intervals extending from ten to twenty-five years, with a maximum diameter ranging from 16–24 in. (41–61 cm) (Table 10-2). Curves delineating the diameter distributions in all of these stands have a similar form, but with different levels of residual stocking and different maximum diameters (see Fig. 10-6).

Silviculturists have also proposed a structural guide for ponderosa pine (Fig. 10-7) in the Rocky Mountains of North America. It has a maximum diameter of 24 in. (61 cm) and a residual density of 98 ft²/ac (22 m²/ha) (Lexen 1939, after Alexander and Edminster 1977a). It has a fixed-q structure. Another for Norway spruce in Britain (Fig. 10-7) includes trees as large as 20 in. (51 cm) dbh, and a distribution based upon changing q values (Osmaston 1968). In other cases, the authors utilized stocking levels from even-aged yield tables to delineate an appropriate residual

TABLE 10-2
ALTERNATIVE RESIDUAL STRUCTURES TO PRODUCE LARGE SAWTIMBER WITH DIFFERENT LENGTH CUTTING CYCLES FOR NORTHERN HARDWOOD STANDS UNDER SELECTION SYSTEM (AFTER HANSEN AND NYLAND n.d.; NYLAND 1986, 1987).

	Target residual basal area for cutting cycles of different lengths		
dbh	15 yrs.	20 yrs.	25 yrs.
(In)		(ft^2/ac)	
2–5	10	10	10
6–11	25	20	30
12–16	35	30	25
17+	15	10	–
Total	85	70	65
(Cm)		(m^2/ha)	
5–13	2.3	2.3	2.3
14–28	5.7	4.5	6.8
29–40	8.0	6.8	5.7
41+	3.4	2.3	–
Total	19.4	15.9	14.8

density, and calculated *q*-based structures to approximate a residual diameter distribution. Examples include the following:

- eastern spruce-fir (Frank and Bjorkbom 1973; Frank and Blum 1978)
- western spruce-fir (Alexander 1986a; Alexander and Edminster 1977b)
- grand fir–western redcedar–western hemlock (Graham and Smith 1983)
- ponderosa pine (Alexander 1986c; 1987)
- Sierra Nevada mixed conifers (Guldin 1991)
- loblolly and shortleaf pine (Brender 1973; Baker 1991b; Baker et al. 1995)
- lodgepole pine (Alexander 1986b)
- northern hardwoods (Leak et al. 1987; Anderson et al. 1990; Majcen et al. 1990; Marquis et al. 1992)

These guides suggest more than one possible residual structure, each based upon a different single-*q* value. Some link residual conditions to site quality. They do not suggest how to select an appropriate *q* distribution to fit differing stand conditions or management objectives.

For community types where research has not yet identified a sustainable structure through cutting trials, *q*-type distributions must suffice as a convenient interim means for controlling selection cutting. Periodic remeasurement will eventually indicate if deficiencies or excess numbers develop in the smaller size classes. How-

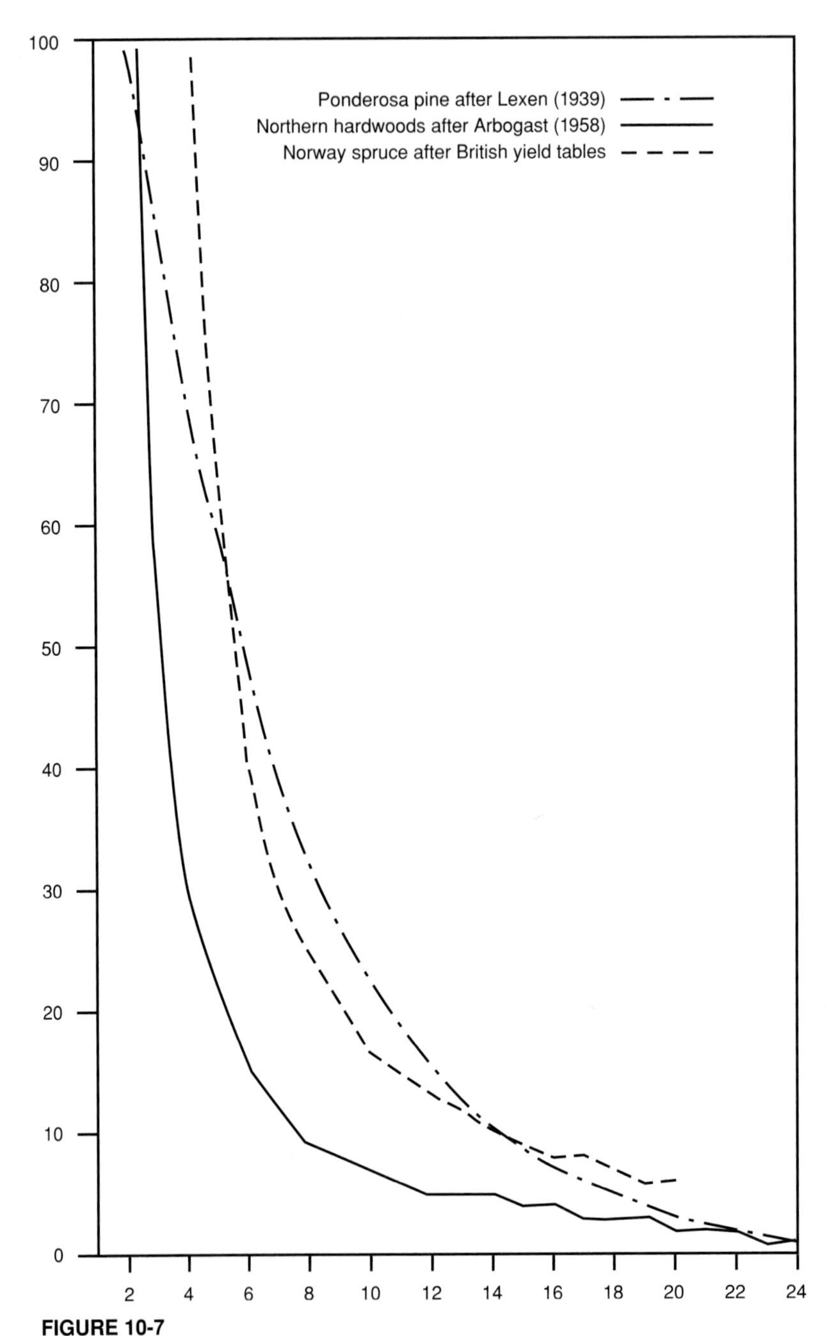

FIGURE 10-7
Recommended residual diameter distributions for uneven-aged northern hardwoods (after Arbogast 1957), ponderosa pine (after Lexen, 1939, reported in Alexander and Edminster 1977a), and Norway spruce (after Osmaston 1968), based upon empirical studies of managed stands.

ever, in the short run foresters must use this judgment in deciding how to make necessary adjustments.

APPLYING SELECTION SYSTEM

For early selection system cutting in Europe (the *check method*), foresters inventoried each stand to enumerate the volume by size class (Knuchel 1953; Kostler 1956; Assmann 1970; Ford-Robertson 1971). They usually measured all the trees present, and compared the existing volume distribution to that in an ideal stand. Then they wrote a prescription to remove excess trees across the distribution and leave the desired residual structure. At fairly frequent intervals thereafter (e.g., five to six years) they made another detailed inventory, and once more compared the current and ideal volumes for each size (age) class. The difference represented the amount added by accretion and ingrowth (the *production*). They then programmed another treatment to remove the growth and reestablish an ideal residual structure once again. Each successive cutting also established a new age class to help maintain the desired structure over repeated cutting cycles, and insure a regular and sustained yield of timber products for the landowner.

Even today, foresters start the process with an inventory to determine the nature of an existing diameter distribution. For this, they usually need to tally the trees by 1- or 2-inch (2-5 cm) diameter classes to detect the deficiencies and excesses in a stand adequately. In addition (after Trimble et al. 1974; Trimble and Smith 1976; Hansen 1987; Nyland 1987; Majcen et al. 1990; Guldin 1991) they must also

1 determine the diameter for maturity;
2 ascertain the residual density and cutting cycle length to best fit the management objectives for a stand (e.g., Table 10-2);
3 compare the existing diameter distribution with one of appropriate structure, density, and maximum tree diameter for financial maturity;
4 make a marking guide to remove excess trees from across the structure; and
5 mark and cut the stand to re-create the desired structure and density.

Traditionally, landowners have set the maximum diameter by financial criteria (see Chapters 2 and 9). With this, they determine the time it takes a tree to grow from one diameter class to another, estimate the expected increase in average volume and value over a cutting cycle, and calculate the compound rate of return. Then they find the diameter where expected returns from holding a tree for another cutting cycle will no longer equal to the required rate of return. This identifies the diameter for financial maturity (Duerr et al. 1956; Mills and Callahan 1981; Murphy and Guldin 1987), or the maximum diameter for the stand structure. It may vary by species depending on any differential rates of growth and market values.

Since the required rate of return and site conditions differ between ownerships, so will the diameter for financial maturity. Generally, enterprises demanding high rates of return (e.g., a commercial property) will use a smaller diameter for financial maturity, and owners having lower financial requirements (e.g., public lands) will grow trees to larger sizes. Where landowners seek primarily nonmarket values and services (e.g., recreation opportunities, or wildlife habitat enhancement) they

TABLE 10-3
ILLUSTRATION OF PROCEDURES FOR EVALUATING STAND STRUCTURE AND DEVELOPING A MARKING GUIDE FOR SELECTION CUTTING BASED UPON THE ARBOGAST (1957) RECOMMENDATIONS FOR UNEVEN-AGED NORTHERN HARDWOODS.

Cuyler Hill Single-Tree Selection System Plots
About 24 acres   Cuyler Hill State Forest

		Original Stand				Arbogast Guide				Excess		Mark
	Dbh	No/ac		BA/ac		No/ac		BA/ac		No/ac		No/Ac
Seedling-Sapling	1	–	61 trees	–	6 sq. ft.	–	223 trees	–	11 sq. ft.	–	Deficient	Mark None
	2	–		–		118		2.6		-118		
	3	8.2		0.4		53		2.6		-45		
	4	41.7		3.6		31		2.7		11		
	5	11.5		1.6		21		2.9		-10		
Pole	6	11.9	44 trees	2.3	15 sq. ft.	15	57 trees	2.9	21 sq. ft.	-3	Deficient	Mark None
	7	10.0		2.7		12		3.2		-2		
	8	9.1		3.2		9		3.1		–		
	9	4.5		2.0		8		3.5		-3		
	10	5.0		2.7		7		3.8		-2		
	11	3.1		2.0		6		4.0		-3		
Small Sawtimber	12	7.1	33 trees	5.6	36 sq. ft.	5	26 trees	3.9	29 sq. ft.	2	Cut 7/33 = 21%	Mark 1/5 of trees
	13	7.2		6.6		5		4.6		2		
	14	5.2		5.6		5		5.3		–		
	15	4.8		5.9		4		4.9		1		
	16	4.8		6.7		4		5.6		1		
	17	3.4		5.4		3		4.7		–		
Large Sawtimber	18	3.9	17 trees	6.9	37 sq. ft.	3	14 trees	5.3	32 sq. ft.	1	Cut 3/17 = 18%	Mark 1/6 of trees
	19	4.4		8.7		3		5.9		1		
	20	3.2		7.0		2		4.4		1		
	21	1.6		3.8		2		4.8		–		
	22	1.0		2.6		2		5.3		-1		
	23	1.8		5.2		1		2.9		1		
	24	.8		2.5		1		3.1		–		
	25+	.6		2.3		–		–		1		ALL
	ALL	156		96		320		92				

usually make subjective judgments rather than use strict financial criteria. In these cases, foresters must often describe the options (differences in yield and stand character), both financial and intangible, and ask the landowners how the different alternatives might fit their ownership goals. Historically, these landowners have often opted for a larger maximum diameter, like that built into the selection system guidelines for northern hardwoods (Arbogast 1957), ponderosa pine (after Lexen, see Alexander and Edminster 1977a), or Norway spruce (Osmaston 1968).

PREPARING A MARKING GUIDE

Table 10-3 and Fig. 10-8 illustrate the process of stand assessment and prescription making, using the Arbogast (1957) guideline summarized by 1-in. (2.54 cm) classes for a balanced uneven-aged northern hardwood selection system stand. Under this guideline, an ideal target structure for an eight to twelve year cutting cycle would have a residual of 92 ft^2/ac (21 m^2/ha) among immature age classes, using 24 in. (61 cm) as the maximum residual diameter (see the column labeled *Arbogast guide*, Table 10-3). For this simplified case, the data reflect only tree size, and not character or species as well. The cut would remove all trees larger than 24 in. (61 cm). Also, to balance the residual structure, the marking crew would remove excess numbers from the smaller classes (see Table 10-3, under the column marked *Excess*).

In this example, some size classes less than 12 in. (30 cm) dbh have an excess of trees (see Fig. 10-8). Yet as a whole, the saplings and poles show an important deficiency. To accommodate this, foresters will leave extra trees in some 1-in. (2.54-cm) diameter classes to help compensate for shortages in the next larger ones (after Eyre and Zillgitt 1953; Trimble and Smith 1976). Here, it would mean leaving excess numbers of well-spaced 6- and 8-in. (15- and 20-cm) trees to compensate for deficiencies in the 9- and 11-in. (23- and 28-cm) classes, respectively. This prescription would remove 21% of trees in the 12- through 17-in. (30–43 cm) classes. In practical terms, that means cutting about one tree in every five, or 1/5 of the small sawtimber. The marking guide would include similar instructions for the larger sawtimber classes as well.

Note that while foresters normally need fairly detailed inventory information to assess the stand structure adequately, the final field marking guide consolidates the 1- or 2-in. (2.54- or 5.1-cm) inventory information into broader classes to simplify implementation. Besides identifying the specific needs for treatment, data for the more narrow initial classes often help a forester to set boundaries for the broader ones. Those can be quite large, but should not likely exceed more than 5 or 6 in. (12–15 cm). Using too broad a class to control marking often proves ineffective. So would a guide based upon a complex array of many 2- or 4-in. (5–10 cm) classes. So to make a cutting plan useful to their marking crews, silviculturists must reduce the prescription to no more than three or four simple instructions. These need include only the size classes where a crew must make some cut, and where the intensity of marking would differ from that used in adjacent diameter classes. For example, the stand depicted in Table 10-3 has an overall deficiency of trees less than 6 in. (15 cm) dbh, and a reasonable balance for the 6–11 in. (15–28 cm) dbh classes (Fig. 10-8). Thus, the crew would do no marking among trees less than sawtimber size,

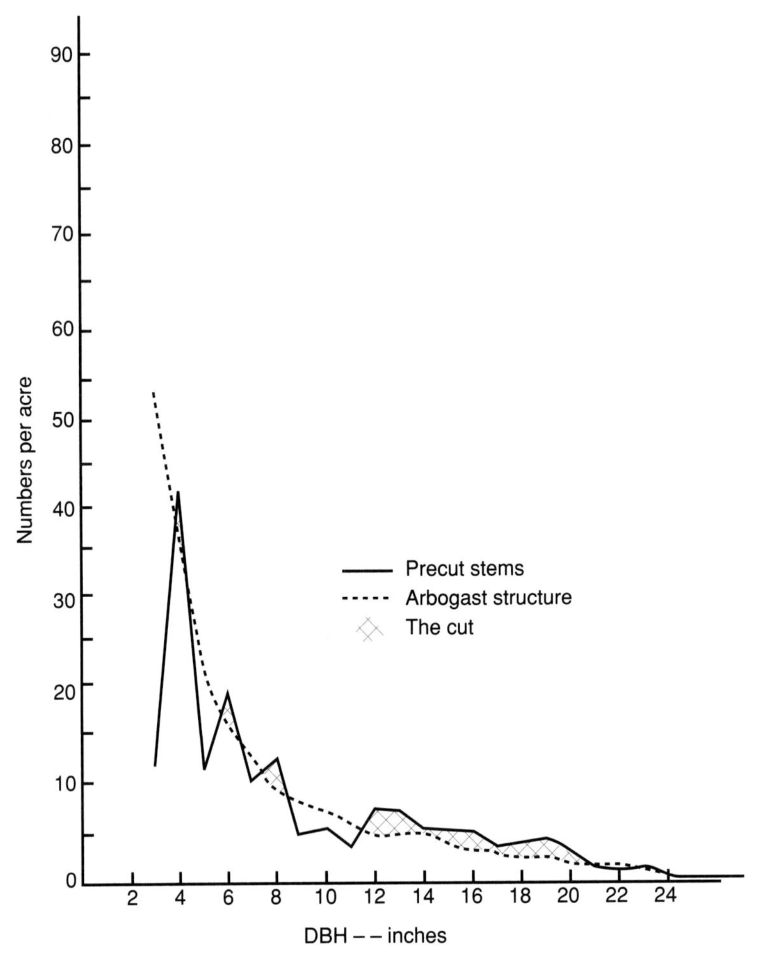

FIGURE 10-8
Comparison of precut stand structure with the residual diameter distribution recommended by Arbogast (1957) for the uneven-aged northern hardwood stand described in Table 10-3.

except to reduce localized crowding in some places. The actual marking guide would look like this:

DBH CLASS	MARK
Sapling	None
Pole	None
SST	1/5
LST	1/6
Mature[a]	ALL

[a] Trees 25 in. (64 cm) and larger.

The ratios describe the relative intensity of marking in different size classes. They also help a crew to focus on those parts of a structure needing particular attention during marking. Note that in this example, conditions within the sapling and pole classes did not require cutting of trees less than sawtimber size. In other stands, the abundance of small trees may require adjustment, and foresters must prescribe a cut for the lower diameter classes. Evidence indicates that failure to reduce these excesses leads to intense intertree competition, and a slowing of diameter growth (McCauley and Trimble 1975). This would affect the stability of the structure and the potential for long-term sustained yields.

ENHANCING QUALITY AND VALUE

When they have commodity production goals landowners normally want to grow trees to fairly large diameters, and produce high-quality sawlogs and veneer products. The silviculture would favor fast-growing trees with straight, sound, round, branch-free, and unblemished trunks. To enhance nonmarket values (e.g., habitat for particular plants and animals, or special visual qualities) they may use other attributes in picking trees to favor. Even so, many nonmarket objectives seem best served by growing at least some trees to large diameters, and having a good interspersion of size classes. Some fauna depend upon cavities for nesting and protection, and decadent trees as sites for foraging insects. This tends to conflict with commonly held sawtimber objectives, and requires resolution where landowners hope to integrate timber and wildlife management on the same area.

In marking for timber management goals, most foresters use a priority system for choosing trees to leave or cut among the immature classes, within confines of the residual stacking goals (diameter distribution and stand density). This would (after Arbogast 1957; Trimble et al. 1974; Matthews 1991) include:

1 *risk*—keep trees of greatest vigor, and the least chance of dying or declining during the ensuing cutting cycle;

2 *vigor*—keep full- and deep-crowned (high-vigor) trees of each age class, and regulate intertree spacing to enhance crown volume and diameter increment;

3 *soundness*—focus the growth on stems lacking unacceptable internal decay, or structural weakness;

4 *stem form, crown size,* and *branching habit*—maintain trees with quality attributes and appearance of interest (a long clear bole, straightness of form, a deep crown that encircles the main stem, and evidence of a uniform diameter growth rate around the stem);

5 *species*—retain those that optimize values of interest, and have the potential for long-term growth and development;

6 *crown position*—insure adequate spatial interspersion of high-vigor trees of different age classes, and maintain a discontinuity of canopy closure to insure a deep vertical distribution of live foliage among trees growing near one another; and

7 *maturity*—adhere to financial or other maturity criteria to open sufficient space for a new age class, and help to concentrate the growth potential onto younger trees of desirable attributes.

By adhering to these standards over repeated cutting cycles, foresters can progressively improve stand quality and maintain a consistency of desirable attributes into the future.

Foresters can use the precut inventory to determine whether each sample tree has acceptable or unacceptable characteristics. Simple classifications might only separate those of high risk (low vigor, poor health, or unacceptable internal decay) from the better ones. Then during the stand analysis, they can determine if a stand has sufficient acceptable trees for the diameter distribution and density of choice, or how to accommodate existing quality conditions. This assessment may even show that some stands lack adequate acceptable growing stock for meaningful management under selection system. If so, foresters will look for some alternative that addresses the management objectives to the fullest extent possible.

For the first treatment in communities not previously under deliberate management (but simply cut), foresters cannot necessarily remove all less-than-perfect trees in a single cut. That would reduce stocking in some diameter classes below target levels, and might reduce residual density too much. Instead, they would retain sound, high-vigor, and low-risk trees without strict regard for bole quality or species. To favor these, they would remove any neighboring tees with holes in the main stem, rot, fungal fruiting bodies, lean, low vigor, major die-back in the crowns, or other conditions that make them a risk for continued management. Later as successive treatments improve the overall quality of a stand, foresters would place more emphasis on quality of the bole, and particularly that of the butt log.

Even in cases where an inventory shows considerable numbers of high-risk trees, the prescription must still keep a residual density appropriate to the intended cutting cycle, and work toward a diameter distribution that will remain stable over time (balanced). For some stands, this may require an adjustment period of two (possibly three) conditioning cuts to progressively upgrade stand conditions and character, and develop a balanced structure. Frequent prescriptions for these conditioning cuts include measures like the following:

1 opting for a lower residual density, and accepting a longer cutting cycle to match;

2 lowering the maximum residual diameter when the largest trees have poor vigor, high risk, and little potential for future value growth;

3 keeping some healthy immature trees of suboptimal grade, form, or species to fill in the structure for one or two additional cutting cycles.

Such compromises help foresters to gradually develop a reasonably balanced diameter distribution, and insure fairly consistent yields during future selection cuts.

Table 10-4 illustrates the prescription-making process based upon tree quality and diameter. In this case, *unacceptable growing stock* (often called *UGS*) means high-risk trees as described above. The example also uses 2-in. (5 cm) inventory classes. Some experienced foresters prefer these to smaller ones for field inventory, since they can often make ocular estimates of diameter to within 2 in. (5 cm) with an acceptable degree of accuracy. Also, the 2-in. (5-cm) classes will suffice for effective structure analyses in most stands. To target the trees of acceptable quality

TABLE 10-4
DEVELOPING A MARKING GUIDE FOR SELECTION CUTTING IN AN UNEVEN-AGED NORTHERN HARDWOOD STAND, TAKING TREE CONDITION INTO ACCOUNT WHEN FORMULATING THE PRESCRIPTION, AND FOLLOWING THE ARBOGAST (1957) RECOMMENDATIONS FOR A RESIDUAL STRUCTURE.

Stand Heiberg #69 # Points ⊠⊠ 20 Date Apr 83

	Acceptable				Unacceptable		Total Stand			
DBH	# Detected	BA/AC	Guide	Excess	# Detected	BA/AC	BA/AC	Guide	Excess	
2-3	··	1.0	5.2	−4.2	⁖	2.5	3.5	5.2	−1.7	−1.5 ft²
4-5	·:	1.5	5.6	−4.1	·:	1.5	3.0	5.6	−2.6	
6-7	·⌉	3.0	6.1	−3.1	⊠·	5.5	8.5	6.1	2.4	
8-9	⊠··	6.0	6.6	−.6	··	1.0	7.0	6.6	.4	
10-11	⊠□	9.0	7.8	1.2	□	4.0	13.0	7.8	5.2	40%
12-13	⊠⁖	7.5	8.5	1.0	□	4.0	11.5	8.5	3.0	
14-15	⊠⊠⊠·:	16.5	10.2	6.3	□	4.0	20.5	10.2	10.0	
16-17	⊠⊠··	11.0	10.3	.7	⊠·	5.5	16.5	10.3	6.2	
18-19	⊠⊠·	10.5	11.2	−.7	⊠··	6.0	16.5	11.2	5.3	49% UGS
20-21	⊠⁖	7.5	8.2	−1.7	⧄	4.5	12.0	9.2	2.8	
22-23	⁖	2.3	8.2	−5.7	·⌉	3.0	5.5	8.2	−2.7	
24+	⊐	3.5	3.1	.4	⊠⊠	10.0	13.5	3.1	10.4	
All	–	79.5	92.0	−12.5	–	51.5	131.0	92.0	39.0	

DBH Class	ALL BA/AC	AGS BA/AC	(Arbogast) Guide	Excess	%Cut	Mark
⩽9	22	12	24	−2	0	None
10-11	13	9	8	5	40%	2/5
12-17	49	35	29	20	41%	2/5
⩾18	48	24	32	−8 AGS	49%	1/2*

*To Reduce UGS

UGS = Unacceptable Growing Stock

NOTE: Due to excess UGS in LST, cut 1/2 instead of 1/3 as called for in the Arbogast Guide.

during the precutting inventory, they would record the trees by diameter class in either the *AGS* (*acceptable growing stock*) or *UGS* column on a tally sheet.

Using quality criteria to assess the distribution of basal area or numbers per size class for the sample stand illustrated in Fig. 10-9 reveals an important deficiency of acceptable large sawtimber. Since experience has shown that foresters can safely truncate a structure (using a smaller maximum diameter) under these circumstances (Eyre and Zillgitt 1953), the prescription would call for removing one half of the large sawtimber. This differs from a recommendation of cutting one third if the

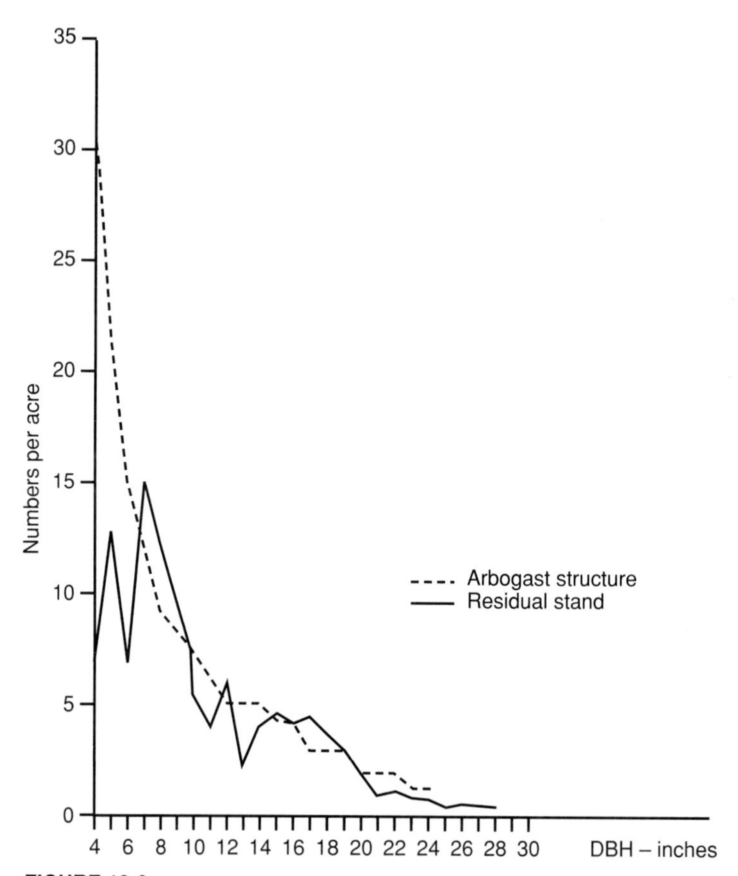

FIGURE 10-9
Truncating the residual structure by removing high-risk trees of large diameters will not disrupt stand stability, as long as the relative numbers in smaller classes are reasonably well balanced.

treatment conformed strictly to the chosen structural guideline without making adjustments for risk and defect. For other diameter classes, the stand also has some deficiencies or excesses. The final prescription reflects these conditions. In actual marking, a crew would strive to remove the high-risk trees from all age (size) classes, but also maintain a uniform spacing among the residual trees, and an interspersion of those having different diameters.

The residual stand shown in Fig. 10-9 illustrates one where foresters need two cuts to bring the stand into conformity with the selected structural guideline. Removing one half of the sawtimber larger than 18 in. (46 cm) will reduce the basal area below the recommended level, but the poor condition (high risk) of trees in this size class makes their salvage necessary. The cut would also leave some shortage in pole-size trees, but a good structure in other diameter classes. Future cuttings would adjust any irregularities not ameliorated by growth of the residual trees or by ingrowth of the new trees following the first selection cutting.

11

UNEVEN-AGED REPRODUCTION METHODS

THE ROLE OF A REPRODUCTION METHOD IN SELECTION SYSTEM

All selection system stands must have at least the requisite minimum number of residual trees per size class at the end of each cutting cycle, and sufficient excesses for a commercially operable cut (Adams and Ek 1974). In such stands, upgrowth among residual trees and ingrowth of new ones will allow foresters to maintain the distribution by periodic cuttings. In essence, the residual structure determines the patterns of upgrowth to sustain the structure. The reproduction method serves the complementary purpose of recruiting new stems to maintain the left-hand tail of the reverse-J distribution. In this context, the *selection method* involves the removal of financially mature trees for the singular purpose of regenerating a new age class. Since these large-diameter trees also have value, landowners can merchandise them to generate revenues.

The selection method allocates the space previously occupied by mature trees to the new age class. It must provide at least the minimum amounts of light, nutrients, and moisture to insure survival and development of the new age class following seed germination. Yet because selection method takes out only part of the stocking during any single entry, it increases only *somewhat* the levels of solar energy near the ground, and reduces only *somewhat* the withdrawal of moisture and nutrients from the soil by transpiration. As a result, selection cutting often regenerates species with at least a moderately high tolerance for shading.

Historically, foresters have recognized two distinct reproduction or regeneration methods under selection system. They differ essentially in the spatial distribution of financially mature trees removed, and the distribution of new trees that regenerate.

The methods (after Ford-Robertson 1971; Smith 1986; Soc. Am. For. 1989; Matthews 1991) include

Single-tree Selection Method—a method of regenerating new age classes in uneven-aged stands by removing individual mature trees more or less uniformly across a stand

Group Selection Method—a method of regenerating new age classes in uneven-aged stands by removing mature trees in small groups or clusters

These reproduction or regeneration methods remove trees that have reached a specified stage of maturity (financial or otherwise). In some cases, foresters may simply use the term *selection cutting* or *selection system* to imply the use of single-tree method to recruit a new age class. Otherwise they will specify *group selection cutting* or *group selection system.*

Foresters often find both spatially dispersed and spatially clustered mature trees in the same stand, particularly if past cutting practices did not balance the age classes. Then they may need to integrate both methods. In addition, for group selection stands, foresters must tend (thin) the existing immature groups to reduce crowding within each single-age cluster.

While the spatial distribution of regeneration differs between the two methods, each will allocate a fixed proportion of space to a new age class. They will also remove excess trees to balance the structure. With single-tree selection method, the regeneration area consists of small discontinuous openings uniformly distributed across the stand area. With group selection method, foresters create fewer but larger openings, reflecting the spatial dispersal of mature groups. For comparable cutting cycles and residual structures, total regeneration area will not differ between single-tree and group selection methods.

SINGLE-TREE SELECTION METHOD

With single-tree selection method the *regeneration openings* cover an area equivalent to the crown spread of a single mature tree. These openings improve environmental conditions sufficiently to promote a new age class within the space. The openings gradually narrow because of lateral branch extension by surrounding trees, and the space becomes increasingly shaded. As a consequence, seedling height growth may diminish by the end of a cutting cycle, except at the center of an opening, and the most shade-intolerant species fail. Gradually the small group of trees becomes stratified in heights, and the most shaded seedlings and saplings decline. Over time, many die, and subsequent tending removes others. Eventually only one remains to fill the space originally occupied by a single mature tree.

Some foresters have considered the regeneration dynamics of single-tree selection method similar to the concept of *gap-phase replacement* in unmanaged old-growth (Watt 1947; Bray 1956; Runkle 1981, 1982, 1990, 1992; Pickett and White 1985; Baker et al. 1995). According to the gap-phase hypothesis (Chapter 8 and Fig. 8-1), the death of an individual old-growth tree creates a bright spot that fills with new trees. In cases of blowdown, a falling tree may break off smaller ones as it crashes to the ground—leaving a circular opening where the old tree once stood, a narrow linear lane where the falling trunk broke side branches off other trees, and another smaller circular opening where the crown impacted the ground. When the opening diameter equals less than one half the height of surrounding trees, regeneration normally fails (Hibbs 1982).

In many regards, selection method differs from gap-phase replacement in old-growth forests. First, selection cutting creates more gaps per unit area and with a more regular distribution than normally occur in a single year by natural gap-phase replacement. Further, foresters apply selection method at regular intervals, creating a more uniform age class distribution over time. Also, the concurrent tending of immature age classes between the mature-tree openings creates additional limited-size gaps (*thinning gaps*) that generally close before the end of a cutting cycle. Yet during early years after a cutting, sunlight penetrates the canopy between the mature-tree and thinning gaps, and this triggers a fairly uniform pattern of germination and early survival of at least shade-tolerant species. Seedlings under the thinning gaps decline after a few years. Those inside the larger mature-tree openings develop to form a series of discontinuous but regularly dispersed clusters of new trees that fill the lower end of a diameter distribution.

By contrast, gap-phase replacement in unmanaged old-growth stands stimulates regeneration in a spatially irregular and randomly distributed pattern. Further, new trees develop in only one spot at a time, and only sporadically over time. In fact, during some years no mature trees may die, while two or more may succumb at other times. Overall, limited numbers may regenerate in any period of years, and a long time may pass for gaps to open an appreciable proportion of the stand area. This makes the patterns of recruitment unpredictable.

Overall, single-tree selection stands have greater numbers of seedlings and saplings per unit of area, less distance and more regular spacing between the regeneration openings, added understory brightening due to the periodic tending, and regularly scheduled cutting to recruit a new age class across a fixed proportion of the stand area at predictable intervals. Also, crowding in unmanaged old-growth stands slows the growth of surviving trees. As a result, old-growth and other unmanaged stands may have a truly uneven-aged character, but with an irregular distribution of age classes (*unbalanced*). The trees often exhibit only a weak correlation between diameter at breast height and total age (Gates and Nichols 1930; Blum 1961; Gibbs 1963; Betters and Woods 1981).

In an ecologic sense, any selection method essentially manipulates the crown canopy to alter its character as an *active surface*, in which the leaves directly intercept incoming solar energy. Cutting breaks the horizontal continuity of this active surface, so that leaves at all heights receive direct sunlight. Concurrent tending maintains the best trees in a vigorous condition. After a series of cutting cycles, the age classes become uniformly intermixed. Due to differential heights of adjacent trees, green, functioning leaves extend in a continuous vertical profile from near the ground to the top of the tallest trees (Fig. 11-1). This gives the stand great *vertical structural diversity*—vertically dispersed foliage due to trees of disparate heights growing interspersed in the same community. In fact, single-tree selection system maximizes this attribute. The stands have a persistent dense appearance, with poor visibility through the understory.

In most applications, foresters use single-tree selection method cuttings with relatively short cutting cycles (e.g., 10–15 years for structures depicted in Table 10-1). They traditionally leave relatively small and dispersed openings, and a fairly high residual stand density. This results in a characteristic environment with:

- only limited reduction of root competition and relatively small decreases in the withdrawal of moisture and nutrients from the soil
- an improvement of diffuse light throughout the understory after each treatment, but with only limited direct solar energy reaching ground level
- some increase in soil temperature and nutrient release (from added litter decomposition) after cutting, but these changes do not persist

The cuttings leave fairly uniform shading throughout the stand, and favor primarily shade-tolerant species (Trimble 1965, 1970; Frank and Blum 1978; Gottschalk 1983; Johnson 1984; Crow and Metzger 1987; Leak et al. 1987).

ALTERNATIVE APPROACHES TO SINGLE-TREE SELECTION METHOD

Assessments following cutting intensity experiments show that to increase light levels by 30% in northern hardwoods, the treatment must remove about 60% of the stand basal area. To increase them by 50%, the cut must remove 80% of the stocking (Marquis 1972). This suggests that traditional single-tree selection cutting has a limited effect on understory brightness. Further, some species have a more dense foliage mass than others, and these trees cast a heavier shade. Thus, it takes about one-third less residual basal area of beech and oak to provide about the same canopy cover as conifers, ash, and basswood (Kelty 1987). Within pure stands of southern pines, a residual stocking of 60 ft^2/ac (13.8 m^2/ha) allows about 60% full sunlight within the understory. Adding 30 ft^2/ac (6.9 m^2/ha) of hardwoods for a total stocking of 90 ft^2/ac (20.6 m^2/ha) would reduce understory light levels to only 25% full sunlight (Shelton and Baker 1992). For a 60% light level, a stand could have only 30 ft^2/ac (6.9 m^2/ha) of pine, and 15 ft^2/ac (3.4 m^2/ha) of hardwoods (Baker et al. 1995). Foresters must accommodate these differences to make conditions favorable to shade-intolerant species.

Other research (Logan 1965, 1966a, 1966b, 1969, 1973) indicates that the seed of many hardwood species will germinate and the seedlings survive for at least short periods under heavily shaded conditions (e.g., 13% light). On the other hand, seedlings develop optimal and balanced tops and roots only with 45% or more light. The most shade-tolerant species can grow reasonably well in a shaded understory. Still, most of these also develop better tops and shoots with light at or above 45%. Cutting only a small proportion of the total stocking would inhibit the less shade-tolerant species.

FIGURE 11-1
Selection system stands have a great vertical distribution of foliage, extending from near ground level to tops of the tallest trees.

Past tests of single-tree selection cutting in eastern hardwoods have shown that given an adequate seed source, foresters can secure a representation of the less shade-tolerant species by lowering the residual stand density and lengthening the cutting cycle (Metzger and Tubbs 1971; Leak and Solomon 1975; Crow and Metzger 1987). These steps will not insure as high a stocking of shade-intolerant trees as results from an even-aged reproduction method (Fig. 11-2). They will improve opportunities to regenerate species of intermediate shade tolerance (Leak and Solomon 1975; Crow and Metzger 1987).

Table 10-2 illustrates some alternatives for maintaining a balanced structure while still lowering the residual stand density among northern hardwoods. These options leave 60 to 85 ft^2/ac (13.8 to 19.5 m^2/ha) of residual stocking, with cutting cycles that range from 10 to 25 years (after Hansen 1987; Hansen and Nyland 1987; Nyland 1987). Conceptually, each alternative should leave different degrees of shading and create varying levels of brightness in the understory. Given an adequate seed source, the stocking of more shade-intolerant species should increase with each progressively lower level of residual basal area. This should help foresters to satisfy a broad spectrum of landowner interests using single-tree selection system.

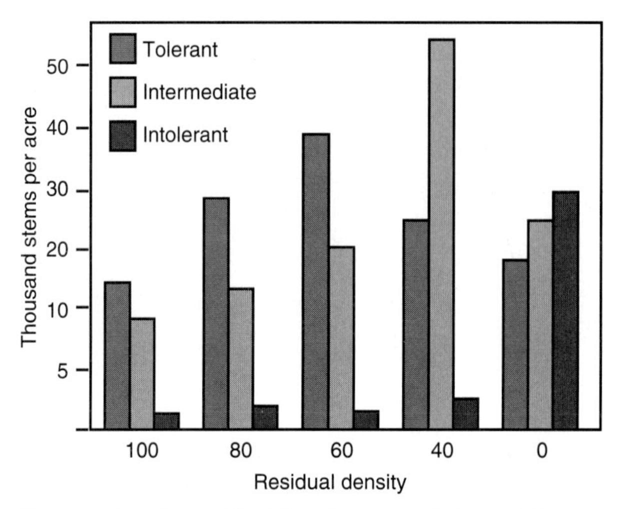

Regeneration (by residual basal area and shade tolerance groups) 10 years after selection cutting in second–growth northern hardwoods in the Bartlett Experimental Forest. Beech, red maple, and paper birch dominated the preharvest stand. Treatments included individual tree selection cuts to 100, 80, 60, and 40 ft^2/ac of basal area in trees 4.5 in dbh and larger (Leak and Solomon 1975). The zero residual density represents 3–year results from group selection on the forest (Marquis 1965a). Shade tolerance groups: tolerant– beech, sugar maple, hemlock, red spruce, balsam fir; intermediate– yellow birch, red maple, white ash; intolerent– paper birch.

FIGURE 11-2
Effect of residual density on species composition of regeneration in partially cut northern hardwood stands (after Crow and Metzger 1987).

GROUP SELECTION METHOD

Because of difficulties in regenerating shade-intolerant species by traditional single-tree selection cutting, some foresters use group selection method. For this to succeed, the mature age class must occur in scattered clusters of at least two to three adjacent trees, and not singly intermixed among the younger age classes. These *family groups* regenerated as contiguous clusters, following some event (cutting or natural disturbance) that created sizeable openings in the stand. This mosaic of single-age clusters gives the stand its overall uneven-aged character (Johnson 1972). In balanced-group selection stands, the age classes still cover similar amounts of space. Additionally, each cutting cycle foresters concurrently tend (thin) the immature family groups, and regenerate the mature ones.

In both group and single-tree selection methods, the treatment opens the same fixed proportion of stand area of regeneration ($A = r / cc$). Even so, the spatial arrangement of the regeneration area differs, with group selection method concentrating the removals into fewer gaps of larger sizes. In addition, cutting groups of adjacent trees substantially brightens the area under each opening. The amount depends upon the characteristics and configuration of a group (the opening), its size, the slope steepness and aspect, and the height of adjacent residual trees. These can vary from one group to another (Champion and Trevor 1938; Marquis 1965b; Law and Lorimer 1989). For example, removing three trees at a triangular spacing will create a fairly round opening, with high levels of direct solar energy reaching ground level at the center. If the mature trees occur in a line or arc, the opening will have a long, narrow shape. Then the levels of light inside a group space will not increase as dramatically. In addition, linear openings close sooner due to lateral branch extension on adjacent residual trees. Further, openings as wide as one time the height of adjacent trees on south slopes will have light levels similar to those twice as wide on north aspects. Because of the variation of conditions created by cutting naturally formed groups, the new age class will usually include shade-intolerant, shade-tolerant, and mid-tolerant species. The abundance of each depends upon the nature of the group openings, and the dispersal of viable seed into them.

While the degree to which shade-intolerant and intermediate species develop and persist within the group openings depends substantially on the shape of the gaps, it also depends on the group size as well. In general, openings created under the group selection method range in size from about one-fifth to one-half acre (0.40–0.80 ha). Evidence from old-growth northern hardwoods indicates that among otherwise undisturbed stands, regeneration will not succeed in openings with diameters of less than one-half tree heights (Hibbs 1982). Further, the effect on species composition depends upon the silvical attributes of the species. As a good rule of thumb, the diameter of an opening created by the group selection method need not exceed 1–2 times the height of surrounding trees in order to recruit trees having a wide range of shade tolerance (Minckler and Woerheide 1965; Smith 1981; Smith 1995). This often equates to the area covered by three to four adjacent mature trees.

Group selection method creates certain characteristic conditions, particularly where the cutting leaves fairly circular openings. Those include the following:

- Levels of light, nutrients, and moisture appreciably increase *within* the group space.
- Surrounding trees modify environmental conditions *somewhat* within the group space (particularly at the periphery), making the changes less dramatic than with the removal of all trees over a large area (e.g., from 1 ac [0.4 ha] or more).
- Increase of direct solar energy at the surface, coupled with higher soil moisture in the upper horizons, promotes more rapid decomposition of the organic litter, releasing nutrients *within* the group space.
- Between the group openings, conditions do not change sufficiently for a new age class to form and grow at good rates.

Light will not limit survival and development *within* a group space (especially at the center). As a consequence, the new age class may include a higher proportion of the less shade-tolerant species (Crow and Metzger 1987), given an adequate seed source. Between the group openings, tending of the immature trees will have similar effects to those of tending under single-tree selection cutting.

These features often make the group selection system appear ideal for regenerating shade-intolerant or mid-tolerant species. Yet the method has some important limitations (after Roach 1974; Murphy et al. 1992; Smith 1995), including the following:

1 Trees within uneven-aged stands do not always occur in well-arranged (e.g., circular) family groups that occupy comparable proportions of area within a stand.

2 Group openings have lots of edge for their size, and adjacent trees temper environmental conditions within most of the group space.

3 Naturally formed groups often differ in shape and size, such that surrounding trees quickly overgrow some openings and make them unsuitable for shade-intolerant species.

4 The fairly small sizes and irregular dispersion of family groups or openings in uneven-aged stands make them difficult to map and record, and to locate during marking and at later entries to a stand.

5 Standard inventory methods do not provide information about the numbers, sizes, ages, and spatial arrangement of family groups within a community.

6 The difficulty of clearly seeing from one mature group to another, or between immature groups of a common age class, complicates marking and other means of on-the-ground control.

7 The existing skid trail system may not lead to the mature groups scheduled for removal, forcing foresters to continually expand the network with each successive entry.

The lack of adequate methodology for inventorying the spatial dispersion and sizes of family groups across a stand confounds the marking for group selection method and system. In fact, past trials suggest that after about two cutting cycles the arrangement of family groups often becomes confusing, and foresters cannot visualize the dispersal and juxtaposition of them during marking (Roach 1974b; Leak and Filip 1977; Smith 1995; Murphy et al. 1993).

The irregular dispersal of family groups also complicates planning for an access system through the community. In fact, in order to effectively remove the dispersed clusters of adjacent mature trees, foresters must open a skidding corridor to each separate group as it matures. Otherwise, a contractor will have difficulty in moving the heavy skidding machines to and from the groups of fallen trees, or in preventing damage to many immature residual trees that grow between an existing skid trail and the group center.

To overcome these problems, foresters might use groups so small that they become almost indistinguishable as distinct multitree openings. Or they could cut more regular-shaped openings large enough to control by methods for even-aged area regulation (Murphy et al. 1993). Then a community would eventually become a composite of small even-aged stands, perhaps interspersed in a standard systematic fashion. Foresters might accomplish this by first setting up an elaborate access trail system having a fixed geometric design, and then locating the openings by some regular pattern along the trail system.

Where a forester wants to create circular openings, the marking crew will often need to cut some immature trees along with the group of mature ones. This often leads to overcutting unless they compensate in some way. If landowners balk at felling unmerchantable trees as a tending operation within the immature family groups, the treatments become *diameter-limit cuttings* that leave an irregular dispersal of residual growing stock (Roach 1974b). Neither of these cases balances the age classes, nor tends the immature (unmerchantable) size classes. Such procedures really amount to *selective cuttings*, rather than *selection system* (see Chapter 23).

HYBRID UNEVEN-AGED METHODS

In reality, many uneven-aged stands have the mature age class arranged both as scattered individual trees and in family groups, often due to past diameter-limit or species-removal cuttings (see Chapter 23). In these classes, foresters may apply single-tree selection throughout, delaying the removal of some sound and vigorous trees within the mature groups until later entries. This will leave a uniform spatial distribution of single-tree openings. Subsequent cuttings will eventually create a fairly regular interspersion of age classes throughout. In addition, they could disrupt some large immature family clusters (groups) by cutting several adjacent trees to create openings comparable to the crown space of single mature trees. This will facilitate regeneration or break up the homogeneity of the large family groups. Further, they will tend the other immature groups to develop more balanced diameter distribution.

As an alternative, foresters could use a combination of group selection and single-tree selection methods, and do the tending by groups or individual trees. This approach would perpetuate the heterogeneous character of a community, but remove the entire mature age class at each entry. It would require careful control over the total area opened by the combined single-tree and group methods. In fact, foresters would regulate the level of residual stocking to fit the planned cutting cycle, and the numbers of age classes within a stand. Such a combination of single-tree and group selection cutting would give them more flexibility than forcing some stands into a totally single-tree or grouped arrangement.

A third hybrid approach would add some larger openings within single-tree selection stands to increase the potential for regenerating shade-intolerant species. This *patch-selection method* combines single-tree selection method across most of the stand area, with some small fixed-size patches at widely scattered locations (Leak and Filip 1977). Each patch would have a regular shape (e.g., circular), and cover sufficient area to create suitable environmental conditions for the indigenous shade-intolerant species (Marquis 1965a). Each patch might cover one-fifth acre in size (0.08 ha). A marking crew could disperse the patches by a random or geometric arrangement, or place them at convenient locations adjacent to seed-bearing trees of a desirable shade-intolerant species. The total area covered by cutting mature individual trees and the fixed-area patches would not exceed the total space allocated to an age class. The community would have an interspersion of single trees of different ages, but with an occasional large patch of a single cohort.

In marking for *patch-selection system*, a crew would locate places with two or more adjacent mature or large-diameter trees, or with a cluster of trees of poor condition or high risk. They would use these as the center of a multitree patch. To make certain that each patch had the minimum requisite size, the marking crew might remove some smaller trees along with the mature or high-risk trees. They would make the openings circular or rectangular to control the patterns of light at the ground (after Marquis 1965b). However, the width of an individual patch would not exceed more than one to two times the height of a mature tree.

When laying out these supplemental patches the marking crew would select locations for a predetermined number of patches, perhaps at a preset spacing, and mark the intervening area by single-tree procedures. To keep from creating too large a regeneration area at any entry, they might keep some sound and vigorous financially mature trees between the fixed-size openings, rather than take all mature trees throughout an entire stand. Those left will increase in size and volume until a later cutting cycle. In the interim, they will control the space open to regeneration and preserve the overall balance within a stand.

To implement a patch-selection system in uneven-aged stands, foresters must

1 select an appropriate residual stand structure, and prepare a marking guide for single-tree selection cutting;

2 determine the numbers of trees (per unit of area) larger than the diameter for maturity;

3 decide how many patches to cut per unit of area, and the size for each one;

4 determine what proportion of the regeneration area to allocate to these patches; and

5 ascertain the number of single mature trees to retain per unit of area to compensate for the space opened up by the patch openings.

To illustrate, mature trees with an average crown radius of 20 ft (6.1 m) occupy about 1257 ft^2 (117 m^2) of space. If each patch covers one-fifth acre (0.08 ha), it takes 8712 ft^2 (809 m^2) of space. This equals the space occupied by

$$8712 \text{ ft}^2 / 1257 \text{ ft}^2 = 6.9 \text{ trees}$$
$$(809 \text{ m}^2 / 117 \text{ m}^2 = 6.9 \text{ trees})$$

Then:

1 Cutting one patch per acre (2.4/ha) opens a regeneration area equivalent to seven single mature trees/ac (17/ha); and

2 if a community has seven age classes (15-yr cutting cycle, and a life span of 105 years for each age class), each one would occupy 43,560 ft^2 / 7 = 6223 ft^2 per ac (10,000 m^2 / 7 = 1429 m^2 per ha); so

3 8712 ft^2 per patch / 6230 ft^2/ac per age class equates with the regeneration area on 1.4 ac (809 m^2 per patch / 1429 m^2/ha per age class equals 60% of the regeneration area per hectare).

To compensate for cutting each patch, the crew must mark five fewer single mature trees per acre (15/ha) for each patch opening. Otherwise they will exceed limits of the regeneration area prescribed by the residual stand guideline, and upset the age class balance. Exact requirements vary among stands, depending upon the preferred patch size and the average crown radius of trees that have reached maturity.

PATCH AND STRIP CUTTING METHODS

Foresters face unusual challenges in converting even-aged stands to an uneven-aged condition. This usually means entering the community at frequent intervals to create small dispersed but persisting gaps in the upper canopy. This will trigger regeneration similar to that of single-tree selection method, and mostly of shade-tolerant species (Trimble and Smith 1976; Erdmann 1987; Lamson and Smith 1991). In addition, they must tend the stand to keep the older trees alive and vigorous over a protracted life span. Further, for optimal sawtimber growth in northern hardwoods subjected to this method, the residual stand should have 70–80 ft^2/ac (16.1–18.3 m^2/ha) of residential basal area, with about 45% of it in sawtimber-size trees (Solomon 1977). Conceptually, other community types would also need some minimum level of residual sawtimber stocking to maintain good levels of volume growth through the transition period.

As one alternative, foresters might apply a fairly uniform cutting of this kind to most of the stand area, but cut a predetermined number of fixed-size openings per unit area (Erdmann 1987; Matthews 1991). Cutting in the overstory will maintain a high degree of vigor among the oldest (original) trees, and the dispersed patch openings enhance chances to regenerate the less shade-tolerant species. Some of the shade-tolerant seedlings may develop between the patches, grow slowly, but persist as advance regeneration. The added light under the patch openings will promote more rapid height development of the new cohort in those spaces. Patch size could vary, but normally would not exceed the minimum size for rapid development of the sought-after species. In essence, these *patch-thinning* treatments mimic patch-selection system, and allow foresters to create multi-age conditions within even-aged communities.

As another alternative, foresters could maintain a closed canopy over most of the area in an even-aged stand, but remove all the trees from fixed-area patches at selected places. To promote shade-intolerant species, these patches would have a circular shape, and a diameter of at least one to two times the height of surrounding residual trees (Hawley 1921; Troup 1928; Smith 1986; Law and Lorimer 1989; Matthews 1991; Murphy et al. 1993; Smith 1995). Foresters would locate the patch openings (after Law and Lorimer 1989) around

- existing advance regeneration of desirable species
- clusters of low-vigor trees or trees of less desirable species
- pockets of high-risk trees likely to die or deteriorate
- excess numbers of trees with similar diameters
- locations in the path of natural seedfall from desirable species
- places where cutting would enhance wildlife habitat or other nonmarket values

When repeated over several cutting cycles, the cuttings would gradually create small even-aged clusters of spatially dispersed openings of regular shape and size, and suited to eventual group selection system.

These methods would also work effectively in stands degraded by past selective cuttings having no balance of age classes. Such stands often require several cutting cycles to adjust the age class distribution, or to permit single-tree selection silviculture. During the long conversion period, landowners must commit to considerable noncommercial cultural work to maintain good vigor in the younger age classes. Particularly with patch cutting systems or modifications of group selection method, foresters must assure themselves (after Smith 1995) of the following:

1 The existing species composition, stand health, quality, and residual tree vigor serve the stated long-term management objectives.

2 With mixed-age stands, trees of the different age classes occur in interspersed clumps or clusters that facilitate patch or group cutting.

3 A substantial portion of the growing stock will remain alive and healthy for several decades.

4 Site conditions allow easy road and skid trail development, and use of heavy harvesting equipment at regular intervals.

With small groups or patches (e.g., less than 1–1.5 times the height of border trees) mostly shade-tolerant species will regenerate. In larger openings where the center gets full sunlight, more shade-intolerant and midtolerant species may regenerate. Yet with large openings, epichormic branching will most likely increase on border trees of susceptible species (Smith 1985).

The length of time before the first cohort of seedlings reaches maturity represents the conversion period. Foresters can determine the area to regenerate at each entry by the following formula:

$$A = r/cc$$

with A equivalent to the area in each age class,
cc the interval between successive cuttings, and
r the life span for an age class, and time to complete the conversion process.

They can use area control methods like those applied to even-aged forests to regulate the numbers and dispersion of openings within a stand (Murphy et al. 1993). This will insure an eventual balance of age classes at the stand level. They can put the openings at a geometric or regular spacing to facilitate harvesting operations, or in a random pattern to satisfy special nonmarket objectives.

Instead of using dispersed circular or rectangular patches, foresters could combine the area of several "patches" into a series of narrow strip cuts that run across the entire width or length of a stand. They call this approach the *strip-selection method* (Hawley 1921; Troup 1928; Smith 1986; Matthews 1991). The strip width and its orientation should fit silvical attributes of the target shade-intolerant species. Foresters would disperse the strips in a geometric pattern, using the same distribution to locate new strips each cutting cycle, and moving the strip cuttings progressively across a stand with each successive entry.

These hybrid methods draw heavily upon historic even-aged techniques for regeneration, but modified to (1) recruit new trees over only part of the stand area at any time; and (2) extend the process by stages over a protracted period exceeding 25%–30% of life span for an individual age class. The hybrid methods will eventually convert an even-aged community to an uneven-aged one. Also, they leave a peculiar spatial dispersion of age classes, and that offers some unique benefits (after Troup 1928; Marquis 1965a, 1965b; Smith 1986; Matthews 1991):

1 Strips and patches of proper orientation and width provide side shading or screening that modifies surface and light conditions during the establishment period, and protects the new trees from wind and exposure during juvenile stages of development.

2 Side lighting from recently cut strips and patches often promotes advance regeneration along edges of the bordering uncut areas, and these seedlings develop rapidly when eventually released by overstory removal.

3 Removal of adjacent strips and patches in a progressive series reduces shading over the previous regeneration area, thereby encouraging the continued development of each previously established age class.

4 By using directional felling and confining skidding to uncut parts of a stand, contractors protect the young trees from logging damage.

5 The well-organized arrangement of age classes facilitates subsequent tending and harvesting operations.

6 Environmental conditions inside the patch and strip openings usually favor shade-intolerant and intermediate species, increasing the tree species richness in a stand.

7 The patterns of development within each strip resemble those in even-aged communities, which may enhance some nonmarket values (e.g., early dominance of herbaceous vegetation and a high initial stem density that provide important habitat for some wildlife).

8 Creating strip and patch openings reduces interception of snow and rain so that more of the precipitation reaches the ground surface within the regeneration areas, even to the point where deeper snow protects the seedlings against winter exposure and animal browsing.

Conversion to a multi-aged condition by patch and strip cutting takes a long time to fully implement, and the cutting strategies create highly artificial conditions (Smith 1986; Matthews 1991). Further, trees kept well past a normal rotation age may become decadent and decline in value. They also lose vigor due to senescence, reducing the return on the investment of holding them past financial maturity, and increasing the risk of defect and mortality. For these reasons, the concept of balancing the age classes by strip and patch cutting may serve more as a conceptual model for planning, than as a practical means of silviculture (Hawley 1921).

EXAMPLES OF UNEVEN-AGED SILVICULTURE IN A SYSTEMS CONTEXT

When used as part of a selection system, each reproduction method serves only to establish a new cohort of seedlings over a fixed proportion of the stand area. Ideally, the new age class forms uniformly under small openings created by cutting individual mature trees, or in larger spaces previously occupied by two or more mature trees. The treatment must also reduce crowding among immature trees between the regeneration openings, concentrating the growth potential onto the best of each age class, and stimulating their diameter increment. By creating and maintaining a balanced structure, foresters can promise a stable structure of well-growing trees to provide regular and consistent long-term yields of products or other values from a stand.

Northern Hardwoods Foresters in North America have used selection system mostly in forest community types having desirable shade-tolerant species. These include northern hardwoods in the Upper Lake States (Eyre and Zillgitt 1953; Arbogast 1957) and northeastern regions (Gilbert and Jensen 1958; Leak et al. 1969) of the United States; and in Ontario (Anderson et al. 1990), Quebec (Majcen et al. 1990), and the Maritime Provinces of Canada (Lees 1978). Both single-tree and group selection methods have worked successfully, and research trials with groups and patches increased the species diversity in the Upper Lake States, Quebec, and New England (Eyre and Zillgitt 1953; Marquis 1965a, 1965b; Leak and Filip 1977; Majcen et al. 1990). Among stands with abundant advance regeneration of sugar maple, that species tends to dominate the new age class following all of the selection methods (Crow and Metzger 1987).

Single-tree selection cutting to residual densities common to ten-to-fifteen-year cutting cycles (at least 80 ft^2/ac, or 18.3 m^2/ha) will yield mostly shade-tolerant species and particularly if they occur as dense advance regeneration (Blum and Filip 1963; Leak and Wilson 1950; Seymour 1995). The less shade-tolerant species will increase when foresters use group-selection or patch-selection method with concurrent site preparation to reduce the shade-tolerant advance regeneration (Eyre and Zillgitt 1953; Arbogast 1957; Trimble and Fridley 1963; Leak and Filip 1977; Tubbs 1977; Leak et al. 1987). Compared with normal single-tree selection cuttings where shade-tolerant species comprise 90%–95% of regeneration, group selection method has yielded 35%–40% shade-intolerant and intermediate species (Leak and Wilson 1958). Research trials also indicate that leaving low residual stocking and using long cutting cycles (e.g., at least 20 years and less than 65 ft^2/ac, or 15m^2/ha) will increase the proportion of shade-intolerant and midtolerant species to at least some degree with single-tree selection cutting as well (Sander and Williamson 1957; Blum and Filip 1963; Trimble and Hart 1961; Trimble 1973; Crow and Metzger 1987).

If foresters cut across the size classes to maintain a balance among the age classes, the diameter distribution will remain stable (Eyre and Zillgitt 1953; Mader and Nyland 1984; Crow et al. 1981; Nyland 1987). This insures the chance to restructure a stand at the end of each cutting cycle, and to stabilize structural attributes and related conditions over long periods. Further, a stable structure insures consistent environmental conditions of importance for many nonmarket values, including the habitat for some wildlife and the hydrologic characteristics of the soil.

Western Spruce-fir Foresters have effectively used uneven-aged silviculture in multistoried stands of western spruce-fir in North America, and particularly at sites with a limited risk of postcutting blowdown (Alexander 1971, 1974; Long 1995). Engelmann spruce on these sites regenerates best in partial shade, and on seedbeds with a relatively thin undecomposed litter layer and at least 40% exposed mineral soil (Alexander 1977, 1987). Many stands have marked deficiencies among sapling and pole classes. Many also have advance regeneration of Engelmann spruce and subalpine fir, including small seedlings that have survived for a long time in the understory (Alexander 1974). Both this advance regeneration and even fairly old understory trees grow well following release by cutting (Alexander 1974; Adams 1980).

Foresters must limit uneven-aged silviculture in western spruce-fir to fairly protected sites (with a low risk of windfall) and stands with a fairly regular interspersion of age classes. Then (after Alexander 1977; Alexander and Edminster 1977a, 1977b; Alexander 1986a) they should

1 reduce the basal area by about 40% using single-tree selection procedures;
2 keep trees of the highest vigor and best condition;
3 balance the diameter distribution across the age classes; and
4 not create large openings in the canopy.

These cuttings often favor regeneration of subalpine fir rather than Englemann spruce. Few shade-intolerant pines will regenerate even if initially present in a stand. Spruce seedlings will become established as advance regeneration, and subsequent cutting can release these for better growth.

Because of past disturbances, many of these uneven-aged spruce-fir forests have a group arrangement of age classes and species. Some have an interspersion of age classes by both single trees and groups. The groups tend to cover too little area to operate as separate stands, so that group selection system, patch cutting, or a combination of single-tree and group selection appears appropriate (Alexander 1974, 1986b; Daniel 1980; Adams 1980; Long 1995). Each cutting must create a suitable residual diameter distribution. Foresters should take account of the risk for blowdown, and leave higher residual densities in stands having the greatest chance of windfall. Each treatment will normally remove up to 40% of the basal area, and not create openings across more than 30% of the stand area. Where mature trees occur in scattered small clusters, a combination of group and single-tree selection allows flexibility in addressing a greater variety of management objectives (Alexander 1971, 1987).

Since partial shading enhances the survival and development of spruce advance regeneration, foresters must plan the size and shape of a group or patch opening carefully. Most important, new seedlings develop best when not exposed to high levels of insolation, or on deep organic seedbeds (Alexander 1971, 1987). Thus, a square or circular opening covering more than 0.5 ac (0.21 ha) would often leave too much area exposed and unsuitable for the regeneration (Daniel 1980). For most cases, the widths of patches should not exceed more than about two times tree height. Partial scarification will also enhance new seedling establishment. In some cases, foresters can also locate group or patch openings to release already-established advance regeneration. This makes the regeneration process more certain, and speeds age class development (Alexander 1971, 1987; Long 1995).

Eastern Spruce-fir Spruce-fir stands of eastern North America often have a two-aged character, as a result of past diameter-limit cuttings or spruce budworm damage (Baskerville 1975; Seymour 1992, 1995). In fact, the highly shade-tolerant red spruce and balsam fir develop as advance regeneration following even minor overstory disturbance, and this characteristic facilitates application of selection system (Frank and Bjorkbom 1973). Light partial cutting will favor balsam fir, and the species may dominate a new age class (Barrett 1980). Because of balsam fir's susceptibility to insect and disease problems, and its lower commodity value, foresters generally prefer spruce. To insure an abundance of the species, they could use periodic tending to reduce the proportion of fir, leaving spruce as the principal seed source (Frank and Bjorkbom 1973; Frank 1977). Once established as advance regeneration and at least 6 in. (15.2 cm) tall, red spruce will develop rapidly if released by partial cutting treatments (Frank and Bjorkbom 1973).

The eastern spruce-fir community types often occur on shallow soils with a deep organic layer, or with a shallow depth to an impervious layer. These conditions make the trees susceptible to blowdown, and all except light partial cutting inappropriate. On the better-drained slopes, balsam fir and red spruce develop deeper roots, and foresters can use selection system more freely (Frank and Bjorkbom 1973; Seymour 1995). Hardwood species also regenerate in profusion on these sites, so the community composition will shift unless foresters take steps to release the conifers, and preferably spruce due to its higher market value (Frank and Bjorkbom 1973). Besides providing an opportunity to concentrate the growth potential onto spruce in favor of other species, regular tending also serves to maintain good rates of diameter increment among residual trees of desirable stem qualities (Frank and Bjorkbom 1973; Frank and Blum 1978). When marking, the crew should maintain a uniform spacing among the residuals, and favor those of best vigor and quality.

Observations after conditioning treatments in spruce-fir suggest that a cutting cycle of 10 years and a residual basal area of 100 ft^2/ac (23 m^2/ha) well distributed among the size classes would provide the best opportunity for controlling stand quality and composition. Also, residual trees in stands cut to either 80 or 115 ft^2/ac (18.3 or 26.4 m^2/ha), with a cutting cycle of 20 and 5 years, respectively, reportedly grew at favorable rates. Further, stands over this full range of stocking produced good levels of volume growth. Once balanced by carefully executed conditioning cuts, the diameter distribution should remain stable over the long run (Frank and Bjorkbom 1973; Frank and Blum 1978). Even so, in some cases spruce-fir stands developed deficiencies in middle parts of the structure because of poor recruitment into the pole classes as a result of high-stand densities and the short cutting cycles

(Seymour 1995). The lighter partial cuts have reduced losses to wind. Most blowdown should occur within 3 to 5 years after logging, and large fir seems most susceptible (McLintock 1954; Grisez 1954).

Ponderosa Pine Foresters in the southwestern United States have long recognized the suitability of selection cutting among ponderosa pine communities growing in areas of limited summer precipitation. The species has a well-developed and uniformly branched tap root, making it well suited to dry soil conditions common to low elevation sites (Berndt and Gibbons 1958; Schubert 1974; Long 1995). The trees grow at wide spacing and many stands have an open-grown and irregular uneven-aged character. A grassland community persists beneath the widely dispersed trees, and ranchers graze cattle in these forests. Available soil moisture proves more limiting than light in regenerating this shade-intolerant species (Pearson 1950; Schubert 1974).

Uneven-aged silviculture tends to emulate the open-grown and uneven-aged character of old-growth stands. Early cuttings removed part of the growing stock. As a consequence, many current stands have multiple age classes, and a diameter distribution common to uneven-aged communities. Trees of different age classes often occur in a group arrangement, with separate groups covering 0.5–1.0 ac (0.2–0.4 ha). Apparently, periodic natural fire kept the understory of presettlement stands free of competing woody plants, and grasses proliferated. Grazing has similar effects today.

In areas with dwarf mistletoe, foresters must remove all infected trees of the older age classes, and prune lower branches from the smaller ones (Alexander 1974; Schubert 1974). Practical considerations often dictate even-aged silviculture in these cases. Where conditions permit, foresters can partially cut ponderosa pine stands to maintain their open-grown character (Adams 1980; Daniel 1980; Alexander 1974, 1986c). In these cases, abundant natural regeneration (Schubert 1974) depends upon

1 a plentiful supply of viable seed;
2 a well-prepared seedbed;
3 no competing understory vegetation;
4 few seed losses to birds and mammals;
5 sufficient moisture to support germination and early seedling development; and
6 no browsing or critical insect pests.

Regeneration mostly fails when foresters do not adequately prepare the seedbed or reduce interfering vegetation. They must prohibit grazing until the seedlings become several feet (meters) tall (Larson and Schubert 1969; Pearson 1923, 1950; Schubert 1974; Alexander 1974, 1986c).

Group cutting, linked with scarification and removal of interfering understory vegetation, will effectively regenerate ponderosa pine (Harrington and Kelsey 1979). The groups may cover 0.5–2.0 ac (0.2–0.8 ha), depending upon the spatial arrangement and clustering of mature trees. Foresters must also tend the immature family groups to maintain good inter-tree spacing and balance the age classes. Single-tree selection system also works effectively with a residual stand density of about 60 ft²/ac (13.8 m²/ha) in trees at least 4 in. (10.2 cm) dbh. The residual diameter class distribution should look like that in Table 10-3. During the conditioning period, reentry can occur at about 10-year intervals. Later, foresters can extend the cutting cycle to 30 years, or when volume growth permits another commercial logging operation (after Schubert 1974; Alexander 1974, 1986b).

Upland Oak Communities Selection methods have not worked well among oak-dominated and other mixed mesophytic hardwood forests in the east-central portion of the United States. Much of this region supports even-aged communities with an important component of shade-intolerant and midtolerant species. The small understory trees have ages equivalent or similar to the larger trees of upper canopy positions, and diameter has no clear relation to age. Early single-tree selection cuttings failed to perpetuate the valuable shade-intolerant species, and the composition shifted (Watt et al. 1979; Sander et al. 1981; Nyland 1988; Sander and Graney 1993). Experiments with circular patches maintained a component of shade-intolerant species (Minckler and Woerheide 1965; Smith 1980, 1981), except where a dense understory of shade-tolerant trees and shrubs interfered with their development (Sander et al. 1983; Loftis 1990; Sander and Graney 1993). Each opening must cover 0.5–1.0 ac (0.2–0.4 ha), with a minimum width of one to two times the height of surrounding residual trees (Hawley 1921; Troup 1928; Watt et al. 1979; H.C. Smith 1981; D.M. Smith 1986; Law and Lorimer 1989; Matthews 1991; Murphy et al. 1993; Parker and Merritt 1995; Sander and Graney 1993). Subsequent cuttings should enlarge the circles along their east-west axis. However, border trees often develop epichormic branches that reduce their quality (Parker and Merritt 1995; Smith 1995).

The center of circular openings of at least 0.5 ac (0.2 ha) will receive the equivalent of about one-third full insolation. That should prove optimal for oak seedlings. Actual patch diameter should vary between slopes and flats, and with aspect as follows (after Law and Lorimer 1989):

	Opening diameter	
Aspect	Tree heights[a]	Area
South and west slope	1	0.09 ac (0.04 ha)
Level areas	1–1.5	0.20 ac (0.08 ha)
North and east slope	2	0.35 ac (0.14 ha)

[a]Based upon an average residual tree height of 70 ft (21 m).

Within smaller canopy gaps, residual trees along the borders shade too much of the opening, and light levels remain too low for good growth of oaks (Sander and Clark

1971; Sander et al. 1983; Smith 1981; Murphy et al. 1993; Parker and Merritt 1995). Pre- or postcutting site preparation must also reduce understory and midstory competition, or regeneration of oaks and some less shade-intolerant species will fail (Hannah 1987; Loftis 1990; Murphy et al. 1993).

Group or patch cutting methods have not consistently established new seedling of difficult-to-regenerate oak species (Beck 1970; Sander and Clark 1971; McQuilkin 1975; Carvell 1979; Sander et al. 1984; Parker and Merritt 1995). For best results, foresters should center the openings over clusters of well-developed advance oak regeneration at least 3–4 ft (0.9–1.2 m) tall, and having well-established root systems (Watt et al. 1979; Smith 1981; Parker and Merritt 1995; Sander and Graney 1993). Further, to convert a community to an uneven-aged condition, foresters must recut the stand periodically to establish new age classes, regulating the total area covered by patch openings and their dispersal at each entry. Exact procedures depend upon the intended residual stand diameter distribution, cutting cycle length, slope steepness and aspect, average heights of surrounding residual trees, and a landowner's management objectives (Law and Lorimer 1989; Murphy et al. 1993).

Southern Pines Long-term research and demonstration trials with loblolly, longleaf, and shortleaf pines in southeastern North America have demonstrated that selection system (Reynolds et al. 1984; Barnett and Baker 1991; Baker 1991b, 1992; Baker et al. 1995) must include

1 intensive site preparation to control interfering trees and woody shrubs;
2 single-tree selection with low residual densities, or group selection or patch cutting systems;
3 short cutting cycles; and
4 control of the diameter distribution through concurrent tending.

This combination has created and maintained balanced uneven-aged stands of shade-intolerant loblolly, shortleaf, and longleaf pines over multiple cutting cycles. Specific guidelines (after Reynolds et al. 1984; Farrar et al. 1989; Barnett and Baker 1991; Baker 1991b, 1992; Baker et al. 1995) include the following:

1 Cut at approximately 3–10 year cycles to maintain a fairly open canopy and a bright understory.
2 Leave no more than 45–60 ft^2/ac (10.3–13.8 m^2/ha) in single-tree selection system stands.
3 With group selection or patch cutting, create openings two to three times the height of surrounding trees.
4 Regulate the diameter distribution, leaving two-thirds to three-fourths of the basal area in sawtimber-size trees.
5 Remove interfering hardwoods from understory and midstory positions, and in the group openings.

With group selection or patch cutting the openings should not exceed 250 ft (76.2 m) in width (Willett and Baker 1990), and site preparation should remove the hardwoods from these spaces. If the hardwood component increases to any degree, pine regeneration and volume production will decline. This often necessitates repeated control at regular intervals (Cain and Yaussy 1983).

Research has yet to provide effective site preparation methods for selection system in Southern pine stands. Foresters currently use intensive and frequent herbicide application, since prescribed burning and traditional mechanical techniques destroy the small pines as well as the hardwoods. Also, among previously managed stands, overtopped pines at least 40–60 years old may not grow well following release (Shelton and Murphy 1990). This complicates efforts to balance the diameter distribution, even when the regeneration method succeeds. Even so, experience with these shade-intolerant pines suggests that intensive methods might also work in other communities of trees with low shade tolerance.

12

GROWTH AND DEVELOPMENT IN SELECTION SYSTEM STANDS

ADVANTAGES AND DISADVANTAGES OF SELECTION SILVICULTURE

When properly implemented (Matthews 1991), the component treatments of selection system silviculture allow foresters to continuously maintain

1 an appropriate array of age classes in correct proportions;
2 an acceptable admixture of species;
3 good growth rates on residual trees of all ages; and
4 a periodic upgrading of stand quality through judicious tending.

Cuttings that remove only large-diameter trees (selective cuttings) simply exploit available growing stock for its current value. This usually degrades a forest and compromises the opportunity for stability of stand and ecosystem conditions, and consistent yields of products or other values at regular intervals.

Even though stand conditions differ somewhat between single-tree and group selection methods, both provide (after Hawley 1921; Troup 1928; Champion and Trevor 1938; Cheyney 1942; Kostler 1956; Nyland 1987; Farrar and Boyer 1990; Baker 1991a; Matthews 1991) the following:

1 an interspersion and balance of age classes in perpetuity within each stand;

2 a well-distributed tree cover that continually moderates environmental conditions within the soil and near the ground surface, making selection system ideal for protection forests;

3 spatially uniform and vertically dispersed foliage that slows wind movement and reduces blowdown on soils with good rooting depth;

4 well-tended residual trees with large and well-developed crowns and good rates of radial increment among all diameter classes;

5 an abundance of reproductively mature trees that insure an adequate source for regeneration, even with heavy-seeded species that serve as mast for wildlife;

6 partial shading that protects seedlings from extremes of light and temperature, and against wide variations of soil moisture;

7 moist and cool understory conditions, and the dispersed nature of logging slash, limit the risks of forest fire, except in years of drought;

8 a continual upgrading of quality in an age class, so it contains high-value trees at maturity;

9 an intermixing of size classes, making selection system stands picturesque to many viewers, and well suited to many nonmarket objectives;

10 even-flow sustained yield from each stand, offering the opportunity for income at frequent intervals;

11 a component of large-diameter trees to insure high levels of sawtimber volume growth, and a steady supply of accessible sawtimber; and

12 an interspersion of size classes and uniform distribution of canopy cover that provide good habitat for many plants and animals, including several adapted to old-growth stands.

Foresters can adjust the cutting cycle length, residual density, and maximum tree size to accommodate different kinds of landowner objectives. Poor markets for small-diameter and low-quality trees may discourage tending of the submerchantable classes, except by noncommercial methods. Often contractors will fell these stems during the harvesting operations. Otherwise, work crews can return to a stand after logging to fell the unwanted trees or kill them with herbicides (Fig. 5-7).

Selection system has some clear disadvantages. Common ones (after Hawley 1921; Troup 1928; Champion and Trevor 1938; Cheyney 1942; Kostler 1956; Nyland 1987; Farrar and Boyer 1990; Baker 1991a; Matthews 1991) include the following:

1 Foresters must inventory a stand once each cutting cycle to secure timely data for prescription making, and this adds to the cost.

2 It takes a skilled marking crew to maintain a balance among the age classes.

3 Because of the interspersion of different age classes, at least some of the residual trees suffer logging damage, even with special harvesting methods.

4 Logging will destroy many small trees, and damage other residuals.

5 Shade-intolerant species commonly fail unless foresters cut fairly large group openings or keep only a low density of residual stocking.

6 The interspersion of small and large residual trees precludes some methods of site preparation.

7 Failure to regenerate shade-intolerant species or those with special seedbed requirements eventually reduces plant community diversity.

8 Overbrowsing by cattle or wildlife will damage or destroy the regeneration, leading to an imbalance of age classes.

9 Surface fires will kill the younger age classes, reduce the habitat for many plants and animals, and alter other attributes having important nonmarket value.

10 For stands with an excess of small or otherwise unmerchantable trees, landowners must invest in tending, and this reduces short-term profits.

11 Periodic tending enhances radial growth, but lower branches remain alive longer, reducing the quality of upper logs.

12 Frequent entry for selection cutting requires an elaborate network of carefully planned and permanent skid trails and access roads.

13 Contractors incur high logging costs to remove the widely dispersed sawtimber trees, and this affects the revenues to landowners.

14 Short cutting cycles increase the frequency of site disturbance, particularly along main skid trails and at the landings.

15 Selection system stands provide poor habitat for animals that depend upon early successional plant communities, or high-density stands at early stages of development.

16 To convert even-aged stands, landowners need several cutting cycles to establish multiple age classes, and create a balanced condition.

Foresters address these shortcomings by altering the cutting cycle length and residual density, using an alternative reproduction method, and through special practices to contain the damage to soil and residual vegetation. Also, contractors must install water bars and take other measures to stabilize the soil along skid trails and at the landings (Fig. 12-1). Even so, selection system provides the only alternative for keeping uneven-aged stands balanced and vigorous.

REQUIREMENTS FOR SUCCESS IN SELECTION SYSTEM

Any silviculture presumes that a landowner wants to control stand development and composition for some special reason, and to sustain the desired values over the long run. This requires careful planning to control the composition, condition, density, and age structure. Foresters must also adjust the silviculture to the species and community conditions of interest. In the case of selection system, this means working primarily with (1) shade-tolerant species, except where foresters can modify the reproduction method or stand density to recruit a component of other trees and plants; (2) long-lived species that will grow well even at advanced ages and produce regular seed crops; and (3) species with seed that germinates in undisturbed litter and a partly shaded environment.

Species that become established, persist, and grow reasonably well as advance regeneration often fill the lower part of a diameter distribution (Leak and Wilson 1958; Metzger and Tubbs 1971; Mader and Nyland 1984). With regular tending, these move up through successive diameter classes on time to eventually become the mature trees removed by another reproduction method cutting (Knuchel 1953; Eyre and Zillgitt 1953; Kostler 1956; Osmaston 1968).

Figure 12-2 shows the effects of selection cutting on individual tree development

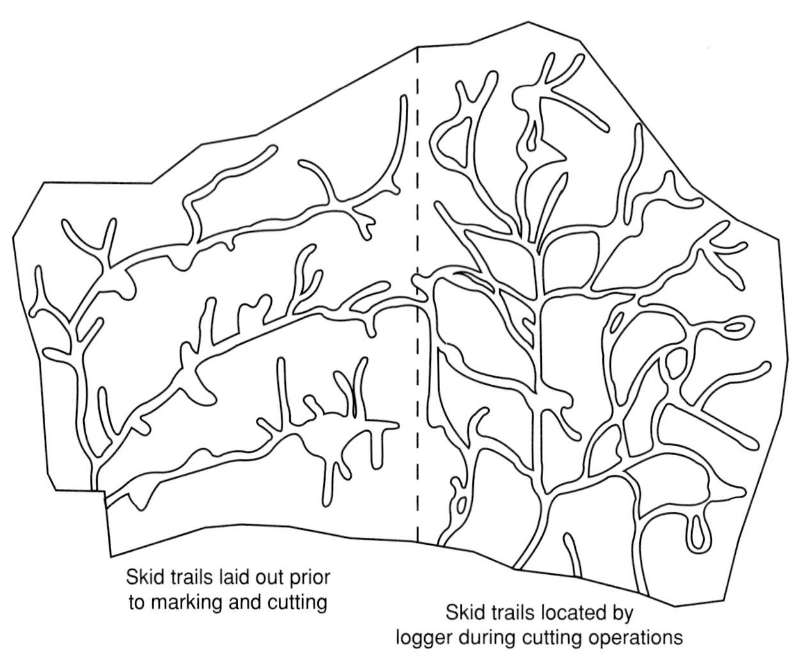

FIGURE 12-1
Selection system requires a well-planned network of permanent skidding corridors within each stand, and postcutting maintenance of the trails and landings to insure soil stability following each entry.

and stand production in northern hardwoods (after Eyre and Zillgitt 1953). Similar patterns would appear likely in other uneven-aged communities dominated by fairly shade-tolerant species. These data indicate that

1 radial increment generally increases in proportion to the intensity of cutting (Fig. 12-2a), but with little benefit from only light cutting;
2 intermediate and heavy cuttings better stimulate the growth of trees less than 15 in. (38 cm) dbh, and especially poles and saplings growing in the shade of taller trees (Fig. 12-2a);
3 appropriate tending insures good rates of volume growth per tree, especially within the small sawtimber classes (Fig. 12-2b);
4 matching the residual density to the cutting cycle length insures peak levels of production (Fig. 12-2c); and
5 balanced stands remain structurally stable over repeated cutting cycles (Fig. 12-2d).

Generally, selection system provides a steady and consistent yield of large-diameter and high-quality sawtimber in perpetuity. It also maintains a stable habitat for indigenous plants and animals, and a consistency in vegetal attributes important to many other nonmarket values.

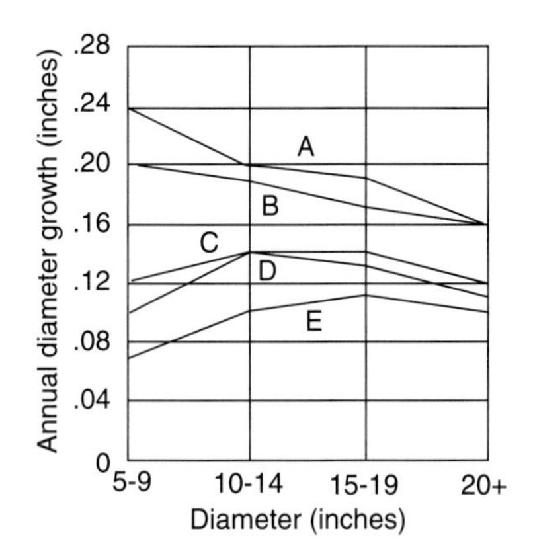

a. Average diameter growth on different cuttings over 20-year period. A, Heavy; B, Intermediate; C, Light; D, Group selection; E, Reserve.

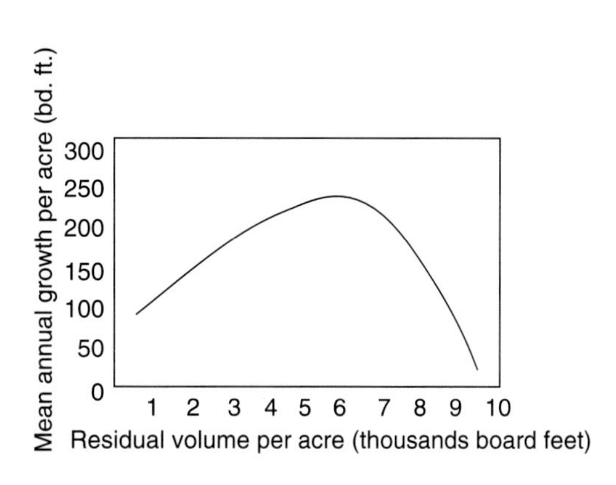

c. Twenty-year average annual net growth per acre with different residual volumes.

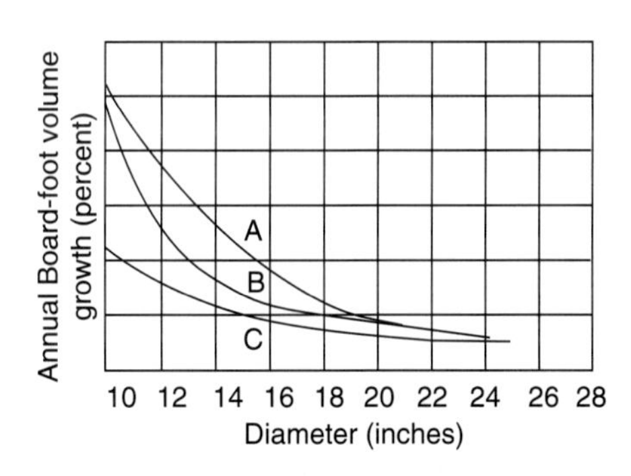

b. Annual board-foot volume growth in percent for sugar maple trees of different diameter classes (20-year average). A, Moderate cutting (Overmature and Defective No. 1); B, Light cutting (Light improvement); C, Reserve area (No cutting).

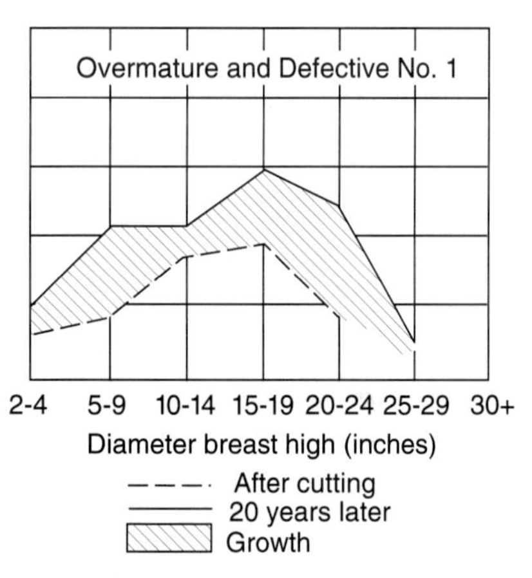

d. Basal area per acre by diameter groups immediately after cutting and 20 years later.

FIGURE 12-2
Effects of selection cutting on individual tree development and stand production within uneven-aged northern hardwood communities (from Eyre and Zillgitt 1953).

EFFECTS OF STAND STRUCTURE CONTROL ON TREE GROWTH RATES

Evidence from Europe suggests that small trees may grow poorly in high-density selection stands operated on short cutting cycles (Knuchel 1953; Assmann 1970). In these cases, trees in middle and lower canopy strata grow in a partially shaded environment, and this limits the photosynthetic potential and the radial increment. Then as a tree ages, it eventually becomes part of the uppermost canopy stratum, and free of shading by adjacent crowns. Crown size, vigor, and radial increment then increase. At some point, radial growth rate slows again, perhaps due to inadequate side lighting to insure continued crown development and photosynthetic efficiency.

Some silviculturists link these growth phases to the three basic legs of the rotated-S curve in Figure 10-6. Figure 12-2a (line E) shows this feature quite well (Eyre and Zillgitt 1953), at least for the unmanaged and lightly cut stands. In the unmanaged one, radial increment increased with tree size through large sawtimber, then leveled off. As trees surpassed about 18 in. (46 cm) dbh, radial growth declined. Light partial cutting (lines C and D) improved the rates of growth across all diameter classes, but did not change the general pattern observed in the uncut stand. By contrast, moderate and heavy cutting (lines A and B) profoundly stimulated the growth of small trees, and they grew most rapidly in diameter. At least when moderate and heavy cutting regulated spacing throughout the diameter classes, the smaller trees (e.g., less than 15 in. or 38 cm) did not show the signs of oppression common to unmanaged stands. Similar effects showed up in other selection system experiments as well (e.g., Leak and Graber 1976; Mader and Nyland 1984).

Such evidence indicates that appropriate control of density and spacing across the diameter classes elevates light levels throughout the multilevel crown canopy of selection stands, and brightens the understory appreciably. This maintains long and wide crowns on trees of all sizes, and insures good vigor in trees of all ages. Due to the uniform spacing, both the short (young) and tall (older) trees receive more direct and indirect solar energy. Crown vigor and volume improve, increasing the rate of diameter growth. Keeping only trees with the best crowns and vigor, and favoring trees of the best stem quality and height, insures that the growth will accrue at the best possible rates on the most valuable trees.

CONTRASTS BETWEEN SELECTIVE AND SELECTION CUTTING

Regulating a diameter distribution and maintaining a residual stocking matched to the cutting cycle length should temper the amount and distribution of regeneration that develops, and also maintain stand production over repeated cutting cycles. In a conceptual sense, each age class within a well-balanced stand would have just the number of trees needed to fully capture a proportionate share of the incoming solar energy. By contrast, exploitive cuttings that remove only large trees do not maintain an appropriate diameter distribution, do not regulate the spacing, and do not control stocking at optimal levels. Generally, they lead to reduced production, structural instability, and an irregular age arrangement (Roach 1974b). This may take up to three cutting cycles to become obvious on the ground.

Figure 12-3 illustrates some effects of diameter-limit cutting in an uneven-aged community. The first exploitation removes much of the standing sawtimber, leaving only small residual trees (perhaps one-third of the original stocking). The many large openings brighten the understory, and dense regeneration develops across the stand (assuming adequate seed and few losses to animals). Eventually, the new age class forms a dense understory subcanopy, much like a young even-aged stand. The

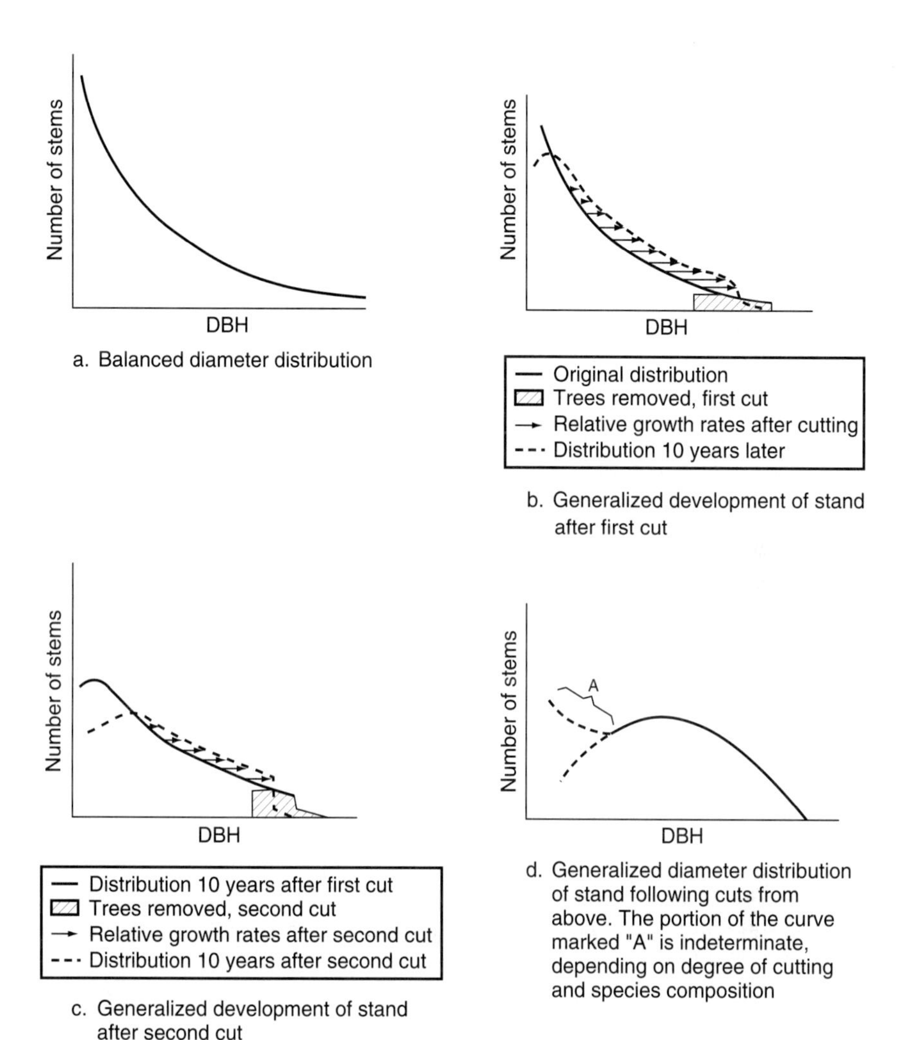

FIGURE 12-3
Expected changes in an uneven-aged stand diameter distribution following repeated diameter-limit cutting (from Roach 1974b).

shade inhibits subsequent regeneration, even of fairly shade-tolerant species. This limited recruitment of seedlings and the slow growth of the crowded young trees become manifest as a drop in the left-hand tail of the diameter distribution (Fig. 12-3b through d). At the same time these diameter-limit cuttings (Fig. 12-3b and c) often release the larger pole classes, and the trees grow into sawtimber at good rates to support another diameter-limit removal within a normal cutting cycle (Fig. 12-3c). Due to the continued lack of tending within the small pole and large sapling classes, large numbers of slow growing trees accumulate just below the threshold cutting diameter (Fig. 12-3c and d). As this glut of pole trees eventually reaches minimum sawtimber size, landowners can make another diameter-limit cutting. It will leave only a low-density stand (Fig. 12-3d), and regeneration will fill the open spaces (given a seed source). A long period will pass before the landowner can make another commercial harvest.

Contrary to selective cutting, selection system stabilizes the structure and optimizes volume production over repeated cutting cycles. Periodic cutting to a balanced structure maintains a good vertical distribution of leaf surface area flooded by sunlight, and good growth in all the age classes. The interspersion of age classes maximizes the vertical dispersion of foliage area (*vertical structural diversity*) due to the intermixing of trees with different heights, and having high live crown ratios. This stability of the stand structure sustains many ecosystem conditions at consistent levels, enhancing essential habitat for a characteristic community of animals and nontree plants. In that sense, by realizing sustained and even-flow yields of timber, landowners also insure a steadiness in many other values from an uneven-aged forest ecosystem.

REGULATING STAND STABILITY AND GROWTH

The concept of a selection system incorporates some basic ideas about regulating timber harvests from stands so that landowners recover financially attractive volumes on a regular schedule. These concerns normally fall within the discipline of *forest management*, where foresters plan how to manage a property as a whole. The forest management plans delineate patterns and timing of treatments at this larger geographic scale, while silviculture designates how to treat individual stands to create and maintain the requisite vegetal conditions within each one. By carefully scheduling the forestwide program of annual silvicultural treatments, forest managers insure regular and fairly uniform yields of products (Osmaston 1968; Davis and Johnson 1987), and a continuity of nonmarket values. With timber management objectives, harvests pay operating costs and overhead, or provide raw materials for a firm's manufacturing operations. For business enterprises, recovered annual yields must provide an adequate return on the investment of ownership.

In addressing traditional timber management goals, different stand treatments (after Osmaston 1968; Davis and Johnson 1987) must:

1 regenerate new age classes in a regular and timely fashion;
2 continually tend immature classes already in a forest;
3 protect investments in the growing stock and insure the yield of quality materials in perpetuity; and
4 annually harvest mature and excess younger trees to maintain vegetal conditions essential to continued production and regular revenues.

Where owners have noncommodity objectives as well, the forest management plans program any requisite ancillary treatments to maintain the desired conditions and values of interest.

From this perspective, an owner's management plans will include a cutting budget or treatment schedule that delineates how to sequence the silviculture for different stands on a property, and what volume to recover in the process. This requires some measure of growth and change to delineate potential treatment needs and the associated harvest yields. Particularly in volume control schemes, the accuracy of a forestwide management plan and cutting budget depends upon how well a forester can predict present growth and future yields, and regulate cutting accordingly (Osmaston 1968; Davis and Johnson 1987). In fact, when regulating a forest or working circle by volume control, foresters strive to harvest the expected long-term sustained growth, or some preset proportion of it. Poor growth data increase the risk of overcutting, or of removing less than the full potential yield from a property.

With selection system, foresters use area-volume regulation schemes. They can formulate a simplistic area-based cutting budget without precise growth information, but cannot develop adequate stand-level prescriptions unless they know how each community changes during a cutting cycle. The growth data help them to

1 judge stand responses, and to set the interval between successive treatments (the cutting cycle);
2 forecast the products available during periodic entries;
3 insure consistent yields from one year to the next across an ownership; and
4 determine the potential for implementing the treatments through commercial sales.

Harvesting removes the growth and controls losses to intertree competition. Even so, some trees die, others grow larger, and new trees regenerate. The residuals also improve or decline in condition or quality (Wenger 1989). As a consequence, the character of each stand changes. So will the resources across an entire property.

ASSESSING STAND STRUCTURAL CHANGE UNDER SELECTION SYSTEM

With selection system, foresters use growth assessments to evaluate the cutting cycle lengths, and the residual stand structure and density. They want to know

1 whether the structure remains stable (sufficient ingrowth and adequate upgrowth) between cuts;
2 if they can re-create the target structure each time; and
3 if the volume in mature trees and excess immature trees will sustain a commercial harvest at each entry.

Mostly, foresters reinventory stand conditions before each cutting to delineate the numbers of trees present by size class. Using techniques described in Chapter 10, they compare the existing structure with an ideally balanced one. Then they prescribe a treatment to restructure the stand again. This keeps the residual trees vigorous, and regenerates a new cohort to replace the mature age class.

These procedures comprise the *check* or *control method* initially devised for European selection system forests (Knuchel 1953; Kostler 1956; Osmaston 1968). They provide no direct information about growth per se, but delineate how to maintain some idealized structure. Still, by remeasuring the same stands before and after a series of treatments, or by using information from other selection system stands on similar sites, foresters can determine the average patterns of net change over time, and the volume they can recover by periodic selection cutting. They can also apply growth information from permanent sample plots and other sources to project how different stands change over time.

Access to affordable personal computers has triggered much research and development work to simulate forest stand development and change (Moser 1976; Adams and Ek 1974; Moser et al. 1979; Buongiorno and Michie 1980; Haight et al. 1985; Hansen and Nyland 1987). This has increased opportunities for foresters to evaluate various management options before actually implementing them by cutting. Even fairly simple stand table projection methods allow foresters to test alternative prescriptions in advance, identify an approach most suited to the conditions at hand, and refine the marking guides to better fit the needs.

Foresters could use a similar approach in assessing treatments for nonmarket values. They need only delineate the stand structural attributes that enhance the values of interest, and determine if and how these conditions change over time. In fact, once foresters develop an appropriately balanced structure for selection system stands under short-to-moderate length cutting cycles, they can assure landowners that the vegetation attributes and associated ecosystem values will remain fairly stable throughout a cutting cycle, and over longer periods of time. Combining these assessments with an area-based regulation scheme identifies the specific stands to treat in any given year, as well as how many and what kinds of trees to cut from a stand.

AN EXAMPLE OF SELECTION SYSTEM

Uneven-aged stands put under deliberate silviculture for the first time usually do not have fully balanced structures. Often the largest trees have considerable defects and other attributes that make them poor candidates as residuals. Unmanaged stands may also have deficiencies within the smallest diameter classes, and a combination of excesses and deficiencies throughout the structure. Foresters must make some adjustments to mitigate these irregularities, and to make the stand a reliable unit of sustained yield silviculture.

Fifteen-year remeasurement data (Fig. 12-4a through 12-4c) from one experimental northern hardwood stand show the changes that follow selection cutting in such a community (Nyland 1990a). Following its first treatment, the residual structure conformed reasonably well to the target condition (see Eyre and Zillgitt 1953, or Arbogast 1957). The stand had some deficiencies in poles and large sawtimber. Upgrowth of excess residual trees deliberately left in selected classes attenuated adjacent deficiencies, and supplemented earlier shortages of trees greater than 18 in. (46 cm). As a result, the stand grew into a balanced condition within fifteen years.

Comparing the fifteen-year stand with the same residual guide shows what to do for a second treatment (Fig. 12-4d). First, salvaging so much large sawtimber during the initial conditioning cut means that contractors will remove few larger-saw-

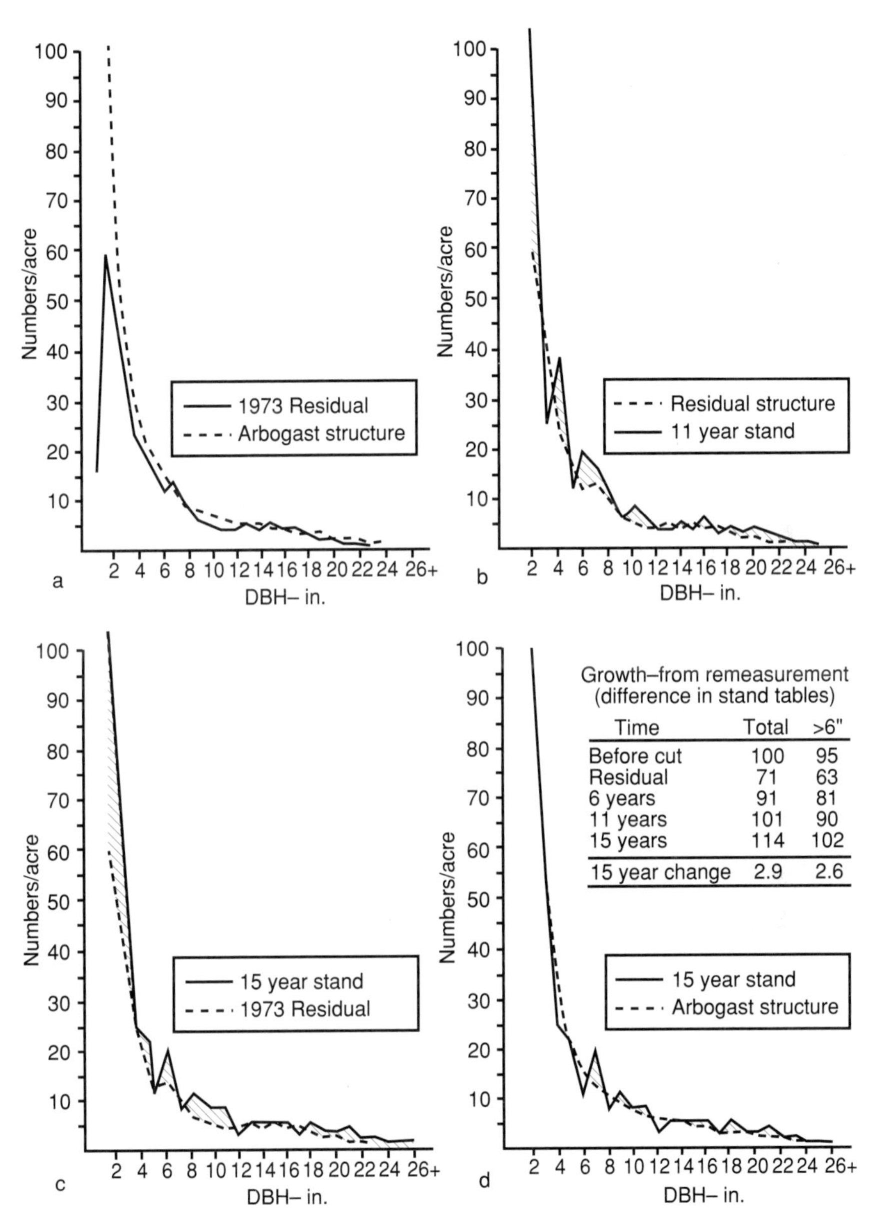

FIGURE 12-4
Fifteen-year development of the uneven-aged northern hardwood stand treated by selection cutting under the prescription depicted in Table 10-3 and Fig. 10-8.

timber trees during the second treatment. Even so, the stand has sufficient excess volume for a commercially viable cut (at least 2,500 board/ac [26.9 m^3/ha] in the region). Ingrowth of new trees brought the stocking of saplings to a desired level, but without excesses to cut during the second entry. The second treatment will remove some poles and small-sawtimber trees, once more re-creating the target residual structure. Further, since the first cut removed most of the high-risk and unacceptable trees, the second cut will focus future growth on the best individuals in each diameter class. This will further upgrade the growing stock quality, as well as leave the requisite numbers of trees in all residual classes. As a result, recoverable yields should increase to a peak level by the third entry.

In essence, the first cutting left a truncated but balanced structure that remained stable throughout the ensuing fifteen-year cycle. Similar smoothing and rebuilding occurred in other stands with a diameter distribution close to a balanced one (Eyre and Zillgitt 1953; Crow et al. 1981; Mader and Nyland 1984). These experiences suggest that with careful planning and marking during a first entry, landowners need not wait several cutting cycles for a balanced structure. Also, truncating the diameter distribution to salvage high-risk and defective trees of larger sizes should not deter long-term success if the stand has a good balance among the younger age classes. A prescription can leave some excess trees in selected diameter classes to compensate for shortages in the next larger one. This will insure a reasonable balance within a relatively short period of years (Eyre and Zillgitt 1953). Further, if foresters must overcut the sawtimber, they can extend a cutting cycle until sufficient regrowth makes another commercial harvest feasible (Hansen and Nyland 1987; Nyland 1987).

When a stand contains few acceptable large sawtimber trees, the marking crew could ignore any strict diameter limit for financial maturity, and retain all the sound and vigorous large sawtimber trees for another cutting cycle. This will improve the harvestable yield at the next entry. Foresters may also leave some trees beyond the normal age of financial maturity (after Hawley 1921) if they (1) have unusual thrift and vigor, and will increase in value at an acceptable rate of return; (2) would serve as seed trees adjacent to group or patch openings; (3) would protect the soil or small seedlings, or help to maintain some uniformity of canopy cover; (4) would enhance visual qualities of a stand; or (5) provided some component of requisite habitat and other nonmarket value. Foresters must often compensate for these extra large-diameter trees by cutting some others that would not reach financial maturity until the next cutting. This preserves a structural balance across the broad size-age classes, and in the space for a new cohort. Experienced marking crews can easily make these adjustments, as long as they understand the philosophy of management.

INTEGRATING CUTTING CYCLE LENGTH, RESIDUAL DENSITY, AND STRUCTURE

As the time lapse between successive cuttings lengthens, crowns of the taller trees expand. Shading increases over the smaller trees and crowding intensifies among older residuals. Diameter increment begins to slow, and height growth may decline among heavily shaded young trees. Where foresters have properly set the length of a cutting cycle, they will have entered a stand before the growth slows perceptibly, to once more open space around crowns of an appropriate number of the best trees in each age class. As a result, diameter increment will continue at fairly consistent rates.

Appropriately matching the cutting interval and residual density also allows sufficient regrowth of volume to sustain another commercial treatment at a predictable time. It will control mortality by sustaining individual residual tree vigor, and also insure optimal accretion by maintaining an adequate complement of well-distributed growing stock. To illustrate, Fig. 12-5 shows the relationship between stand density at the midpoint of a cutting cycle and gross basal area growth of stands under selection system (after Leak et al. 1969). These data suggest that maximum production will occur in northern hardwood communities having a midcycle stocking of about 90 ft^2/ac (20.6 m^2/ha). Sawtimber accretion peaks with residual density between 60 and 80 ft^2/ac (13.8 and 18.3 m^3/ha) in trees at least 5 in. (12.7 cm) and larger dbh, and 25–35 ft^2/ac (5.7–8.0 m^2/ha) of sawtimber (Leak et al. 1987). Thus, stands cut back to 60 ft^2/ac (13.8 m^2/ha) of residual poles and sawtimber trees, or about 70 ft^2/ac (16.1 m^2/ha) of total basal area, would increase in stocking at a rate of about 2.9 ft^2/ac (0.67 m^2/ha), and reach 90 ft^2/ac (20.6 m^2/ha) in about 10 years. For optimum production, the cutting cycle would extend to about twenty years.

Table 10-2 illustrates one set of alternative structures and attendant cutting cycles to optimize large-sawtimber production in northern hardwoods (after Hansen and Nyland 1987; Nyland 1987). These have 85%–90% of the basal area in trees at least 6 in. (15 cm) dbh, and at least 60% in sawtimber. Simulation trials indicate that all provide similar levels of production. They give foresters some alternatives for addressing alternative management objectives for different uneven-aged stands. In choosing an appropriate strategy, they might consider factors (after Gilbert and Jensen 1958; Adams and Ek 1974; Buongiorno and Michie 1980; Martin 1982; Haight et al. 1985; Hansen and Nyland 1987) like the following:

1 the importance of maximizing either volume or value growth
2 the preference for producing large sawtimber versus other kinds of products
3 the effects of site quality or product potential between stands
4 the importance of optimizing return on investment, compared with annual value growth
5 the need to adjust for variations in operating costs based upon access or the difficulty in logging

A landowner's interest in managing the habitat for different wild animals or plants, influencing visual qualities to enhance recreational values, or addressing a range of nonmarket ecological opportunities also affects decisions. All suggest a need for flexibility in selection system, and temper the choices about selection cutting in each stand.

Figure 12-6 illustrates expectations from several stand structures and densities within northern hardwoods (Hansen 1987; Hansen and Nyland 1987). The curves

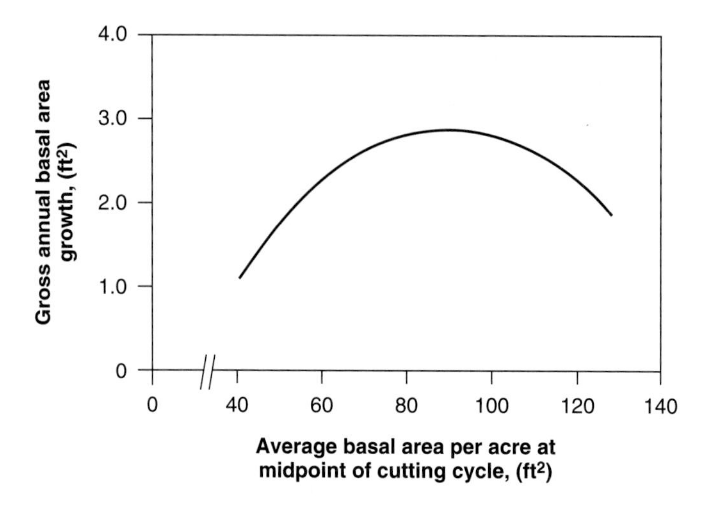

FIGURE 12-5
Relationship between stand density and growth to determine an appropriate cutting cycle length for stands under selection system (from Leak et al. 1969).

depict output (e.g., volume or value growth) as a function of residual density, with separate curves for each different structure and maximum residual tree size. Some distributions have a high stocking of large-diameter trees (e.g., q=1.2), and others proportionately more poles and saplings (e.g., q=1.8). For these cases, differences in stand structure and maximum tree size have little effect on total volume production among trees 6 in. (15.2 cm) and larger dbh. For maximum wood fiber production, foresters need only use a residual density appropriate to the cutting cycle. By contrast, best large-sawtimber or value growth comes from stands with an abundance of sawtimber trees (e.g., a q=1.2), though not necessarily larger than 20 in. (50.8 cm) dbh. Again, foresters would temper the residual density to fit the cutting cycle (see Table 10-2). To maximize the return on investment, they would cut to a low residual density, use a long cutting cycle, and set the maximum residual diameter at 16 in. (40.6 cm). The structure would need to insure high ingrowth to large-sawtimber sizes (e.g., like that proposed by Eyre and Zillgitt 1953; or Arbogast 1957).

Simulation of selection cutting in uneven-aged loblolly-shortleaf pine stands suggests similar effects (Hotvedt et al. 1989). Commonly recommended residual densities range from 60 to 75 ft²/ac (13.8–17.2 m²/ha), and cutting cycles cover four to ten years. Yet within these ranges, optimal simulated sawtimber volume production came from stands with 60–65 ft²/ac (13.8–14.9 m²/ha) of residual basal area, a high proportion of basal area in sawtimber-size trees, and a cutting cycle of four to six years. With longer cutting intervals (e.g., nine to ten years), foresters would lower the residual stocking to 45 ft²/ac (10.3 m²/ha), but maintain the same basic stand structure. Stands with 55 ft^2/ac (12.6 m^2/ha), limited numbers of residual sawtimber trees, and operated at five-year intervals would provide the best return on investment (e.g., 10% interest rate). In general, the greater the residual value, the more a stand must produce to insure the requisite rate of return. Low residual stocking translates into lower levels of residual value, and better chances for higher returns on the investment. At the same time, stands with low residual stocking take longer to regrow sufficient volume for another operable cut, and this extends the investment period. In fact, the short cutting cycles deemed optimal for high returns in southern pines (Hotvedt et al. 1989) may not sustain commercial logging under some circumstances. Instead, landowners must often extend the cutting cycle and accept lower returns to maintain consistent levels of production over repeated entries to a stand.

Experience and research emphasize that landowners will not stabilize product yields and other opportunities by simply extracting salable trees of large diameters at periodic intervals (e.g., diameter-limit cutting). They must use each cutting to regulate crowding, stimulate the radial increment, and concentrate the growth on trees of the best qualities. Also, they must integrate an appropriate reproduction method to recruit adequate seedlings of acceptable species. In combination, these practices will maintain a balance across the age classes, and sustain the values of interest from uneven-aged communities.

APPLICATIONS FOR NONCOMMODITY OBJECTIVES

Historically, foresters have used selection system primarily for wood production goals. In the process they also created and maintained fairly stable vegetal and physical environment conditions, and conserved many important ecosystem values. These include long-term consistency in water yield and quality, of visual characteristics within stands, and of habitat for indigenous and introduced plants and animals. Further, by controlling the stand density and structure, foresters concurrently maintain fairly consistent levels of organic matter production and accumulation; growing season soil moisture withdrawals and temperature fluctuations; understory brightness, light quality, and air temperature; and vertical dispersion of green foliage, with a resultant dispersal of primary net production. These environmental conditions sustain a characteristic community of plants, ranging from microscopic organisms to conspicuous herbs and woody species. They function as loci of photosynthesis, and substrata for insects and fungi. Collectively, these organisms support an array of invertebrates and larger animals that find the unique physical environment of selection system stands a suitable habitat.

Periodic stand maintenance under single-tree selection system regularly adds considerable amounts of downed woody debris, keeps light levels within certain threshold levels, and recruits new members to the plant community. The latter preserves the high degree of vertical structure. Further, with short-to-moderate cutting cycles, environmental conditions normally favor shade-tolerant species of both woody and herbaceous plants like that common to climax communities. Using group selection method, or cutting fixed-area patches in conjunction with single-

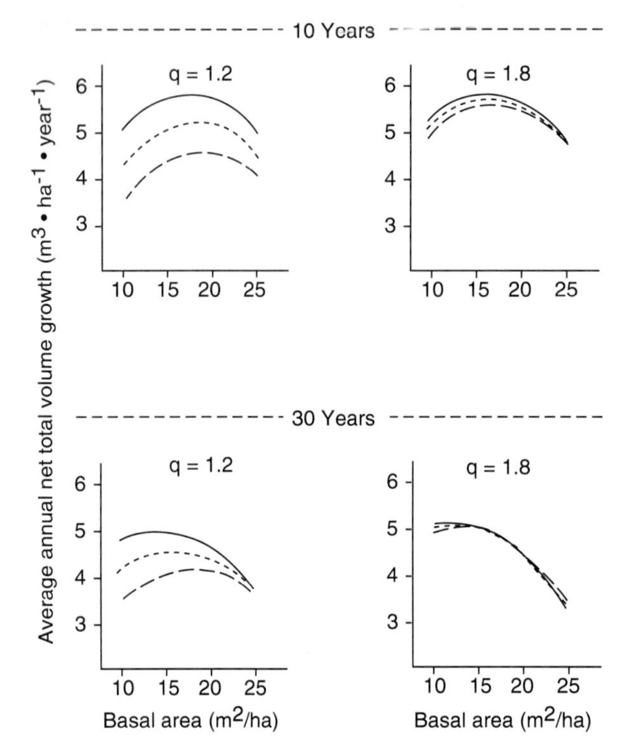

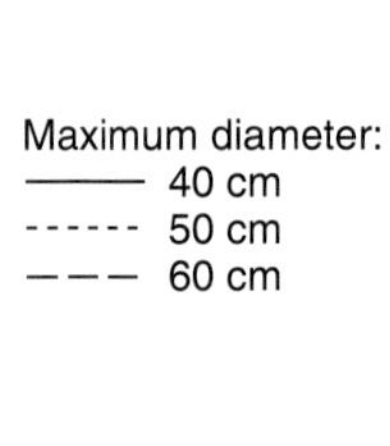

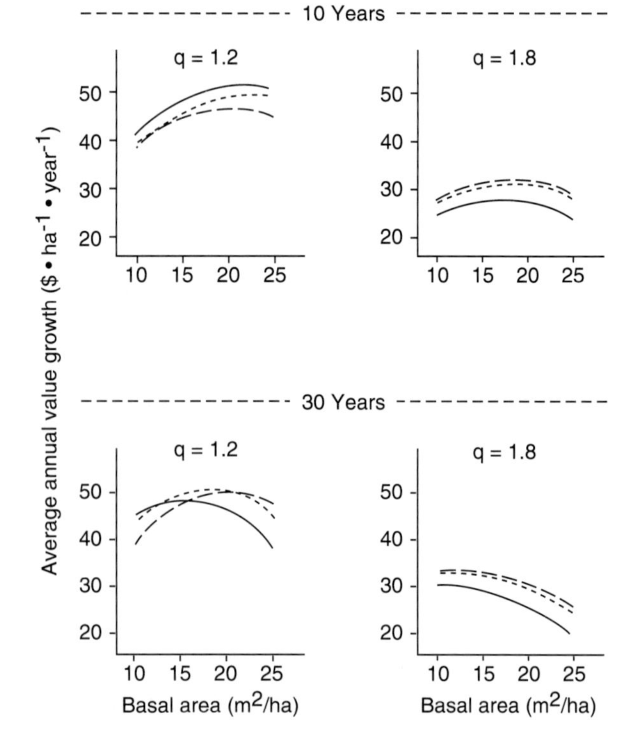

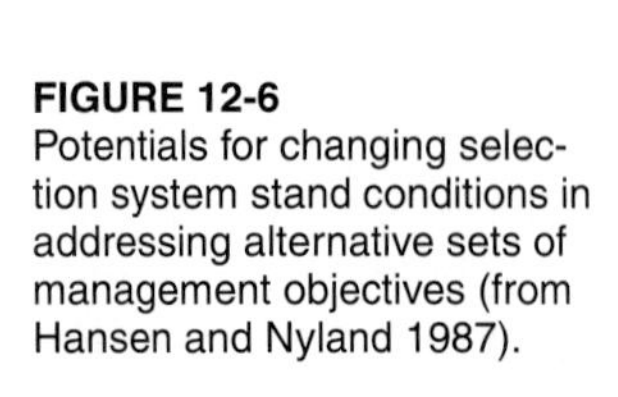

FIGURE 12-6
Potentials for changing selection system stand conditions in addressing alternative sets of management objectives (from Hansen and Nyland 1987).

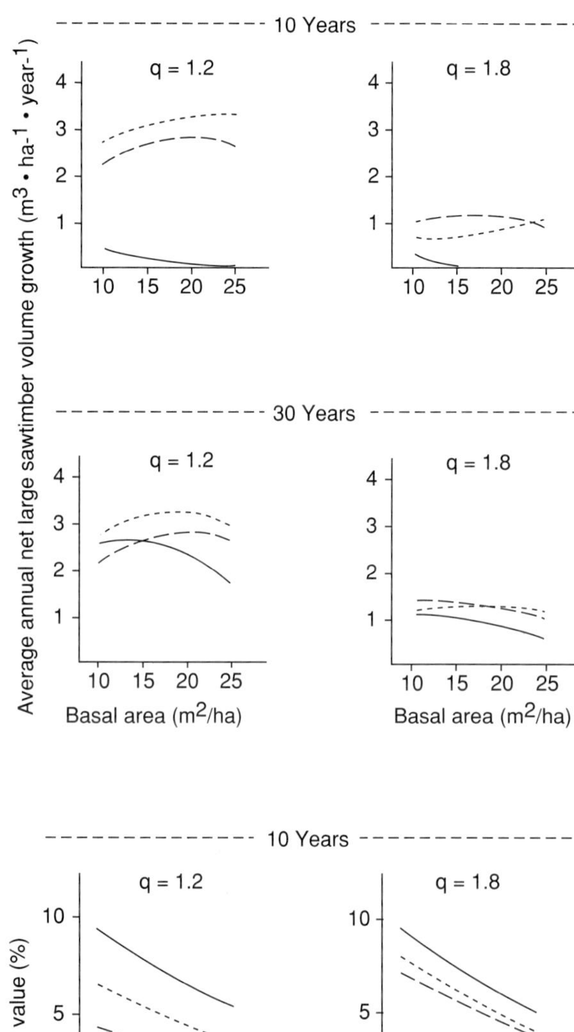
10 Years
q = 1.2
q = 1.8
30 Years
q = 1.2
q = 1.8
Average annual net large sawtimber volume growth (m3 • ha-1 • year-1)
Basal area (m2/ha)
Basal area (m2/ha)

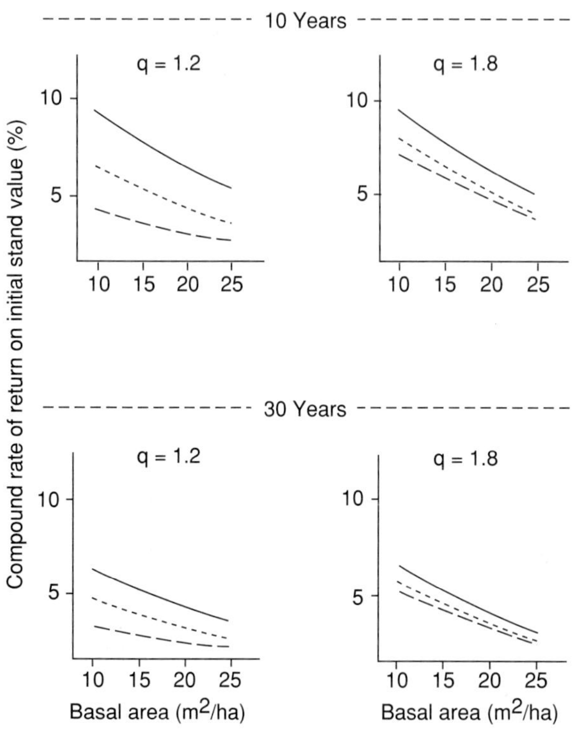
10 Years
q = 1.2
q = 1.8
30 Years
q = 1.2
q = 1.8
Compound rate of return on initial stand value (%)
Basal area (m2/ha)
Basal area (m2/ha)

tree removals, improves opportunities to regenerate a component of less shade-tolerant species (communities of an earlier successional seral stage) in the larger openings. This may enhance the health and reproduction potential for different animals, and increase the faunal community richness within a stand. Similar benefits would accrue from using a low residual density and long cutting cycle in conjunction with single-tree selection system.

Available evidence suggests that well-managed uneven-aged stands contain more herbs, browse, and cover (a higher small-stem density) than old-growth stands or fully stocked even-aged communities of at least the pole stage (Healy 1987). Repeated selection cutting will sustain these features over time, as well as the characteristic high vertical structural diversity (Fig. 12-7). This provides a food base for both insects and insectivores at different canopy levels. The short trees provide protection, cover, and nesting sites for avian species that function at or close to the ground surface, as well as suitable habit for midcanopy feeders. Perpetual supplies of green leaves, twigs, and buds at multiple levels also enhance opportunities for several forest-dwelling herbivores.

The height profile of vegetation and density of foliage at various elevations above the forest floor will influence the bird community composition substantially. Maximum avian species diversity will occur in stands with a balance of short (under 2 ft, or less than 0.6 m), medium (2–25 ft, or 0.6–7.6 m), and tall (over 25 ft, or more than 7.6 m) vegetation. This gives the community a high degree of vertical structural diversity (MacArthur and MacArthur 1961; Noon and Able 1978; DeGraaf 1987; Noon et al. 1979). Plant community composition has a lesser effect, except for some birds that depend upon a particular species for food or other life essentials. Such findings suggest that most selection cuttings would maintain conditions suited to an array of forest-dependent birds, and cause few lasting changes in avian community composition. At the same time, adjacent stands have similar structures and vegetal conditions, giving a selection system forest little diversity from one stand to another (Fig. 12-7).

Like a variety of other partial cuttings, selection system will increase the diversity and frequency of nesting birds heard during the first few years following a treatment (Webb and Behrend 1970; Nyland et al. 1976; Webb et al. 1977). Most avian species common to precut conditions remain abundant. Rare or infrequently seen species may temporarily decline, and several that utilize fairly open habitats move into treated areas during at least early parts of a cutting cycle. With fairly light partial cutting, the community reverts to a pretreatment status within a few years, making the effects fairly short-lived (Webb et al. 1977). This probably reflects the increase of crown canopy density, and the filling of the small gaps caused by the last cutting. As a consequence, at least the avian community composition will probably vary little over repeated cutting cycles, except for the influx of some additional species during a few years immediately following a selection treatment.

The small mammal community may also not change much over the course of a cutting cycle. This reflects the consistency of conditions important to them, at least in stands with a perpetual high vertical structural diversity. Large mammals generally roam across areas larger than individual stands, and move between different ar-

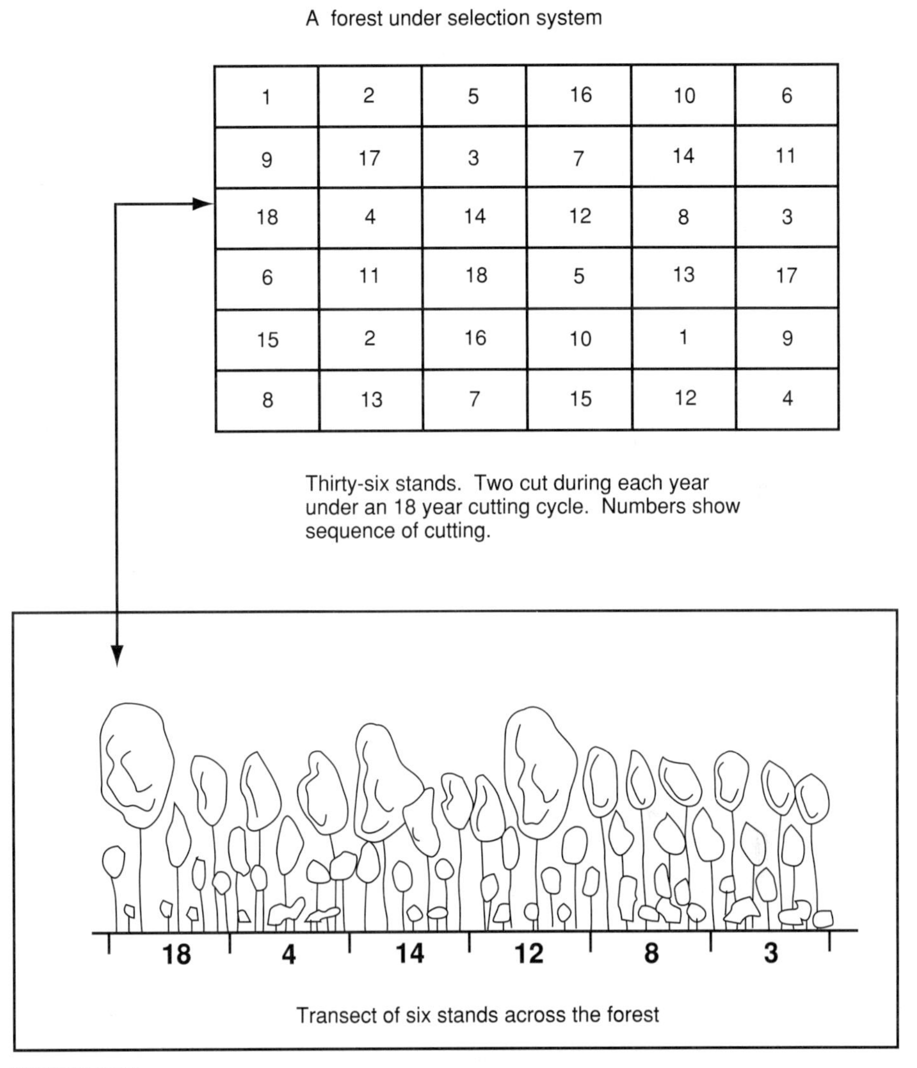

FIGURE 12-7
Intermixing of trees having different ages and heights gives selection system stands a high degree of vertical structural diversity, but lends little variation in vegetal attributes between adjacent stands or even across an entire selection system forest.

eas to derive life essentials from a broader landscape. Compared to recently regenerated even-aged stands, ones under selection system have less total usable materials for browsing animals (woody browse, succulent shoots, and herbs). However, supplies remain more consistent over long periods, and selection system stands never have prolonged periods with little or no understory browse (Pruitt and Pruitt 1987).

Selection cuttings based upon long cutting cycles and low residual densities would trigger more drastic short-term changes of conditions, and this could substantially affect the habitat for rarely occurring birds or those with particularly narrow niche requirements (MacArthur and MacArthur 1961; Noon and Able 1978; Noon et al. 1979). It would alter the habitat for many mammals as well. The longer period between successive entries may reduce disturbance of wild creatures with an apparent sensitivity to human activity. On the other hand, it would add slash and downed logs less frequently, and this may prove less desirable for some other animals.

The temporal continuity of an animal community within selection system stands provides many opportunities for viewing and hearing a fairly predictable array of wild creatures. Further, the permanent trail system allows recreationists easy access for a variety of uses. The intermixing and diversity in tree sizes (diameters and heights) add a special visual quality to the *standscape* in selection stands, and many lay viewers cannot readily discern cut from uncut areas (Rutherford and Schafer 1969). The high degree of vertical structure blocks visibility over all but short distances. This screens most stumps, logging slash, and exposed mineral soil that many lay viewers consider visually obtrusive (Nyland et al. 1976). Foreground images and those along skidding trials become the most dominant, and these often have positive characteristics. The lanes add patterns of openness and brightness that alter visual texture, extend the lines of sight, and interrupt the visual monotony common to most intertrail spaces. Each cutting temporarily increases vertical and horizontal visibility through the trees. However, rapid regrowth of the vegetation reduces the effect after just a few years.

The wide spacing between residuals and elevated understory light in selection systems stands under a long cutting cycle (e.g., twenty to twenty-five years) somewhat lengthens the period of improved visibility. After ten to fifteen years a dense low-level canopy will have formed and lines of sight become short again, except along the major skidding trails. The lower density schemes will recruit trees having a wider range of shade tolerance, and this adds visual interest. Also, the more open canopy and higher understory brightness make the trail surfaces good seedbeds for herbaceous plants and woody shrubs. These enrich the plant community composition with species of added visual appeal (e.g., in foliage colors and textures, flowers, and fruits), and that provide succulent food and small seeds otherwise absent or limited in closed tree communities. Animals using these foods also add interest for viewers, and enhance the recreational values of selection stands.

SETTING THE MAXIMUM DIAMETER FOR NONCOMMODITY OBJECTIVES

Figure 12-6 suggests that owners who wish to maximize absolute volume or value growth (sawtimber volume or dollars) gain little by retaining really large trees (e.g., over 20 in. or 50.8 cm dbh). Yet they would keep 60% of the total basal area in large sawtimber, maintain a residual density of 75–85 ft^2/ac (17.2–19.5 m^2/ha), and use a cutting cycle of not more than fifteen years. Conversely, to maximize return-on-investment criteria, foresters must use a maximum residual tree size of not more than

16 in. (41 cm), a residual density of 60–65 ft²/ac (13.7–14.9 m²/ha), and a cutting cycle of twenty years or more. The treatments must remove excess trees across the diameter distribution to maintain a balance within the structure (Hansen and Nyland 1987).

Financial maturity criteria usually encourage foresters to truncate the diameter distribution well below that of old-growth forests. They could keep larger trees without important negative biologic consequences where other objectives dictate. They must maintain a fixed proportion of area in each age class, including the oldest one, but could use a flexible maximum diameter for large sawtimber. Thus, with a fifteen-year cutting cycle for northern hardwoods (see Table 10-2), a landowner would still retain about 80–85 ft²/ac (18.3–19.5 m²/ha) of residual basal area, including 15–20 ft²/ac (3.4–4.6 m²/ha) in trees larger than 17 in. (43.2 cm) dbh. The large sawtimber might include scattered trees as big as 25–30 in. (63.5–76.2 cm), if they have good vigor and a high probability of surviving through another cutting cycle.

Foresters may also retain larger-than-usual trees within lower-density stands operated at extended cutting cycles, as long as they select fairly long-lived species that do not deteriorate when exposed by heavy cutting. In northern hardwoods this largely means sugar maple and red spruce, since yellow birch and hemlock often decline in vigor after heavy release. For selection systems employing short cutting cycles and high residual densities, foresters could retain almost any tree species. That gives them greater flexibility in addressing landowner interests.

Leaving only one *oversized* tree per acre (3/ha) adds considerable visual diversity to a stand, and enhances other noncommodity needs as well. These trees do not represent a great investment. Foresters could maintain them in perpetuity (or as long as they live). In other cases, they could replace some of the oversized trees each cutting cycle, insuring a continuing stocking of large trees in good health and vigor and allowing a landowner to periodically use some of the old-growth stems to supplement revenues from a stand. Contractors must protect these residuals against logging wounds to the crown and basal portion of the stems, or to the root systems. Damaged trees have a high probability of deteriorating over the ensuing cutting cycle, especially if skidding causes deep ruts and severs part of the root system (Shigo 1966; Anderson et al. 1990; Anderson 1994).

Both the diameter growth and the crown size of even fairly old trees of many species will increase following heavy cutting, as long as those trees initially have full crowns and good vigor. They will increase substantially in volume over time, and in the absolute value per tree. For example, in the Huntington Wildlife Forest in New York, one experimental cutting to a residual of 50 ft²/ac (11.5 m²/ha) released twenty-three 200-year-old sugar maple trees. The diameter growth averaged about 2 in. (5.08 cm) per 10 years during a 33-year period following cutting, compared with 1.2 in. (3.05 cm) per 10 years prior to release. Crown radius increased by an average of about 5–7 ft (1.5–2.1 m) over the period. At other sites, many residual sugar maple trees with ages up to 250 years or more survived and grew at similar rates once released by cutting. This evidence suggests that at least for sugar maple, trees of good health and vigor will readily survive for long periods in managed stands.

SOME OTHER CONSIDERATIONS FOR SERVING NONCOMMODITY GOALS

In some cases, landowners may have stringent requirements for visual qualities of logged areas, such as insuring minimal litter disturbance during harvesting. Under these circumstances, foresters working at high latitudes can limit logging to wintertime periods when the soils freeze and snow blankets the surface. Then landowners will see little sign of litter disturbance after the snow melts. Also, confining logging to periods when soils readily support the machinery without rutting leaves the trails in a more trafficable condition, both for pedestrians and for skidding at later cutting cycles. Further, preventing rutting protects root systems of residual trees, and this helps to insure good stand health and vigor.

Some landowners may require slash reduction by lopping within proximity to a pathway, and removal of obstacles along trails. Use of directional felling will usually suffice for keeping debris away from the access corridors, and operators can easily drag off unused stems and tops from the main skidding trails. Top lopping may cost as much as 15% of the value in logs taken from the cut trees (Nyland et al. 1976). Further, fine slash usually breaks down naturally within about three to five years. When confronted with such high costs for such a temporary effect, many landowners decide to forgo the lopping. They may also recognize that the slash piles enhance habitat conditions for small mammals, and come to consider the woody debris as a benefit to the ecosystem. When contractors produce pulpwood and other fiber products as well as sawlogs, crews must cut more limbs from the felled trees. This leaves smaller unused tops that satisfy many landowners' aesthetic interests without the extra expense of lopping.

When addressing traditional commodity interests, foresters normally remove trees with physical defects and disease. This includes trees with deformities that lend a unique visual diversity to the *standscape*, and trees with cavities and fruiting bodies. Foresters could just as easily preserve scattered defective stems if they serve special needs. For example, some birds and mammals depend upon trees with cavities as nesting sites and shelter. To help to maintain these wildlife species, landowners could keep some *den* trees in the managed stands (Cunningham et al. 1980; Scott and Oldemeyer 1983; Moriarty and McComb 1983; Healy 1987; Healy et al. 1989; Welsh et al. 1992). Also, logging breaks large branches from the crowns of a few trees at each entry (Nyland and Gabriel 1971; Nyland et al. 1976). These injuries may eventually become cavities. Other badly damaged trees may die, becoming snags. Thus, if foresters retain even a few badly damaged trees over successive cutting cycles, they enhance cavity and snag formation and help maintain an important element of wildlife habitat. The damaged trees do represent an investment equivalent to the potential production lost, but not as much as commonly thought. Natural compartmentalization around the wounds contains the extent of rot and defect that develops in living trees (Shigo 1966; Anderson 1994). Normally, wood laid down afterward will not become discolored or decayed. As a result, trees with upper stem wounds may continue to produce high-quality wood on the lower logs. The high marketplace value of many old-growth trees with defective centers witnesses to the effectiveness of this compartmentalization process.

13

CLEARCUTTING

CLEARCUTTING AS A REPRODUCTION METHOD

With even-aged silvicultural systems, foresters completely remove the entire tree community over a relatively short period at the time of economic maturity. This allocates total *ecologic space* to the new cohort, and creates a new even-aged community. In this context, *clearcutting* removes the entire overstory (the mature trees) in one operation (Fig. 13-1) to establish a new cohort of desirable species across the site (Hawley 1921; Troup 1928; Champion and Trevor 1938; Cheyney 1942; Smith 1986; Matthews 1989). Landowners may use either natural or artificial means to establish the new trees.

In describing clearcutting as a silvicultural method, foresters differentiate (after Ford-Robertson 1971; Soc. Am. For. 1989) between the following:

Clearcutting—clean felling, complete cutting, complete exploitation . . . *strictly* removing all the trees in one operation . . . a harvesting method

Clearcutting method—a method for regenerating an even-aged community, where the new seedlings become established in fully exposed microenvironments after removal of most or all of the mature trees

Clearcutting system—a silvicultural system incorporating the clearcutting method to remove (clear) the mature community over a considerable area at one time

The terms *clearcutting method* and *clearcutting system* identify clearcutting as a reproduction method to enhance the regeneration of a new even-aged community. Clearcutting may also improve food resources for wild animals or increase water yields from a site.

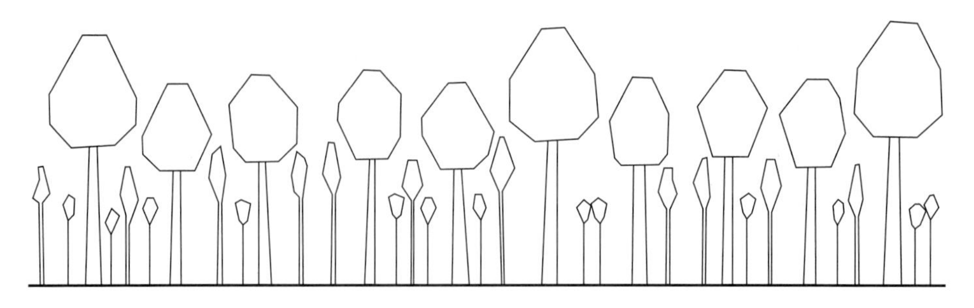

Mature stand ready for reproduction method

After clearcutting

Seeding stage

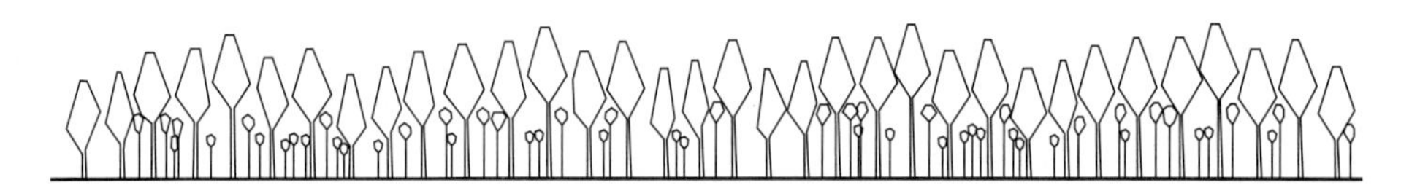

Early stand development
(early pole stage)

FIGURE 13-1
Clearcutting removes the entire overstory (the mature trees) in one operation to establish a new cohort of desirable species across the site.

The sizes of clearcuttings have historically differed widely from one ownership to another, and from one clearcut to the next. Actual size depends upon the objectives for its use, site conditions in different stands and on different ownerships, and characteristics of different forest communities under management. Actual arrangements will reflect both ecologic and economic considerations (Matthews 1989). Yet as a reproduction method, clearcutting removes the entire standing crop *over a considerable area* (Ford-Robertson 1971; Soc. Am. For. 1989). Placing clearcutting within the context of a silvicultural system fixes that area as the space occupied by a *stand*.

In a technical sense, each stand has distinct ecologic and operational attributes. These determine the minimum amount of area that clearcutting must cover. To illustrate, common definitions (Soc. Am. For. 1950; Ford-Robertson 1971) identify a stand as an aggregation or community of trees *occupying a specified area*. The contiguous nature of this space facilitates record keeping and planning. Second, each stand has *sufficient uniformity* in composition, age condition, spatial arrangement, and other conditions *to distinguish it from adjacent communities*. As a result, foresters can easily locate stand boundaries at the time of some programmed use or management activity, and readily delineate each separate stand on maps. Third, the distinctive conditions that separate one stand from others justify unique silvicultural treatments for each one, or require their application at different times. Consequently, foresters will *treat each stand as an entity* (Davis 1954). Fourth, a stand must cover sufficient space *to operate efficiently as a unit*. This usually means yielding adequate volume and value to support commercial operations during both tending and regeneration phases of a silvicultural system. So from a silvicultural perspective, a stand has four essential characteristics:

1 recognizable in character
2 covering a contiguous unit of space
3 treated as an entity
4 large enough to manage efficiently

Within these constraints, landowners can subdivide their forests into a mosaic of management units based primarily upon operational or use considerations, or ecologic factors.

Silviculturally, stands represent the minimum land area treated by a reproduction method. When a landowner elects to remove all the standing trees in a single operation, the treatment qualifies as a clearcutting. If done to regenerate a new even-aged community, it becomes the clearcutting method (Hawley 1921; Champion and Trevor 1938; Cheyney 1942; Smith 1986; Matthews 1991). Further, foresters can use clearcutting method to regenerate a group of target species by natural reproduction, or make ecologic space for artificial regeneration by planting or direct seeding. They may supplement the clearcutting with treatments to control interfering vegetation or improve seedbed conditions for germination and seedling development. Removing all the standing trees in a single operation facilitates this site preparation, and often allows greater flexibility in the methods used (Smith 1986; Klinka and Carter 1991).

In contrast with the clearcutting method, some landowners use selective cuttings they loosely call *commercial clearcutting* or *economic clearcutting*—removing all the merchantable trees, leaving the unsalable trees standing (Ford-Robertson 1971; Wenger 1984; Soc. Am. For. 1989). These simply cut and extract the usable trees for immediate commercial value. They remove choice species and the large-diameter trees of upper canopy positions, leaving the low-vigor and smaller trees as residuals. These often interfere with the regeneration and its development, and frequently develop into branchy trees of limited value (Troup 1928; Matthews 1991). New seedlings may become established between the irregularly spaced residuals, but exploitive cuttings do not control the species that regenerate nor the abundance and spacing of seedlings that develop. As a consequence, foresters cannot readily forecast the long-term character of the community that might regenerate, how soon it will become established, nor what value it will have to a landowner.

ATTRIBUTES OF CLEARCUTTING AS A REPRODUCTION METHOD

Stands regenerated by any even-aged reproduction method have certain recognizable characteristics of both ecologic and operational importance. These include:

1 The time to sexual maturity determines the minimum rotation for silvicultural systems based upon natural regeneration from seed.
2 A long period will elapse until many species begin producing abundant seed.
3 Healthy even-aged stands will produce abundant seed after reaching sexual maturity.
4 After that time, adequate numbers of well-distributed seedlings should regenerate across the stand, given no important interference or herbivory.
5 After crown closure, the even-aged community will have a reasonable degree of homogeneity throughout.
6 Main canopy trees will grow at similar rates, so foresters can apply a common cultural treatment throughout the entire community, and tend it periodically to maintain stand health and vigor.
7 Trees in the entire community will reach financial maturity together.

Variations of topographic and other physical site attributes may influence the initial tree composition and density, and their rates of development. Subsequent tending will concentrate the growth potential into the best of the cohort. Then when the community reaches a desired stage of maturity, foresters will once again use clearcutting and site preparation to regenerate a new stand of sought-after species, and once more set the stage for that stand's long-term development.

Aside from knowing that trees in a community have reached sexual maturity, foresters must also consider the mechanisms for seed dispersal and time of dissemination. Concerns include the distance and spatial uniformity of a seed rain relative to the parent trees, reliability of those patterns, and chances of seed loss during and after dissemination (see also Chapter 9). Since clearcutting removes all standing trees in a single step, it also takes away any in-stand source of new seed following the treatment. Consequently, to secure adequate seed-origin regeneration, (Hawley 1921; Troup 1928; Smith 1986; Matthews 1991) foresters must depend upon

1 viable seed already stored within the litter layer;
2 seed stored in unopened cones or fruits attached to branches of recently felled trees; and

3 seed disseminated by wind, animals, and water from adjacent stands after clearcutting.

Some seedlings derive from each source, but seed dispersed prior to cutting, or off the logging slash, usually provides the most uniform coverage (Hawley 1921; Troup 1928). With many deciduous species some new trees also may arise vegetatively particularly after clearcutting of fairly young even-aged communities or species that do not regenerate easily from seed (e.g., basswood).

Early studies in lodgepole pine showed the importance of seed stored in cones on uniformly dispersed logging slash, especially when released by prescribed burning following clearcutting (Bates et al. 1929). By contrast, in northern hardwoods the seed stored in the organic layer serves as the major source for new germinants (Morash and Freedman 1983; Lees 1987). For species as diverse as loblolly pines and western hemlock most seed germinates during the first growing season after dispersal (Haig et al. 1941; Little and Somes 1959; Barnett and McGilvray 1992). So the importance of stored seed depends on when it fell and how much animals, insects, and fungi destroy. In addition, forest floor disturbance by site preparation or logging displaces portions of the seed bank and can alter the spatial patterns of seedling establishment. This disturbance may also create more favorable seedbeds for newly dispersed seed of small-seeded species (Roberts and Dong 1991). In some community types, skid trails and other areas scraped bare during logging remain largely devoid of tree regeneration following clearcutting (Walters and Nyland 1989), apparently due to the lack of viable seed. These factors vary widely among species, and with site conditions.

IMPORTANCE OF SEED SOURCE AND ADVANCE REGENERATION

For most species, wind, water, and animals will not transport seed far from the source. In fact, safe distances may not exceed more than one to five times the height of adjacent parent trees (Smith 1986). To illustrate, with western larch, Engelmann spruce and subalpine fir in the Rocky Mountains of North America, only about 40% of the seed rain dispersed as much as 100 ft (30 m) downwind into a clearcutting, about 10% drifted as far as 300 ft (91 m), and only 1% fell as far as 600 ft (183 m) from the edge (Fig. 13-2). The actual distances depend on the total supply on parent trees, and the force of the wind. With heavy seed crops, new seedlings will arise for distances of 500–600 ft (152–183 m) downwind from the border trees. For most years, little falls more than about 300 ft (91 m) into a clearcutting (Schmidt et al. 1976; Noble and Ronco 1978; Alexander and Edminster 1983). Likewise, wind-disseminated loblolly, shortleaf, and slash pines in southeastern United States disperse effectively no more than 200–300 ft (61–91 m), and longleaf pine less than 100 ft (30 m) (Wenger and Trousdell 1958; Crocker and Boyer 1975; Boyer and White 1990; Barnett and Baker 1991; Baker 1992). With ponderosa pine in the Black Hills, effective dispersal does not exceed more than about 130 ft (40 m) (Boldt 1974). Even with light-weight paper birch seed, only 10% may glide up to 175 ft (53 m) from the edge of a clearcutting (Bjorkbom 1971). In fact, most seed of commercially important North American trees will not travel more than 100–300 ft

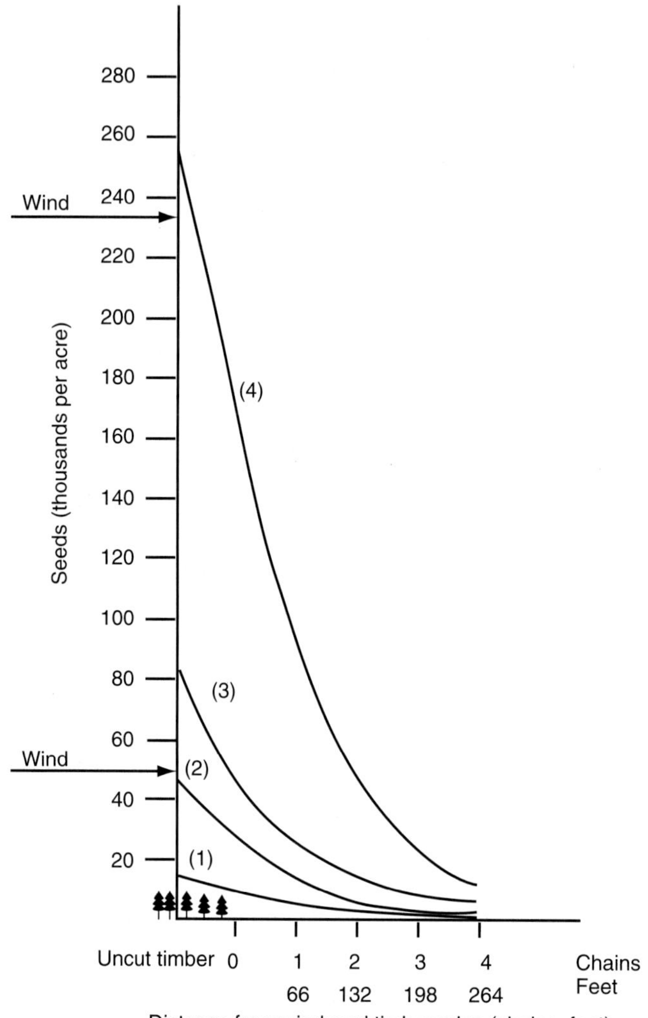

Engelmann spruce-subalpine fir seed dispersal from windward timber edge to 3.5 chains into the opening, averaged over all years and areas, illustrates (1) poor, (2) fair, (3) good, and (4) heavy seed crop years

FIGURE 13-2
Seed dispersal patterns from Englemann spruce and subalpine fir trees into clearcuts during years with varying levels of seed production (after Noble and Ronco 1978).

(30–91 m) downwind (see Table 13-1). These represent maximum safe widths for clearcuts.

Technically, clearcutting initiates a new age class *after* overstory removal. Yet in many community types having a major shade-tolerant component, a cohort of advance regeneration often forms during latter parts of a rotation—especially when some cultural treatment reduces the density of the overstory canopy. These seedlings often become well established prior to overstory removal, and grow rapidly following release. They may prove critical to success, or provide an insurance against failure. To illustrate:

1 For northern hardwoods and spruce-fir communities in northeastern North America, clearcutting usually succeeds in stands with abundant well-developed advance regeneration (Metzger 1980; Grisez and Peace 1973; Marquis et al. 1975; Bjorkbom and Walters 1986; Chamberlain 1991; Davis 1991; Ruel 1991; Smith 1991).

2 In eastern North America complete overstory removal often fails to perpetuate oaks in stands with fewer than 430–440/ac (1062–1087/ha) well-distributed advance oaks at least 4.5 ft (1.37 m) tall, or where stump sprouts will not supplement the advance regeneration in adequate amounts (Sander et al. 1976, 1984; Johnson 1977).

3 In old-growth red fir in the Cascade Range of northern California, well-spaced advance saplings at least 5–12 ft (1.5–3.7 m) tall lessen the need for site preparation, and shorten the time for stand establishment following complete overstory removal (Oliver 1986).

4 In the northern Rocky Mountains, one-third to three-fourths of stocked plots not subjected to site preparation had advance regeneration prior to clearcutting (Ferguson 1984).

5 Among western hemlock and mixed conifer clearcuts in the mountains of eastern Oregon, advance regeneration supplements the distribution and density of new germinants, and its abundance often determines whether foresters can rely upon natural regeneration (Seidel 1976b).

6 Conifers in western North America almost exclusively develop best in stands having abundant advance trees with at least a 50% live-crown ratio (Seidel 1980; Carlson and Schmidt 1989).

In cases like these advance trees supplement the distribution of new germinants, and could make the difference between adequate and insufficient stocking of natural regeneration.

Most shade-intolerant species germinate almost entirely following the reproduction method cutting. The more shade-tolerant species emanate mostly from stump sprouts and advance regeneration (Trimble et al. 1986; Wang and Nyland 1993). The actual spatial distribution of a species within a new community depends upon the variability of physical site conditions. Given adequate sources for regeneration different species will dominate areas with conditions optimal to their growth and development (Leopold and Parker 1985). Further, at least among eastern hardwoods where regeneration normally follows promptly and develops rapidly after

TABLE 13-1
APPROXIMATE MAXIMUM DISPERSAL DISTANCES FOR THE BULK OF SEED BLOWN DOWNWIND INTO OPENINGS FROM ADJACENT PARENT TREES OF SEVERAL COMMERCIALLY IMPORTANT NORTH AMERICAN SPECIES (AFTER INFORMATION COMPILED BY FOWELLS 1965; HARLOW ET AL. 1978; BURNS AND HONKALA 1990A, 1990B).

Effective dispersal distance for most seed in an average year[a]								
Adjacent to parent tree[b]	Only short distances	Within 50–100 ft (15–30 m)	Up to 100–120 ft (30–36 m)	Up to 120–150 ft (36–46 m)	Up to 200 ft (61 m)	Up to 300–330 ft (91–101 m)	Up to 400 ft (122 m)	At least 500 ft[c] (152 m)
American beech	Pacific silver fir	Balsam fir	Engelmann spruce	Nobel fir	California red fir	Blue spruce	W. white pine	Fraser fir
Black cherry	White spruce	Subalpine fir	Ponderosa pine	White fir	Grand fir	Black spruce		E. white pine
Butternut	Red pine	Sitka spruce		Tamerack	Lodgepole pine	Red spruce		Atl. white cedar
Walnut	Green ash	Longleaf pine		Western larch	Redwood	Jeffrey pine		Western hemlock
Black tupelo	Honey locust	Sugar pine		Shortleaf pine	N. white cedar	Pitch pine		Paper birch
Oaks		Jack pine		Slash pine	Sweetgum	Scotch pine		Yellow birch
Hickories		Eastern hemlock			Tulip-tree	Loblolly pine		Aspens
		Blue gum				Douglas-fir		Poplars
						W. redcedar		Cottonwoods
						Sugar maple		Red alder
						American elm		

[a] Some seed will disperse farther into a clearcutting, especially in strong winds, but the bulk will fall within the average distances shown.
[b] Birds and mammals will carry some seed over much longer distances, and this serves as an important dispersal mechanism within species with heavy seeds that normally land in close proximity to the parent tree.
[c] Birches often blow long distances over crusted snow. The silky-haired seeds of aspens, cottonwoods, and poplars may float for several miles before landing.

overstory removal, the dominant species generally become established soon after clearcutting, and later germinants often fail (Connell and Slatyer 1977; Leopold et al. 1985; Wang and Nyland 1993). Similarly, most black spruce regeneration develops within three years of strip clearcutting, probably because seedbed conditions deteriorate after that time (Gagnon et al. 1991). Within mixed-species subalpine stands in the Cascade Range of Oregon, natural regeneration has sometimes appeared insufficient in early years, only to increase by ten to fifteen years after clearcutting (Minore and Dubrasich 1981).

Foresters who use the time of seedling origin to separate clearcutting from other even-aged reproduction methods call the release of advance regeneration by complete overstory removal *one-step overstory removal*, or *one-cut shelterwood method* (Smith 1986; Matthews 1989; Loftis 1990). This strict interpretation does help to differentiate the even-aged reproduction methods. Still, as long as the method secures adequate desirable regeneration in a timely fashion, it makes little practical difference whether the new community arises from postcutting germinants, or includes trees from advance regeneration. In either case, the overstory removal promptly establishes desirable species in adequate numbers. Otherwise foresters follow the cutting with tree planting or direct seeding, or use an alternate reproduction method that will succeed with natural regeneration.

In the absence of advance regeneration, clearcutting the seed source forces foresters to rely upon seed stored in the litter layer and on logging slash, or delivered naturally from adjacent stands. These sources must provide abundant seedlings or sprouts throughout the regeneration area, including the center. With species that produce good seed crops annually, or where the seed remains viable in the litter for two or more years, seed source may not limit success if conditions favor prompt germination and early survival of new seedlings. For other species, foresters must carefully time their clearcuttings to follow dispersal of a good seed crop. Otherwise, they use artificial methods to insure prompt and adequate stocking of the desired species. For hardwood species, stump sprouts or root suckers add some new trees (Johnson 1977). These may supplement the seedling regeneration substantially, or even provide the only reliable source for some species (e.g., the oaks).

EFFECT ON ENVIRONMENTAL CONDITIONS NEAR THE GROUND

Tree cutting of any kind changes the physical environment within a stand. The degree depends upon the amount of growing stock that remains, the completeness and density of canopy cover, and the general nature of a site. Yet at least from a biologic perspective (after Hart 1961; Marquis et al. 1964; Lull and Reinhart 1967; Hornbeck 1970; Aubertin 1971; Pierce 1971; Aubertin and Patric 1972; Marquis 1966, 1972; Reinhart 1972; Martin et al. 1986; Hornbeck and Leak 1992), whenever a cutting removes an appreciable portion of the tall vegetation:

1 temperatures of the air near the ground and within upper soil horizons increase;

2 humidity near the surface decreases, and the potential for surface evaporation increases;

3 moisture of the upper soil increases, partly due to less transpiration after overstory removal;

4 interception of rain and snow decreases, and more precipitation reaches the surface;

5 more water infiltrates and percolates through the soil, increasing subsurface flow;

6 organic decomposition increases in the warmer and moister microenvironment, releasing nutrients; and

7 some elements not bound to soil particles or taken up by vegetation leave the site through subsurface flow.

Clearcutting also adds considerable slash to the surface, and its shade reduces the extremes of temperature at the forest floor. The degree depends upon the completeness of slash cover, and if it still has needles or leaves attached. Yet even with this protection, the soil becomes warmer and wetter, and more rich in nutrients. In turn, the altered environment supports a rapid regrowth of new vegetation.

Clearcutting potentially shifts at least temperature and light to a biologic extreme, often creating widely varying microenvironment conditions within a regeneration area (Tesch and Mann 1991). This may prove critical to regeneration success, since conditions of microenvironment determine if germination occurs, where new germinants become established, and how well they grow. On a standwide scale, the balance among environmental conditions determines the potential density and spatial uniformity of regeneration, and the relative abundance of different species. Even with artificial regeneration, postcutting macroenvironment conditions profoundly influence the species and seed sources that will develop best following clearcutting.

Complete removal of the mature trees abruptly allocates total ecologic "space" to a new age class, dramatically changing the macroenvironment (standwide) and the microenvironment (more localized). In fact, observations in redwood forests of California and among Rocky Mountain conifer stands indicate that mean daily maximum soil temperatures could increase by more than 18° F (10.0° C) (Li 1926), and mean growing season soil temperatures at a 1-in. (2.54-cm) depth by as much as 10° F (5.6° C) (Bates 1924), following clearcutting. Soil temperatures in clearcuts may exceed those of particularly cut sites by 24%, or about 3° F (1.7° C) (Haig 1936; Haig et al. 1941). Temperatures of unshaded duff surfaces may reach 150° F (65.5° C) among Sitka spruce-western hemlock stands in the Pacific Northwest of North America, depending upon moisture, exposure time, and sun angle (Ruth and Harris 1979). Along south aspects in the central Rocky Mountains, maximum temperatures may average 140° F (60° C) inside clearcuts, with peak levels of 176° F (80° C). Partial shading can reduce the extremes by 8°–12° F (4.4°–6.7° C) (Alexander 1984). Surface temperatures within clearcuts in the Lake States region of North America may increase by 18°–30° F (10.0°–16.7° C) at 8 in. (20 cm) (U.S. For. Serv. 1935), and those inside clearcuts in the southwestern ponderosa pine region by 15° F (8.3° C) (Pearson 1928). Duff temperatures have risen to as much as 130°–160° F (72.2°–88.9° C) (Jemison 1934; Hallin 1968) following clearcutting, and this could prove lethal to new germinants (Seidel 1986; Tesch and Mann 1991).

Moisture in the upper soil horizons can compensate for the higher temperatures to a considerable degree. In balance, the soil moisture depends upon three factors (Kittridge 1948; Coleman 1953): (1) the quantity delivered to the surface through precipitation; (2) the proportion that infiltrates the surface and percolates through the soil; and (3) the rate of subsequent loss by evapotranspiration and subsurface runoff. Following clearcutting, soil moisture will often increase sufficiently to increase subsurface runoff. Also, precipitation reaching the surface may increase by 100% for showers delivering less than 0.1 in. (0.25 cm) of water, and by 10%–40% for showers dropping more than 0.4 in. (1.02 cm). The effect varies with forest community characteristics, season of the year, and duration of precipitation (Horton 1919; Mitchell 1930; Kittridge 1948). Among conifer forests in western North America, clearcutting increases precipitation at the forest floor by as much as 25% (Krutilla et al. 1983). It also increases the water held in upper soil horizons, and the subsurface runoff (Bormann and Likens 1979; Krutilla et al. 1983; Mitchell et al. 1986; Hornbeck and Leak 1992).

Clearcutting has decreased evapotranspiration by as much as 20% among conifer stands in the Pacific Northwest of North America (Klock and Lopushinsky 1980). Soil moisture also has increased after tree removal in conifer forests of Utah (Croft 1950), hardwoods in the Piedmont of North Carolina (Ferrell 1953), pine-hardwood mixtures of the southern coastal plain (Throusdell and Hoover 1955), and western white pine of the Northern Rocky Mountains (Larson 1922, 1924). In fact, during summer when evapotranspiration exceeds precipitation, soils in 0.1–0.4 ac (0.04–0.16 ha) openings within mature upland hardwood stands in North Carolina had a significantly greater water content than under the surrounding forest. Soil water increased dramatically for 30–40 ft (9.1–12.2 m) into the openings, then leveled off farther from the edge (McNab 1991).

Following clearcutting of northern hardwoods in the White Mountains of New Hampshire, nutrient outflow peaked during the second year (Pierce et al. 1972), and returned to precut levels by four years (Pierce et al. 1972; Martin et al. 1986). Apparently, warmer soil temperatures and greater moisture in the upper horizons accelerated decomposition, releasing nutrients from accumulated organic materials. Since little vegetation remained to take up nutrients from the soil solution, the freed elements either became bound chemically to soil particles, or moved from a site via the increased subsurface flow. Yet under most circumstances, the hydrologic and nutrient budgets should return to precutting levels within ten years (Hornbeck et al. 1987; Hornbeck and Leak 1992). Additionally, limiting the proportion of a watershed cut at any time, reducing the size of individual clearcuts, and leaving uncut streamside buffer strips will reduce the magnitude and duration of nutrient outflow (Martin and Pierce 1980). In fact, clearcutting only 15% of a northern hardwood watershed in New England, and 13% of an aspen-covered one in Utah, did not affect the water chemistry of major streams flowing from them (Johnson 1984; Martin et al. 1986).

Slope and aspect may amplify the temperature changes and affect internal soil

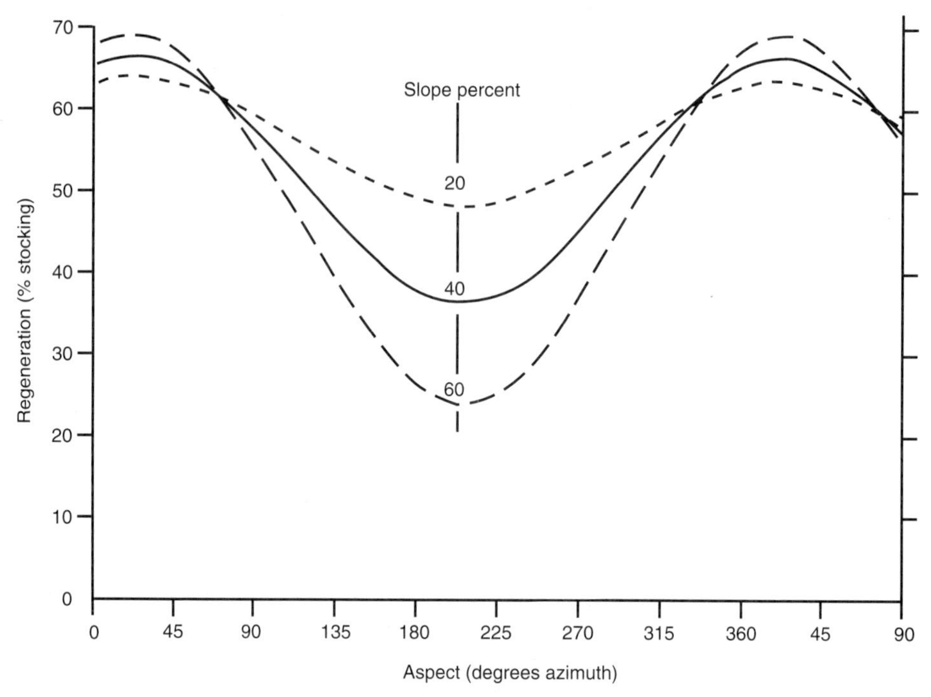

FIGURE 13-3
Effects of aspect and slope steepness on success of regeneration following clearcutting in southwestern Oregon in North America (from Graham et al. 1982).

drainage to the detriment of the regeneration. Particularly in geographic areas having potentially limiting growing-season climatic conditions and highly variable physical site features, foresters cannot easily predict how clearcutting will affect the microenvironment. To illustrate, along the Pacific Coast of Oregon, clearcutting succeeds best at sites with certain herbaceous indicator plants (suggesting favorable inherent site conditions), and on sites facing from northwest to east (azimuth 315° through 90°), independent of slope steepness (Graham et al. 1982). On the more southern-facing sites, success decreases on slopes greater than 20% (Fig. 13-3). With Engelmann spruce in the central Rocky Mountains, even well-prepared seedbeds prove unsuitable on unshaded south slopes. By contrast, prepared surfaces in clearcut openings as wide as 200–350 ft (61–107 m) will restock adequately within five years on north aspects (Alexander 1974, 1984, 1986d). Western white pine communities show similar effects of slope orientation (Haig et al. 1941). Also, mixed-conifer communities in mountainous areas of eastern Washington and Oregon regenerate best on north and east aspects, and within about 330 ft (101 m) of an edge having western larch or lodgepole pine trees of seed-bearing ages (Seidel and Cochran 1981). This suggests that even where foresters limit the size of a clearcut to fit the seed dispersal patterns of the target species, the method will often fail as a natural reproduction on steep and dry sites in mountainous regions of northwestern North America.

As a new community of herbs, shrubs, and tree seedlings develops, moisture and temperature gradually shift back toward their original levels. In fact, by the time a tree canopy closes over the site, environmental conditions at the ground once more resemble those of a precut stand. Shading reduces surface temperatures, and helps to increase humidity at the forest floor. Transpiration once more reduces moisture in the upper horizons, and decreases the water and nutrients that leave a site. Decomposition decreases in the cooler understory environment, and the new community adds litter to the surface. The coarse woody debris left during overstory removal continues to decompose into humus, replacing some of the surface organic material that decomposed immediately after clearcutting. These processes continue as a stand matures, stabilizing the understory environment throughout most of the life of an even-aged community.

ACCOMMODATING SPECIES TOLERANCES

From an ecologic perspective, clearcutting must alter light, moisture, and other physical site resources in favor of prompt germination and sustained growth by a new plant community of desired floristic composition. Light and other critical environmental conditions must remain within safe limits, and match the optimal range of the target tree species. Otherwise the regeneration will fail altogether, not have adequate density or spatial uniformity, or lack the species sought by a landowner.

As long as a clearcutting improves essential environmental conditions, its maximum size depends largely upon the available seed supply and the seed dispersal patterns of desired species. In most cases, the safe maximum width should not exceed 300 ft (91 m) (see Table 13-1). Exceptions include stands where the litter layer or logging slash contains sufficient viable seed to meet seedling density requirements, or where advance regeneration, stump sprouts, or root suckers will supplement the new germinants. Particularly among heavy-seeded species (e.g., oaks and hickories), short seed dispersal distances may preclude clearcutting in favor of other even-aged reproduction methods. With artificial regeneration, clearcuttings could conceptually cover whatever space a landowner can reasonably replant or reseed in a single operation, as long as the physical and biological environments remain suitable for the long-term growth and development of a new plantation.

While many species have a fairly wide ecologic amplitude and a considerable biologic potential to regenerate under a wide range of environments, others grow well only under more restricted conditions. Foresters must know these silvical requirements. To illustrate, one classification system for conifers in northwestern North America separates three broad groups of species based upon protection requirements (Klinka et al. 1990; Klinka and Carter 1991):

1 shade-intolerant species that require bright conditions, and need little protection;

2 shade-tolerant and midtolerant species that withstand or benefit from high light levels, and also need little overhead protection; and

3 shade-tolerant and midtolerant species that survive and grow poorly if not protected to some degree during early stages of development.

Clearcutting would work with the first two groups, given no limiting site conditions (particularly of soil moisture) and an adequate seed source. With the last group, well-distributed and adequately developed advance regeneration may withstand complete overstory removal, whereas new germinants fail. For stands lacking advance seedlings of these species, planting must often follow clearcutting, or foresters should use an alternative reproduction method. They must follow clearcutting by site preparation and planting in any stand lacking an adequate source of seed, or where competing vegetation would impede natural seedling establishment and development.

To shelter seedlings for a few years, foresters normally leave widely spaced overstory trees to partially shade the site. Even logging slash mitigates surface temperatures to an important degree, if uniformly distributed and of adequate density (Edgren and Stein 1974; Ronoco et al. 1984; McInnis and Roberts 1991). This *dead shade* (Place 1955; Ryker and Potter 1970; Edgren and Stein 1974; Geier-Hayes 1987) will (1) reduce peak surface temperatures by as much as 70° F (38.9° C); (2) reduce temperature and moisture fluctuations within upper soil layers; (3) retard wind movement near ground level; and (4) slow surface evaporation and transpiration losses. This reduces mortality during succulent stages of seedling development. Dense slash may also protect small advance seedlings during skidding operations (McInnis and Roberts 1991).

ALTERING THE CONFIGURATION OF CLEARCUTTING

Foresters use clearcutting primarily in community types having desirable species that readily regenerate, adequately tolerate, and rapidly develop in environments with full brightness, high surface temperatures, and elevated soil moisture and temperature. They may also clearcut to facilitate intensive site preparation in stands with appreciable ground vegetation or pest problems that might preclude success with partial cutting. They have no other option for silvicultural systems incorporating tree planting to regulate spacing, species, or genotype.

Sites with a frequent growing-season precipitation and soil water storage capacity will support higher rates of surface evaporation, and this helps to modify temperatures at and near the surface. At some drier sites, the conditions could prove lethal to new seedlings, and foresters must often use another reproduction method, or alter the clearcutting. Also, clearcutting on poorly drained soils will reduce transpiration and might raise the water table. The saturated surface could inhibit germination and early survival of many tree species (Lees 1970). This condition would also force foresters to alter the clearcutting to keep environmental conditions within safe limits. In addition, they must consider how a clearcutting might enhance or detract from a variety of nontimber values, including the habitat for wild plant and animal communities, and other ecologic values of importance.

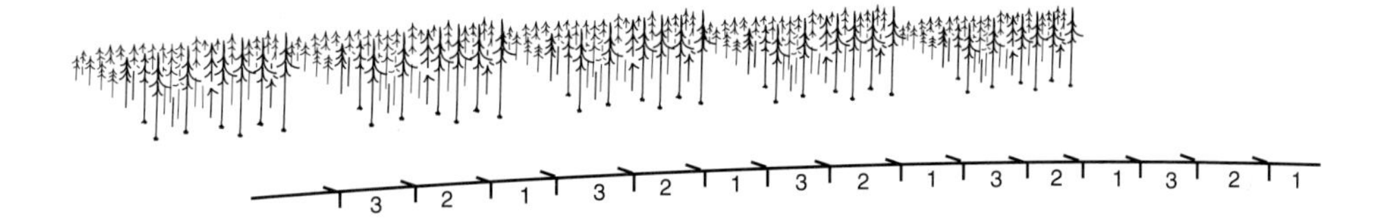

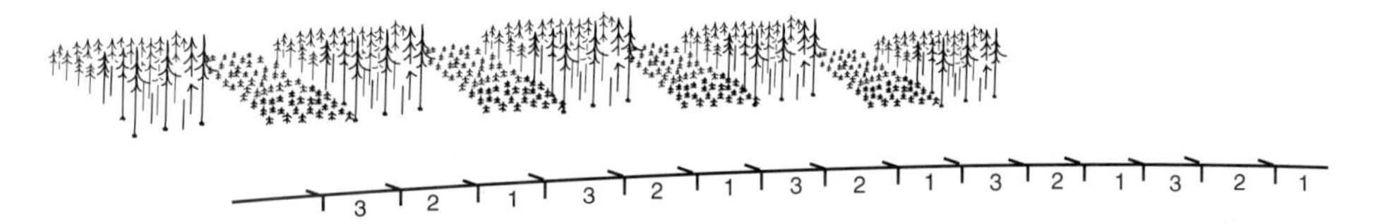

FIGURE 13-4
Foresters can apply clearcutting by alternate or progressive strip or patch methods to create a new even-aged community over the entire stand area within a period of time not exceeding 20% of the intended rotation.

In addressing these concerns, foresters have several options available. These include applying clearcutting in either large blocks that cover an entire stand at one time, and conform to its existing size and shape; or strips or patches that cover only a proportion of the stand area at one time, with the remainder removed by subsequent cuts extending over no more than 20% of intended rotation length (Hawley 1921; Troup 1928; Cheyney 1942; Daniel et al. 1979; Smith 1986; Matthews 1989). Foresters can clearcut strips and patches covering one-half the stand during the first entry and the remainder at an appropriately timed second cutting (Fig. 13-4). They call this the *alternate-strip* or *alternate-patch method.* They could also clearcut a progressive series over three or more entries, covering an equal area on each occasion. Foresters call this the *progressive-strip* or *progressive-patch* clearcutting method. Foresters regulate strip or patch size to control ecologic conditions within the clearcut areas (e.g., seed dispersal, or side shading of the surface), but also to accommodate operational requirements of the logging. For shade-intolerant species, each opening must cover sufficient area to increase light and moisture adequately at the center. Then foresters must expand these openings after a few years (e.g., with alternative and progressive strip or patch clearcutting). Otherwise they lose the species that require high levels of light for long-term survival, growth, and development.

By regulating strip or patch size and orientation foresters can simultaneously use side shade from the adjacent uncut area to reduce temperatures at and just above the forest floor. This often proves critical to seedling establishment. Foresters can even cut narrow strips or patches to maintain shading across the entire open area. To il-

FIGURE 13-5
With progressive strip clearcutting, foresters can open fairly narrow areas to maintain partial shading over the regeneration area, or to protect against blowdown, and move across an entire stand in a period of time not exceeding 20% of the intended rotation. First-cut strips opened one-third of the area to regeneration in the three-step process depicted here. Once regeneration becomes established in the first-cut strips, foresters remove a second series, and then return to complete the process after the second strips regenerate successfully.

lustrate, at noon-time on level ground at 44° N latitude the summer shadow of border trees extends about 2 ft (0.6 m) for every 5 ft (1.5 m) of their height, and for about 7 ft (2.1 m) in morning and afternoon. Shadow lengths from south-bounding trees increase by about 1.5 ft (0.5 m) for each added 10% of slope on north aspects, and decrease by similar amounts on south slopes (Marquis 1965). Once the seedlings no longer need the protection, foresters clearcut a new strip or patch along the south side of each original opening (alternate and progressive strip or patch clearcutting). This increases light around the established seedlings and promotes their development (Fig. 13-5). Simultaneously, it sets the stage for another band of regeneration in the shaded environment of each newly cut strip or patch (Marquis 1966; Gordon 1979). The design must simply allow contractors to remove sufficient volume for a commercial harvest, and permit efficient operation in the process.

Figure 13-6 demonstrates effects of patch size, shape, and orientation on light patterns (from Marquis 1965). These diagrams also suggest some effects of extending fairly short openings into long strips. To illustrate, orienting a fairly narrow clearcut strip (e.g., not wider than the height of adjacent trees) with the long axis running east and west would maintain continuous shade over the southern portion, but provide full sunlight along the northern half during the hottest part of a summer day. Orienting a narrow strip north-south leaves more of the area with no shading. Even so, north-south strips cut to only 25% or 50% of the height of adjacent trees increase daily illumination to 21% and 40%, respectively, of full sunlight (Table 13-2). In strips equivalent to the average height of adjacent trees, sunlight increases to 75% of that in the open (Berry 1964).

Experiments with openings of different widths from one-quarter to two times tree height showed differing effects on light, soil moisture, and regeneration development. Within patches no more than three-quarters tree height, the northern part will receive the most light, followed by the center, west, east, and south portions. Openings of at least one tree height have greatest light levels at the center, followed by north, west, east, and south parts. Illumination toward the middle will increase as the growing season progresses. By contrast, light will not change much in openings of only one-half tree height. Patches on south slopes receive the highest insolation. Also, soil moisture will increase from its lowest level under uncut parts of the stand, to intermediate levels just inside an opening, and become highest at the center. Conditions for tree growth will vary from least favorable just inside small openings, to best at centers of openings equivalent to two times the heights of border trees (Minckler et al. 1973).

Tests using clearcut strips in mixed hardwood-conifer stands on coarse-textured soils in northeastern North America demonstrate the effect (Smith and Ashton 1993). Within east-west strips slightly wider than the height of border trees, most shade-intolerant species (e.g., pin cherry, paper birch, and grey birch) developed poorly along the shaded southern edges. These same species became dominant at fully lighted parts of the strips. Longer-lived and more shade-tolerant species (e.g., red maple, yellow birch, and black birch) persisted in the subcanopy within brighter

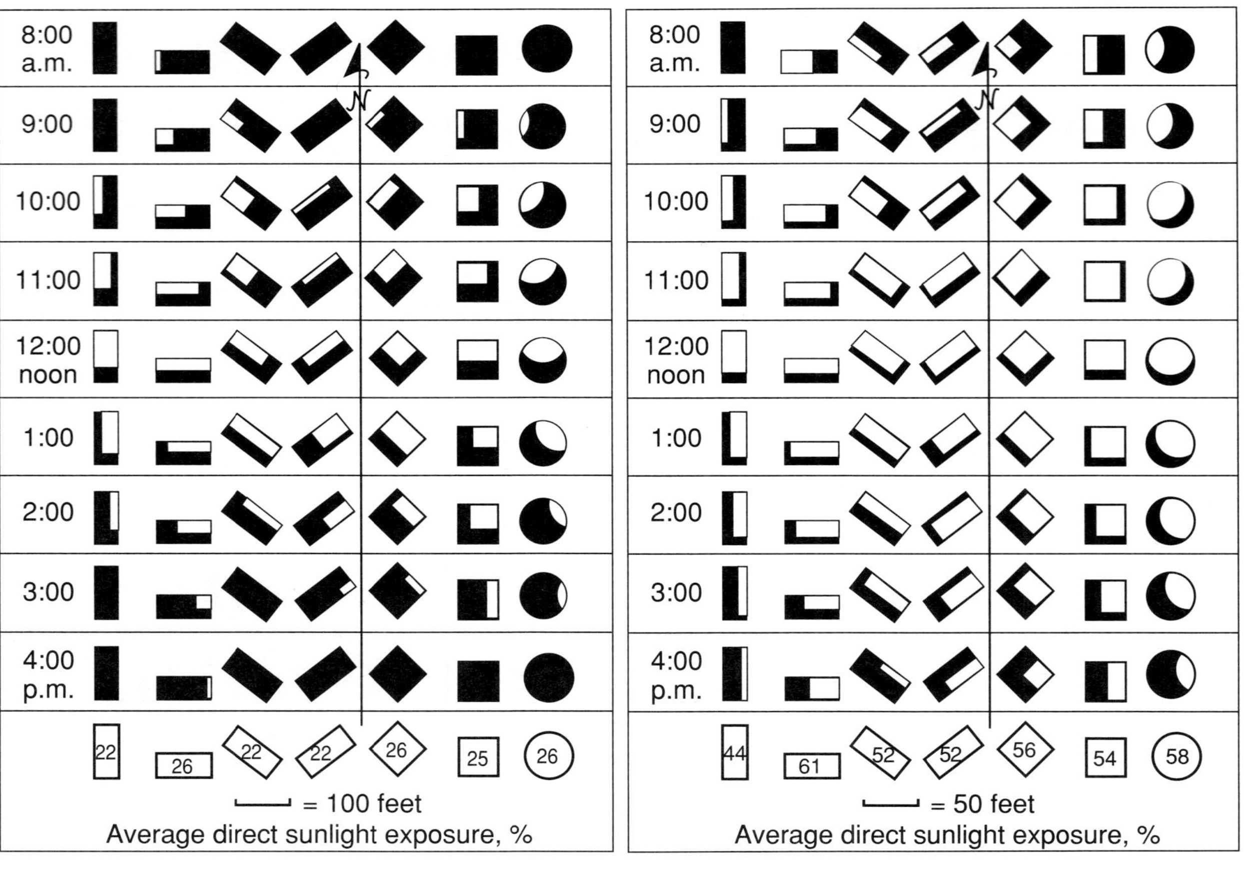

Patterns of light and shade in various 1/10-acre patches - June 7, latitude N. 44°, at different times of day

Patterns of light and shade in various 1/2-acre patches - June 7, latitude N. 44°, at different times of day

TABLE 13-2
EFFECT OF STRIP WIDTH ON THE PROPORTION OF DAILY SUMMERTIME ILLUMINATION REACHING THE GROUND IN CLEARCUT STRIPS AT 46° N LATITUDE, AND ORIENTED NORTH-SOUTH (BERRY 1964).

	Width as % of adjacent tree heights		
Position in the strip	**25**	**50**	**100**
		—% full light—	
Center	21	41	97
Intermediate	21	40	76
Edge	21	38	63
Average across strip	21	40	74

portions of the strips. These same species dominated only the side-shaded zones. Similarly, clearcut strips in the upper Great Lakes region of North America had more shade-intolerant species in the central portions, and in those cut 122 vs. 66 ft (37 vs. 20 m) wide. Yet neither strip orientation nor distance from the uncut edge, per se, affected individual species' growth rates (Metzger 1980). Likewise, in northern hardwood stands of New Hampshire's White Mountains, shade-intolerant pin cherry comprised only 5% of the stems at age 10 within progressive strips made 1.25 times the height of dominant trees, compared with 22% in the block clearcuts. The strips also had generally lower stem densities, and larger diameter trees (Martin and Hornbeck 1990). Such findings suggest a potential for using relatively narrow strips to temper the overall species composition and minimize the relative importance of fast-growing pioneer species.

Narrow strips or rectangular openings oriented at right angles to the prevailing winds would also optimize seed dispersal from wind-disseminated species (see Table 13-1). In fact, experiments among western white pine in the northern Rocky Mountains showed that strips not exceeding 300–450 ft (91–137 m) wide regenerated promptly, and with abundant seedlings, at all but exposed south-facing slopes. With larger openings, regeneration did not succeed more than 400 ft (122 m) from a seed source (Haig et al. 1941). Foresters have also used the progressive strip and patch clearcutting methods to safeguard against blowdown at windthrow-susceptible sites (e.g., exposed slopes, or sites with poor root anchorage due to shallow, poorly drained, or rocky soils). They use fairly narrow strips or patches with the long axis oriented 90° to the direction of prevailing winds, and make each progressive cutting on the upwind side of the previous opening (Hawley 1921; Troup 1928;

FIGURE 13-6
Patterns of light and shade in clearcut openings of two shapes and sizes (from Marquis 1965).

Ruth 1976; Smith 1986; Matthews 1989). Additionally (Alexander 1986a, 1986b, 1987b) they can keep clearcut edges away from areas with

1 topographic features that funnel the winds and accelerate the velocity (e.g., in saddles along ridges, or indentations on slopes);
2 ridge lines in hilly or mountainous country;
3 poorly drained and shallow soils within a stand;
4 signs of old blowdown or pronounced hummock and hollow surface topography; and
5 concentrations of unsound trees with decayed roots and trunks.

In general, clearcutting works better than partial cutting in areas with a high risk of blowdown. Use of strips allows foresters to accommodate other concerns of equal or greater ecologic importance.

In practice, the choices between strip, patch, or large block clearcutting, and their configuration and sizes, often reflect a compromise between the desire to manage shading patterns, safeguard against blowdown, and insure adequate seed dispersion across a regeneration area. Openings up to two times tree height generally have more moderate microclimates and better coverage by wind-disseminated seeds (Metzger 1980; Ffolliott and Godfried 1991; Smith 1995). They also provide better security against losses on windthrow-susceptible sites. Even so, strip or patch cutting must remove sufficient volume for efficient harvesting, and the widths often depend upon the kinds of machinery that a contractor will use for the logging.

EFFECTS ON SOME OTHER ECOSYSTEM ATTRIBUTES

Where landowners want to address certain ecologic or other nonmarket interests, foresters must often regulate the extent of clearcutting and find appropriate ways to disperse them across a landscape in time and space. Not clearcutting in setback zones along streams, ponded water, special plant and animal habitats, and difficult-to-regenerate sites often accommodates many special ecologic needs. Between the clearcuts, foresters can also leave areas of tall-tree cover to (1) serve as refuges for sensitive plants and animals; (2) provide habitat for animal species that derive critical needs from tall forests; (3) serve as travel corridors for birds and mammals that move long distances in fulfilling their basic needs; or (4) influence visual qualities of frequently seen portions of a property. Ecologically, these residual communities must cover adequate area (minimum widths, and adequate overall sizes) to provide true tall-forest environments, or with adequate edge to sustain plant and animal species that depend upon those conditions. Areas of tall-tree cover also diversify the patterns of color and texture from critical viewing points. Usually, needs like these require planning at a landscape scale, rather than at the stand level.

For many ownerships, and particularly public lands, foresters will often alter the shapes of block clearcuttings for visual effects, and disperse them to create a mosaic of tall-forest and newly regenerated stands. This makes the total area treated by clearcutting at any time less apparent. For large clearcuttings, they avoid straight sides and square corners, and blend the shapes with topographic features (Hough,

FIGURE 13-7
For large clearcuttings, foresters can avoid cutting straight edges and making square corners, since these features would have a stark and unnatural appearance in most forested landscapes (courtesy of U.S. Forest Service).

Stansbury & Assoc. Ltd. 1973). Otherwise the clearcuts will have a stark and unnatural appearance in most forested landscapes (Fig. 13-7). In some cases, they might prefer to limit the size of clearcuts in highly visible areas (e.g., smaller than 10-15/ac [4-6/ha]), and disperse them to reduce the apparent cutting intensity (Schroeder et al. 1993; Palmer et al. 1995).

Often, strip and patch clearcutting allows foresters to address a broad array of ecologic, visual, commodity, and other concerns. These methods extend the time to convert a tall forest stand to a new young community. This also prolongs the recovery of products from limited-area ownerships while still capitalizing on the operational and biologic characteristic of clearcutting. Foresters can also curve the strips to shorten lines of sight, or fit them to topographic conditions. They can use irregularly shaped edges rather than straight ones, and disperse the cut areas in either systematic or irregular patterns across the stand. They can also wait for regeneration on one strip or patch to reach rather large sizes (e.g., eye level) before cutting in other parts of a stand. These approaches soften the visual starkness of recent clearcuts.

Foresters have often used clearcutting for wildlife habitat management. It produces a dense and species-rich assemblage of vegetation, and abundant leafy biomass close to the ground. This provides a diversity of food sources for many creatures, and dense cover and concealment for some. Yet the size, shape, and widths of clearcuts influence the distance to escape cover, and their configuration and area tend to affect the ways mammals use the area for feeding and other purposes. In fact, evidence suggests that female black bears with cubs will not venture far from trees large enough to climb (Rogers 1987, 1988). Deer and elk also tend to avoid the center of a large opening when feeding, and will not benefit from food resources in the middle of large clearcuts (Lyon 1977; Hunter 1990). By contrast, moose freely use the entire portions of even large clearcuts, so opening size neither limits their feeding nor proves effective as a means for reducing any harmful browsing effects (Hunter 1990). On the other hand, the deeper snow inside clearcuts may impede the movement of both deer and moose, limiting their access to winter browse (Crawford and Frank 1987).

Factors like these force foresters to make careful choices about the ways they use clearcutting, and how they arrange them across a forest. Large clearcuts may prove advantageous where browsing would threaten the success of a reproduction method. Concentrating a series of clearcuts within a limited area also helps to reduce browsing impacts by providing more food resources than the creatures can use. On the other hand, with strip and patch clearcutting, foresters can maintain nearby escape cover and make the cutting area more accessible to the animals. The uncut strips also preserve tall-tree travel corridors and provide visual screening until the new communities develop. Further, until the final cut, stands will have areas with tall and young trees adjacent to each other, providing a diversity of habitat in close proximity.

Some other values accrue from openings of specific dimensions and shapes. In these cases, foresters seek prompt regeneration to insure a long-term continuity of forest cover, but often capitalize on the clearcutting for other purposes. To illustrate, in mountainous regions of western North America, watershed managers make strip or irregularly-shaped patch cuts to trap snow during a winter (Figure 3-1). Opening width varies with the community type, elevation, aspect and slope, and other physical site conditions (Ffolliott et al. 1965; Troendle and Leaf 1980; Troendle 1983; Kattelmann et al. 1983; Krutilla et al. 1983). These range from about two times tree height among ponderosa pine communities in Arizona (Ffolliott et al. 1965), to five times tree height for subalpine stands of mixed conifers in Colorado (Troendle and Leaf 1980; Troendle and Meiman 1984). Managers orient the strips or long axes of the patches at right angles to the prevailing winds to create an eddy at the clearcut edge. Blowing snow collects in these calm places. The cuttings also reduce the amount of snowfall intercepted by tree crowns. Protection and shading by border trees reduce sublimation of the snow pack, increasing its water content, and retarding snow melting in springtime. In composite, this increases total water yield, extends the period of runoff during spring, and reduces the peak flows and potential for flooding of adjacent streams. However, among conifer stands of the Rocky Mountains, the water yield will decline to about one third of the postclearcutting level within 25 years. Among aspen stands the effect diminishes within 5 years (Troendle 1982).

SITE PREPARATION AND OTHER ANCILLARY TREATMENTS

With some forest community types, natural seedbeds and logging slash will not inhibit natural regeneration of desirable species. In fact, the debris of branches and other unused parts of trees may actually discourage the movement of deer and other browsing animals, and help to protect emerging seedlings from browsing (Grisez 1960). Shade cast by the slash, a small object like a stump or rock, or herbaceous plants also helps to improve the environment on sites with limited moisture and high surface temperatures (Ruth and Harris 1979), even for advance regeneration. Under these circumstances, foresters may opt for no, or only partial, site preparation.

In other cases, site preparation may actually prove critical. Then foresters must follow the clearcutting by one or more necessary ancillary treatments. For example, red and white pine germinate best in a mineral soil or humus seedbed, so that scarification improves chances for success (Haig et al. 1941; Blum 1990; Wendel and Smith 1990). Also, in the northern Sierra Nevada of California, natural regeneration of mixed-conifer communities depends upon site preparation to reduce slash, expose mineral soil, reduce competition from broadleafed species, and control seed-eating mammals (McDonald 1983). Likewise, pales weevil will damage new seedlings of several conifer species in western North America. This insect uses slash and downed branches as a habitat for its reproduction, and the population often builds in the dense slash left during logging. Site preparation to reduce the density and distribution of downed branches and promote their decay helps to control the weevil and protect the seedlings during susceptible stages. Slash reduction may also reduce the habitat for other harmful insects, and even some mammals that eat tree seed or feed on young trees.

Vegetation control also often proves critical to the success of a reproduction method. Undesirable tree species, shrubs, and herbs may delay or interrupt recruitment and dominate the new age class. For example, within northwestern North America, interference by weed species has resulted in a loss of conifers on about 10% of the commercial forest area (Walstad et al. 1987). The problem reflects a capacity of the weeds to (1) better survive disturbance; (2) resprout if broken off or killed back; (3) germinate from large quantities of buried seed; and (4) derive from new seed disseminated from nearby sites. These plants often slow the regeneration of desirable conifers unless vegetation control treatments (site preparation) reduce the interference.

Postharvesting weed control has become important in other regions of North America as well. The reasons differ from one area to another, but the need seems fairly widespread. To illustrate, in Southeastern North America, fire control has led to a buildup of hardwood understories, and these often dominate a site following removal of the overstory. As a consequence, low-quality hardwoods now occupy many areas that previously supported pine forests. Also 40%–70% of managed forest lands have weed control problems. Herbs often deter the initial establishment of pines, but hardwood competition intensifies by about the fifth year. Where pines regenerate in the absence of site preparation or postcutting hardwood control, the stands have low stocking and the trees grow slowly (Gjerstad and Barber 1987). By contrast, loblolly and shortleaf pine seedlings kept free of competition for a five-year period will have greater heights, better live-crown ratios, and more volume per tree than those on untreated sites (Cain 1991a).

Across the Northeast and Lake States regions as well, hardwoods now occupy vast areas that once supported conifers. These attest to problems resulting from interference by hardwoods, raspberries, grasses, and sedges. Landowners must often use site preparation to control existing plants, followed by postestablishment release to keep the conifers free to grow (Newton et al. 1987). Even among hardwood communities, species like pin cherry, striped maple, American beech, some ferns, and some grasses interfere with the commercially desirable species. Foresters can often identify the critical sites in advance (Marquis et al. 1992; Tubbs 1977; Farnsworth and Barrett 1966). Application of herbicides to the understory has promoted a more diverse and more favorable species composition (Metzger 1980; Kelty and Nyland 1981; Sage 1987; Horsley and Marquis 1983; Loftis 1985).

Programs to control interfering vegetation should span all phases of an even-aged silvicultural system (after Walstad, Newton, and Gjerstad 1987; Stewart 1987; Tappeiner and Wagner 1987). They may begin with early tending to insure full site utilization by healthy trees of desirable species, and protection against harmful agents. Later treatments may reduce any problem vegetation, insure dominance of desirable trees, and reduce interfering plants (Fig. 5-2). At stand maturity, foresters must use an appropriate reproduction method and good logging practices, necessary site preparation, and prompt reforestation (natural or artificial). In many cases they must also return during early stages of community development to keep the desirable species in a free-to-grow position. This involves more than simply killing undesirable vegetation in conjunction with the reproduction method. In fact, site preparation comprises only one part of comprehensive vegetation management plan that foresters initiate with the even-aged reproduction method, and continue through subsequent stand development. It must fit the ownership objectives, and needs only to suppress the vegetation that would interfere with the species of interest.

CLEARCUTTING WITH ARTIFICIAL REGENERATION

Natural regeneration methods usually cost less than artificial techniques. They often improve the profitability—given a suitable species and appropriate stocking across the regeneration area. Nevertheless, the long-term financial and nonmarket benefits depend upon the species that regenerate, and the density and distribution of trees in the stand. If sufficient desirable seedlings seem likely to become established, foresters will not need to intervene by tree planting or direct seeding. However, when owners must have some minimum stocking of a particular set of species, and the prospects for securing them by natural methods seem poor, foresters will often opt for artificial methods.

The silviculture of longleaf and shortleaf pines in southeastern United States illustrates cases where foresters often incorporate tree planting as part of a reproduction method following clearcutting. Longleaf pine has heavy seed that does not disperse far from a parent tree, remains viable in the duff for only a short period, and germinates best on a mineral soil seedbed. Abundant regeneration usually becomes established no more than 100 ft (30 m) from a seed source, and during years with bumper seed crops and low predator pressure (Crocker and Boyer 1975; Boyer and White 1990). By contrast, shortleaf pine seed disperses up to about 200–300 ft (61–91 m). It regenerates better on a mineral soil seedbed than in organic litter. Germination may fail in a poor seed year, or during hot and dry weather. Good seed years come at three- to ten-year intervals (Baker 1992; Shelton and Wittwer 1992). With both species, these critical factors make natural regeneration risky, and foresters frequently recommend planting or direct seeding when landowners use clearcutting as a reproduction method.

Linking artificial regeneration with clearcutting in these circumstances has certain economic or ecologic advantages. It provides foresters with the most certain means for securing prompt regeneration of desired species (after Hawley 1921; Cheyney 1942; Brender 1973; Daniel et al. 1979; Matthews 1989), even where environmental conditions might prove unsuitable for new germinants. In some large clearcuts, they might rely upon natural regeneration along edges of a cutting area where seed dispersal and environmental conditions favor natural regeneration, and plant the central portions to insure adequate stocking throughout. Even if indigenous species proved suitable, a landowner could introduce trees from selected sources to enhance the production potential, or the resistance to pest problems. Planting or direct seeding would best serve these interests. Similarly, a landowner might add a missing species by enrichment planting or spot seeding at relatively wide spacing, or supplement the natural regeneration by reinforcement planting of nursery-grown trees. This could include areas devoid of natural regeneration. In these cases, planting stock of adequate quality and size should become part of the main crown canopy (Needham and Clements 1991). In fact, planting two years after clearcutting with site preparation by slash burning has increased the proportion of Douglas-fir within treated areas. Thirty to 40 years later the planted trees also had larger diameters and 40% more volume than the natural regeneration (Miller et al. 1993).

In some regions and ownerships (e.g., with production forests) foresters use mechanized tending operations, and elect planting to carefully control spacing and arrangement (see Chapters 6-8). They also clearcut the stands at maturity, and use intensive site preparation (including fertilization) and tree improvement to maximize fiber yields. They usually convert from indigenous species to those better suited to local markets, or to genotypes with better growth or resistance to local pests and diseases. This, too, necessitates artificial rather than natural regeneration methods.

In circumstances like these, planting or direct seeding allows foresters to better control regeneration success or influence the characteristics of stands that develop following clearcutting. They add flexibility to capitalize upon special landowner interests (after Cheyney 1942; Brender 1973; Daniel et al. 1979; Smith 1986), including

1 altering species composition and genetic constitution to serve selected markets, or enhance tree growth;
2 planting nursery-grown seedlings to overcome site constraints to natural regeneration, and insure prompt stocking of desirable species;
3 introducing genotypes resistant to indigenous pests and diseases;
4 regulating spacing to promote more rapid diameter growth;
5 controlling spacing and arrangement to facilitate eventual tending, harvesting, and other operations; and
6 introducing well-developed seedlings to insure early tree domination at the site.

Especially with direct seeding, foresters must prepare the seedbed, and sometimes also reduce competing vegetation (see Chapter 8). With tree planting they must often reduce the logging slash to facilitate the work or reduce pest problems, and may also need to control interfering vegetation. In fact at droughty sites like those on upper slopes in the Cascade Range in Oregon and Washington, success depends upon *prompt* control of herbaceous vegetation (Dimock 1981). Prompt action also often reduces the colonization of interfering vegetation, or its regrowth, following overstory removal (Matthews 1989). However, these measures increase the costs, and the resulting community has a structural simplicity and much less ecologic complexity than naturally regenerated stands.

OPERATIONAL CONSIDERATIONS IN CLEARCUTTING

Clearcutting removes large volumes of biomass that accumulated in a stand over many years, and converts that volume into revenues. This makes clearcutting commercially attractive. In some timber types, harvesting merchantable trees actually costs about the same for clearcutting and other reproduction methods. Cost differences may really reflect the amount of time spent felling unmerchantable trees, or time lost to other nonproduction activities (Filip 1967; Nyland et al. 1971). Landowners must pay the costs of these treatments, and for site preparation and other regeneration practices, from the value of a sale, or carry them as an investment until the new stand will support an operable cut of useful products. On the other hand, important cost savings accrue from needing only simple and fairly inexpensive steps to control the cutting (Anderson et al. 1990). Foresters must only identify the stand boundaries, mark the location of skid trails, and delineate edges of buffer strips along water bodies and other sensitive habitats. Most contracts also require the contractor to fell unusable stems, protect soil and water resources, and fulfill other requirements to protect advance regeneration. By recovering high yields from a concentrated area, and handling a large volume of logs from a single setup of equipment, contractors can often pay higher stumpage prices, or even operate in inaccessible areas that cost more to log. Yet to optimize revenues, contractors must find markets for trees of different sizes and qualities.

For any clearcutting, foresters must protect soil and water resources, the viability of habitats for threatened and endangered plant and animal species, the integrity of unique environments, and several other noncommodity values associated with forests. They must not let skidding disturbances accelerate base erosion to unacceptable levels, nor cause soil instability. In fact, except where a regeneration plan calls for special site preparation, they will protect litter layer to

1 guard against mineral soil compaction;
2 preserve the stored seed;
3 maintain the nutrient pool represented by the undecomposed organic material;
4 maintain high moisture-holding and infiltration rates; and
5 preclude surface erosion by overland flow.

Proper design, layout, and maintenance of the skid trails and haul roads will guard against these potential risks. In addition, prompt restocking will also help to stabilize the site by increasing transpirational use to reduce subsurface flow, and replenishing the litter by adding new organic material.

VALUES AND LIMITATIONS OF CLEARCUTTING AS A REPRODUCTION METHOD

When applied appropriately, clearcutting will promptly restock a mature even-aged stand with desirable species at adequate density. The new community will develop rapidly, and normally contain a wide variety of species. Altogether, removing the mature stand in a single step has some distinct advantages (after Hawley 1921; Troup 1928; Knuchel 1953; Kostler 1956; Daniel et al. 1979; Seidel and Cochran 1981; Smith 1986; Matthews 1989; Anderson et al. 1990; Barnett and Baker 1991), including the following:

1 high yields per unit of area potentially lower the harvesting costs;
2 setup and control require few technical skills, except for the skid trail system;
3 brightness of the area will sustain even the most shade-intolerant species, and promote the rapid growth of most species;
4 cutting all trees facilitates site preparation to control pests and competing vegetation, improve seedbeds, and ameliorate soil deficiencies (e.g., by cultivation, drainage, or fertilization);
5 easy access by machines simplifies artificial regeneration;
6 clearcutting controls pests that damage older trees left by partial cutting (e.g., dwarf mistletoe);
7 clearcutting facilitates natural regeneration of species with serotinous cones;
8 clearcutting precludes blowdown of residual trees, and removes decadent trees from a site;

9 the density of the new community promotes early lower branch mortality, formation of long clear boles, and less taper on survivor trees;
10 herbaceous vegetation and woody shoots close to the ground provide abundant food and excellent cover for many birds and small mammals;
11 limiting the regeneration period to a small part of the rotation facilitates later uses, such as grazing or recreation; and
12 when applied systematically across a forest ownership, clearcutting creates well-defined age classes in distinct stands, simplifying the management for even-flow sustained yield from a forest.

To capitalize on its benefits, foresters must limit clearcutting to sites suited to its use, and with abundant supplies of seed. They must determine in advance that nothing will preclude regeneration establishment or development.

Clearcutting also has certain potential limitations, particularly when used as a natural regeneration method. These (after Hawley 1921; Troupe 1928; Cheyney 1942; Eyre and Zehngraff 1948; Kostler 1956; McDonald 1976; Daniel et al. 1979; Seidel and Cochran 1981; Smith 1986; Hannah 1988; U.S. Corps Eng. 1988; Matthews 1989; Anderson et al. 1990; Barnett and Baker 1991; Tesch and Mann 1991) include the following:

1 landowners must depend on stored seed, and that dispersed into the site from adjacent sources;
2 any shortage of seed on site limits regeneration to light-seeded species, barring an unusual dispersal mechanism;
3 the abundance and uniformity of any particular species, and the spacing and species composition of a new stand depends on an uncontrollable seed supply;
4 dependence on seed trees in adjacent stands and seed already stored on site reduces the chance to control the seed source for genetic improvement of the new community;
5 cutting during poor seed years may lead to failure or irregular stocking, and particularly with species having a distinct periodicity for seed production;
6 the open environment may inhibit some species, and will also favor many herbaceous plants that impede the regeneration of desirable trees;
7 dense competing vegetation or harsh soil conditions may require costly site preparation;
8 soils with a shallow depth to the water table may become saturated or waterlogged due to reduced transpiration, inhibiting seed germination and reducing seedling survival;
9 reduced transpiration increases percolation and subsurface flow, accelerates nutrient leaching until a new vegetation cover develops, and increases the chances of mass soil movement on steep slopes;
10 in flattened or concave topography, the lack of overstory protection may increase the chance of freezing temperatures early in a growing season, killing or damaging all but frost-resistant species;
11 at dry sites, the unshaded surface may become unsuitable for many species;

12 disturbance of the surface litter during logging displaces stored seed and increases chances for surface erosion on hillsides, at least until new plants colonize the site;

13 overstory removal precludes a second chance for regeneration if unusual conditions cause failures immediately following clearcutting;

14 prolonged litter decomposition in areas that do not regenerate promptly may change the soil moisture balance and nutritional status;

15 removing all the mature trees, leaving abundant logging slash and fresh stumps, and exposing soil across the area degrades the visual quality for many forest users;

16 abundant dry logging slash increases the fire danger during dry periods, and provides ideal habitat for some harmful insects and small mammals;

17 resulting even-aged communities have less resistance than uneven-aged stands to snow and wind damage; and

18 removing all the large trees eliminates essential habitat for some wildlife.

Foresters must take these potentially adverse factors into account, weigh their importance, and find ways to ameliorate them in favor of the target species and landowner objectives. Otherwise, they will opt for another reproduction method that controls the critical conditions more effectively.

Aside from the potentially important biophysical concerns, clearcutting suddenly changes the visual qualities of a treated area. Removing the large and tall trees alters visual characteristics from those of a closed forest to those of an open area, improving visibility through a stand and increasing the distances people can see. In most cases, the cutting leaves considerable amounts of logging slash or unmerchantable stems and branches. After a short period, needles and leaves on the slash dry and turn brown, lending an unnatural image of devastation to the area. By seeing longer distances—and a greater proportion of the total area within a clearcut—viewers often notice more readily that logging-exposed soil along skidding and yarding corridors, and this may increase their concern about soil erosion. Along the skidding and yarding trails people also see large numbers of fresh stumps that contrast in brightness with the darker slash and exposed soil. Some people equate this with destruction and devastation, at least until a new age class reestablishes the natural greenness that most people associate with forests and undisturbed natural areas.

14

SHELTERWOOD AND SEED-TREE METHODS

A RATIONALE FOR ALTERNATIVE EVEN-AGED REPRODUCTION METHODS

Foresters can devise numerous ways to accommodate potential concerns associated with large clearcuts. Common adaptations (Kostler 1956; Smith 1986; Klinka and Carter 1991; Barnett and Baker 1991) include reserving some mature trees as a seed source; and retaining sufficient residual trees to tamper the physical environment, or the structure and visual qualities of a community. At some point they usually remove the old trees to free the new community from shading. Otherwise, the residuals may oppress the new age class, and keep it from developing at optimal rates (Cheyney 1942; Kostler 1956; Smith 1986).

Historically, foresters have used shelterwood and seed-tree methods to provide these alternatives to clearcutting. Each has unique characteristics. Yet all of the even-aged methods produce a new community of trees, and the spread of ages does not exceed 20% of the intended rotation. Clearcutting may include artificial reproduction methods or natural seeding. Foresters use shelterwood and seed-tree methods primarily with natural regeneration.

CHARACTERISTICS OF SHELTERWOOD METHOD

Foresters use shelterwood method where an inadequate seed supply or a sharp change of environmental conditions might preclude success following clearcutting. Shelterwood method also allows them to temper visual characteristics within regenerating stands, and to maintain essential habitat conditions for selected animals and nontree vegetation. It differs from clearcutting in three principal respects:

1 Foresters leave a relatively low density residual of vigorous seed-bearing trees of good phenotypic character as seed source, and for protective cover.
2 The residual overstory trees provide sufficient canopy cover to mitigate sensitive environmental conditions.
3 Foresters remove the reserve trees once new regeneration of adequate size and density forms, and no longer needs protection.

Some foresters call the residual trees the *overwood*, meaning the same as the terms *residual overstory* or *residual shelter*. Their large size and superior condition, and their numbers per unit area, permit landowners to eventually remove the residual trees commercially, spreading potential revenues from the stand over two closely timed cuttings.

Shelterwood method makes up one part of a shelterwood system for even-aged stands. Later tending will nurture the new community as it develops, and help to concentrate the growth potential onto trees that interest a landowner. In describing this method, foresters commonly use the following terms (after Ford-Robertson 1971; Soc. Am. For. 1989; Smith 1986):

Shelterwood method, or *shelterwood cutting*—any regeneration cutting to establish a new even-aged community under the protection of older trees
Shelterwood system—an even-aged silvicultural system where the reproduction method removes the mature community in two or more successive cuttings, temporarily leaving some of the old trees to serve as a source of seed and to protect the regeneration

The shelter trees remain on site only until a new community of adequate density forms and no longer needs their protection.

Foresters may apply as many as three component treatments in the shelterwood method. These (after Ford-Robertson 1971; Soc. Am. For. 1989; Smith 1986; Hannah 1988) include

Preparatory cutting—an optional initial treatment to increase tree vigor and seed production within a mature stand
Seed or *seeding cutting*—a treatment to establish regeneration throughout the stand area
Removal or *final cutting*—a harvest to take away the overwood so the new community can develop uninhibited

The optional preparatory cutting serves only to improve the vigor of potential reserve trees and to stimulate seed production. Well-tended stands of rotation age usually have adequate numbers of good-vigor trees for abundant seed production. Then foresters can bypass this step. For any preparatory cut, they will retain sufficient stocking to inhibit seed germination and development, except possibly for quite shade-tolerant species.

Seed cutting creates permanent openings in the main crown canopy (Ford-Robertson 1971), and a void in the ecosystem (Fig. 14-1). It elevates light levels near the ground and reduces the withdrawal of soil moisture sufficiently to support

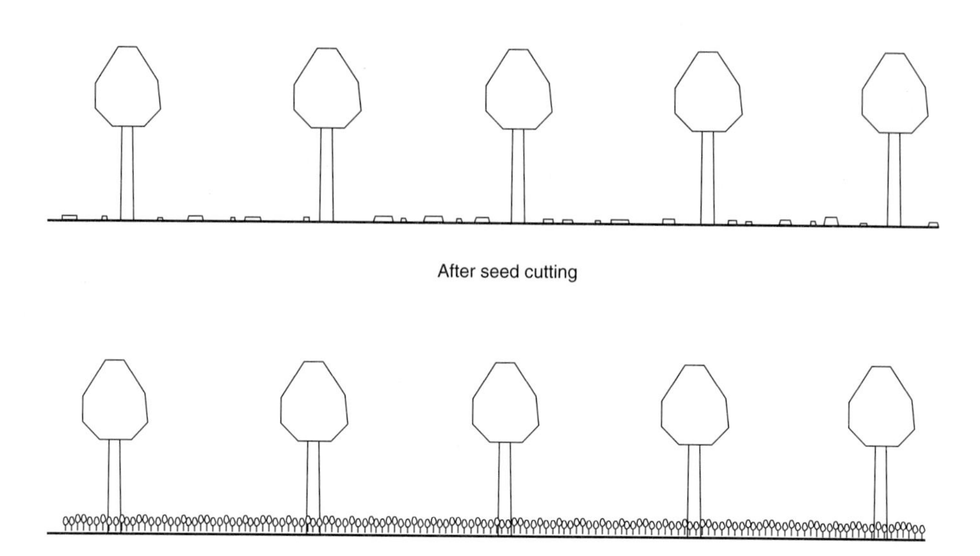

Prior to removal cutting

After removal cutting
(seedling stage)

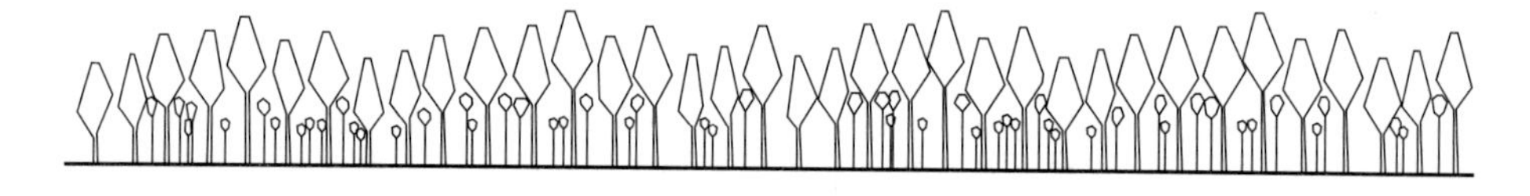

Early stand development
(early pole stage)

FIGURE 14-1
A shelterwood method seed cutting creates permanent openings in the main crown canopy, and establishes natural regeneration underneath the reserve (overwood) of mature trees. The later removal cutting takes away the remaining mature age class, leaving a new cohort of regular constitution that develops into an even-aged community.

a new age class. It serves as the principal mechanism to promote seed germination and seedling establishment underneath the reserve trees. For this seed cutting, foresters must (1) appreciably modify the environment; (2) retain sufficient residual cover of tall trees to modify conditions in favor of the target species, and supply seed for the new community; and (3) keep the environment conducive to seedling development. They may return for a second seed cutting if the new age class develops too slowly or shading becomes too heavy.

Once a new cohort has adequate numbers of desirable trees and they have grown to some minimum threshold size, foresters will remove the overwood by a removal cutting. This takes away the overwood, leaving the new even-aged stand free to grow (Fig. 14-2). When the shelterwood method includes only a seed cutting and a removal cutting, foresters call the process a *two-cut shelterwood method.* In communities where they find an adequate density and suitable advance regeneration to skip the seed cutting, they call it a *one-cut shelterwood method*, a *one-cut overstory removal,* or a *simulated shelterwood cutting* (Alexander 1987). These cuttings resemble clearcutting, except that most of the new age class already grows at a site prior to overstory removal. With other forest community types, the regeneration may take twenty years or more to reach a size and condition suitable for release. Then foresters may make a second cutting to reduce overstory density after a period of years, and return a third time for the removal cutting. Foresters call such a program a *three-cut shelterwood method.*

The actual intensity of shelterwood cutting may vary greatly between stands and localities to accommodate variations in the physical environment, silvical characteristics of the target species, and purposes of an ownership. In all cases the residual trees must (1) adequately control environmental conditions in favor of the target species; (2) have sufficiently wide spacing to brighten the understory; and (3) produce enough seed to completely cover the regeneration area. If foresters retain high-vigor trees of upper crown positions as the residual overstory, a stocking that adequately tempers the physical environment will also usually produce abundant seed. In addition, leaving the residual trees well spaced across a site will insure a uniform distribution of seed, freeing foresters of any biologic requirement to control the size or configuration of a regeneration area.

Seed cutting does elevate the light and temperature near the ground surface, but not as much as after clearcutting. Soil moisture will increase within the upper soil horizons, but to a lesser degree than following a one-step removal of a mature stand. In fact, continued transpiration by the residual shelter trees limits the change, and in proportion to the amount of growing stock left. The cutting will promote decomposition of the soil organic layer and a release of nutrients, but the residual trees take up a portion of those nutrients. This reduces losses through subsurface flow.

CHARACTERISTICS OF SEED-TREE METHOD

Seed-tree method maintains only a few widely spaced residual trees as a source of seed on site, insuring its uniform distribution across the regeneration area (Cheyney 1942; Schmidt et al. 1976; Daniel et al. 1979; Smith 1986; Barnett and Baker

(a)

FIGURE 14-2
Once sufficient regeneration of the required minimum size and consisting of desirable species develops across a site (a), removal cutting takes away the mature age class, leaving a new even-aged stand of regular constitution (b).

1991). Foresters normally use the method with fairly slight-seeded and wind-disseminated species that produce abundant seed and regenerate easily in an open and unprotected environment. It largely frees them from constraints over the size, shape, and orientation of a regeneration area. Further, seed-tree method allows them to retain only trees of the best phenotypes as a seed source. It leaves insufficient residual stocking to mitigate changes of environmental conditions (Fig. 14-3). In fact, levels of light, temperature, soil moisture, and other physical site attributes change similar to clearcutting (Cheyney 1942; Schmidt et al. 1976; Smith 1986.)

(b)
FIGURE 14-2 (*continued*)

Standard definitions pertinent to the seed-tree method reflect these conditions (after Ford-Robertson 1971; Soc. Am. For. 1989; Smith 1986) as follows:

Seed-tree method—removing most mature trees in one cut, except widely spaced seed-bearing reserves (singly or in groups) as a seed source for the new even-aged community

Seed tree—a tree left as a source of seed for regeneration

Seed-tree system—an even-aged silvicultural system where the reproduction method removes the mature community in two or more successive cuttings, temporarily leaving only a few widely-scattered old trees to serve as a source of seed

With seed-tree method, *preparatory cutting, seed* or *seeding cutting,* and *removal cutting* have meanings identical to those for shelterwood method. The two approaches differ only in the amount of residual stocking left during seed cutting (Figure 14-4), and its effect. With the seed-tree method, residual trees serve only as a seed source.

Most seed tree cuttings leave only 2-12 seed trees per acre (5-30/ha) (Cheyney 1942). The number and distribution depends upon seed dispersal mechanisms and patterns within a target species, and to some degree upon stand conditions. To illustrate, in forests of western United States and Canada foresters may leave only three seed trees/ac (7/ha) for larch (Schmidt et al. 1976), and 6-8/ac (15-20/ha) with Douglas-fir (Seidel 1983). For eastern white pine and red pine, a seed-tree seed cutting

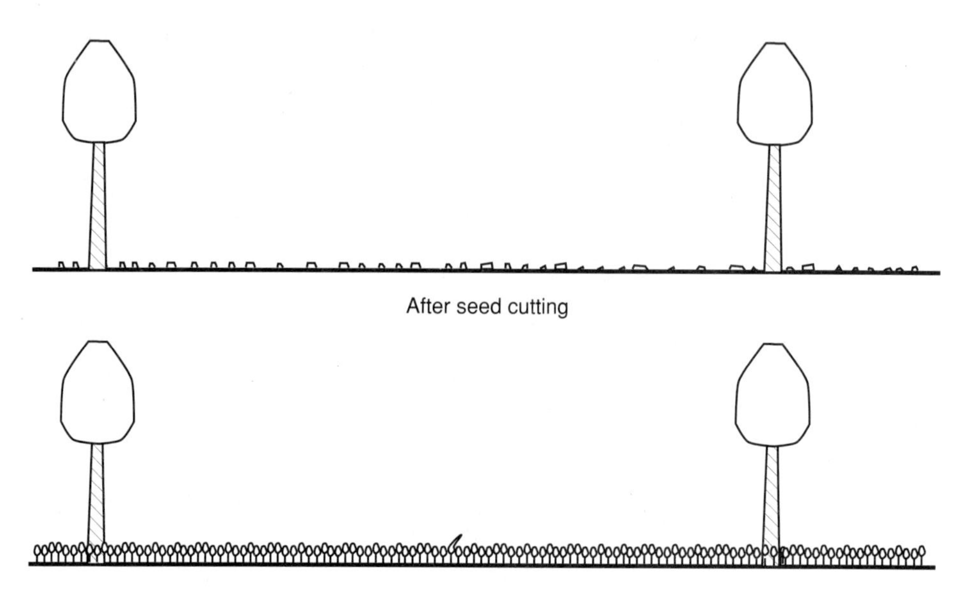

After removal cutting
(seedling stage)

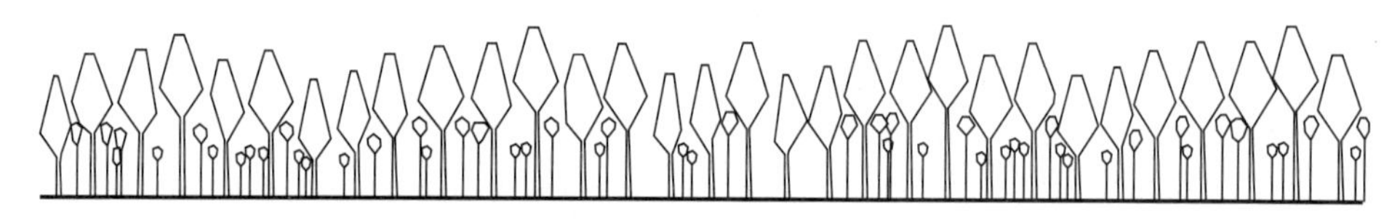

FIGURE 14-3
Seed-tree method serves primarily as a means to insure an adequate seed supply across the regeneration area, but does *not* leave sufficient residual cover to appreciably modify conditions of the physical environment.

may leave as few as 3/ac (7/ha), to as many as 10/ac (25/ha) (Horton and Bedell 1960). For loblolly and slash pines in southeastern United States they leave about 30 trees/ac (74/ha) when the diameters average about 10 in. (25.4 cm), but keep only 10/ac (25/ha) in stands with seed trees as large as 16 in. (40.6 cm) dbh. By contrast, it takes approximately 50 and 20 trees/ac (124 and 49/ha), respectively, for shortleaf and sand pine of comparable diameters (Barnett and Baker 1991). So few seed trees per unit of area may not support a commercial removal cutting, but they also have only a nominal effect on the physical environment. As a consequence, foresters can often leave them indefinitely without harming the new age class (Cheyney 1942; Smith 1986). In some cases, foresters may leave more than the minimum number of seed trees just to make a normal removal cutting commercially feasible.

In some cases, foresters may elect to delay cutting particularly good seed trees for many years into the new rotation, until the trees reach extra-large sizes. This may also help to make an early intermediate treatment commercially feasible (Cheyney 1942). Foresters call this a *reserve cutting*, and the seed trees *reserve trees*. They may also keep reserve trees as a part of shelterwood system, but normally to satisfy some special management objective.

Because of the extra-wide spacing and exposure of residuals with any seed-tree method (reverse cutting or not), some species deteriorate in quality or vigor. Then foresters would need to use shelterwood method. For other species, the vigor will improve and the diameter increment increase, making seed-tree method feasible. Whether the change enhances or harms a tree depends upon its silvical characteristics, and its health and condition at the time of seed cutting. For example, in conifer stands of Canada's western British Columbia, foresters can safely use seed-tree method with species that do not suffer from exposure, as long as they leave vigorous trees of upper canopy positions. They can also opt for a shelterwood method with as little as 20% cover of vigorous upper-canopy trees with exposure-tolerant species. For species requiring protection from exposure, the seed cutting must leave at least 40% canopy cover (Klinka and Carter 1991). As an alternative for some community types, foresters may leave scattered groups or strips of seed trees to provide mutual protection within these clusters (Cheyney 1942; Smith 1986). This will not necessarily insure good postcutting growth of the seed trees. In fact, only shelterwood and seed-tree cuttings that leave uniformly dispersed seed trees insure adequate crown release to effectively stimulate diameter increment.

SELECTING SEED TREES FOR SHELTERWOOD AND SEED-TREE CUTTING

Foresters must carefully select each seed tree based upon both economic and biologic factors. Keeping the best trees as overwood insures a good financial return from the removal cutting. From a genetic perspective, it also enhances chances for passing desirable attributes to the new stand and future generations of trees that will occupy the site (Maynard et al. 1987). This has a long-term economic benefit for the landowner.

Generally, foresters use readily observed characteristics in selecting the seed trees (after Hawley 1921; Haig et al. 1941; Cheyney 1942; Smith 1986; Pait et al. 1991) and commonly opt for trees they consider

1 phenotypically superior, to insure a good growth and form in the new community, as well as resistance to local insect, disease, and other pest problems;
2 prolific in seeding and flowering to sustain a high level of seed production and to include both sexes among dioecious species; and
3 sturdy and healthy enough to withstand wind and exposure and remain alive until the removal cutting.

Seed-tree method allows foresters a high degree of selectivity over the residual seed source. They must find good trees at regular spacing to insure a uniform seed dispersal. Generally, most of the most dominant members of a well-managed even-aged community will serve as good seed trees. In the course of different silvicultural treatments, foresters would have selected them as representing the best of the age class, at least relative to the management objectives. In some cases they might retain selected trees with actively used upper-stem cavities to maintain essential nesting and denning sites for a variety of birds and mammals. Yet even those trees should have good vigor, and generally good phenotypic characteristics. They, too, contribute seed to the new cohort, and also must remain alive and healthy throughout the regeneration period.

Seed trees must withstand blowdown and exposure following a seed cutting, and this often precludes use of seed-tree method at sites with shallow soil or among shallow-rooted species. With shelterwood method, foresters can leave a higher residual stocking (closer spacing between the seed trees) and this provides a reasonable degree of protection. Otherwise, they might opt for progressive strip or patch clearcutting, or artificial reforestation.

Windfirmness depends upon several physical site factors, as well as inherent attributes of different species and individual trees. At the community level, rooting depth and the cohesiveness of a soil influences the risks of blowdown. So do the prevailing direction and velocity of wind, as well as other factors like:

1 phenotypic characteristics of a species
— rooting depth and branching patterns influence anchorage
— crown form, size, and density amplify or resist the stress by wind
2 physical site characteristics that influence anchorage
— soil texture, drainage, and depth affect rooting depth and patterns
— texture and moisture influence cohesion, and how easily a soil pulls apart
3 physiographic features that affect wind velocity
— aspect and slope influence the forces and direction of winds
— shape of topography may funnel winds and intensify velocity
— landforms may block prevailing winds and lessen exposure
4 individual seed-tree characteristics that affect sturdiness

— size, density, and length of the crown affect the degree to which a tree intercepts the wind
— the height-diameter ratio influences the degree of bole flex under stress by wind, and the resistance to breakage
— tree bole characteristics as related to size and distribution of dead branches (knots) influence its strength

Foresters cannot alter topographic and edaphic conditions. They can consider the hazard in choosing between shelterwood and seed-tree methods, or in selecting some other alternative. For the most part, stands along ridge tops with shallow soil, and on poorly drained flats along the base of slopes and in the valleys, have the highest risk of blowdown.

In well-managed stands, trees of upper-canopy positions usually have stocky boles for their height, and well-balanced crowns. Compared with less vigorous trees of poorer positions, or trees that grew in a crowded environment for long periods, they do not sway as much when loaded with snow or water. That reduces the risk of blowdown or breakage substantially. Of course, neither the sturdy or weak trees withstand severe storms like hurricanes or tornadoes, even at sites with deep and cohesive soils. In fact, not even well-designed shelterwood seed cuttings afford much protection against such severe storms. Instead, the winds often flatten a forest across a large geographic area, independent of the silviculture or the intensity of past management.

DETERMINING AN APPROPRIATE LEVEL OF RESIDUAL STOCKING

Several factors affect natural seeding from residual trees (see Chapters 8 and 9). These depend in part upon the species, and the seed trees themselves. Key characteristics (after Cheyney 1942; Daniel et al. 1979) include

1 durability of seed trees to exposure, their resistance to windthrow and breakage, and how long they continue to produce abundant seed;

2 frequency of anticipated good seed crops, and amounts produced during subsequent years;

3 quantities of seed produced per tree based upon its health, vigor, age, and species;

4 direction and velocity of prevailing winds;

5 heights of seed trees, and the effect on seeding distance downwind;

6 likely seed dispersal distance from each seed tree, and the composite seed rain across a regeneration area;

7 length of time that seed of the target species remains viable in the forest floor after dispersal;

8 characteristics of the seedbeds, and their suitability for germination and early survival;

9 length of time that seedbeds remain suitable after a seed cutting and the site preparation that may follow;
10 expected germination rates, and subsequent survival of new germinants; and
11 density of seedlings for effective site utilization, including the advance regeneration.

In determining the average numbers of seed trees to leave, foresters can use a version of the formulae for calculating sowing rates with direct seeding, and in a tree nursery:

$$\text{\# seed trees/ac} = (43{,}560 / F) * (D / G * N * Y)$$
$$[\text{\# seed trees/ha} = (10{,}000 / F) * (D / G * N * Y)]$$

where

43,560 = the number of ft^2/ac
10,000 = number of m^2/ha
F = the area of ground space (ft^2 or m^2) covered by seed falling from a single tree
D = the density of seedlings needed per unit of area (ft^2 or m^2) to exactly occupy the site
G = the germination percent expected among the fallen seed
N = the number of seeds cast per tree
Y = a survival factor

Y represents a forester's best guess about the proportion of seed that will germinate and produce thrifty seedlings of some predetermined minimum size.

The residual stocking for a shelterwood method seed cutting can vary appreciably from stand to stand, even for a single community type. Conceptually it must include the minimum number of trees for effective seed supply and also maintain sufficient canopy cover to mitigate environmental conditions. At the same time, shade from the overwood canopy must not preclude germination or inhibit seedling development. The nature of a seed cutting may also depend upon factors (after Cheyney 1942; McDonald 1976; Laacke and Fiddler 1986; Tesch and Mann 1991) like

1 the minimum volume to eventually sustain a commercial removal cutting;
2 the volume that accumulates on the seed trees between the seed and removal cuttings;
3 how rapidly crowns of the residual trees close, and light near the ground drops below acceptable levels;
4 the ease and cost of logging during the seed and removal cuttings; and
5 the extent that logging will damage the regeneration during removal cutting.

The latter varies with the sprouting capacity of the target species, and the likelihood that damaged seedlings will break off at ground line or higher up the stem. Since most conifers do not sprout, sufficient seedlings of these species must escape damage to insure full site occupancy in due course.

In northern hardwood forests of eastern North America, foresters use shelterwood method in communities lacking abundant and well-developed advance regeneration. Where stands have abundant interfering plants, they must precede seed cutting by site preparation, and normally using herbicides (Kelty and Nyland 1981; Sage 1987; Horsley 1994). The seed cutting can have as little as 30 or as much as 80 ft^2/ac (7 or 18 m^2/ha) of residual basal area, but 40–50 ft^2/ac (9–11 m^2/ha) in large sawtimber trees provides the best residual canopy shading for optimal seedling development. Where interfering plants like ferns and other herbs might proliferate following an overstory reduction, foresters may need to leave as much as 90 ft^2/ac (21 m^2/ha). A removal cutting can follow once at least 5000 new trees/ac (12,350/ha) reach free-to-grow positions (e.g., 2–4 ft or 0.6–1.2 m tall) (Jacobs 1974; Kelty and Nyland 1981; Kelty 1987).

For oak communities, foresters may need a three-cut shelterwood method, plus site preparation to reduce the non-oak understory vegetation. Site preparation with prescribed fire may begin even several decades prior to any cutting treatment, and include several successive burns. With herbicides, they could treat the understory just prior to a reproduction method cutting. Often abundant oak seedlings become established following a good seed year, and these new trees remain alive for several years (Nyland et al. 1982; Loftis 1990a, 1990b). In stands having these advance seedlings, foresters can reduce the overstory to about 60%–70% of the stocking found in unmanaged stands, removing trees of lower crown positions and keeping the overstory largely intact. This will promote oak advance seedling development, but inhibit many undesirable understory species (Loftis 1990). Once the oak seedlings reach about 3 ft (0.9 m) tall, foresters would reduce the stocking to about 50% of that in unmanaged stands. This helps to stimulate the understory oaks even more. A removal cutting will follow after at least 435–440/ac (1074–1087/ha) well-distributed oak saplings reach a minimum of 4.5 ft (1.37 m) tall. Overall, this sequence of treatments may span at least twenty-five to forty years (Sander 1979; Sander et al. 1984).

Among conifer forests in western North America, foresters should maintain 50%–60% canopy cover, and leave no more than 50% of the basal area found in undisturbed (unmanaged) mature stands (Williamson 1973; Tesch and Mann 1991). Actual numbers of seed trees depend upon their crown dimensions, and the silvical characteristics of each species. Within Douglas-fir communities, leaving about 20 old-growth trees/ac (49/ha) will produce the best seedling stocking. Commonly, silviculturists recommend leaving 10–20/ac (25–49/ha). This converts to a residual basal area between 50 and 150 ft^2/ac (11.5 and 34.4m^2/ha) (Tesch and Mann 1991). With mixed-conifer types, foresters may leave up to 35/ac (86/ha) (Seidel 1983), or an overwood stocking of about 50–60 ft^2/ac (11.5–13.8m^2/ha) (Seidel 1979a, 1979b). For spruce-fir forests of the Rocky Mountains, foresters remove about 30% of the basal area. With ponderosa pine, they take out 40%–60% (Alexander 1986a, 1987). For red pine in the Lake States region of North America, a seed cutting should remove 40%–60% of the basal area in trees at least 10 in. (25.4 cm) dbh (Eyre and Zehngraff 1948). Exact requirements vary with the species, site conditions, and special interests of the landowner.

In many cases, logging for both the seed and removal cuttings exposes mineral soil along the skid trail surfaces, and somewhat in the intertrail space. This creates a seedbed that herbaceous species readily colonize, and where little or no tree regeneration may develop (Kelty and Nyland 1981; Walters and Nyland 1989). Consequently, foresters must limit the trails to a minimum number and distribution for efficient logging, and place them to divert traffic away from sensitive microsites (e.g., wet or thin soils), or highly erodible areas.

After logging, contractors must install waterbars and grade the trails to minimize surface runoff. By using the same trails for both the seed and removal cuttings, foresters can protect much of the new age class from skidding damage. At the same time, many different herbaceous plants colonize on the mineral soil along the skid trails, and these plants provide food for many wild animals until the tree canopy closes over a trail. The herbs also help to stabilize the soil until a new litter layer forms.

RESERVE, GROUP, AND STRIP SHELTERWOOD METHODS

Seed-tree method leaves too few residual trees to harm the new community, even if foresters do not return for a removal cutting. By contrast, the overwood with shelterwood method will eventually inhibit seedling development. The oppression becomes manifest (Cheyney 1942; Smith 1986; Tesch and Mann 1991) as (1) slower rates of regeneration height growth; (2) losses of some young trees to overstory shading (particularly among shade-intolerant species); (3) reduced foliage density of even the tallest trees of the new age class; and (4) increased variation in the heights and density of trees within the new stand. Oppressed trees often develop flattened tops due to the slow rates of height growth, and many of the less shade-tolerant species die or become overtopped by more shade-tolerant trees. These symptoms indicate that foresters waited too long for a removal cutting.

Foresters may sometimes deliberately retain all or some of the old trees for more than 20% of the new rotation to accommodate special management objectives (Hannah 1988). With these systems, foresters keep only the minimum number of trees for adequate seed production *and* shelter, or remove only part of the overwood once the new age class develops to an appropriate condition. Both approaches create a two-aged community, with two distinct canopy layers (Fig. 14-5). These schemes allow foresters to regenerate a new cohort of even fairly shade-intolerant species, and concurrently to accommodate some special nonmarket (nontimber) objectives as well.

Foresters call these specialized schemes the *irregular shelterwood method, delayed shelterwood method, extended shelterwood method*, or *reserve shelterwood method* (after Ford-Robertson 1971; Smith 1986; Wenger 1984; Klinka and Carter

FIGURE 14-4
Seed-tree seed cutting leaves only widely scattered trees of excellent phenotypes to serve as a seed source, but the sparse canopy cover has little effect on conditions of the environment near the ground.

1991). Though recognized by silviculturists for decades (Hawley 1921; Cheyney 1942; Kostler 1956), foresters have not historically used reserve shelterwood schemes to any great extent, particularly for timber management. Still, the method offers some potentially important nonmarket values where a landowner also chooses to use an even-aged reproduction method. For example, reserve shelterwood method may allow landowners (Hannah 1988; Tesch and Mann 1991; Nyland 1991) to

1 leave the older age class to maintain some protection for the new community over an extended period;

2 realize additional growth on the overwood trees, and maintain them to extra-large diameters for specialty purposes or products;

3 enhance scenic values;

4 maintain the habitat for particular plants and animals; or
5 address special ecological or social needs, consistent with the management objectives.

In all of these cases, landowners will need special silviculture to maintain the two-aged arrangement, or to eventually convert the community to an even- or multi-aged condition (see Chapter 23).

Foresters must choose the reserve trees carefully, since these will remain well into the next rotation. Appropriate criteria include

1 long-lived species;
2 deep-rooted and durable species not prone to blowdown, breakage, or dieback from exposure;
3 trees largely free of structural defects that might make them susceptible to breakage;
4 high-vigor and healthy trees with full and well-balanced crowns that encircle the trunk; and
5 trees free of logging injuries that might serve as entry courts for disease organisms, or root damage that might make the trees more susceptible to blowdown.

Foresters must keep the reserve trees at a uniform and wide spacing to insure high levels of insolation across the lower strata.

Though attractive in many respects, the reserve shelterwood system has some potentially important operational and financial disadvantages (Hawley 1921; Cheyney 1942; Daniel et al. 1979; Smith 1986). These include:

1 higher logging costs to remove only part of the seed trees;
2 greater chances of damaging some reserve trees during the initial and follow-up cuttings;
3 higher risks of reserve tree loss due to exposure, blowdown, or old age;
4 difficulties of maintaining an open overstory canopy cover to insure good development of the younger age class;
5 heavy losses of the younger (shorter) trees during eventual reserve tree removal, unless contractors use special harvesting practices; and
6 higher logging costs to recover the widely spaced reserve trees at the time of their maturity.

Landowners must accept these limitations or use an alternative approach, including *group-* or *strip-shelterwood systems*.

Strip-shelterwood method resembles strip clearcutting. In fact, the two approaches may look alike in many cases. For strip-shelterwood cutting, foresters remove the mature age class over a series of entries, cutting narrow parallel strips not exceeding the height of adjacent standing trees. These do not create a clearcut-like environment in the center, but brighten the forest floor considerably. The adjacent residual strips provide seed and partially shade the openings. To maximize seed dispersal and reduce chances of blowdown on sensitive sites, foresters orient the cut strips at right angles to the prevailing winds. Where side shading

FIGURE 14-5
Reserve shelterwood method creates a two-aged community having two distinct canopy layers, with the older trees remaining for at least 20% of the rotation for the younger age class.

seems more important than seed dispersal, they align the strips in an east-west direction to minimize direct insolation (Hawley 1921; Cheyney 1942; Smith 1986).

Foresters might space the cut strips at intervals equivalent to three to four times their widths, and return at regular intervals to remove a progressive series of additional strips. They would cover the entire stand area in three or four entries. Progressively cutting into the wind or toward the south (or northward in the southern hemisphere) will increase the brightness over each previously cut strip, but still maintain the critical patterns of seed dispersal or shading for the newly opened strips. Ideally, regeneration will form quickly, allowing successive removals at fairly short intervals, and limiting the spread of ages to no more than 20% of the intended rotation (Hawley 1921; Cheyney 1942; Smith 1986). Foresters must remove each progressive strip before shading oppresses seedling development, particularly of desirable shade-intolerant species. In fact, the narrowness of each cut strip makes good light management critical with the strip-shelterwood method, and forces foresters to return at fairly short intervals to complete the series of progressive strips and produce a new even-aged community with little difference in the heights, species composition, and rates of development of component trees.

For some applications, foresters might prefer group-shelterwood method. In these cases, the seed cutting creates well-dispersed openings, each with a diameter not exceeding one time the height of adjacent trees. This improves environmental conditions, but not to the extremes of clearcutting. Once seedlings become established, foresters remove a band or ring of residual trees around all or part of each initial opening. Cutting along the south sides of previous patches (north sides in the southern hemisphere) lets full sunlight into the regenerated areas, but maintains partial shade over the new openings. At each entry, foresters might cut additional patches to create new pockets of regeneration throughout the stand. Eventually the progressively wider openings coalesce, leaving no mature trees within the stand (Hawley 1921; Ford-Robertson 1971).

Foresters can sometimes center the original group cuttings around pockets of advance regeneration. Side lighting from these first-cut openings will promote additional advance regeneration under edges of the remaining uncut areas. Strip-shelterwood method will also have the same effect, and this usually helps to shorten the overall regeneration period. Each successive cutting band will then extend far enough into the mature community to release the existing advance regeneration. As a result, the total regeneration period depends on how rapidly advance regeneration develops adjacent to each opening or strip. In most cases the procedure regenerates a new even-aged stand over a period equivalent to that of strip- or uniform-shelterwood methods. However, with really slowly regenerating species, group- or strip-shelterwood methods may produce a multi-aged stand (Hawley 1921; Ford-Robertson 1971).

REMOVAL CUTTING AND SEEDLING DAMAGE

Foresters cannot predict exactly how rapidly regeneration will develop at each site, but can use past experience and observations from research to indicate when to schedule a removal cutting. Conceptually, they must leave the overstory in place at least until adequate numbers of desirable tree seedlings regenerate and no longer require any protective cover. Conversely, they should remove the seed trees before shading and competition interfere with the young trees. Between these points in time, they must remove the seed trees, and that often gives foresters a new problem to deal with. The logging inevitably damages or destroys a portion of the new cohort. In some cases, it has left only a poorly stocked new stand. In fact, many foresters consider this problem the most important shortcoming of shelterwood method, and the major deterrent to its more widespread use as a reproduction method (Hawley 1921; Cheyney 1942; Smith 1986; Tesch and Mann 1991).

With fairly large conifer saplings that break off when pushed over during skidding or by a falling tree, the damaged trees will not resprout. As a consequence, the removal cutting normally reduces the density and alters the distribution of regeneration. Having large numbers of thrifty seedlings and releasing these trees before they become inflexible helps to minimize the losses. With hardwoods, broken and bent over seedlings will resprout if not uprooted. Here, too, making the removal cutting while the seedlings still remain flexible reduces the losses. With all species, seed remaining in the litter layer will often germinate if logging opens gaps in the new vegetation layer and exposes the forest floor. Foresters cannot necessarily count on these additions, particularly with species that develop best underneath the protection of an overstory shelter, or where an established herbaceous community readily spreads into the regeneration gaps and overtops the new germinants. Rather, as a common preventive measure, they must schedule a removal cutting before the new trees get too tall and inflexible to resist breakage, and before the community density decreases appreciably due to normal intertree competition. Then sufficient numbers will usually escape to fully occupy a site.

Prerequisites for overstory removal differ among forest community types, with variations of site, and with the species. For conifers in western United States and Canada, seedlings should grow to 1.5–2.0 ft (0.5–0.6 m) tall before overstory removal, but not beyond 3.0–3.5 ft (0.9–1.1 m) tall (Schubert 1974; Tesch and Mann 1991). With pines in southeastern United States, overstory removal can occur when stands have 1000/ac (2470/ha) of seedlings at least 1 ft (0.3 m) tall. It takes 6000/ac (14,820/ha) for loblolly, shortleaf, slash, and sand pines. Normally this requires 3 to 6 years (Barnett and Baker 1991; Boyer and White 1990). With conifers of eastern United States and Canada, the removal cutting should occur when seedlings reach between 0.5 and 1 ft (0.2–0.3 m) for eastern spruce-fir (requiring as long as 25 years on some sites), and 2–5 ft (0.6–1.5 m) tall for eastern white pine (Hannah 1988). Smaller seedlings do not survive the release, and logging will destroy many larger ones. Losses during removal cutting range from 10% to 100% among conifer communities along the west coast of North America (Tesch and Mann 1991), and 35% to 70% among longleaf pine stands along the southeastern coastal plain of the United States (Boyer and White 1990).

With hardwoods, most small seedlings bend over or break off close to the ground, and resprout from the root collar. The new shoots often reach the prebreakage height within two to three years, filling gaps caused by damage, and overtopping any new germinants that arise after the logging. By contrast, many saplings at least 1 in. (2.54 cm) in diameter became uprooted. Others snap off well above the ground. They sprout at the point of breakage, and become forked and of poor form (Nyland 1992). Logging may break up to 50% of the new trees and leave as much as 25% of the area initially unstocked. Yet within a few years of a well-timed removal cutting, replacements will occupy all but a small proportion of the area between skidding trails (Kelty and Nyland 1981; Kelty 1987).

Foresters can contain the losses of regeneration by careful logging practices, and by appropriately timing a removal cutting. Timing includes both the degree of seedling development, and the time of year. Particularly at high latitudes and with conifers, real benefit accrues from limiting the removal cutting (Eyre and Zehngraff 1948; Seidel 1979a) to winter months when a snow cover protects the seedlings, seasons when the trees lack tender shoots, and before the regeneration has grown very tall. The harvesting plan should include provisions (Seidel 1979a, Boyer and White 1990; Tesch and Mann 1991) to

1 keep landings outside the regeneration areas;
2 limit ground skidding to a minimum number of well-marked trails, and preferably those established for the seed cutting;
3 carefully clear skid trails of old logging debris and windfalls without pushing them into the adjacent regeneration;
4 require directional felling to minimize turning and twisting of the logs during skidding;
5 keep tractors on the trails to the degree possible, and use winches to move the logs from stump to the skidding corridor;
6 restrict skidding to fully limbed pieces with no large side branches; and
7 remove unmerchantable logs by the same procedures required for the original slash disposal, or leave them in place to decay.

Foresters must also help the logging crews to understand the importance of careful work, and how to control the logging to accommodate the silvicultural objectives.

Logging will break off or push over many of the understory trees during the eventual removal cutting in a reserve shelterwood system as well. In these cases, the falling overstory trees will crash down on many in the younger age class, breaking off the tops, and pushing others over. This leaves fairly large openings that will not subsequently regenerate due to the lack of a seed source (except along stand edges). If foresters delay the removal cutting until the new age class grows up to the lower level of the reserve tree canopy, the logging may only thin the younger community (see Chapter 23, and Nyland 1992).

SITE PREPARATION WITH SHELTERWOOD AND SEED-TREE METHODS

Site preparation serves the same purposes in shelterwood and seed-tree methods as with clearcutting. On the other hand, foresters must take special measures to safeguard the health and condition of seed trees, the seed stored in the litter layer, the stumps and root systems that would contribute sprouts or root suckers to the new age class, and any desirable advance regeneration (Alexander 1974; Walstad and Seidel 1990). Since most foresters use these two reproduction methods with natural regeneration, the site preparation will primarily improve seedbed and soil conditions, kill or set back interfering vegetation, and reduce hazards from fire and pests. Doing this work prior to seed cutting (no logging slash) often reduces the costs and improves the results. In addition, prescribed fires often become too hot and too difficult to control on sites covered with dry logging slash. Besides, logging slash will partially shade the new seedlings, and protect them from desiccation and heat damage. In addition, the skidding may disturb and mix the litter sufficiently to leave good seedbeds over an adequate area, making additional scarification unnecessary (Alexander 1974; Seidel 1979a; Nyland 1989; Walstad and Seidel 1990).

With many forest community types, interfering understory plants may preclude success. In hardwood associations of the eastern United States, these include grasses and ferns, as well as several woody shrubs and undesirable tree species. Applying mistblown herbicides prior to seed cutting has effectively cleared the understory, and prevented postcutting interference with the new tree community (Kelty and Nyland 1981; Horsley and Marquis 1983; Sage 1987; Hannah 1988; Horsley 1994). In many areas, foresters must also control browsing by several forest-dwelling mammals (e.g., white-tailed deer and moose) to protect the new seedlings until they grow beyond heights that these herbivores can reach (Kelty and Nyland 1981; Horsley and Marquis 1983; Kelty 1987; Hannah 1988).

With conifers of northwestern North America, foresters often use shelterwood method for natural regeneration. Prescribed burning enhances the process (Alexander 1974b, 1986a; Walstad and Seidel 1990) by disposing of accumulated fuels to lessen the risk of subsequent fires, reducing slash and interfering vegetation, providing protection against potential insect and disease problems, and improving seedbed conditions for desirable species. Postlogging slash burns will destroy advance seedlings of shade-tolerant conifers, as well as many broadleafed species that interfere with the conifers. The fires also often damage some residual seed trees (Isaac 1963). To avoid this, foresters must move the slash away from the seed trees before burning, or substitute mechanical treatments. Both add to the costs of broadcast burning (Alexander 1974, 1986, 1987). In contrast, carefully planned prescribed burning before seed cutting (Marlega 1981) will successfully reduce the litter and duff layers, kill back advance understory vegetation, and leave a more favorable seedbed for the conifers. These advance underburnings may also promote the germination of fire-adapted shrubs, necessitating a second burn about three years after the first one (Martin 1982). Periodic underburnings have also successfully controlled undesirable vegetation in pine forests of southeastern United States (Van Lear and Waldrop 1991).

Conceptually, both shelterwood and seed-tree methods establish and promote new regeneration coincident with the seed cutting, precluding any dependence on advance regeneration. Yet in several hardwood forest community types of eastern North America, foresters often utilize the existing advance regeneration as part of the new community (Sander et al. 1976; Clark and Watt 1971; Marquis 1987; Nyland et al. 1981; Loftis 1990a). The seed cutting helps to fortify the density and distribution of this natural regeneration, or enrich the species composition. In these cases, they may use site preparation only to enhance some critical site conditions, but use a method that preserves the advance regeneration. Oak forests within the central United States illustrate this circumstance. Foresters must establish 400–600/ac (988–1482/ha) of saplings at least 4.5 ft (1.37 m) tall before any overstory removal (Clark and Watt 1971; Sander 1972; Sander et al. 1976). Otherwise an array of interfering woody plants will overtop the more slowly growing oak seedlings and inhibit their development. Also, oak regeneration takes many years to form, and often develops only in years with (1) a heavy seed crop; (2) low predation by seed-eating insects and animals; (3) seedbeds of only thin leaf litter; (4) favorable weather conditions; and (5) freedom from interference by understory vegetation or faster-growing competing plants. The seed cuttings serve to encourage the height growth of existing advance oak regeneration, and to enhance seedling density (Loftis 1990b). The site preparation would maintain favorable seedbeds and reduce understory competition, often by periodic understory burning beginning well in advance of stand maturity (Hannah 1988; Nyland 1989).

SOME NONCOMMODITY ASPECTS OF SHELTERWOOD AND SEED-TREE METHODS

Available evidence indicates that young even-aged communities established by seed-tree and shelterwood methods create a special kind of habitat. In particular, they benefit wildlife that thrive in dense stands of small trees, feed upon fruits and other parts of the herbaceous and shrub species that often dominate early stages of community development, or browse the shoots, bark, and buds of woody vegetation. Such favorable conditions characterize all young naturally regenerated even-aged communities. At the same time, the overwood in shelterwood or seed-tree stands provides additional habitat for species that operate within the elevated canopy of tall trees. As a consequence, as long as the older age class remains in place, regenerating stands will have a greater songbird diversity than either young even-aged communities, or closed-canopy stands of at least intermediate ages (Webb and Behrend 1970; Thomas et al. 1975; Medin 1985; Deisch and Sage 1989, 1990; Tobalske et al. 1991).

The actual effect probably depends upon the numbers and dispersion of the residual trees, and how long they remain in place. Particularly, reserve or group shelterwood methods that retain mature trees for extended periods provide a gradual transition from conditions found in mature communities, through a regeneration period, and to a new stand containing trees of substantial sizes. The shelterwood seed cutting maintains large trees used for nesting, denning, and food production critical to many faunal species. Retaining some dead snags as well as living trees with cavities insures both continued seed production through early stages of community development, and den or nest sites for many creatures (Alexander 1977; Medin 1985; Tobalske et al. 1991; Nyland 1991). When foresters eventually remove the seed trees, they reopen the skid trails, re-creating conditions for new herbs to colonize the exposed seedbeds. This prolongs the usefulness of an area for creatures that depend upon a herbaceous community for many life requirements (Nyland 1991).

For a long period following clearcutting and after overstory removal, even-aged stands have a nominal vertical structural diversity. The green leaves essentially occupy only a narrow vertical layer that rises higher above the ground as the trees get taller. At least initially, shelterwood stands have a two-layered canopy. Though present for only a short portion of a rotation, this two-layered structure makes shelterwood method preferable to clearcutting under some wildlife management scenarios, even where conditions would make clearcutting feasible. This same two-story condition gives shelterwood and seed-tree methods better visual qualities than clearcutting, both for near and far viewing. At a distance, residual tall trees add interesting shadow patterns and visual textures, particularly after new vegetation *greens* the understory. For near viewing, large- and small-diameter trees provide a sharp visual contrast that adds visual diversity to the *standscape*. This illusion continues for long periods in reserve shelterwood stands. Gradually, as the size differential between the older and younger trees decreases, the community comes to resemble a more simple maturing even-aged stand (Nyland 1991).

15

EARLY STAND DEVELOPMENT

EVEN-AGED COMMUNITY ESTABLISHMENT AND FORMATION

During the first few years, new even-aged communities often contain a rich mixture of herbaceous, shrub, and tree species. Initially, the nontree vegetation may make up much of the total accumulating biomass, and dominate the site (Bormann and Likens 1979). As the community develops, it transpires increased amounts of moisture, taking up nutrients from the soil and the decomposing organic materials. Gradually, moisture within upper soil layers and the volume of water (and dissolved nutrients) leaving a site by subsurface flow also diminish. Shade from the new vegetation reduces temperatures at the forest floor and within the upper soil layers. This slows organic decomposition and reduces the amounts of nutrients released to the soil solution.

In time, the tree seedlings become taller than the herbs and shrubs. Their crowns grow together, shading the ground and the plants of shorter stature (Fig. 15-1). Darkened understory conditions, coupled with the limited amounts of ungerminated and viable seed remaining in the litter, reduce or stop new germination. This initially stabilizes the numbers of plants present. Then overtopped plants of the less shade-tolerant herbaceous plants begin to die. Total plant density and species diversity decline, and the floristic composition once more begins to resemble the precutting community, with most of the biomass accumulating on the trees. Thereafter, trees dominate a site until the next reproduction method cutting. Some shrub and herbaceous species persist in the understory, and new species may even enter the community as it matures. Still, as long as the upper canopy remains dense and tight and casts a heavy shade over the surface, most new germinants will not live long.

FIGURE 15-1
Once the crown canopy closes in a developing even-aged community, the forest floor becomes darkened and the rich community of herbs and shade-intolerant shrubs that regenerated after the reproduction method cutting declines, leaving an understory similar to that of more mature forests.

The bulk of photosynthesis occurs in the upper canopy, with tree species producing most of the biomass.

The timing of these events depends upon the promptness of germination and seedling establishment, the rates of community development, and the time until canopy closure. All vary with conditions of the physical environment. Yet clearcutting, seed-tree, and shelterwood methods tend to trigger certain characteristic trends in community development. These (after Bormann and Likens 1979; Leopold and Parker 1985; Leopold et al. 1985; Lees 1987; Martin and Hornbeck 1989; Oliver and Larson 1990; Hornbeck and Leak 1992; Wang and Nyland 1993) include the following:

1 the proportion of early-succession and shade-intolerant species increases, given a seed source and a suitable physical environment;

2 advance tree regeneration usually develops rapidly and becomes an important component of the new tree community;

3 a complement of broadleafed species sprout or develop from root suckers, including from trees of relatively young to moderate ages;

4 viable seed in the organic horizons germinates, adding other species that proliferate and grow rapidly;
5 species previously absent from a site may appear, due largely to lightweight seed from adjacent areas;
6 herb and shrub biomass increase rapidly in the first few years, and then decline as trees dominate the site; and
7 once the trees form a closed canopy, total stem density decreases due to competition-induced mortality and other factors.

Among eastern hardwoods and southeastern pines, trees dominate a site within ten to fifteen years (Leopold and Parker 1985; Martin and Hornbeck 1989; Walters and Nyland 1989; Cain 1991a; Hornbeck and Leak 1992; Wang and Nyland 1993). For other conifer communities, tree domination may take longer.

EARLY DEVELOPMENT OF EVEN-AGED COMMUNITIES AT MESIC SITES

Early silviculture research (Jensen 1943; Wilson and Jensen 1954; Gilbert and Jensen 1958; Marquis 1965, 1967), as well as more recent modeling work (Adams and Ek 1974; Botkin et al. 1972a, 1972b), has provided information about even-aged stand development (see Chapter 9 and Figs. 9-6 and 9-8), including general kinds and patterns of change following an even-aged reproduction method.

Among hardwood stands of eastern North America (Kelty and Nyland 1981; Walters and Nyland 1989) tree stem density remains fairly stable for the first ten years after both shelterwood and clearcutting methods. Shade-intolerant species and sprouts off a wide array of trees grow the most rapidly in height and diameter, while large numbers of shade-tolerant seedlings grow slowly and remain in understory positions. By about the fifth to seventh year, the shade-intolerant tree species emerge through the dense *Rubus* that formed across each site, and the berry bushes begin to decline. Subsequently, height growth of the more shade-tolerant trees growing underneath the taller *Rubus* also increases. By the tenth year, a closed tree canopy forms, lower branches die on the survivor trees, and many herbaceous species disappear or decline significantly. This leaves the understory largely devoid of live foliage, except in open lanes along the old skid trails.

Other studies also documented dramatic temporal shifts of species composition and herb domination among newly developing even-aged communities. The timing of events differs by region and site-community type, and the dominant species often shifted throughout the reorganization phase. To illustrate, in Douglas-fir communities of northwestern North America herbaceous cover increased from 35% in the first year to 79% by the third year following clearcutting and burning, and then declined to 75% by the fifth year. At a coverage of 25%–85%, these interfering plants slowed Douglas-fir development. Greater herbaceous densities prevented new seedling establishment (Isaac 1940). Among slash pine plantations in Florida, herbaceous cover peaked in the fourth year after different degrees of site preparation (chopping and bedding vs. stump removal, shearing and piling, burning, disking, and bedding), and then declined. Numbers of species and density of cover increased. Values for woody competition and the planted pines increased through five years. Even so, the biomass in herbs declined after one to three years, and woody competition lessened after four years. That for pine increased steadily through five years (Swindel et al. 1983; Swindel et al. 1989). In naturally regenerated loblolly pine stands on the Piedmont of southeastern North America, herbaceous plants comprised 85% of the first-year biomass, while pines and hardwood sprouts accounted for only 1% and 13%–15%, respectively. By five years the pines accounted for about 70%–85%, hardwoods about 10%–20%, and herbs less than 10%. Further, the standing biomass had accumulated most nutrients in quantities that equaled or exceeded the precut levels, mostly in the pines (Cox and Van Lear 1985). Studies following strip and block clearcutting among northern hardwoods in New Hampshire showed similar patterns. The plots had advance tree regeneration. Herb density increased over a ten year period following cutting, but the biomass in herbaceous species decreased from a level of 3%–28% of the total at the first year, down to less than 0.1% by year ten. Shrub biomass went from a first-year high of 19%–50%, down to 2%–10% by ten years. At the same time, tree density reached a peak of about 400,000/ac (988,000/ha) between the second and fourth years, and the canopy closed by the tenth year. Tree biomass increased from 40%–60% of the total at year one, to 90%–98% by the tenth year (Martin and Hornbeck 1989; Hornbeck and Leak 1992).

The period of years when germination and other sources add new trees to a site, the relative abundance of herbs and woody species, and the rates that trees grow through and overtop the early-developing nontree vegetation vary among community types (species composition) and conditions of the physical environment. Important determinants to community composition, the rapidity of tree development (including crown closure), and the number of consecutive years when new trees continue to initiate (after Noble and Slatyer 1977; Oliver 1981; Oliver and Larson 1990) include

1 the intensity of a disturbance, and its effect on soil, seedbed, and microenvironmental conditions;

2 the degree that logging and site preparation reduce or damage seed in the litter and on slash, the advance regeneration, and the parent sources for suckers and sprouts;

3 the means of transport to and persistence of propagules at a site after the disturbance;

4 the relative contribution of different sources of new plants, and how source of origin influences the early rate of development (e.g., seedlings vs. suckers) and proliferation (e.g., through rhizomes and other underground parts);

5 the timing of a reproduction method cutting relative to the abundance of seed on residual and adjacent vegetation (including seed trees), and the occurrence of weather conditions favorable to germination and early development of different species;

6 the size and shape of area treated, and how these factors affect seed dispersal from adjacent communities;

7 the amounts of seed lost to predation (e.g., birds and mammals) and environmental damage (e.g., desiccation), and when these losses occur in relation to the timing of a reproduction method cutting;
8 the rates of growth among different species that make up the new community, and particularly that of trees compared with interfering herbs and shrubs; and
9 the longevity, density and comparative rate of development by interfering plants, and the tolerance of different tree species to the competition.

Particular advantages accrue from having advance regeneration or securing prompt postcutting germination. These early trees (or trees that initiate at sites free of preexisting herbaceous competition) seem to better withstand interference from other plants, and grow more rapidly than do trees initiating after the herb community has become established (Isaac 1940; Oliver 1979; Cain 1991b). Early establishment of adequate trees at uniform spacing should also reduce the time to crown closure, and more rapidly move a community out of the stand initiation or reorganization phase.

In some cases, trees may not dominate for several years, extending the reorganization phase. If empty space persists during this time, new plants (trees and other vegetation) may continue to become established until some critical environmental factor limits additional recruitment (Oliver 1981). In fact, dense interfering vegetation may slow tree community development appreciably. Even so, rapidly forming herbs and shrubs serve some highly beneficial ecologic functions that include (after Walstad et al. 1987)

1 inhibiting soil erosion by effects of the root systems, protective canopy, and newly deposited litter;
2 reducing nutrient losses from a site by taking up and storing large quantities of available elements through photosynthesis, and later recycling them through leaf fall and eventual plant mortality;
3 adding organic materials to the soil surface through deposit of leaf litter and other dead plant parts;
4 reducing surface evaporation and heating by shading the ground and shorter plants, thereby decreasing environmental stress on small tree seedlings;
5 protecting small seedlings from harm by some animals; and
6 making the habitat less favorable to some soil-borne pathogens.

Ideally, the herbs and shrubs would regenerate in sufficient numbers to contribute these desirable effects, but not in sufficient density to suppress tree seedling development.

In some regions, herbaceous competition may often slow initial tree seedling height growth and crown closure. Actual effects differ with the species present, and probably also reflect the nature of any limiting physical site conditions. Though not necessarily detrimental in northeastern North America hardwood or spruce-fir communities (Kelty and Nyland 1981; Wall 1982; Kelty 1987; Walters and Nyland 1989), *Rubus* may dominate for ten to twenty-five years following clearcutting of conifer stands that lack adequate advance regeneration. Also sedges and grasses may impair survival and development in conifer plantations on some sites (Newton et al. 1987). Similarly, Douglas-fir that becomes established soon after clearcutting and prescribed burning will survive and dominate a site, even though a dense cover of interfering herbs and shrubs also regenerates. Most trees from later germination will fail (Isaac 1940). Naturally regenerated loblolly and shortleaf pine seedlings that germinate beneath herbaceous vegetation also tend to fail, even if abundant (Cain 1991b). Likewise, ponderosa pine planted in California on plots with established woody shrubs had diameters equivalent to those attained three years sooner in site-prepared areas. Yet pines growing in dense communities of newly germinated woody shrubs showed no significant effect (Oliver 1979).

Even when the trees do not die from interference, they may grow more slowly. This delays the onset of stand aggradation and volume accumulation. For example, among pine communities of the southeastern coastal plain of North America, herbaceous plants may reduce early above-ground biomass accumulation by as much as nine to ten times below that of plantations kept free of interfering plants (Gjerstad and Barber 1989). In fact, by age ten, loblolly pine plantations given good weed control may have twice as much volume as those left untreated (Burkhart et al. 1987). This has such financial consequences for timber production that foresters include intensive site preparation to inhibit herb and shrub establishment, or do a postestablishment release of the tree seedlings before interference threatens their survival or development.

Once trees overtop the herbs and shrubs, the lesser vegetation will have little detrimental effect. In due course, a closed tree canopy forms. By that time, foresters also see a distinct differentiation of heights within a tree species, largely reflective of inherent genetic attributes of those present. Early differences also often result from unequal rates of height growth among various species, with the less shade-tolerant species growing into dominating positions. Coincident with canopy closure, a stand passes into the aggradation or stem exclusion phase, marked by the onset of sharp losses among trees of poor crown position (Oliver 1981; Peet and Christensen 1988). Early mortality includes slower-growing individuals of both the shade-intolerant and shade-tolerant species, although the latter group persists longer in the shaded understory environment. Generally, the best trees of more shade-intolerant species grow most rapidly, and soon form an upper canopy layer of varying density. Those shade-intolerant saplings that remain in understory crown positions normally die after a few years, and this narrows the range of height classes among the more shade-intolerant trees. By contrast, shade-tolerant trees of poor crown positions survive for longer periods, so that even at a fairly young age a community has considerable vertical stratification (differentiation) in the distribution of foliage within the tree canopy, and a spread of diameters. Many overtopped trees of even the most shade-tolerant species continue to decline in vigor, the poorest eventually die, and total numbers of trees begin to decline.

As a stand matures through the aggradation or stem exclusion phase, trees of even main crown positions also succumb to intertree competition, disease, and insect attack. Early upper-canopy losses often include short-lived, shade-intolerant species (often called pioneer species) that initially grew rapidly into dominant positions. They commonly flower and bear fruit at early ages, although the unfavorable understory environment may limit germination or preclude survival of any progeny. Their death (from the overstory) decreases tree species richness within the community, and removes one wildlife food resource during early stages of stand development. Also, loss of the dominant pioneers frees other species from full or partial overhead shade. These then grow better, and become the new upper canopy species. In fact, the process may continue throughout much of the aggradation phase, resulting in a long-term series of shifts in species within the upper canopy (Oliver 1981; Oliver and Larson 1990).

Whether the remaining community includes diverse species and appreciable numbers of each one really depends on events that affect stand initiation. In turn, the numbers of species that regenerate and whether they survive and develop depend largely on (1) the nature of a reproduction method with its ancillary site preparation; (2) the diversity and abundance of different sources of reproduction; and (3) conditions of the physical environment after a cutting. This probably applies to herbs and shrubs as well, although many of the new herbaceous plants survive only in unshaded environments, and die off by the time of canopy closure. By contrast, species remaining from the precut stand persist, giving some stability to the nontree communities. In fact, fifty-year-old southern Appalachian mixed-hardwood stands initiated by clearcutting tend to have herbs and shrubs similar to that of nearby undisturbed communities with comparable aspect and elevation (Leopold and Parker 1985; Leopold et al. 1985).

As the crown canopy forms, lower branches of the trees become shaded, photosynthesis diminishes, and the radial increment begins to slow. Eventually weak trees of subordinate crown positions die. Each death leaves a space where insolation penetrates deeper into the canopy, illuminating lower branches of adjacent survivors. Their photosynthesis and growth improve. Branch elongation soon fills these small gaps, maintaining a continuous closed canopy. As low vigor trees continue to die throughout the aggradation phase, the woody biomass accumulates on fewer trees that become bigger faster. Over time, this mortality reduces gross biomass production substantially. Tree species richness may also diminish somewhat with each loss, until a mixture characteristic of mature communities within the region remains. Even so, living biomass increases annually throughout the aggradation phase, but not at a fixed rate (Fig. 9-6).

Eventually biomass peaks, and then begins to decrease. During this transition or understory reinitiation phase, mortality of large physiological mature trees with big crowns creates persistent canopy openings. Light penetrates to the understory (see Fig. 8-1), changing the environment near the ground, and making conditions conducive to germination and survival of a new cohort (Runkel 1981, 1982, 1990). Gradually more and more old trees die, and new age classes of predominantly shade-tolerant species form beneath the old survivors. This gradually transforms the community into a multi-aged condition (Watt 1947; Bormann and Likens 1979; Oliver 1981; Peet and Christensen 1988; Oliver and Larson 1990).

A complex array of physical site conditions and biotic factors influences the patterns of change in species richness over time, and actual membership in the flora at a site. Soil fertility and pH, particularly, seem to affect the temporal trends for herbs and shrubs (Christensen and Peet 1988), yet most tree communities tend to develop consistent with the general model described here and in Chapter 9. That knowledge helps foresters and ecologists to explain probable ecologic changes that will occur as an even-aged community matures into an old-growth condition. Most landowners interested in commodities would not keep a stand as long. Normally their silvicultural systems would include a ten- to twenty-year period of reorganization, and another seventy to one-hundred years of the aggradation phase before a stand reaches financial maturity. To foster more rapid rates of diameter development and influence the economic worth of a stand, many landowners intervene at intermediate ages with one or more cultural treatments. These alter the patterns and rates of stand development to favor financial, institutional, and social interests of landowners and other users.

INTERMEDIATE TREATMENTS FOR EVEN-AGED STANDS

The *regeneration phase* of an even-aged silvicultural system coincides with the *reorganization phase* of stand development. For planning purposes, foresters may define this period as beginning with an even-aged reproduction method; extending through an establishment (initiation) period; and ending once new trees of adequate numbers, size, distribution, and health cover a site. With shelterwood and seed-tree methods, the regeneration phase ends at the time of overstory removal. The tree crown canopy may not yet have closed, but eventual site occupancy by the new community seems assured.

By the time of tree crown closure, foresters know if the reproduction method succeeded. Exact criteria depend upon the objectives, but they usually evaluate (1) the abundance of different species present; (2) the relative sizes of each one; (3) the distribution (spatial uniformity) of acceptable tree seedlings; and (4) the overall stem density. Given satisfactory results, the silvicultural system enters its second (tending) phase. Thenceforth, the treatments serve to keep a new community growing at an acceptable rate. They usually include measures to protect the stand, to reduce intertree competition, and to concentrate the growth potential onto the most desirable trees and species.

Most intermediate treatments occur during the aggradation phase, until financial maturity. With noncommodity objectives, the rotation ends when a community no longer provides the values of interest. By definition, an *intermediate treatment* or *intermediate cutting* includes (after Ford-Robertson 1971) any removal of trees between the time of stand initiation and the reproduction method cutting that ends a rotation. This implies that managers can discern when stands contain (1) appropriate species; (2) trees large enough to survive into the future; and (3) an adequate distribution and spacing for full site utilization. Such questions have relevance in deciding whether to make a *release cutting*. This means freeing young trees *not past sapling stage* from undesirable competing vegetation that overtops or closely surrounds them (Ford-Robertson 1971; Smith 1986; Soc. Am. For. 1989). To *release* a young tree (not past sapling stage) means to kill or cut undesirable trees or other vegetation that overtops it (Smith 1986). Further, these definitions imply another, namely *isolation* (after Ford-Robertson 1971)—removing surrounding trees or other vegetation to give selected trees permanently free positions. In this context, release cuttings isolate selected young trees to insure their survival and enhance their long-term growth. A release cutting controls the composition of a young community in favor of a landowner's interests.

From this perspective, *composition* means both species composition and quality composition as represented by the form and condition of trees in an emerging stand. Ideally, comprehensive silvicultural programs to improve species composition should include

1 site preparation before or coincident with the reproduction method that initiated a community;
2 follow-up measures (like release treatments) to influence stand composition once the trees become established; and
3 continuing control of undesirable vegetation in all intermediate treatments through the aggradation phase.

This usually includes a complex of cultural measures that foresters implement throughout an even-aged silvicultural system (after Lotan 1986; Walstad et al. 1987b; Stewart 1987; Tappeiner and Wagner 1987), including (1) well-planned tending operations to insure full site utilization by healthy trees of desirable species; and (2) integrated health management to protect the trees from harmful agents. A comprehensive program also (after Walstad and Kuch 1987) incorporates measures to

1 inhibit and retard the establishment of weed species through appropriate site preparation;
2 control existing interfering plants to reduce their numbers and sizes;
3 maintain a plant species diversity that will minimize long-term pest problems and protect essential ecological conditions;
4 accept all species that have value to a landowner, and not control those having no important negative effects; and
5 insure compatibility between the vegetation control practices and other components of the silvicultural system, and a landowner's objectives.

In this sense, the vegetation management would include site preparation with an even-aged reproduction method, and continual attention to community composition and health throughout a rotation. It should fit the long-term ownership objectives, benefit the preferred species at critical stages of their development, and ignore plants that do not interfere with those of economic interest or that have a beneficial ecologic effect.

Release treatments free desirable seedling- and sapling-stage trees from oppression and suppression, and insure that they become the dominant vegetation within a stand. They reduce both interference and competition. In this context, the term *interference* (Peterson 1986; Radosevich and Osteryoung 1987) means any interaction between neighboring plants including positive (stimulation), negative (depression or antagonism), and neutral effects. Interference may result from production and consumption of resources, excretion or leaching of growth stimulants or toxins, or parasitism, predation, and protection.

By contrast, *competition* (Zedaker 1983; Peterson 1986; Radosevich and Osteryoung 1987; Kimmins 1987) means negative interactions related to the shortage of resources in and around the physical space occupied by adjacent plants, or interference with use of even abundant resources. Competition may involve a single resource (e.g., light), or multiple resources (e.g., light and soil moisture). It includes interference by plants of the same species (intraspecific competition), and different species (interspecific competition).

Interference and competition often become adverse when logging or site preparation physically disturbs high proportions of a regeneration area, removing the litter or mixing it with mineral soil. Especially in the brighter environment created by an even-aged reproduction method, and with the sudden release of nutrients due to postcutting litter decomposition, the regeneration area often becomes ideal for germination and subsequent survival of many herbs and shrubs. If they interfere with tree seedling survival and development, foresters must often release the trees to insure success (Walstad et al. 1987). The treatments increase essential site resources by (1) making conditions less suitable for the interfering vegetation (e.g., by preregeneration site preparation); or (2) reducing the numbers and density of competing and interfering plants to economically safe levels (e.g., by release treatments). Even-aged reproduction methods allow foresters to use highly mechanized methods of site preparation and release treatments. This often increases the options and reduces costs compared with those for uneven-aged stands. The absence of residual trees and the distribution and abundance of slash even make clearcut areas suitable to prescribed burning.

JUDGING SUCCESSFUL ESTABLISHMENT

A decision to do release treatments of any kind presumes that the oppressed trees have promise and warrant an investment of early release. In making these judgments, foresters must first determine if sufficient numbers of desirable new trees have become established, and if their quality will serve a landowner's long-term interests. Actual criteria may differ with the intended uses, and particularly in evaluating commodity and other values (after Daniel et al. 1979). Yet in most cases, foresters consider a seedling or stand of seedlings *established* when they are safe from normal adverse influences common to the site (e.g., frost, drought, weeds, and browsing) and they no longer need special protection or tending (Ford-Robertson 1971). The actual criteria may differ from one site to another, but common features that distinguish established trees (Daniel et al. 1979; Tappeiner and Wagner 1987) include: good health and form; high vigor; a threshold minimum size (height, diameter, and crown volume); and freedom from suppression or probable damage by weed species and pests. Release treatments address the latter criteria by freeing otherwise suitable seedlings and saplings from interference by less desirable plants.

Ultimately, foresters must judge seedling-sapling development against the landowner's purpose for a stand, and the different species' tolerance to interference.

They must also account for a landowner's willingness to invest in vegetation management, or to accept the long-term consequences of not taking some kind of action. Overall, each evaluation may involve

1 identifying the objectives, and whether a stand will satisfy these without special release treatments;
2 evaluating whether interfering vegetation threatens the desirable trees, or the probable benefit of investing in the control;
3 setting characteristics of established trees, and determining their numbers and distribution in the community;
4 estimating probable future mortality of established trees (with and without control of interfering plants); and
5 determining the minimum numbers needed for adequate future stocking, commensurate with the objectives for management.

Foresters must decide that a stand will contain the requisite stocking of desirable trees *after* normal mortality. Otherwise, they need to increase the numbers (reinforcement planting), or release the trees to reduce losses as a stand develops.

As depicted in Fig. 15-2, the interference (Tappeiner and Wagner 1987) might (1) slow crop tree development, rather than reduce survival appreciably; (2) suppress crop tree development, and trigger eventual mortality; and (3) reduce both survivorship and rates of growth to an economically important degree. In sorting out these possibilities, foresters must judge (1) the probable growth and development of important competing vegetation; (2) how those species might temper crop tree survival and growth; and (3) how to effectively reduce existing interference, and suppress reinvasion from surrounding areas or residual seed. They often rely on general experience with similar sites and community types in making final decisions about when and what kinds of vegetation control to apply.

With timber production goals, the released crop must yield useful products of reasonable value, and repay the initial treatment costs. This includes an appropriate balance between species and quality composition, the numbers and spacing of acceptable stems, and the distribution of potential crop trees across a site. Realistically, a commercially viable stand might have only the minimum numbers of suitable trees for full site utilization by the time a crop matures. Less desirable trees may fill spaces between the crop trees during intermediate stages, as long as the fillers do not damage or interfere with the crop trees. In this context, the term *crop tree* (Ford-Robertson 1971; Soc. Am. For. 1989) means any tree forming part of a mature stand or age class, or a tree selected at an intermediate age to become part of the final crop when a stand or age class matures. For nontimber objectives, foresters must decide if a release treatment would enhance the values of interest, or improve opportunities for some uses. In either case, appropriate treatments (after Tappeiner and Wagner 1987) will

1 suppress the interfering vegetation and prevent its reinvasion;
2 enhance long-term crop tree survival and development;

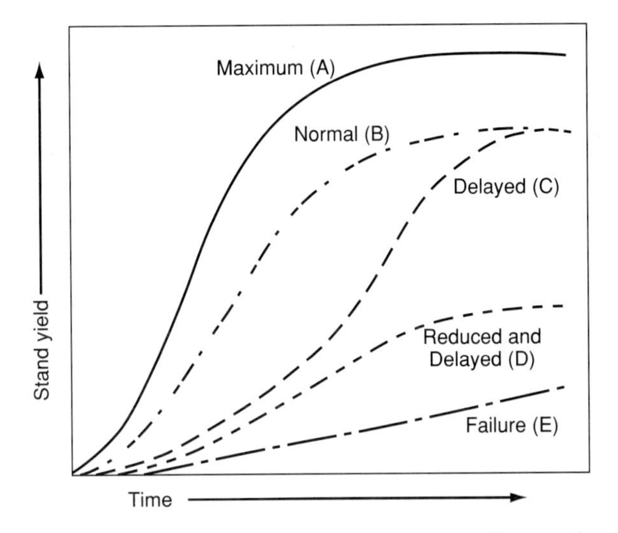

Five hypothetical scenarios of stand development under different kinds of vegetative competition: (A) maximum site productivity under low levels of vegetative competition; (B) normal rate of stand development and final yield as determined by commonly used yield tables or growth models; (C) effects of competition that reduce the rate of stand growth but not the final yield as compared to normal stands; (D) effects of competition that reduce both the rate of development and the amount of final yield; and (E) severe effects of competition that cause high mortality of desired species, thereby precluding development of a manageable stand.

FIGURE 15-2
Possible effects of interfering vegetation on tree community survival and development (from Tappeiner and Wagner 1987).

3 prove cost effective in the long run; and

4 protect ecologic resources and desirable nonmarket values from loss or degradation.

Otherwise foresters would let an existing community develop naturally.

VEGETATION AND REGENERATION SURVEYS

In designing a regeneration and vegetation survey, foresters take into account any special management objectives and unique conditions of a site. The findings might help them (after Tappeiner and Wagner 1987) to

1 select an appropriate reproduction method in advance, and identify any preharvest and postcutting site preparation requisite to success;

2 determine any postcutting conditions to rectify by site preparation, including slash reduction and scarification;

3 evaluate the status of regeneration and look for threats to seedling development;

4 inventory seedling numbers and distribution, as well as the status of undesirable or interfering species;
5 monitor or detect potential health problems;
6 look for site conditions that might deter germination or seedling establishment; or
7 evaluate the status of forage for grazing.

The surveys might also include features like the habitat suitability for indigenous wild creatures.

Objective assessments of this kind might lead to one of three possible follow-up actions:

1 *None*—because the status and condition of regeneration seem adequate to the need, or nothing seems likely to interfere with the new age class (e.g., no harmful amounts of interfering vegetation present or expected)
2 *Normal measures*—because conditions should support a new age class of adequate composition and stocking, given an appropriate reproduction method and standard site preparation (e.g., interfering vegetation below a critical level)
3 *Special measures*—because existing conditions portend a probable failure unless controlled by special treatments (e.g., interfering vegetation dense, and widely distributed across a site)

In many cases, foresters must judge if seedlings or saplings of the acceptable species have sufficient vigor and capacity to actually grow better following a release, and to surpass the undesirable species (Wenger 1955; Zutter et al. 1985; Burkhart et al 1987; Tappeiner and Wagner 1987; Cain 1989). For this, they can judge each seedling's growth potential based upon factors like (1) its position and height relative to surrounding vegetation; (2) the average numbers of side branches with healthy foliage on the seedlings; (3) past rate of height growth; (4) freedom from past insect and disease damage; and (5) the probable future growth of nearby competing vegetation. The assessment criteria must indicate whether the trees will remain free to grow, and if a release treatment might prove beneficial.

The methodology used must allow foresters to make an efficient, simple, accurate, unbiased, and reproducible assessment appropriate to the sizes and kinds of vegetation present (after Wagner 1982; Zutter et al. 1985). With objective surveys, foresters can delineate the numbers, sizes, and characteristics of species present. This would portray the patterns and degree of vegetation development and the community composition, or characterize attributes of the vegetation. Alternatively, foresters could simply use a *stocked quadrat survey,* where they install sample plots to determine what proportion of them has at least some minimum number of seedlings of a requisite size (Wenger 1955; Doucet 1991), or what proportion of them would need some special remedial treatment to correct a problem. This method requires only a yes-or-no decision to indicate if each sample plot satisfies the conditions of interest (e.g., stocked or not, or having too many interfering plants or not). The proportion of plots satisfying different threshold conditions indicates if foresters must include special release treatments.

The time to make a regeneration survey depends upon the silvical characteristics of a target species, and whether a forester will rely upon artificial method or natural regeneration. For mature even-aged stands having an important shade-tolerant component that will develop from advance regeneration, the assessment must precede the reproduction method. In fact, with difficult-to-regenerate species like the oaks, foresters may begin their assessments after the last thinning. Findings would indicate if foresters should increase the numbers and size of advance seedlings before the time of financial maturity. With species that regenerate after or coincident with a reproduction method, or with artificial regeneration, a survey would indicate any special site preparation needs. For all species, postcutting assessments show if an age class has become established in adequate numbers, and whether it requires release.

SOME SAMPLE GUIDELINES FOR JUDGING REGENERATION SUCCESS

Research has provided several field-applicable guidelines to aid foresters in assessing recently regenerated areas in common forest community types. The criteria reflect commercial timber production requirements, and often the minimum conditions for long-term success. Foresters have traditionally considered the regeneration successful when 60%–70% of the sample plots satisfy the minimum conditions of stocking. For example, within the Allegheny hardwood stands of eastern North America, successful stocking has developed when at least 70% of the 6-ft (1.83-m) radius regeneration plots have at least two acceptable seedlings at least 2 ft (0.6 m) tall (Marquis et al. 1984). Foresters can consider such stands *established.* For even younger regeneration (e.g., after the seed cutting and before overstory removal in shelterwood method), the plots should contain at least twenty-five acceptable seedlings having two or more postjuvenile leaves, or five or more acceptable trees at least 3 ft (0.9 m) tall, or a minimum of five acceptable trees at least 5 ft (1.5 m) tall. When at least 70% of the plots have any of these conditions, a commercially viable stand will probably develop, even in areas with fairly high levels of deer browsing.

Among eastern oak forests, threshold conditions for overstory removal, or likely development into an acceptable oak stand, differ markedly from those necessary in northern hardwoods. While most northern hardwood species grow rapidly in height if free of overtopping vegetation or repeated deer browsing, the oaks grow more slowly than their common competitors. In fact, oak seedlings take many years to reach free-to-grow positions, and usually need an advance start to remain in the upper canopy of a new age class. As a result, foresters must apply more stringent criteria in judging the adequacy of young oak regeneration. This usually means postponing removal of the entire mature overstory until a stand has a minimum of about 60% of 4.3-ft (1.3-m) radius plots with at least one 4.5 ft (1.37 m) tall sapling, or an oak at least 2.0 in. (5.1 cm) in diameter at the ground line, or at least one tree 1.5 in. (3.8 cm) at breast height (Sander et al. 1976; Sander et al. 1984). With stocking below this threshold, the stand must contain sufficient numbers of sproutable oak stumps to compensate for the shortage of tall saplings (Johnson 1977). Large saplings of these minimum sizes, and stump sprouts, normally remain in upper canopy positions within a new community, assuring at least a minimum acceptable future stocking of oak.

16

RELEASE CUTTINGS

EARLY RELEASE TREATMENTS

Release treatments salvage a situation deemed untenable in light of ownership goals when stands have adequate regeneration, but other vegetation threatens its long-term survival and development. Before committing to the work, foresters should recognize that the poor position of sought-after seedlings *may* reflect ecological conditions suboptimal for the chosen species. These might include circumstances such as:

1 inhospitable or limiting site conditions;
2 species not suited to the physical environment;
3 high population levels of uncontrolled destructive agents that limit tree development or threaten their survival; or
4 inadequate past site preparation practices to alter physical or chemical soil conditions in favor of good germination, early survival, and long-term growth of choice species.

Under any or all of those circumstances, the species of choice may not grow well, even if released during early ages. Then landowners must reassess their objectives, often accepting a secondary species better suited to the environment that prevails.

Release treatments alter the species composition in favor of the trees best suited to a landowner's interests. Common objectives (after Kostler 1956) include

1 protecting young trees from injuries and suppression by shrubs, vines, and other interfering plants;
2 improving the species and quality composition of trees in a young community;

3 reducing interference by more rapidly developing or older trees along the margins (edges) of a stand;

4 reducing crowding of desirable species within too densely stocked young growth; and

5 enhancing the growth and development of selected elite trees or those of a particular species.

In any of these cases, foresters will eliminate or suppress herbaceous vegetation, shrubs, or undesirable trees that (1) germinated immediately following a reproduction method cutting, and grew more rapidly in height; (2) occurred as advance vegetation, and grew rapidly enough to overtop new seedlings of more desirable species; or (3) remained from a previous rotation, and overtop the new trees.

They must schedule the treatments before suppression by interfering vegetation becomes irreversible, or before the trees become too large for some kinds of mechanized operations. On the other hand, they must usually wait until the new stand differentiates sufficiently to make the potential crop trees apparent, and their potential for rapid future growth and development reasonably certain. Generally, these early treatments accelerate stand development (shorten the reorganization phase, hasten crown canopy closure, and hasten aggradation). They also alter the wildlife habitat diversity by eliminating important food plants, or moving stands more quickly into the aggradation phase. In particular, the diversity of avian species may decrease sooner, compared with areas not given early tending (Hunter 1990).

Foresters classify different release treatments based upon the age and nature of vegetation removed, relative to the sought-after species or crop. In some cases, the treatments also differ in the age of trees released, as follows (after Hawley 1921; Ford-Robertson 1971; Daniel et al. 1879; Soc. Am. For. 1989; Smith 1986):

Cleaning

• a treatment *during the sapling stage* to free selected trees from competition of overtopping trees *of comparable age*, and to favor the trees of better species and quality

• removing woody vines and shrubs that overtop or seem likely to suppress desirable trees not past sapling stage

Weeding

• a treatment during the seedling stage to eliminate or suppress mainly herbaceous plants and shrubs that overtop or interfere with desirable young trees

In natural stands, foresters normally do cleanings following crown canopy closure (early in the aggradation phase of stand development), and after the crop trees become overtopped or oppressed. In some cases they anticipate a need and treat a stand before serious problems develop, even before the crown canopy closes. Weedings occur during the reorganization phase, to insure timely emergence of the tree seedlings from herbaceous and shrub competition. For the most part, foresters use these practices in even-aged silvicultural systems. Within uneven-aged stands, cleanings reduce the density of young trees and maintain a balance of saplings in a stand.

WEEDING

Most commonly, foresters do weeding to follow up artificial regeneration at sites with a dense herbaceous and shrub community. Weeding has also proven necessary after reforestation of open sites formerly occupied by a dense herbaceous community. In both cases, the release may extend earlier site preparation treatments. It may also remedy problems where landowners skipped any earlier vegetation control, or where it proved ineffective. Weeding may include either chemical or mechanical methods, as spot treatments around individual seedlings, or by broadcast applications across the entire stand.

Spot weeding includes both mechanical (e.g., grubbing or hoeing) and chemical (foliar spray and ground application) methods. For the former, workers cut or dig out other plants within a fixed circular radius of selected tree seedlings. They might also cover the surface with a paper or plastic mulch to kill the herbs or prevent reestablishment of obtrusive plants after their removal. Spot chemical treatments normally involve foliar sprays (Fig. 16-1) or ground applications to all vegetation within a fixed radius of the selected trees (e.g., 3–6 ft, or 0.9–18 m). The extent of area treated depends upon the kind, density, and size of interfering plants in relation to the preferred tree seedlings. To reduce costs in natural stands, foresters usually release only a limited number of trees (e.g., at a spacing similar to plantations), or release only the potential crop trees at even wider intervals.

For broadcast methods, foresters apply sprays, mists, or granular herbicides using specially designed aircraft or ground vehicles. Backpack mistblowers and small fertilizer spreaders have also proven effective for treating relatively small areas. Normally, broadcast chemical applications of these kinds kill most of the herbaceous and broadleaved woody plants. With many conifers, foresters can choose a selective herbicide that will not harm the trees. They may also apply foliar active chemicals over the seedlings during dormant periods (after winter buds set in late summer, or before shoots elongate at the beginning of a growing season), but after the other species have sprouted. Herb and shrub reinvasion will normally occur during the first season following weeding. Yet the tree seedlings benefit from even a temporary reduction of competition. In fact, one study in eastern Texas showed that a broadcast spray of foliar-active herbicides applied over dormant loblolly pine reduced herb competition for only about four months. Nevertheless, it increased first-year seedling volume (size) by about 300%–340% (Schoenholtz and Barber 1989).

Other research showed a long-term cumulative effect upon tree size and volume as well. Most experiments removed both woody and herbaceous plants. To illustrate, controlling hardwood competition in a young loblolly pine plantation in Louisiana resulted in greater heights and diameters, and 151% more volume by ten years of age (Harwood 1986). In Virginia, release treatments in one-year-old plantations at numerous sites increased the volume of loblolly pine after seventeen years by 123% to 203% (Dierauf 1989a, 1989b, 1989c, 1990a, 1990b, 1990c, 1990d, 1990e, 1990f). Also in Virginia, weed control in one-, two-, and three-year-old loblolly pine plantations increased tree volume growth the most when done during the second growing season—and where weeding removed all the herbaceous vegetation and two-thirds of the hardwoods. Complete hardwood control did not

FIGURE 16-1
Spot treatments around recently planted seedlings will often reduce the competition and interference by herbaceous plants, improving survival and speeding development of the new plantation.

seem necessary, nor prove desirable (Bacon and Zedaker 1985). Other tests to control both herbaceous and woody competition at fourteen different sites across southeastern United States showed best results with:

total vegetation control > herbaceous control > woody control > no control

Over five years, individual tree volume increased by 67% following woody species control, and 171% after herbaceous control (Miller et al. 1991). Similarly, using herbicides to remove herbaceous and hardwood competition from around balsam fir saplings in Maine increased the two-year posttreatment volume increment by 1.6 times on the better-drained soils, compared with 1.3 times among imperfectly drained soils (Briggs and Lemin 1991). With southern pines, appropriate herbaceous control may shorten the rotation by as much as three years. This would make the treatment financially attractive (Busby 1989).

Aerial application of herbicide over Norway spruce in the central Appalachians of North America effectively reduced hay-scented fern, blackberries, and goldenrod. New plants colonized the site by the second year, and Norway spruce survival did not change over two years following treatment. Height growth did increase, and numbers of trees in a free-to-grow position improved from 38% to 68% (Wendel and Kochenderfer 1984). Releasing young black spruce and jack pine in northeastern Canada increased terminal shoot growth, but only in areas with hardwoods twice as tall as the conifers (Richardson 1982). Similarly, longleaf pine in Georgia showed greater height growth following early release on sites that had only low-intensity site preparation, compared with more intensive treatments (Boyer 1989). In California, survival also did not improve following a one-time chemical application over a 4-ft (1.2-m) area around planted ponderosa pine at year two, or mechanically grubbing around them during the first (4-ft or 1.2-m radius) and third (6-ft or 1.8-m radius) years for the two treatments. Released seedlings had 40% and 53%, respectively, greater six-year heights, and 93% and 62% larger stem diameters (McDonald and Fiddler 1990). In another California test, spot treatments of herbicide in a 3-ft (0.9-m) radius around Douglas-fir seedlings during the third and fifth growing seasons significantly improved survival (73% vs. 42%). Stem diameter growth also increased by about 55%, and height growth by about 170% (McDonald and Fiddler 1986).

Experience with different forms of vegetation management in western North America has demonstrated that foresters should deal with herb competition through site preparation. Even so, when herb cover exceeds 50% of the surface within established stands, or even less on droughty sites, early weeding will double seedling survival. Established seedlings will probably not succumb to shrub competition, but may grow substantially slower at sites with at least 30% shrub cover (20% on drier sites). For best results, foresters must apply the treatments before the trees become overtopped, and before shrubs become well established and large in size (Miller 1986a). Among pine stands in southeastern United States, competition from herbs, shrubs, and vines has greater effects than that from hardwood trees. Postrelease improvements of diameter and height growth depend upon the nature of an interfering plant community, the timing and effectiveness of treatment, and stand age. Also, the degree of competition will vary with (1) the abundance and species composition of available herb and shrub seed stored in the litter; (2) the density and size of residual plants of these species; (3) the degree of surface disturbance during logging and site preparation; and (4) the environment at a site. Benefits accrue primarily in diameter growth, but height increment may improve as well (Minogue et al. 1991).

CLEANING

Figure 16-2a illustrates a stand needing cleaning, and the results. In the top diagram, an unwanted species partly covers the sought-after trees, and dominates the stand. Unless released, the overtopped trees will continue to decline, and eventually die. This often happens, for example, when pin cherry overtops a species of low or intermediate shade tolerance (e.g., yellow or paper birch) in northern hardwood stands of eastern United States (Marks 1974; Safford and Filip 1974; Heitzman and Nyland 1994). If the overtopping species casts a low-density shade, and if the overtopped species has high shade tolerance, the smaller trees may remain alive for decades. This can occur in mixed-species stands where a rapidly growing tree like

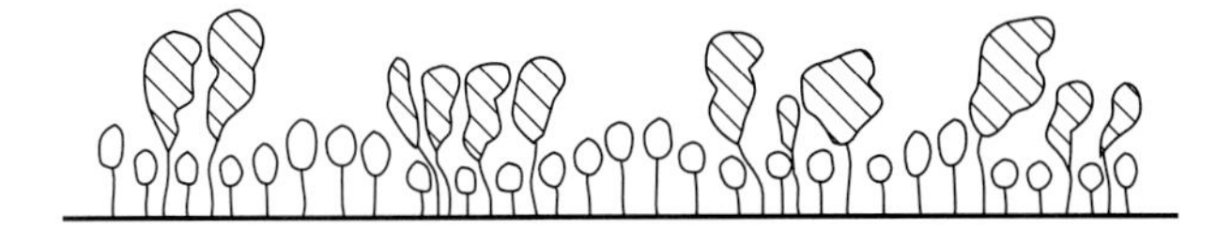

Before cleaning

In some stands, undesirable trees may partly cover the more desirable trees, requiring a release cutting to free the oppressed saplings of good quality or species.
a

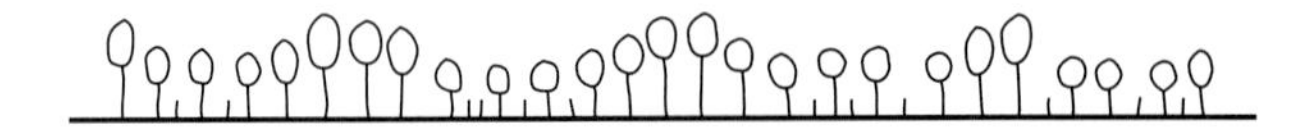

After cleaning

Once in a free-to-grow position, the saplings develop rapidly to form the new even-aged community.
b

FIGURE 16-2
Cleaning removes undesirable trees from sapling stands to free selected trees of desirable species or quality from overtopping trees of comparable age.

trembling aspen overtops a fairly shade-tolerant one such as sugar maple and beech (Roberts 1992), or western spruce fir (Alexander 1974a). The understory saplings persist for many years. Cleaning would free these desirable trees, insuring their development into upper canopy positions (Fig. 16-2b).

Benefits of release vary with species, age, degree of suppression, and completeness of a treatment. Generally, cleaning works best in young stands, while the preferred trees still have sufficient vigor to respond to release (Fig. 16-3). One survey of cleaning and weeding in the upper Lake States region of North America indicated that released conifers had 43% better survival, 120% greater height development, and about 815% more above-ground biomass (Perala 1982). In North Carolina, releasing young shortleaf pine by mechanical or chemical methods did not improve survival or height growth, but diameter increment increased by about 20% (Lloyd et al. 1991). With shortleaf and loblolly pine, cleaned stands may produce 4.5% more volume by age 18. At current stumpage prices this would provide landowners with a favorable return on the investment (Guldin 1984). For greatest effect, cleaning should favor trees of upper canopy positions. Consequently, in natural stands foresters must wait until the trees differentiate by crown position, and the dominant stems become apparent within the species of interest (Anderson et al. 1990).

Common methods for cleaning include (after Miller 1986b; Smith 1986; Miller 1991; Minogue 1991):

FIGURE 16-3
Cleaning releases saplings of desirable characteristics by removing overtopping trees of the same age class, as with this spruce-fir community in northeastern North America.

1 mowing, using tractor-mounted brush cutters (brush hogs) to create cleared lanes adjacent to the crop trees;

2 directed foliar sprays, mists, granules, and pellets applied from tractor-mounted equipment; and

3 broadcast application of sprays, granules, and pellets from aircraft.

With broadcast sprays, crews must follow exact protocols to prevent drift into adjacent areas, to insure uniform coverage of the treated stand, and to regulate herbicide concentrations and application rates (see Chapter 5).

Trials that completely controlled hardwoods in young naturally regenerated longleaf pine stands illustrate the biologic gains from early cleaning (Boyer 1985).

Researchers removed all hardwoods beginning at ages one, two, three, four, or eight years, and re-treated the plots to maintain full control. At age four, they also equalized the pine seedling density across all plots to minimize competition between pine trees. By age thirty, pine density ranged from 600–800/ac (1482–1976/ha) in released plots, compared with 256/ac (632/ha) in untreated plots. Pines in untreated areas had significantly shorter average heights and smaller average diameters, and 7.9 to 9.6 times less merchantable volume. Delaying the release for at least a few years did not diminish long-term volume production, at least at reasonably good sites with adequate pine stocking.

With ground methods, workers can direct the sprays, mists, and solid herbicides into lanes between rows or lines of crop trees. This protects the crop trees to some degree, at least in plantations. Foresters might also use directed sprays to create narrow lines or bands of residual trees at regular intervals, similar to that of plantations. Mechanical methods (mowing with brush hogs or rotary mowers) also allow foresters to reduce competition along two sides of the rows in plantations. In natural stands they can leave narrow strips of saplings between the mowed lanes (Fig. 16-4). For intensive release, crews with brush saws could also cut competitors in the unmowed strips, leaving the residual stand with a spacing similar to that of plantations (Seymour et al. 1984).

FIGURE 16-4
For early cleaning workers can use brush hogs or brush saws to cut narrow lanes, leaving rows or bands of saplings at regular intervals and with open space on two sides of the crowns.

Operational strip treatments in southern pines in southeastern North America have involved early entry (e.g., ages two to five years) using heavy-duty rotary mowers to cut swaths through the regeneration. Foresters can probably justify this method in naturally regenerated or direct-seeded southern pine stands having 1500/ac (3705/ha) or more seedlings (Balmer and Williston 1973). With loblolly pine in Arkansas, cutting 12-ft (3.7-m) wide lanes to leave 1-ft (0.3-m) strips of pine regeneration decreased the density by 89%, leaving about 1920 residual pine trees per acre (4742/ha). After two years, diameters of the released trees exceeded those of untreated stems by 130% (Cain 1983). Prescribed burning after three growing seasons reduced the fuel hazard and the nonpine vegetation on both sets of plots without killing the pines. By ten years, the average annual diameter growth of pines in treated plots averaged 0.51 in. (1.3 cm), compared to 0.31 in. (0.8 cm) for trees in untreated plots (Cain 1993).

Experimental strip treatments in a dense stand of direct seeded three-year-old loblolly pine have proven as effective as removing selected individual trees when both reduce the stem density to comparable levels. Researchers cut 6.6-ft (2.0-m) wide swaths, leaving equally wide bands of seedlings between them. Or they cut 6.6-ft (2.0-m) swaths at right angles to each other, leaving a checkerboard of uncut squares. They also cut 7.5-ft (2.3-m) wide swaths, leaving 3.5-ft (1.1-m) uncut between them. For other plots they selected the best trees as residuals, and reduced the pines to densities of 750 to 4350/ac (1853–10,745/ha). The average diameters and merchantable volumes at age sixteen inversely correlated to the posttreatment residual density. For residual levels of 1450–2900/ac (3582–7163/ha), strip treatments proved as effective as the individual tree removals (Lohrey 1977). The treatment proved best at early ages when the seedlings still had good vigor, and when foresters could still use fairly light equipment (e.g., a rotary mower) to reduce seedling density to as low as 500–750/ac (1235-1853/ha) (Nebeker et al. 1985). Likewise, prescribed burning under a 17-yr-old loblolly pine stand of 10,000/ac (24,700/ha) reduced stem numbers by 65%. It left most stems greater than 2.5 in. (6.4 cm) dbh (NcNab 1977). Burning would also reduce hardwood competition in poorly-stocked stands.

Experiments among dense stands of 25-yr-old lodgepole pine saplings in the Rocky Mountains of Canada produced similar results (Bella and DeFranceschi 1977). Workers cut parallel 10-ft (3-m) wide swaths, leaving 4–6 ft (1.2–1.8 m) wide lanes of residual trees between them. This enhanced the second five-year mean increment of the biggest trees of upper-canopy positions by 65%, compared to untreated trees. Trees within the most dense areas showed the most dramatic improvement in growth. Also, since the strip treatments freed trees on only one side, a second release seemed appropriate within five years. Further, scheduling the first treatments before crop trees reach 6 ft (1.8 m) tall (age five to ten) would allow foresters to use tractor-mounted choppers or other brush cutters.

For broadcast chemical methods, foresters must kill or suppress the unwanted vegetation without harming the preferred species. If undesirable saplings completely overtop the preferred species, delivering a fine herbicide spray or mist from above may reduce penetration through the canopy, protecting the saplings in understory positions. At least with conifers, many species survive low-dosage sprays that will kill broadleaved competitors. For example, among spruce-fir stands in northeastern North America, triclopyr and glyphosate compounds will kill broadleaved species such as maples, aspens, and birches without damaging the conifers. Phenoxy herbicides leave more hardwood diversity and cover, and also do not kill the spruce or fir. Many conifers also have less susceptibility to these formulations during the late growing season, or prior to shoot elongation in spring. Altering the dosage will affect the degree of hardwood control, allowing foresters to even maintain some scattered hardwoods for a more diverse tree community (Newton et al. 1992a, 1992b).

For all chemical methods, users must carefully control the timing, dosage, and rate of application in accord with registration labels (see Chapter 5). Even so, vapors volatilizing from the herbicides during hot weather may damage some conifers, and this limits the timing of application (Miller 1991; Minogue et al. 1991). Broadcast methods do not allow crews a high degree of selectivity in choosing individual trees and shrubs to remove, and require caution in treating sensitive areas. Individual tree treatments give workers a greater degree of control, but due to the great inputs of labor they usually prove inefficient for large areas (after Mahoney 1986; Miller 1986b).

Figure 16-2 suggests another situation, where the competing vegetation does not overtop an entire stand of desirable saplings. Even this may reduce the vigor and growth of many desirable trees, as suggested by their condition in the figure. Here, too, cleaning would remove the taller undesirable saplings to improve growing conditions for the preferred trees, and to promote their long-term development. Because many of the desired saplings have exposed crowns, broadcast methods might prove infeasible unless foresters can use herbicides that have a selective effect, or apply them at times when the preferred species would not suffer damage (e.g., after winter buds set late in a growing season).

Spot treatments often prove effective and less costly where unwanted trees overtop only part of the desired crop. These include both mechanical and chemical methods (after Miller 1986b; Smith 1986; Miller 1991; Minogue et al. 1991) as follows:

1 tree cutting using axes, machetes, brush hooks, brush saws, small chainsaws, and similar devices;

2 tree cutting, plus spraying stumps with a herbicide to prevent sprouting;

3 injecting a herbicide into cuts in the bark;

4 selectively spraying all or parts of individual tree and shrub crowns, usually limited to vegetation less than 6 ft (1.8 m) tall;

5 spraying a herbicide mixed with a carrier (e.g., kerosene, diesel fuel, or mineral oil) onto the bark within 1–2 ft (0.3–0.6 m) of the ground; and

6 applying granular, pellet, or concentrated liquid forms of herbicide to the soil surface around the base of the target vegetation.

Simply cutting most broadleaved trees and shrubs will normally stimulate sprouting, and often with increased vigor. To prevent this, workers might apply a herbicide spray or granules to the stumps, or distribute a granular herbicide around the base.

Ground-spread granular imazapyr and hexazinone will kill hardwoods, but imazapyr can also reduce the growth of loblolly pine (Loveless and Page 1991). Soil treatments may also kill pines if the roots extend into the treated area. Small hardwoods with limited root systems may escape (Minogue et al. 1985). Basal sprays and stem injection allow greater control. With the former, workers direct the flow of herbicide onto the bark at the base of each treated tree. The compound enters through the bark, and translocates to the foliage. Depending on susceptibility of the species and its size, the treatment can involve

1 a full basal spray to cover the lower 1–2 ft (0.3–0.6 m) of bark on all sides of a stem, and run down onto the root collar;

2 a low-volume basal spray that covers only the lower 1–2 ft (0.3–0.6 m) of bark, but does not run off; or

3 a streamline basal spray of concentrated herbicide applied horizontally across or vertically up and down the target stem.

Alternatively, workers make axe or hatchet cuts through the bark at a convenient height, and squirt a measured volume of herbicide into each cut. They can also use the same technique, but with hypo-hatchets. These have a hollow handle fitted with a tube that runs to a supply of herbicide, and an impact-activated valve that releases a measured volume of herbicide into the cut through openings in the face of the hatchet blade (Fig. 16-5). For some operations, they might also use tubular tree injectors consisting of a long hollow tube fitted with a chisel-type blade at the base. The tube holds herbicide. After making a cut at the base of a tree, the worker opens a valve to release a measured amount of chemical that flows into the incision (Miller 1991).

Tree cutting, stem injection, and basal spray methods often prove necessary for cleaning in hardwood stands, or with any species that has a high susceptibility to foliar sprays. Each method kills individually treated stems, but not adjacent crop trees. Chainsaw felling of hardwoods less than 3–4 in. (7.6–10.2 cm) dbh may amount to only about one-half the cost of killing the trees by basal spray, and two-thirds that of stem injection (Miller 1984). With early cleaning, the small size of many saplings may preclude cutting or stem injection, leaving basal sprays or soil surface application as the only practical methods.

The choice of a broadcast or spot method often depends upon the density of trees to remove, and their sizes. These spot methods may prove cost effective in stands with only sparse competition, and that require no more than about one day of labor per acre (2.5 days/ha) (Perala 1982). In fact, experience in northern hardwoods indicates that when removing less than 500/ac (1235/ha) of trees at least 3 in. (7.6 cm) dbh, individual stem treatments usually cost less than broadcast methods (Sage 1987). Similarly, in southern pine communities, individual stem herbicide injection costs less than broadcast methods where foresters must release no more than 250–300/ac (618–741/ha) of trees at least 3–4 in. (7.6–10.2 cm) (Williamson et al. 1976; Nelson and Miller 1991). They have little choice in hardwood communities

FIGURE 16-5
Hypo-hatchets and tree injectors have an impact-activated valve that releases a measured volume of herbicide into the cut through openings in the face of the blade (courtesy U.S. Forest Service).

where foliar sprays and herbicides absorbed through root systems damage or kill all species. Factors like size of the area treated, access to an adequate and affordable labor force, and limitations imposed by environmental regulations may also influence the decision.

RESPONSE OF CROP TREES TO RELEASE

The posttreatment response of released trees depends upon the permanency of release, the species, and the age and condition of the favored saplings (Trimble 1973, 1974; Smith and Lamson 1986; Wendel and Lamson 1987; Lamson and Smith 1989; Voorhis 1990; Rathfon et al. 1991). Degree of release also often proves important. To illustrate, cleaning 11-yr-old second-growth hardwoods in the upper Lake States of North America by removing competitors within 2.5 or 5.0 ft (0.8 or 1.5 m) increased annual diameter growth to 0.25 and 0.30 in. (0.64 and 0.76 cm) over eight years, compared with 0.24 in. (0.61 cm) for unreleased trees. With the heavier release, 64% of crop trees increased 2 in. (5.1 cm) or more in diameter over eight years, compared with only 38% for untreated trees (Stoeckler and Arbogast 1947). Similarly, clearing competing trees from a 11–12 ft (3.4–3.7 m) radius around selected yellow birch saplings more effectively stimulated their growth than less intensive cleaning (Erdmann et al. 1981).

Cleaning 8-yr-old northern hardwoods in New England also significantly increased the diameter growth, vigor, and crown width of crop trees (Voorhis 1990). The degree varied with species, and lasted longer following heavy release. For paper and yellow birch, crop trees had significantly better diameter and crown growth over a 16-yr period following cleaning. For sugar maple, cleaning enhanced diameter growth, but not significantly. It significantly increased crown width only following heavy release. Cleaning generally kept the lower limbs alive longer, and crop trees given heavy release had significantly shorter heights to the base of the live crown and the first major branch. By contrast, cleaning 7-yr-old yellow-poplar and black cherry in the central Appalachians did not effectively enhance the diameter or crown position of crop trees (Trimble 1973). Over five years, sprouts of cut trees and grape vines overtopped the released trees and largely nullified effects of the treatment. In another Appalachian study, releasing 9-yr-old sugar maple, red oak, black cherry, and yellow-poplar also did not significantly improve diameter growth, crown position, or length of clear stem (Lamson and Smith 1987). In fact, five years after release only 41% of the dominant and 24% of the codominant trees remained in those same crown classes. Others had regressed to intermediate or overtopped positions, or had died. Overall, about one-quarter of the released trees had adequate crown position, vigor, and stem form to remain as crop trees. Generally, delaying cleaning until age fifteen seemed more appropriate (Trimble 1971).

Releasing 12-yr-old planted Douglas-fir in the Pacific Northwest of North America by removing sprouts of madrone, tan oak, and all other vegetation within 9 ft (2.7 m) also stimulated diameter development. After nine years, released trees averaged 1.2 times greater in diameter at 12 in. (30.5 cm) above the ground, but their heights did not differ significantly from those of nonreleased trees (McDonald and

Tucker 1989). With 7-yr-old western hemlock, cutting all trees within 22 ft (6.7 m) increased the diameters by 5.3 in. (13.5 cm) after seventeen years at one location, and by 5.5 in. (14.0 cm) after eleven years at another site. Intertree competition became apparent among trees left at a 16-ft (4.9 m) spacing within twelve years at the first location, and nine years at the other (the better site). At closer spacings (down to four feet), diameter growth slowed sooner (Hoyer and Swanzy 1986). Even cleaning a dense natural stand of fifty-five year old sapling-size ponderosa and Jeffrey pine advance regeneration improved the growth. Researchers selected crop trees at about an 8-ft (2.4 m) spacing, and cut all others. Though growth slowed after twelve years, crop tree diameter and height growth, respectively, increased by 167% and 62% during the first five years, by 43% and 38% during a twelve-year period, and by 91% and 39% over thirty years (Lilieholm et al. 1989). Released trees also had a 31% greater live crown ratio.

General areawide treatments have often proven less effective. In one case, researchers reduced central hardwood sapling stands in Ohio by a series of treatments at ages eight, ten, and seventeen. Each time they cut sufficient trees to maintain 30%, 50%, and 70% residual stocking. By age twenty-two, none of these treatments had substantially enhanced the growth or quality of potential yellow-poplar, red maple, and oak crop trees (Hilt and Dale 1982). Better results have followed cuttings that removed all adjacent trees whose crowns touched that of the crop trees—the *crown touching method* (Smith and Lamson 1986). The treatment must favor saplings of upper-canopy positions (dominants and codominants).

In eleven-year-old cove hardwood stands of eastern North America, completely releasing the crop trees did not significantly increase the numbers of defects in butt logs at age twenty-five (Della-Bianca 1983). Red oak lower bole quality improved because the better diameter growth healed over the branch stubs more rapidly. By contrast, released red maple and white oak had more large live limbs on the lower bole. Likewise with many other hardwood species, lower branches of released trees remain alive longer, increasing low forking and reducing the clear bole length and potential value. Even so, early cleaning may prove critical in stands where rapidly growing pioneer species of little commercial potential overtop high-value trees (Heitzman and Nyland 1991). In fact, an untreated 7-yr-old paper birch-aspen stand in northern New England converted to essentially pure aspen after twenty-four years. The trees averaged 7 in. (17.8 cm) dbh. Cleaned plots had dominantly paper birch that averaged 6 in. (15.2 cm) dbh, with some trees as large as 10 in. (25.4 cm) (LaBonte and Nash 1978). In cases like these, overtopped trees of shade-intolerant species may succumb unless released, substantially altering both the commercial value and the species richness of a stand (Heitzman and Nyland 1994).

LIBERATION CUTTING

Liberation cutting differs from weeding and cleaning in the kind and age of vegetation removed, and the nature of the stand treated. *Liberation* or *liberation cutting* (after Ford-Robertson 1971; Smith 1986; Soc. Am For. 1989) means

• freeing trees not past sapling stage from competition of older overtopping trees

• freeing young growth not past the sapling stage from overhead competition

Foresters often use liberation cuttings where past logging removed the good trees, leaving culls and other unmerchantable trees behind. These residuals have little value and suppress the development of young understory saplings.

In many cases *liberation cutting* corrects a problem not addressed by earlier site preparation. It serves to promote the vigor of an oppressed sapling understory, and to salvage an untenable situation. For example, one study of grand and Shasta red fir in the Pacific Northwest of North America showed that poisoning the residual overstory tripled the diameter increment of the younger saplings within five years. Further, radial growth increased significantly more during the second five-year period. Height growth also doubled during the first growing season following release (Seidel 1983). In other studies, red fir height growth did not improve for the first five years (Gordon 1973), but trees with at least a 50% live crown ratio responded best to release (Seidel 1980). For hot and dry sites, some advantage accrues from a two-stage liberation treatment that maintains some overstory protection until the understory trees improve in vigor and condition (Ferguson and Adams 1979).

Figures 16-6 and 16-7 illustrate liberation cutting. As suggested by the diagram,

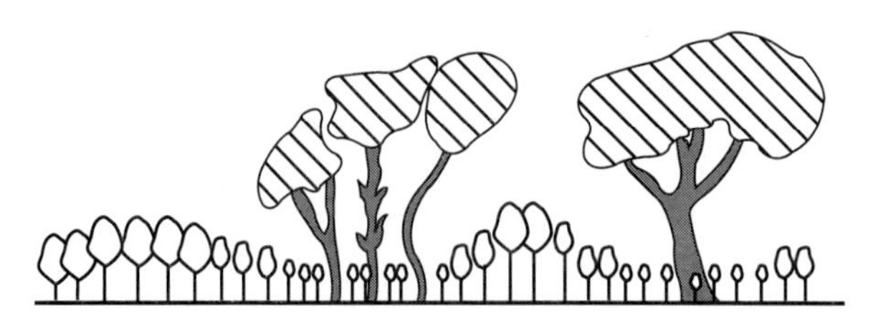

Before liberation cutting

After liberation

FIGURE 16-6
Liberation cutting removes older trees of poor quality to release a desirable understory not past sapling stage to become the dominant community at a site (courtesy of U.S. Forest Service).

FIGURE 16-7
Liberation cutting eliminates undesirable residual trees left during past mismanagement, as in this case where stem injection techniques killed poor-quality residual red maple to free a new age class of trembling aspen.

in some cases, the poor-quality older trees might form a complete canopy. Most of the time they occur at irregular spacing, and with a discontinuous elevated canopy. The drawing emphasizes the poor quality of these older trees, and that they interfere with the development of younger and shorter saplings of acceptable species and condition. This clearly separates liberation cutting from an overstory removal (Fig. 16-8) in shelterwood or seed-tree methods. In the latter cases, cut trees have excellent phenotypic characteristics, and foresters remove them as a part of the reproduction method. By contrast, a liberation cutting takes off undesirable residual trees left during past logging, and converts the space to more productive growth (Fig. 16-7). It usually yields limited volumes of poor quality trees, whose value (e.g., pulpwood or firewood) may only marginally cover costs of the operation. In fact, for many liberation cuttings, landowners pay for cutting or poisoning the cull trees. In these cases, foresters often call the liberation cutting a *cull removal.*

Most liberation cuttings remove only scattered individual trees, and broadcast treatments prove unnecessary or inappropriate. Common techniques include

1 tree cutting using chainsaws;
2 tree cutting, plus spraying the stumps with a herbicide to prevent sprouting;

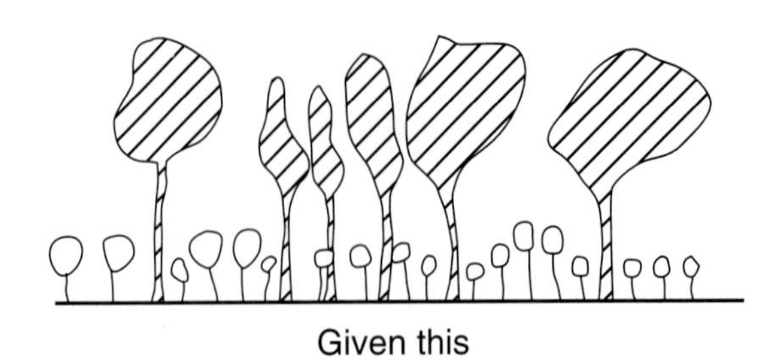

Is this liberation or removal in shelterwood method?

FIGURE 16-8
Liberation cutting differs from a removal cutting by virtue of the poor condition of the overstory, and the lack of intent in previous logging to regenerate a new community from seed off selected trees of desirable phenotypes.

3 injecting a herbicide into cuts in the bark; and

4 applying granular, pellet, or concentrated liquid forms of herbicide to the soil surface close to the base of the target tree.

Felling culls with large crowns will damage saplings in the impact area. This could substantially reduce the stocking, or leave many flattened areas within a sapling stand. Likewise, if machine operators use bulldozers to push over or uproot the cull trees, they also damage many desirable young trees. Even carefully planned mechanized logging will cause much damage, and the low-value products may not compensate for losses from the harvesting. Instead, foresters often use stem injection techniques. The dead trees eventually fall down in pieces, doing little harm to the sapling stand.

To kill trees of only moderate diameters, workers can often save time by using spaced cuts, and filling them with herbicide. Reduction of the crown by more than 95% has followed injection of herbicides in cuts spaced 4 in. (10.2 cm) apart among a variety of southern hardwoods as large as 12 in. (30.5 cm) dbh (Ezell et al. 1993). With larger trees of many species, they usually make complete axe or chainsaw girdles around the trunk, and then inject herbicide into all of the cuts. For soil treatments, workers must keep the herbicide near the base of each treated tree in order to protect adjacent saplings. In fact, tests with white pine in the Appalachians of the eastern United States indicated that treating the soil closer than 3–4 ft (0.9–1.2 m) from a 3-ft (0.9 m) tall tree might kill it. In stands with larger saplings, workers should keep the herbicide spots even farther from the pines (Wendel and Kochenderfer 1988). Otherwise, sapling losses will leave the stand with an irregular intertree spacing and incomplete site utilization, much like that caused by felling damage.

Broadcast applications often prove impractical unless the overtopping trees form a complete and closed canopy. Otherwise the spray drifts into the understory and damages the saplings. It can work effectively for stands with a susceptible species in the overstory, and an understory of herbicide-resistant saplings (e.g., scattered hardwoods over conifers). In fact, broadcast spraying overtopping hardwoods in the southeastern United States has significantly increased the diameter growth of 5-yr-old loblolly and shortleaf pine for five to ten years (Guldin 1985).

SOME BROAD ECONOMIC CONSIDERATIONS

Herbicides have gained widespread use for many types of vegetation control. They work more effectively than mechanical means, and often cost less if applied by broadcast methods over fairly large areas. In fact, one assessment based upon 1984 costs and values indicated that backpack and aerial spraying of herbicides for release treatments would cost only one-third to three-quarters as much as brush cutting. Also, landowners realize greatest cost efficiency in treating tracts larger than 50–100/ac 124–247/ha) (Row 1987). Further, assessments for loblolly pine suggest that where landowners control the hardwoods by site preparation and timely post-planting release treatments, stands have more pine volume and provide higher returns (Fig. 16-9). Vegetation control helps to concentrate the production on the more valuable pines. As a consequence, stands sustain a commercial thinning earlier, and contain more trees in sawtimber size classes by age thirty (Brodie et al. 1987).

Release cuttings primarily address commodity production interests by insuring adequate stocking of valuable trees. The treatments frequently reduce species richness in a stand, and almost always alter the relative importance of those that remain. In many cases, species removed by a cleaning include fast-growing shade-intolerant species with relatively short lives. Often called *pioneer species*, these have little commercial value and usually disappear from a stand fairly early in the aggradation phase. While present, the pioneers interfere with slower trees of greater value. Cleaning may also remove long-lived species of little worth to a landowner, thereby affecting long-term successional trends in a stand.

Both cleaning and weeding have implications for several nontimber values. The most obvious pertain to habitat for mammals and birds. For example, in eastern hardwood forests the fast-growing and short-lived tree and shrub species include several that produce both soft and hard *mast*—fruits used as food by animals. Cleaning may remove these, leaving longer-lived trees that do not flower until much older ages, and affecting early mast production and the seed sources for fu-

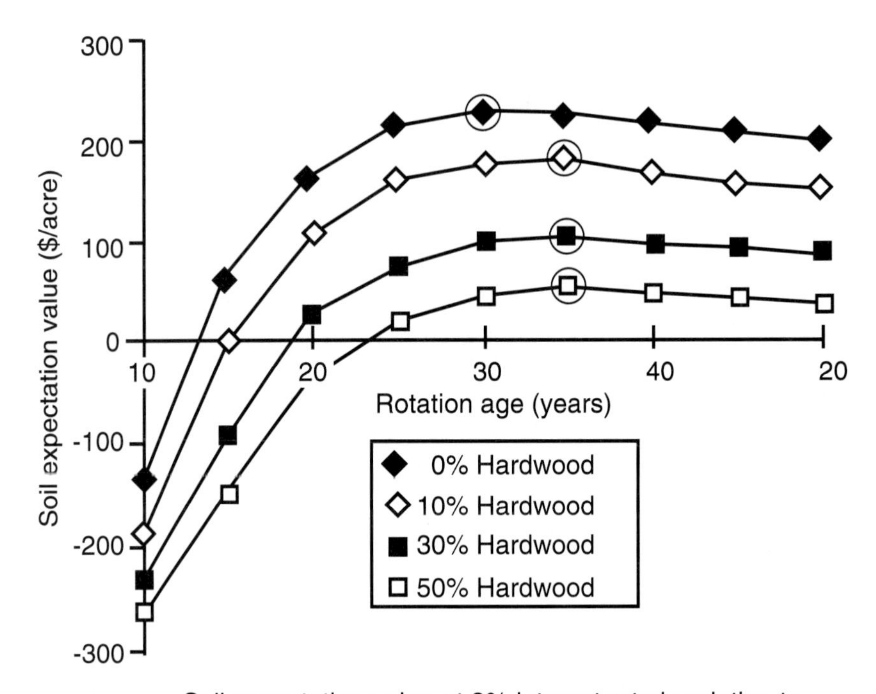

FIGURE 16-9
Where landowners control the interfering hardwoods in loblolly pine communities using site preparation and timely postplanting release treatments, the new stand will contain more pine volume, and provide a higher return on the investment of management (from Brodie et al. 1987).

ture communities at a site. In addition, removing all of the short-lived species with relatively soft wood (like aspens and grey birch) also reduces cavity formation when these trees begin to decline and die during pole stages. Limiting the release to individual crop trees at wide spacing helps to maintain a component of early-fruiting and short-lived species between the crop trees, and lessens the effects on community composition and wildlife habitat diversity (after Hunter 1990).

Weeding speeds tree development, promotes earlier crown closure, and shortens the reorganization period. This reduces the numbers of years when a dense community of herbaceous plants provides cover, concealment, and food for small mammals and songbirds. Weeding also removes early-fruiting herbs and reduces total plant species diversity and early community structure. As a consequence, bird species density and diversity will decline following a treatment (Santillo 1987). Further, if the chemical also kills invertebrates used as food, avian density will decrease (Slagsvold 1977). The earlier the application, the greater the potential effect.

Cleaning may reverse the effect. In 7–9-yr-old loblolly pine plantations in southeastern United States, reducing pine density from about 600 to 450 trees/ac (1482–1112/ha) delayed canopy closure and increased deer forage by 42% during the first growing season, and 360% during the second winter. Benefits continued at least through the third winter (Hunter et al. 1982). Once the crown canopy closed, total forage declined. However, creative use of herbicides to maintain areas in herb cover can help to prolong or increase plant species diversity as a stand develops (Hunter 1990).

Liberation cutting also has implications for wildlife, particularly for birds that feed in the upper canopy and for an array of cavity-nesting birds and mammals. By removing the tall trees, it reduces the vertical structure and eliminates essential habitat for creatures that depend upon the tall trees. The songbird community will subsequently decline in diversity, leaving only those species that function close to the ground and in young (short) tree cover. Also, the trees removed by liberation cutting often contain cavities used as nesting sites. Poisoning leaves skeletons of these tall trees, but animals often abandon holes in dead snags. If they do not find substitutes nearby, the populations of den-dependent fauna decline. As an alternative, workers could mechanically girdle the culls so that they die slowly, or wound them sufficiently to reduce the vigor and promote their decay (Hunter 1990).

Cleaning and weeding will not affect recreational values unless the shift in animal species lessens a user's interest in an area. Cleaned and weeded stands look similar to untreated stands, except for their canopy density. By speeding crown closure, and decreasing herb density, weeding shortens the time before people can comfortably walk through a recently regenerated area. This facilitates recreational use. With liberation treatments, the poisoned trees look bad, and branches falling from them pose a hazard to recreationists. These effects of release cuttings diminish after two or three decades. By then, the stands look like any other even-aged community.

Cleaning and weeding treatments require an investment that landowners cannot recover for long periods. By contrast, with liberation cutting they can sometimes sell the cut trees to pay the costs. This often makes liberation cuttings more economically attractive than other release treatments. Further, it concentrates the productive capacity of a site on young trees of good quality and potential. Even so, all release treatments involve some investment at the beginning of a rotation. Landowners must wait long periods to repay the costs. That often makes cleaning and weeding financially unattractive, except to salvage an otherwise hopeless situation. In fact, foresters may more easily justify

1 combining site preparation with a reproduction method cutting to favor desired species and discourage others;

2 tailoring a reproduction method to optimize conditions for the target species and promote their early rapid development; and

3 using a variety of tending operations to eliminate regeneration sources for undesirable species, and reduce their numbers in the next rotation.

These measures often cost less than release treatments. Further, foresters can use them as part of their routine silviculture, and cover the costs from proceeds of a commercial sale.

17

THINNING AND ITS EFFECTS UPON STAND DEVELOPMENT

SOME DYNAMICS AMONG INTERMEDIATE-AGED EVEN-AGED STANDS

Even during the reorganization or stem initiation phase, differences in growth between individual plants and the several species present begin to determine how a stand will develop, and what trees will eventually dominate a community. Perhaps most important, differential rates of height growth within and between species result in a stratification of trees as the crown canopy forms (Fig. 17-1). Poorly developing trees become overtopped or partly shaded, and foliage on the lower branches dies. Intertree interference (physical obstruction and shading) within closed-canopy stands also limits lateral shoot extension among all but the upper most branches. As a consequence, crowns of individual trees decrease in length and expand little in diameter. Tree vigor declines and radial increment slows, especially among the most heavily shaded trees. Eventually, many overtopped trees become suppressed and die (Baker 1950; Assmann 1970; Smith 1986; Oliver and Larson 1990).

Heaviest losses occur during the first two to three decades, and the rates of mortality decline as a community ages (Fig. 9-7). Actual losses depend upon the shade tolerance of species in a community, with greater rates of mortality in stands of shade-intolerant species. Also, the time of crown canopy closure and onset of stem exclusion vary with tree density, spatial distribution, and rates of growth. Generally, the trees grow more rapidly at the better sites, and the crown canopy closes earlier (Baker 1950; Assmann 1970; Smith 1986; Oliver and Larson 1990).

Over time, unequal growth of survivor trees increases the spread of heights and

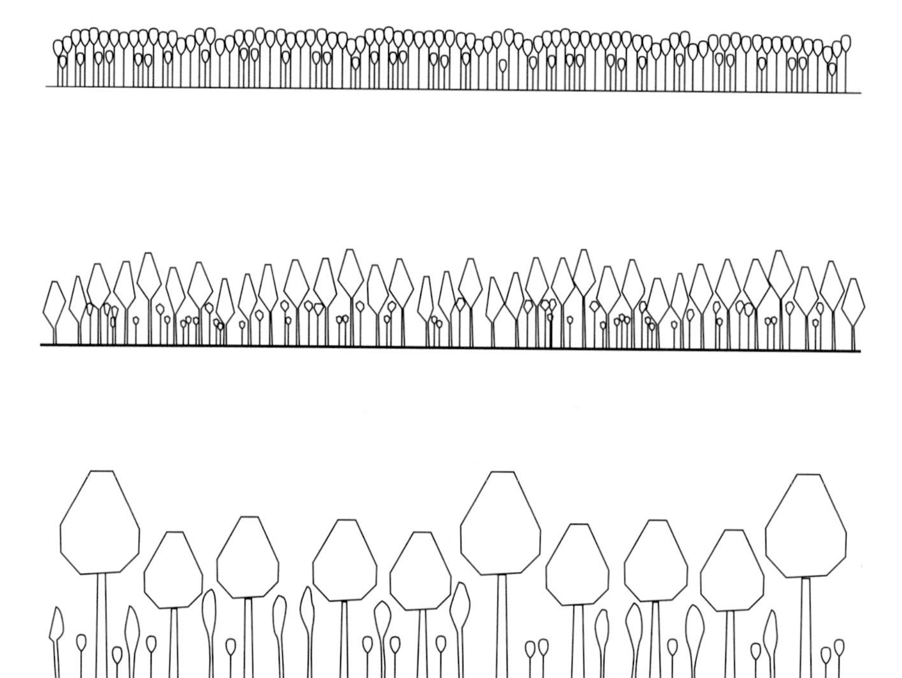

FIGURE 17-1
Differential rates of height growth within and between species result in a stratification of trees as the crown canopy forms, as shown by this series showing stand development from seedling through early sawtimber stages.

diameters within a stand (Fig. 17-2), particularly when a community has species with a wide range of shade tolerances. The shade-intolerant species grow more rapidly, but the shade-tolerant species develop a wider spread of heights and diameters, giving the stand a greater degree of differentiation. To illustrate, within new even-aged communities in northeastern North America, shade-tolerant sugar maple trees begin to differentiate in heights soon after establishment. The more rapidly growing trees overtop others by the time of canopy closure. Those in understory positions decline in vigor and grow slowly, but remain alive for decades. Maples in the overstory develop more rapidly, so that differences in tree diameters and heights increase with time. Paper birch also differentiates in height, but overtopped trees of this shade-intolerant species soon die. As a consequence, the spread of paper birch diameters and heights becomes more narrow with time (Gilbert 1965). Where both species occur in the same stand, the birch will overtop the maple, further exaggerating the spread of heights and diameters in a single stand. Differences between species, as well as the spread of sizes among the shade-tolerant sugar maples, tend to increase as the stand matures (Gilbert 1965; Shugart et al. 1981). Where both

FIGURE 17-2
Over time, unequal rates of growth among survivor trees increase the spread of heights and diameters within an even-aged community, particularly stands with an important component of shade-tolerant species like these northern hardwoods.

species groups regenerate in abundance, the shade-intolerant trees may completely overtop the shade-tolerant cohort at an early age, and oppress its development. Then a two-storied stand forms (Smith 1986; O'Hara and Oliver 1992; Oliver and Larson 1990; Roberts 1992). A stratified mixture may also develop among two or more species having similar growth potentials, but where environmental or biological factors delay the germination of one. Then the slightly older trees partially shade the late-developing species, at least somewhat oppressing the height and diameter increment. The effect increases with time, and a vertically stratified mixed-species stand develops (Palik and Pregitzer 1991). The complexity of stratification depends upon the mixture of species that regenerate, their abundance and growth rates, and time when they initiate following a disturbance (Oliver and Larson 1990).

In general for even-aged stands, canopy position influences the radial increment of individual trees, and shade tolerance determines the survival of slow-growing trees in heavily shaded and overtopped positions. Further, communities of species with disparate growth rates show the greatest heterogeneity of diameters and heights. As a consequence, even-aged tree communities may have at least three distinctly different kinds of diameter distributions (after Baker 1950; Marquis 1968; Smith 1986):

1 bell-shaped distributions for mixtures of largely shade-intolerant and midtolerant species, where trees of poor canopy position die, reducing the spread of diameters;

2 reverse-J distributions for stands of shade-tolerant species, or a mixture of species with a wide range of shade tolerances, and where many slow-growing understory trees persist for several decades; and

3 bimodal or multimodal distributions where the species have widely different degrees of shade tolerance or rates of height and diameter growth, and form distinctly layered canopy strata.

Figure 17-3 illustrates these basic structural types. Actual stands might have other kinds of distributions that represent an intergrading of those described. Exact attributes depend upon the numbers of species, their relative abundance and spatial distribution, and differences in growth rates (Shugart et al. 1981; Oliver and Larson 1990). Though similar in general form to those in uneven-aged stands, the reverse-J distributions in even-aged communities reflect differential growth rates among trees of a single cohort. The small trees grow poorly compared with others, due to their low vigor (oppressed or suppressed).

During early stand development, death of the shortest and smallest trees has no appreciable or lasting effect upon the density of the upper canopy, nor the understory shading. Each death narrows the range of diameters present. If the loss opens space within the upper canopy, light increases along one side of neighboring survivors, stimulating their radial increment. Lateral branch extension soon fills the canopy space, limiting the diameter growth of all but the largest trees in upper-canopy positions (Oliver and Larson 1990). As time passes, larger trees in main canopy positions also die, creating wider and more lasting openings. As a result, lower branches of neighboring trees remain exposed for extended periods, maintaining higher rates of radial increment for a longer time. Then as a stand approaches physiological maturity, even large trees of uppermost crown positions begin to die. Lateral branch extension fills the opening only slowly, leaving more lasting but randomly located gaps in the canopy. A new cohort forms beneath each opening, with species composition dependent upon the interaction of canopy openness, regeneration source, and microsite physical conditions. This process gradually transforms the community into a multi-aged condition that may persist for many decades (Watt 1947; Baker 1950; Bray 1956; Bormann and Likens 1979; Oliver 1981; Woods and Whittaker 1981; Runkle 1981, 1982, 1990, 1992; Pickett and White 1985; Oliver and Larson 1990).

CROWN CLASSES AND THEIR SILVICULTURAL IMPORTANCE

From initiation to physiological maturity of an individual tree, a complex of interacting biophysical factors determines its rate of growth, and longevity (Fig. 17-4). Localized environmental conditions (microsite attributes, density of adjacent trees, and residual tree canopy shading) affect seedling establishment and the initial rates of individual tree development, at least during the reorganization or stem initiation

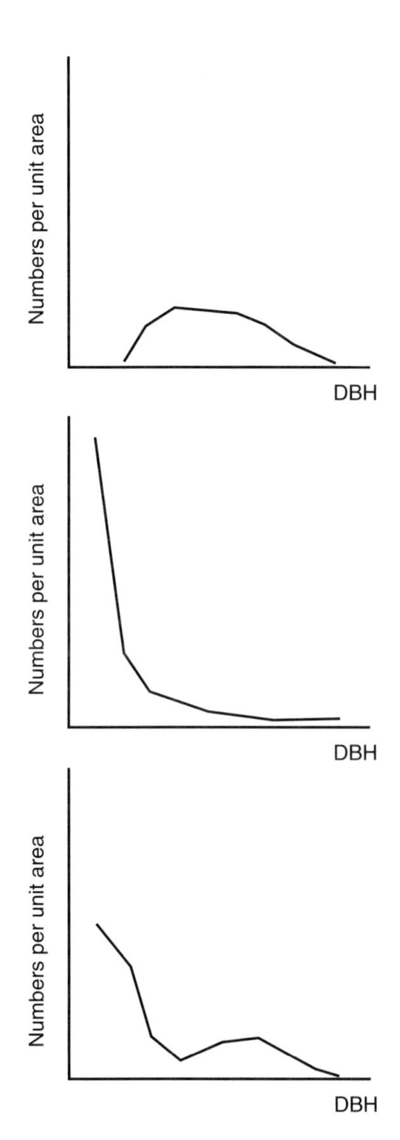

FIGURE 17-3
Common alternative diameter distributions for even-aged stands (after Marquis 1986).

phase. Yet as the root systems spread laterally, the trees integrate edaphic conditions over an increasingly larger area, and small-scale differences in site have less effect. Instead, the differentiation into size classes more probably reflects inherent individual tree genetic factors, as well as differential rates of species growth. Together, site and genetic influences result in a greater crown canopy differentiation. The tallest trees intercept greater amounts of solar energy, photosynthesize at higher levels, and grow better, amplifying the size differences and increasing the crown canopy stratification even more (Assmann 1970; Woods and Whittaker 1981; Oliver and Larson 1990).

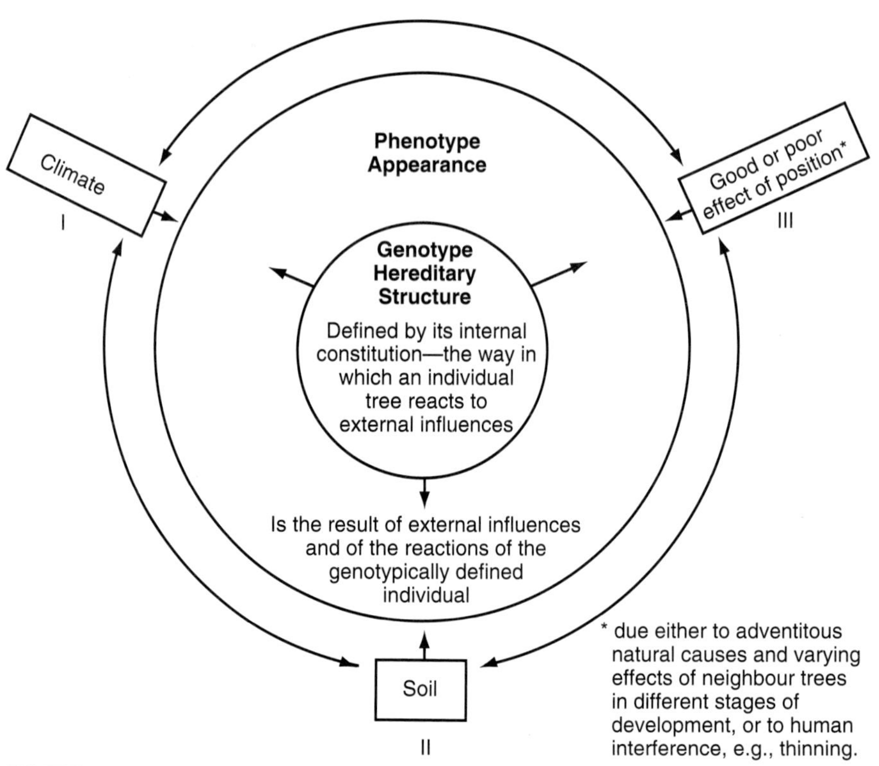

FIGURE 17-4
A complex of interacting biophysical factors determines the growth rate and longevity of individual trees (from Assmann 1970).

Historically, foresters have attempted to assess implications of crown canopy stratification by rating individual trees according to their relative crown canopy position, and using the classification to judge the potential for future tree growth. The commonly used systems apply only to even-aged communities. They segregate the trees into four classes based upon position relative to the general crown canopy, and the proportion of crown surface area exposed to direct sunlight. These classes (Fig. 17-5) (after Kraft 1884; Hawley 1921; Baker 1950; Kostler 1956; Assmann 1970; Ford-Robertson 1971; Daniel et al. 1979; Hocker 1979; Spurr and Barnes 1980; Smith 1986; Oliver and Larson 1990) include

Dominant

- —crown extends above the general canopy layer for the stand
- —crown intercepts direct sunlight across the top and along sides of the upper branches
- —crown well developed and large, though usually somewhat crowded along lower branches
- —tree diameter usually among the largest in the stand

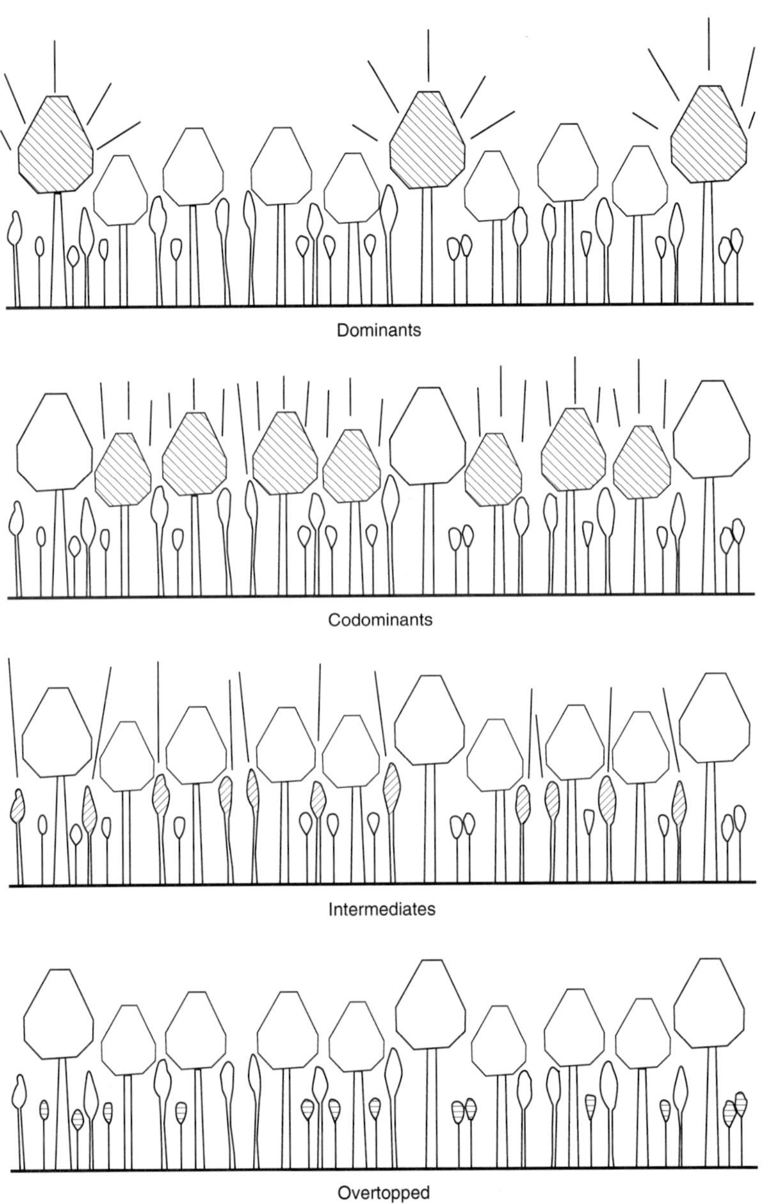

FIGURE 17-5
Tree crown classes reflect their relative position in an even-aged stand crown canopy, as well as their general vigor and rates of growth. Lines above and around the crowns indicate the direct solar energy intercepted by trees of different crown classes.

Codominant

—crown within and helping to form the main crown canopy for the stand
—crown intercepts direct sunlight across the top, but only at tips of the upper side branches
—crown well developed, but of only medium size and crowded at the sides
—tree diameter among the upper range of those present, but not the largest

Intermediate

—crown extends somewhat into the lower part of the main canopy
—crown intercepts direct sunlight only at a limited area on the top, and none at the sides
—crown narrow and short, with limited leaf surface area and a low live-crown ratio
—tree diameter within the lower range of those present, but not necessarily the smallest

Overtopped

—crown entirely below the main canopy, and covered by branches of taller trees
—no direct sunlight strikes at any portion of the crown
—crown small, often lopsided, flat-topped, and sparse
—tree diameter among the smallest in the stand

Foresters can usually identify overtopped and most intermediate trees fairly readily, but often cannot clearly differentiate between the best of the codominants and poorest of the dominants (Smith 1986). An important overlap often makes the poorest codominants and best intermediates difficult to separate as well.

Some classifications subdivide the overtopped and weaker intermediate trees into oppressed and suppressed trees, and this may prove difficult. *Oppressed* trees have small and poorly developed crowns, but a potential to recover and enlarge if released. *Suppressed* trees have declined to a point that they will not recover (Toumey and Neethling 1924; Baker 1950). Where foresters can recognize this condition, they can make better judgments about uncovering overtopped saplings by release treatments, or removing the overstory from older stratified communities. Oppressed trees might include a shade-tolerant species overtopped by taller shade-intolerant trees (in a stratified species mixture), or intermediate trees in communities of shade-tolerant species. In the former case, many of the oppressed trees have a genetic potential to grow well. By contrast, intermediates in a single-species even-aged stand grow better following release, but never as well as upper-canopy trees.

Other classification systems might include other variables. Exact factors depend upon whether foresters want to make relative judgments about the condition and status of individual trees (e.g., when selecting those to retain or remove during intermediate treatments), or to predict future growth with a fairly high degree of precision (Hocker 1979). To illustrate, for growth predictions of longleaf pine, foresters must add site index and the live crown ratio (Chappell 1962). Similarly, to accurately predict the radial growth of sugar maple, they must know the initial tree diameter and its crown position (Nyland et al. 1993). For sessile oak, foresters

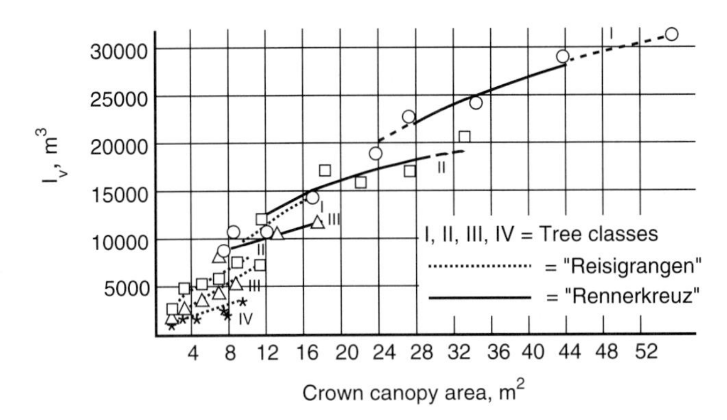

Increment by Kraft tree classes, plotted over the crown canopy areas for the two strands of sessile oak (R. Mayer.)

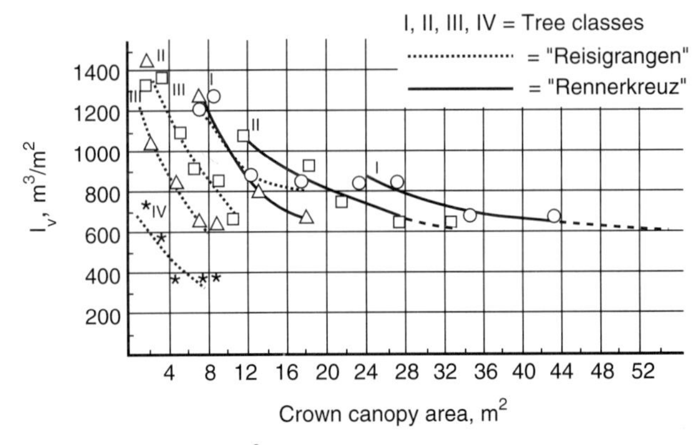

Increment per 1 m^2 crown canopy area for the same tree classes.

FIGURE 17-6
Crown classes indicate relative past rates of development among trees in a community, and provide a basis for selecting ones having the greatest promise for future growth (from Assmann 1970).

should consider crown diameter, degree of development, and canopy position (Assmann 1970). In any case, they should use discrete attributes related to crown condition, tree vigor, or health. Also, foresters will have difficulty classifying the trees if they use elaborate systems based upon only subtle differences, and having too many classes. In fact, time has proven the practicality of the simpler classification depicted in Fig. 17-5 (Baker 1950).

Most crown classification schemes have little relevance in communities with multiple age classes. There, both height and diameter reflect age, rather than just the relative rates of growth among neighboring trees. Thus, for uneven-aged stands, an evaluation would have to assess tree vigor and height in comparison with others of similar age. To that end, one system for uneven-aged ponderosa pine separates the trees into four age classes and then assesses dominance, crown length and width, and tree vigor *within age groups*. Foresters would use features like foliage color and density, bark characteristics, and absence of disease in judging the vigor of individual trees (Dunning 1928; Keen 1936). For hardwood species, the classification might include the presence or absence of large dead branches in the upper crown, condition of the main stem or bole (freedom from rot and cracks), and degree of exposure to direct sunlight from above (Eyre and Zillgitt 1953).

Since all trees in an even-aged community regenerated at the same time (or nearly so), the crown classes indicate relative rates of development since stand initiation. They also imply much about differences in tree vigor and future growth. To illustrate, the curves in Fig. 17-6 depict the correlation between tree crown volume, crown position, and volume increment of sessile oak in Europe (from Assmann 1970). The crown position classification by Kraft (1884) incorporates a four-class system with I the best, and IV the poorest. Similarly, for shade-tolerant sugar maple growing in northeastern North America, while trees of all diameters had higher rates of radial increment during nine years following release, those of poor initial crown position grew much more slowly than the codominant trees (Nyland et al. 1993). In these examples, overtopped trees were probably oppressed rather than suppressed, since they grew better following release. The evidence suggests that within even-aged stands, trees that grew the poorest will grow the slowest in the future. Further, ones of upper canopy positions will have the highest rates of radial increment following release by thinning.

SOME IMPLICATIONS OF TREE AGING

Figure 17-7 illustrates two characteristics of individual tree development over long periods. The left-hand curves depict a classic pattern where individual trees initially grow slowly in height. Then height increases rapidly among a relatively few elongating branch tips. As time passes, the shoot elongation potential becomes dissipated over increased numbers of growing branch tips along both the sides and top of the crown. The rate of height growth slows appreciably, until at old age total height may not change much over long periods. Foresters call these the *juvenile, full vigor* or *intermediate*, and *senescence* or *decline* phases of growth (Assmann 1970).

Shade-intolerant species pass through these growth phases more quickly than do shade-tolerant species, reaching a culmination or peak of height growth at younger ages. For all species, culmination comes sooner on the poorer sites (after Baker 1950; Assmann 1970). Also, at least in hardwood trees at northern latitudes, environmental forces (e.g., ice and snow damage to crowns) may periodically break the upper stem, distort the bole, and fix or limit the overall merchantable length of the main stem. Under these conditions, middle-aged and older trees will often have similar usable bole lengths, and volume increment depends primarily upon an increase of diameter along the merchantable portion of the main stem.

The right-hand curves in Fig. 17-7 depict a similar growth pattern for radial increment. Generally, diameter increases rapidly during young ages, and declines as a

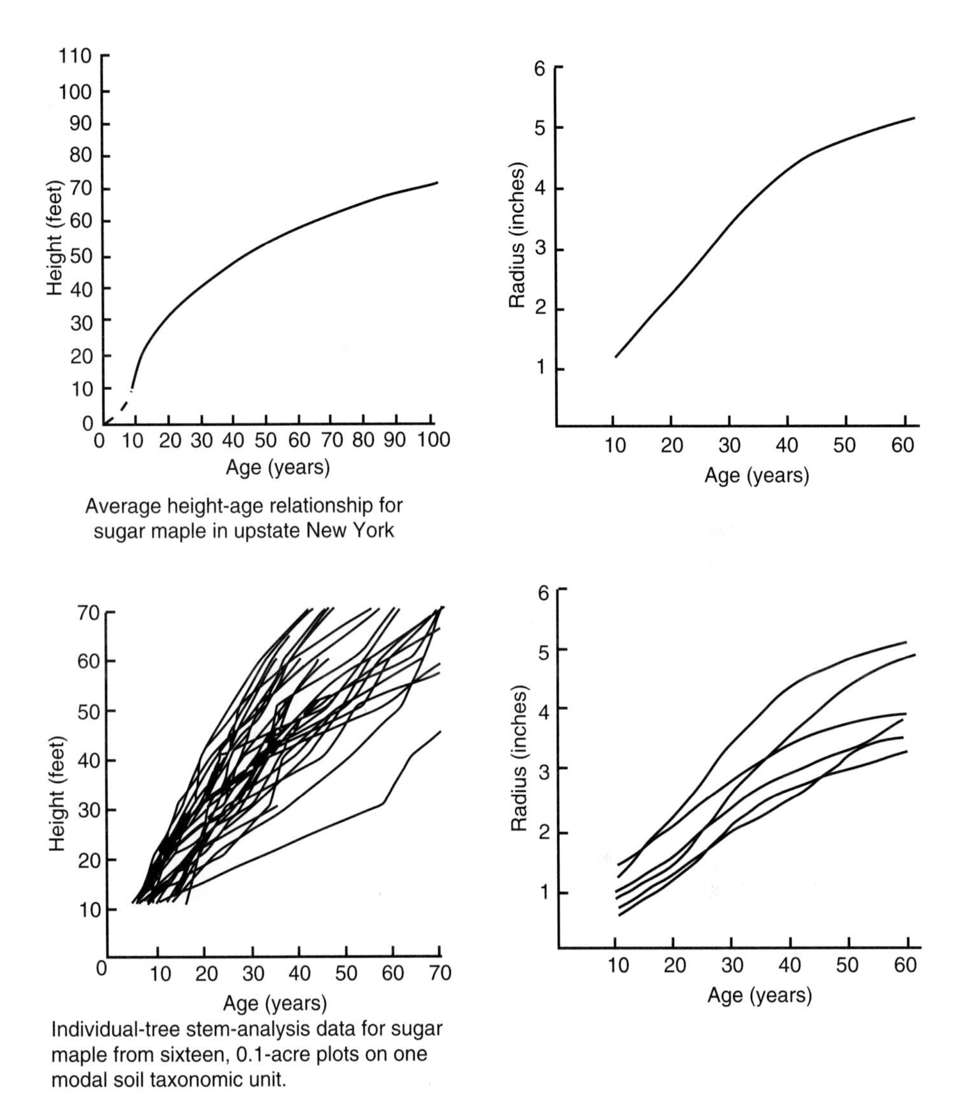

FIGURE 17-7
Differential patterns of growth between individual trees over long periods result in a stratification of heights and diameters within an even-aged community.

Left-hand curves illustrate patterns of height development for one and several sugar maple trees in a single even-aged stand (from Berglund 1975).

Right-hand curves illustrate patterns of radial growth for one and several yellow birch trees in a single even-aged stand.

As even-aged communities mature, the spread of heights and diameters increases, particularly for communities with both shade-tolerant and intermediate species.

tree matures. In fact, old trees may increase in diameter by only a small amount from year to year. A complex of factors probably controls radial increment and the time of peaking, including effects of stand density (degree of competition) on crown volume, and a species' growth characteristics. Generally, radial growth culminates sooner for shade-intolerant species, and vigorous trees peak later than weak ones (Baker 1950; Assmann 1970; Oliver and Larson 1990).

Figure 17-7 probably also reflects a reduction in photosynthesis as crown volume and tree vigor decline following prolonged intertree competition, and an increase in maintenance respiration as the mass of bole and branches becomes greater over time (after Baker 1950; Kimmins 1987; Oliver and Larson 1990). Generally, as a tree matures, continued intertree competition and a complex of physiological factors limit lateral branch extension and height growth, making any increase in total leaf surface area more dependent upon a thickening rather than a widening of the crown. Also, lower branches continue to die due to prolonged shading, limiting crown depth or length. Further, leaves on the lowest branches may remain continually shaded, diminishing their contribution to total production. Photosynthesis may not exceed respiration in some branches, or it becomes negative and they die. As a consequence, leaf surface area stabilizes or begins to decrease, and the amount of carbohydrate material produced annually by photosynthesis levels off, or may diminish. Also, as the tree becomes larger and larger, metabolism and respiration consume proportionately more of the photosynthate, leaving less each year to form new wood and maintain the rate of radial increment (Daniel et al. 1979; Oliver and Larson 1990).

The actual size of an individual tree also affects its potential rate of diameter growth. This becomes particularly evident among trees with a stable leaf surface area. In these cases, as individual trees become larger over time (even at slowing rates) with a bigger main stem and more branches, the wood produced annually must cover an ever-increasing total surface area. Thus, if the amount of woody material laid down each year does not increase over time (or increases only by small amounts), the thickness of the new woody sheath will necessarily diminish annually. This becomes apparent as a slowing rate of radial increment at breast height. In severe cases, the wood production may diminish to a point where growth rings do not extend all the way down the trunk to the stump (Daniel et al. 1979). Then as trees reach physiological maturity and biological senescence, diameter growth diminishes for reasons not related to the increasing mass of bole and branch surface area alone (Baker 1950; Daniel et al. 1979). At some point, a complex of factors causes respiration to exceed photosynthesis, and the tree dies. This may occur among low-vigor trees of any age, and old trees that reach a stage of physiological maturity.

Overall, the photosynthetic output per tree must increase annually to sustain even a constant rate of diameter growth. This means using different silvicultural practices to maintain an increasingly larger leaf surface area on the crop trees, and to increase light around their foliage (the active surfaces). Otherwise, crown volume declines or remains unchanged, photosynthesis levels off or diminishes, and the rate of annual wood volume production slows or declines (after Baker 1950; Assmann 1970). Further, once the growth of appreciable numbers of individual trees slows, total stand-level volume production (p.a.i.) also levels off, or falls.

THE PRODUCTION FUNCTION AS A MODEL OF EVEN-AGED STAND DEVELOPMENT

Figures 9-5 and 9-6 depict the pattern of basal area or volume development for fully stocked stands left unmanaged over long periods. It suggests several important ecologic and economic factors (after Baker 1950; Osmaston 1968; Assmann 1970; Smith 1986; Davis and Johnson 1987; Oliver and Larson 1990) including:

1 Volume rises rapidly during the early aggradation phase due to rapid individual tree height and diameter growth, much ingrowth of trees into minimum measurable (economic) sizes, and little mortality among trees with measurable volume.

2 As a community ages, slowing tree diameter and height growth, plus the death of trees with measurable volume, cause p.a.i. to peak and then decline, slowing volume accumulation within a stand.

3 As height growth slows, or merchantable length becomes fixed by damage or other factors, volume increases depend largely on the diameter growth of survivor trees (accretion), and the sizes of trees that die.

4 Over time, tree crowns stabilize and greater numbers of merchantable trees die, and standing volume or basal area reflects the balance between growth (p.a.i.) and mortality.

5 Among untended stands, mortality eventually equals production for a relatively short period (at physiological maturity), beginning the transition phase.

6 Subsequently, losses of the large trees, compensated by only slow growth of the survivors, reduce standing biomass or volume, until the community eventually passes into a steady-state stage.

The rates of change and the amounts of volume that accumulate depend upon site conditions and the species composition, and how soon a stand regenerates after the reproduction method. Delays in stand establishment and canopy closure, differences in the extent and rate of stem differentiation, and variations in the degree of site occupancy often make the patterns less regular than those portrayed in Figs. 9-5 and 9-6 (Peet 1981; Oliver and Larson 1990).

Curves like these suggest that to maximize total standing sawtimber volume, foresters should set the rotation at physiological maturity. Total volume will increase up to that age, but then decline. Conversely, to maximize average total annual volume production they will grow a stand only until m.a.i. peaks. The *zone of rational action* for rotations serving other objectives occurs between these two points in time, with the reproduction method coming no sooner than the culmination of m.a.i., and no later than physiological maturity (after Osmaston 1968; Smith 1986; Davis and Johnson 1987). For sawtimber products they extend the rotation at least until trees grow to the minimum sizes for sawlogs, and normally look for the rotation that optimizes the rate of financial return. To speed individual tree growth they would remove some trees to reduce the crowding. This will enhance sawtimber volume production and value (Smith 1986).

A management rationale to maximize volume production using rotations set at the culmination of mean annual increment assumes that landowners can find markets for all the trees containing measurable volume, that their value covers the logging costs, and that the stumpage prices provide an acceptable return on the investment. Foresters balance these factors in large measure by deciding how small a tree to include in the volume assessment, and the units of measure to apply. For example, with traditional pulpwood markets in North America they would tally the total volume in trees of at least minimum merchantable size for local markets. With access to chip markets or those that take total fiber independent of tree size, the production function curves more closely reflect the total accumulation of woody above-ground biomass. In all cases, p.a.i. and m.a.i. peak sooner for total biomass than pulpwood, and sooner for pulpwood crops than sawtimber.

Uncontrolled insect and disease episodes or other natural agents may reduce stem densities or kill most of the trees at any time. Low-vigor trees often succumb more readily than more thrifty trees, and management provides opportunities to protect against many natural hazards (see Chapter 21). If a disturbance kills only some of the trees, it effectively reduces stand density and intertree competition. Thereafter, crowns of surviving trees expand into spaces formerly occupied by the killed trees, and they grow with renewed vigor. Competition-induced mortality may temporarily decline, and p.a.i. increase. However, as the crown canopy closes again, tree vigor once more decreases and p.a.i. (or c.a.i.) and m.a.i. eventually diminish.

INTERMEDIATE TREATMENTS TO TEMPER STAND DEVELOPMENT

Natural agents reduce stand density at irregular and unpredictable intervals, and with uncontrolled intensity. Also, they may harm trees and species that a landowner prefers, causing a major economic loss. Early foresters recognized that they could use deliberate cutting to reduce stand density, and also favor the trees that best serve a landowner's objectives. Foresters call these *intermediate stand treatments,* and *intermediate cuttings* (after Hawley 1921; Assmann 1970; Ford-Robertson 1971; Daniel et al. 1979; Wenger 1984; Smith 1986; Soc. Am. For. 1989):

- silvicultural operations in closed-canopy stands of intense intertree competition
- any removal of trees from a stand following its formation (after a reproduction method) and before the end of a rotation
- any manipulation to modify or guide the development of an existing community of trees, but not replace them with new trees
- treatments to improve the composition, quality, spacing, and growth of a developing stand or age class, and to recover values that a landowner would otherwise lose

These intermediate treatments include cleaning, liberation cutting, improvement cutting, thinning, pruning, and sometimes also salvage and sanitation cuttings. They allow foresters to substitute timely tree cutting for natural mortality, and to stimulate the vigor and diameter growth among residual trees of choice. They concentrate the growth potential of a site on trees and species of highest value, and control mortality by reducing intertree competition.

Foresters apply release cuttings during sapling stages (Chapter 16). For older even-aged stands, they commonly do one or more *thinnings* (after Hawley 1921; Assmann 1970; Ford-Robertson 1971; Daniel et al. 1979; Smith 1986; Soc. Am. For. 1989)—a treatment to increase the diameter increment of residual trees, improve stand quality, and increase stand-level production by cutting excess and potential mortality trees *without permanently breaking the crown canopy*. Generally, thinnings remove the less desirable trees to reduce intertree competition around better trees. They improve residual tree vigor, and promote higher rates of diameter increment. Thinning also maintains a uniformly dispersed canopy cover with no lasting openings. This helps to prevent a new age class from prematurely regenerating beneath the immature trees.

Thinning increases light around crowns of the residual trees. It also reduces the transpiration and nutrient uptake from the soil solution, increasing the amount available to residual trees. They eventually develop larger crowns, and radial growth increases. This (Hawley 1921; Daniel et al 1979; Smith 1986)

1 shortens the time it takes an individual tree to reach some specified diameter, and favors trees with the best growth potential;
2 prolongs the time until p.a.i. peaks;
3 improves the quality and value by focusing the growth on the most valuable trees;
4 increases product yields by harvesting trees that would otherwise succumb to intense intertree competition;
5 strengthens the bole and branches of residual trees, making them more resistant to breakage;
6 maintains high tree vigor, and removes trees susceptible to local insects and diseases; and
7 provides early income to help pay investments and operating costs.

For the most part, thinnings enhance the potential to produce high-quality trees of large diameters in shorter rotations (Smith 1986).

If foresters control intertree spacing and reduce canopy crowding, they will not measurably affect height growth across a stand. In fact, stand density (degree of crowding) will not affect height growth, except at extremely close or wide spacing or at sites with limiting physical resources (nutrients or water). Likewise, thinning to common levels of residual stocking neither improves nor reduces height increment on trees in upper crown positions (Hawley 1921; Hummel 1948; Bennett 1956; Braathe 1957; Munzel 1962; Smith 1986; Oliver and Larson 1990). Rather, thinning keeps the lower branches of residual trees alive and increases light around the entire crown. Lower branches remain alive longer, and the crown expands into spaces opened by the cutting. Gradually, the trees develop longer and wider crowns, and foliage thickens within the old crown area. Photosynthetic output increases, as does the amount of woody material produced annually. Radial growth also increases, or at least does not diminish as readily as the tree ages. In composite, the improved growth, lower mortality, and harvest of excess trees increase the realized yield by at least 25%–40% of gross growth (Baker 1950; Daniel et al. 1979).

A PRODUCTION FUNCTION FOR THINNED STANDS

Foresters represent thinned stands by a specialized production function. The shape (Fig. 17-8) shows how thinning periodically draws down the volume by removing some of the growing stock, and then how growth rebuilds the stocking over time. Further, as the crown canopy closes again and before intertree competition intensifies, foresters do another thinning. This gives the production function curve a saw-toothed appearance, with the exact shape depending upon the numbers, intensity, and timing of the thinnings throughout a rotation. Major differences compared with unmanaged stands include the following:

1 The reduction of stand density controls mortality and maintains higher tree vigor, so that numbers per acre do not change appreciably between well-timed thinnings.

2 Improved growth on the residual trees compensates for the increment a landowner would have realized from the cut trees, insuring full net production (maximum p.a.i.) in stands with adequate residual stocking.

3 The standing volume within appropriately thinned stands regrows above the previous precut level, given a proper interval between successive thinnings.

4 The removals during thinning, subsequent growth by residual trees, and controlled mortality give the production function its saw-toothed shape, and an upward positive slope.

A landowner can alter the shape of this curve by changing the timing, frequency, and intensity of the thinnings. As long as a forester balances the thinning intensity and interval with the p.a.i., the saw-toothed curve should have a positive upward slope, representing increasing standing volume throughout a rotation.

A program of well-timed thinnings of appropriate intensity increases the yield of usable sawtimber volume, compared with unmanaged stands (Daniel et al. 1979; Smith 1986; Davis and Johnson 1987). This results from

1 removing and marketing potential mortality trees to increase the long-term realized yield;

2 cutting the excess trees not needed for full net production; and

3 maintaining sufficient leaf surface area to convert available insolation to biomass at the highest feasible level (insure full site utilization).

As long as the thinning does not reduce the total leaf surface area below the amount needed to fully intercept insolation and convert it to biomass via photosynthesis, the post-thinning p.a.i. will remain higher than in unthinned stands of a comparable

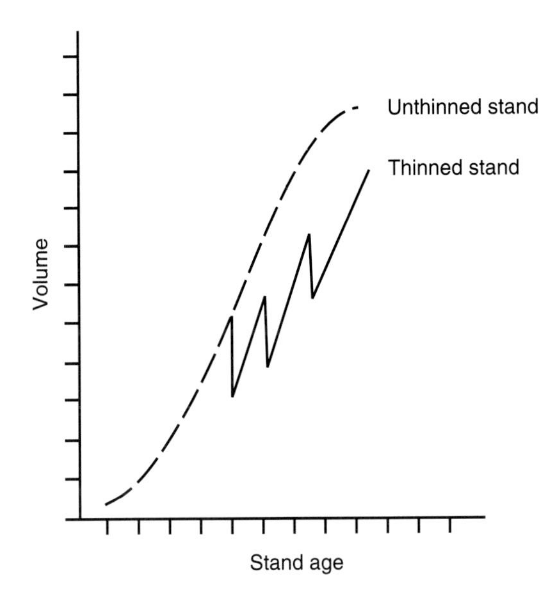

FIGURE 17-8
Production function for a thinned even-aged stand, showing effects on the levels of stocking and patterns of regrowth across a sawtimber rotation.

age. As individual tree growth and p.a.i. drop during latter parts of a rotation, the recovery time to rebuild the standing volume to the precut level increases. Consequently, foresters usually increase the time between thinnings as the rotation progresses (see Fig. 17-8).

Even with the decreasing p.a.i., foresters can extend the thinning interval to allow the volume to regrow to the precut level. Then the slope of the production function will remain at least level. If they also leave a higher level of residual stocking with each consecutive treatment, the slope will remain positive. Further, by periodically stimulating individual tree growth, foresters can produce sawtimber-sized trees much sooner. Nevertheless, when landowners want to maximize fiber production (independent of tree sizes), they will realize more yield from unthinned stands grown to the culmination of m.a.i.

STOCKING AND RELATIVE DENSITY AMONG UNMANAGED STANDS

Foresters have many different terms for conditions in even-aged forest stands, and some apply to uneven-aged stands as well. Important ones include:

Stocking (Soc. Am. For. 1950; Ford-Robertson 1971)

- the number of trees per unit area compared with the desired number for best growth and management

- the amount of any measurable attribute on a given area, particularly in relation to what is considered optimum

• a subjective indication of the numbers of trees, compared with the desired number for best results

• the proportion of area actually occupied by trees, as distinct from stand density

Stand Density (Ford-Robertson 1971)

• the degree of crowding within stocked areas, using various growing space ratios based on crown length or diameter, tree height or diameter, and spacing

• the numbers, basal area, or volume expressed relatively or absolutely per unit area

Fully Stocked Stand (Soc. Am. For. 1950)

• trees effectively occupy all growing space, but have ample room for development

• synonymous with *normal stand*

Normal Stand or Forest (Soc. Am. For. 1950; Ford-Robertson 1971)

• the best-producing fully stocked area for a given species, site, and method of treatment

• a forest or stand that has reached and maintains a particularly attainable degree of perfection relative to the objectives for management

• a standard for comparing other stands to bring out their deficiencies

Overstocked (Soc. Am. For. 1950)

• trees occupy the growing space completely, so that growth has slowed

• a condition where many trees are oppressed or suppressed

Relative Density (after Curtis 1970)

• stand density expressed as a proportion of some reference level

• the absolute density (e.g., basal area/acre) expressed as a percent of the average maximum common to stands at a similar stage of development and species composition

These definitions suggest a degree of subjectivity that foresters can avoid by using some kind of easily applied analytic technique to express stocking, density, or other stand conditions in quantitative terms.

Currently, foresters use the term *relative density* to describe the degree of crowding in even-aged stands, and in judging needs for thinnings and other intermediate stand treatments. A major breakthrough came with a field-expedient stocking guide (Fig. 17-9a) based upon easily obtained measures of stand conditions (Gingrich 1967; Roach and Gingrich 1968; Leak et al. 1969; Roach 1977; Rogers 1983; Lotan et al. 1988; Schmidt and Seidel 1988). These use tree-area ratios (after Chisman and Schumacher 1940) to show the expected maximum level of some mathematically derived density measure in previously undisturbed stands of a particular forest community type. They use easily measured stand attributes to predict how conditions change as a stand develops. The line for maximum density (100%) portrays the expected change of conditions in undisturbed stands. To account for differences in species composition, researchers have developed separate guides for each community type, or they include adjustment factors that correct for effects of species composition on stocking (MacLean 1979; Marquis et al. 1984; Ernst and Knapp 1985; Stout 1987; Stout and Larson 1988).

Common stocking guides utilize changes in measures like basal area and numbers of trees per unit area to portray the development of undisturbed closed-canopy even-aged stands during the aggradation or stem exclusion phase (Fig. 17-9a). In essence, the curved line in Fig. 17-9a represents a transformed version of the classic production function (Fig. 9-5). The vertical axis portrays the amount of basal area present as a stand develops, and the horizontal axis the change in numbers of trees per unit area. Though not explicitly linked to time in these stocking or relative density charts, numbers of trees follow the survivorship curve depicted in Fig. 9-7 and the change in number of trees can serve as a surrogate for stand age. Time moves from right to left on the diagram. Thus, the curve depicts how basal area normally accumulates over time in undisturbed communities, and the maximum level of basal area associated with differing numbers of trees at different stages of development. Foresters often call this *A-level stocking*, or *100% relative density*. Some stocking guides also add lines for the change in average tree size, and commonly the *quadratic stand diameter* (called QSD). QSD means the same as the diameter of the tree of average basal area, derived by

1 dividing the total basal area by the numbers of trees to determine the average basal area per tree; and
2 determining what size (diameter) tree would contain that average amount of basal area.

QSD gives an index to even-aged stand development, but does not reveal the actual sizes of trees present.

For community types that form closed canopies, stands with a relative density above 90% have some common attributes. These include

1 much mortality evident, particularly of trees in inferior crown positions;
2 intermediate and overtopped trees have low vigor and slow growth;
3 most trees have a relatively long dead length along the lower bole;
4 trees of most crown positions have low live crown ratios and narrow crowns, except perhaps for the dominant trees;
5 direct solar radiation does not reach ground level, so the stand has a dark understory;
6 advance tree regeneration sparse and short-lived if present; and
7 the stand looks crowded.

Figure 17-10 highlights the classic production function for unthinned stands, with a second one to depict cumulative gross production (standing crop, plus cumulative volume of trees that died). The diagram dramatizes the amounts of volume lost to mortality over long rotations, and suggests that foresters can increase the realized yield appreciably through effective silviculture. The treatments must periodically reduce stand density to maintain the vigor of residual trees, and concurrently remove potential mortality trees to merchandise as roundwood products.

Even-aged stands with a species composition appropriate to the stocking guide will follow the pattern of development shown by the diagram, independent of site quality. The change in numbers of trees and basal area occurs more rapidly at the

a. From Roach and Gingrich 1968.

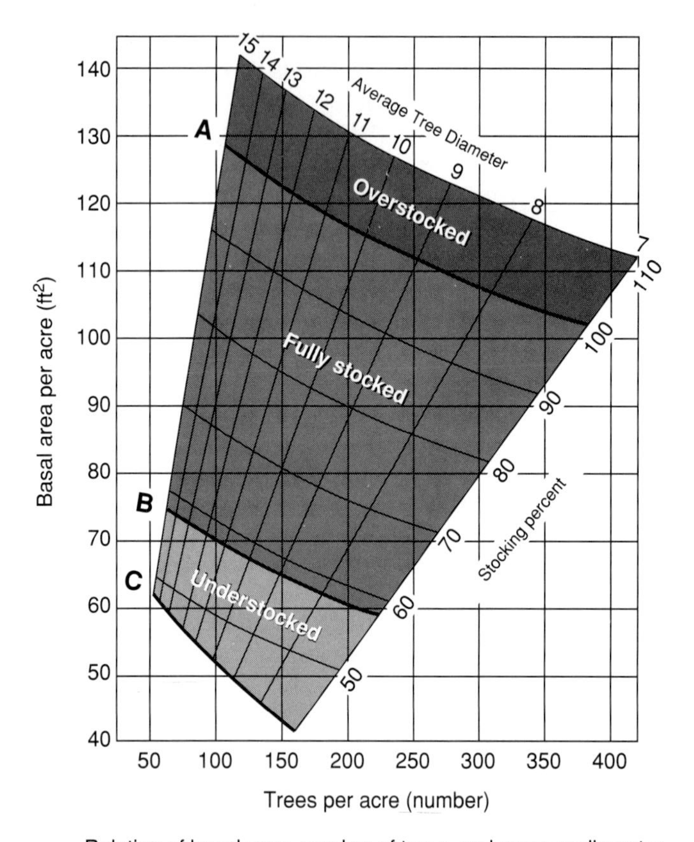

Relation of basal area, number of trees, and average diameter to stocking percentage for upland central hardwoods

FIGURE 17-9
Commonly used stocking guides portray patterns of stand development using easily measured attributes, and provide recommendations for minimum and maximum stocking of managed stands (a. from Roach and Gingrich 1968; b. from Drew and Flewelling 1979).

best sites, and those stands arrive at a given point along the 100% relative density curve sooner. Even so, communities with the same QSD usually have similar basal area and numbers of trees (after Reineke 1933; Chisman and Schumacher 1940; McArdle et al. 1961; Reukema and Bruce 1977; MacLean 1979; Stout and Larson 1988; Goelz 1990). Differences in site conditions also affect the species composition, and this may alter the relationship between basal area and numbers of trees in different stands. In addition, sites with better soil conditions have taller trees, and the community develops faster. This affects the rate at which volume accumulates, and the amount of volume associated with different combinations of basal area and

b. From Drew and Flewelling 1979.

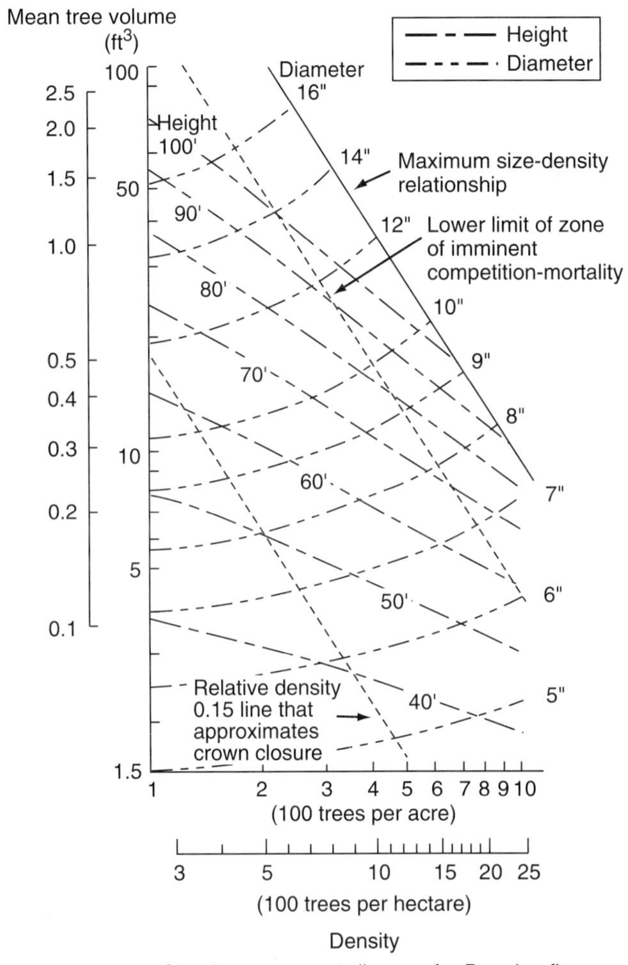

Stand management diagram for Douglas-fir with estimates of diameter and height.

FIGURE 17-9 (Continued)

tree density (taller trees with more volume on the better sites). To account for this, some density-management guides combine numbers of trees, QSD, and site index or tree height (Fig. 17-9b). These use the Reineke stand density index (SDI) as a measure of space allocation per tree (Reineke 1933), and portray relationships between easily measured stand characteristics and volume. Like other guides, they do not indicate the rates of development (Drew and Flewelling 1979; McCarter and Long 1986; Long 1988; Dean and Jokela 1992).

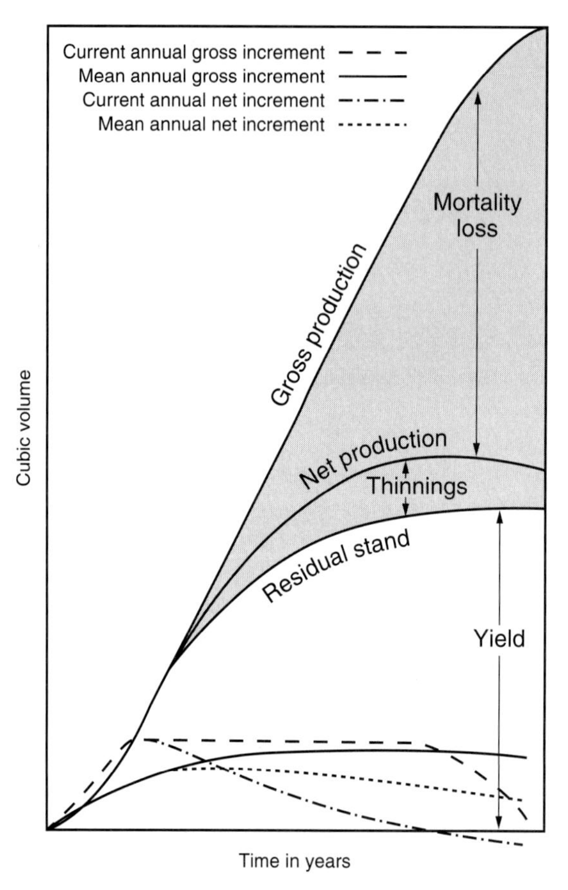

A hypothetical relationship between gross and net increment in an even-aged stand. Note that the net production curve is the production of an unthinned stand, and the area between the residual stand and gross production curves is the volume of thinnings.

FIGURE 17-10
The change of standing volume and cumulative gross production in managed and unmanaged even-aged stands over long periods of time (from Hall 1955).

The first stocking guides did not recognize effects of species composition, per se. They applied to a single community type in a specific region (Roach and Gingrich 1968; Leak et al. 1969), or to single-species stands (Philbrook et al. 1973; Benzie 1977; Reukema and Bruce 1977). Later research demonstrated differences in *tree-area ratios* between separate groups of species (Chisman and Schumacher 1940; Stout and Nyland 1986), and showed the benefits of adjusting relative density to reflect species composition (Roach 1977). Some guides account for this by presenting different curves for each particular forest community type (Tubbs 1977). Figure 17-11a illustrates another approach (Roach 1977). It presents five different A-level curves, each representing 100% relative density for stands with differing proportions of basal area in the black cherry, white ash, and yellow-poplar species group (called *CAP*'s). Stands with high levels of these species support greater amounts of basal area and volume for any given number of trees per acre. By enu-

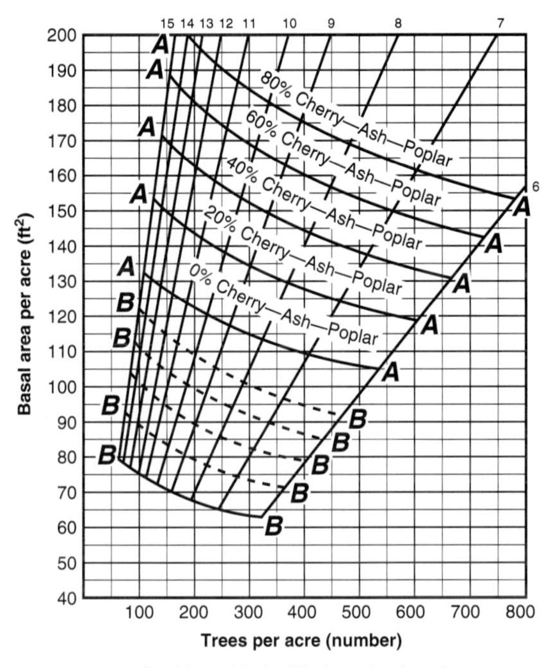

a. Stocking guide for Allegheny hardwoods.

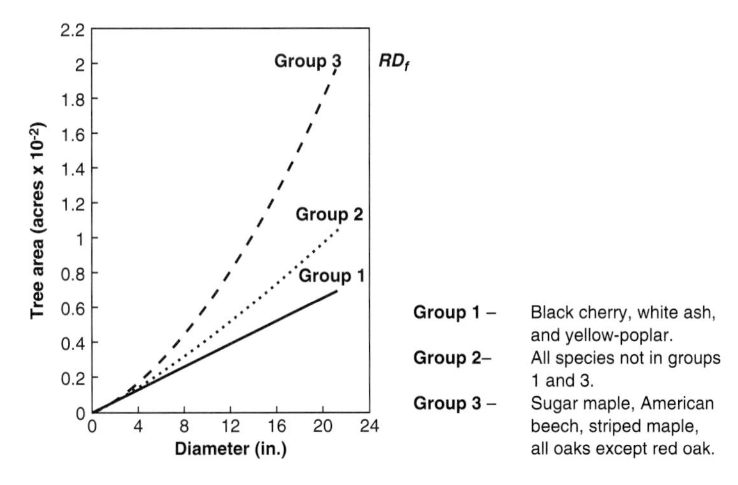

b.

FIGURE 17-11
Research indicates that among some fairly complex community types a forester may benefit from accounting for as many as six different species groups, but field-expedient guides generally use only two or three general groups for simplicity (from Roach 1977; Stout and Nyland 1986).

merating basal area into two species groups as shown, foresters can more accurately assess stand relative density. A more recent version (Marquis et al. 1992) uses the curves in Fig. 17-11b to derive a series of conversion factors rather than presenting a graphic stocking guide, per se (Marquis et al. 1992). Foresters use point sampling to enumerate basal area in three species groups, and five diameter classes within each group. They also separate the tally into acceptable (AGS) and unacceptable (UGS) growing stock. Applying the conversion factors, labeled RD_f, to the basal area by species and size class produces an estimate of stand relative density reflecting the exact species composition in each separate stand.

Personal computers and electronic data recorders simplify the use of increasingly complex techniques for assessing relative density in stands with diverse species mixtures. Even so, simple graphic guides suffice for the practical management of many even-aged stands. All the methods define relative density in quantifiable terms, and give foresters a new definition of normality.

CONTROLLING RELATIVE DENSITY THROUGH THINNING

By linking relative density to easily measured stand attributes, the stocking guides give foresters a consistent means to assess relative density, independent of personal judgment and experience. They provide a standardized reference point to use in comparing stands to bring out their deficiencies, at least relative to unmanaged stands in the aggradation or stem exclusion phase of stand development. They also help foresters to anticipate how even-aged stands should develop if not appreciably disturbed by insects, diseases, other natural agents, or human intervention. In addition, the information gives them a starting point for determining how much to reduce relative density to control mortality and still fully utilize site resources to keep wood volume production at an optimal level.

Conceptually, by removing a substantial number of trees from an even-aged stand through thinning, foresters stimulate the growth of remaining ones. This added increment compensates for the production that would have accrued on cut trees. Further, by regulating intertree competition and interference, thinning reduces mortality, so that all of the increment accumulates on trees that stay alive between subsequent treatments. Thinning will not reduce volume production below the full site potential if foresters leave sufficient numbers of the better-growing trees. In fact, by harvesting potential mortality before it occurs and concentrating the growth potential of a site on trees that remain alive, net or realized growth will increase over the course of a rotation (Assmann 1970). Further, at an appropriate level of relative density, net growth will equal gross growth, as depicted in Fig. 17-12.

Empirical studies of the relationship between residual stand density and post-thinning periodic annual increment suggest much about the optimal stocking. Generally, if foresters installed thinning plots having a range of residual stocking from none to maximum levels, *gross* volume production (p.a.i.) would remain fairly constant from 100% down to about 50%–60% relative density (Fig. 17-12, after Mar:Moller 1954). Below this threshold, gross production decreases steadily, suggesting that full site utilization occurs in stands kept at or above this critical mini-

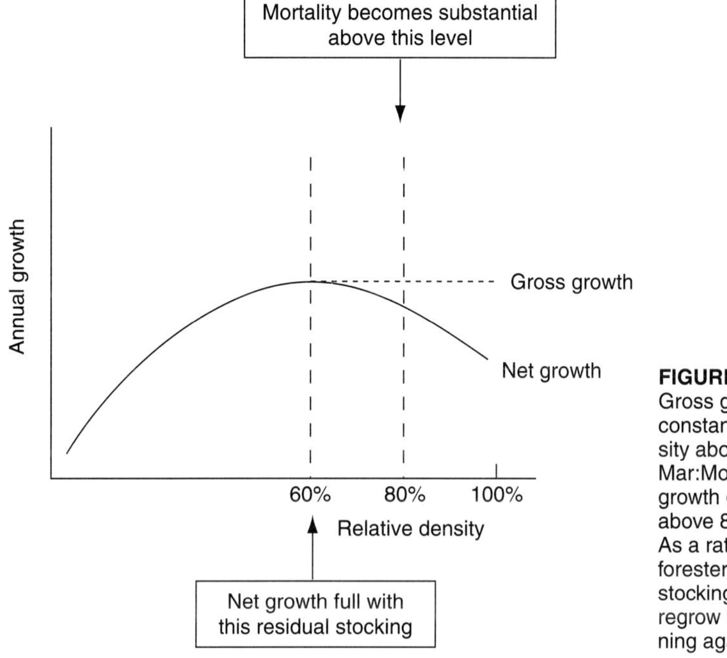

FIGURE 17-12
Gross growth remains fairly constant with relative density above 60% (after Mar:Moller 1954), but net growth drops substantially above 80% relative density. As a rationale for thinning, foresters would reduce stocking to 60%, and let it regrow to 80% before thinning again.

mum threshold. Other studies have shown that individual tree radial increment continues to increase with progressively lower relative density (Braathe 1957; Seidel 1977). In fact, among hardwood stands in northeastern North America, maximum individual tree growth occurs at about 30% relative density (Marquis 1986). This reflects the added light, moisture, and nutrients available per tree in low density stands, and a second threshold where competition begins to limit individual tree diameter growth.

Figure 17-13 portrays an additional threshold of importance to foresters. It shows curves for both gross and net cubic volume periodic annual increment based upon thinning studies in oak communities of central United States (after Dale 1968, 1972), and aspen in the eastern plains of Canada (after Steneker and Jarvis 1966). Those data show an increase in mortality among stands with relative density above 55%–60%, and losses of about 45% of total production within stands at 100% relative density. A similar pattern appeared among western larch thinning studies in Oregon. The ten-year net cubic volume growth increased with progressively higher stocking levels from 22% to 66% of initial basal area. Plots cut to 81% of prethinning levels had a lower periodic annual increment than those reduced to 67%. Mortality slightly reduced net growth at the two higher levels of stocking (Seidel 1977). Apparently, the higher-density plots reach an upper threshold sooner, and require a shorter interval between successive thinnings. Other research has shown that with stocking up to about 80%, the mortality occurs primarily among overtopped and intermediate trees. Codominants may also die when stands reach a higher relative

Relation between stocking percent and annual gross growth and net cubic volume growth in trees 2.6 inches d.b.h. and larger in Iowa and Kentucky.

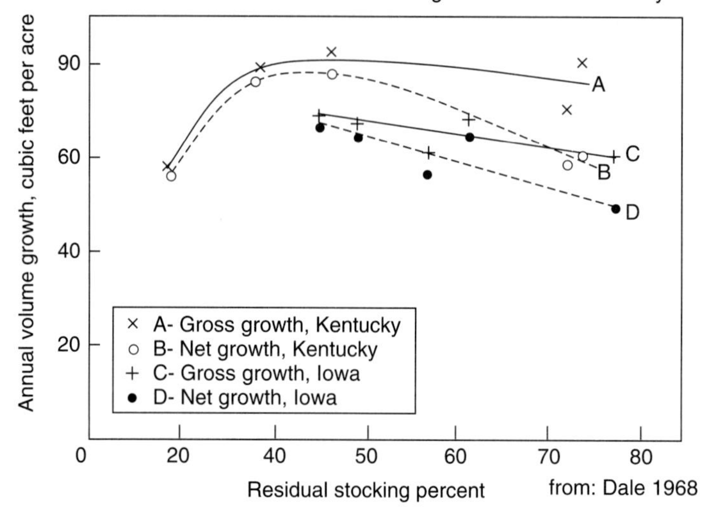

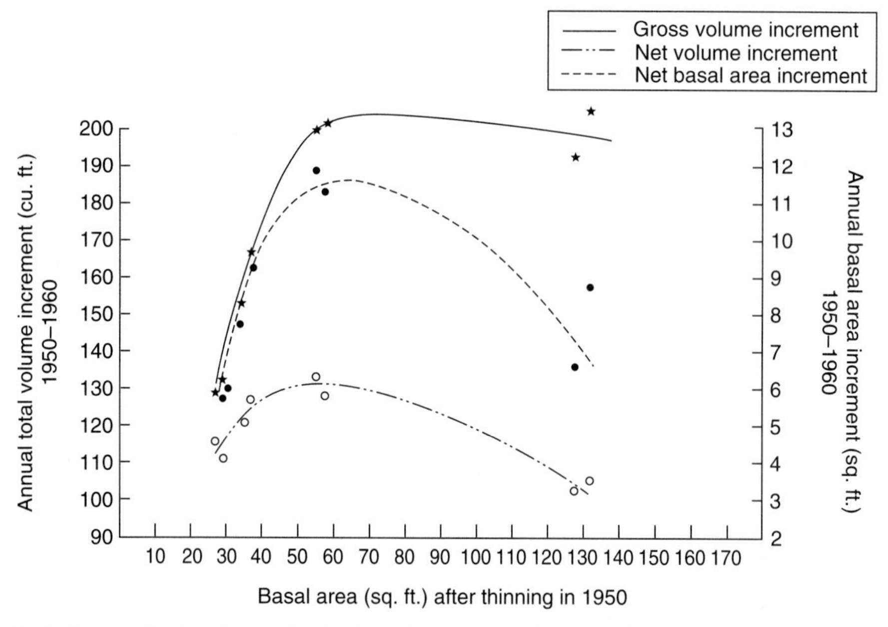

Periodic annual net and gross total volume increment and net basal area increment per acre related to basal area per acre after thinning. Study 5: a 22 year-old stand thinned in 1950.

FIGURE 17-13
Net growth will equal gross growth in appropriately thinned even-aged communities (from Dale 1968; Steneker and Jarvis 1966).

density (Marquis 1986). This suggests a need for repeated thinning once stands regrow to about 80% relative density, and a zone of rational action as depicted in Fig. 17-12. The figure also validates an important axiom of silviculture: that within certain wide limits, stand density does not influence the gross volume increment (Langsaeter 1941; Wiedemann 1950; Heiberg 1954; Mar:Moller 1954; Braathe 1957; Smith 1986). However, it will affect net accumulation.

In recognition of these concepts, many stocking guides include a B-level line (Fig. 17-9a). While somewhat different between community types, the B line is generally set at 55–60% relative density. Guides that recognize differences in species composition (e.g. Roach 1977) include a family of B-level curves, each paired with a different 100% line (Figs. 17-9a and 17-11a). In these cases, the guides suggest leaving a higher residual basal area with each successive thinning, and different amounts for stands with different species composition. The sawtoothed production function depicted in Fig. 17-8 portrays current thinking about an appropriate residual stocking to maximize full net annual production within stands under long-term management.

18

METHODS OF THINNING

COMPARING METHOD AND INTENSITY OF THINNING

Thinning will substantially influence natural patterns of growth, development, structure, and yield among stands past sapling stage. It allows foresters to regulate the levels of site occupancy and subsequent stand production, and to bring a tree community into conformity with many kinds of relevant management objectives. In general, thinning allows foresters (Daniel et al. 1979; Smith 1986) to

1 salvage potential mortality trees before they die, thereby recovering as much as 25%–35% of gross production as marketable yields;
2 concentrate the growth potential onto the fewer trees so they become bigger faster;
3 influence the composition for quality and species, thereby enhancing value over a rotation;
4 realize earlier and more regular income by regulating the growing stock density throughout a rotation;
5 insure good seed sources for the next rotation; and
6 shorten the rotation by speeding tree development.

Though foresters most commonly thin stands to enhance wood production, they can also use thinning to control conditions of essential plant and animal habitats, or to enhance other nonmarket values. They need only know the kinds of structural and density conditions that enhance those values. Then they can determine how best to alter stand attributes to address the objectives.

For many nonmarket and commodity values, foresters look for ways to trigger different kinds and magnitudes of tree growth by the way they concurrently control the amounts of growing stock left in place, the spatial distribution of residuals, and the kinds of trees cut. Of these three components, the first denotes the *intensity* of thinning, or its severity. The last reflects the *method* of thinning, or really the kinds of trees kept as residuals. Integrating these elements in a single prescription recognizes that simply regulating the numbers of residual trees and their spacing (the intensity) may not provide the best results. Rather, foresters must also leave trees with desirable characteristics, and having full and vigorous crowns. This will insure good health, and good rates of post-thinning diameter growth. Further, they usually must maintain an appropriate degree of crown canopy cover (intercept sufficient insolation) to keep biomass production at a full level (Cheyney 1942), or to influence the kinds and amount of understory vegetation that develops.

When foresters control the intensity of a thinning, they regulate the degree of site utilization, the potential for lateral branch elongation and survival, and the effect of intertree competition on individual tree diameter growth. The amount of residual cover (determined by the intensity of cutting) also controls understory brightness. This affects the establishment, composition, and development of low vegetation as well. In fact, fairly dense residual stands (few large gaps in the upper-canopy layer) may suppress understory plants, and keep low vegetation sparse. Conversely, an open crown canopy will favor the proliferation and growth of herbs, shrubs, and tree regeneration. In large measure, thinning intensity determines how much volume a stand will produce afterward, what proportion of it is lost to mortality, at what rate individual trees will grow, and how the total vegetation community changes.

From a silvicultural perspective, thinning intensity translates into controlling the level of residual stocking, rather than designating the amount to remove. It follows the notion that silvicultural practice either tends an existing community, or regenerates its replacement at an appropriate time. Usable materials accrue by removing excess or mature trees during a tending or regeneration operation. The residual stand or new age class represents the productive machinery (usually called *growing stock*) that captures incoming solar energy and converts it to biomass. The yield from thinning represents a *byproduct* of management, and to the degree that foresters adequately control the level of residual stocking for full site utilization, they insure high levels of future revenues to pay off investments in land ownership and management.

Besides selecting an intensity for thinning, foresters must also choose a method. These two differ. Conceptually, foresters can apply each method of thinning at a range of intensities, thereby controlling post-treatment stand production and the levels of individual tree growth in each case. Whereas the intensity designates the amounts of growing stock retained, the method reflects the sizes and kinds of trees kept and cut. Further, for even-aged stands, where tree size reflects its past performance and potential future growth, the method essentially prescribes the crown positions of trees to retain or remove. To the extent that a method of thinning concentrates the growth potential on trees having full and well-developed crowns, and the intensity maintains an appropriate residual relative density, both the stand and its component trees should develop at good rates. If they cut the dominant and better codominant trees, foresters can still maintain full net production per unit area. However, the residual trees will grow slower than if they had left trees of better crown positions, and the stand will not have as many large-diameter trees of high value at the end of the rotation.

Historically, foresters have recognized five distinct methods of thinning (Ford-Robertson 1971; Daniel et al. 1979; Smith 1986):

1 low, or thinning from below
2 crown, or thinning from above
3 selection, or diameter-limit thinning
4 mechanical, or geometric thinning
5 free

For the first three, foresters select trees to cut or leave based largely upon crown position. With mechanical thinning, they use fairly strict spacing criteria. In a free thinning, they often incorporate both of these elements, so the method may represent a hybrid approach. Foresters can control the intensity of thinning with all five methods, thereby regulating the degree of site utilization as well as the sizes of trees retained as growing stock.

LOW THINNING

Foresters may call *low thinning* an *ordinary thinning,* a *German thinning,* or a *thinning from below* (after Hawley 1921; Cheyney 1942; Assmann 1970; Ford-Robertson 1971; Smith 1986; Soc. Am. For. 1989), meaning that it

- removes trees of lower crown positions, leaving the larger and more free-to-grow trees more or less evenly spaced over the area;
- removes overtopped trees and those in subordinate canopy positions, and creates few gaps in the main crown canopy except when applied at a high intensity; and
- favors upper crown trees by removing varying proportions of the others.

This method simulates natural mortality by cutting smaller trees that would die first. It retains the largest and most vigorous trees with the best-developed crowns. As the intensity increases, the treatment removes some trees of better crown positions. Even so, it focuses the growth on trees with the greatest promise of growing at maximum rates (those of upper-canopy positions).

Figure 18-1 depicts different historically recognized intensities of thinning from below, called *grades* or *degrees* of thinning. Specifically, to increase the intensity of thinning, foresters take increased numbers of trees from below the main crown canopy, thereby opening more space around crowns of the dominant and upper codominant trees (after Cheyney 1942; Kostler 1956; Assmann 1970; Smith 1986) as follows:

Thinning from below

Grade	Trees to cut
A - very light	Overtopped
B - light	Overtopped + intermediates
C - moderate	Overtopped + intermediates + some codominants
D - heavy	Overtopped + intermediates + most codominants

Grade A and B thinnings from below most closely parallel natural mortality by removing the weakest and bent-over trees with the highest probability of dying. The photosynthetic output per residual tree will not increase to a great degree, and individual tree growth may not improve appreciably. A and B grades of thinning from below primarily allow landowners to create a single-layer crown canopy or utilize the small-diameter trees before they die and become useless (Assmann 1970).

Some classifications do not recognize a D-grade thinning from below (Kostler 1956; Assmann 1970), or they subdivide B-grade thinnings into two classes to recognize the proportion of intermediates removed (Hawley 1921). Even then, C- and D-grade thinnings from below remove trees from the main canopy (Hawley 1921; Cheyney 1942; Kostler 1956; Assmann 1970; Smith 1986). They create gaps within the canopy and elevate the light around crowns of the residual trees. As lateral branches extend, the gaps close. The degree that these higher grades of thinning from below increase understory brightness depends on how soon crown expansion closes the canopy gaps. This affects what herb, shrub, and tree seeds germinate, and if any new plants persist and develop. At least in stands that produce viable seed, a D-grade thinning from below often stimulates regeneration similar to that after a shelterwood method seed cutting (Smith 1986). In fact, all methods of thinning should trigger regeneration in maturing stands, given an adequate source of viable seeds and adequate brightness in the understory.

Foresters regulate the residual density for thinning from below by setting an approximate minimum diameter for trees to keep in the residual stand, and prescribing the proportion of trees to remove from each crown class. Well-trained work crews can then apply the treatment without having a forester first mark specific trees to remove. This reduces costs of supervision and layout. At the same time, thinning from below recovers only small trees, and may prove impractical to apply by commercial logging. Especially for early thinning, foresters often use only an A- or B-grade thinning from below. Then they must either find markets for small-diameter trees, or cut and leave (or kill) any that they cannot sell. This may cost too much for enterprises that require a high rate of return on the investment. Thinning from below may prove attractive as the first treatment when landowners accept precommercial treatments, or do the work independent of a timber sale. Subsequent thinnings from below will necessarily shift to at least a C grade. After a series of them, only the final crop trees of uppermost canopy positions remain. As an outcome, stands have little vertical structure, with live foliage confined to only upper

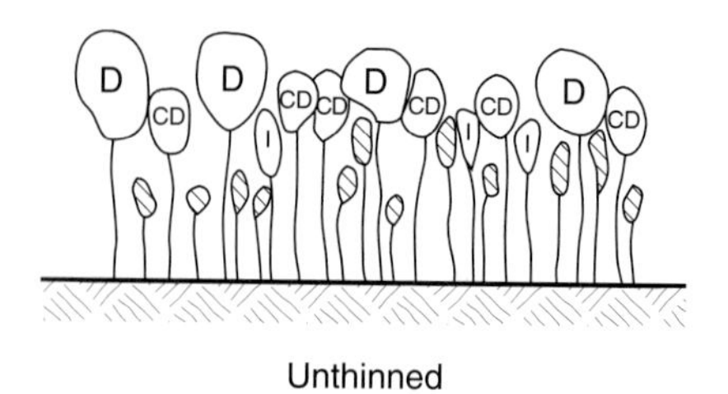

Unthinned

Grade A thinning from below
(Cut overtopped trees)
. . . A very light thinning

Grade B thinning from below
(Cut overtopped and intermediates)
. . . A light thinning

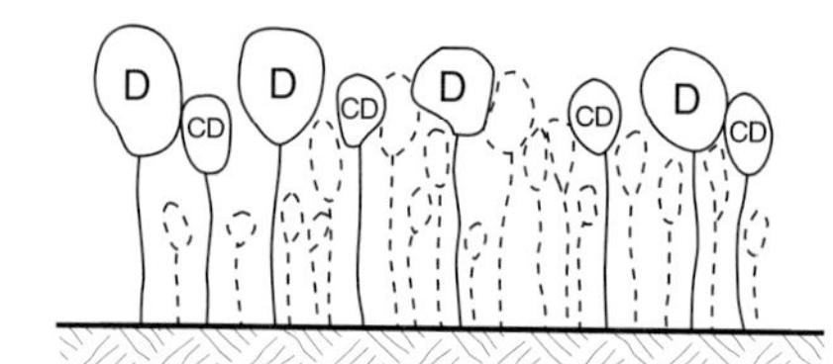

Grade C thinning from below
(Cut overtopped, intermediates,
and *some* codominants)
. . . A moderate thinning

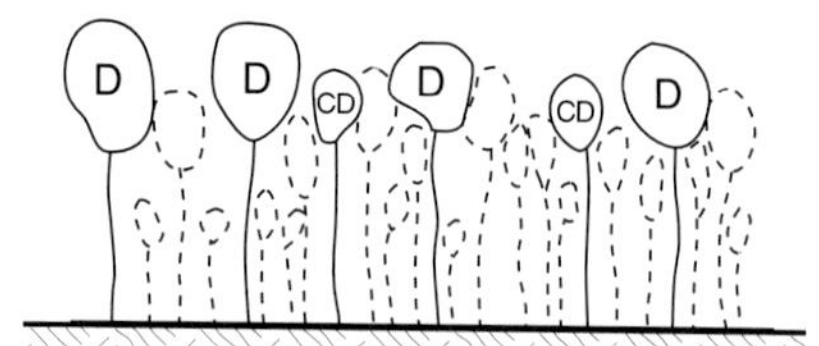

Grade D thinning from below
(Cut overtopped, intermediates,
and *most* codominants)
. . . A heavy thinning

FIGURE 18-1
Low thinning removes trees of subordinate crown positions to favor those of upper-crown classes, with little effect on the main crown canopy for intensities of less than C-grade thinning from below.

portions of the trees' crowns. Thinning from below also leaves stands with little visual diversity, and eliminates short trees that would interrupt the lines of sight through a stand.

CROWN THINNING

Crown thinning reduces crowding within the main crown canopy and increases the light around foliage of the residual trees. Sometimes called *thinning from above,* the

French method, or *high thinning,* it has these characteristics (after Hawley 1921; Cheyney 1942; Kostler 1956; Assmann 1970; Ford-Robertson 1971; Smith 1986; Soc. Am. For. 1989):

- removes some of the codominant and dominant trees to favor the best trees of those classes
- removes trees from middle and upper crown positions (and diameter classes) to open the canopy and favor the most promising dominants and codominants
- leaves overtopped and intermediate trees that do not interfere with those of better canopy positions
- favors the most promising trees of upper-canopy positions (though not necessarily all dominants) by removing varying proportions of other stems, giving due regard to residual tree spacing

Crown thinning may remove some dominants of undesirable characteristics, including trees of noncommercial or low-value species. It may take out dominants of poor form and quality if better codominants would develop into trees of higher value. Primarily, crown thinning favors selected trees of high vigor and removes other competing trees, regardless of crown class. It ignores understory trees that do not affect the better trees, and retains the overtopped and lower intermediate trees as part of the residual stand.

Figure 18-2 depicts crown thinning. It shows that the thinning maintains uniform spacing between residual trees, and retains the best and most vigorous trees as growing stock. It reduces crowding around their crowns to foster improved diameter increment (after Hawley 1921; Cheyney 1942; Kostler 1956; Assmann 1970; Smith 1986). If parts of a stand lack stems of ideal quality, foresters select the best tree available and remove poorer neighbors that interfere with the selected trees. This stimulates the growth of all main-canopy residuals. Keeping the overtopped and lower intermediate trees maintains any vertical structure in the community. The overtopped and intermediate trees may also partially shade lower branches of better trees, and help to promote lower branch mortality. Their shade may also retard the establishment of understory plants, particularly early in a rotation. Understory trees of high shade tolerance often improve in vigor following crown thinning, and grow better than before thinning. However, they don't develop as rapidly as those in upper-canopy positions.

Crown thinning improves the diameter growth of upper-crown residual trees. It concentrates the growth almost entirely on the *crop trees.* These will form the final tree community at the time of stand maturity (Ford-Robertson 1971; Soc. Am. For. 1989). Characteristics of crop trees depend on the objectives of a landowner, but they generally have high quality and value. The number of crop trees depends upon their expected average crown radius at the end of a rotation, and the spacing that will insure full canopy cover between the last thinning and the end of a rotation. During intermediate ages foresters will keep additional trees between the crop trees to insure full site utilization as the stand develops. Foresters call these extra trees *fillers.*

The numbers of filler trees left after each thinning depends upon the target residual relative density. In crown thinning, foresters deliberately thin around the filler

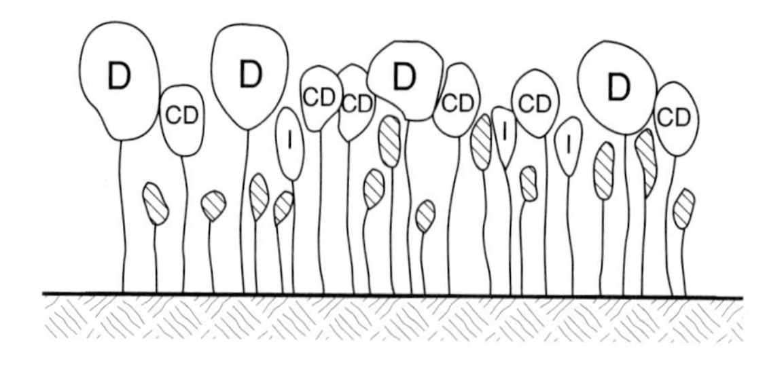

Unthinned

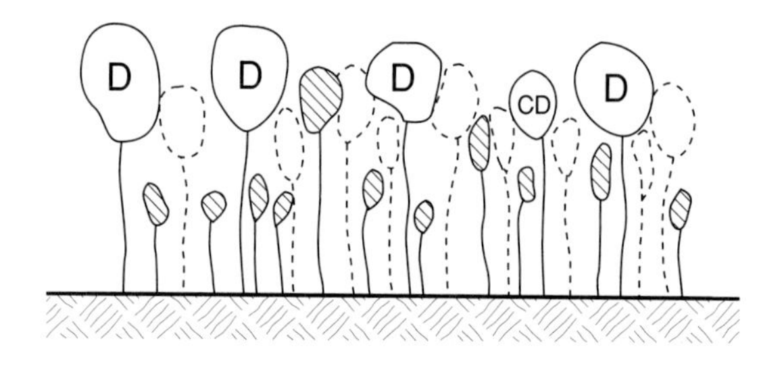

Crown thinning

FIGURE 18-2
Crown thinning favors the best trees available among those of upper-crown positions, removing competitors from the main crown canopy.

trees to maintain their vigor and improve their growth, at least until the next thinning. This insures spatial uniformity of production across a site, and controls the mortality of filler trees. With each thinning, foresters remove some of the filler trees to open space around the crop trees. Ideally, after the last thinning only the crop trees remain.

Crown thinning increases light levels around the foliage of residual trees, including their lower limbs. Lateral branches throughout the crown will elongate into the open space without obstruction, and the crowns gradually become wider. Since the lower branches remain alive longer, total crown length also increases as the trees grow in height. This two-dimensional crown expansion (length and width) increases the photosynthetic output within a tree, and the diameter increment. Crown thinning delays self-pruning (falling off) of branches at the base of the crown. This may contradict objectives to produce knot-free wood and to keep the knots small. Keeping judiciously spaced filler trees maintains at least partial shade on lower branches of the crop trees, and minimizes this side effect of crown thinning. Once branches have died along an acceptable merchantable length of main stem, foresters will want to keep the lower branches alive and maintain as much live crown length as possible. This will enhance the crop trees' diameter growth.

To promote self-pruning during early stages of stand development, foresters sometimes leave neighboring trees of shorter heights to interfere with lower

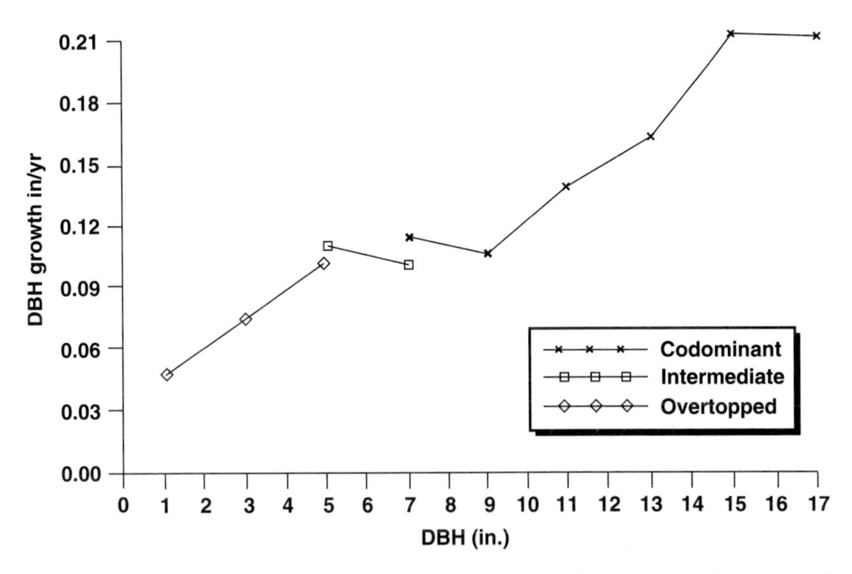

FIGURE 18-3
Effects of thinning on the growth rates of residual shade-tolerant sugar maple trees of different initial crown positions in a previously unthinned 75-yr-old even-aged northern hardwood stand (from Lareau 1985; Nyland et al. 1993).

branches on the crop trees. These *trainers* hasten lower branch mortality by their shading, or through abrasive action against the crop tree (Soc. Am. For. 1989). The shorter trainers also benefit from a crown thinning, but primarily serve to enhance the quality of adjacent crop trees. They make crop tree crown expansion more one-dimensional (largely by extension of the terminal shoot and uppermost branches) than following a thinning that opens space all around the crop trees.

For a crown thinning, foresters mark no overtopped trees, and will leave any intermediates that do not interfere with the crop trees. The small stature of these inferior trees probably reflects their limited genetic potential for height and diameter increment, at least compared with upper canopy trees of the same species (Fig. 18-3). On the other hand, with shade-tolerant species that became overtopped by more rapidly growing shade-intolerant trees, the smallness may simply reflect oppression. Then crown thinning may promote their growth, at least to a minimum merchantable size. For suppressed understory trees (e.g., in single-species stands), thinning simply keeps them alive (but growing slowly), and forces foresters to remove them as a site preparation treatment at the end of the rotation. In the interim, they enhance stand vertical structure, and contribute to the habitat of several indigenous animals. They also alter visual qualities important to some nonmarket values.

In planning a crown thinning, foresters must deal with a combination of factors. These include (1) selecting the crop trees; (2) controlling the intensity of each cutting; (3) determining the numbers and positions of filler trees; and (4) judging the need for trainers. As a result, many foresters consider crown thinning a fairly complex method. By removing trees mostly from the middle-diameter classes, it usually provides sufficient volume for a commercial sale (Hawley 1921, 1929; Cheyney 1942; Smith 1986), at least in areas with a demand for fiber products. Crown thinning usually proves useful after an earlier thinning from below. Then foresters need not worry about promoting continued self-pruning of the lower branches.

SELECTION THINNING

Selection thinning has no technical relationship to selection cutting or selection system. Sometimes called *thinning of dominants* or the *Borggreve method* (after Hawley 1921; Kostler 1956; Ford-Robertson 1971; Smith 1986; Soc. Am. For. 1989) it

- removes trees larger than a prescribed diameter limit to favor smaller trees of good *growth form* and condition
- removes dominant trees to favor smaller trees of at least reasonable vigor

The term *growth form* means the life form or habit (Ford-Robertson 1971), and suggests nothing about a tree's size or rate of growth, per se. Selection thinning deliberately removes the most vigorous and largest trees from a stand. Those released may have good stem form and even a balanced and reasonably developed crown. Yet they generally have a limited growth *potential.* Foresters justify selection thinning primarily when the dominant trees consist entirely or mostly of low-value species or have poor stem quality. Selection thinning also removes more volume during early entries to an even-aged community.

Figure 18-4 illustrates selection thinning, and that it essentially amounts to a diameter-limit cutting. The smaller trees improve in vigor and diameter growth only if they have sufficient crown (e.g., at least 30%–40% of total height) to ensure good long-term development at reasonably fast rates (Smith 1986). Usually, shade-tolerant species offer the best promise for such an improved post-thinning growth, yet their growth will not match the past increment of larger codominant and dominant trees taken out in a selection thinning (Fig. 18-3). In fact, within loblolly pine, the post-thinning rates of growth decline consistently from dominants, to codominants, to intermediates (Little and Mohr 1963). Similarly, among Allegheny hardwood stands in Pennsylvania, diameter growth averaged (Marquis 1991):

	Stand age		
Crown position	10–20 yrs	30–40 yrs	40–50 yrs
	(Growth as a % of dominant trees)		
Codominants	69	63	61
Intermediates	49	35	32
Overtopped	41	13	16

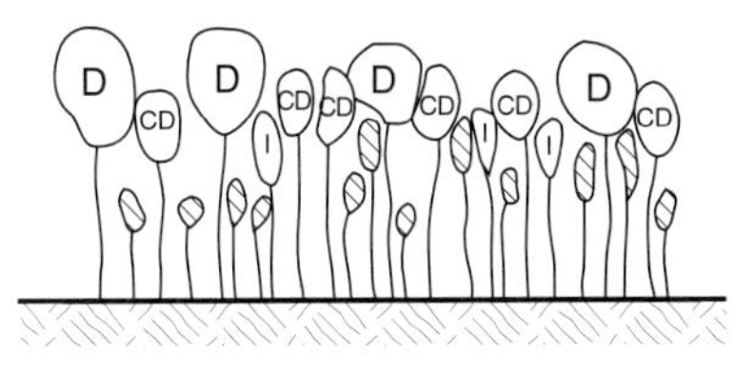

Unthinned

Selection thinning, first cut
(Remove trees marked with //)

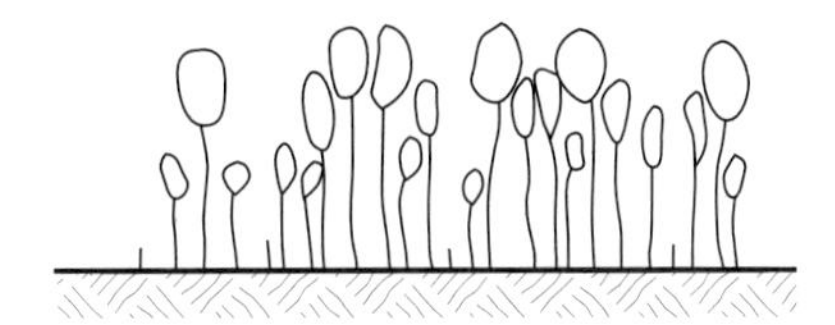

Residual stand after
first selection thinning

Selection thinning, second cut – after growth
(Remove trees marked with //)

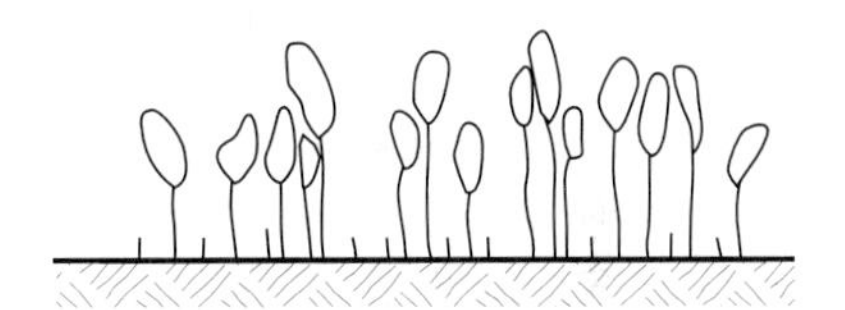

Residual stand after
second selection thinning

FIGURE 18-4
Selection thinning removes the largest trees in upper crown positions, uncovering smaller trees for future growth and development.

Such data indicate that the relative rates decline as stands age, and most dramatically among trees of lower crown classes. Foresters can minimize the loss by removing only the dominant trees, and retaining the better codominants as residuals. Those trees should grow reasonably well following release. When selection thinning removes a fast-growing and low-value species that overtops more valuable shade-tolerant trees, it may enhance the long-term value growth for stands dedicated to commodity production.

Selection thinning yields considerable volume of larger-diameter trees, and may make early thinning feasible in areas with poor markets for small-diameter trees. If judiciously controlled to an appropriate intensity, selection thinning should not greatly diminish cubic volume or bulk fiber production. Yet removing the best growing trees early in a rotation reduces long-term sawtimber volume production (Daniel et al. 1979; Smith 1986; Beck 1989; Nyland et al. 1993). In fact, one study in Appalachian hardwoods of eastern North America showed that selection thinning would result in only 81% of the total yield produced with regimes employing thinning from below (Beck 1989). This difference may prove important for ownerships dedicated to producing large-diameter logs in the shortest possible period of time. The difference becomes magnified over time if foresters use selection thinning as the sole method for an entire rotation.

Figure 17-4 emphasizes the importance of both the genetic quality and relative crown position of trees in even-aged stands (after Assmann 1970). It highlights how climate, soil, and crown position interact with the genotype to become manifest as the phenotype. Since foresters cannot influence either soil or climate, they work with tree appearance and crown position in selecting those to favor or remove during thinning. Where they apply a selection thinning, the residual stand contains trees that developed under the negative influence of shading and crowding by better-positioned trees. Further, the inherent genetic potential tempered by the interaction with environmental factors apparently contributed to their poor growth and crown position prior to thinning, and would likely limit their growth in the future as well.

In effect, taking the largest diameter trees from an even-aged stand through selection thinning removes the best growing stock (Daniel et al. 1979; Maynard et al. 1987). It decreases rates of development for a stand and prolongs the time before residual trees reach a predetermined minimum tree diameter. In addition, when foresters repeatedly remove the tallest trees, those left at the end of a rotation have shorter average heights, further reflecting the disgenic character of this method. Clearly, if followed by a *natural* reproduction method at the time of financial maturity, selection thinning could degrade a stand genetically, particularly among species with considerable inherent variability in tree diameter and height growth.

With plantations of some shade-tolerant species, foresters have occasionally used selection thinning and artificial regeneration as part of a long-term approach called the Borggreve silvicultural system (Hawley 1921; Daniel et al. 1979, after Dengler 1944). They let the planted trees grow until dominants reach merchantable sizes, and then harvest them. They also cut the overtopped trees, as in a Grade A thinning from below. As the residual codominants reach merchantable sizes, the second selection thinning also removes the larger trees, leaving the poorer codominants and intermediates for continued growth. Landowners repeat this process over consecutive cycles until a stand contains only the minimum number of trees for full crown closure through the remainder of a rotation. Then when the last of the original stand finally reaches merchantable size, they clearcut and replant (Daniel et al. 1979). Note that by using tree planting as the reproduction method, foresters need not worry about degrading the genetic capacity of a stand. Nevertheless, they will not optimize long-term sawtimber production by this system, and must extend a rotation until the original intermediates grow into usable sizes.

MECHANICAL THINNING

Mechanical or *geometric thinning* uses some spacing or spatial arrangement criteria to identify the trees to cut or leave. Technically, mechanical thinning (after Ford-Robertson 1971; Smith 1986; Soc. Am. For. 1989)

- removes trees within a fixed spacing interval, or by strips with fixed distances between them; and
- removes trees in plantations by rows.

Foresters separate mechanical thinnings into two broad groups. These include *spacing thinning* applied to natural stands, and *row thinning* for plantations. Row thinning, in particular, gives little regard to crown position or condition of the cut and residual trees. Foresters usually take tree vigor into account with spacing thinnings. Generally, mechanical thinnings leave a residual stand of some predetermined spatial arrangement, and often prove arbitrary in other respects. They remove much of the decision making from workers, so a crew can often thin a stand without the investment in tree marking by a forester.

A spacing thinning often works best as a first thinning early in a rotation. Foresters first select an appropriate spacing between the residual trees, in essence mentally superimposing a preset grid across the stand (Fig. 18-5). The interval between grid intersections increases as a stand matures and the trees become larger. In fact, spacing should reflect the area needed for adequate crown expansion until the next thinning. For an early precommercial thinning, foresters may not even mark the stand. Instead, they may simply designate the spacing interval and rely upon a well-trained crew to fell or poison trees within the intervening space. For stands in poletimber or later stages of development, foresters may take both spacing and crown position into account. In these cases they designate an appropriate preset spacing interval, and then look within a certain maximum distance of the designated point for a tree of upper-crown position and high vigor. They may consider stem quality as well. As a consequence, the stands look similar to those given a Grade D thinning from below (all foliage concentrated in a narrow horizontal band), except that the crown canopy will not have large open spaces between the residual trees.

Row thinnings often work well in plantations. Foresters simply designate an interval for the rows to keep, and then cut all trees between those rows (Fig. 18-6). They regulate the intensity of thinning by removing either every second row or every third row. The former releases each residual tree on two sides and reduces the stocking and relative density by 50%. Foresters often must then wait an extended time until the crown canopy closes and they can do another thinning. The prolonged crowding along two sides of trees within the residual rows flattens their crowns into an oblong shape. With third-row thinning, the residual trees remain crowded on three sides, and the crowns become distorted if foresters wait too long between thinnings. Removing only every fourth or fifth row leaves three-fourths or four-fifths of the trees without any release. The interior unreleased trees will show hardly any response to release (Little and Mohr 1963). This does not qualify as a thinning.

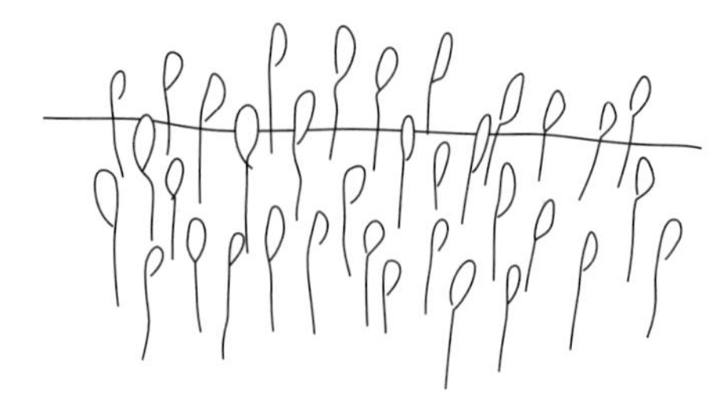

Unthinned

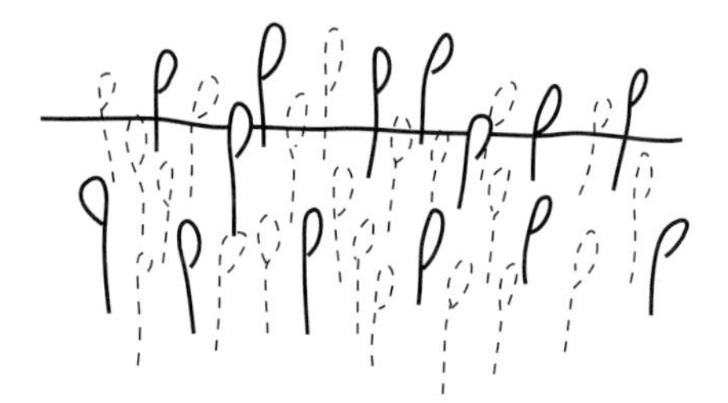

Mechanical thinning by spacing

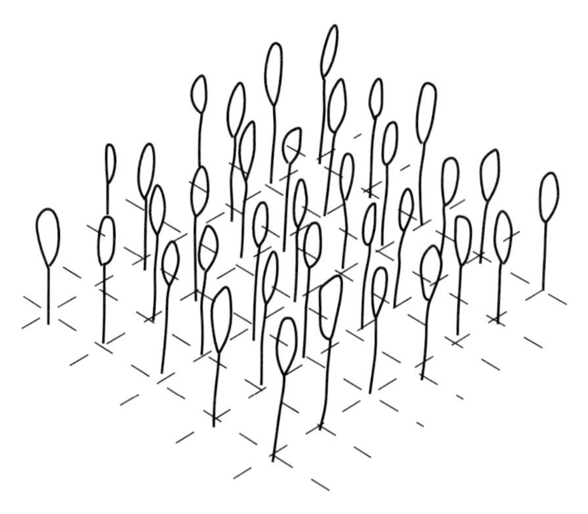

Residual stand after spacing thinning (Mechanical thinning)

FIGURE 18-5
Spacing thinning releases trees at a fixed spacing interval, removing others to favor the development of residual trees, and adjusting the spacing interval to reflect the degree of stand development and anticipated growing space needed for good residual tree vigor and development.

Row thinnings remove different sizes of trees in proportion to their frequency in a stand. They do not alter the quadratic stand diameter (QSD) or average tree height. Row thinning does take sufficient large trees to make an earlier commercial sale feasible (one-third or one-half of the standing volume), but does not degrade the growing stock. It fits well with mechanized logging systems (Fig. 18-7), and that often makes it particularly attractive for a first treatment in young stands. It does not reduce the numbers of crop-tree-quality stems below the minimum for

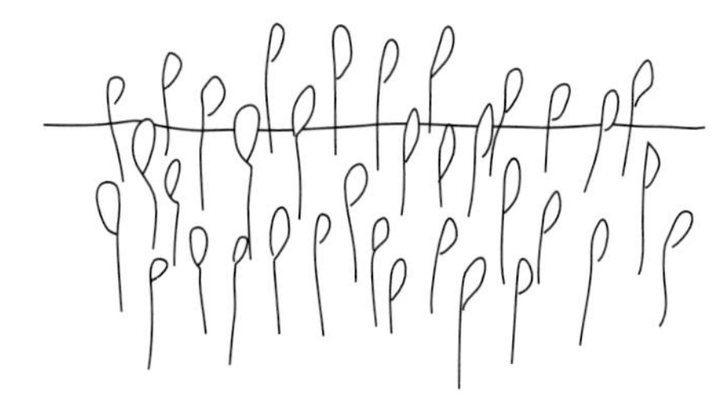

Unthinned

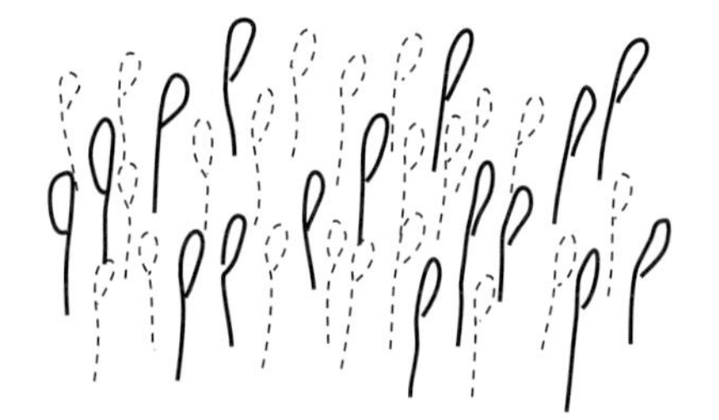

Mechanical thinning by rows

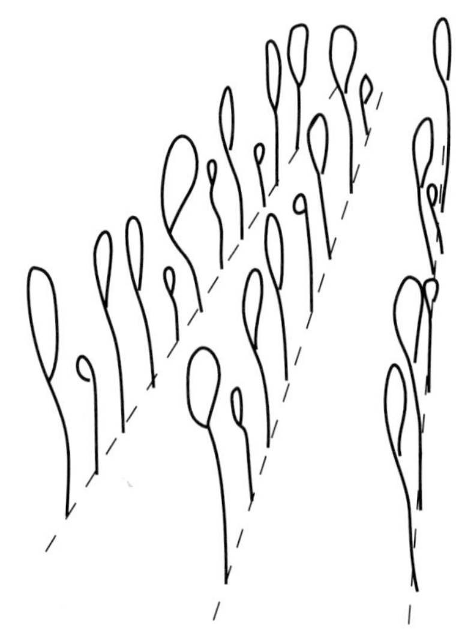

Residual stand after row thinning (Mechanical thinning)

FIGURE 18-6
Row thinning removes all trees in designated rows without reducing crowding among trees within a row, freeing the residual trees on one or two sides for third-row and alternate-row thinning, respectively.

full site occupancy at the end of a rotation. In fact, in many plantations, foresters could even do a second row thinning at right angles to the first one, and still have sufficient crop-tree-quality residuals for full site utilization at the end of a rotation (Day and Rudolph 1972). By retaining trees of all sizes (diameter and height) within the uncut rows, row thinnings do not alter the vertical structure important to

FIGURE 18-7
Row thinning in this Norway spruce plantation removed two rows to facilitate mechanized harvesting, leaving two-row bands of residual trees, each released on one side.

some wildlife. Second-row thinnings open the crown canopy appreciably, and may brighten the understory sufficiently to recruit shrubs and herbs that contribute to wildlife habitat and alter the visual qualities of a stand. Cutting the trees in straight rows creates long lines of sight and changes a stand's appearance. Foresters must consider such factors for ownerships with important noncommodity objectives.

Among natural stands, foresters may emulate row thinnings by a technique called *strip thinning* (Hawley 1921). This method creates a rowlike arrangement by removing trees in continuous strips running at preset intervals from one side of a stand to the other, similar to the arrangement depicted in Fig. 16-4. The individual residual trees do not necessarily line up in a single row, or in a line of single stems. Instead, they remain in narrow strips or bands, with a continuous strip of open space along two sides. Foresters regulate residual stocking or relative density by keeping the width of each cut strip and the interval between residual strips fairly narrow. Generally, strip thinning lends itself to mechanized precommercial thinning, and to mechanized harvesting systems for thinnings in older stands. Strip thinning affects wildlife habitat and visual qualities in ways similar to those of row thinning.

FREE THINNING

Free thinning primarily releases especially selected crop trees (after Ford-Robertson 1971; Soc. Am. For. 1989; Smith 1986). As such it

- favors only the crop trees, leaving the remainder of a stand unthinned;
- favors desired trees using a combination of thinning criteria; and
- releases crop trees without strict regard to the crown position.

Free thinning may combine elements of two or more thinning methods to adequately free designated crop trees for better growth. It has no precise definition, except for the emphasis on crop tree release.

In applying a free thinning, foresters must first set the criteria for selecting crop trees based upon a combination of crown position, bole quality, and spacing interval. A crop tree may simply represent the best main canopy tree available at some preset interval. Then the treatment resembles either a spacing or a crown thinning, depending upon the distance between crop trees. If foresters decide to select only perfect trees for release (e.g., dominants or upper codominants with a straight bole and small branches), the actual spacing may vary widely from one part of a stand to another. Then free thinning has a distinctive character that resembles no other method.

In marking for a free thinning, foresters work through a stand to find the crop trees. They normally maintain some minimum distance between them, and mark only neighbors that interfere with the crown of each crop tree. Areas lacking perfect trees receive no treatment under a free thinning method. This distinguishes free thinning from crown thinning, in which foresters free the best tree at some designated spacing interval, and strive to keep a uniform relative residual density throughout the stand. Free thinning may leave places with reduced density around the crop trees, and other unthinned spots where the marking crew found no appropriate crop trees. That makes exact control of relative density difficult. Foresters may simply prescribe how much to free each individual crop tree (e.g., freed on two, three, or all sides). Under these circumstances, stands treated by either free or crown thinning would have similar attributes for wildlife habitat and visual qualities. Free thinning might appear more disorganized due to the spatial inconsistency in residual stand density.

Foresters often call free thinnings a *crop tree release* (Fig. 18-8). They may keep only the minimum number for full site utilization at rotation age. This allows a high degree of selectivity for crown condition and main stem qualities, permitting foresters to favor only elite trees. In most of these cases, crop trees have upper-crown canopy positions (usually the best codominants or dominants). The thinning frees space around all sides of the crowns, and promotes rapid diameter increment. It may remove or leave overtopped trees, and even keep some intermediates as trainers. In this sense, the term *crop tree release* better describes the intent.

Crop tree release (free thinning) fits well into fairly extensive silvicultural systems. It allows landowners to free a minimum number of choice trees per unit area

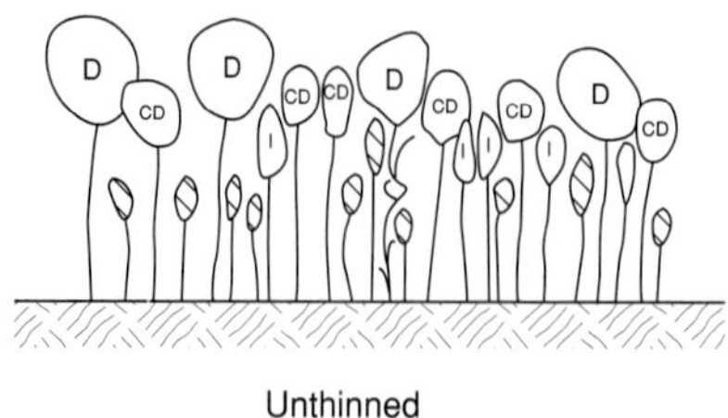

Unthinned

Unthinned
(Marked trees indicated by ⫽)

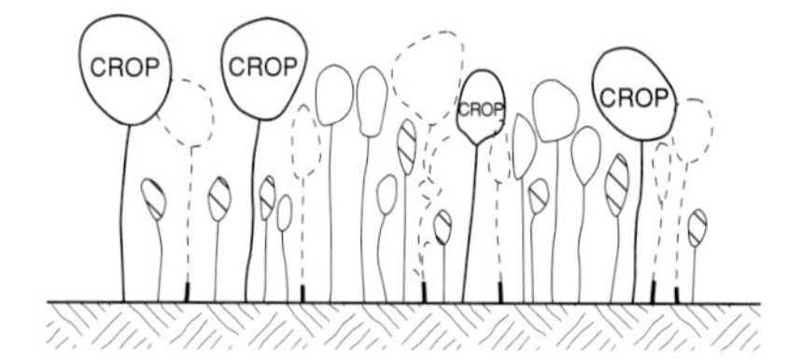

Free thinning
(Crop trees released)

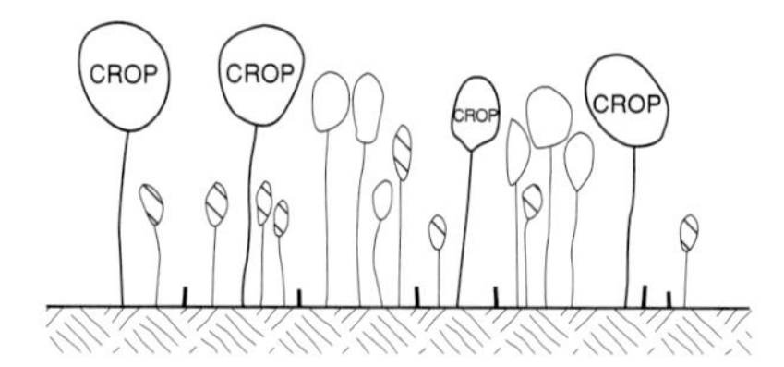

Residual stand after free thinning
(Crop tree release)

FIGURE 18-8
Free thinning releases selected crop trees of ideal characteristics without necessarily improving the growing conditions for others of lesser quality and condition.

without investing in less perfect trees as well. To illustrate, if the rotation-age stand would have residual trees at a 30-ft (9.1-m) spacing, foresters would select crop trees at that interval. They would free each one on an appropriate number of sides, and leave the area between crop trees unthinned. Also, if a stand contained fewer good-quality trees, a forester would select only those good stems and thin around them, leaving most of the stand untended. This would reduce costs, and might improve chances for a good return on the investment.

With an extensive silvicultural system involving only a single entry, or with long intervals between successive treatments, foresters might use a *crown-touching crop tree release* (after Smith and Lamson 1986; Lamson and Smith 1987). It has also worked well for first treatments in fairly young stands under both intensive and extensive systems. These thinnings remove any tree with branches touching the crown of a crop tree, thereby opening space on all sides. Foresters would mark only the crop trees, thereby reducing the costs. Later, work crews can easily determine the trees to cut. They may also mark some extra trees of poor condition to increase the recoverable yield and make the work more attractive for a commercial sale.

Foresters most commonly use free thinning as a fairly informal approach to stand management. With a crop tree release they may not make a pretreatment stand assessment, nor develop any exact prescription for regulating residual relative density. They simply enter the stand with a sense of what to look for in selecting crop trees, and locate those trees. Freeing the crop trees on an appropriate number of sides maintains good diameter growth, but does not necessarily control other aspects of stand development as well as with crown thinning or thinning from below.

SELECTING A METHOD

Each thinning method has distinct characteristics, and foresters use different combinations of them to serve different kinds of landowner objectives. Primary factors to weigh in selecting a method include

1 stand condition, including the diameters of trees present, their quality, and the arrangement or spacing of potential crop trees;

2 species mixture, with concern for tolerances to crowding and shading, their potential to grow better following release, their stratification into height classes, and their compatibility with a landowner's objectives;

3 stand age, particularly relative to the time left in a rotation, the degree of self-thinning that has already occurred naturally, the degree of stratification within the main crown canopy, and the potential for residual trees to increase in radial increment following release;

4 purposes of a landowner, including the product goals and noncommodity objectives, the financial constraints and requirements, the planned intensity of management, and the intended frequency of treatments; and

5 end use possibilities, as reflected by market demands for different kinds of products, a landowner's nontimber objectives (e.g., recreational uses or as habitat for diverse plants and animals), and the relative value expected from different potential uses for a stand.

The age of a stand at the time of first thinning, the condition and species of trees in it, and the financial constraints of a landowner usually prove most decisive. One method might seem biologically ideal, but financially impractical to apply. Particularly when landowners will do the work only by commercial sales, some approaches may become impractical. Ultimately, foresters must balance the ecologic-biologic potential against a landowner's economic objectives in recommending the best method for thinning at any given stage of stand development.

Often foresters change the method of thinning as a rotation progresses. In fact, past thinnings may have changed conditions sufficiently to make some methods useless. For example, a thinning from below for the first entry eliminates the trees of subordinate crown positions. For subsequent tending a forester would necessarily use crown or free thinnings. With a mixture of low-value shade-intolerant and valuable shade-tolerant species, foresters might prescribe selection thinning for the first entry, followed by crown or free thinnings. They generally have greater flexibility for early thinnings, and usually must shift to methods that reduce crowding within the main canopy as a stand becomes older.

CONTROLLING SPACING AND SERVING DIVERSE LANDOWNER VALUES

Once foresters have prescribed the intensity and method of thinning, they prepare a marking guide with instructions for field application, and arrange for the marking. In some cases, they do the marking themselves, and at other times they have specially trained technical assistants do the work. Either way, marking gives them the means to control the cutting (or poisoning), and to manipulate the growing stock effectively.

Except for free thinning, the marking should leave a well distributed residual stand, and an appropriate stocking. Landowners realize little production at places not occupied by trees. In fact, by creating large openings within the main canopy layer, foresters lose control over plant community dynamics, often provoking untimely regeneration within stands of seed-bearing ages. They may want to lower the stocking sufficiently to promote regeneration for some nonmarket objectives, but irregular stocking does not usually satisfy a landowner's commodity interests. In fact, foresters can optimize production only by leaving a residual stand of proper residual relative density (e.g., 60%), and with the growing stock uniformly distributed across the site.

Marking crews must work carefully to maintain uniform spacing between the residual trees, and avoid creating large openings. They can best control the conditions by focusing on the residual trees, rather than looking for trees to cut. This usually insures the most effective thinning by creating a residual stand with trees at uniform spacing, and receiving an appropriate degree of release around the crown. To fulfill these goals, foresters must select residual trees at some appropriate spacing, and then mark others that interfere with the crowns of these well-spaced residual trees. This often requires extra effort in stands of poor quality, or with an irregular initial spacing.

Stands with this uniformly distributed growing stock have an organized and orderly appearance. This often enhances the visual qualities that many landowners deem essential to their recreational pursuits. It conveys a sense of careful and deliberate tending, and creates uniform patterns of sunlight and shadows in the understory. By selecting one method of thinning in preference to others, foresters may change the images that people see in a stand. In that respect, foresters can sculpt the appearance of a stand and influence how viewers react to it.

19

THINNING REGIMES IN THE EVEN-AGED SILVICULTURAL SYSTEM

CHARACTERISTIC RESPONSES AMONG THINNED STANDS

Thinned and unthinned stands differ in absolute and relative density, and individual tree vigor and growth. Thinning also affects general structural attributes of a stand (size-class distribution, spacing among individual trees, and volume or basal area), and how these change with time. Readily available information describes these largely from a financial or commodity perspective. Yet in a biologic or ecologic sense, thinning really serves two functions: (1) to temper *successional* trends, bending succession in some desired direction over the short and long run; and (2) to temper *development* of a stand, eliminating some size classes, and enhancing the growth of selected individuals. Effects on succession come from eliminating undesirable species, altering the relative abundance of different species present in a stand, and controlling the species that will serve as seed sources for the next rotation. Foresters can also use thinning to maintain slow growing species that might otherwise become overtopped and die. They temper stand development by removing some particular diameter classes (e.g., thinning from below vs. selection thinning), and by controlling stand density and residual tree spacing to speed diameter growth. In these cases, thinning hastens the reduction of stem density and shortens the time for individual trees to reach certain threshold diameters. Generally, release cuttings provide the best control of species composition and can have the greatest effect on successional trends. Thinnings have more effect on the rates and patterns of individual tree and stand development, and the structural attributes of a vegetation community. If properly executed, thinnings maintain sufficient crown cover to insure full standwide net growth for the site, species composition, and stage of development (Assmann 1970). A uniform canopy cover also mutes understory development, at least during most of a rotation.

Thinning trials in a 60-year-old Douglas-fir stand in Pacific Northwest North America illustrate the general expectations. Three successive treatments reduced the basal area by approximately one-third to one-half by cutting trees mostly of poorer crown positions to improve spacing and stand quality. Moderately thinned plots had the highest net 30-year increment, followed by unthinned and heavily thinned plots. Thinning reduced mortality and increased individual-tree diameter growth. Heavy thinning increased diameter increment the most, and trees of the largest diameters grew the best at all levels of residual density. Thinning reallocated the growth potential to fewer and larger trees that grew better. It also reduced losses to mortality, and provided earlier revenues to the landowner (Worthington 1966).

Thinned aspen stands in the eastern plains of Canada showed similar patterns. For stands up to about 40 years of age, maximum net basal area increment occurred with about 60% residual stocking. Older stands peaked at progressively higher levels. In a twenty-three-year-old thinned stand, gross growth remained relatively constant from about 60–130 ft^2/ac (13.8–29.8m^2/ha), but the ten-year net basal area and cubic volume increment decreased above and below these levels. Unthinned stands produced only about 70 percent of the potential increment. Individual-tree radial growth increased progressively with reductions of a residual stocking down to about 26%. Larger, upper-canopy trees grew the best. In thirty-five-year-old stands, removing 36% of the basal area also improved the radial growth, and the fifty largest trees grew 1.4 times better over the next thirty-four years (Steneker and Jarvis 1966). These findings reinforce the concepts depicted in Figs. 17-6 and 18-3.

If the overstory canopy fully captures incoming solar energy and utilizes available moisture and nutrients to the highest possible sustainable level, no voids will develop within the plant community. This optimizes usable wood volume production, and limits the regeneration of trees or shrubs underneath a stand. To promote an understory (e.g., for wildlife habitat), foresters must create more lasting openings in the canopy. In fact, in the southern Appalachian Mountains of eastern North America, thinning twenty to fifty-six-year-old cove hardwood stands increased the combined weight of understory trees, shrubs, and herbs by 1.7 times during the first year, and 2.3 times during the third year. Then the understory declined. Both the amount of understory vegetation and its longevity depended upon the intensity of thinning (Beck 1983). Similarly, removing 60% and 80% of the basal area from a sixty-five-year-old Rocky Mountains lodgepole pine stand increased annual understory forage production (herbs and woody shrubs) by about 40% and 50%, respectively, over five years after thinning. Reducing stocking by about 50% did not appreciably affect total understory production (Crouch 1986). Such heavy removals stimulate residual tree growth rates as well, but reduce total wood production. Also, a new age class may form underneath the open understory, creating a two-aged community at the site.

Effects of thinning on habitat attributes differ with the method employed, as well as the intensity (after Hunter 1990). Taking trees from the lower part of a diameter distribution only (thinning from below) will reduce the already limited vertical structural diversity of closed-canopy even-aged stands. Except with C and D grades, thinning from below would not open the canopy sufficiently to promote much understory development. By contrast, selection thinning also reduces the spread of heights, but it disrupts the continuity of a crown canopy, and brightens the understory. To the degree that different species have become stratified by height classes, these two methods reduce tree species richness. Crown thinning maintains the diversity of tree heights and species richness, unless the prescription discriminates against a particular set of species. Also, where the marking crew deliberately maintains trees suitable for wildlife nesting and foraging sites, the abundance of cavity-dependent birds will not decline following crown thinning (Welsh et al. 1992). Crown and selection thinning will open the main crown canopy, improve the vigor of understory trees, and initially increase the mass of low foliage, forage, and browse. By increasing the density of tree foliage at midstory levels and initiating a new age class underneath the older main canopy, these methods eventually darken the understory sufficiently to reduce long-term forage production (Blair and Enghardt 1976). Exact effects vary with stand age at the time of a first thinning, the intensity of each treatment, and the character of a community (species composition and stratification by height and diameter classes).

From a business perspective (commodity goals), foresters thin stands when available markets place a premium on high-quality trees and logs of large sizes (higher value per unit of volume). Landowners may also realize better prices if they grow only one or two particular species, and produce logs free of cavities and other defects that make the logs less valuable for lumber and veneer products. Then foresters prescribe silvicultural systems incorporating long rotations and periodic thinnings. This allows a landowner to capitalize on the full productive capacity of a site, to concentrate the production on trees of high value, and to get the trees to optimal market-place sizes in the shortest time. Where landowners can market trees only for fiber products (e.g., pulpwood, or chips for fuel), they realize greater yields over the long run by regenerating a stand using an appropriate even-aged reproduction method (natural or artificial), protecting it from destructive agents, and making another reproduction method cutting at the culmination of mean annual increment (when m.a.i. peaks). This insures maximum fiber yield over the long run.

With noncommodity objectives that do not depend on tree sizes and stand density, thinning may also prove unnecessary. It might even detract from the primary values. The decision to thin depends upon stand structural attributes that best provide the values of interest, and whether some intermediate treatment would enhance those conditions. In general, thinning reduces the numbers of trees, makes the spacing more regular, decreases the variability in tree size and lower bole characteristics, and dampens understory development (Fig. 19-1a). Visually, thinned stands have a more regular and orderly appearance, and seem more open and organized. They contain few trees with cavities, deformities, and dead branches in the main crown, unless foresters leave them for some particular purpose.

INFLUENCING YIELD BY THINNING

Besides recognizing that the intensity and method of thinning influence the kinds and magnitudes of stand and tree development, foresters also know that soil quality and other environmental conditions affect the levels of production in a stand (after Nebeker et al. 1985). First, site quality (e.g., as measured by site index) integrates a variety of environmental conditions that regulate the transformation of solar energy into biomass. The better the site, the more rapidly that volume accumulates, and the higher the potential stocking. Second, because site quality affects height growth, communities on the best sites develop a closed canopy and differentiate into crown classes sooner. This triggers competition-induced mortality earlier, so that the best sites have fewer trees at any age (at least through physiological maturity). Third, the diameter growth depends upon stand density and crown position. The dominant and codominant trees grow the best, and subordinate trees decline in vigor. Many die early in a rotation. Trees in low-density stands grow most rapidly, and those in upper-canopy positions have the best radial increment at any density. Fourth, as a consequence of mortality, the density of natural stands reflects the carrying capacity of a site. The change in the numbers of surviving trees and the stocking (basal area or volume) follows a predictable pattern over time. The better sites develop more rapidly (fewer trees and more stocking for a given age). Fifth, stands on the best sites have more volume because of better tree heights. For unthinned stands of comparable site quality, the total cubic volume does not differ appreciably with even a wide variation in the numbers of trees. However, sites with fewer trees will have a higher average diameter, and more sawtimber volume at an earlier age.

At least four main factors influence the subsequent long-term volume growth following thinning (after Oliver and Larson 1990). These include

1 the timing of each thinning, particularly with reference to effects of prolonged crowding on crown size and tree vigor, the species' tolerance to interference and competition, and ways that site conditions temper the rates of tree and stand development;

2 the crown condition and position of residual trees, recognizing that dominants and codominants with well-developed crowns grow the best;

3 the post-thinning spacing, and its influence over crown vigor and development; and

4 the interval between successive thinnings, reflecting the decline of periodic annual increment as a stand matures.

Most important, removing trees from the upper canopy reduces long-term growth and development, and the sawtimber volume production. Delaying the first entry until the thinning will remove sufficient volume for a commercial sale might reduce the tree vigor, so that some trees never rebuild an adequate crown for good rates of growth. Further, weak and slender trees in high-density stands of advanced ages may break off or bend over under the weight of snow once released by thinning (Fig. 19-1b). Many also sway and whip back and forth in heavy winds, damaging the crowns of adjacent trees or tipping over (Kramer 1975; Braastad 1979; Groome 1988; Oliver and Larson 1990). At least for Norway spruce plantations in Europe, the most resistant stands have a wide initial spacing, and more sturdy trees (Johann 1980; Kramer 1980; Mracek 1980). Also, favoring the stronger trees of upper-canopy positions (i.e., thinning from below) reduces the losses following a delayed first entry (Kramer 1980). However, early thinning (even precommercial) proves best in reducing these potential losses to snow and wind (Chroust 1980).

Researchers often portray the way that thinning alters stand development by a series of *yield tables* (Table 19-1, after Gingrich 1971). The tables outline a program of treatments for long-term management to enhance certain commodity values, and usually include (1) the time and intensity for each treatment; (2) the amount of volume that a landowner can expect to remove, and the probable numbers of trees to cut; and (3) some key attributes of the residual stand following each treatment (e.g., basal area, and average tree size). The tables delineate the growth by showing the expected change in numbers of trees, basal area, and volumes for both the residual stand and cut trees. They also enumerate the cumulative yield at various stages of a rotation (standing trees plus those removed during past thinnings).

Some yield tables list the diameter of the tree of average basal area (called *QSD*, or *quadratic stand diameter*), the average height of cut and residual trees, the p.a.i., and the m.a.i. An *empirical yield table* reflects the development actually measured on managed plots (Table 19-2). These may take the form of computer growth and yield models where users can input various combinations of thinning treatments to simulate the effects and compare possible alternative outcomes (Ford-Robertson 1971; Wenger 1984; Davis and Johnson 1987). Such computer models can depict rather complex patterns of stand and tree growth in readily accessible software that produces easily interpreted results. In other cases, biometricians have experimented with the models and summarized the output into yield tables that portray several management scenarios (Alexander et al. 1975; Reukema and Bruce 1977; Curtis et al. 1882; Farrar 1985; Solomon and Leak 1985; Baldwin and Feduccia 1987).

Table 19-1a and Fig. 19-2a illustrate one possible thinning program for oak-dominated stands, beginning at age 40 and continuing at ten-year intervals thereafter. The upper curve in Fig. 19-2a also shows the expectation for unmanaged stands. In the region where these tables apply (eastern United States), contractors normally require at least 5–7 cds of pulpwood per acre (44.8–62.7 m^3/ha) to sustain a commercial operation. Thus the proposed schedule includes a *precommercial thinning* at age 40, before the cut trees have sufficient merchantable volume to sustain a commercial operation (Ford-Robertson 1971). Under these circumstances, a landowner must carry the investment until later revenues repay its compounded value. By contrast, the program listed in Table 19-1b delays a first entry until age 50, when a contractor can recover sufficient volume to make the operation profitable. Such a *commercial thinning* covers a landowner's costs for marking and supervision, and may also return some profit.

The thinnings portrayed in Table 19-1 leave B-level relative density, with subsequent entries at 10-year intervals. This maintains full site utilization, and also:

1 protects tree quality from degradation due to epicormic branching and other damages resulting from exposure of the bole;

2 controls mortality losses within practical limits; and

3 stimulates individual tree diameter growth to the extent possible without reducing stand-level p.a.i.

a

FIGURE 19-1
Timely thinning keeps the upper-canopy trees vigorous and strong (a) and less prone to breakage, but slender and weak trees in stands not thinned until an advance age may break under the weight of ice and snow (b), as with this red pine stand.

Figure 19-2b overlays the two different thinning programs onto the traditional stocking guide for upland oak stands. This version of the saw-toothed production function (called the *stand development trajectory*, after Kershaw and Fischer 1991) begins on the right-hand side of the diagram, with stand development progressing to the left. Vertical increases of the stand development trajectory following each treatment reflect an assumption that few trees die between thinnings. Changing heights of the sawteeth also show the gradual decrease of basal area regrowth that results from maintaining a constant thinning interval, even while p.a.i. declines with stand age. Simply lengthening the time between thinnings would insure a regrowth to 80% relative density after each entry, and maintain an upward positive slope to the peaks of the trajectory depicting expected stand development.

EFFECTS ON INDIVIDUAL TREE GROWTH

Figure 19-3 shows how different management programs affect average tree size, with periodic thinning starting at ages 20 and 50, respectively. Overall, the diagram shows two important outcomes: (1) for a given age, trees in thinned stands have

b

FIGURE 19-1 (continued)

larger diameters, as evidenced by the height of the curves for QSD; and (2) the rates of change in tree diameter do not drop off as early in managed stands, as witnessed by the steepness and shape of their QSD lines. The earlier the thinnings commence, the greater the long-term effect on tree size. Also, early thinning allows a landowner to better control mortality, given a market for the small trees and a cost-effective means of extracting them from the stand. Otherwise landowners must weigh the option of having larger trees at rotation age, against the compounded cost of precommercial thinning.

The slope and height of curves for thinned stands reflect two kinds of change. First, most thinnings (except selection thinning) retain the larger trees as residuals. Each thinning increases QSD, and elevates the curve by a series of stair-step jumps. The steepness or angle of the lines between the stair steps (between subsequent thinnings) reflects the rate of individual residual tree growth. Thinnings in oak stands in the Ozark region of North America illustrates this effect. Those cuttings released 30–35-yr-old upper-canopy crop trees by removing neighbors with interlocking crowns, interlocking plus touching crowns, with crowns within 8 ft, or all but the best 100 trees. After twenty-eight years, the average crop tree diameter exceeded that in the unthinned stand by 0.75, 1.02, 1.38, and 2.16 in. (1.9, 2.6, 3.5, and 5.5 cm), respectively, across treatments (Mitchell et al. 1988). Within a sixty-three-year-old baldcypress stand in the Atchafalaya River Basin in Louisiana, crown thinning to 82, 64, and 45% of the initial basal area increased the five-year diameter growth by 1.4, 1.6, and 2.5 times, respec-

TABLE 19-1a
THINNING REGIME FOR UPLAND OAK STANDS—SITE INDEX 65 (from Gingrich 1971).

	Residual stand					Cut stand				Cumulative total yields (cut stand plus residual stand)		
Age (years)	Basal area	Average tree diameter	Yield			Basal area	Yield					
	Square feet	Inches	Cubic feet	Cords	Board feet	Square feet	Cubic feet	Cords	Board feet	Cubic feet	Cords	Board feet
40	69	6.5	1,600	15.9	440	27	240	2.3	-	1,840	18.2	440
50	66	8.5	1,910	17.7	1,800	28	410	4.0	200	2,560	24.0	2,000
60	70	10.4	2,200	20.7	4,200	18	400	3.6	280	3,270	30.6	4,680
70	74	12.4	2,485	23.1	7,210	16	420	3.7	710	3,955	36.7	8,400
80	77	14.5	2,720	24.8	8,960	15	410	4.0	1,050	4,600	42.4	11,200
90	79	16.5	2,925	26.6	10,710	13	460	4.0	1,630	5,265	48.2	14,580

[a] Yields per acre for upland oak. First thinning at age 40.

(*continued*)

TABLE 19-1b
(Continued)

Age (years)	Residual stand					Cut stand				Cumulative total yields (cut stand plus residual stand)		
	Basal area	Average tree diameter	Yield			Basal area	Yield					
	Square feet	Inches	Cubic feet	Cords	Board feet	Square feet	Cubic feet	Cords	Board feet	Cubic feet	Cords	Board feet
50	75	8.0	2,130	19.6	1,850	30	670	7.3	300	2,800	26.9	2,150
60	68	9.6	2,130	19.5	4,090	29	470	4.4	210	3,270	31.2	4,600
70	70	10.4	2,240	20.6	6,160	18	400	3.7	330	3,780	36.0	7,000
80	74	12.2	2,480	22.8	8,240	14	300	2.6	520	4,320	40.8	9,600
90	77	14.8	2,745	23.2	10,305	12	275	2.7	935	4,860	45.9	12,600
100	79	17.0	3,000	28.5	10,700	10	235	1.8	1,905	5,350	51.0	14,900

[b]Yields per acre for upland oak. First thinning at age 50.

(Site Index 65) based upon a ten-year thinning interval and a relative density control, beginning at age 40[a] (with a precommercial thinning) or at age 50[b] (after Gingrich 1971).

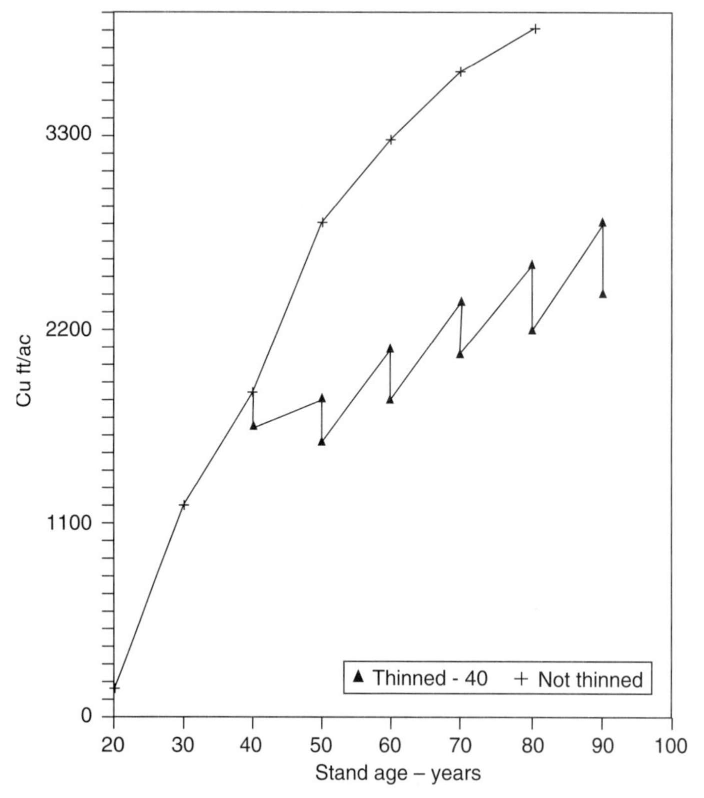

FIGURE 19-2a
Long-term development of cubic-foot volume in upland oak stands (Site Index 65) for a thinning involving a ten-year thinning interval and a relative density control, beginning (A) at age 40 with a precommercial thinning, and (B) plotted on the oak stocking guide (after Gingrich 1971).

tively. Removing two-thirds or more of the stocking stimulated epicormic branching on many trees, but crown thinning to all levels increased sawtimber volume and increment (Dicke and Toliver 1988).

Such findings reinforce the notion that releasing upper-canopy trees increases their radial increment, and the steepness of change in QSD. Actual responses depend upon the residual stocking, and the condition of individual trees. To illustrate, thinning lodgepole pine in western North America indicated that (1) young trees respond more than old trees; (2) dominant and codominant trees grow better than those of poorer crown positions; and (3) the rate of individual tree radial increment increases in proportion to the degree of release (Johnstone and Cole 1988). Thinning to low residual stocking stimulates diameter increment more than light or modest intensity release. As a rule of thumb, the increase may vary from 1 to 6.5 times the previous increment (Baker 1950). This improvement makes the curves for change in diameter more steep, and provides larger trees sooner.

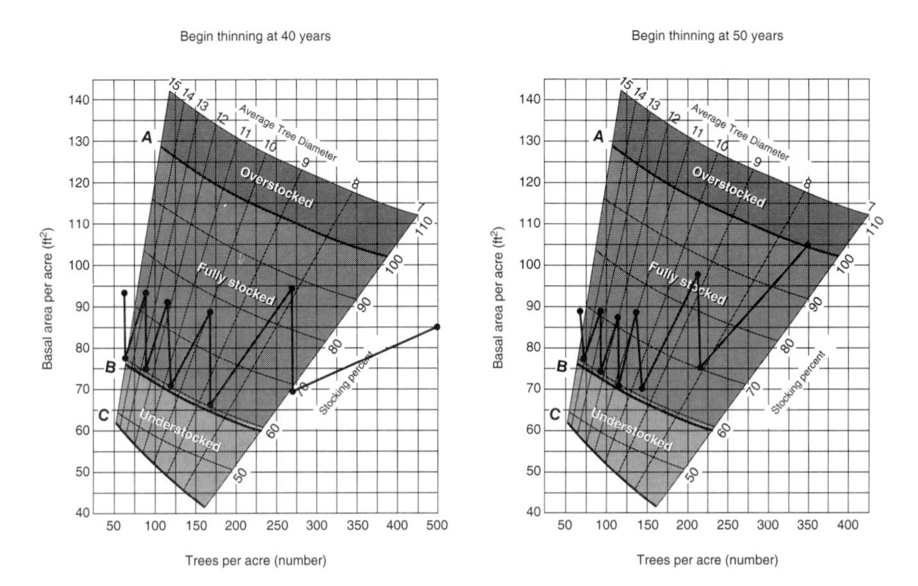

b. Relation of basal area, number of trees, and average tree diameter to relative density for upland oak stands. The tree diameter range (QSD) is 7–15 inches in the chart. QSD means the diameter of the tree of average basal area. The area between the A and B lines represents the range of stocking where trees can fully utilize the growing space. Curve C shows the lower limit of stocking necessary for a stand to reach the B level in 10 years on average sites.

FIGURE 19-2b (continued)

In natural ponderosa pine stands in southwestern North America (Ronco et al. 1985), thinning left residual basal areas ranging from 19–80 ft²/ac (4.4–18.3 m²/ha). Trees at the lowest levels grew 2.3 times faster than those in highest-density plots during the first five years, and 3.0 times faster during the second five-year period. Crowns increased 3 times more in length and 2.5 times in width after the first five years, and 4 times in length and 3 times in width during the second period, respectively, for these high- and low-density plots. Within sixty-five-year-old ponderosa pine in central Oregon, ten-year radial increment increased progressively with decreasing residual stand density between 99 and 31 ft²/ac (22.7 and 7.1 m²/ha), but not much between 99 and 144 ft²/ac (22.7 and 33.0 m²/ha) (Barrett 1983). Rates of radial increment increased with time after thinning to 99 ft²/ac (22.7 m²/ha) or less, but not for residuals of 113 or 144 ft²/ac (25.9 or 33.0 m²/ha). Among 30-year-old ponderosa pine stands, trees at 35 ft²/ac (8.0 m²/ha) grew 2.7 times or more in ten years than those at 92 ft²/ac (21.1 m²/ha). The rates increased with time in the low density plots, but not at the higher level (Barrett 1972). Within a 50-yr-old unthinned western larch stand in the northern Rocky Mountains, crop tree release significantly increased the radial growth over twenty-five years, but the rate slowed by about one-half during the last fifteen years. Thinning at this late age improved diameter growth by about 1% over that among dominant and codominant trees in unthinned plots (Cole 1984). In younger stands, larch grew appreciably better after thinning, with no decline in the growth for fifteen years following release (Seidel 1982). Similarly, thinning young longleaf pine significantly increased the diameters

TABLE 19-2
MULTIPRODUCT YIELDS IN MANAGED NATURAL SLASH PINE STANDS (from Bennett 1975)

	Initial stocking				Cut				Residual stand			
			Products				Products				Products	
Age (years)	Basal area	All trees	Sawtimber	Cordwood[1]	Basal area	All trees	Sawtimber	Cordwood[1]	Basal area	All trees	Sawtimber	Cordwood[1]
	Sq ft	Cu ft	Bd ft	Cu ft	Sq ft	Cu ft	Bd ft	Cu ft	Sq ft	Cu ft	Bd ft	Cu ft
SITE 80												
20	75	1,558	0	1,558	25	478	0	478	50	1,080	0	1,080
30	76	2,189	1,862	2,034	26	689	199	672	50	1,500	1,662	1,361
40	68	2,350	5,765	1,870	18	572	832	503	50	1,778	4,933	1,367
20	100	2,016	0	2,016	33	612	0	612	67	1,404	0	1,404
30	92	2,607	3,177	2,342	25	652	115	642	67	1,955	3,062	1,700
40	85	2,861	6,338	2,333	18	557	723	497	67	2,304	5,615	1,836

	Projected stand				Periodic annual growth				Total yield					Estimated average log diameter[4]
			Products				Products				Products			
Age (years)	Basal area	All trees	Sawtimber	Cordwood[1]	Basal area	All trees	Sawtimber	Cordwood[1]	All trees	Saw-timber	Slab and edging[2]	Kerf[2]	Cord-wood[3]	
SITE 80														
30	76	2,189	1,862	2,034	2.6	111	203	94						
40	68	2,350	5,765	1,870	1.8	85	410	51						
50	64	2,452	7,256	1,847	1.4	67	232	48	4,191	8,287	435	242	2,823	9
30	92	2,607	3,177	2,342	2.5	120	234	101						
40	85	2,861	6,338	2,333	1.8	91	328	63						
50	81	3,025	8,282	2,335	1.4	72	267	50	4,846	9,120	570	281	3,235	8

[1]Contains slab and kerf volumes required to produce the board-foot yield, if any.
[2]Bennett, Frank A., and F. Thomas Lloyd, 1974. Volume of saw-log residues as calculated from log rule formulae. USDA For. Serv. Res. Pap. SE-118. Southeast, For. Exp. Stn., Asheville, N.C.
[3]Includes nonsawtimber trees and topwood from sawtimber trees.
[4]Estimated average log length = 14 feet

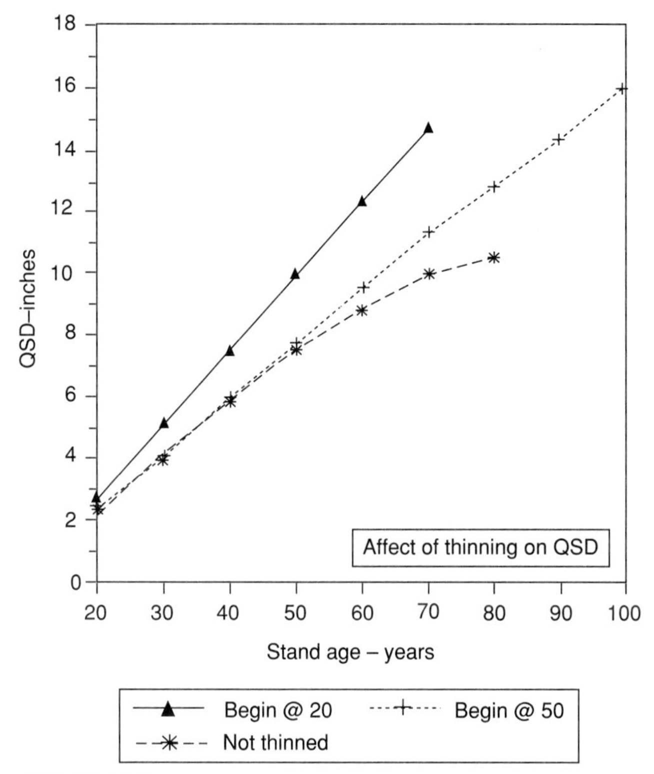

FIGURE 19-3
Effect of thinning on the diameter of the tree of mean basal area (QSD) for thinning regimes commencing at ages 20 and 50 in Site Index 65 upland oak stands (after Gingrich 1971).

of crop trees fifteen years later, even though radial increment did not differ across treatments after the eleventh year (Sparks et al. 1980). Collectively, such observations indicate that the improvement of radial increment depends upon the open space around the crowns, and the crown position. Relatively small improvements may occur among trees in uppermost canopy positions (already relatively free to grow), particularly when first released at advanced ages. Further, without additional release, growth rates decline as the canopy closes again.

Generally, experience and research indicate that the rate of individual tree radial growth will continue to increase with reductions of relative density down to about 30% (Daniel et al. 1979; Marquis 1986; Sonderman 1986). This suggests a potential to shorten a rotation dramatically by entering a stand early and keeping the stocking at low levels. To illustrate, a series of thinnings from below at ages 12, 15, 24, 27, and 30 significantly increased the 40-year mean diameters of planted loblolly pine. Further, each 20 ft^2/ac (4.6 m^2/ha) reduction of stocking from 90–30 ft^2/ac (20.6–6.9 m^2/ha) resulted in a dbh increase of about 1.5 in. (3.8 cm). Trees maintained at about 30 ft^2/ac (6.9 m^2/ha) averaged about 1.5 times larger in diameter than those in unthinned portions of the stand (Wiley and Zeide 1992). Yet despite such gains, a series of thinnings to low stocking eventually reduces the numbers of trees below that needed for complete site utilization, and persistent large openings remain in the main canopy. Then total volume production drops below that of stands at appropriate levels of stocking, and even below the production of unthinned stands (Assmann 1970). Repeated heavy thinning also promotes understory development, potentially creating a two-aged structure in community types that regenerate fairly readily. Further, the lower branches remain alive longer, producing larger branch diameters and larger knots. This could lower the eventual grade of the lower logs. Trees grown in low-density stands also have greater main stem taper, with greater losses to slabs and edgings when sawn into boards. However, after the lower branches have died up to the expected merchantable length for a site, individual tree quality will not suffer following heavy thinning, at least among conifers.

Differences in thinning affect the rates of diameter increment substantially, and the stand-wide volume production. Foresters must weigh the impulse to grow residual trees in low density stands from early ages to realize maximum rates of diameter growth, against accepting more modest rates of radial increment to maintain high-value logs (Abell 1954; Jagd 1954; Smith 1986). Delaying thinning reduces the potential tree sizes at rotation age, and opting for higher residual stocking (e.g., at least 60% relative density) to maintain full volume production requires more frequent entry to a stand. Thinning early and to low residual densities gets trees to threshold diameters sooner, and extends the thinning interval, with some loss in tree quality and stand-level volume production over a rotation. The method of thinning also affects the response, even at a wide range of residual stocking. To illustrate, among Appalachian hardwood stands in West Virginia, the five-year radial increment increased by 0.25, 0.40, 0.59, and 0.80 in. (0.6, 1.0, 1.5, and 2.0 cm), respectively, for trees released on one, two, three, and four sides. Yet thinning from below to 60% relative density (using the stocking guide by Gingrich 1967; Roach and Gingrich 1968) freed 75% of the crop trees only on one or two sides. Thinning to 45% relative density released only 60% on more than two sides (Lamson et al. 1990). By contrast, crown or free thinning would increase the degree of release at all levels of residual density, and better enhance the radial increment of crop trees.

Even when they delay a first thinning well into the rotation, foresters can improve the growth of high-quality trees in upper crown positions. To illustrate, a crown-touching release (freed on all sides) among 75–80-year-old Appalachian hardwoods in West Virginia increased the five-year radial increment by an average of 1.6 times. Rates improved for all species, but the difference proved statistically significant for only six of ten species. Epicormic branching would not degrade upper-canopy trees lacking such shoots before release (Smith and Miller 1991), particularly after thinning to an appropriate residual density (Brinkman 1955; Smith 1977; Marquis 1987, Sonderman 1987). In fact, data from thinned and unthinned loblolly pine plantations across southeastern North America indicate that thinning increases the proportion of veneer-quality peeler logs by concentrating the growth on the best trees. Using different intensities of thinning will not alter the overall proportion of sawtimber by diameter class (Burkhart and Bredenkamp 1989).

Even accounting for effects of thinning intensity, differences in curves like those in Fig. 19-3 reflect both the improved radial increment and the increase of average diameter due to tree cutting alone. Thinning has its greatest effect in young stands, and diameter growth generally declines as a stand matures. With late thinnings, increment may also improve appreciably compared to the pretreatment growth, but not to the same degree as for appropriately thinned young trees (Braathe 1957). The potential for response differs with species, crown position, and age. To illustrate, slash pine trees grown in even moderately dense stands for 25–30 years do not show an appreciable improvement in radial increment following even heavy thinning (Johnson 1961). By contrast, a first thinning that removed 36% of the basal area increased the 8-year radial increment of 52-year-old dominant Norway spruce trees by 1.8 times. The average growth of codominant trees did not change (Drew et al. 1985).

Overall, appropriately thinned stands (except following selection thinning) contain larger trees, and have more uniform crop-tree diameters at the end of a rotation. With selection thinning, average diameters may not exceed those in unmanaged communities. In fact, removing mostly dominant and codominant trees from Douglas-fir stands in western North America did not appreciably increase the QSD, and after 18 years the thinned plots had fewer sawtimber-sized trees than the unthinned plots. Also, removing 12%–31% of the volume at a first entry and cutting 90 percent of the increment during subsequent thinnings (at 3–9 year intervals) reduced gross volume growth by 20% over the 18-year period (Reukema 1972). Other studies of Douglas-fir showed better growth of upper-canopy trees than among those in subordinate positions (Staebler 1956; Reukema 1961). At the same time, repeatedly removing the largest trees from red pine stands in eastern North America did not reduce total volume increment, perhaps due to the thinning intensity, or the general homogeneity of growth within the species. Yet selection thinning limited sawtimber production, since no trees reached larger sizes (Buckman and Wambach 1966). Similarly, about 46% of the designated crop trees had to come from the intermediate class, and 39% from codominants during selection thinning of loblolly pine in the southern United States. After ten years only 27% remained suitable for crop tree status, largely because the intermediates grew too slowly to keep pace with the codominants. By contrast, 50% of the selected trees continued to serve as crop trees in a stand thinned from below (Kennedy 1961).

Such experiments highlight the importance of retaining trees of upper-canopy positions, and carefully regulating the intensity and frequency of thinnings. In fact, European trials clearly indicate that few intermediate and overtopped trees ever grow well following release, and they do not develop at rates comparable to those of upper-canopy positions (Braathe 1957). North American thinning trials in northern hardwoods (Marquis 1991; Nyland et al. 1993), aspen (Steneker 1964; Steneker and Jarvis 1966), and red pine (Day and Rudolph 1972) reinforce these findings.

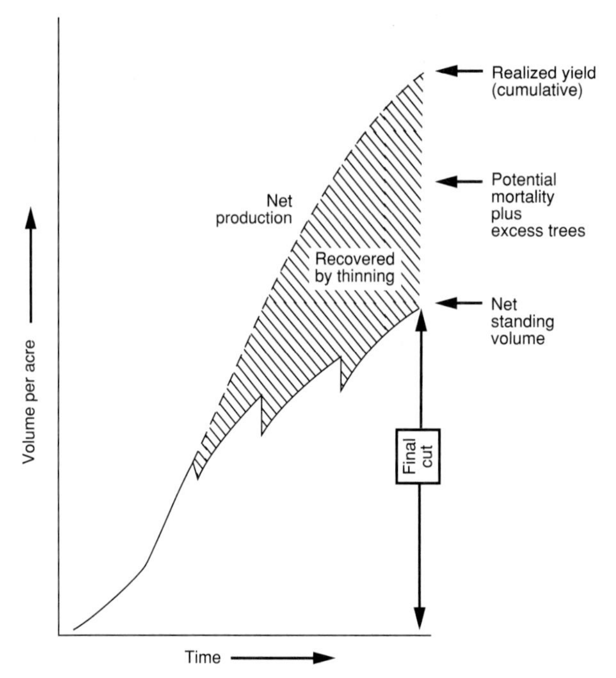

FIGURE 19-4
Thinning recovers the yield in potential mortality trees and excess growing stock, increasing the total realized over an even-aged rotation for large-diameter sawlogs.

All indicate that trees of subordinate crown positions (smaller diameters) may grow better following release by thinning, but none grow as well as the larger trees in dominant and upper codominant classes.

EFFECTS OF THINNING ON STAND VOLUME PRODUCTION

Figure 19-4 summarizes the long-term potential of thinning programs. The upper line depicts cumulative or realized yield (standing crop, plus amounts removed in all past thinnings up to that age). The lower curve shows how the thinning controls standing volume over time, allowing sufficient regrowth to maintain an upward positive slope to the saw-toothed production function. Adding a middle line for unthinned stands would isolate the proportion of cumulative yield attributable to the cutting of excess trees (those not needed to maintain full site utilization, as represented by B-level relative density). Exact differences between thinned and unmanaged stands vary with conditions like site quality, species mixture, growing season length, and several other variables.

Figure 19-4 illustrates that over long rotations landowners can recover more usable volume by periodically thinning to control stand density, and by merchandising the excess and potential mortality trees. Figure 19-5 shows the effects of adding a precommercial treatment (after Reukema and Bruce 1977). In both the upper and lower diagrams, the area above the dashed line represents the gain from thinning over a rotation that ends when QSD reaches 18 inches. Both regimes maintain sufficient growing stock for full site utilization. An extra stippled area in the lower diagram depicts how much more volume a landowner would realize by adding a precommercial thinning to help control early mortality losses.

Growth and yield studies of loblolly pine in the southern United States further illustrate the effects of thinning. Generally, they show that a series of well-designed treatments will increase recovered yield over sawtimber rotations. To illustrate, predictor equations developed from 85 unthinned plots and 167 others thinned from below indicate that tended stands (Site Index 70) would have 55% as much standing volume at age 40. Yet two thinnings at younger ages would recover sufficient extra volume (potential mortality, plus excess trees) to bring total yields to 102% of the unthinned stand (Baldwin and Feduccia 1987). In other experiments, yields varied with residual density as well (Wiley and Zeide 1992). In this case, thinning from below reduced stocking to 40, 60, 80, and 100 ft^2/ac (9.1, 13.8, 18.3, 22.9 m^2/ha) at age 12, and then to 30, 50, 70, and 90 ft^2/ac (6.9, 11.5, 16.1, 20.6 m^2/ha) at 15, 24, 27, and 30 years. Plots periodically thinned to 90 ft^2/ac (20.6 m^2/ha) produced 1.5 times more cubic volume than the unthinned stand, and also yielded more wood than did thinning to lower residual densities. Thinning to the lowest density produced the greatest cumulative sawtimber yields by age 15. Thereafter, sawtimber yields peaked with residuals of 50, 70, and 90 ft^2/ac (11.5, 16.1, and 20.6 m^2/ha) at 19, 24, and 27 years. By age 30, periodically thinning from below to 70 ft^2/ac (16.1 m^2/ha) provided 1.5 times more sawtimber volume than the unthinned stand. The 90-ft^2 (20.6 m^2) plots yielded 99% of the volume grown at the 70-ft^2 (16.1 m^2) level, but the trees had a 1.4-in. (3.6 cm) smaller mean diameter. The 30-ft^2 (6.9 m^2) level produced only 97% of the volume in the unthinned plots, and two-thirds that produced at the 70-ft^2 (16.1 m^2) level. Gains from the best-producing plots represent the added volume from controlling potential mortality and removing excess trees while still maintaining sufficient stocking for full site utilization.

One growth model for thinned ponderosa pine stands in western North America shows that site conditions profoundly affect both basal area and volume growth. Also, volume production will decline significantly at residual densities below levels recommended for the site (Myers 1967). Plots at 40 ft^2/ac (9.2 m^2/ha) produced only two-thirds to three-fourths of those at 110 ft^2/ac (25.2 m^2/ha), with differences increasing with improvements in site quality. On the other hand, individual trees annually grew 0.1 in. (0.25 cm) better at the low level of stocking (Oliver and Edminster 1988). Likewise, experiments with lodgepole pine in the same general region indicated that thinning significantly increases the long-term merchantable yields, stands on the best sites produce more volume for a given level of residual density, and low density stands sustain the best rates of individual tree growth (Johnstone and Cole 1988). Predictor equations for white fir in the Pacific Northwest of North America indicated that even-aged stands past 30 years of age and reduced to 75%,

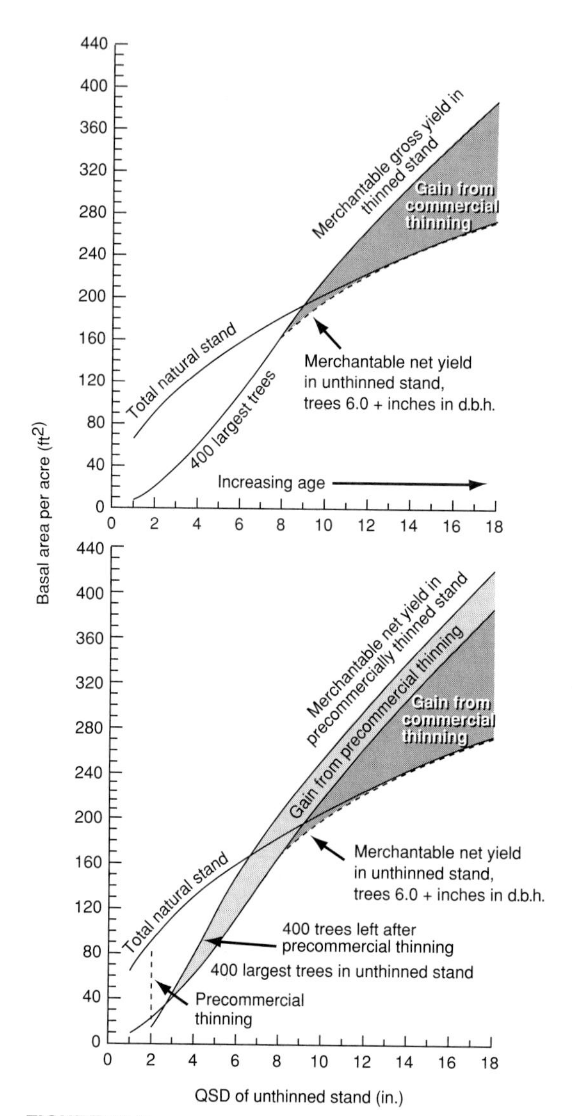

FIGURE 19-5
Sources of added volume realized by thinning Douglas-fir stands in western North America (from Reukema and Bruce 1977).

50%, and 25% of normal density would produce 93%, 80%, and 53% of gross periodic annual cubic volume growth of untreated stands (Cochran and Oliver 1988.)

Research has indicated that some methods of thinnings provide only limited or modest gains if applied fairly late in a rotation. Actual effects may depend upon the condition of residual trees, and the potential for crown development after treatment. To illustrate, among mixed conifer stands in the northern Rocky Mountains of North America, beginning a thinning regime after stem density has naturally decreased to 1000 or fewer trees per acre (2470/ha) has little effect on yields (Foiles 1956). Consistent with this, a single thinning from below during the sixth decade (approximately 1000 trees per acre, or 2470/ha) increased the sawtimber volume only modestly. Diameter growth of the 100 largest trees did not differ appreciably between thinned and unthinned stands. Three thinnings from below at 10-year intervals beginning at age 55 (about 1300 trees/ac, or 3211/ha) also only slightly increased the realized sawtimber yield. By contrast, a first thinning after the age when western white pine still responds well increased the average diameter of the 100 tallest trees by nearly one inch. Further, a single late thinning of moderate intensity (at 87 years of age, with <200 trees/ac or 494/ha) increased the diameter growth of both western white pine and grand fir. Compared with selection thinning, crown thinning produced trees (5.5 in. or 14 cm and larger) that averaged 5 in. (12.7 cm) greater in diameter for the pines, and 2.5 in. (6.4 cm) for fir. Further, thinnings that concentrated the growth potential on trees of upper-canopy positions also enhanced stand and tree quality and value (Foiles 1956; Foiles 1972; Graham 1988).

Comparisons of row thinning and thinning from below in loblobby and slash pine in southeastern North America also show differences related to method. Even when reduced to comparable residual densities, row thinning resulted in slightly less average diameter increment, and less basal area and volume yield over a 10-year period. Similar findings among radiata pine plantations in New Zealand and Australia, and red pine in the upper Great Lakes region of North America, suggest that differences reflect residual tree characteristics, rather than residual stand density (Day and Rudolph 1972; Wright 1976; Cremer and Meredith 1976). Thinning from below concentrated the production entirely on trees of upper-canopy positions, whereas row thinning left trees of all crown positions (greater range of diameters, crown sizes, and vigor classes). The row-thinned plots had more residual trees, but the smaller trees did not accumulate volume or basal area as rapidly as those in upper-canopy positions. In all three studies, the largest trees grew comparably following both methods of thinning. (Wright 1976; Cremer and Meredith 1976; Baldwin et al. 1988). Similarly, within red pine plantations in Michigan (Day and Rudolph 1972), the more dominant the tree and the denser its crown, the greater the post-thinning growth.

MANAGEMENT SCHEMES FOR FIBER PRODUCTION

Data from Table 19-1 provide a basis for comparing yields from thinned and unmanaged stands when a landowner elects to produce bulk quantities of fiber, independent of tree sizes. In this case, the volume includes only trees of merchantable sizes for fiber markets, so it incorporates some concern for minimum diameter. The mean annual increment for unmanaged stands peaks at about 55 years of age. Stands at that time contain 3100 ft^3/ac (217 m^3/ha). A second unmanaged fiber crop would produce about 2320 ft^3/ac (162 m^3/ha) during the next 45–50 years, increasing the total to 5420 ft^3/ac (380 m^3/ha) over a 100-year period. Where thinning begins at age 50 with follow-up treatments each 10 years thereafter, a stand yields 5150 ft^3/ac (361 m^3/ha) over that same 100-year period. This includes 2115 ft^3/ac (148 m^3/ha) taken out during the five thinnings, and 3000 ft^3/ac (210 m^3/ha) harvested as part of the reproduction method cutting at the end of 100 years. Under these commercial schemes, landowners who produce fiber crops gain more by regenerating a stand each time m.a.i. reaches a peak, and protecting stand health until m.a.i. culminates again about 55 years later.

Table 19-2 depicts this scenario for natural slash pine in the southeastern United States (after Bennett 1975). It includes two management options, as represented by different tree densities and residual basal areas. Both incorporate thinnings at 10-year intervals. The higher density option consistently shows higher cubic volumes at the end of a 50-year rotation, but with smaller trees. By contrast, stands at the lower density consistently yield more sawtimber volume, and larger trees. Landowners could apply stumpage price information to those yields to weigh financial benefits of producing more total wood volume from the high-density stands, against the value from holding fewer trees and growing them faster to sawtimber sizes. For commercial ventures, a landowner would use the regime offering the greatest net revenues.

MANAGEMENT SCHEMES FOR SAWTIMBER

The concept of a silvicultural system implies not just applying appropriate types of treatments throughout the life of a stand, but also arranging them in a proper sequence, frequency, and intensity to fulfill the management goals. In essence, thinning programs arrange the treatments to insure continuous optimum growth of the trees, and the stand as a whole. They also provide for periodic harvest of intermediate products. Landowners normally want to integrate these silvicultural plans with those for other stands, and to carefully orchestrate the different thinnings into a coordinated program of annual work and cutting. Otherwise the commodity goals of an enterprise do not come to fruition, and the management provides an irregular flow of income to balance expenditures. Owners who manage for nonmarket values also frequently want to create and sustain some specific stand and forest attributes of importance, or to maintain existing conditions that seem valuable to their objectives. In many cases, management can create an appropriate spatial balance of conditions at the property and landscape scale by appropriately arranging the time and sequence of the reproduction method cuttings, and integrating these with the intermediate treatments in immature stands not ready for replacement.

Generally, *thinnings* (Ford-Robertson 1971; Winters 1977; Smith 1986) serve the purpose of

1 stimulating the growth of trees that remain;

2 recovering the volume in trees that would normally die due to intertree competition;

3 improving growth and form of the crop trees;
4 maintaining a particular stand density for optimal growth; and
5 increasing total yields of large-diameter products over the life of a stand.

Foresters must not emphasize the option of increasing intermediate yields, lest they detract from the primary purpose of nurturing stand development toward some desirable future condition (Braathe 1957). They want to accelerate diameter increment *without permanently breaking the canopy* (Ford-Robertson 1971), or reducing relative density below full site occupancy. Thinnings concentrate the growth potential on fewer trees of choice species and quality, so that the residuals grow to bigger sizes at a faster rate (after Heiberg 1954).

The term *intermediate cutting* incorporates important related concepts (after Hawley 1921; Cheyney 1942; Ford-Robertson 1971; Smith 1986). It means any deliberate cutting within a regular stand between the time of its formation and the final harvest, including release cuttings, thinnings, and other tending operations. These improve the condition of an existing community, influence its development, and provide intermediate financial returns. Usually, a silvicultural system incorporates different kinds of intermediate or tending operations to manage intertree competition throughout the life of an age class. From this perspective, thinning clearly removes useful material from a stand, as well as managing competition within the residual community.

The definition for *timber stand improvement* includes another important related idea (Ford-Robertson 1971; Soc. Am. For. 1989). It includes all intermediate cuttings to improve the composition, constitution, condition, and increment of an even- or uneven-aged stand. This infers a selectivity to concentrate the growth potential onto trees best suited to the ownership purposes. Where commodity interests prevail, foresters select the trees with a long merchantable length, straight bole, few surface blemishes, no internal defects, and high growth potential. With other management goals, they need only determine what kind of trees would best fit the objectives, what method of thinning would best create a stand of desirable characteristics, and what residual relative density would best promote its development.

Generally, in choosing trees to leave or cut, foresters try to control the phenotypic (and genetic) constitution of a stand (Maynard et al. 1987; Bassmann and Fins 1988). With thinning, they repeat the selections over successive treatments, realizing considerable improvement over the course of a rotation. The gain varies with the method and intensity of thinning. In a relative sense, the different methods rank in potential for genetic improvement as follows: low > crown > mechanical > selection. The latter method has clear disgenic effects, while those of free thinning depend upon how a forester uses the method in any set of circumstances (Bassmann and Fins 1988). Especially with low and crown thinning, foresters keep sufficient numbers of the best trees for long-term full site utilization, consistent with a landowner's objectives. Over time, this reduces the growing stock to a fairly small elite population of carefully selected trees from among the large numbers that initially regenerated at a site (Schadelin 1942; Assmann 1970).

While many benefits accrue from each individual thinning, an owner will realize the fullest long-term value only by programming an appropriate series of consecutive thinnings that amplify the gains throughout an entire rotation. Foresters call this sustained program of tending the *thinning regime* or *thinning schedule* (after Ford-Robertson 1971). It means the types, degrees, and frequency of thinnings applied over a rotation, generally including when they begin, and sometimes when they end. As one element, a thinning regime delineates a *thinning cycle* (Ford-Robertson 1971; Soc. Am. For. 1989). It means the planned intervals between successive thinnings in a single stand, either regular or irregular. Regimes also specify the target residual basal area or relative density. This, too, frequently changes over a rotation, with higher levels of residual basal area left during each subsequent thinning. The sequence of treatments delineated in Table 19-1 and Fig. 19-2b illustrates two possible thinning regimes or thinning schedules for even-aged oak stands.

By regulating the periodicity of successive thinnings (regular, or with a changing lapse time), the regimes provide for the deliberate control of stand density throughout a rotation. This insures *continuous stimulation* of final crop trees, and regular maintenance of filler trees that remain for only part of a rotation. This maintains uniform growth rates among the crop trees (Fig. 19-3) and lends an important element of quality to the lumber, in that boards with fairly uniform annual rings remain more dimensionally stable. Further, the uniform growth rates also bring the trees to the requisite threshold sizes without delay, shortening the rotation. In addition, timely thinning gives foresters the opportunity to recover potential mortality trees, and to sell them to generate revenues. In fact, experiments with one growth and yield model for slash pine plantations in southeastern United States indicated that stands (Site Index 70) given three thinnings over a 40-year rotation would yield 8111 ft^3/ac (568 m^3/ha), including 3204 ft^3/ac (224 m^3/ha) removed during thinning, compared with 7371 ft^3/ac (516 m^3/ha) for unthinned stands (Zarnoch et al. 1991). Foresters must also control the intensity as well as the timing of thinnings to realize the full potential benefit. Thus, removing 29% of the volume at age 13 and 25% at age 24 from slash pine plantations would increase total realized yields by 12% over a 29-year period. On the other hand, taking 56% at age 13 and 35% at age 26 would reduce yields by 10% (Keister 1972).

For ownerships dedicated to nonmarket values, the planned reentry maintains essential ecosystem attributes that may depend upon periodic opening of the crown canopy layer, or getting trees to large sizes in the shortest possible time. Some objectives may encourage thinning heavy enough to deliberately provoke an understory response, or to sustain shrubs and herbs of only moderate shade tolerance. In this sense, a thinning regime can guard against the loss of noncommodity values important to a landowner or enterprise, similar to controlling mortality of trees that have financial worth for commodity goals.

SOME CONSIDERATIONS FOR TIMING A FIRST THINNING

Generally, the sooner the regime commences, the greater the cumulative effect, at least in a biologic sense. Yet financial limitations may make precommercial thinnings impractical. For many enterprises all treatments must pay for themselves, or insure sufficient *added* value yield to make an investment financially attractive. Then the value at some time in the future must pay off the initial unpaid costs compounded at the prevailing rate of return for the ownership, plus yield any additional required profit. Foresters who manage for noncommodity objectives will also need to promise adequate additional intangible values to make any investment worthwhile. This may mean demonstrating a sufficient improvement in nonmarket benefits to encourage a landowner to spend the money to alter natural stand development in the way proposed. As a tangible example, the thinning regime depicted in Table 19-1 begins with a precommercial treatment at age 40. If the ownership required an interest rate of 6%, every $50 invested compounds to $89.54 in 10 years (the time for a second thinning in Table 19-1). Then a landowner would recover 410 ft^3 of salable material from each acre (28.7 m^3/ha). This would just marginally sustain a commercial sale, and would need to return a stumpage value of $22.39 per cord ($6.17/m^3) to repay the compounded value of the earlier $50 investment. At the same time, this regime produces 16% more sawtimber volume by age 90 (see Table 19-1). Likewise, among ponderosa pine stands in northwest North America, precommercial thinning would reduce the age when landowners could do a first commercial thinning by 70% (24 vs. 87 years), and shorten the rotation age by 90% (114 vs. 201 years). Even then, the stand given a precommercial thinning would yield about 7% more sawtimber volume (Sassaman et al. 1977).

Available evidence for southern pine plantations indicates that early intervention increases the proportion of large-diameter trees. This has especial value for sawtimber schemes involving short rotations (Nebeker et al. 1985; McMinn 1965; Langdon and Bennett 1976). To illustrate, retaining 20%, 11%, and 4% of the trees in a 7-yr-old natural slash pine stand (3500 trees/ac or 8645/ha) increased the average diameter at age 35 by 106%, 113%, and 123%, respectively. Sawtimber volume (removed in thinning, plus that left standing) increased by 202%, 256%, and 287%, respectively, for the same plots (McMinn 1965). Yet to enhance sawlog quality, stands should remain unthinned until lower branches die for about 25 ft (7.6 m) (Wahlenberg 1946), but not after the live crown length of upper-canopy trees drops below 35%–40% of total height (Wahlenberg 1946; Mann 1952; Wakeley 1954). This coincides with the time when 4–5 in. (10.2–12.7 cm) diameter trees begin to die due to intertree competition (Mann 1952). The actual time comes earlier at the better sites, and among plantations of initially closer spacing (Nebeker et al. 1985).

Foresters can sometimes adjust the thinning to insure a break-even operation. They might change the method of thinning, use a lower residual relative density, or postpone the treatment. For example, the forester responsible for this same oak stand depicted in Fig. 19-2a might opt to reduce stand density to fairly low levels of residual stocking. Just delaying the thinning until age 50 would yield sufficient volume to make the work commercially viable (Table 19-1b), and possibly provide some profit. The decisions reflect choices of *immediate* financial benefit, but at a biologic cost. To illustrate, reducing stand density to extremely low levels might threaten individual tree health and quality, and brighten the understory sufficiently to regenerate herbs, shrubs, or trees beneath the existing stand. Cutting the largest trees would remove the best genotypes, at least to the degree that inherited genes control the rates of past growth and development. Further, by delaying thinning a landowner forgoes the opportunity to grow the trees to the required minimum merchantable sizes in the shortest time (Table 19-1).

Among close-canopy communities, clear boles form on trees only after lower branches die and fall off, and diameter growth covers over the imbedded knots. This dead length increases at a fairly rapid rate during early stages of the aggradation phase, when the stand has a dense and closed canopy and individual dominant and codominant trees increase in height at fairly rapid rates. At least through pole stages, the base of the canopy rises fairly steadily above the ground over time following crown closure, and little live foliage remains in the understory. As a stand ages, height growth peaks and slows. Then the top of a crown canopy rises more slowly. Continued lower branch mortality reduces the live crown ratio, and the radial increment declines, even among upper-canopy trees (Holsoe 1947). Landowners who want to produce high-quality sawtimber over the shortest possible rotations must strike a balance between investing in early thinning to maximize long-term diameter development, and delaying thinning to promote optimal clear lengths or to insure that the first treatment will sustain a commercial sale.

Early thinning keeps the lower branches alive longer, maintains a longer and wider live crown, and insures better rates of radial increment. The crop trees will have live limbs on the lower logs for an extended period, develop larger branches (knots) along the lower bole, and have knots that extend outward much farther from the center of the trunk. This reduces the quality and value of the lower sawlogs. Also, once a first thinning in older communities begins to control stand density and reduce lower branch mortality, additional dead length will form only slowly on the crop trees. This limits additional dead length to an important degree. The already dead branches continue to fall off the residual trees, increasing the length of bole where radial increment eventually adds clear wood. In this sense, the first thinning has a more profound effect than later treatments in influencing tree condition and character. To balance these opportunities, one rule of thumb suggests delaying the first thinning in hardwoods until height growth of upper-canopy trees begins to slow, and the crown starts to form a domelike shape. After that time, clear length develops more slowly in both thinned and unthinned stands, and landowners will increase upper-log quality only by keeping the stand unthinned for a disproportionate number of additional years (Holsoe 1947). At high latitudes, ice and snow damage often distorts the form sufficiently to limit the usable length. Then decreases in height growth or change in crown form give only a poor indication of when to schedule the first thinning. Rather, foresters may simply delay a first entry until branches die to the expected average merchantable height for the site. They willingly sacrifice some potential to maximize early diameter growth, or to control early mortality, in favor of financial or quality limitations imposed by an owner.

In delineating an optimal relative density to maintain standwide net production (e.g., 60% relative density), most stocking guides imply that the residual canopy intercepts sufficient insolation to maintain net photosynthesis at or near the full potential of a site. This may not happen in even-aged stands that grew at or near 100% relative density for long periods (Roach 1977). Then tree crowns must first expand to fill much of the open space. The recovery rate depends upon the age and vigor of the trees at the time of release, the growth characteristics of the species, and the rates of lateral crown extension into the open spaces. Lateral branches of low-vigor pole trees extend at surprisingly slow rates, while crowns of larger trees in upper-canopy positions expand more over a thinning cycle. For example, the crown of a 40-year-old 6-in. (15.2 cm) red oak tree would expand by about 9 ft (2.7 m) in 15 years, and that of a 12-in. (30.5-cm) tree by about 19 ft (5.8 m). Crowns of comparable tulip-poplar trees would increase by about 8 and 12 ft (2.4 and 3.7 m), respectively (Trimble and Tryon 1966). This suggests a difference in the recovery rate and production between thinnings that leave the larger vs. the smaller trees as residuals. Also, species composition may influence the rate of canopy closing for stands at a common residual relative density.

Guidelines for hardwoods in eastern North America commonly recommend removing no more than 35% of the relative density during any single thinning. Especially for stands much beyond 50 years of age, a lighter thinning will better maintain full site utilization until the crowns of residual trees recover from a protracted period of intense intertree competition (Roach 1977; Marquis et al. 1984). At the same time, the lighter thinning mandates a shorter thinning interval in order to sustain the benefits, and will not stimulate individual tree radial increment to the fullest degree (Oliver and Larson 1990). Especially with shade-intolerant and midtolerant species, crowded trees have narrow and short crowns, and radial increment may not improve appreciably for several years after thinning. By contrast, shade-intolerant trees (like tulip-poplar and red pine) with good crowns may increase measurably in diameter during the first growing season (Stephens and Spurr 1948; Holsoe 1951). In fact, early studies in northeastern United States detected a 40% increase of radial increment within twenty-four hours of thinning around vigorous red pine trees in young stands (Stephens and Spurr 1948). Among shade-tolerant trees, considerable foliage persists on branches at mid- and low-canopy levels, even in crowded stands. Shading limits the level of photosynthesis in this foliage. Once exposed by thinning, the previously shaded foliage photosynthesizes more efficiently, and radial increment increases appreciably during the first season. Growth continues to increase as the crowns expand, lengthen, and become more dense.

The actual improvement in radial increment depends upon the extent of lateral crown extension, and the degree that foliage density increases within the existing crown. Thinning keeps the lower branches alive longer, so that the crowns of residual trees become longer as they grow taller. Crowns of young to moderate-aged trees increase noticeably in a few years. For trees where height growth has slowed appreciably, the live crown ratio will not increase much over even long periods. Instead, diameter growth responses reflect the rate of lateral crown expansion and how much foliage density increases (Daniel et al. 1979). Generally, radial increment increases or remains steady over long periods following thinning. In fact,

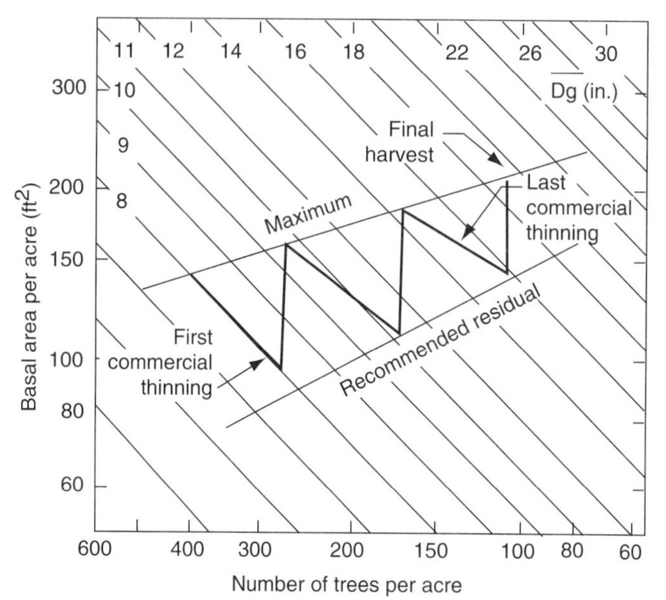

FIGURE 19-6
Application of a stocking guide to control the intensity and frequency of thinnings of even-aged Douglas fir stands in western North America (after Reukema and Bruce 1977).

maximum crown volume and photosynthetic capacity often peak just prior to an appropriately sequenced second thinning (Mar:Moller 1945).

SETTING A THINNING INTERVAL

In some cases, a stand may reach financial maturity before conditions warrant a second thinning, particularly if the first one occurs late in a rotation. Other regimes may include two or more thinnings, and then foresters must decide how long to wait between successive entries. In principle, these should (1) maintain the full growth potential of individual crop trees at a sustained and steady level; (2) remove filler trees before competition intensifies and they die or seriously deteriorate; and (3) provide sufficient time for volume to rebuild and sustain another operable cutting. Generally, the rate that stocking rebuilds depends upon the intensity of a previous treatment with reference to p.a.i. and the rate of lateral branch extension. In addition, as trees become larger they become more valuable, allowing contractors to operate efficiently by cutting fewer trees.

Foresters can use an available stocking or relative density guide to monitor when stocking rises above some predetermined threshold. Figure 19-6 illustrates this concept based upon a stocking guide for Douglas-fir in western North America

(Reukema and Bruce 1977). The upper line delineates a maximum level of basal area for managed stands, and the lower one the target residual stocking following thinning. Generally, thinning reduces stand basal area only to the lower line, and this insures an appropriate degree of site utilization. Because thinning controls mortality, stocking increases without appreciable changes in the number of trees (shown by the vertical lines on Figure 19-6). When basal area regrows to the upper line, foresters should once again reduce the stocking to the recommended residual level (Fig. 17-12). The residual basal area should increase with each successive thinning, similar to other stocking guides. In concept, the guide parallels recommendations for keeping relative density between 60% and 80% (e.g., Roach and Gingrich 1968; Leak et al. 1969; Roach 1977; Marquis et al. 1992). In fact, in thinned longleaf pine stands in the southern United States, leaving increased levels of residual stocking with each subsequent thinning provides greater sawtimber yields than cutting back to a constant or declining level of residual basal area (Ferrar 1985).

The time when stocking reaches the recommended upper threshold determines the thinning interval (e.g., keeping the stand basal area between the upper and lower lines). Since thinning controls mortality, p.a.i. essentially represents the composite increment among residual trees. This diminishes with stand age, and largely in proportion to the decline in individual tree radial increment. In fact, as height growth declines, the importance of diameter growth increases. And though periodic thinning can enhance the growth of even fairly old trees, diameter will not change as rapidly as it does in young stands (Braathe 1957). As a consequence, p.a.i. declines with stand age, and the time required for basal area to increase from the residual level to a recommended upper limit will increase. To compensate, foresters need to lengthen the interval between successive thinnings if they elect to optimize the rate of volume accumulation within a residual stand.

Once a stand becomes operable (sufficient removals to sustain a commercial thinning), cutting to the target relative density or stocking usually allows sufficient regrowth to sustain a series of commercial operations throughout the remainder of a rotation. Most methods of thinning fit these techniques, giving foresters considerable flexibility in designing a thinning regime. In fact, they can change methods over the course of stand development, and still use a relative density or stocking control to simplify long-term planning. Where they know the likely growth rates in advance, foresters can anticipate the approximate ages for each successive thinning. If they lack good growth data, they simply monitor stand conditions and wait for stocking or relative density to increase to the recommended upper limit. Then they thin again.

Under less formal modes of control, foresters may simply wait until a stand regrows sufficient volume for another operable cut, and then reduce stocking to some level of interest based upon experience. These more subjective and casual approaches generally lack the means for monitoring stocking to insure full site utilization and full net current increment. They may suffice when landowners primarily seek nonmarket values. In some cases, foresters base subsequent treatments on changes of individual tree characteristics, like watching for a decline of radial increment or a reduction of live crown ratio among the crop trees. This may include such simple rules of thumb as keeping the live crown ratio above 35%–40%, or never letting the lower branches of crop trees interlock and die. Reductions of live crown ratio below 30% usually indicate that tree vigor has declined sufficiently to retard radial increment and a slow recovery following release (Daniel et al. 1979; Smith 1986).

20

MANAGING QUALITY IN FOREST STANDS

MARKETING AND QUALITY FACTORS IN SILVICULTURAL PLANNING AND OPERATIONS

Landowners with commodity goals commonly opt for programs offering the highest returns that deliberate management can sustain indefinitely at a specified level of investment. They may not grow trees for a particular kind of product, but simply view their forests as profit centers that must yield some minimum level of return. Then foresters may capitalize on a wide variety of revenue-generating operations, and manage for a balance of activities that collectively fulfill the financial requirements. This may include selling commodities like timber and minerals, and leasing opportunities for recreation uses such as hunting and fishing. A firm may use its forests to grow specific products as raw materials for its manufacturing operations, or that command high prices as roundwood products.

For ownerships with strong revenue objectives, foresters must insure that stands capture the full productive capacity of the land. Also, they must market products and leases on a regular schedule, in the fullest quantities that accessible markets will absorb, and consistent with the maximum sustainable productive capacity of the land. This insures regular revenues to pay costs of operations and sustain the financial interests of a landowner. Additionally, where they do not control the primary or secondary manufacturing, landowners must continually attend to the quality of their offerings, knowing that markets normally accept and willingly pay more for goods and services that best satisfy the exact desires of a buyer or user. In fact, history shows that to succeed as a business, a forest enterprise must aggressively identify the market opportunities that pay best prices, and develop a long-term working relationship with those buyers to sustain a market share based upon trust and fairness. Further, they must deliver the agreed-upon quantities and quality of products at specified times. Of course, this strategy presumes a relative stability in the kinds and quantities of commodities and services that landowners can sell at an acceptable price, and that consumer demand will maintain or increase the values over the long run.

At least four special conditions affect forestry valuation decisions, and the marketing strategies to maintain a desirable flow of revenues. These add a high degree of uncertainty to resource management planning, and often affect a landowner's willingness to make long-term investments in silvicultural operations (after Davis and Johnson 1987). First, a relatively small number of buyers or sellers may monopolize local markets, or control the way they function. That gives them a clear advantage in negotiating prices, or determining who can exchange what quantity of goods and services. This works to the disadvantage of those not in a dominant negotiating position. Second, even in historically active areas with ongoing sales of timber and other forest-based commodities, no one can actually know what the markets will demand when young age classes eventually become merchantable. They can use information from recent sales with a high degree of certainty in negotiating a fair market price for currently available timber or use opportunities. At the same time, the long wait for young trees to grow into marketable sizes adds a risk to investments in silviculture, and to choices about how long to hold different age classes before regenerating them. Third, many different natural agents (e.g., insects, diseases, and fire) and human activities can damage or destroy trees in a stand, or change the pattern of their development. These may reduce the value, or reduce opportunities for renting or leasing uses of a stand or forest. This, too, adds great uncertainty in forecasting the kinds and qualities of trees and general forest conditions available at some distant time, and what the markets will pay for them. Fourth, for many ownerships, foresters must make investments to increase the array of use opportunities and to maintain many kinds of ecological conditions that have no traditional market value. This could include extending the life of an age class beyond normal financial maturity to maintain some important visual qualities. That further increases the long-term risks of loss or damage, and commits an owner to a lower rate of return on the investment of growing the trees. Collectively, such factors often force foresters to rely on rather indirect and subjective methods for estimating and projecting values. This reduces the precision of their forecasts.

As with other kinds of agricultural production, forest owners also cannot often adjust quickly to changes in market conditions, or immediately capitalize on new demands by consumers. Instead, they may need to reorganize their silvicultural operations to regenerate substitute species and grow them to marketable sizes. They may also need to develop new facilities to support changing in-forest use opportunities. On the other hand, landowners who own the species or kinds of trees that match consumer interests can often shift in their management operations and work schedules, concentrating on stands containing the desired products or use opportunities. This may prove highly profitable in the short run. In most cases landowners need to adjust gradually over long periods, often accepting reduced prices for commodities that they have already grown over long periods based upon past investment decisions (Kohls and Uhl 1985). Some may gamble that future shifts in consumer demand will once more make their forest more valuable by the time much of the immature growing stock reaches a marketable condition.

Large firms that both own the forest and manufacture consumer products can often influence supply and price, or even create new demands for the kinds of products available on their lands. Similarly, public agencies that administer large blocks of land may play a major role in controlling the timber supply across a broad region, and the goods offered for sale. This, too, can affect demand and price. On the other hand, noncorporate individuals who manage limited areas must often rely on aggressive marketing to complement their silvicultural operations. Important components (after Kohls and Uhl 1985) would include

1 deciding what to grow or what use opportunities to develop, and at what rates to harvest or otherwise utilize them over the long run;

2 determining when to offer different species or stands for sale or use in response to periodic fluctuations (both increases and declines) in price and demand, and what to hold for the future;

3 ascertaining how much to control the actual harvesting and use (e.g., doing the logging, or selling stumpage), when to cut the timber, and what kinds and sizes of roundwood pieces to offer to different potential buyers;

4 judging how aggressively to develop unique marketing opportunities, compared with offering available products to traditional buyers who regularly operate in an area; and

5 devising an active advertising program to announce ongoing timber and use opportunities, and to attract more potential buyers and create competition among them.

In many respects, the potential to keep well-managed growing stock for protracted periods without a decline in condition and value gives forest owners considerable flexibility in their marketing. Some may do the harvesting themselves, or with their own employees. Then they can often better control production and delivery, including log sizes and quality. This may influence price and increase profitability. Where landowners or the foresters who represent them can develop new and special markets, or offer unique harvesting and use opportunities to buyers, they may enhance prices or insure a steady demand for their offerings. All of these choices shape the strategy for merchandising available commodities and use opportunities, for generating revenues to pay costs of operations, and for realizing the profits that encourage continued ownership and management.

When landowners can offer a particular kind of product that buyers generally find in short supply, or consistently make available products of above-average quality, they can usually command higher prices and capture a better share of the market. This generally holds true as much for nontraditional offerings like recreational leases, as it does with sawlogs and other commodities. Foresters must recognize that people (buyers and fee users) willingly pay more for the things they want the most. Then they can produce those commodities or services in limited supply or high demand, and realize the best prices. Further, they will often look for specialty forest products and noncommodity uses that an enterprise can offer in quantity. This, too, will make future sales easier and more profitable.

In some cases, landowners may have mostly large volumes of relatively low-value growing stock (e.g., trees containing mostly pulpwood vs. high grade sawlogs), and can offer only common products. Then they survive financially by marketing large amounts of limited-value materials at low costs. They function primarily as price takers, and have little control over the amount of goods purchased or the prices paid for them. To succeed, they must cut costs of production, and compete aggressively for opportunities to sell their goods. In some cases they have banded together with other forest owners in marketing cooperatives to capture a large collective market share, or control the amount of materials offered for sale at any time (after Kohls and Uhl 1985). Some individuals also offer adjunct services that neighbors cannot as readily provide (e.g., dependability of supply, and year-round access). This makes their goods more attractive to potential buyers. Usually, these situations give little incentive to invest in high-intensity silviculture aimed at growing specialty products over long periods of time. Instead, success depends upon efficient extraction and marketing of available products, and producing large quantities of salable products at low costs. Silviculture serves primarily to insure a continuing supply of naturally growing trees through timely and effective reproduction method cuttings, plus adequate measures to protect the growing stock from damage by indigenous destructive agents. Circumstances might also force landowners to shift to extensive management if their properties contain supplies of high-quality trees, but local markets demand only low-value bulk products. Conversely, buyers may find so much of the sought-after products that they can offer low prices and still satisfy their quotas. In cases of limited competition or monopolies, buyers may elect to suppress prices to increase their own profits.

When landowners control an uncommon supply of scarce products (e.g., high-value sawlogs and veneer logs, or opportunities for unique recreation experiences), and anticipate continuing and steady opportunities to access those markets, they more willingly invest in managing their growing stock. Further, if landowners can access markets that take a variety of products (e.g., bulk products, as well as those that command higher prices due to their scarcity), they have more flexibility to sell a wide variety of different products, and this usually encourages more intensive silviculture. Then the marketing strategy might include scheduling a regulated harvest of high-grade products, plus selling considerable quantities of pulpwood from the immature age classes. In this way, a firm may profitably implement a management scheme to enhance growing stock quality and rates of growth, and also create and maintain ecosystem conditions desirable for other fee uses like recreation or water production.

DETERMINING QUALITY IN TREES AND STANDS

Foresters often view quality as the goodness or measurable condition of a product or opportunity in its existing state, while awaiting some future use. Common examples include the characteristics of a log lying at roadside ready for transport, or the suitability of a property as the setting for a particular recreation activity. In most forestry situations, however, the price offered does not necessarily reflect what the seller considers valuable, or even how a seller rates the quality. Rather it represents the buyer's judgment about how well the logs or forested areas will eventually satisfy some anticipated use or provide some value of interest. In the case of logs or trees, it depends upon how much of a particular kind of lumber a sawmill can eventually cut from the pieces (Lockard et al. 1950, 1963; Campbell 1951; Mitchell 1961; Ostrander and Brisbin 1971; Rast et al. 1973).

As a general rule, buyers look for logs that should yield the highest proportions of desirable lumber, and they will generally offer landowners higher prices for those logs. This holds for noncommodity uses as well. People normally should pay higher fees for unique use experiences, or those that promise the greatest benefit for the time and money invested. In this context, quality depends upon three elements: (1) how a buyer intends to use an item or opportunity; (2) what a buyer defines as desirable for the intended end use; and (3) how well a buyer thinks an item or experience will provide the sought-after values. Quality depends upon *purpose,* and represents a buyer's expectation of how well an item or opportunity offered for sale will provide the values sought.

From a technical or engineering perspective, quality depends upon how well the physical and chemical characteristics of the wood in a tree will meet the physical property requirements of different end products and uses. Accessibility, operability, and costs of processing or utilization have no bearing on intrinsic quality. Instead, characteristics like soundness, size, and clearness of the logs affect their end product use for wood, and its quality as a material. Wood density, fibril orientation, chemical composition, and distinctive grain patterns also affect its usefulness and desirability in many manufactured products (Mitchell 1961). Knowing this, landowners can systematically evaluate their stands using specific measurable quality characteristics to judge the end-use potential. This means broadening their inventory procedures to include estimates of quality as well as an enumeration of the quantities of goods and nonmarket features of each stand. This information will (1) show landowners how much quality they control; (2) indicate how best to invest available funds and resources in managing their lands and growing stock to perpetuate or enhance those values; and (3) suggest how much leverage they have in marketing the things that they offer for sale. Further, landowners can use the quality information to project the probable future value of many goods and services, and assess the long-term profitability of many investment opportunities.

As a prelude to prescription making, silviculturists evaluate various physical characteristics of the trees and stands that they intend to put under management, and identify those that promise the best payoff for investments in deliberate management. Then they prescribe practices to concentrate the growth potential of a site on those stands and trees offering the highest market value to a landowner. This reasoning infers *selectivity,* and suggests that both sellers and buyers can differentiate between trees and logs of high and low end-use value. Ideally, they can also identify the stands that offer the best chances for providing valuable noncommodity opportunities as well. It also suggests that they can judge when individual trees and stands have reached an optimal stage of development to harvest or rent at the best price.

Conceptually, landowners could develop their own grading systems, and use them to determine potential values. In fact, as long as buyers and sellers can agree on a price, they can negotiate sales without actually sharing information about the quantity and quality of goods or opportunities offered. Using a standard system of grading and measurement will facilitate the exchange of information, and often reduce the costs of arranging a transaction. As with the sale of other kinds of commodities, buyers and sellers of forest products in different regions have developed an agreed-upon market language (e.g., standardized grades) to simplify communication, avoid confusion, and facilitate the marketing process. These standardized grades also provide an ethical basis for buying and selling (Kohls and Uhl 1985). This proves particularly important in dealing fairly with landowners who lack the technical knowledge to make independent assessments and judgments about their sales programs.

QUALITY REQUIREMENTS FOR DIFFERENT WOOD PRODUCTS

When foresters have information about the quality of the growing stock and various marketable use opportunities that a forest offers, they can usually make more informed decisions about where to invest resources. In essence, adding quality ratings to their assessment procedures allows them to stratify information about physical inventory of goods and opportunities on the land into standardized grades (quality- or use-classification groups). These reflect how well each tree fits the anticipated end-point market for particular commodities or use opportunities based upon

1 the intended use that buyers and lessees plan for a particular commodity or area (e.g., trees and stands);

2 the desirable characteristics that consumers look for in the final product or use opportunity; and

3 the expected yield of desirable goods and services to consumers once they process the trees and logs or take advantage of an opportunity offered for sale or lease.

From this perspective, quality is a function of purpose, and depends upon the ways that buyers intend to put a log or tree or stand to use.

Table 20-1 lists several common wood products, and the general attributes that make a tree valuable for each one. Though not all inclusive, the list illustrates typical products, generally arranged in order of increasing grade requirements. These include

- fuelwood, or parts of trees intended for burning as round pieces, chips and sawdust, or in a pelletized form;
- pulpwood, or wood intended primarily for wood pulp or chips for secondary manufacturing into reconstituted fiber-based products like paper, fiberboard, or chip board;

TABLE 20-1
SOME IMPORTANT FOREST PRODUCTS AND THE FACTORS CONSIDERED IN GRADING TREES AND LOGS FOR THOSE USES.

Type of product	General grade requirements
Fuelwood	Sound, reasonably dense
Pulpwood	Sound, good density, suitable fiber characteristics
Posts	Durable, straight, and sound
Poles and pilings	Straight, sound, and strong
	Minimum diameter, length, and taper
Timbers	Straight, sound, and strong
	Minimum length and small-end diameter
Tie logs	Sound and straight, but knots acceptable
	Minimum length and small-end diameter
Factory sawlogs	Sound and straight, and with minimum knots or other defects
	Minimum length and small-end diameter
Veneer logs	Straight, round, and minimum taper
	Sound, and free of defects
	Minimum length and small-end diameter

- roundwood, or tree trunks (or parts thereof) used as solid round pieces like posts, poles, and pilings;
- timbers, or wood in forms suitable for heavy construction, and usually squared by some primary processing;
- tie logs, or logs suitable for use as railroad ties and blocking;
- factory sawlogs, or logs for manufacture into lumber; and
- veneer logs, or logs intended for slicing, sawing, or rotary cutting into thin sheets.

Most trees have value for pulp and fuel, but many have little value for veneers and high-grade lumber. Foresters must determine the grade requirements for buyers who operate in their area, and learn to recognize the indicators of quality in standing trees. Then they can prescribe specific silviculture to enhance the desirable characteristics, and favor the most promising trees during different kind of intermediate treatments.

RECOGNIZING PRODUCT QUALITY IN STANDING TREES

Any physical characteristic that limits the use of a tree for the intended product reduces the potential value. Thus for sawlogs, any feature that will degrade or reduce the usefulness of the boards sawn from them constitutes a grading defect (Lockard et al. 1963; Ostrander and Brisbin 1971; Rast et al. 1973). The same characteristics will also leave defects in the sheets of veneer, and limit its usefulness for that purpose as well. Likewise, wood attributes that increase the costs of production or reduce the goodness of a fiber-based product (e.g., paper or fiberboard) reduce the quality and value of the logs or bolts. A feature that might substantially degrade a tree for use in one set of products might not affect its value in others. For example, crookedness may make a log useless for sawn products, but not as a raw material in manufacturing paper. Likewise, species with short fibers may not fit some pulping and paper-making processes, but they may have a high value as fuel, or for other uses that do not require long-fiber wood. To optimize their marketing, foresters must know something about the characteristics of wood in different species, and learn how to detect features that affect the quality of individual trees and logs for a particular end use.

Generally, the most valuable trees have large diameters and straight logs, and also minimum numbers of knots and other physical defects that reduce the usefulness of wood in them (Lockard et al. 1950, 1963; Campbell 1951; Mitchell 1961; Ostrander and Brisbin 1971; Rast et al. 1973). Wood strength, consistency, and other engineering characteristics based upon density and annual ring width also temper wood quality for some uses (Mitchell 1961). But in most cases, buyers tend to correlate value with size, given no unusual amount of rot and defect in the trees offered for sale (after Baker 1950).

At least two kinds of defects may degrade trees. Both have importance, but affect the trees in different ways (Anderson et al. 1990) as follows:

- Value-affecting defects, like mechanical injuries, knots, and deformities that reduce wood quality and usefulness, do not increase or spread, and do not cause any deterioration of the wood.
- Decline-causing defects, like damage by pathogens and wood-rotting fungi that enter through mechanical injuries, trigger a deterioration of wood, and continually spread and rot more of the solid wood in a tree.

Damages that affect the form of the main stem also limit its usefulness, even if the wood remains sound and free of internal defects. Further, some mechanical injuries (natural or other) may affect a tree so severely that the vigor declines, and pathogens may colonize in the tree as an aftermath.

In most cases, foresters can readily separate value-affecting and decline-causing defects. Important ones include (after Rast et al. 1973) the following:

- knots
- pitch
- holes/cavities
- splits
- ring shake
- gum and pitch pockets
- grub holes
- unsound wood (decay)
- bumpy surfaces
- undesirable color
- dead and living limbs
- sweep and crook
- surface wounds
- cracks and seams
- ingrown bark
- imbedded objects
- bird peck holes
- fruiting bodies (decay)
- burls
- twisted grain

For all of these except color of the wood and twisted grain, foresters can usually see external indicators that show up as branch stubs, swellings or other abnormalities on the surface, in bark patterns, or holes and other surface openings (Fig. 20-1). Also, internal defects of a decline-causing character may become obvious from fruiting bodies on the trunk or around the base of a tree. In addition, the usefulness of a tree or log for many kinds of solid wood products also depends upon the lengths, widths, and thickness of pieces that a mill can saw or otherwise manufacture from it (Lockard et al. 1950, 1963; Campbell 1951; Mitchell 1961; Ostrander and Brisbin 1971; Rast et al. 1973). Straightness and size affect these options.

In general, the value of a tree for furniture, cabinets, and structures depends upon wood characteristics of the species, as well as attributes of the component logs. High-value trees have both suitable intrinsic wood qualities, and general physical attributes that match requirements of the intended product. For example, those species most highly prized for furniture normally have wood with sufficient hardness to withstand abrasion and denting, good machining and gluing characteristics, and attractive grain and color. Generally, broadleafed species with a hard wood best serve these requirements. Then within the suitable species, trees that will yield the highest proportions of wide and long lumber free of knots and other defects have the greatest value to the furniture industry (Hanks 1971; Rast et al. 1973). In selecting trees to grow for these uses, foresters will look at the tree trunk for external characteristics that usually indicate defects within the logs.

Among both hardwoods (Vaughan et al. 1966; Hanks 1971; Carmean and Boyce 1973) and conifers (Heiberg 1942) most of the high-quality boards come from the butt log. The exact length of this choice piece varies with site quality and stem form. Except with extremely long rotations, foresters cannot develop the upper stem into top-grade logs (see Fig. 20-2), so they generally favor desirable species lacking signs of internal defects within the butt log (17–33 ft, or 5–10 m). This has proven possible even among fairly young growing stock. In such cases, foresters would evaluate readily apparent tree characteristics like crown class (dominant and codominant) and size, potential length of a branch- and defect-free butt log, soundness of the tree, straightness of the butt log, lack of lean to the tree, and the kinds and distribution of surface features on the butt log. These attributes give an index to the eventual lumber grade yields when a tree reaches sawlog size. The straight ones with a full-length butt log having few large surface defects (sometimes called overgrowths) will eventually yield greater proportions of the highest quality lumber (Boyce and Carpenter 1968; Sonderman 1979). Foresters should favor these trees in marking for thinning.

By contrast, lumber for construction uses must have lightness of weight, strength as an entire piece, and remain free from splitting when fastened. It may contain relatively small-diameter and tightly held knots for some applications, but the best lumber has relatively few of these defects and contains a high proportion of knot-free surface area (Ostrander and Brisbin 1971). Generally, conifer species have the best characteristics for construction uses, both as boards or when manufactured into plywood and similar kinds of building products. The most valuable trees have no branches on the lower part of the bole, few knots in the lumber sawn from those

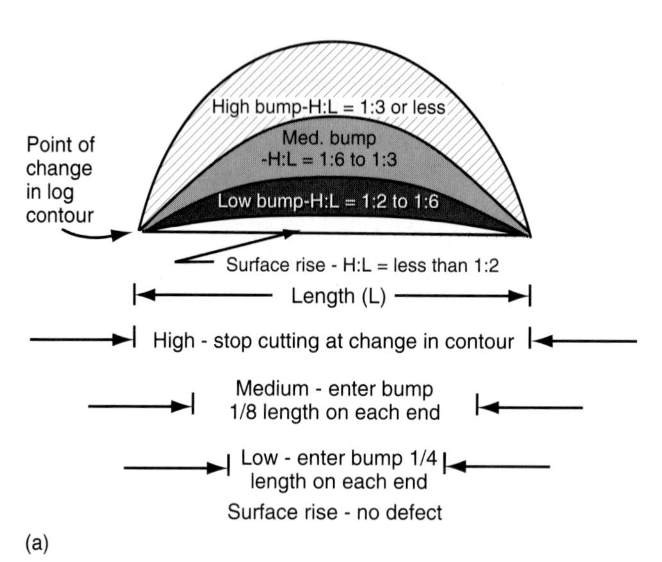

FIGURE 20-1
For most internal defects except color of the wood and twisted grain, foresters can usually see external indicators that show up as branch stubs, swellings or other abnormalities on the surface, in bark patterns, or holes and other surface openings. A. The height of a bump in relation to its width indicates the closeness of a defect to the surface. B. Heavy bark distortions indicate a defect fairly close to the surface of this white oak tree. C. The first board has a clear outer face but a distinct grain. D. The first small defect appears at 0.7 in. (1.8 cm) below the log surface. E. A major defect appears at 1.2 in. (3.0 cm) below the surface (from Rast et al. 1973; Rast et al. 1989).

logs, or only small-diameter branches and long internodal distance between the whorls of branches. Usually, foresters find these trees in closed-canopy stands, and among individuals of large diameter with straight boles.

With both hardwood and softwood lumber, manufacturers can use long and wide boards with knot-free surfaces for a great variety of products. They can work with these as whole boards, or cut them into smaller defect-free pieces for special applications. As a consequence, buyers willingly pay higher prices for such lumber, and for the long and large-diameter logs that will yield it. By contrast, a manufacturer has few options for using narrow and short boards, and those with knots distributed across much of the surface. This lumber may have good strength, but will not machine well and often lacks aesthetic qualities valued in many products. Builders might use it in framing and hidden parts of houses, but not for the molding around exposed areas like windows and baseboards. As a consequence, they place a lower value on knotty lumber and pay less for knotty boards or for logs that will yield high proportions of knotty lumber.

As a general rule, the characteristics that users consider valuable in lumber also determine the worth of a standing tree, or of a log ready for sale. Thus, trees and logs that mills can convert into solid wood products have greater value than those they can use only for fiber products. Trees that will yield considerable volumes of defect-free lumber of wide widths and long lengths have the greatest value for

(b)

(c)

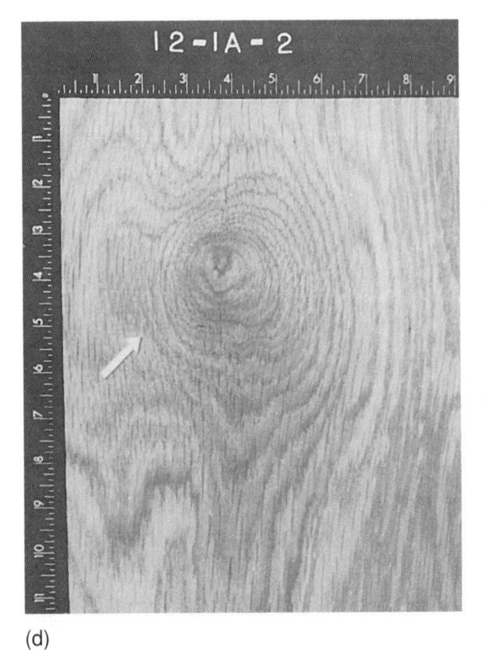

(d)

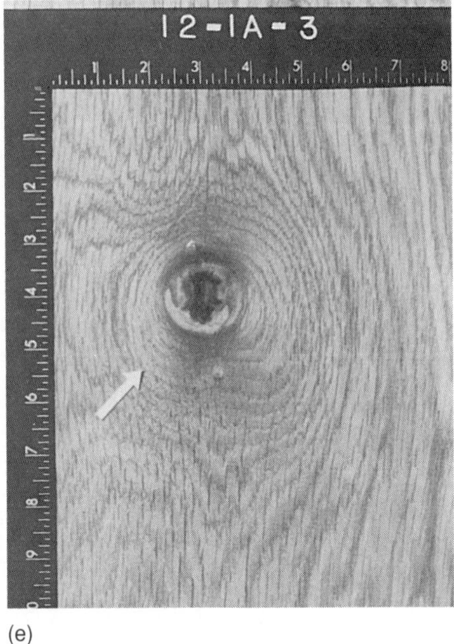

(e)

FIGURE 20-1 (continued)

sawlogs. Veneer producers must have the straight logs of large diameter and minimum taper, and with much defect-free wood of attractive grain and fairly uniform intrinsic qualities. Buyers pay especially high prices to get those logs. In principle, then, just as long and wide lumber offering the potential for use in the greatest variety of products commands the highest prices, so will fairly large-diameter trees that grew from an early age with no branches on the lower bole, and that remained free of injuries causing physical defects to the trunk. The larger the tree and the straighter and more branch-free its main stem, the greater the value for high-quality wood products (after Smith 1986).

FACTORS THAT PROMOTE QUALITY AMONG TREES AND STANDS

Exact factors that determine the lumber quality differ between hardwood and conifer species, reflecting the end uses. Even so, some fairly common factors influence the potentials for producing high-quality trees and logs of both species groups. Important factors (after Anderson et al. 1990) include

1 tree diameter, with the largest trees yielding wider and longer boards, and more of them;
2 site conditions, with regions and sites of the most favorable physical environment for tree growth and development generally producing larger-diameter and more defect-free trees and logs of better form;
3 damage and internal defect, with undamaged, sound, and healthy trees having more usable wood of better quality;
4 tree growth rate, with healthy and fast-growing trees with large crowns healing over wounds and branch stubs more quickly, and adding more wood over enclosed defects in a shorter period of time; and
5 species, with some species shedding lower branches more quickly, growing taller and faster, and proving more resistant to damage and disease.

Site conditions influence the species present, the density of stems that regenerate, the time when the crown canopy closes, the rates of individual tree growth, and the kinds and prevalence of naturally occurring damage. Particularly at high latitudes, ice and snow loading during storms frequently breaks the crowns, especially among hardwoods. This opens potential entry courts for pathogens, fosters discoloration of the wood in damaged trees, and also may affect the form of lower boles (Shigo 1966; Shigo and H. Larson 1969). Further, trees in stands on poor-quality sites in all regions generally grow more slowly, and take longer to heal over branch stubs and wounds. They tend to have a poorer form, and shorter branch-free lengths along the lower boles. As just one example, a dominant or codominant black oak growing in an even-aged central hardwood stand in North America may develop a clear bole along the butt 17 ft (5.2 m) within 25–30 years on Site Index 80 land, but take 70–75 years at Site Index 40. These trees would have 18 and 54 knotty annual growth rings at 17 ft (5.2 m) above the ground, respectively, for the two site classes (Carmean and Boyce 1973).

Figure 20-2 illustrates the patterns of wood formation in a growing tree, empha-

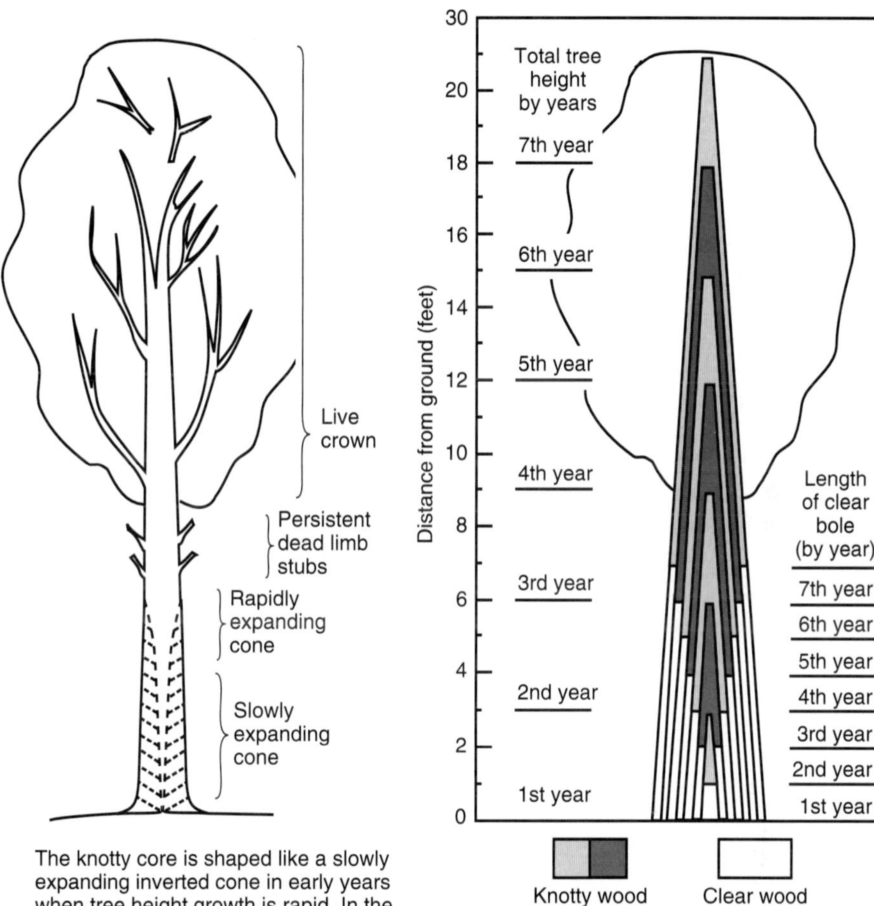

The knotty core is shaped like a slowly expanding inverted cone in early years when tree height growth is rapid. In the later years height growth slows, the base of the live crown moves upward more slowly, and the knotty core is shaped like a rapidly expanding inverted cone. Eventually the knotty core emerges at the base of the crown where persistent dead limb stubs occur.

Schematic diagram showing the first 7 years of tree-height growth, the clear wood formed on the lower one-third of the bole after healing of the dead limb stubs, and the knotty core formed while live limbs and dead limb stubs are on the upper two-thirds of the bole.

FIGURE 20-2
Patterns of wood formation in a growing tree, and distribution of clear and knotty annual rings with height above the ground (from Carmean and Boyce 1973).

sizing the distribution of clear and knotty annual rings with height above the ground. In essence, wood laid down onto the branchless, injury-free portion of a tree trunk will lack knots and other defects. Thus, any set of natural conditions or management strategies that promotes early mortality and prompt shedding of the lower branches from crop trees, and protects them from subsequent mechanical injury, also enhances the potential to produce quality logs in a stand. Further, if grown to large diameters, those trees will contain greater proportions of high-quality wood in the lower logs, as long as they remain free of damage during any logging and other activities within a stand. In addition, if the management includes timely tending to maintain high vigor and fairly uniform growth rates within the crop trees, an age class will yield usable logs of high quality within a shorter period of time. The shorter time to merchantability also reduces the duration of exposure to potential injurious agents, or the extent to which internal decay might spread through a tree. As an added benefit, lumber sawn from the financially mature trees with fairly consistent past growth rates will not warp and twist as readily as the boards cut from trees with erratic growth rates.

Fast growth in conifers or hardwoods does not translate into poor quality for most uses. Instead, thinning promotes the formation of more wood later into summer (called *latewood,* or *summer wood*). This later-formed material has better wood qualities than that produced at the beginning of a growing season (van Lear et al. 1977; Saucier 1981; Mitchell 1961; Smith 1986). Among hardwoods, latewood also has higher density than wood produced early in a growing season (called *earlywood,* or *spring wood*), and trees with higher proportions of it yield wood of better technical quality. Further, trees growing on high-quality sites will produce more latewood of higher density (Zahner 1970). Where a silvicultural system provides timely tending of stands on good sites, concentrates the growth potential on trees of desired external characteristics, and maintains consistent rates of growth on them it will increase the yield of higher-quality wood.

The major determinants of quality take shape at the time of a reproduction method cutting, and continue to influence tree value development until the time of financial maturity. During this time, foresters have several options for securing good numbers of the desired species, and enhancing crop tree quality such as the following:

1 using site preparation to enhance the germination and survival of sought-after species at desirable stem densities;

2 selecting a reproduction method suited to conditions of the physical environment and target species, to secure sufficient numbers of new trees for full site utilization in a reasonable time;

3 selecting species and seed sources with a known tendency for good natural pruning and small diameter branches when planning artificial regeneration;

4 using spacing to promote timely crown closure and early lower branch mortality within plantations;

5 considering tree and wood quality attributes when evaluating seed sources or genotypes for artificial regeneration;

6 delaying the first thinning until dead limbs or a clear bole develops to a merchantable height appropriate to the site;

7 selecting crop trees with branch-free lower boles, small-diameter limbs, and high quality butt logs;

8 regulating residual relative density and leaving trainers during early thinnings to promote continued natural pruning along the main stem;

9 thinning to increase the production of clear and late summer wood on the crop trees once clear boles form; and

10 designing a thinning regime to maintain uniform growth rates on the crop trees, bring them to merchantable sizes over the shortest rotation, and enhance wood qualities for lumber uses

Although more easily recognized in the case of even-aged management, foresters also apply these techniques within uneven-aged stands as well.

Foresters must carefully assess the real long-term advantages of keeping stand density low from an early age to maximize individual tree diameter development. In developing a planting plan, they must weigh the benefit of wide spacing to delay crown closure and promote rapid diameter growth, versus closer spacing that insures earlier crown closure to promote *natural pruning* (natural branch shedding). The close spacing leads to slower diameter growth and the wide spacing brings the trees to minimum operable sizes in a shorter period. With wide spacing, lower branches stay alive longer and develop larger diameters. This increases knot size in the lumber, and also lengthens the rotation to produce clear wood after the branches fall naturally from lower boles. It might force investments to remove the branches artificially.

From a financial perspective, *indirect* measures to promote crop-tree quality often cost far less than artificial measures (e.g., cutting the branches off). Yet costs accrue to owners due to the reduced radial increment on trees in high density stands during early stages of development, the delay for an age class to develop sufficiently to sustain a first commercially viable tending operation, and the need to retain an age class until clear (knot-free) wood forms on crop trees after lower branches die and fall off. In essence, the silvicultural system controls both the quality and value of production, and the balance of costs and returns from management.

PRUNING

Most well-stocked stands with fairly old trees of large diameters contain high-quality logs that yield abundant high-value lumber. This generally holds for both the hardwoods and conifers. When European settlers began clearing old-growth forests across North America centuries ago, or as people have more recently cut old forests in the tropics, they have recovered logs containing high volumes of clear wood. Even today, landowners who wait for extended periods (perhaps centuries) could grow similar high-value trees without doing much more than protecting the growing stock from damage by natural and human agents. Yet most enterprises and private landowners demand a quicker payoff, and foresters react by reducing stand density to promote more rapid diameter increment and shorten the time to grow the trees to a minimum merchantable size. To the degree that this strategy delays lower branch mortality and natural pruning, foresters must use artificial means to promote quality development.

While natural pruning occurs fairly readily among most hardwood species growing in closed stands, dead branches persist for long periods on most conifers. This reduces opportunities for producing knot-free conifer lumber unless landowners invest in *pruning*—removing the lower limbs from standing trees (Ford-Robertson 1971; Soc. Am. For. 1989). If workers cut living branches from the trunk, they do *green pruning*. When removing dead branches they do *dry pruning*. The minimum height for pruning varies with the management objectives, but must extend up the trunk at least to the top of one merchantable log. In most cases, workers using pole-mounted saws cannot reach more than 17–18 ft (5.2–5.5 m) from the ground. Removing branches to this height gives adequate allowance for stump, plus trim at the top end of the butt log. Furthermore, pruning costs increase greatly when workers use ladders or climbing spurs to reach greater heights. In one evaluation, pruning white pine to 34 ft (10.4 m) using a hand saw and ladder took three times longer than pruning to 17 ft (5.2 m) with a pole saw (Horton 1966). Also, spurs damage the trees, and degrade the wood. More important, the value of lumber in the 16-ft (4.9 m) butt log represents two-thirds of the total value in the entire sawlog portion of an eastern white pine tree (Heiberg 1942). So pruning only a single log provides the greatest payoff to a landowner.

Generally, workers do a better job of pruning when they use saws, compared with axes or other devices that slice through a branch. Further, to do the least damage and promote the most rapid healing, workers should make the cut flush and parallel with the trunk, leaving only the basal enlargement (branch collar or shoulder) or callus ridge at the base of a side branch (Arvidsson 1986; Shigo et al. 1979; Shigo et al. 1987; Shigo 1989). For species lacking a distinct branch collar or callus ridge, they should make the cut flush with the main stem, but without damaging any bark on the trunk. This technique (Fig. 20-3) has some distinct benefits. First, cutting off the branch collar or callus ridge creates a serious wound prone to fungal infection and subsequent decay. In addition, new growth to rapidly close over the branch stub initiates from cells along the outer segment of the callus ridges or branch collar, so leaving those tissues speeds healing. The new wood that forms directly over the stub of a well-pruned branch has a distinct grain pattern, but remains clear of knots as the tree grows larger. Landowners just wait until clear boards form outside of the tree. The time depends upon the growth rates of the crop trees, and their size when pruned.

Workers can prune dead branches at any time of year, but must remember that the bark of some species may rip and peel early in the growing season (Cline and Fletcher 1928; Ralston and Lemmien 1956; Funk 1961). Workers often cannot prevent this bark peeling when they prune live branches except by making an undercut beneath each one. Generally, foresters minimize these risks by not pruning early in a growing season, and pruning live branches only during dormant periods. Also, delaying pruning until the branches have died to 16–18 ft (4.9–5.5 m) indirectly reduces the chance of bark damage. In Sweden, researchers have linked autumn pruning of Scotch pine to an increase of *Phacidopycnis pseudotsuga,* a harmful fungus (Arvidsson 1986). Such possibilities warrant monitoring with other species as well.

Cutting living branches will not affect radial increment unless workers remove more than 50% of the crown, or reduce the live crown ratio to 40% or less (Daniel et al. 1979; Smith 1986). In fact, removing the lower inefficient living branches

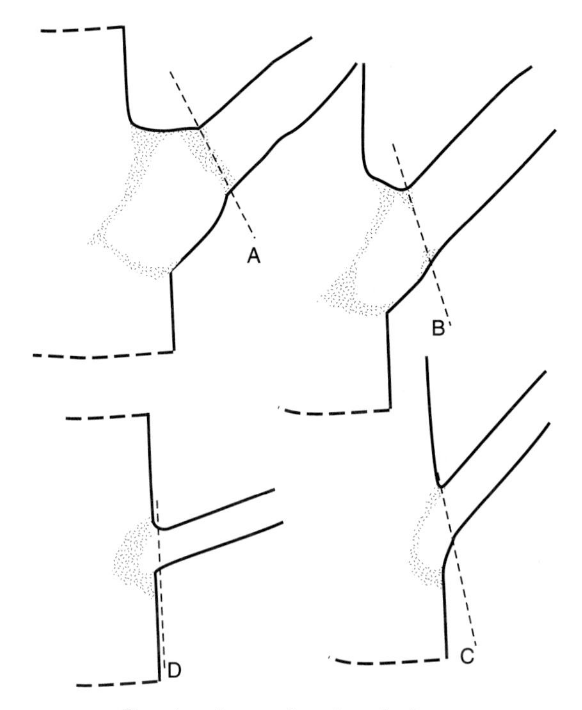

Branch collars and angles of cuts

Proper pruning of a living branch is a cut as close as possible to the branch collar. There is no set angle for a proper cut. Cuts A, B, C and D are proper cuts.

from Shigo 1989

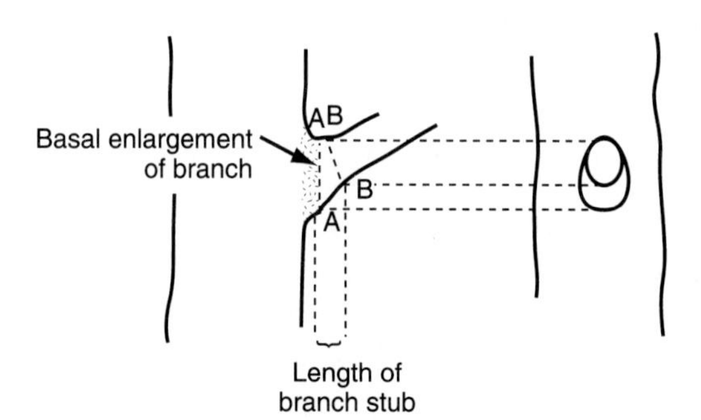

How to cut the branches when pruning.

from Arvidsson 1986

FIGURE 20-3
The appropriate technique for pruning branches from trees (from Arvidsson 1986; Shigo et al. 1979; Shigo et al. 1987; Shigo 1989).

while still maintaining a 50% live crown-ratio on longleaf pine in Louisiana increased radial increment by 14% over a 15-year period. By contrast, reducing the live crown to 38% and 25% decreased diameter growth by 5% and 12%, respectively (Sparks et al. 1980). Similarly, reducing the live crown ratio of loblolly pine in Arkansas to 25% and 40% at ages 12 and 15 significantly lowered the sawtimber volume by age 30. Pruning to a 55% live crown ratio increased the volumes, but not significantly. By age 30 all thinned plots had about a 50% live crown ratio, irrespective of the initial pruning (Wiley and Zeide 1992). Likewise, removing up to 50% of the live crown from black cherry in low-density stands of eastern North America did not reduce diameter growth over 10 years, but pruning 75% of the crown both reduced the growth and promoted epicormic branching. Most wounds healed in 4 years, with no differences between cutting live or dead branches (Grisez 1978). For red spruce in New England, removing one-third or one-half of the live crown reduced annual radial growth for up to 9 years, but taking one-sixth did not. Later growth rates did not differ across pruning treatments. In fact, by the eighteenth year the live crown ratios had become nearly equal (Blum and Solomon 1980). Pruning living branches from open-growth trees will expose the bark, and this could lead to sun scale on young (thin bark) trees of some species. Cut branches also add to the fuel load around pruned trees, increasing the fire hazard in some regions (Daniel et al. 1979; Smith 1986).

PRUNING TOOLS AND EQUIPMENT

Pruning requires only simple tools. Any that will sever the branches and leave a smooth and closely cut branch stub will suffice. Even so, foresters mostly opt for saws with 16–18 in. long (41–46 cm) curved blades, about 5–8 teeth per in. (2–3/cm), and that cut on a pull stroke (Fig. 20-4a). Another kind has a replaceable 36-in. (91-cm) long narrow blade mounted in a triangular frame. This saw cuts on both the pull and push strokes, and its design allows workers to make undercuts more easily. Large pruning shears have also proven efficient. They make a smooth cut, but workers have difficulty getting the cutting blade flush against the callus ridge. This leaves an extended branch stub that takes longer to heal. Attaching the saw blades or shears to telescoping or sectional poles lets workers reach to the required heights. For pruning lower branches, foresters have often mounted the saws onto axe handles (Fig. 20-4b). These Meylan saws allow workers to efficiently remove limbs up to 9–10 ft (2.7–3.0 m) from the ground. In fact, foresters will often organize the work by having some crew members use short saws to cut lower branches, while others follow with pole saws to extend the pruning to a desired 17–18-ft (5.2–5.5 m) height.

To gain efficiency, engineers have developed several kinds of power pruners. Some have hydraulic shears, and others small chainsaws. One has a drive shaft that runs through the center of the pole handle, powering an oscillating saw blade. All increase pruning efficiency by 25%–100% over manual tools (Arvidsson 1986). Other pruners include machines that workers clamp onto each crop tree. A gasoline engine powers drive wheels that move the pruner around and up the tree in a spiral

a.

b.

FIGURE 20-4
Pruning requires only simple tools to cut the branch close to the trunk. a. Long-handle sectional pruning saws allow workers to reach at least 17 feet (5.2 m), insuring clear lumber from the entire 16-foot (4.9 m) butt log. Adding or removing pole sections extends the range from about 8 to 20 feet (2.4-6.1 m). b. Shorter saws make low pruning more efficient, as with this Meylan saw which allows workers to reach branches up to about 8 feet (2.4 m).

pattern. Small saws cut off the branches as the device climbs around and up the tree to some fixed height, and then the device reverses back to ground level. This kind of mechanization has reduced the costs of large-scale pruning operations. Advances in engineering will certainly make other kinds of mechanized pruning possible. The new tools and machines will gain acceptance based upon efficiency of operation and effectiveness in safely pruning branches with a close and smooth cut.

MANAGING PRUNING AS AN INVESTMENT

Pruning requires an investment that landowners will not recover until the end of a rotation. As a consequence, foresters usually incorporate it only into intensive silvicultural systems to produce high-quality conifer logs or to enhance some special noncommodity use (e.g., the visual quality). They may also prune to protect against diseases that enter trees through dead branches, as with blister rust in eastern and western white pine (Weber 1964; Hungerford et al. 1981). Even so, they mostly prune to produce knot-free wood over shortened rotations, and when they anticipate that the long-term value growth will repay the investment at the required rate of return. In large measure, this depends upon the prices paid for the pruned logs, and the time to grow the pruned trees to the requisite merchantable sizes.

Foresters have only two options for containing the costs of pruning. First, they can limit the kinds and numbers of branches cut from individual trees, and prune only the minimum number of crop trees for full site utilization at the end of a rotation. Second, they can delay pruning until late stages of stand development, and thin to maintain rapid diameter growth on the pruned trees. Foresters can also choose the crop trees carefully by looking at the following:

1 the form of the main stem, including its straightness, roundness, taper, and size of its branches;

2 the vigor and health of a tree, as manifest in the size of the crown and density and color of the foliage;

3 the probable rate of future diameter growth; and

4 the species' value and silvical characteristics, including its potential to grow rapidly under the planned scheme of management.

Within even-aged stands, trees of dominant and upper codominant positions usually exhibit all or most of these qualities. Other trees generally have a limited growth potential, and require an extended rotation to reach minimum useful diameters. That prolongs the investment period, and reduces the payoff from investment in pruning.

To an important degree, site quality affects the growth and branching characteristics of trees, and indirectly the costs of pruning. This relationship derives from the annual height growth, and the time until crown closure. To illustrate, conifers growing among high-quality sites and the tallest individuals in upper-canopy positions have a greater distance between the annual whorls. As a consequence, workers must cut off fewer total branches within the designated pruning height. Also, early crown closure causes mortality of the lower limbs at smaller diameters, as witnessed by red pine planted at 5-, 7-, 9-, and 11-ft (1.5-, 2.1-, 2.7-, and 3.4-m) spacings. The branch diameters at 23 years of age increased significantly with each additional 2-ft (0.6-m) interval of initial spacing, except between 7 and 9 ft (2.1–2.7 m) (Laidly and Barse 1979). In eastern white pine, stands at 1500 trees/ac (3705/ha) had average knot sizes of 0.53 in. (1.3 cm), while branches in stands with only 500 trees/ac (1235/ha) measured 0.79 in. (1.6 cm) (Cline and Fletcher 1928).

Besides pruning only individual trees of appropriate characteristics and restricting the treatment to high-quality sites, foresters should prune only the trees that will comprise the final crop under the intended thinning regime (Fig. 20-5). This requires some knowledge about the probable crown radius of well-growing crop trees following the last thinning, and the number that will fill the growing space without crowding. Crews will cut the other trees during intermediate treatments, so many will not grow large enough to yield clear lumber. This means avoiding the tendency to prune insurance trees, just to provide a few extra in the event of loss or damage to some crop trees. To illustrate, crop trees of species with a 20-ft (6.1-m) crown diameter at financial maturity occupy 400 ft^2 (37.2 m^2) of horizontal growing space. This includes the actual crown area, plus the open gaps around their crowns. The final stand will have:

$$\text{Number of crop trees} = \frac{(43{,}560 \text{ ft}^2 \text{ per ac})}{(400 \text{ ft}^2 \text{ per tree})}$$

$$= 110 \text{ trees/ac}$$

$$\text{Number of crop trees} = \frac{10{,}000 \text{ m}^2/\text{ha}}{37.2 \text{ m}^2/\text{tree}}$$

$$= 270 \text{ trees/ha}$$

When foresters lack good examples of maturing managed stands, they can use the average crown radius of dominant trees in unmanaged stands at rotation ages, or somewhat older. Since these trees grew for long periods with restricted space for lateral branch elongation, their crowns give a conservative picture of growing space requirements. Alternatively, foresters could measure the crowns of open-grown trees of crop-tree sizes. These indicate the probable maximum growing space for trees in managed stands, and might somewhat underestimate the numbers needed for full site occupancy under deliberate management. The real crown size for managed stands occurs between these limits, so foresters can use an average of the two intervals in designating a crop tree spacing.

Whenever foresters invest in pruning, they should implement an aggressive and well-planned thinning regime as well. It might use lower-than-normal residual densities to insure rapid individual tree diameter growth. Otherwise, thinning might not shorten the investment period sufficiently to insure an adequate return on the investment (Smith and Seymour 1986). They must also use a thinning method that favors trees of upper canopy positions to insure both rapid radial increment and to speed the healing of pruning wounds. Then they must return to the stand periodically to maintain the growth rate. Otherwise, the investment period lengthens, reducing the rate of return to a landowner.

FIGURE 20-5
To limit the cost, foresters can prune only the better sites where the trees have long internodes, and fully stocked stands where early crown closure helped to limit branch diameter. They can also prune only the number needed for full site utilization at the end of a rotation.

JUDGING THE PROFITABILITY OF PRUNING

Many analysts have studied the profitability of pruning. Their findings vary widely. For Douglas-fir in northwestern North America, pruning returned in excess of 4% at high-quality sites with low-density stands having only 150 trees/ac (370/ha) (Fight et al. 1992). In another trial with eastern white pine, researchers pruned 7–8 in. (17.8–20.3 cm) diameter trees and grew them in low-density stands for 30–35 years (to 22 in., or 55.9 cm dbh). The investment returned 6% (Smith and Seymour 1986). In another case, yields ranged from 5.5% for an 80-year growth period, to 15.5% from pruning larger trees and growing them only 20 years (Horton 1966). The best return seemed possible by pruning a limited number of trees per unit of area, and growing them rapidly by periodic heavy thinning (Smith and Seymour 1986). Even so, foresters cannot simply assume that pruning will prove profitable, or not. It will prove reasonable for some silvicultural systems, and unsuitable to others. Much depends upon the landowners' requisite rate of return, and other demands. If they want to maximize the volume of knot-free wood, over short- to modest-length rotations, they must prune. On the other hand, if a firm demands a high return on investment, pruning might not pay.

Most likely, with either maximum-volume or maximum-return-on-investment criteria, pruning will pay only under limited circumstances. In the former case, foresters would program pruning at the first thinning, but only when they can do the first thinning commercially. The lower branches will have died for at least 15–20 ft (4.6–6.1 m) along the bole. The pruning would not affect radial increment, and the thinning will increase the growth of crop trees. At the other extreme, landowners might delay the pruning until well into a rotation to grow only 1–2 in. (2.5–5.1 cm) of clear wood on the butt log. This probably would not take more than 20 years with upper-canopy trees having a 40% live crown ratio (Horton 1966). Foresters might also keep these stands at a lower-than-usual relative density to keep the crop tree growing at maximum rates.

Time studies in Sweden indicate that skilled workers using conventional saws take about 5–6 minutes to prune Scotch pine trees to 17–18 ft (5.2–5.5 m). For plantation-grown red pine and white spruce in eastern Canada, pruning to 17 ft (5.2 m) took 4.2 and 9.9 minutes, respectively. Time to walk between trees and other activities reduced production rates to 8 minutes per pine tree, and 15–20 minutes per spruce tree (Berry 1964). Pruning red spruce in natural stands to 18 ft (5.5 m) took about 9 minutes per tree (McLintock 1952). Because of the high stem densities and earlier crown canopy closure, these trees had small branches. Also, red spruce has fewer branches per lineal foot than white spruce (Berry 1964). Further, pruning branches above 10–12 ft (3.0–3.7 m) takes twice as long as removing the lower limbs (Ralston and Lemmien 1956; Funk 1961; Berry 1964). With Douglas-fir, rates for pruning trees in a thinned stand ranged from 5.25 minutes for a 5-in. (12.7-cm) tree to 11.02 minutes for trees at 14 in. (35.6 cm) dbh. Those for ponderosa pine ranged from 9.8 minutes for 6-in. (15.2-cm) trees, to 17.4 minutes for those 16 in. (40.6 cm) in diameter (Cline and Fletcher 1928). Generally, workers can remove small and dead limbs faster than large or live ones, and limb size has an important effect. In fact, for eastern white pine, workers will spend an average of 1 second to cut off a 0.25-in. (0.6-cm) diameter limb, and 12 seconds for a branch with a base diameter of 2.5 in. (6.4 cm) (Cline and Fletcher 1928). As a means to control these costs, one rule of thumb suggests limiting pruning to trees with branches no larger than 2 in. (5.1 cm) at the base (Funk 1961).

Actual production rates vary with species and stand conditions: including density, tree diameter, branch diameter, number of branches per whorl, number of whorls to remove, worker skill and motivation, pruning height, terrain and ground cover, weather conditions, and kind of tool used (Shaw and Staebler 1950; Funk 1961; Arvidsson 1986). Besides charging the actual time to saw branches from the crop trees, foresters must also add the costs of buying and maintaining the tools, supervision and administration of the operation, tree selection and marking not associated with the thinning, time for workers to travel to a site, and time for resting and other miscellaneous activities (after Shaw and Staebler 1950). Further, changes in

TABLE 20-2
AN EXAMPLE OF THE COMPOUNDED COSTS FROM PRUNING A 10-IN. CONIFER, AND GROWING IT FOR ANOTHER 30 YEARS TO A 20-IN. MERCHANTABLE SIZE.

Assumed time to prune a tree	Cost compounded for 30 yrs. @6%[a]	Stumpage value to repay costs[b]
Unpruned	$ 0.00	$20.00
5 min.	3.33	23.33
10 min.	6.75	26.75
15 min.	10.51	30.15
20 min.	13.38	33.38

[a]At a wage rate of $7.00/hour, it will cost $0.58, $1.17, $1.75, and $2.33, respectively, when pruning takes 5, 10, 15, and 20 minutes per tree.
[b]The price required above the expected $20 realized from an unpruned tree of comparable size.

community and tree conditions as a stand ages tend to reduce time per tree to prune to 17 ft (5.2 m). In one study, actual costs went from 4.3 minutes/tree for 35-yr-old natural white pine, to 3.4 minutes with 55-yr-old trees, to 2.7 minutes with 80-yr-old ones (Horton 1966). Amount of natural pruning, improved ease of access through a stand, and similar factors affected the costs. Species like the spruces with numerous branches between the whorls take longer than the pines at all ages, even though the spruce might have smaller-diameter branches. The persistent branches also make walking through young to middle-aged spruce stands particularly difficult, slowing the workers and reducing their productivity. Total times for pruning to 17–18 ft (5.2–5.5 m) might extend to 15–20 minutes per tree in some circumstances.

The case summarized in Table 20-2 portrays a range of investments based upon pruning times of 5 to 20 minutes per tree, with labor costs ranging from about $0.60 to $2.30 per tree. One assessment of Douglas-fir reported a total cost of about $4.00 per tree (Fight et al. 1992), so costs shown in Table 20-2 do not exaggerate conditions. The example assumes that a tree at financial maturity might contain about 500 board-feet (1.74 m^3) of lumber. At a value of $40/mbf ($11.50/m^3), buyers would pay $20 for an unpruned tree. If immature trees grow in diameter at a rate of 1 in. (2.54 cm) in three years, a newly pruned 10-in. (25.4 cm) tree would increase to 20 in. (50.8 cm) in 30 years. During this period, every dollar that a landowner invested in pruning would compound at 6% to $5.74. So to repay the cost of pruning, the added value of lumber in a pruned tree must increase the stumpage price to $23.33 if pruning takes 5 minutes, or $33.38 if it takes 20 minutes. If landowners feel that the logs will bring a stumpage price at least this high, they consider pruning a good investment.

PRUNING LANDSCAPE TREES

Landscape trees along streets and in yards and parks often require pruning, but for a different purpose. Primary reasons (after Wenger 1984) include

1 altering the form and directing shoot elongation of young trees to enhance their shapes, preventing them from interfering with wires and structures, and containing their sizes;

2 inducing flowering and fruiting, or promoting sprouting within the crown (pollarding and topiary); and

3 removing dead or diseased branches, and those prone to breakage, to improve safety and tree health.

Generally, best results accrue from pruning at early ages, before the trees and the branches become too large. As with forest trees, workers should make smooth cuts flush with the callus ridge or branch collar, and never leave protruding stubs. In removing large branches, the branch collar may form a large swelling around the base. For these cases, workers should make the cut on the outside face of this enlarged shoulder (Fig. 20-6), and not damage the collar itself (Shigo et al. 1987; Shigo 1989; Harris 1992).

When removing large branches, workers often need to prune a limb in two stages. First, they undercut the branch 1–2 ft (0.3–0.6 m) from its base, and cut the branch off just outside the undercut. This reduces stress at the underside of the branch, and minimizes the risk of ripping the bark as the limb falls. Finally, they make a normal pruning cut to remove the stub (Fig. 20-6). As an added insurance against damage with large branches, workers can also first undercut the stub at the callus ridge before making a follow-up cut from above. They must often lower the largest limbs using ropes to prevent damage to other parts of a tree. After cutting diseased branches, they should disinfect the saws by dipping them in 70% denatured alcohol, or household bleach.

Hand and pole saws or pole-mounted pruning shears usually suffice for all but large-diameter branches. Contractors often use power pruners to gain efficiency, and rely on chain saws for the largest branches. They can facilitate wound healing by cutting away damaged bark and wood, and minimizing irregular edges along the pruning wounds. Severing the branches just outside the callus ridge or branch collar will promote the most rapid regrowth over pruning wounds and deter rotting. Wound dressings do not facilitate healing or prevent decay. They primarily have cosmetic value in making the pruning wound less noticeable. Some dressings actually stimulate rotting, or promote *wound wood* that rolls over the edge of a pruning wound and inhibits healing (Shigo 1989; Harris 1992).

PRUNING AND NONMARKET VALUES

Pruning has little ecologic value, except to remove entry courts for disease organisms through dead branches. Removing the lower branches would also reduce the ladder fuels at fire-prone sites, and lessen the chance of crowning when surface fires move through a stand. Otherwise the lower branches provide several ecologic benefits. They serve as perches and nesting sites for some birds, and contribute to vertical structural diversity important to many songbirds (Hunter 1990). Densely interlocked lower branches in young to middle-aged conifer stands also interfere

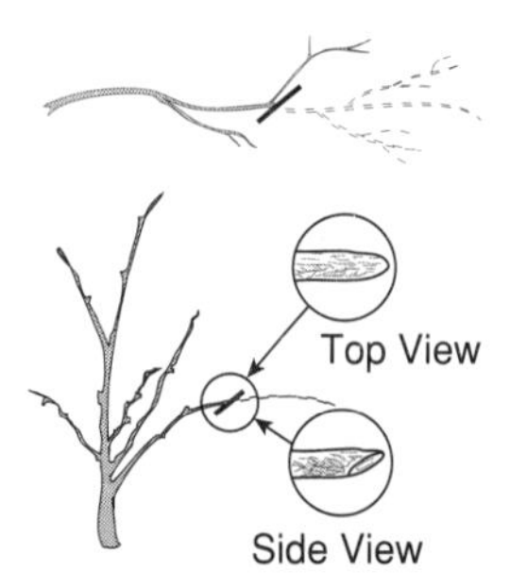

On low-growing shrubs, you can often hide pruning cuts by cutting back to a horizontal lateral growing from the top of the branch (top) or cutting to a bud so the cut surface is toward the ground and away from the viewer's angle of vision (bottom).

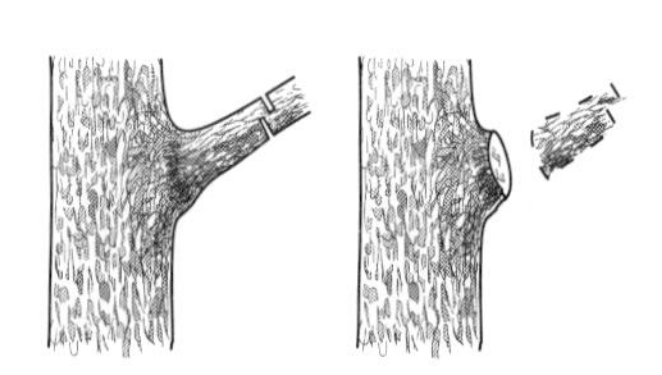

Remove a large limb by making three cuts. Make the first cut on the bottom of the branch about 300 mm (12 in.) from the branch attachment (left). Make the second cut on the top of the branch within 25 mm (1 in.) of the under cut. Make the final cut just beyond the outer portion of the collar and the branch bark ridge (right).

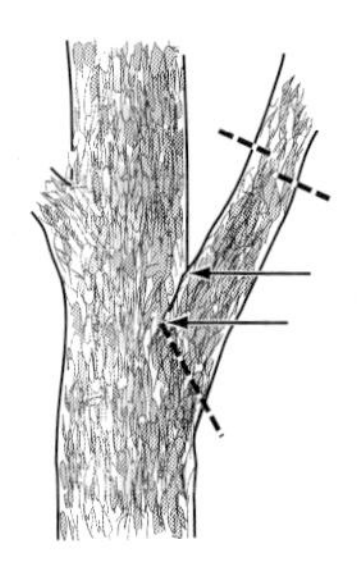

Remove a branch with a sharp "V" crotch in a similar three-step process being aware that the actual union of the two branches is often much lower than the apparent junction. The cut should be at a 40° to 50° angle from the horizontal.

from: Harris 1992

Tree topping injures mature trees.

Young trees can be topped to regulate size and shape.

Proper early pruning can regulate tree height and make later topping unnecessary.

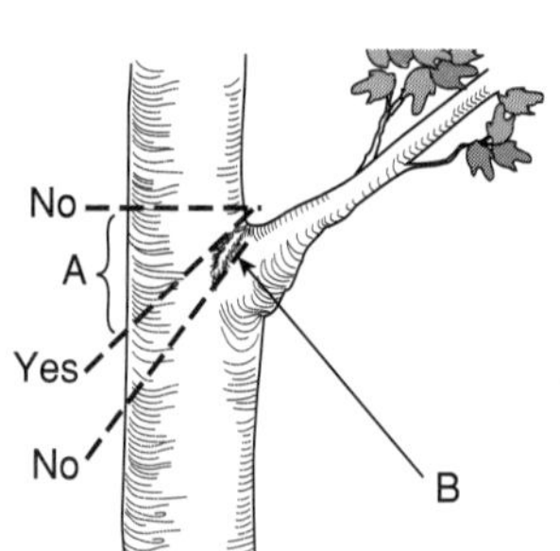

If you think your mature tree needs topping; maybe you need a new tree, especially if it is under a power line.

A = stem stub; B = branch bark ridge

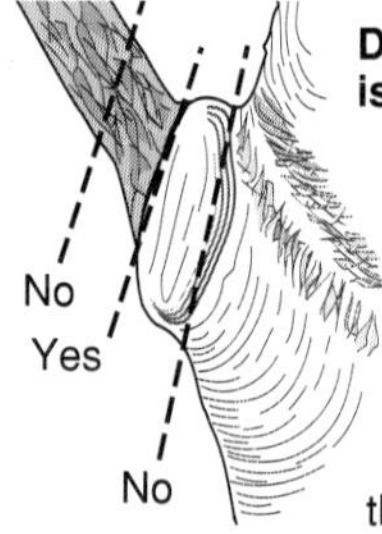

Dead branch removal is a health treatment!

But, do not remove the ring of living wood that surrounds the dead branch. Dead wood is an energy source for the fungi that grow into trees.

Injections, done properly, can benefit trees.

Holes should be small, shallow, and at the tree base.

If holes are not closed after one growing season, do not continue to inject.

from: Shigo 1987

FIGURE 20-6
Pruning techniques for urban and shade trees (from Shigo 1987, Harris 1992).

with the flight of hawks and reduce the risk to prey species. Even dead lower branches in thick conifer stands also impede wind movement, making those communities better winter protective cover for many animals. Lower branches also serve as microsites for plants like lichens and mosses, and contribute to the habitat of wildlife that eat or otherwise depend upon those plants. Removing the low live branches reduces the forage for browsing wildlife like deer and snowshoe hare

(Hunter 1990), and that may reduce the habitat value of some stands. Yet even in unmanaged conifer stands, the lower branches eventually fall off, and the habitat conditions change. The branches die and break off at fairly young ages among hardwood communities.

Many landowners who place a high value on the visual qualities often prefer pruned stands. Pruning may also make conifer stands more useful for many kinds of recreation (added safety, better understory visibility, and easier access). In these cases, landowners might even prune the trees at early ages, and market the cut branches for Christmas boughs. This gives them early revenues to help pay the compounded costs of site preparation and planting for plantations. Pruning living branches from no more than 50% of the tree height does no harm. In such cases, foresters would probably prune all the stems, rather than just the crop trees. The benefit accrues from intangible values to the landowner, rather than as a financial payoff in higher lumber prices.

21

STAND PROTECTION AND HEALTH MANAGEMENT

SILVICULTURE AND STAND HEALTH

Silviculturists perform four primary functions in planning and applying silvicultural treatments to stands, and across forests—*controlling, facilitating, protecting,* and *salvaging* (see Chapter 1). This includes protecting the site by the following:

1 insuring continuous vegetation cover
—establish trees after trees
—promote a tree community suited to the management objectives
—minimize premature tree losses
2 maintaining stable soil
—prevent erosion
—contain nutrient depletion
—sustain basic productivity
3 safeguarding against disruption of landforms and drainages
—protect natural drainages
—preserve general topography

To reduce chances for accelerated erosion following harvesting, resource managers can take simple steps (after Pierce et al. 1993) like:

1 appropriately siting and designing trails and logging roads to reduce the erosion potential;
2 utilizing machines and harvesting methods that minimize soil disturbance;
3 scheduling activities when soils have greatest resistance to compaction, deep rutting, and other disturbance;

4 smoothing the trails and roads following use, and installing water bars on slopes; and

5 reducing activities that remove the litter on slopes and erosion-prone soils.

These steps safeguard water quality. Further, maintaining continuous tree cover in riparian zones adjacent to streams and ponded waters, and appropriately locating and installing stream and wetland crossings by temporary trails and permanent roads, also contribute substantially to water quality maintenance (Pierce et al. 1993).

The objectives of optimizing production and safeguarding investments also require a carefully devised plan for protecting the trees. These health management programs include the following:

1 monitoring and evaluating potential health problems
—changes in pest habitat
—fluctuations in pest populations
—declines in natural control agents
—emergence of problems in nearby areas

2 keeping trees vigorous and healthy
—low-risk species
—well-managed stands
—control of damage

3 preventing deterioration of the growing stock
—vigorous trees
—regenerated on time
—grown under adequate hygiene

These represent some important elements of an approach that forest protection specialists call *pest management*, or more broadly *integrated pest management (IPM)*. Put in a broader context this could convert to *integrated health management* (*IHM*).

Historically, health management programs assumed that problems arise when harmful agents cause economically important damage. They sought to lower pest reproduction or immigration rates, or increase mortality or emigration rates (Reese 1979), to lower the long-term mean density, reduce the amplitude fluctuations, or reduce the frequency of pest and disease outbreaks. In this context, a *pest* includes any organism that prevents optimal development, use, or management of an economically important resource. Examples include pathogens, insects and other invertebrates, vertebrates (including people), and higher plants (after Stark 1977).

By current definition, health management integrates knowledge of ecosystems, pest population dynamics and genetics, and economic considerations to devise a program to keep the threats below unacceptable damage thresholds. It integrates knowledge about all aspects of the pest-host system to provide information about the options for protecting the resources of interest. Since eradication of a naturally occurring pest or disease complex might unbalance an ecological system and cause undesirable site effects, and would usually prove nearly impossible, modern health management simply attempts to keep the damage at economically acceptable levels

(Stark 1977; Zodoks and Schein 1979; Allen 1987). In many respects, the degree of control required and the amount of time and money invested depend upon the objectives for management and use, and a scale of a potential problem.

Definitions for IPM (or IHM) suggest four important ideas (Coulson 1981):

1 It has foundations in the principles of ecology.
2 It involves a combination of tactics.
3 It strives to reduce threats to economically and socially tolerable levels only.
4 It serves as one component of some broader resource management system.

Integrated health management deals with all potential health problems of economic or social consequence, including effects upon many nonmarket benefits. It accounts for the inherent community and landscape complexity associated with modern schemes of multivalue resource management, and includes a strategy that transcends the life of a single tree community or age class (Walters 1974). It consists of systematic decision-making to identify alternative and complementary cultural, biological, and direct control procedures to reduce economic losses while minimizing damage to beneficial and benign organisms. Further, IHM balances what health management specialists consider theoretically possible and what landowners regard as financially viable (after Way 1977; Reese 1979; Neisess 1984; Allen 1987; Speight and Wainhouse 1989). More generally, IHM entails active measures and timely action (after Thatcher et al. 1986) to keep the trees vigorous and healthy (so they more readily recover from damage), and to control important damaging agents (before they cause unacceptable losses). A combination of measures usually proves more satisfactory than relying on a single technique (Horn 1988). Also, prevention proves far more effective and less expensive than trying to control a problem after it develops (Hawley and Stickel 1948).

Integrated health management does not mean making choices between natural and chemical control measures. Rather, IHM incorporates a variety of activities (after Stark 1984), such as

1 articulating the forest management objectives, including the desired future conditions and values sought;
2 determining the long-term potential risks and where damage might occur geographically (e.g., hazard rating);
3 identifying ways to contain the risks, and optimize the long-term outcome of those measures;
4 formulating and periodically updating plans for coping with the problems through pertinent silviculture (e.g., adaptive management);
5 scheduling deliberate and timely action to minimize or reduce susceptibility and make the plant communities more resistant to damage; and
6 monitoring the forest and surrounding area for early warnings of potential problems.

The latter step insures timely and prompt action against potentially damaging agents before they devastate a stand or render it useless to the interests of a landowner (Allen 1988).

Generally, IHM involves using silviculture and other measures to reduce susceptibility or vulnerability to common harmful agents. For this reason, foresters actively work to prevent potential problems from developing, and to improve resistance and resiliency of a community and the trees in it. They intervene with direct control measures when important unforeseen problems develop and appear likely to cause unacceptable damage (Neisess 1984). Primarily they attempt to manage the risks by a holistic program of environmental management to realize the desired level of long-term control and protection at the landscape level (Horn 1988). Important benefits (after Waters and Cowling 1976; Croft 1985; Shea 1985; Allen 1988) include the following:

1 It integrates both direct (short-term) and indirect (ecological) options for control and prevention.

2 It identifies the resources and values warranting protection, the geographic boundaries of potential threats, and the objectives for managing these problems.

3 It fosters advance planning to discern the critical biotic and abiotic factors that temper the potential damage they cause, and to manage these in a specific time frame.

4 It encourages integration of strategies by adjacent owners to manage the problems at scales ranging from individual trees to landscape, or larger levels.

5 It continuously monitors conditions for early warning of potential problems before they reach critical economic thresholds.

6 It draws upon several technical disciplines in devising ecologically based management strategies, and arranging an appropriate response when problems develop.

As an end point of IHM, landowners strive for healthy trees and healthy forests, minimum disturbance to the environment, a greater assurance that management will provide the goods and services sought, and a potential to recover or use a forest without interruption over time (sustained at the requisite levels). It means that foresters make each silvicultural decision and implement each treatment with the goal of minimizing both short- and long-term potential threats to the trees and stands under their care.

While common to the management systems of many commercially important crops, IHM programs for forests must deal with some unique factors. These complicate both the planning and implementation (after Horn 1988). For example, forests serve as the natural habitat for many insects, microorganisms, and other potential pests. These include introduced agents with few local natural regulators over population dynamics. The complexity of most forested systems also makes precise control by sanitation, resistant or tolerant genotypes and species, and intensive regulation of pest habitat impractical. The remoteness of many areas further complicates monitoring and forecasting of potential problems before they develop. In addition, a wide variety of different harmful agents may injure a stand during the long time it takes for the trees to mature economically. Further each different approach to forest stand management, each potentially different combination of species (one- to multispecies stands) and structure (simple to complex), and each mix of products and uses may require unique approaches to long-term health management. In fact, a selected strategy may need to change over time with even-aged stands as they mature (Thatcher et al. 1986). Foresters must deal with all of these complexities in planning a silvicultural system, and putting it to use.

INTEGRATED HEALTH MANAGEMENT AND SILVICULTURE

In general, by accepting silviculture as a critical aspect of health management, foresters hope to avoid direct control measures. From that perspective, effective health management (after Thatcher et al. 1986) would consider

1 the character of potential and existing pest populations, and how they might change over time;
2 the susceptibility of different plant and tree communities, and how it changes over time
3 the suitability of different communities as habitat for an array of economically important pests, and how the conditions change over time;
4 the way and degree to which different pests affect resource values of importance, and at various stages of stand and age class development,
5 how these threats influence the management objectives for a stand and across a forest, and the potential to satisfy those purposes; and
6 the alternative strategies for prevention and resource utilization in relation to the potential for damage as an age class or stand develops.

This entails watching for signs that particular damaging agents might increase to economically harmful levels, and recognizing that many become important only during specific stages of stand and age class development. IHM also assumes that timely silvicultural treatment helps to minimize the threats.

Pine plantations of southern North America provide a useful illustration of how silviculture serves as a critical element of integrated health management. Within those communities, reproduction weevils, leaf-cutting ants, pine tip moths, and fusiform rust may damage and kill trees during the regeneration stage. To anticipate problems with these, foresters can prescribe site preparation (e.g., prescribed burning) to reduce the habitat and inoculum, tree planting to insure adequate stocking in areas lacking abundant viable seed or to introduce more resistant genotypes, and postestablishment weed control to insure good seedling vigor and early growth. Later, insect problems intensify as stands become crowded and tree vigor begins to decline. Thereafter, problems with *Annosus* root rot, plant parasitic nematodes, and some particular fungi may increase. In fact, thinning may magnify the problem by promoting the incidence of *Annosus* root rot through the stumps and damaged roots. Trees damaged by logging also become more susceptible to wood decay organisms and a variety of insects. To address these threats, silviculturists would use a wide initial spacing to delay the onset of active competition and reduce the need for thinning, careful logging to minimize the damage during any thinning in a stand, stump treatment to reduce the spread of root rot, and salvage of trees severely infected with diseases or most susceptible to insect attack (Thatcher et al. 1968,

Nebeker et al. 1985). In some cases, management schemes that incorporate no thinning might prove the most feasible. For other kinds of forests, the potential for damage and the exact silvicultural measures depend upon the nature of particular problems, the site and stand conditions, the tree species in a community, and the objectives for management and use.

Despite the good intentions of any health management plan, most crops suffer periodic threats from a variety of natural agents, and from human activities. Some outbreaks may actually kill the forest over vast areas, but the majority only alter its character by reducing tree vigor and rates of growth, causing defects, promoting rot, and creating holes in the main stem. This reduces many economic values, but may actually enhance the habitat for some wild animals and plants and serve other essential ecological functions. In fact, insect and disease outbreaks have historically proven ecologically important to forest renewal and development. For example, activity of mountain pine beetle may kill individual trees within maturing lodgepole pine stands in western North America. This natural thinning reduces the numbers of serotinous cones containing viable seed, and contributes dead fuel. The latter increases the potential for natural fires that kill the remaining live trees and release seed for a new cohort. Yet by reducing the seed supply, earlier insect-caused mortality indirectly influences the density of later regeneration, and lessens the potential for early stagnation due to overcrowding (Peterman 1978). Natural outbreaks of spruce budworm in the boreal coniferous forests, southern pine beetle in the southeastern coastal plain, and jack pine budworm in the Great Lakes region of North America also all have occurred periodically at a landscape scale, often when large areas of their host tree species reached advanced ages (e.g., Peterman 1978; Belanger 1981; Blais 1985). Even so, most landowners want to control the time of stand replacement to their own advantage, to protect their investments against untimely losses, and to safeguard nonmarket benefits of interest. They would consider these natural events as unacceptable, and look for ways to reduce the risk through preventive measures.

In many respects, the strategy depends upon what might cause a pest population or disease to reach harmful levels. Managers would use silviculture to change conditions that might foster a buildup, primarily by disrupting some essential habitat condition for the organism. To illustrate, entomologists have forwarded two alternative hypotheses to explain mountain pine beetle infestations in lodgepole pine forests of western North America (Amman 1978). The classic hypothesis suggests that a stress external to the trees themselves (e.g., drought or a pathogen) weakens the community, creating a favorable habitat for insects to proliferate. Under the *stress* hypothesis, management ignores stand age per se. It keeps trees healthy and more resistant to stress by periodic tending, and ends a rotation when the trees serve the economic objectives. This precludes a tree community decline due to senescence, and keeps the habitat suboptimal for the insect. Alternatively, some hypothesize that physiological maturity of a community leads to a general decline in tree vigor, and this improves the habitat to support a pine beetle population increase. If managers accept the *physiological maturity* philosophy, they would also use periodic thinning to keep trees vigorous, and extend the rotation only to some preset age (Amman 1978). This minimizes the concentration of weak trees that would provide a good habitat for the beetles. Available guidelines suggest a maximum age of 80 years for mountain stands, or alternatively not beyond the culmination of mean annual increment. For low elevation sites (the most susceptible), the rotation would end at about 60 years, when the trees average about 7 in. (18 cm). Additional benefit accrues from growing mixed-species stands, removing large and high-risk trees from stands not yet ready for replacement, and creating a mosaic of interspersed age classes at the landscape level. The latter reduces chances of simultaneous outbreaks over large areas (Safranyik et al. 1974; Amman et al. 1977).

Even early researchers cited tree vigor as the single-most important factor to good tree health (Boyce 1948; Hawley and Stickel 1948; Graham 1952). Their work led to an important principle for management: while landowners often wish to grow large trees to optimize many commodity and nonmarket values, they must regenerate an age class before senescence, since old trees gradually lose vigor and succumb more readily to attacks by insects and diseases, or the stresses of abiotic agents (Spurr and Barnes 1980). More recent evidence substantiates that high tree and stand vigor provides the best resistance to bark beetles among conifers in western and southeastern North America. Further, regulating stand density by thinning reduces the risks of bark beetle infestations (Wood 1988; Belanger and Malac 1980; Belanger 1981; Schowalter and Turchin 1993). Based upon evidence like this, foresters usually assume that by enhancing crop tree vigor through silviculture, and favoring high-value trees, they improve both tree health and value. Also, by removing low vigor trees and others that might serve as a host to support harmful insects and diseases, they hope to keep potential problems in check (e.g., See Knight and Heikkenen 1980; Neisess 1984; Speight and Wainhouse 1989).

Prevention silviculture in pine forests of southeastern North America incorporates three primary strategies (Belanger and Malac 1980; Belanger 1981):

1 promoting individual tree resistance
—favoring the most resistant individual trees and species
—removing weakened and damaged trees that have a high risk

2 promoting stand resistance
—maintaining an appropriate stand density by regulating initial spacing, and periodic thinning after canopy closure
—keeping the most vigorous trees of upper canopy positions as residuals
—maintaining a mixture of pines and hardwoods
—minimizing damage from skidding and tree cutting
—limiting the rotation age

3 protecting the site
—minimizing soil compaction and erosion that might stress the trees
—avoiding practices that reduce internal soil drainage
—maintaining or supplementing soil fertility

Quite important, pine-hardwood mixtures, and stands dominated by longleaf pine, provide a less favorable habitat for southern pine beetles. For plantations, foresters must also properly match the species to the site to insure good growth and development, and preclude stress due to unsuitable growing conditions. Also, prescribed burning serves as a useful precommercial treatment to kill weak understory trees and to reduce the severity of root rot before and following later thinnings (Schowalter and Turchin 1993).

Exact strategies will differ with community type, geographic location, and the management objectives. They also take account of pest dynamics and requirements, and attempt to prevent ideal habitat from developing over large areas. Thus, for single-species even-aged stands, foresters might control tree density at early ages (e.g., a wide planting interval, or early cleaning), and use relatively short rotations (e.g., set at the culmination of m.a.i.). Further, to keep the growing stock healthy and vigorous throughout longer rotations, they would periodically thin a stand to remove weak individuals and maintain large crowns on the residual trees. They would also favor species suited to the site conditions, and pick crop trees promising the best growth (dominants and codominants) and long-term economic benefits to a landowner. They would regularly monitor stand and tree health, as well as the populations of indigenous damaging insects and diseases. Then if a harmful one appears likely to increase to outbreak levels, they could institute timely control. Also, before a stand declined physiologically, they would regenerate it, using necessary site preparation to control the pests and insure prompt regeneration of a suitable mixture of species. In combination, these measures would reduce the probability that harmful pests and diseases become a problem, and minimize a need for direct control to protect stand health.

For some special noncommodity needs, landowners may retain trees well past the age of financial maturity, even keeping them beyond *physiological maturity* (Ford-Robertson 1971). This represents the time when a tree reaches full development, particularly in height and seed production, and when the vigor, health, and soundness begin to decline or fail. Foresters also use this term to describe an entire even-aged stand of fairly advanced age, when losses to mortality exceed growth, and the standing biomass begins to steadily decline over long periods. Many foresters apply the same meaning to old age classes within uneven-aged stands as well (e.g., in old-growth communities), since these ancient trees often show symptoms of reduced health, vigor, and soundness. Generally, physiological maturity implies that the trees have less resiliency, and recover more slowly when damaged.

Some pathologists hypothesize that old and large trees decline and show signs of dieback as a cumulative effect of recurring injury over time by the actions of people or a variety of natural agents. Old age, genetic make up, site conditions, and similar factors *predispose* or weaken a tree, permanently stressing it and increasing the vulnerability to actions of other harmful agents called *incitants*. These inflict injuries at a later time, and then disappear. They include single short-term or intermittently occurring factors like insect defoliation, weather damage, and mechanical injury. Since the already-stressed trees have lower resistance and a reduced potential for recovery, the inciting agents make them weaker or less healthy. Finally, a third group of *contributing* agents may follow, become established in the trees, and continue to affect them over long periods (e.g., wood-decay fungi, and bark beetles).

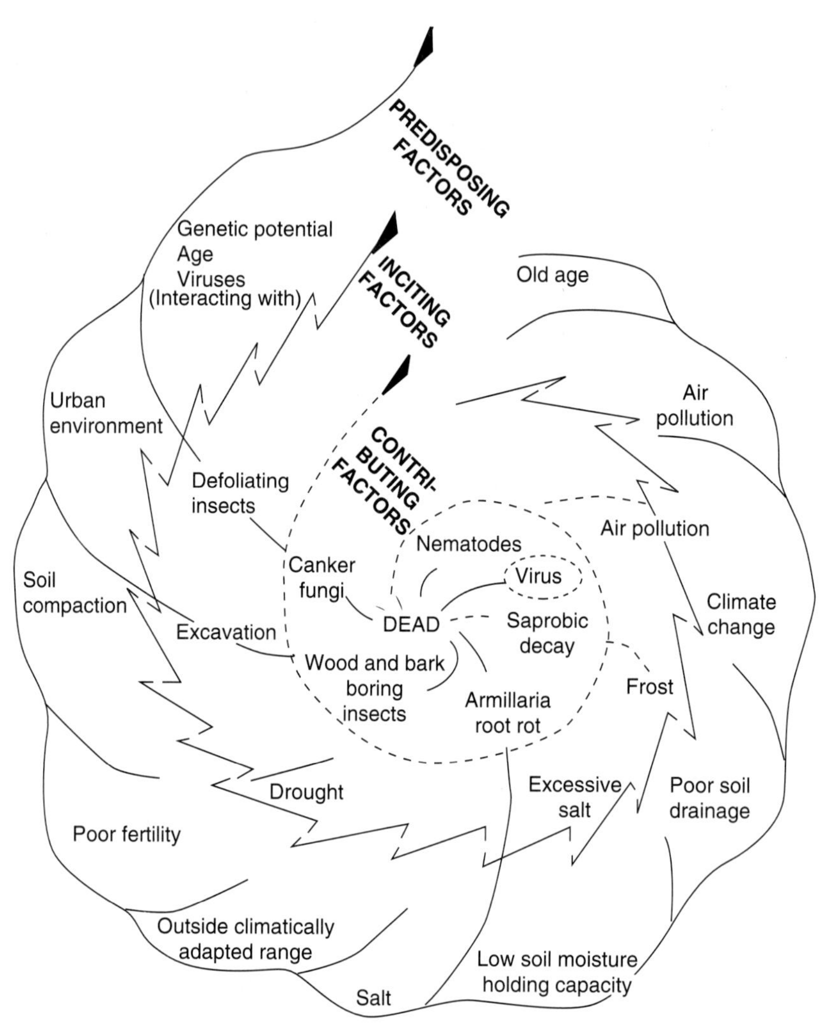

FIGURE 21-1
The role of predisposing, inciting, and contributing agents to tree decline and mortality (from Manion 1991).

These further weaken the hosts through their persistent action. In combination, all three factors stress trees through a continuing spiral or progression of increasingly important interactions over long periods, eventually leading to death (Fig. 21-1). The predisposing agents (outer spiral) precondition the trees for accelerated decline following the occurrence of inciting agents (middle spiral) that may follow over time. Finally, the contributing factors (inner spiral) add stresses that weaken the trees to a point of mortality (Manion 1991). Trees that escape these different kinds of damaging agents remain healthy longer.

HEALTH MANAGEMENT THROUGH SILVICULTURE

Past research and experience have shown that a particular array of silvicultural practices may actually increase the abundance of a pest, or the frequency of economically important damage to the trees. Similarly, practices that effectively contain one problem may even stimulate another (Waters and Cowling 1976). Further, management practices that disrupt balances of ecologic conditions providing a natural regulation over pest populations may increase the susceptibility to damage and loss (Schowalter et al. 1981). To prove effective, silviculture must manage the vegetation community and environmental conditions (after Nebeker et al. 1984) to

1 reduce existing levels of pest populations and other inciting and contributing agents that eventually would reduce tree health and push damage to economically unacceptable thresholds;
2 decrease the potential for pest populations to fluctuate from sparse levels to outbreak status;
3 increase the time between possible pest outbreaks related to normal population cycles, or the frequency of other damaging agents, thereby limiting tree injury and mortality or other forms of economic loss;
4 decrease the duration of pest outbreaks that occur, thereby reducing the impacts; and
5 maintain pest populations and the occurrence of other inciting and contributing factors at sparse levels, reducing the potential for future problems.

To this end, foresters will infuse their silviculture with measures to safeguard the health and condition of the immature growing stock, and to insure its timely replacement before the trees reach physiological maturity. They often also develop contingency plans for reacting to unexpected events that might unfavorably alter the patterns of development in a stand or age class. In this planning, foresters usually assume that vigorous trees and stands better withstand disease and insect attack, and recover better if affected. Further, they argue that indirect protection afforded through appropriate silviculture serves as the principal means for maintaining tree health and thriftiness.

This premise suggests that well-managed stands consisting of species suited to the site will have fewer health problems, and recover better if damaged or attacked by some harmful agent. It reflects an assumption that

- appropriate site preparation makes environmental conditions suboptimal for harmful pests and other damaging agents, thereby reducing the potential health problems during the regeneration and juvenile stages of stand development;
- control of the species composition and age class distribution insures adequate stocking of trees well suited to the growing conditions, and provides an ecological diversity to help keep the growing stock healthy;
- intermediate treatments like release cuttings and thinning maintain tree vigor and eliminate diseased and unhealthy trees to prevent initial infestations that might spread to healthy trees as well; and

• replacement of an age class before it becomes physiologically overmature (senescent) to reduce losses to many pests and diseases, and replenish the stand with a new community for continued long-term growth and development.

The protection measures actually used reflect experience with past pest and disease problems. Important considerations in planning a strategy (after Manion 1991, 1992) include

1 the logical ecological role of harmful organisms, including the beneficial effects in a long-term ecologic sense;
2 the implication of planned management activities on the agents that regulate, kill, and recycle plants and plant parts in a forested ecosystem;
3 the acceptable damage levels of these agents; and
4 the opportunities to salvage and utilize the trees that would otherwise die and eventually deteriorate and decompose (naturally recycle) in the forest.

It must deal with the potential for damage by people and their machines during various uses of a stand, as well as the biotic and abiotic natural agents common to a site or region.

Despite careful planning and protection, forces beyond human control sometimes devastate a stand or threaten it with unacceptable damage. This may force a drastic change of management in some instances, or at least reduce the options for using the trees and stand as originally intended. In many regards, the necessary and appropriate action depends upon the landowner's objectives, the character of a harmful agent, the extent of its damage, and the degree to which a landowner can stop the loss by prompt intervention.

FACTORS THAT INFLUENCE FOREST TREES AND STANDS

Intertree competition and interference weaken trees in forest stands, and those suffering long periods of suppression may photosynthesize too poorly to survive. This kind of mortality occurs so predictably in closed tree communities that foresters do not classify competition and interference as pests or natural destructive agents. Insects, diseases, animals, and a variety of abiotic factors also damage and kill trees as a natural part of ecosystem dynamics. The latter group includes many kinds of environmental stresses like lightning, wind, and fire (Spurr and Barnes 1980). They all serve critical functions in long-term ecosystem maintenance and renewal (Fig. 21-2), but may occur at unexpectedly high levels from time to time and disrupt the planned use (economic) of a forest community. When the change occurs abruptly and over large areas, and destroys entire stands of trees, foresters call it a *catastrophe* or *natural disaster*. Examples include large wildfires, extensive blowdown from hurricanes and tornados, damage from widespread flooding, bark beetle outbreaks, major episodes of defoliation and disease, and killing effects of drought. The action of less spectacular natural agents does not destroy an entire stand, and only causes injuries to individual trees. This may include damage by animals or disease organisms, from nutrient deficiencies and extremes of temperature, and by the actions of people. In the long run, both the natural disasters and the agents that function primarily as endemic irritants generally have greater economic than ecologic importance.

As defined, the harmful agents do not include intertree competition or interference, nor any internal physiological dysfunction triggered by a genetic abnormality. In that sense, foresters cannot necessarily guard against or eliminate them by simply regulating stand density or species composition. In fact, while good stand vigor and health can often reduce impacts of the endemic irritants or facilitate tree recovery after an attack, silvicultural measures will not usually stop regional-scale outbreaks and disasters of many kinds. In fact, when environmental conditions lead to an outbreak of some insect or disease over a widespread geographic area, the status of stand health may not have much effect in preventing a loss. Foresters can only react to those catastrophes by prompt remedial measures to remove the dead and damaged trees, and by rehabilitating a devastated area through special regeneration programs or other measures.

HOW TREES BECOME UNHEALTHY

Foresters can classify the causes of plant damage or disease into two basic groups. The *biologic* ones result directly from an infection by or the action of a biotic agent other than intertree competition for light, moisture, and nutrients. Examples include:

- insect feeding
  —defoliation
  —inner bark boring
  —feeling on other plant parts
- disease
  —leaf diseases
  —root diseases
  —wood rotting organisms
  —vascular diseases
  —viruses
- animal activity
  —browsing
  —feeding on bark
  —breaking the stem or branches
- allelopathy of one plant to another
  —preventing germination
  —inhibiting growth
  —killing established plants

Mechanical and physical agents also inflict some physical injury not associated with the feeding or development of another organism. These injuries may themselves prove important, or they may set the stage for secondary insects and microorganisms to invade the damaged sites. Examples include:

- mechanical breakage
  —ice and snow damage
  —blowdown
  —wind damage
  —machine use damage
  —trampling
  —landslides/avalanches
- fire
  —partial or entire defoliation
  —kill cambium or buds
  —wound main stem
  —consume entire plants
- flooding
  —inundate
  —saturate soil
  —uproot
  —break off or bend over
- pollution
  —contaminate soil
  —kill tree
  —damage tissues

Though not necessarily comprehensive, this list illustrates the array of damaging agents and how they affect tree health. In many cases, trees of good vigor survive the damage or infestation, and recover from it. In some cases the attack or infestation may trigger a plant reaction or set the stage for secondary organisms that damage a tree even more, or even kill it.

EFFECTS OF NONCATASTROPHIC AGENTS

Injurious agents having a noncatastrophic effect usually affect individual trees, damaging only parts of a stand or a forest. They have three basic effects:

1 damage or alter the form, vigor, or soundness (by defoliating, debarking, breaking, rotting, or bending over)
2 kill or otherwise destroy (by eating, trampling, defoliating, debarking, uprooting, poisoning, breaking off, and destroying cells and tissues)
3 transform the environment, thereby disrupting physiological functions or growth patterns (by impairing nutrient or moisture uptake, or polluting the environment with toxic substances)

The first two groups directly affect individual trees and other plants, and change their character. The latter group alters conditions essential to plant and tree growth, thereby indirectly affecting tree health and condition. This may include effects of soil compaction, flooding, and toxic chemicals. In many cases, agents of the first group degrade a tree sufficiently to reduce its economic worth for commodities or to support some nonmarket value of interest.

Since most noncatastrophic agents occur naturally in forested ecosystems, foresters must plan deliberate preventive measures before they seriously damage the values desired from a stand. Protective steps might include the following:

1 direct attack on a predator or disease by
—chemical control
—hunting and trapping
2 reducing habitat for a pest or disease by
—site preparation to alter the essential physical environment
—selectively eliminating critical food sources
—removing unhealthy trees that might serve as infestation sites
—removing hosts and alternative hosts essential to pest or disease survival and reproduction
3 protecting trees from direct attack by
—surrounding them with special structures (e.g., fences)
—discouraging attack with repellents
—creating diversions to attract pests away from the trees of interest
—using resistant genotypes and species

To insure success, foresters must first identify the agents that will most likely affect a stand. Then they can select the prevention measure that appears the most financially and ecologically efficient.

SPECIES AND GENETIC SELECTION AS A CRITICAL PREVENTIVE MEASURE

In large measure, the process of selecting a species or group of them for regeneration involves at least five steps. These include:

1 identifying the species suited to the intrinsic site conditions;
2 determining what local pests or diseases might preclude success under the intended scheme of management, or for the desired products and uses;
3 considering if (and how) various ancillary measures might effectively suppress or prevent a recognized problem;
4 excluding species susceptible to economically important damage; and
5 selecting species most likely to succeed at the least cost.

Figure 21-3 portrays many important considerations for species selection related to conditions of the site, the species, and the objective for management (from Savill and Evans 1986). Foresters must weigh similar factors in making choices about seed sources and other matters of genetic importance as well. Examples include seed source and family resistance or susceptibility to problems like needle cast of lodgepole pine, western gall rust on lodgepole pine, pitch nodule moth on lodge-

FIGURE 21-2
Many ecological processes like the bark beetle infestation that killed this lodgepole pine community serve as a natural means for restoration and renewal in unmanaged forests, but may also cause important economic losses to a landowner.

pole pine, blister rust on western white pine, and budworm on western spruce (Hoff 1984). Particularly for artificial regeneration, foresters could deliberately mix several resistant sources to insure genetic diversity. This might also increase standwide resistance to a broader array of health problems, and give assurance of an adequate representation of sources fitted to the physical environment. The decisions become particularly important when the management objectives demand use of a susceptible species. In some of these cases, foresters may foresee a greater promise from growing other species that will not require important pest-control investments, and then using the revenues to purchase the required species on the open market.

As a critical determinant in selecting a species either for tree planting or to encourage through a natural reproduction method, foresters must consider a variety of biological factors indigenous to the site, and related to the species of preference. Where conditions indicate a high hazard from local insects, diseases, and abiotic agents, foresters discount a species or discourage it by the design of a reproduction method and associated site preparation. Figure 21-4 illustrates the nature of this assessment, using information about the biology of root collar weevil and the susceptibility of different pines as determinants in selecting between species commonly

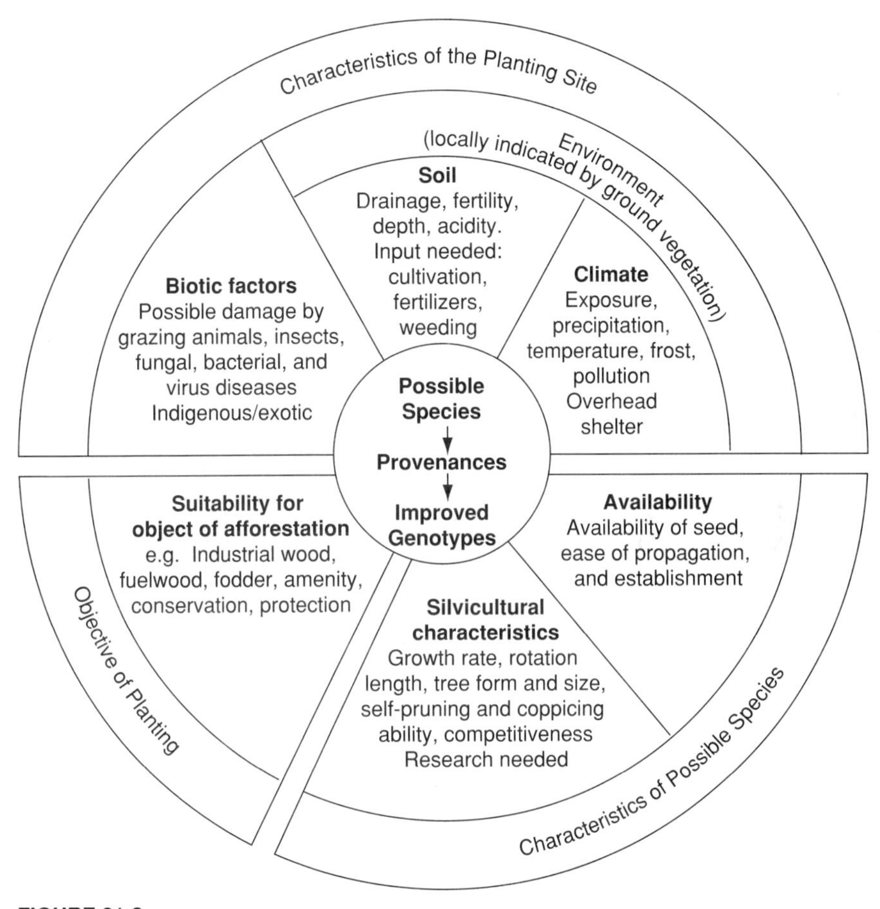

FIGURE 21-3
Some major factors to weigh in making choices about species, seed sources, and other matters of genetic importance in planning a forest regeneration program or reproduction method (from Savill and Evans 1986).

grown in the Lake States region of North America (after Wilson and Millers 1983). Under high-risk conditions, foresters reject susceptible pines (e.g., Scotch pine). For moderately susceptible species, landowners would commit to prevention measures like low pruning or application of insecticides. With white pine (low susceptibility) they must only separate a planting by 1400–1500 ft (427–457 m) from known infestations of the pest (Wilson and Millers 1983). Foresters use similar guidelines to judge the probability of damage from a variety of important damaging agents in other regions (Hedden et al. 1981), and to make choices about the tree species for planting or other reforestation efforts.

Note that even in relatively safe zones, a site must include the requisite balance of physical and biotic elements for a species to grow well over the long run. To a large degree, foresters first consider the *intrinsic site conditions*, and then select a species suited to that environment (see Figure 21-3). This recognizes an important ecologic reality: Most species survive under a fairly wide range of site conditions, with some having a greater ecologic amplitude than others, and most species grow well under a more limited range of site conditions than necessary for survival. Limitations of a site often become manifest at middle ages, but the relative importance of different critical environmental conditions may differ as trees grow in stature and change in physiological condition. Ecologically, the primary measures of species suitability include the potential for germination to occur, new seedlings to survive and become established, established trees to grow at good rates to maturity, and biologically mature trees to flower and set seeds. As long as the physical environment has adequate accessible supplies of basic resources within the growing space of an individual tree, it will grow and survive. Where sites have more than minimal amounts of the requisite resources, the trees grow better, have higher vigor, and remain in good health as long as some damaging agent does not interfere. Trees unsuited to a site senesce and decline earlier. Besides selecting species well suited to the environment, foresters must also use care to maintain an adequate tree species diversity across the landscape.

Generally, pathologists and entomologists argue that vast areas containing simplified forests of only a single species provide an overabundance of a host, and of conditions where a pest or disease can spread rapidly if environmental conditions favor its multiplication. In fact, the reforestation literature includes many references to the potential health risks of creating large-scale monocultures (Gibson and Jones 1977; Zobel and Talbert 1984; Zobel et al. 1987). By definition, *monocultures* include both natural stands and plantations containing a single species, and commonly of an even-aged character (Ford-Robertson 1971). Even early writings about forest protection caution against creating and maintaining these communities over large areas (Boyce 1948; Hawley and Stickel 1948; Graham 1952; Knight and Heikkenen 1980) and recommend

1 mixing species within each stand;
2 limiting the size and extent of any single-species community;
3 concentrating on species well adapted to the site;
4 planting a species only within its natural range, or under conditions essentially similar to its native habitat;
5 maintaining age class diversity within a forest or at the stand level; and
6 implementing timely silviculture to maintain tree vigor and regenerate age classes before senescence.

Since serious threats of outbreaks occur primarily in simplified monocultures extending over large areas, maintaining structural-age class and community composition diversity at the landscape level helps to reduce the risks.

Genetic diversity also plays an important role in forest health. In fact, many naturally occurring forest communities that cover expanses of area actually have limited tree species diversity, or none, but survived long periods without devastation (Gibson and Jones 1977; Chou 1979; Bain 1979). Apparently, trees in these naturally evolved species-simple communities have sufficient genetic diversity to afford a reasonable degree of resistance, at least until stands approach old ages. By contrast, stands and forests having a restricted genetic base appear more vulnerable to indigenous pests and diseases that may erupt to damaging levels from time to time (Zobel et al. 1987). Plantations of a single species and having a limited genetic diversity, including stands or exotic species propagated from a limited number of parent trees growing in a restricted geographic area, tend to suffer the most devastating health problems, even if established on sites well suited to the species (Zobel et al. 1987). In fact, foresters may increase their risks more by creating highly simplified artificial forests (plantations) established from genetically uniform sources, than in maintaining monoculture communities having natural structural and genetic diversity at both the stand and landscape levels (after Zobel and Talbert 1984; Zobel et al. 1987; Speight and Wainhouse 1989).

DAMAGE, DISCOLORATION, AND DECAY IN LIVING TREES

Landowners who manage their forested lands for sawtimber and veneer products usually want to produce high-quality logs of large diameters. They normally prefer fast-growing trees of valuable species, and periodically tend their stands to insure that the crop trees develop large crowns and grow rapidly in diameter. To further enhance their investment, landowners usually want to have growing stock made up of trees with good lower-bole characteristics (straight, round, branch free, and unblemished). That part of a tree has the greatest market value (Heiberg 1942; Vaughn et al. 1966; Hanks 1971; Carmean and Boyce 1973). They also want to protect the crop trees from damage by agents that would slow the growth or degrade the butt logs due to physical defects, discoloration, or decay.

Past studies have shown that major wounds from both logging and natural agents can cause physical defects, discoloration, and decay within standing trees. The injuries open trees to a complex biochemical process that discolors the wood present at the time of the damage, and increases the probability that wood decay organisms can infect the discolored tissues (Shigo 1966; Shigo and vH. Larson 1969; Shortle 1987). Whether decay actually develops depends upon a wood-rotting fungus inoculating and colonizing the wound. The extent of discoloration depends upon tree vigor, the severity of an injury, the presence of old wounds in the tree, and the time that has lapsed since the injury (Shigo 1966). Subsequent wounding close to an older injury may cause the discoloration and any decay to spread to healthy tissues laid down since a previous wounding (Shigo 1985).

Normally, injuries to living trees trigger an internal chemical response leading to the formation of special cells by the cambium. These compartmentalize the wounded tissues, isolating them from the rest of the tree, and restricting discoloration and decay to the wood present in a tree at the time of injury (Shigo 1966; Shigo and vH. Larson 1969; Shortle 1987; Anderson 1994). Compartmentalization may also reduce the internal space for storage of energy, thereby causing stress within a tree (Shigo 1985). Because of this internal compartmentalization, discoloration does not spread outward. Instead, new wood laid down in subsequent years will not discolor, nor will decay develop in the new tissues (Shigo 1966; Shigo and vH. Larson 1969). As a result, previously injured trees usually have a darkened or discolored center, surrounded by light-colored wood that developed following the wounding (Fig. 21-5). The size of the discolored cylinder generally reflects the diameter of the tree at the time of an injury.

Even skilled pathologists cannot precisely predict the degree of decay and discoloration based upon external wound characteristics alone (Shigo 1966; Shortle 1987). Generally, injuries to the crown have a greater effect than injuries to the trunk, especially in trees damaged repeatedly throughout their life (Shigo 1966; Shigo and vH. Larson 1969). Also, observers found a high probability of measurable decay in hardwood trees (e.g., sugar maple) from broken branches with a basal diameter of at least 3 in. (7.6 cm) (Silverborg 1954, 1959; Hesterberg 1957). Further, logging injuries that remove the bark from more than one-third of the circumference of the butt log often lead to discoloration and decay (Shigo 1966), and wounds covering 150 in.^{2} (968 cm^2) or more on sugar maple have at least a fifty-fifty chance of developing decay within one decade (Hesterberg 1957; Ohman 1970). With yellow birch, measurable decay may develop from bark wounds as small as 90 in.^{2} (581 cm^2) (Lavalle and Lortie 1968), suggesting that some species have greater sensitivity to these kinds of injuries. The effect also varies with the nature of a new wound and other factors, but trees subjected to repeated injury generally decline in health and quality. The more frequent the logging in a stand, the greater the probability of repeated wounds to any one tree, and the greater the proportion of trees that receive logging injuries during the life span of an age class. Further, repeated entry for uneven-aged silviculture will likely maintain some basic level of defect and degradation in a stand (Nyland and Gabriel 1971).

In addition to affecting the quality and volume of usable sawlogs, logging injuries may have other adverse effects. The most severe even cause mortality. For example, felling one tree may damage the crown of others, reducing their leaf surface area, and diminishing the growth rates. Some smaller trees may also be broken off, bent over, or knocked into leaning positions by either the felling or skidding operations (Nyland and Gabriel 1971). Also, when logging machines sink into saturated soil leaving deep ruts, they cut the roots on one side of the adjacent trees. In severe cases this removes sufficient root area to reduce moisture and nutrient uptake, as well as the energy storage capacity of the root system (Shigo 1985). Equally as important, cutting the roots around one side of a tree makes it more prone to windthrow. Further, damage to roots during management activities opens infection courts for root diseases like *Annosus* root disease (*Heterobasidion annosum*), black stain root disease (*Leptographion wageneri*), and red-brown butt rot (*Phaeolus schweinitzii*) (Reaves et al. 1989). Yet assessments following thinning among loblolly pine stands in southeastern North America indicated no negative effect on the radial increment of trees adjacent to trails with ruts less than 6 in. (15 cm) deep, and having no exposed or severed roots. Growth did decline adjacent to trails with deeper ruts, and where skidding damaged lateral roots of the trees (Nebeker and Hodges 1985; Reisinger et al. 1994).

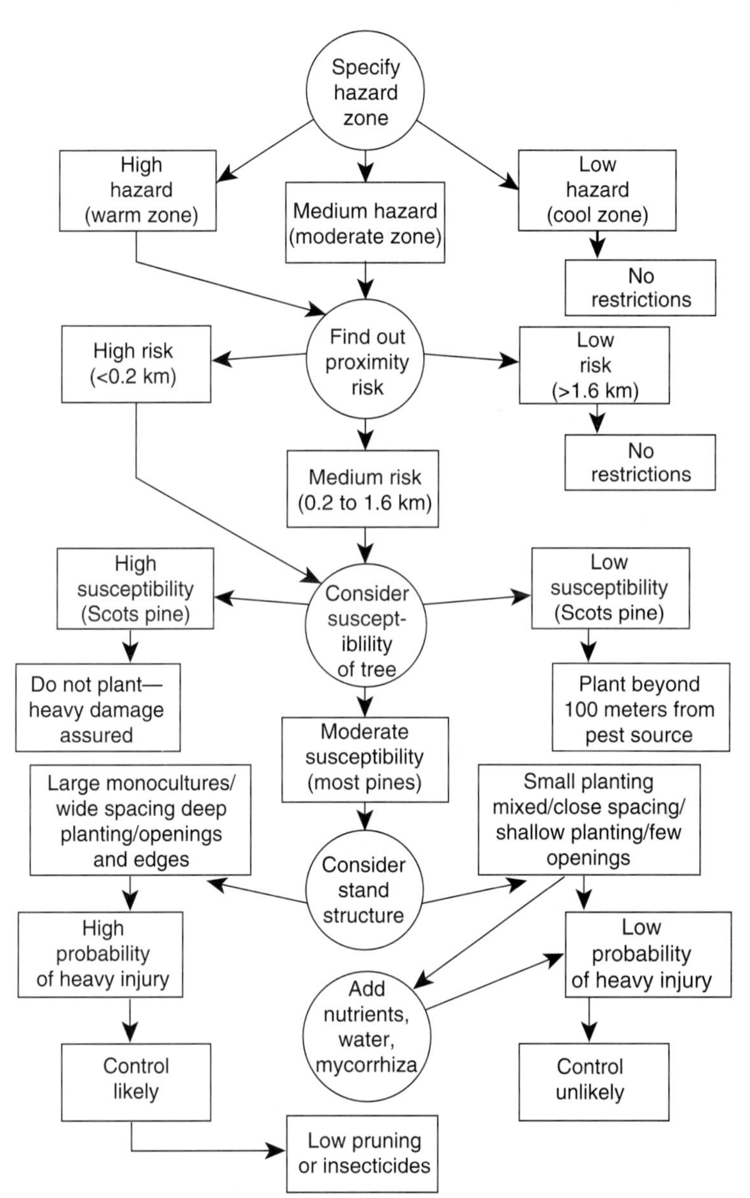

Decision-making guidelines and the probable consequences of management of *Hylobius radsicis.*

FIGURE 21-4
Risk assessment serves a critical role in many kinds of management decisions, including choices of species in a reforestation program (after Wilson and Millers 1983).

DIFFERENCES BETWEEN SELECTION SYSTEM AND THINNING

Studies of partial cutting indicate that logging will damage crowns of trees in all sizes, break off and bend over many residual saplings and small pole-size trees, uproot some of the advance regeneration, and leave basal wounds on an unpredictable number of other trees (Zillgitt 1945; Weitzman and Holcomb 1952; Trimble and Hart 1961; Arbogast 1950; Engle 1947; Nyland and Gabriel 1971; Nyland 1986). The frequency of wounds among the size-classes depends upon the number of residual stems in each one. In uneven-aged stands, trees with major injuries may comprise about one-fifth of the residual basal area. Further, logging may destroy sufficient numbers of saplings and poles to cause an imbalance in the age classes (Nyland and Gabriel 1971; Nyland 1990a, 1991).

Most management guides for selection systems recommend freedom from damage and disease as a major criterion for selecting residual trees (Arbogast 1957; Trimble et al. 1974). Wounds within 10–15 ft (3–5 m) of the ground have the greatest economic consequence, since the butt log has the highest value. By watching for signs of damage when marking for later treatment, foresters can periodically remove many previously injured trees (Tubbs 1977), but they must use caution not to leave irregular spacing, reduce the growing stock below acceptable levels, or distort the balance of age classes. Also, as many as 60%–70% of trees within 10 ft (3 m) of the primary skidding trails may receive basal wounds during skidding (Nyland and Gabriel 1971; Davis and Nyland 1991). Removing all of these previously injured trail-side trees would widen the skidding corridors excessively (Ohman 1970; Davis and Nyland 1991), and eventually destroy the integrity of a stand. Instead, the marking crew may need to leave a band of damaged trees along each side of the skid trails. If operators exercise reasonable care during skidding, they should contain most of the basal damage to these same trees during each entry to a stand. The majority of trees growing within the intertrail space remain free of skidding injuries, and will yield high quality logs when they reach sawtimber sizes (Davis and Nyland 1991).

The extent and importance of damage to even-aged stands during intermediate treatments depend upon the method of thinning used, and the intensity of a treatment. With most thinnings, felling and skidding break off or bend over mostly trees of small sizes, and may destroy one-third to one-half of the sapling class (Lamson et al. 1984; Miller et al. 1984; Nyland 1989). Actually, these trees have no value in stands treated by crown thinning, free thinning, and thinning from below. In these cases, foresters reserve the largest stems as crop trees, and destroying the small ones of inferior crown position during logging may actually save a landowner from removing them at the end of a rotation. Loss of the overtopped and smaller intermediate trees may lower the target relative density by up to 10%, but will not appreciably reduce the residual basal area or volume (Lamson et al. 1984; Lamson and Miller 1983; Miller et al. 1984; Nyland 1986). By contrast to other methods, selection thinning cuts the biggest and largest-crown trees. This causes felling damage like that to the smallest size classes in partially cut uneven-aged stands (Nyland 1986), breaking off and bending over high portions of the overtopped and lower intermediate trees, and heavily damaging many of the better intermediates and lower codominants. This may leave too few uninjured trees to comprise a satisfactory residual stand.

With all the thinning methods, injuries from skidding tend to vary greatly from one site to another (Lamson et al. 1984; Miller et al. 1984). The extent of injuries depends more on the numbers of trees felled and skidded than the residual relative density. Skidding may injure trees of any size and inflict both large and small wounds. In fact, it may leave 15%–20% of the residual trees with basal injuries (Nyland 1986), and cause both new and repeated injuries with each entry. Whether a high proportion of the crop trees escape serious injury depends upon the carefulness of each logging. Also, while only the larger injuries generally lead to substantial decay and discoloration, even small wounds to the lower bole degrade the lumber in a tree and reduce its value. Actual patterns vary from one stand to another, depending upon worker habits and the logging practices used.

MEASURES FOR CONTAINING LOGGING DAMAGE

Past experience indicates that the amount of logging damage during partial cutting and thinning varies greatly between jobs, suggesting that work practices substantially influence the patterns and extent of injuries. Logging crew training, use of special work practices, proper supervision, and the choice of equipment appropriate to the job help to reduce logging damage. Also, limiting logging to times when soils have drained to field capacity or have frozen sufficiently to support the machinery helps to prevent deep rutting and root cutting. Good skid trail layout and design minimize the numbers of trails in a stand, and keep the machines away from poorly drained areas where deep rutting would most likely occur (Nyland 1989, 1990a, 1991; Davis and Nyland 1991). Even so, these practices will not prevent damage altogether in stands with moderate to high levels of residual density, and a uniform spatial distribution of residual trees (Nyland et al. 1976; Nyland 1989). At best, foresters contain the numbers and patterns of injuries at an acceptable level, and reduce the financial impacts from damaging trees intended for high-quality timber and veneer products.

Most important, foresters who want to contain logging damage during partial cutting must work closely with the logging contractor and work crews to explain the need and identify ways to prevent injuries. These might include measures (after Nyland 1989, 1991; Davis and Nyland 1991) like

1 scheduling logging for times when the bark will not peel off easily, nor when the branches most readily break from residual tress;

2 limiting skidding and forwarding to times when the soil will support the machinery;

3 using the smallest size equipment possible without making the operation unnecessarily inefficient;

4 developing a well-planned access system of straight or gently curving trails and corridors that accommodate the equipment, minimize turning and surface disturbance, and avoid problem areas with wet soil or difficult topography;

5 laying out the skid trails in advance of marking, and leaving space for appropriately placed corridors of adequate width when selecting the residual trees;

FIGURE 21-5
The wood present in trees injured by natural causes or by people will discolor and can decay, but that laid down onto the trunk after the injury will remain free of these effects. (courtesy of U.S. Forest Service)

6 promoting careful work habits among tree fellers and machine operators, and providing incentives for good work (or penalties for poor performance);

7 using directional felling to safeguard trees of high value, and to align the logs for efficient skidding with minimum turning;

8 cutting out forks and removing large branches from felled trees so the logs track well and fit within the skidding corridors; and

9 limiting skidder loads to the effective capacity of the machine under conditions of the site.

The tendency for logging damage to vary in extent and severity from stand to stand underscores the importance of using an appropriate mix of machines and good logging practices to protect the residual trees. Above all, it suggests that practices and attitudes of the logging crew importantly influence the patterns of damage during partial cutting in either even- or uneven-aged stands.

22

IMPROVEMENT, SALVAGE, AND SANITATION CUTTINGS

SILVICULTURAL RESPONSES TO EFFECTS OF INJURIOUS AGENTS

Most environmental factors contributing to the buildup of insects and diseases across a region, or even maintaining them at endemic levels within a particular stand, generally remain outside reasonable human control. In fact, populations of some insects, diseases, and animal pests rise and decline in often unpredictable cycles over time. They generally reach epidemic or outbreak stages during periods when the different life elements essential to reproduction-propagation, survival, and development also become optimal. At some later point, the balance of habitat conditions eventually changes again, making the environment less ideal. Disease, predation, and parasitism may also increase, and then once more the population will collapse to endemic levels. Such cycles typify the dynamics of several economically important native defoliating insects and bark beetles, a few destructive mammals and birds, and some common indigenous diseases. Several introduced pests (e.g., gypsy moth, white pine blister rust, and the beech bark disease) have few indigenous regulators to naturally control their proliferation. In other cases, hunting, habitat loss, or a variety of control measures may have eliminated or reduced the predators or parasites that regulate the population of a potentially destructive agent. As a consequence, it may persist at outbreak levels for long periods, or explode to high population levels more frequently than the common native insects, diseases, and other pests.

Well-conceived and properly applied silviculture and management often improve tree community resiliency and resistance, and help to reduce the risk of heavy losses against recurring problems. Fairly simple approaches provide an important degree of protection. These include the following:

• well-timed reproduction method cuttings to renew a tree community prior to the onset of physiological maturity

• well-controlled regeneration and tending operations to maintain species suited to the site under management

• well-timed tending operations to keep the growing stock vigorous

• appropriate control of composition during young ages to eliminate species most susceptible to known injurious agents

• careful control of the size, placement, orientation, and configuration of stands as a preventive measure against potential environmental threats like strong winds and other kinds of weather-related damage

• continued scrutiny over population levels of potentially harmful agents, and threatening physical environment conditions

• use of silvicultural methods fitted to the forest community type, topographic form, other ecological characteristics of the land

• timely prevention and control to contain the populations of specific injurious agents of economic importance

• prompt salvage and sanitation cuttings when problems develop

These exemplify critical silvicultural components of a well-conceived health management program aimed at limiting the risk of potentially harmful external agents (see Chapter 21). They will not necessarily protect an individual stand or even an entire ownership when some insect or disease explodes to an outbreak level across a fairly large region. Nor will routine silviculture eliminate the diverse community of potentially harmful organisms that stay at endemic levels over long periods, often causing a base level of economically important damage to some individual trees in even well-managed forest stands.

While silviculture cannot prevent all harm to individual trees, foresters do routinely remove the damaged, diseased, and weak trees as a recurring part of thinning and tending operations. Besides upgrading the condition and quality of the growing stock, these measures may also

1 remove infected trees and host plants;

2 take out trees and other vegetation that would serve as essential habitat to a variety of harmful pests and diseases; or

3 manipulate stand density, character, and composition to maintain tree vigor and make the physical environment less optimal for a destructive agent.

These measures help to safeguard the growing stock and make it more resilient to recovery. To the degree that the treatments remove poor trees and favor good ones as well, they also help to make a stand more valuable to a landowner.

THE ROLE OF IMPROVEMENT CUTTING

Among forests not previously under silvicultural treatment, component stands often contain many damaged and defective trees intermixed with trees of acceptable condition and quality. Under these circumstances, foresters often give a high priority to improving the general condition of the growing stock as a first step in long-term management (Cheyney 1942; Hawley and Smith 1954). This usually means focusing on tree health and condition as important criteria in selecting trees to cut and leave. In many cases, damaged, diseased, and defective trees may constrict crown development of better growing stock, so at least a part of the production accumulates onto stems of little or no value to a landowner. *Improvement cuttings* fit this need (Cheyney 1942; Hawley and Smith 1954). Some of these trees contain cavities that serve as nesting sites and shelters for birds and small mammals, and landowners may elect to preserve a predetermined number to sustain the habitat for those creatures.

Common definitions of *improvement cutting* include the following elements (after Cheyney 1942; Hawley and Smith 1954; Smith 1986; Soc. Am. For. 1989; Ford-Robertson 1971):

- applied to stands or age classes past sapling stage by removing trees damaged by an injurious agent to favor better trees within the main canopy
- removes trees of less desirable species, poor form, and poor condition from the main canopy to favor better trees and improve stand quality and composition

Note that trees removed during improvement cutting are part of the main canopy. Removing them releases better trees that also have upper-canopy positions (Fig. 22-1). Further, improvement cuttings remove damaged, defective, or otherwise imperfect trees specifically to enhance the growth and development of better trees that remain. Also, the damage did not result from intertree competition. The cuttings serve as a first step for improving degraded stands and upgrading the commodity potential of a site (Cheyney 1942; Hawley and Smith 1954). Thinning and other kinds of tending operations follow in subsequent years to better manage the growth, and to bring a stand more fully to the desired future condition.

Some definitions specifically characterize improvement cuttings as treatments for uneven-aged stands (Ford-Robertson 1971). Still, the phrase *past sapling stage* implies an even-aged condition, and suggests a role for improvement cuttings in those stands as well. Also, the commonly used definition for *intermediate treatment* links improvement cutting with even-aged silviculture by calling the former a removal of trees from a stand after its formation, and generally including release cutting, thinning, improvement cutting, and possibly salvage and sanitation cutting (after Ford-Robertson 1971). Within even-aged communities, improvement cuttings represent one possible intermediate treatment after a stand passes out of the sapling stage, and before it reaches financial maturity. They serve primarily to upgrade the quality of an intermediate-aged stand, and do not necessarily regulate the relative density to optimize periodic annual increment. Among uneven-aged stands, improvement cutting may include any cutting to release good trees from oppression by damaged trees of comparable age, but without necessarily maintaining a specific stand structure or balance among the age classes.

ASSESSING OPPORTUNITIES FOR IMPROVEMENT CUTTING

Improvement cutting serves primarily as a remedial treatment to correct problems brought about by some past event, or the action of some noncatastrophic injurious

FIGURE 22-1
Improvement cutting removes trees damaged by natural agents like wind and ice, or injured during logging, freeing adjacent good quality trees in the upper canopy for better growth and development.

agent. This may include damage during past logging, or from other uses of a stand. Generally, the poor trees have some kind of readily recognizable deformity or wound caused by a specific injurious agent, or a combination of them. These may have healed over long ago, covering a considerable amount of internal defect and discoloration within the tree. In other cases, the injuries remain open as holes and other abnormalities on the main stem. If left in a stand, most damaged trees remain alive for a long time, and continue to occupy valuable growing space. Removing them does not serve any ecologic purpose. Rather, it improves future economic opportunities.

Before deciding to make an improvement cutting, foresters must do a stand evaluation, just as when planning any other silvicultural treatment. They begin with an inventory, and evaluate each sample tree for its condition, size, and species. In judging tree characteristics they look for signs of major injuries:

- broken tops
- large-diameter broken and dead branches in the upper crown
- basal holes and other openings in the main stem
- fruiting bodies and deformities caused by fungi
- injuries caused by boring insects and other animals

By keeping track of the numbers and sizes of acceptable and unacceptable trees based on criteria like these, they can later use standard analytic techniques to segregate the high-risk and low-value trees from good growing stock (Marquis et al. 1984). Then they can determine if a stand has sufficient numbers of acceptable trees to justify the investment of continued management.

The requisite amount of acceptable growing stock depends upon the objectives of an owner, and the forest conditions to fulfill those purposes. Commodity production goals might require some minimum threshold number of potentially salable trees to sustain a commercial timber harvest when the stand reaches sawtimber status. In other cases, a landowner might require full site utilization and volume production by good trees within a period of time not exceeding a normal cutting cycle. Among even-aged communities of eastern hardwoods, foresters have generally accepted a level of 45% relative density (C-level) as the minimum needed for full site utilization (B-level stocking, or about 55%–60% relative density) within 10 years (e.g., Leak et al. 1969; Roach and Gingrich 1968). Similarly, the stocking guide for ponderosa pine in the Pacific Northwest of North America (Meyer 1961) includes a *minimum stocking line.* Stands below this threshold will not produce volume at the full potential for a site (Fig. 22-2). For loblolly-shortleaf pine communities in southeastern North America, foresters consider 60% of full stocking the minimum acceptable level to justify long-term management. Stands with only 30% stocking in well-distributed trees of good condition and vigor will increase in stocking by 2.4% per year, reaching the 60% level within 12–15 years (Willett and Baker 1990). By inference, foresters could conclude that even-aged communities having at least the minimum amount of acceptable stocking to insure full site utilization by good-quality trees within 10–15 years warrant continued stand management. Those with less than this minimum level in acceptable growing stock (e.g., < 45%) would not likely serve a landowner's commodity production goals. In those cases, foresters should apply an even-aged reproduction method, rather than an improvement cutting.

Some circumstances recommend against heavy improvement cutting, or attempts to completely condition a stand in one operation. Especially for sites with a moderate to high risk of windthrow, foresters need to cut judiciously. To illustrate, management guides for ponderosa pine, spruce-fir, and subalpine mixed conifer communities in the Rocky Mountains of North America recommend against taking more than 10%–20% of the basal area from sites with a high risk of blowdown, depending upon stand type and condition. For well-stocked stands or low-risk sites, a cutting may remove as much as 40%–50%. This will also promote regeneration in

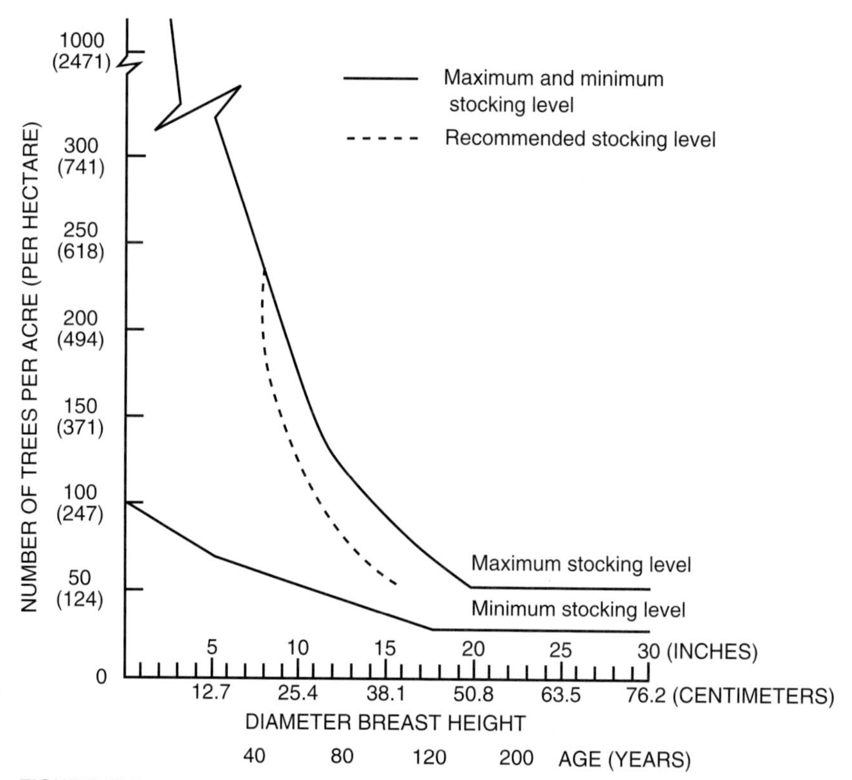

FIGURE 22-2
Foresters can use the minimum stocking line on an appropriate relative density guide like this one for site index IV or better ponderosa pine (from Meyer 1961; Sussman et al. 1977; Barrett 1979) to determine if a stand has sufficient acceptable growing stock to justify improvement cutting or other rehabilitation treatments, or if they should regenerate the community instead.

communities of seed-bearing ages (Alexander 1974, 1986c, 1987b). Even greater restrictions might apply to communities of shallow-rooted trees on poorly drained sites, those on shallow soil, or stands in exposed areas having a high risk of damage from wind (e.g., many lodgepole pine sites in the central and southern Rockies). For these cases, foresters may decide to take no action, or to regenerate the community using progressive strip or patch clearcutting (see Chapter 13). These techniques also prove feasible for other community types on wind-prone sites where management or operational constraints do not permit light partial cutting.

Research has not yet yielded explicit criteria for the minimum acceptable stocking in most uneven-aged community types, probably reflecting the more complex character of those stands. Available evidence does suggest that to apply selection system adequately foresters must maintain at least three age classes of acceptable trees (after Smith 1986), and keep some minimum level of stocking in each one. Some guidelines recommend at least 45 ft^2/ac (10.3 m^2/ha) of residual basal area to insure a minimally acceptable level of production. Also, they suggest not removing more than 40 ft^2/ac (9.2 m^2/ha) in any single cutting (Leak et al. 1969). Consistent with this, simulation experiments of uneven-aged sugar maple stands in New York indicated that for a 30-year growth period, production of total volume would drop with a residual basal area less than about 45–50 ft^2/ac (10.3–11.5 m^2/ha) (Hansen 1987). Analyses of 20-year growth measurements from uneven-aged northern hardwood stands in Ontario set the minimum level at about 60 ft^2/ac (13.8 m^2/ha) in trees at least 9 in. dbh (23 cm) (Ont. Min. Nat. Resourc. 1983). Opening the stand more might promote epicormic branching, sunscald, and dieback due to exposure.

Recommendations like these lead to a conclusion that foresters might consider improvement cutting in degraded uneven-aged hardwood stands with at least 45 ft^2/ac (10.3 m^2/ha) of acceptable trees (about 35%–40% of maximum possible basal area), and where they would not remove more than 40 ft^2/ac (9.2m^2/ha) in a single operation. With uneven-aged loblolly-shortleaf pine communities in southeastern North America, stands cut back to 20% of maximum potential basal area regrew to 60% in about 25 years, and those reduced to 30% reached that level in about 16 years (McLemore 1983). Thereafter, landowners realized good levels of production using uneven-aged silviculture. Epicormic branching would not develop in low-density stands of these species, but dense hardwood understories would form.

Available evidence suggests that uneven-aged northern hardwood stands operated at a 25-year cutting cycle for optimum sawtimber production should contain a residual basal area of 55–65 ft^2/ac (12.6–14.9 m^2/ha) for all trees, appropriately distributed among the age classes (Hansen and Nyland 1987). This would insure full site utilization and a good level of sawtimber production over the quarter-century period. In most cases, cutting to that residual stocking would allow foresters to remove large numbers of poor trees, and improvement cutting would seem appropriate if foresters can also develop a balanced structure within the stand. Otherwise, they might consider clearcutting or shelterwood method, depending upon the abundance and condition of any advance regeneration already growing at the site. Where two-aged silviculture seems appropriate they could retain widely spaced trees of large diameter and good vigor, similar to reserve shelterwood method. Both the amount of acceptable growing stock and its spatial distribution throughout a stand determine the feasibility of working with an existing community or age class, or the need to replace it through an appropriate reproduction method.

The notion of removing damaged and defective trees to favor better growing stock does not necessarily preclude management for noncommodity purposes. In fact, foresters can usually provide multiple values by making only minor adjustments to their prescriptions. For example, to maintain habitat requirements for cavity-nesting species, they can retain a minimum requisite number of healthy cavity trees, while still removing other defective trees of little wildlife value. Where a landowner attributes aesthetic value to trees of unusual form and branching patterns, foresters can favor some of these along pathways and skid trails that landowners routinely use for walking and hiking. If damaged trees of otherwise good health would serve as an important source of mast production by a poorly represented species, they can keep these individuals until appropriate silviculture regenerates adequate progeny and brings them to seed-bearing ages. In all of these and similar cases, improvement cutting removes trees of little value to a landowner, favoring better trees of good health for future development and growth. The exact nature of an improvement cutting may differ from one property to another, but primarily in the ways that foresters characterize unacceptable trees to reflect the management objectives and interests of different landowners.

INTEGRATING IMPROVEMENT INTO OTHER TREATMENTS

Foresters normally integrate at least a component of improvement cutting into other routine silviculture (Hawley and Smith 1954). This way they can address more than a single management objective with each entry to a stand, particularly within previously unmanaged forests. This links growing stock improvement to a target level of residual relative density. Row thinnings, selection thinnings, and thinnings from below do not readily incorporate an element of improvement cutting, since they use spatial juxtaposition, tree size, or crown position as primary criteria for designating the cut or leave trees. By contrast, with crown and free thinnings, foresters commonly remove defective and high-risk trees to allocate more site resources to the good trees of upper canopy positions. They also salvage useful volume in the defective trees before further losses to cull development or mortality, and while the trees still have merchantable value. If they succeed in adequately conditioning the growing stock and also maintaining a uniform distribution of residual trees, foresters can use later thinnings almost exclusively to concentrate the growth potential on the best quality and most valuable trees, and to further stimulate residual tree development.

Though applied to stands not past sapling stage, and thereby distinct from improvement cuttings, release treatments also have the common objective of improving the condition and production potential of younger stands (Hawley and Smith 1954; Smith 1986). This becomes manifest through the selection of trees to remove, and those to keep. Most obvious, liberation cutting takes away an older overstory of defective and low-value trees to release desirable saplings of a young age class (Fig. 22-3). This directly improves the growth potential of younger and better trees, and concentrates the growth potential on trees of value to a landowner. Also, cleanings often remove saplings of poor species that overtop or compete with desirable trees of comparable age. Yet more broadly, they release desirable saplings by cutting taller trees of undesirable characteristics, including those deformed and damaged by injurious agents. Both liberation cuttings and cleanings directly improve the condition of a stand, and enhance the long-term potential for producing high-quality timber. In that sense, they complement improvement cuttings within older communities.

These same ingredients characterize the tending of immature age classes in selection cutting within uneven-aged communities. There the criteria for selecting trees to keep as residuals (Arbogast 1957) include a concern for spacing in relation to trees of similar and different ages, and the potential for individual trees to yield high proportions of valuable lumber at the time of financial maturity. Foresters commonly mark diseased and defective trees, those of high risk, the poor species,

FIGURE 22-3
Liberation cutting removes defective overstory trees to free desirable saplings for better growth and development, substantially improving stand quality as well.

and trees of poor form and quality (Arbogast 1957). Extending tending operations into the sapling classes during uneven-aged silviculture parallels cleaning in even-aged stands and focuses the growth potential on the best members of each young age class. In addition, by carefully evaluating tree condition when deciding what trees to keep and remove in tending the pole and immature sawtimber classes within uneven-aged stands, foresters can usually upgrade growing stock condition and quality over the long run. In fact, trials of selection cutting among Appalachian hardwoods in eastern North America showed that by proper residual tree selection, foresters could bring a two- to three-fold increase in the proportion of volume in grade 1 and 2 sawlogs (Trimble 1970). In reality, the tending component of selection system provides foresters as many opportunities to improve stand quality as during thinnings within even-aged communities.

When taken seriously, improvement cutting offers an opportunity for foresters to do more than simply remove poor trees. By carefully planning the treatments, they can usually maintain residual stands or age classes at an appropriate density for full site utilization. Where necessary to limit the intensity of a treatment to meet other management objectives (e.g., visual quality goals), they can spread the improvement work over two or more cutting cycles, gradually upgrading stand quality through a series of consecutive treatments (after Hawley and Smith 1954). They can also integrate improvement cutting with other treatments to bring a stand into optimal growth and development without delay. By taking time to carefully assess conditions, and by using the evidence to develop a prescription for appropriate management, they can more quickly bring a stand into conformity with the interests of a landowner.

CAVITIES, CULLS, AND DEAD TREES AS HABITAT ESSENTIALS

In addressing timber production goals, silvicultural activities during intermediate stages of age class or stand development serve to reduce the numbers of defective trees, and trees in a poor state of health. Viewed from a wildlife perspective, those trees with holes in the trunks and large branches, and in declining vigor, serve as critical habitat components. The cavities provide nesting and hiding sites that offer a refuge from predators, serve to moderate and buffer fluctuations in air temperature, and protect against rain and snow. In fact, a multitude of vertebrates depend upon them for survival. Further, dead and deteriorating trees, as well as down woody debris, support many kinds of invertebrates and wood-decomposing organisms, and these plants and animals represent a critical food source to a wide variety of higher organisms (Fig. 22-4). Trees in declining vigor become the *snags* (dead, standing trees) of tomorrow, eventually replenishing the supply as older trees fall over, or dry out and become less suitable for the plant and animal organisms that live in and on them. To the degree that improvement cutting or other silvicultural treatments reduce the numbers of declining trees, snags, and trees with cavities they profoundly affect the character of wildlife using a stand and a forest (after Robinson and Bolen 1989; Hunter 1990). Preserving even two to four per acre (5–10/ha) would suffice for most wildlife populations (Hunter 1990). Releasing the cavity trees during tending operations, and protecting them from injuries, keeps them vigorous and alive. This often precludes the need to retain additional cavity trees during later management activities.

Cavity tree retention will not make management infeasible in most circumstances. In fact, landowners may lose relatively little in overall value production. To illustrate, one study in even-aged oak stand sawtimber communities in northeastern North America indicated that cavities occurred in only 4% of live trees. These represented only 8% of the total basal area, and included trees of all diameters. Few of the cull and poor quality trees had a cavity (Healy et al. 1989), so removing them during improvement cutting would have little impact on the cavity-dependent species. Further, natural compartmentalization seals off most wounds, confining the decay to a central core of the stem, and keeping it from spreading outward into wood laid down following an injury (see Shigo 1966; Shigo and vH. Larson 1969; and Chapter 21). The inner sections where rot develops normally yield only low grades of boards (e.g., 2 and 3 common lumber) having little market value (see Fig. 21-5), so losing the central portions of a few trees to decay will not really cost landowners and sawmills much money. In a strange way, those trees might prove more profitable to process, since they would yield mostly boards of better grade and higher value.

Silviculturists have several other alternatives for maintaining these important

FIGURE 22-4
Weakened and damaged trees with internal decay can become part of the habitat for many cavity dependent wildlife species, and eventually develop into snags if left until killed by contributing agent.

habitat elements. They probably contribute more to wildlife habitat by using a variety of measures, rather than concentrating on only a few. For example, maintaining even narrow bands of uncut or lightly cut forest to protect flowing and ponded water or steep slopes adjacent to timber sales provides opportunities for reserving old trees, trees with cavities, and snags. So will scenic buffers along public roads, recreation sites, and buildings. In fact, maintaining about one-quarter to one-half acre (0.1–0.2 ha) of reserve area per 10 ac (4 ha) of well-managed forest would probably serve the needs (Hunter 1990). Retaining trees with large broken branches also increases chances for decay to develop. Then cavity excavators can more readily make the holes they need for nesting sites and protection (DeGraaf and Shigo 1985). Where landowners make improvement cuttings, felling the cull trees puts a large piece of fresh woody debris on the ground, where it slowly decays. Large-di-

FIGURE 22-5
Salvage cuttings recover the timber value in trees harmed by some injurious agent like the fire that killed these lodgepole pine trees over a fairly large area.

ameter upper stems and branches of the merchantable trees also become down woody debris, further improving the habitat (Hunter 1990). Girdling the cull trees instead slowly kills them, so they become an excellent substratum for invertebrates and a host of decay organisms. On the other hand, girdled trees tend to decay from the outside inward, and central portions may not rot sufficiently for the primary excavators to create large holes before the snag topples (Bull and Partridge 1986; Hunter 1990).

SALVAGE AND SANITATION CUTTING

While many different indigenous agents continually damage a small proportion of the growing stock in most forested stands, the majority neither reach outbreak levels nor have a catastrophic effect on any community or forest as a whole. In most cases, foresters can deal with them through judicious improvement cutting, as a part of routine management, or as a separate operation. They can also prevent many kinds of damage through aggressive integrated health management built into routine silviculture. Yet despite these efforts, foresters have little control over catastrophic events like widespread epidemics of insects and diseases, destructive effects of winds from tornadoes and hurricanes, or major conflagrations during periods of extreme fire danger. When some disaster occurs, they can only enter the devastated stands to recover whatever value remains among the fallen, broken, damaged, and dead trees.

Foresters use salvage cuttings as a direct response to some catastrophic event. These disasters normally come without warning, and allow no advance planning. Further, *salvage cutting* usually demands prompt action to deal with pressing economic needs, as suggested in the standard definition (Hawley and Smith 1954; Ford-Robertson 1971; Smith 1986; Soc. Am. For. 1989): removing dead or badly damaged trees, or trees that appear likely to succumb to an injurious agent, *to recover their value.* This definition excludes effects of intertree competition that foresters control by thinning and other tending operations. Salvage cutting primarily harvests the dead and damaged trees for their commodity value. Some foresters use the term salvage cutting more generally, to mean harvesting the volume in any dead or damaged tree, including potential mortality in overmature age classes (those in senescence).

Salvage cuttings address financial rather than ecologic needs (Fig. 22-5). In fact, removing damaged timber has no ecologic value per se. The catastrophe will have abruptly altered ecologic conditions across a large area, but also sets the stage for natural succession to restore the forest in due course. In some cases, remedial measures may seem warranted to control the buildup of insects and other pests that might proliferate in the down timber, or to reduce the fire hazard. In fact, such presuppression measures often provide the only practical means to contain the hazard and prevent damage across a broader landscape. Yet where the catastrophe killed the trees and made them worthless, landowners have no reason for salvage cutting. They will simply let the trees deteriorate in place.

In some cases, ecosystem monitoring may reveal incipient stages of an insect or disease epidemic of known destructive potential. Then foresters may elect to remove susceptible trees before the injurious agent causes much actual mortality or serious damage. Technically, these treatments also qualify as salvage cuttings, although some foresters highlight their anticipatory character by calling them *presalvage cuttings* (Smith 1986). They anticipate losses from imminent damage by some kind of injurious agent.

Foresters also use *sanitation cutting* as an early response to pending pest and disease problems. Standard definitions highlight this characteristic (after Hawley and Smith 1954; Ford-Robertson 1971; Smith 1986; Soc. Am. For. 1989) by describing the purpose as follows:

- removing trees to prevent the buildup or spread of insects and diseases
- removing susceptible and host trees to promote forest hygiene and prevent the populations of harmful insects and diseases from spreading

Sanitation cuttings may remove both merchantable and worthless trees, including those already affected, and others that might serve as good habitat for the spread of a disease or an insect population. Simply cutting low vigor trees from a community does not qualify as sanitation cutting.

Salvage and sanitation cuttings do have some features in common with thinnings and other tending operations. First, they all remove potential mortality trees from forest stands, even though the reason differs from one method to another. Also, all of the cuttings have potential financial value by increasing recoverable yields. Salvage or sanitation cuttings serve primarily to reduce financial losses from a recognizable injurious agent.

REGENERATION AFTER SALVAGE AND SANITATION CUTTING

Because foresters can take time to assess conditions in stands programmed for improvement cutting, they can make deliberate plans to follow the treatment by either natural or artificial regeneration measures to replace a stand or supplement the residual growing stock if that seems appropriate and necessary. By contrast, they use salvage and sanitation cuttings as emergency measures, forcing them to put aside other work to concentrate on recovering salable wood products or protecting nearby communities. The urgency overrides any immediate concern for establishing a new age class to fill voids caused by the death and destruction. In fact, foresters may not assess the effects of a catastrophe and prescribe effective responses until long after the disaster passes. By both definition and function, salvage and sanitation cuttings do not attempt to establish regeneration or maintain any particular minimum degree of site occupancy within the damaged communities.

Any catastrophe may create some important long-term regeneration problems that will take unique solutions. In some cases, the geographic scale exaggerates the operational problems and makes even moderately intensive reforestation programs logistically or financially impractical. The catastrophe may also leave large areas without effective seed production and dispersal, or with conditions that deter germination and survival. In these circumstances, landowners must invest considerable resources to reestablish forest cover throughout an area, or accept the natural deforestation as a condition of continued ownership.

Even when financial or logistical problems do not preclude prompt reforestation, many biologic factors may prevent easy regeneration. These may reflect either direct or indirect effects of the catastrophe. For example, an injurious agent that damages and kills the growing stock often also destroys all or most of the seed source in the affected area. The disaster may also cover an area that makes natural seed dispersal from residual sources unlikely, except along margins of the destroyed stands. In some cases the damage kills young trees before seed-bearing ages, or it affects regeneration as well. Also, some diseases or insects may persist in the damaged area, or new ones may develop in the deadened material. As a result, new losses may continue for long periods, or the devastation may eventually spread to surrounding areas. Losses to persistent toxic agents and continued inundation of forested areas also may continue to kill some or all species after the initial disaster. Some destructive agents may also damage the environment, precluding success with even artificial regeneration across large areas.

Despite these difficulties, silviculturists must look ahead and plan ways to regenerate devastated areas that landowners want put back into forest cover. They may accept suboptimal species that have a natural resistance to the injurious agent, or that survive and grow in the harsh conditions caused by a disaster. They may have such a large area to cover that most techniques prove impractical, and need to use direct seeding to reforest extensive tracts of land. The dead and downed trees may heighten the fire danger, requiring aggressive prevention and presuppression measures. All together, conditions may require extraordinary measures to insure prompt forest replacement and recovery. In some cases, landowners must simply wait long periods while the forest regenerates naturally. This may make continued ownership impractical for firms that depend upon their forests for raw materials, or as a backdrop for other revenue-producing activities.

SOME ECONOMIC CONSIDERATIONS IN IMPROVEMENT AND SALVAGE CUTTING

Besides the destruction of many commodity values, and the financial implication of those losses, many disasters render an area worthless for noncommodity uses such as outdoor recreation. This reflects the loss of visual and other amenity values, and the hazards that dissuade people from entering an area. Loss of tree cover also affects hydrologic regimes by reducing interception and transpiration. If it also destroys the litter and organic layers (e.g., as during a severe fire), overland flow and surface erosion may increase. Later decomposition of the dead organic material also releases large quantities of nutrients that dissolve in the water and move off site at a higher-than-usual rate. The catastrophe may destroy the habitat for many resident terrestrial creatures, or set the stage for new plant species to regenerate across the landscape. All of this brings ecologic change that affects the natural balances and biologic diversity, as well as the opportunities for people who utilize the forest for different purposes. Over the long run, the ecosystem recovers, assuming that human activities do not inhibit natural processes. In fact, natural disasters have shaped the character of forested lands since closed stands first appeared across the landscape. They will continue to recur periodically as long as the forests last. From this perspective, most natural disasters have few long-term ecologic consequences. They mainly upset people because of the immediate economic loss, or the inconvenience caused by an abrupt change in the character of a forest. Where creative management quickly restores lost values, or redirects uses to capitalize on potentials that a catastrophe did not destroy or actually created, landowners may find new opportunities for profit and gain.

Salvage cutting has immediate economic urgency in minimizing commodity losses associated with one of these disasters (Hawley and Smith 1954). Prompt action will not prevent or reduce losses of nonmarket values, nor dramatically influence the natural revegetation of a site. On the other hand, landowners can often recover some of the value if they promptly salvage the dead and damaged trees. Yet, they face at least three major obstacles to financial success. First, the injurious agent or a secondary pest may have destroyed some of the product value, or rendered the logs useless for anything except low-value products. For example, an infestation of wood-boring insects may make even large-diameter trees unsuitable for quality lumber and veneer products, leaving a landowner little recourse except to market them for lower-priced pulpwood. If the region lacks markets for these alternative products, the trees will have no value despite their size and accessibility. Second, some catastrophes cover a fairly large geographic area (e.g., a hurricane causing extensive blowdown across a large region). As a result, many landowners may simultaneously begin salvaging fallen and damaged trees (Fig. 22-6), saturating available markets across a region (Hawley and Smith 1954). Prices may fall, making both the salvage operations and the potential for financial recovery economically marginal. Demands by landowners may also overtax the capacity of available logging crews and trucks, making salvage impossible unless they can attract additional workers and equipment from outside areas. That may increase costs of the logging, also making the operations financially marginal for many landowners. Having a good access system across the property helps to make the work most feasible in a short period, given a market. Third, some disasters also make areas inaccessible in the short run. This prevents the salvage. For example, a major storm system with heavy rains and gale-force winds might cause flooding that washes out roads and bridges, or triggers landslides that block roads or impounds fairly large areas. This might prevent access to much of the damaged timber, unless landowners invested considerable money to repair the roads and build new ones. Then costs may exceed any recoverable value and make the operation impractical.

Prompt action will not overcome all of these obstacles, but will afford at least some access to available markets and better prices. This concern usually forces foresters to immediately discontinue routine and preplanned operations in favor of the salvage cutting. It also necessitates a major new planning effort to ascertain the needs, and to determine how best to begin a program of recovery and rebuilding for the enterprise. In turn, the transformed forest may pose new challenges for future silviculture and use, based upon new ownership goals that better reflect ecologic and economic opportunities prevailing after a disaster has passed.

The need for sanitation cuttings also suggests some important economic implications for a landowner. On the surface, these may seem less pressing, in the sense that a disaster has not yet occurred. Sanitation cutting may even include investments to fell and dispose of unmerchantable trees. On the other hand, by not investing in timely sanitation cutting landowners may allow a serious insect or disease problem to develop, setting the stage for an economic loss that eventually exceeds even the compounded costs of sanitation cutting. By contrast, endemic injurious agents rarely cause or threaten mortality or damage to an entire stand, let alone large geographic areas. More commonly they affect single trees at dispersed locations. The damage may reduce the worth of some individual trees, but not necessarily

FIGURE 22-6
When natural agents destroy forests over a large area, like this tornado blowdown in Pennsylvania, landowners must move promptly to salvage the fallen timber before secondary agents cause more damage or the markets become saturated and prices fall.

kill them or cause a rapid loss of the volume. This makes improvement cutting less urgent, giving foresters considerable latitude to integrate improvement cuttings with the regular silvicultural operations across a forest or management unit.

Landowners will recognize the financial importance of many endemic injurious agents when they try to market the trees after improvement cutting or other regular treatments, and realize reduced stumpage prices for the damaged trees. In many cases, landowners can generate only enough revenues to repay the costs of improvement cutting, with little left for profit. For other stands, damages that have built up within the community over long periods of time may make many trees unmarketable, as with stands subjected to repeated high-grading in the past. Then contractors may not recover sufficient volume to sustain a commercial operation. Landowners must often implement the treatment as a noncommercial venture by poisoning or girdling the unmerchantable trees. Generally, landowners justify the investment for the value that accrues on the acceptable growing stock favored by the improvement cutting. The total area treated often depends upon the amount of low-grade volume that they can sell in available markets, the accessibility of stands needing improvement cutting, and the amount of funds available to support improvement treatments where they cannot accomplish the work by commercial sales (Hawley and Smith 1954).

OTHER ASPECTS OF IMPROVEMENT, SALVAGE, AND SANITATION CUTTING

Improvement, salvage, and sanitation cutting differ from most silvicultural treatments in at least one important regard: They generally put fewer technical demands upon the practitioner, and have less exact requirements for their implementation. Despite their apparent casualness, improvement, salvage, and sanitation cutting qualify as legitimate silviculture. They have far more value than most informal cuttings, in being

- planned in response to a definite need, and implemented in accord with a deliberate plan that calls for decisive action;
- purposeful in satisfying a specific objective, and having a definitive outcome of value to a landowner; and
- controlled when preceded by careful assessment and evaluation, and implemented to fulfill a prescribed plan of action.

In fact, even after salvage and sanitation cutting, foresters can often program follow-up measures to begin a recovery from the loss. This will include subsequent treatments that eventually create and maintain a requisite set of stand conditions to satisfy the prevailing interests of a landowner over the long run.

When faced with problems like those caused by both endemic and epidemic injurious agents, foresters must respond with decisive action. They must act promptly by deliberately planning and carefully executing these measures to gain control over the future as quickly as possible. Further, by intervening aggressively in the face of disaster, they improve chances for reorganizing their silvicultural programs to address the new needs without delay. They may even need to carefully reassess the feasibility of continuing to satisfy the landowner's goals in light of the new environmental conditions that prevail across the forest. If they can reestablish order and regain control promptly after a disaster, foresters can continue to promise the benefits that routinely accrue from a continuity of management in forested ecosystems.

23

OTHER PARTIAL CUTTINGS

DELIBERATE SILVICULTURE AND CONSERVATION

Foresters acting in their role as silviculturists create and maintain forest stands having particular combinations of species, tree sizes, age classes, and other attributes that fit the needs and interests of different landowners. They also evaluate each stand and forest in a broader spatial context to ascertain how any single activity will strengthen or alter several critical ecologic balances on a landscape scale. In some cases, landowners can best accommodate the silvical characteristics of a chosen species and provide the requisite stand and habitat values of establishing new even-aged communities across some of the area, and tending immature stands that already exist. At other times and places, an aggregation of uneven-aged communities seems more appropriate. Landowners might even intermix stands with different age arrangements to increase spatial diversity or take advantage of stand conditions that already exist.

In making these choices, foresters first determine the purposes for the management, the conditions at each site, and the biotic attributes of the stands and forest. They must also understand how these contribute to the character of a broader landscape. In addition, they take cognizance of the financial and other economic limitations and opportunities, product options, and landowner interests. Then they can work with an owner to plan a program of management consisting of biologically-ecologically viable treatments that make the forest more valuable.

In many respects, this process of devising a long-term ecologically based approach to silviculture and management represents the essence of conservation (see Chapter 2). Frequently, resource professionals convey this idea by the phrase *wise use,* with use implying a deliberate program of creating, maintaining, and actually using the resources of concern. In this sense, conservation also implies the idea of *use in moderation*, insuring adequate future access to the readily usable resources covered by the plan. In fact, forests become resources only when people can use them, or otherwise derive value in some tangible or personal way.

To help foresters make decisions, concepts of conservation must account for the renewable nature of plants and animals that make up the tree-dominated communities under their care. They recognize that biologic resources become replenished only following long periods of growth and development. At any point in time, landowners may find that the inherent productive capacity of a site or the biotic community itself limits the opportunities. Given a suitable combination of environmental conditions, the indigenous plants and animals reproduce, and their offspring grow and mature. This gives rise to a never-ending cycle of replenishment, development, and renewal that characterizes all living systems. Yet from an economic perspective, landowners want to know when they can derive some intended use, what quantity and size of plants and animals they will find, and how much and at what rate they can utilize the resources of interest.

As habitat conditions evolve over time, one mix of organisms may flourish, and others decline. If some natural event or human activity alters the environment, the community of living organisms changes in response to the new physical resources. Periodically, some of the large plants die, opening ecologic space within a community, and altering the habitat nearby. Depending upon the nature of that change, new individuals of the existing flora, or even different species, regenerate within the void. The community composition comes to reflect the physical environment at the time of regeneration, and the distribution of viable seed and spores at a site. In fact, in the absence of some irreversible change in the physical environment or the potential for reproduction, a fairly predictable community of plants and animals continues to thrive in the area. When landowners find that mixture of species and their age arrangement suitable, they need not interrupt natural processes. Where the natural communities seem undesirable or unattractive, they intervene with silviculture. The time and intensity of their action depends upon the kind and degree of change that they seek.

This potential represents an important dimension of conservation for silviculturists. They take cognizance of economic interests about when to appropriately capitalize upon the values in a stand, knowing that appropriate silviculture insures a timely renewal with new age classes to sustain future requirements as well. Through judicious tending they can alter the rates of development and shorten the time until some desired condition once more becomes a reality. This becomes possible only with appropriate silviculture of an appropriate intensity, and applied at the proper times.

MAINTAINING AN ECOLOGIC STANDARD

From this perspective, conservation becomes embodied in the silvicultural system. It addresses economic (financial and institutional) demands and opportunities, taking cognizance of the biologic-ecologic limitations and potentials of each site. It applies a minimum intensity of disturbance to maintain an ecologically and economically viable forest, and to insure its timely renewal. At appropriate times it draws economic benefit from the resources, and maintains a basic level of productivity into the future. Simply extracting available resources for some short-term economic purpose (*exploitation*) does not insure ecologic stability over long periods.

Foresters need to intensify their ecologic scrutiny when landowners demand maximum use of a particular resource within a compressed period of time. When faced with an immediate financial need, owners may have little interest in long-term management or a continuing and regular supply of resources. They may want to use maximum amounts in a short time, and at the least cost without much concern for the damage to an ecosystem. In these circumstances, foresters must propose alternatives to preclude irreversible ecological changes, and maintain the long-term integrity and viability of an ecosystem. They need to test each proposed use against some minimum ecologic standard, and accept temporary ecosystem changes over only a finite amount of space. They might hold succession off balance for a period of time or set conditions back to an earlier seral stage. This might also make the yield of desirable goods and services irregular and unpredictable in the future. However, foresters should not knowingly encourage uses that permanently deforest large areas of land, or harm the plants or animals within a landscape.

Economically and politically, these standards reflect personal views of landowners and other people who use natural resources in some way. This often leads to regulations representing a compromise between the preferences of people having different points of view. From an ecologic perspective, the essential minimum requirement should insure that

1 trees and the commonly associated plants regenerate promptly to replace any removed from a site, thereby maintaining essential ecologic balances over time;

2 the soils remain stable, thereby sustaining the potential for plant reproduction and development, and providing a suitable habitat for the many organisms that live in the soil and the biologic communities it supports; and

3 the basic land forms and drainages remain intact, insuring a continuity of the indigenous physical resources into the future.

Especially when faced with a need or opportunity that does not readily fit fairly traditional silviculture, foresters must use creativity and imagination to devise alternative solutions consistent with these standards. In doing so, they must satisfy a landowner's economic interests (financial, institutional, and social), and simultaneously insure long-term ecosystem stability, resilience, and function. Then landowners have the assurance that future generations will also find useful resources and viable forests to serve their needs and interests. Further, the management should keep open some options for landowners to change direction in the future, either at the stand or landscape levels (Grant and Brooks 1992a, 1992b).

Usually, alternatives to standard even- or uneven-aged silviculture take the form of partial cuttings that remove only some of the growing stock. These may actually resemble traditional treatments, or differ substantially. When performed as silviculture, the management serves to tend an existing age class, or regenerate replacements for some of the trees (Kostler 1956). In fact, when a cutting reduces stocking below some critical ecologic level, regeneration usually develops underneath a residual community of seed-bearing age (given no limiting herbivory, or no hindrance by interfering vegetation). This transforms a previously even-aged community into a two-aged condition, and adds a new age class to uneven-aged stands.

Generally, partial cuttings represent one of two distinctly different approaches to resource use. As a part of silviculture, they include thinning and other intermediate cuttings among even-aged stands, salvage and sanitation cutting, liberation cutting, improvement cutting, and selection system. All have a positive and constructive influence over community development and condition. Partial cuttings may also exploit available resources in a fairly uncontrolled manner, except to insure efficient logging. Only the former kinds (silviculture) offer creative approaches to dealing with special interests of a landowner.

TWO-AGED AND TWO-STORIED SILVICULTURE

Reserve shelterwood method represents one clearly recognized form of two-aged silviculture (see Chapter 14). Foresters initiate this method by a seed cutting, keeping sufficient numbers of residual trees to serve recognized ecologic and economic needs. Later they may remove all but selected individuals of the mature age class, leaving only a low density of reserve trees at a wide and uniform spacing (Fig. 23-1). That insures adequate sunlight and other site resources to sustain a rapid development of the new age class that has formed as a consequence of the seed cutting (Fig. 23-2). Generally, for two-aged silviculture like this, each age class comprises at least 20% of the basal area for most of a rotation. The older age may have as few as 20 trees/ac (50/ha), but with generally large diameters (Beck 1987; Smith et al. 1989; Smith 1995).

Foresters must use especial care in selecting trees to retain under a reserve shelterwood or for other two-aged silviculture systems. Primary characteristics (after Smith 1995) include the following:

1 expected to live for at least 50–80 years following the seed cutting;
2 have a single main stem with no strong forking;
3 of dominant or codominant crown position;
4 having no more than 10 degrees of lean from the vertical;
5 with no more than 15% deduction for sweep, crook, and decay;
6 have no dead or dying major branches in the upper crown;
7 showing no signs of developing epicormic branches within the butt log following heavy release; and
8 of a species not prone to dieback or decline following sudden exposure due to heavy cutting.

Foresters should select the reserve trees at uniform spacing, often marking those to keep rather than the ones to cut. Crews must use care to avoid damaging these trees

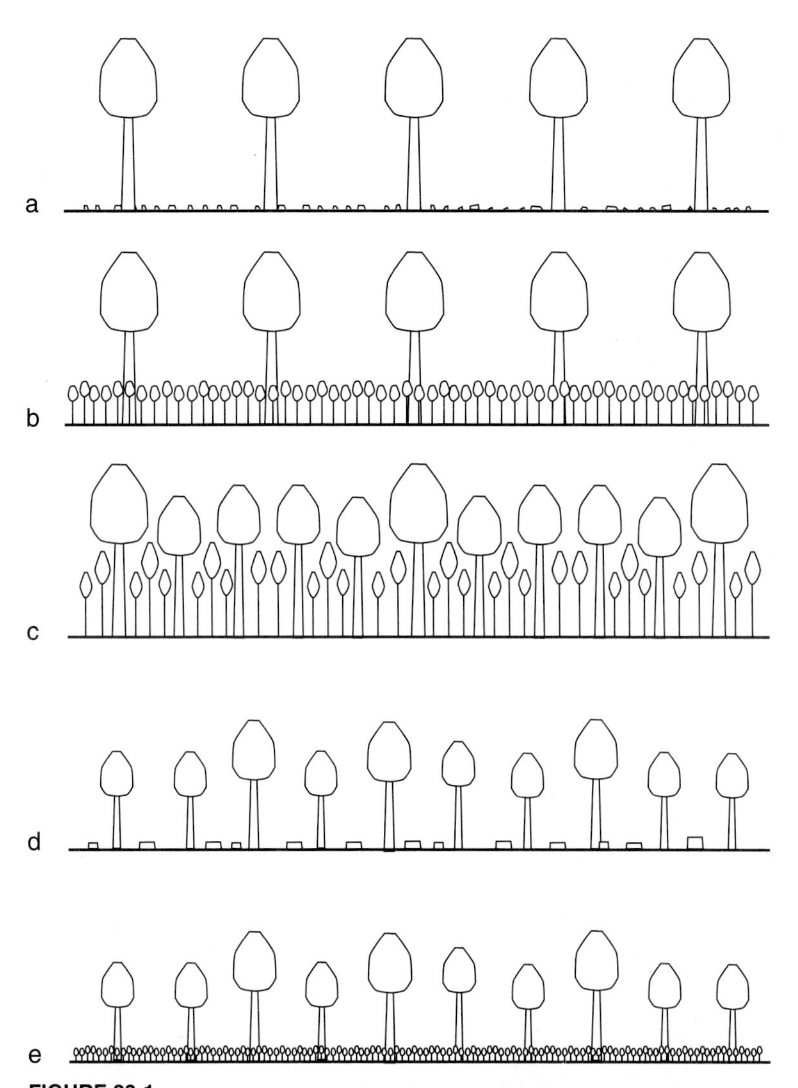

FIGURE 23-1
Reserve shelterwood method leaves a two-aged arrangement with carefully selected individuals of the mature age class left at wide spacing (a), insuring adequate sunlight and other site resources to sustain a rapid development of the new age class (b). If foresters wait until mid-rotation age to remove the reserve trees (c), they can also reduce the understory to widely spaced trees of good phenotypes (d). This will promote a new age class and initiate another two-aged community (e).

FIGURE 23-2
Two-aged silviculture leaves a low density of older reserve trees, enhancing conditions for a new age class to regenerate and develop under the widely spaced overwood (*courtesy of U.S. Forest Service, and David Wm. Smith*).

during logging, and limit the work to seasons when the trails (well drained or frozen) will support the skidding machines without rutting the soil and cutting the roots.

Reserve shelterwood stands have a two-layered structure. Large crowns of widely spaced old trees form a discontinuous overstory, casting spots of shade across a portion of the understory. The understory that develops alters the vertical structural diversity within a stand, enhancing the habitat for several animal species, and giving a stand new visual qualities (Fig. 23-1b). Retention of a biologically mature overwood of high quality and vigor also insures enduring mast production, and high levels of value growth on the large residual sawtimber trees (Beck 1987; Smith et al. 1989; Smith 1995). Gradually, the young trees develop a closed and continuous lower-canopy layer that rises progressively higher above the ground over time (Fig. 23-1c). Eventually, crowns from the new age class reach those of the reserve trees, giving the stand an appearance of one continuous leafy layer, punctuated with widely scattered bumps of foliage that extend above the general canopy level. Throughout all of this period, crowns of the younger trees completely shade the understory, giving it the appearance of a closed-canopy even-aged community. The younger age class develops like that of any other even-aged stand, except that shading by the low-density overwood may somewhat restrict the development of close-by younger trees.

Landowners may keep the reserve trees for an indeterminate length of time, depending upon several biologic and financial conditions. But with two-aged silviculture they would not do a removal cutting for at least one-half of a rotation (Fig. 23-1d). Conceptually, as long as the reserve trees serve the needs, they should remain in place. The decision often reflects financial factors (financial maturity), the visual qualities realized from a two-layered community, or the ecologic benefits from retaining a component of old trees until the new age class grows to some minimum stature. Ultimately, reserve tree health and longevity determine when a two-aged arrangement no longer seems logical. Landowners can recover or harvest the reserve trees before they die or deteriorate, or let all or some of them succumb and become large snags.

Stratified mixed-species communities represent another kind of two-storied stand, but where the trees all have the same or similar ages (Smith 1986; Oliver and Larson 1990). In many respects, their management depends upon having desirable species in both strata, and verifying that height differences reflect unequal growth rates due to species characteristics rather than site conditions. Given suitable conditions, foresters can thin in both strata, as if each comprised a separate stand in itself (Smith 1986). They regulate density and spacing within each layer, but probably maintain the overstory species at a fairly wide spacing to insure sufficient understory brightness for reasonable growth rates of the shorter trees. When the overstory reaches some specified stage of development, they remove that stratum to release the understory species. Overstory stratum removal also creates considerable open space within the residual stand, and a new age class should became established, creating the two-aged structure.

In some cases the understory consists of many deformed, suppressed, and genetically undesirable trees of little promise. This situation requires a special tending operation to eliminate all but the acceptable trees. Otherwise, foresters would eliminate the entire stratum in conjunction with an overstory treatment (Oliver and Larson 1990). Damage to the shorter trees during overstory removal may also make their retention unrealistic. If foresters remove the understory species first or at the same time, and simultaneously reduce the overstory density to brighten the understory and promote regeneration, they can circumvent the problem. The overstory residuals then serve as a seed source, much like seed trees in a shelterwood method.

Foresters can often also create a two-aged community by removing all but really widely spaced (scattered) trees of the overstory species to serve as a seed source, and reducing the understory stratum to a low density of carefully selected trees (Smith 1986). Such a cutting would resemble a shelterwood seed cutting, but with a mixture of interspersed residuals from both height strata. This maintains a degree of vertical structure, and contributes to visual qualities and wildlife habitat needs. The treatment creates sufficient openings to brighten the understory. Reserves include mostly carefully selected trees from the lower stratum, but with a minimum number of uniformly distributed overstory trees to insure an adequate seed source for that species. Where the lower stratum lacks adequate vigor for good seed production, foresters may first thin the community as a preparatory treatment. This also gives them an opportunity to eliminate the poor, defective, and genetically undesirable trees from the lower stratum. After the understory trees become more thrifty, they would reduce stocking to the target level for a seed cutting. Finally, once a new age class forms, they remove the scattered overstory trees, and reduce the second layer to only widely spaced reserves that remain for an extended rotation over the new age class.

At least with shade-tolerant hardwoods in eastern North America, carefully selected lower strata trees left at wide spacing have grown quite well following release. Where left at only a low density, the reserve trees did not interfere with the new age class, and it develops much like any other even-aged community (Marquis 1981a, 1981b). Careful reserve tree selection becomes imperative to success, because limited intertree competition within the understory stratum may not have eliminated the genetically undesirable individuals and those with a potential for only poor growth (Oliver and Larson 1990). Also, closely spaced residual trees would cast sufficient shade for shade-intolerant or mid-tolerant species to become established. Foresters must find a balance between these requirements and the need to retain an adequate volume of reserve trees to insure a commercial operation with each future entry to the stand.

Foresters can reduce the residual density of formerly stratified mixed-species stands to insure adequate regeneration of some species, as well as suitable rates of growth within the new age class. Deliberate control of residual tree density (cutting to control spacing within the reserved species), plus the addition of site preparation measures, separates this approach from *diameter-limit cutting*. The latter removes only large trees without concern for the character of a residual stand, or without making adequate provisions to deliberately regenerate desirable species in adequate numbers. Financial and biologic factors influence how long landowners keep the older trees and maintain the two-aged arrangement. The structural attributes of these stands resemble those of a reserve shelterwood community, except for having smaller and younger trees in the upper stratum.

Foresters can also create a two-aged community by heavily thinning an even-aged stand of seed-bearing ages. In these circumstances, they reduce the relative density below full site occupancy (e.g., to 35%–45%). That would brighten the understory, setting the stage for a new age class to form underneath the scattered overstory trees. Due to the younger ages of the residual trees, epicormic branching might become a major problem within some species (e.g., the oaks, black cherry, and yellow birch). They can leave residuals of all species, thereby insuring a diverse seed source. To insure adequate understory brightness over the long run, the numbers of reserve trees would mimic those retained with reserve shelterwood method.

Foresters can lower the potential for epicormic branching by reducing the stocking no lower than 50%–60% relative density, and thinning more frequently (probably every 8–10 years). To make the treatments commercially feasible, each consecutive cutting would necessarily leave a lower level of residual stocking, reaching the desired residual tree spacing after two to three entries. Foresters would retain only healthy trees of the best crown positions (dominants and better codominants), and control logging to minimize losses within the younger age class. Carefully managed two-aged stands of this kind have a structure like those of converted stratified mixed-species communities. Landowners would keep the older age class until financial maturity. Then they would remove all of the older trees, and reduce the younger age class to a low-density community similar to the initial treatment. This sets the stage once more to regenerate a new age class, repeating the cycle for a second rotation.

As an alternative form of two-aged silviculture, foresters could keep the reserve trees in groups or clusters, rather than uniformly spaced across a site. The clusters could have a special geometric shape (e.g., circular patches), each covering only a fairly small area (Kostler 1956). Foresters could also leave straight or curved strips of reserve trees. The design might reflect ecologic factors such as serving as protective corridors for animals, covering vernal pools and the drainages of intermittent streams, or keeping high shade over areas with unique plants. The reserve clusters or strips might also enhance the visual qualities of a regeneration area, and then the design would reflect that purpose.

Where a landowner has timber production goals, foresters would regulate intertree spacing within the reserve clusters or strips to insure good crown and tree development over long periods. The general structure of these communities would depend upon the arrangement of the reserve clusters, and the residual tree density and spacing within them. Regeneration would also reflect the patterns of shading and light, the status of advance regeneration, the dispersal of seed from the residual trees, and the abundance of seed stored in the litter layer prior to cutting. In fact, the arrangement of reserve clusters and strips might serve a useful purpose in securing regeneration of selected species (see Chapter 13), or keeping scarce species as a long-term seed source for the stand.

While most two-aged silviculture would involve natural regeneration, foresters have created these arrangements by cutting to a low residual relative density, and then planting a moderately shade-tolerant species beneath the widely spaced reserves (Kostler 1956; Matthews 1989). The cutting must control overstory density to insure appropriate light, nutrients, and moisture for the new age class. Also, using a wide spacing for the planted trees lessens chances of crowding over the long term. Eventually, overstory removal would reduce the community to an even-aged stand, and provide an opportunity to thin the former understory plantation. If the younger trees have merchantable value, landowners could harvest both the over- and understory trees in a single operation, and regenerate a new even-aged stand as the replacement (Matthews 1989). Otherwise, they must control logging to insure an adequately stocked stand of regularly spaced residual trees.

Well-managed even-aged stands of many kinds (plantations or natural) may develop understories of natural regeneration, particularly with a thin-crowned overstory species (e.g., red or white pine). Regular thinning often brightens the understory sufficiently for a reasonably shade-tolerant species to become established and develop. This new age class adds considerable vertical structural diversity to such a stand, enhances the habitat for several wild creatures, and alters the visual qualities appreciably. This understory may also form a closed lower canopy that hinders new regeneration at the end of a rotation. Foresters can address this situation by appropriate site preparation in conjunction with the reproduction method cutting. Otherwise, the understory trees may limit a landowner's regeneration options. Logging might destroy much of the lower stratum during a reproduction method cutting. Yet broadleafed species sprout when broken off, and this would make the situation worse.

Many documented experiences with two-aged silviculture come from trials in Europe, India, and Pakistan (Matthews 1989). Overstory trees have included primarily shade-intolerant species, with fairly shade-tolerant ones forming the understory stratum. In some cases, foresters have produced two or three crops of fuelwood from the younger age class, and retained the scattered overstory reserves to old ages for large-diameter sawtimber. These cases often involved understory crops of sprout origin, and reserve trees regenerated from seed. They may provide a good model for community forestry programs in developing countries, including some options for shelterbelt management. Two-aged stands have also proven useful in regenerating sensitive species that develop well only under some kind of nurse crop, and primarily by artificial methods (Kostler 1956).

ADVANTAGES AND DISADVANTAGES OF TWO-AGED SILVICULTURE

Two-storied stands (including two-aged silviculture) have characteristics that foresters should consider in their planning. These include positive and negative features. The advantages (after Matthews 1989) include the following:

1 Understory trees protect the soil after foresters reduce the older reserves to a low stocking.

2 Wide spacing promotes rapid growth of the older trees, and they reach large diameters in a shortened period.

3 The overstory provides long-term protection to sensitive understory trees, particularly in areas subject to freezing temperatures during the early growing season.

4 Where a landowner intends to grow species of distinctly different growth rates, retaining older trees of the slower-growing species offers a potential to bring both crops to merchantable sizes together.

5 Cutting to a low residual relative density increases the revenues from early thinnings, but still retains high-quality trees for long-term value growth.

6 Two-aged silviculture facilitates a long-term conversion to shade-tolerant species.

7 Recently cut two-aged stands have better visual qualities than clearcuts, and long-term appearances of two-layered stands may surpass those of even-aged communities.

8 Retaining a component of old trees provides essential habitat for specialized plants and animals.

Two-storied or two-aged stands also have some disadvantages for landowners (after Matthews 1989):

1 Reducing older age classes to low densities and wide spacing increases the danger of blowdown at sensitive sites.

2 Eventual felling of the overstory trees can heavily damage the understory unless crews use special felling and skidding methods.

3 Two-storied stands require a combination of overstory removal and thinning with each entry.

4 Retaining widely spaced overstory trees limits the use of broadcast or area-wide site preparation involving heavy equipment, aircraft, herbicides, or fire.

5 Partial shading by tall reserve trees may deter the regeneration of many shade-intolerant species, limiting community diversity at any site.

Where the initial cutting regenerates a dense understory, the new age class will develop much like any other even-aged community. Following canopy closure, intertree competition will slow radial increment and many trees of poor crown positions will die. This may justify cleaning or a precommercial thinning to get the new age class to an operable size in time for overstory removal. Also, the potential for heavy losses during overstory removal may require a two-phased operation. The first would regulate spacing of the understory, opening corridors for felling and skidding of the larger trees. This improves chances to protect selected members of the understory for long-term growth (Matthews 1989).

TWO-AGED TREATMENTS AS A PATHWAY TO SELECTION SILVICULTURE

Though usually considered a sustainable system by itself, two-aged silviculture could allow landowners to convert even-aged communities to multi-aged arrangement. The transition requires a minimum of three to four entries at long intervals, with each treatment retaining trees of all age classes. Foresters could start after a stand began producing good quantities of viable seed, by thinning to a low residual relative density. Before the upper canopy closed sufficiently to suppress the regeneration (perhaps 10–15 years), they would make another cutting, reducing the oldest age class to a residual of widely spaced trees. They would need to control logging damage by establishing a fairly elaborate skid trail network, and mandating special felling and skidding practices like those used with selection system. Even so, logging would break down much of the youngest age class, but open sufficient space to sustain a third age class that regenerates following the cutting. Before the third entry, foresters would wait until the second age class would sustain a commercial thinning (perhaps another 30–40 years). Then they would salvage any of the oldest trees that appeared unlikely to survive for another two decades, and thin the second-oldest age class to a wide spacing. They might also do a precommercial treatment in the 30–40-year-old age class, leaving wide spacing between those stems. The treatment would establish a fourth age class, giving the community a truly uneven-aged character, but with the composition shifting toward the more shade-tolerant species.

Foresters have other approaches for converting even-aged communities to an uneven-aged arrangement (see Chapter 11). All take long periods and several entries to effect, and most shift the composition toward a more shade-tolerant species. Even so, where landowners want an intermixing of age classes within individual stands, they have few choices except to practice some form of partial cutting. To the degree that foresters can carefully control residual conditions and promote regeneration, they can bring about a fairly deliberate transition with a reasonably predictable outcome.

NONSYSTEMS HARVEST CUTTINGS

Many landowners forgo deliberate forestry, perhaps due to not understanding about the potentials, an unwillingness to invest in deliberate management, or the motivation of self-serving interests. Instead they commonly exploit the standing timber for short-term financial gain. They became more concerned about efficiently removing available products than growing new ones for future use, or maintaining any set of vegetation conditions essential to a particular set of nonmarket values. They usually talk primarily about *timber harvesting* or making *harvest cuts*. In this context, the terms generally mean removing financially or physically mature trees or using other forest products (Ford-Robertson 1971). By contrast, *management* and *silviculture* involve a determined effort to regenerate mature trees or tend immature ones, and to provide for the future by using harvesting to recover products that become available as a byproduct of the systematic management.

For the most part, nonsystems timber harvesting involves partial cutting of most of the merchantable trees. These harvest cuts often leave the unmerchantable trees standing, regardless of their age or condition. Technically, they fall within the context of *selective cutting* (after Ford-Robertson 1971):

- creaming, culling, high-grading
- removing only certain species and large trees of high value, while ignoring requirements for regeneration, production, or sustained yield

Foresters call any area treated this way a *culled forest or stand.* It usually has irregular dispersion and stocking of residual trees, and little residual value (Heiberg 1942; Hutnik 1958; Jacobs 1966; Trimble 1971b). Where the exploitation leaves a low-density community of unusable or unsalable trees, foresters call the operation a *commercial clearcutting* (Ford-Robertson 1971). This term has a negative connotation.

In a general sense, any cutting that simply extracts timber independent of deliberate provisions to regenerate a replacement age class or tend the residual immature trees qualifies as a selective cutting. This includes the following:

- *high-grading,* removing choice species or trees larger than some specified diameter limit if they fit common utilization standards
- *diameter-limit cuttings,* taking out all trees larger than a specified diameter
- *maturity selection removals,* salvaging individual old trees or groups of them at their senescence
- *salvage cuttings,* removing dead, damaged, and dying trees to recover their worth
- *improvement cutting,* removing poor trees to favor others of better conditions and quality

The order of this list suggests a gradation from poor to more acceptable. In fact, foresters often use the latter three as interim or emergency measures within the framework of a deliberate forest management program.

The exact nature of selective cuttings varies from stand to stand, and between regions. Most have some negative long-term economic impact. At best, they serve as only provisional or stopgap measures pending later silvicultural treatments. Primary concerns (after Troup 1921; Zillgitt 1947) include the following:

1 They do not move communities or forests toward the kind of controlled age or size class distribution that insures long-term sustained yields at predictable levels or intervals.

2 They do not insure adequate regeneration, either in numbers, species composition, or by arrangement across a site.

3 They ignore silvical requirements of component and desired species with respect to regeneration and long-term growth and development, and the importance of maintaining any specific ecologic conditions.

4 They remove many sound and vigorous trees with an excellent potential for good volume and value growth.

5 They leave many defective and unhealthy trees, thereby degrading the health, biologic diversity, and economic potential of a stand.

6 When applied to many stands, they degrade the forest as a whole.

Even so, as long as the logging does not upset soil stability and preclude tree regeneration, a replacement stand of some kind will gradually become established. Eventually, a forest will develop again, and whoever owns the land will once more find it suitable for many uses.

DIAMETER-LIMIT CUTTING IN EVEN- AND UNEVEN-AGED COMMUNITIES

When applied to even-aged communities, selective cuts often take the form of diameter-limit removals that extract the best trees of sawlog sizes, and leave the smaller and slower-growing trees. This maximizes short-term profits compared to treatments that also fell other trees for cultural purposes (Filip 1967). Repeated diameter-limit cutting in mixed-species stands would not control species composition and quality compared with even- or uneven-aged silviculture. Also, cutting only the largest trees does not improve the quality or reduce competition among the smaller size-age classes within uneven-aged communities (Trimble 1971; Smith and Lamson 1977; Murphy and Shelton 1991). In fact, leaving the poor and low-value trees reduces future value production, and limits stand quality over time (Miller and Smith 1991). The immediate financial advantage eventually comes back as reduced future value.

Repeated diameter-limit cuttings may also prove disgenic in even-aged communities (Wilusz and Giertych 1974; Maynard et al. 1987; Zobel and Talbert 1984; Anderson et al. 1990). They take out the faster-growing species and trees of best phenotypic characteristics and growth rates, leaving the poor and slow-growing trees as a future seed source for the stand. Among uneven-aged communities, selective cuts usually remove all the merchantable trees. This leaves an irregular age class distribution, unregulated crowding among the pole and sapling classes, and a reduced production potential (Nyland et al. 1993). Even so, selectively cut uneven-aged stands usually had appreciable numbers of large-diameter trees that deposited good quality seed for the new age class. On the other hand, if much of the seed comes from the residual trees after the exploitation, diameter-limit cutting may have negative genetic effects in uneven-aged communities as well.

Most diameter-limit cuttings in stands of seed-producing ages will initiate a new cohort. In fact, among oak-hickory stands in the Central States region of North America, 11- and 14-in. (28–36 cm) diameter-limit cuttings established abundant regeneration. Twenty years later it averaged 400–550/ac (990–1360/ha). Even so, while the original overstory consisted entirely of oaks and hickory, red maple and other commercial hardwood species dominated the new age class (Heiligmann and Ward 1993). Similarly, in the Appalachian region, three consecutive diameter-limit cuttings in second-growth communities established sufficient regeneration to convert the stands to an uneven-aged assemblage. The composition also shifted toward shade-tolerant species (Miller and Smith 1991). Such observations suggest that ineffective control of shading patterns and lack of deliberate measures to influence the species composition (e.g., regulation of the seed source, or adding no necessary site preparation) often trigger a forest type conversion, as well as degrading residual stand quality.

SELECTIVE CUTS RESEMBLING DELIBERATE SILVICULTURE

Some forms of selective cutting may resemble deliberately planned silviculture, but still degrade a stand. These treatments remove part of the growing stock, and have

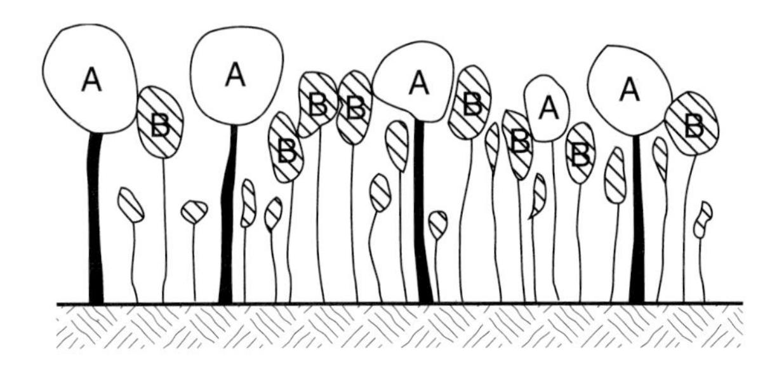

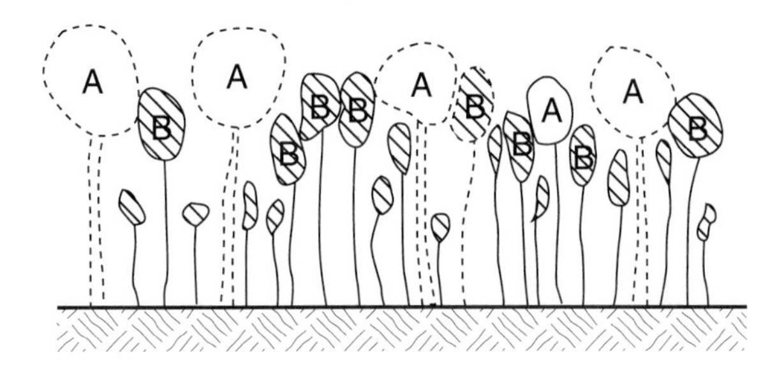

Diameter-limit cutting in this even-aged stand would remove most of species A from the community, leaving a less diverse community comprised primarily of species B.

FIGURE 23-3
Selection thinning and other diameter-limit cuttings in even-aged communities remove the faster-growing trees and more shade-intolerant species that occupy primarily upper canopy positions.

some features common to a regular tending or reproduction method cutting. These include diameter-limit cuts under the guise of selection thinnings among even-aged stands, and sawtimber-only harvests among uneven-aged communities.

All thinnings remove excess trees to reduce intertree competition and concentrate the growth on stems of long-term value. With most methods, foresters select dominants and upper codominants as crop trees. This increases the diameter for the tree of mean basal area (e.g., with crown, free, and thinning from below), or does not reduce it (e.g., with row thinning). By contrast, selection thinning removes trees larger than some specified diameter (Fig. 23-3), making it a type of diameter-limit cutting. In some cases, foresters use a flexible diameter limit to maintain a uniformity of spacing, yet even then a selection thinning removes the best-growing trees. It leaves those of lesser genetic potential as the seed source for future communities. It lengthens the time for volume to regrow to precut levels, and slows the recovery of other important stand characteristics as well. In fact, growth of understory oaks released by overstory removal within even-aged communities on the Cumberland Plateau of central North America depended upon size and condition of their crowns prior to release. Trees increased in height by less than 6 in. (15 cm) per year, and some became shorter as the result of crown dieback after exposure. Epicormic branching increased, degrading the quality of the main stem. After 15 years, only 16% had any promise of developing into quality sawtimber, and most had little potential as future crop trees (Clatterbuck 1993).

To unschooled persons, carefully controlled selection thinnings may look normal. They may have a fairly uniform canopy cover, though a strict diameter-limit control makes this unlikely. The trees usually have smaller than normal live crown ratios (e.g., 20%–30%, rather than 35%–40%), and look poorly developed for their age. They have smaller maximum diameters than expected for even an unmanaged stand of comparable age. The early thinnings remove more cubic volume than with crown or free thinning, yet long-term sawtimber yields may run as much as 30% lower. Exploited stands may also provide only about 10% of the total yield from trees at least 16 in. (41 cm) dbh, compared with 86% under a regime employing crown thinning (Nyland et al. 1993).

Diameter-limit and sawtimber-only cuts among uneven-aged stands may have less dramatic effects in the short run. They distort or destroy the balance of age classes. Further, if repeated at short intervals, diameter-limit cutting leaves a stand structure much like that of even-aged communities (Fig. 23-4). In loblolly-shortleaf pine communities of southeastern North America, mortality has increased and production dropped following heavy diameter-limit cutting. Also, four carefully applied 12-in. (30-cm) diameter-limit cuttings at 5-year intervals did establish adequate regeneration to maintain good stocking of saplings and poles. Upgrowth of residual trees allowed periodic reentry, but with variable yields from one cutting cycle to the next (Murphy and Shelton 1991). Removing all but a low density of residual understory trees from irregular uneven-aged mixed hardwood communities in the southern Appalachians of North America also caused considerable mortality. The residual trees did not increase appreciably in height over 5 years, but radial increment improved due to extensive epicormic branching along the bole (Young et al. 1993). In both cases, findings emphasize the importance of selecting good-quality and thrifty stems to leave following heavy overstory removal, rather than just accepting whatever diameter-limit cutting leaves by chance.

Comparisons of fairly conservative (17-in. or 43-cm) diameter-limit and improvement cutting resembling selection cutting in second-growth Appalachian hardwoods suggest that removing only the largest trees for a limited number of cutting cycles may not necessarily preclude future options. These stands started as predominantly 45-year-old communities with inclusions of older trees left during past selective cutting. The treatments sought to convert them to an uneven-aged condition for long-term sawtimber production of quality timber. Both treatments initially removed the older age class, but the selection-like improvement cut also took undesirable species and poor trees from all sawtimber classes to reduce crowding and upgrade stand quality. Both cuts reduced stocking enough for regeneration to become established, and residual trees grew better. After three entries, the diameter-limit area had 86% as much residual volume as the other treatment. Where cutting extended through the diameter classes, the eventual distribution closely matched the guide for selection system. It became distorted in the stand given the 17-in. (43-cm) diameter-limit removals. Sawtimber volume in Grade 1 and 2 trees

An uneven-aged community prior to cutting

After one diameter-limit cutting

Regrown and ready for a second cutting

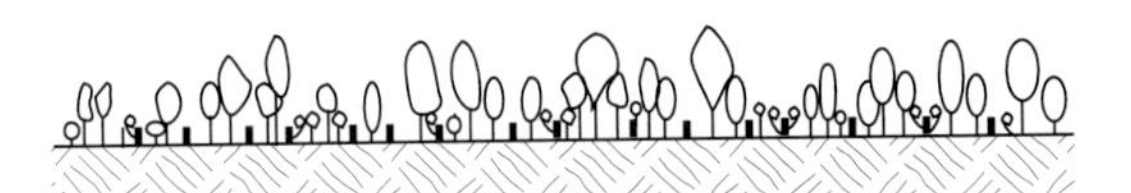

After a second diameter-limit cutting

Eventually the community develops...

...an even-aged-like character

FIGURE 23-4
Uneven-aged stands given repeated diameter-limit or sawtimber-only cuts eventually develop distorted diameter distributions, characterized by glut of trees in the mid-diameter range of size classes, and generally develop a structure more like even-aged communities.

increased by 35% as a result of cutting across the diameter classes, but by only 15%–20% following three diameter-limit cuts. Further, conditions varied from one diameter-limit removal to the next (Miller and Smith 1991; Miller 1993).

Failure to tend the immature size-age classes in uneven-aged stands subjected to cuts with a reasonably high diameter limit usually leaves the residual trees crowded, and at irregular spacing. Older trees may overtop the younger age classes, and interfere with their crown development. This limits the diameter growth of all but the tallest trees, and usually leads to accelerated mortality of the shortest trees. Crowding also darkens the understory, preventing regeneration in all but the most open places (Roach 1974a). Eventually, a dense midheight canopy layer forms, creating a mid-diameter bulge in the diameter distribution, and a decrease of saplings (Fig. 12-3). Stands treated in this fashion eventually develop a naked understory like that underneath middle-aged even-aged communities.

Cutting throughout the sawtimber classes differs somewhat. These cuttings remove all trees larger than some specified diameter and thin the smaller sawtimber classes. Usually, a forester or technician marks the stand to reduce stocking to some target residual basal area. They set the diameter limit based upon value growth factors, and essentially apply an improvement cut to the smaller sawtimber classes. They do no cutting among poles and saplings, and assume that intertree competition will naturally regulate the numbers of survivor trees within these younger age classes.

Regeneration of shade-tolerant species usually develops after a sawtimber-only cut, and the residual sawtimber trees continue to grow at good rates. Smaller trees remain crowded, and their growth rates may decline. This disparity of radial increment reduces upgrowth from the pole classes, and gaps develop in the diameter distribution of sawtimber. Excesses often develop among the poles. This gradually darkens the understory, and dampens regeneration after a second and third cut (Roach 1974). Eventually the mid-diameter hump moves up through the diameter distribution, and continued sawtimber-only removals narrow the spread of diameters in a stand (Fig. 23-4). After three or four of these cuts, conditions resemble an even-aged community (Nyland et al. 1993).

COMPARISON OF STANDS UNDER SILVICULTURE AND CASUAL CUTTING

Selective cutting and other exploitive treatments leave forest stands with an irregular stocking, and poor growing stock. By comparison, well-managed even-aged stands have features (Fig. 17-2) that include the following:

- fast early growth of free-to-grow trees and shade-intolerant species due to adequate sunlight around the crowns
- even levels of intertree competition across the community, and among trees of the same age
- appreciable changes in ecological conditions as a stand ages
- close crowding during early years of stand development, influencing the branching of component trees

- special thinning programs that repeatedly upgrade growing stock quality and control stocking to speed tree diameter growth
- complete removal of the stand at the end of a rotation
- new regeneration containing shade-intolerant species, given an adequate seed source

Release cuttings and thinnings regulate intertree spacing, reducing intertree competition and controlling mortality. As a consequence, managed even-aged stands develop more rapidly, as witnessed by the larger diameters and narrower spread of tree sizes compared with unmanaged stands of a comparable age.

Uneven-aged stands have distinctly different characteristics, and remain more stable over long periods. Important features of selection system stands (Fig. 11-1) include the following:

- uneven levels of competition, as young trees grow under the influence of older trees, and eventually develop into positions of dominance as they mature
- partial shading by taller neighboring trees slows the growth of young trees not released by periodic cutting
- heterogeneous levels and patterns of intertree competition within and between age classes as they mature
- ecological conditions change only nominally over long periods, at least on a standwide scale
- local crowding and tree age influence the branching of individual trees
- live crown length remains relatively high on trees of all ages, and trees grow at good rates throughout their life
- much of the ground surface and understory remains shaded, even after the periodic partial cuts
- mostly shade-tolerant trees regenerate in the stand, except where the silvicultural system makes special provisions to increase light levels at the ground
- periodic tending of immature classes upgrades the quality of the growing stock

Regular tending maintains a uniform spacing among trees within and among age classes in uneven-aged stands. This maintains good rates of diameter increment throughout the life of an individual tree. It also prevents overtopping of intermediate-aged trees by older ones, though partial shading does reduce light levels around the younger age classes. In general, selection system stands change only moderately from the beginning to the end of each cutting cycle.

Two-aged communities have a fairly normal even-aged understory, topped by widely spaced and almost independently growing older trees at wide spacing. These distinctly two-layered communities (Fig. 23-2) have features that include the following:

- fast early growth of free-to-grow understory trees, including shade-intolerant species exposed to direct sunlight
- some oppression within the shadows of overstory trees, leading to somewhat shorter understory heights
- even levels of intertree competition across the understory

• full sunlight around crowns of the widely spaced overstory trees, leading to full and dense crowns, and good rates of radial increment

• unique habitat conditions for wild animals and some unique plants

• appreciable change in ecological conditions within the younger age class as it matures

• branching among component trees influenced by close crowding during early years of age class development, and nominal intertree competition if later left as widely spaced reserve overstory trees

• thinning repeatedly upgrades the quality of the younger age class, improves the quality of the growing stock, and controls stocking to speed tree diameter growth

• complete removal of the overstory trees, with concurrent heavy thinning of the younger age class, when the older trees reach maturity

• the species composition of regeneration depends upon the density and spacing of reserve trees, and the seed source

• careful selection of the reserve trees insures a seed source of excellent phenotypic potential and controlled species composition

If foresters leave reserves at only a wide spacing, the younger age class should develop with only nominal interference. In small areas underneath individual reserve trees, shading might make conditions suboptimal for the most shade-intolerant species, potentially tempering the composition at those places. This shading might also affect the growth of some trees that receive only diffuse light during most of a day.

Characteristics vary widely among stands subjected to selective cuts. Depending upon the nature and severity of cutting, stands have uneven-aged attributes in places, and even-aged conditions elsewhere. For example, common features of selectively cut stands (Fig. 23-5) include the following:

• harvesting leaves trees of unpredictable condition and at irregular spacing and density

• irregular environmental conditions develop across the stand, reflecting the variable density from place to place

• ecological conditions vary greatly, due to the spatial unevenness of residual stocking, tree sizes, and tree conditions

• composition, density, and distribution of regeneration varies widely across the stand, reflecting the intensity of cutting at different places

• tree growth may slow or increase over time due to changes in local crowding and tree size

• lack of tending provides no deliberate improvement in stand quality and condition

Failure to control the intensity and uniformity of cutting leaves a stand of considerable spatial heterogeneity. Some areas have only a low-density residual cover, and regeneration may form there. The more crowded areas may not regenerate at all, or support only scattered seedlings of the most shade-tolerant species. Defective and

FIGURE 23-5
Selectively cut stands have an irregular distribution of residual trees, with a mixture of dense and open places, and many residual trees of poor form and character.

damaged trees accumulate over time, and may oppress better trees of smaller sizes. If applied to even-aged stands, these cuttings degrade the genetic potential and reduce species diversity. Yields and conditions become increasingly less predictable if landowners repeat the diameter-limit cuts, and the options for management become narrower.

EXPLOITATION VERSUS SILVICULTURE

In a general sense, silvicultural systems create and maintain only even- or uneven-aged stands (Mulder 1966; Roach 1974b). These develop in reasonably predictable ways, depending on characteristics of the component species and attributes of the physical environment. Through deliberately planned silvicultural treatments, landowners optimize the values and benefits they seek. A viable forest covers the site in perpetuity, though the sizes and numbers of trees in individual stands vary considerably over time with even-aged systems. Associated nontimber ecosystem values reflect conditions across a forest and on a landscape scale, and the kinds and frequency of treatments applied to them.

Even at its least intense level, silviculture satisfies the basic elements of forest conservation. By contrast, selective cutting and other exploitive uses give landowners no assurance of anything except a short-term payoff from the products extracted. This applies to noncommodity uses as well as timber. In this sense, exploitation can include many activities such as overgrazing of forested rangelands by cattle or wild animals, unregulated human use that degrades recreation sites and damages the vegetation cover, and dumping of pollutants from commercial activities or careless discharge of domestic wastes. These violate even the elementary sense of conservation, and should have no place in forest ownership and use.

24

COPPICE SILVICULTURE

ROLE IN FOREST STAND MANAGEMENT

Clearcutting, shelterwood, seed-tree, and selection methods all depend upon regeneration from seed. This includes new seedlings that germinate directly in the stand, or those grown at a nursery and transplanted into the forest. As a result, the new crop has considerable genetic diversity, reflective of the cross-pollination that occurs between parent trees. To the extent that seed trees have desirable phenotypic attributes, foresters generally expect that a new community originating after a high forest reproduction method or artificial regeneration from seed will have good characteristics as well. In fact, both base-level and special tree improvement programs enhance the genetic potential of a community through careful tree selection during routine silviculture, and in choosing the best parent trees and stands for seed collection (Maynard et al. 1987). In systems employing artificial regeneration, they limit the planting and seeding to carefully selected sources, and use only the seedlings of good quality and vigor. They also maintain the genetic quality by shunning high-grading and other forms of selective cutting.

While the majority of trees in most forest communities arise from seedlings, foresters routinely regenerate some species primarily by vegetative means using *low forest methods*, and *coppice systems* (after Ford-Robertson 1971; Soc. Am. For. 1989)—where the crops originate mainly from shoots and suckers, and are grown for relatively short rotations. In these cases, the new trees have a genetic makeup like the parent trees. In fact, for stands that developed by the vegetative proliferation of a single parent, the progeny will have similar genotypes. They also show common phenotypic characteristics from one generation to another.

Foresters have a unique terminology to describe coppice systems. Important ones include (after Ford-Robertson 1971; Soc. Am. For. 1989) the following:

Coppice
—a stand of shoot origin
—a sprout forest arising from coppice shoots, root suckers, or both
—natural regeneration of stump sprouts or root suckers

Copse
—a small woodland or stand of coppice origin, and managed regularly by coppice methods

Coupe
—a cutting or felling area
—the area cut in a single year, and forming an age class

Coppice method
—regenerating a stand or copse vegetatively by stool shoots, root suckers, or branch layering

Sprout method
—equivalent to coppice method

Coppicing
—cutting trees close to ground level to produce coppice shoots

Coppice shoot
—any shoot arising from adventitious or dormant buds on the stump of a woody plant

Sucker
—a shoot arising from a rhizome or root, also called a coppice shoot

Stool
—a living stump capable of sprouting

Stool shoot
—a sprout or coppice sprout arising from a stool, also known as a stump sprout

Layering
—rooting of an undetached branch lying on or partially buried in the soil, and capable of independent growth after separation from the parent tree

To successfully practice coppice methods, the species and individual trees must have a capacity to sprout or sucker, or to root naturally. That limits coppice systems mainly to stands of broadleaved species, and communities with trees of young to moderate ages (Hawley 1929; Cheyney 1942; Toumey and Korstian 1947; Smith 1986; Matthews 1989). Foresters can establish new stands to manage eventually under coppice systems by planting species capable of sprouting. They often use rooted or unrooted cuttings from clones of known growth and sprouting or suckering potential, and mix several clones to maintain a degree of genetic diversity for protection against insects and diseases, or to accommodate microsite variations in soil conditions across the stand area.

COPPICE METHODS BASED UPON STUMP SPROUTING

Starting a coppice system in fairly young even-aged communities insures both a high stem density and close spacing between the stools, and a potential for multiple vigorous sprouts from each stump. In establishing plantations to manage by coppice methods, landowners would normally plant open fields, reducing the costs of site preparation by using readily available agricultural equipment. Spacing depends upon the species, the desired product, and the rotation length. It may vary from as little as 2 x 3 ft (0.6 x 0.9 m) for energy plantations coppiced on 1-year cycles, to 10- or 12-ft (3.0- or 3.7-m) spacing for poles and sawlogs grown over rotations up to 30–40 years. Exact conditions vary with the species, intended product, and growing conditions.

While many factors influence the outcome of coppice systems, successful programs using plantations begin with careful site selection. Critical soil conditions include good physical properties, abundant moisture availability during the growing season, good aeration, and high fertility. Best yields occur in deep soils with a medium to coarse texture, a loose and porous structure, an organic matter content of more than 2%, suitable levels of pH, and an adequate growing season moisture supply without supplemental irrigation (Broadfoot et al. 1971; Hansen et al. 1993). On these sites, successful establishment usually requires intensive site preparation by disking or plowing, and control of herbaceous growth using herbicides or cultivation. To maximize production, the system must combine planting genetically improved stock, complete and timely weed control, fertilization, irrigation, and protection (Carter and White 1971; White and Hook 1975; Hansen et al. 1993). With short-rotation schemes that involve frequent whole-tree removal, landowners must take particular care to maintain the soil nutrients, especially nitrogen. Otherwise, repeated coppicing reduces fertility to a level that natural nutrient cycling and other ecosystem processes will not sustain, and production declines. The rate of change depends upon the initial levels of available nutrients, the site conditions, and the management strategy (Jorgensen and Wells 1986).

Most broadleaved species produce stump sprouts if cut before the trees reach old ages and large diameters. Some shoots arise from dormant buds that originate at the pith, and grow just under the bark. Sprouts of this origin may appear at the side or base of the stool. Others develop from adventitious buds formed in the cambium. These emerge from callus that forms along the cut surface around the top of the stump, just inside the bark. New shoots can initiate at any height along the stool, and from a combination of these sources (Cheyney 1942; Smith 1986; Matthews 1989). Their origin may shift primarily to adventitious buds after several coppice cuttings (DeBell and Alford 1972).

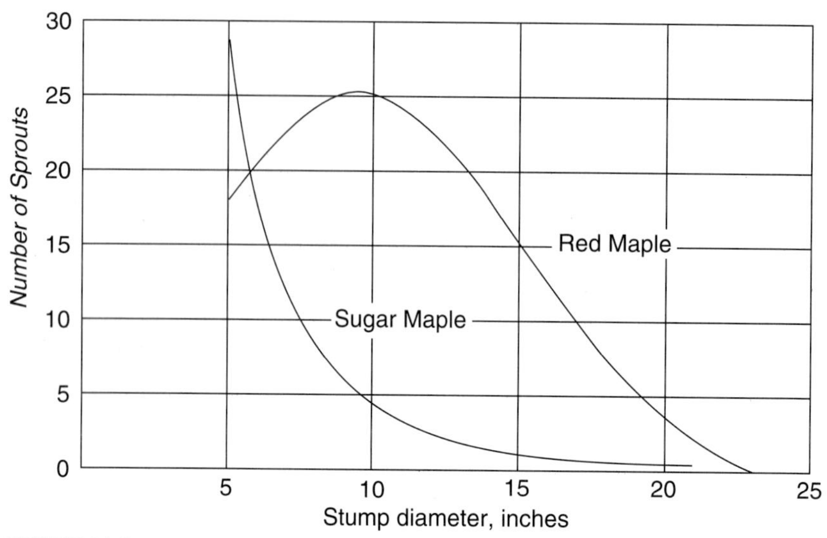

FIGURE 24-1
Effect of tree size on the sprouting capacity of red and sugar maple (from Soloman and Blum 1967).

The exact sprouting capacity differs between species and with tree size. The latter influences the maximum rotation for coppice stands originating by stool shoots. Comparing sugar and red maples (Fig. 24-1), the sprouting capacity of sugar maple peaks early and becomes rather limited by the time trees reach 12–14 in. (30.5–35.6 cm) diameter (Solomon and Blum 1967). By contrast, sprouting increases until red maple trees reach about 10 in. (25.4 cm) diameter, and remains fairly good through 20 in. (50.8 cm). Trials with sycamore indicate that the green weight of sprouts per stool increases through at least 10 in. (25.4 cm) dbh, reflecting the greater carbohydrate storage and nutrient absorption capacities of root systems on larger trees (Schmeckpeper and Belanger 1985). In most species, a reduced sprouting capacity necessitates limiting rotations to about 40 years or less (Hawley 1929; Matthews 1989). The decline apparently results either from a thickening of the bark over dormant buds at the cambium, or from breaking the connection between the dormant buds and the pith. Inhibition due to either factor increases with tree age and size (Smith 1986). Eventually, stools no longer produce sufficient shoots for adequate coppice density, and foresters must replace them with new seed-origin trees established naturally or by planting (Broun 1912).

The most desirable stool sprouts emerge from the root collar. These develop better and appear less prone to decay. To illustrate, stools of sycamore cut close to the ground (less than 6 in., or 15 cm) developed a more extensive taproot and conductive system than those with 12-in. (30.5-cm) stumps. Root collar diameters averaged 3.4 in. (8.6 cm) for close-cut stools, 2.9 in. (7.4 cm) with 6-in. (15.2 cm) stumps, and 2.5 in. (6.4 cm) for 1-ft (0.3 m) high stools (Belanger 1979). Close cutting also improves decay resistance (Cheyney 1942) in that:

1 Low-origin sprouts will more likely develop a root system independent of the old stump.

2 The subsurface portion of a parent tree will not decay as readily as the above-ground stump.

3 Low-origin sprouts have a firmer base and will not break off as readily as those attached to the top of an old stump.

Sprout-origin trees should not develop decay from the parent stump if they emerge close to the ground. By contrast, sprouts off the tops or upper parts of tall stumps have a much higher probability of containing decay from the parent tree (Roth and Hepting 1943; Roth 1956; Berry 1982; Stroempl 1984). Also, sprouts attached to a large mass of decaying wood (a large stump) will more likely develop rot (Lees 1981). Perhaps for that reason alone, sprouts from younger trees (e.g., less than 35–40 years old) do not develop decay as readily. In general, foresters can control these problems by having the logging crews cut the trees close to the ground, and by limiting the rotation length.

Stools repeatedly coppiced decrease in sprouting capacity after several generations (Berry 1982; Lees 1981; Smith 1986; Strong 1989), so that after about three rotations landowners must replace the old stools. This has proven true with red maple in northeastern North America (Lees 1981). Also in India, the yield of *Eucalyptus globulus* coppiced on a 15-year cycle declined by 9% in the third rotation, and 20% in the fourth (Rouse 1984). With *Populus* in North America, stool mortality reached 94% after five rotations of 1 year each, and four rotations of 2 years. It increased to only 30% after the third cutting on a 3-year rotation. Average shoot height and diameter from surviving stools decreased with each successive cycle (Strong 1989).

In many temperate species, the sprouting capacity also differs between the growing and dormant seasons (Cheyney 1942; Belanger 1979; McMinn 1985; Smith 1986). Generally, dormant trees sprout more vigorously than those felled during periods of active growth. This may result from hormonal fluctuations within the trees, or from the greater concentrations of carbohydrates in root systems during periods of dormancy (Kramer and Kozlowski 1960). With sycamore in southeastern North America, coppicing between January and March resulted in at least 50% more dry weight production than after May treatments, and six times more production than following July cutting (Belanger 1979). At least some sprouting will occur throughout the year in temperate regions, but shoots arising late in a growing season do not develop with as much vigor. At high latitudes they may not become dormant before early autumn freezing temperatures kill or damage the tissues (Cheyney 1942; Smith 1986; Matthews 1989).

Coppice method involves *clearfelling* all trees in the original stand by blocks, strips, or patches. This allocates total space to the new age class, and makes maximum resources available to the sprouts and suckers. Also, the new shoots live off well-established and large root systems of the parent trees, with many absorbing and actively growing tips. That increases the potential for moisture and nutrient uptake by the new sprouts and suckers, and supplies do not become limiting in soils with good water-holding capacity and fertility. As a result, new coppice grows rapidly and forms a closed canopy sooner than with even-aged stands initiated by high-forest methods.

Stool shoots originate only from existing stumps, so new trees arise only at locations of the parents. As a consequence, coppicing may produce more than a single sprout off each stump, but does not change the spatial distribution of trees in the community. Gaps may occur due to an irregular stem distribution in natural stands, and due to mortality of some stools in plantations. At least with sycamore, the frequency of cutting affects stool production. Trials have shown that annual coppicing reduces the number of sprouts, their basal diameter, total height, and green weight compared with rotations of 3–4 years (Kennedy 1975; Schmeckpeper and Belanger 1985). Additionally, as many as 15% of the stumps may not sprout due to logging damage (Zobel et al. 1987). With alder and locust, only stools with moderate to heavy damage decline in sucker productivity. Injuries come from skidders running over the stumps, dislodging the stools and separating cambial tissues from the wood (Gibson and Pope 1985). The losses increase with repeated generations of stool shoots within a single stand, so foresters must often limit coppicing based upon stump sprouts to stands of fairly high stem density. After several rotations they increase the numbers of sprouting stumps by reinforcement planting, or from other sources (Zobel et al. 1987).

While coppicing via stump sprouts normally produces a multistem clump, the weaker sprouts die as a clump develops. Numbers per clump decline fairly rapidly during early stages of stand development, leaving only one or two by large pole or sawtimber sizes (Stroempl 1984). With long rotations, this self-thinning leads to a normal-looking community of trees that poses no particular problems during eventual harvesting. In cases of short rotations, logging crews must use equipment suited to cutting multistem stumps close to the ground. Otherwise they leave large and tall stumps that give rise to poor-quality sprouts of high origin.

SHORT-ROTATION COPPICE SYSTEMS

Interest in using wood as an energy source led to several innovations of coppicing throughout the world, generally using plantations of species that reproduce vegetatively by stump sprouts. Other trials include use of woody biomass plantations as components of environmental cleanup programs and for bioremediation (Christersson et al. 1993, White et al. 1993). These usually include production of woody biomass in rotations of one to a few years, and with shrubs as well as trees. As sources of woody biomass, such plantations often prove more attractive (after White and Gambles 1988) because

1 coppicing provides repeated crops of rapidly growing woody biomass without replanting for several rotations;

2 vegetative propagation maintains the genetic integrity of a plantation;

3 high rates of growth allow production of large volumes on limited amounts of land; and

4 returns accrue on short cycles (even annually), providing a quick payoff on investments.

Normally, landowners control clone selection and spacing, capitalize on hybrid vigor in some genera, and fertilize and irrigate these plantations (Carter and White 1971; White and Hook 1975; White and Gambles 1988). Biomass plantations might have anywhere from 440/ac (1087/ha) to as many as 43,560/ac (107,593/ha), respectively, for 10- and 1-year rotations. Landowners measure the output as *biomass*, or weight, and include the total above-ground coppice growth (even bark and buds). That limits its use to applications such as feedstock for energy production.

One early scheme tested in southeastern North America became known as silage cellulose or silage sycamore (McAlpine et al. 1966; Herrick and Brown 1967; Steinbeck et al. 1968). It involved planting unrooted sycamore cuttings at closely spaced intervals (e.g., 1 x 4 to 4 x 4 ft, or 0.3 x 1.2 to 1.2 x 1.2 m). Landowners would harvest the crops at 1- to 4-year intervals using heavy-duty silage harvesters and other kinds of modified agricultural equipment. They would chop the above-ground biomass like silage, and use the material as feedstock for energy production, for different kinds of fiber-based boards and extruded products, or for a variety of other purposes. In early tests, an 11-year-old plantation set out at an initial spacing of 8 x 8 ft (2.4 x 2.4 m) yielded about 13–14 green tons/ac/yr (32.1–34.6 green tons/ha/yr) for 3-year rotations (Schmeckpeper and Belanger 1985). For such short rotations, production actually increases after the initial cycle, when new shoots arise via coppicing rather than off the planted cuttings or seedlings. For example, sycamore set out at 4 x 4 ft (1.2 x 1.2 m) yielded about 45% more biomass in the second 4-year rotation (Dutrow 1971). In *Populus*, it increased 3.4 times in the second 2-year cycle, but did not improve with rotations of 5 or more years (Strong 1989). With other species, yields increased as much as 75% to 170% after the first cutting (Schmidt and DeBell 1973; Rose et al. 1981; Vyas and Shen 1982). The estimates with sycamore range from 3.6 dry tons/ac/yr (8.9 dry tons/ha/yr) for 3-year rotations using a 2 x 5 ft (0.6 x 1.5 m) spacing (Kennedy 1975), to 4.2 dry tons/ac/yr (10.4 dry tons/ha/yr) over 5-year cycles for a 4 x 4 ft (1.2 x 1.2 m) spacing (Dutrow and Saucier 1976). Still, the added yields from closely spaced plantations might not offset the higher initial establishment costs, except with unusually high biomass prices (Dutrow 1971; Dutrow and Saucier 1976; Rose and DeBell 1978).

Other species would also fit this approach, as long as they sprouted profusely over three to four short rotations. In fact, species of *Alnus, Platanus, Populus,* and *Salix* have shown great promise for short-term fiber crops in temperate regions. Common spacings range from about 4.0 x 2.5 to 4.0 x 4.0 ft (1.2 x 0.8 to 1.2 x 1.2 m), using 4–5-year rotations (Matthews 1989). Tests with poplars in the United Kingdom using 4–5-year rotations yielded 4–5 dry tons/ac/yr (10–12 dry tons/ha/yr) (Cannell 1980), while plantings at about 3- or 6.5-ft (0.9- or 2.0-m) intervals produced 5–6 dry tons/ac/yr (12–15 dry tons/ha/yr) over 3-year cycles (Mitchell and Ford-Robertson 1992). Willows planted at about a 3-ft (0.9 m) spacing yielded about 4–5 dry tons/ac/yr (10–12 dry tons/ha/yr) on a 3-year rotation (Mitchell and Ford-Robertson 1992). Other trials with willow at different spacings showed a maximum annual production of 3.2–5.6 dry tons/ac (8–14/ha) using a 4–5-year rotation and a plant density of 8,000/ac (20,000/ha) (Willebrand et al. 1993; Willebrand and Verwijst 1993). Current commercial schemes involve double-row plantings with a spacing of 1 ft (0.3 m) within rows, 1.5 ft (0.5 m) between

FIGURE 24-2
For short-rotation bioenergy plantations of shrub willows, foresters often use the Swedish system of planting two closely-spaced rows (1.5 ft or 0.5 m apart), leaving a distance of 5 ft. (1.5 m) between double rows (courtesy of Daniel J. Robison).

single rows, and 5 ft (1.5 m) between double rows (Fig. 24-2). The double-row arrangement facilitates mechanized harvesting at 3–5-year rotations. Individual sprouts will have grown to 25–35 ft (7.6–10.7 m) tall, but seldom in excess of 3 in. (7–8 cm) diameter. Shrub-form willows generally produce the most (Fig. 24-3), if the clones match soil conditions. Shrub-type willows root readily from cuttings, making large-scale propagation easy. They sprout profusely, and the shoots grow rapidly. Rootstocks may resprout vigorously for five to six rotations before production declines. These systems require high inputs of labor and energy, fertilization, intensive weed control, and protection measures (McElroy and Dawson 1986; White and Gambles 1988; Ledin 1986; Sall 1986; Sennerby-Forsse 1986; Ledin et al. 1992; Christersson et al. 1993; Gullberg 1993; White et al. 1993; Robison et al. 1994).

Trials with willow bioenergy plantations in the United Kingdom indicated that close spacing enhances early yields, although production appears to stabilize eventually over extended periods regardless of the interval between stools. The choice depends upon the intended rotation and the size of material produced, plus compatibility with the intended harvesting equipment (related to spacing between rows). The scheme must balance establishment costs, the need for early returns, and the potential to pay off investments (Stott et al. 1980). Even so, experiments in North

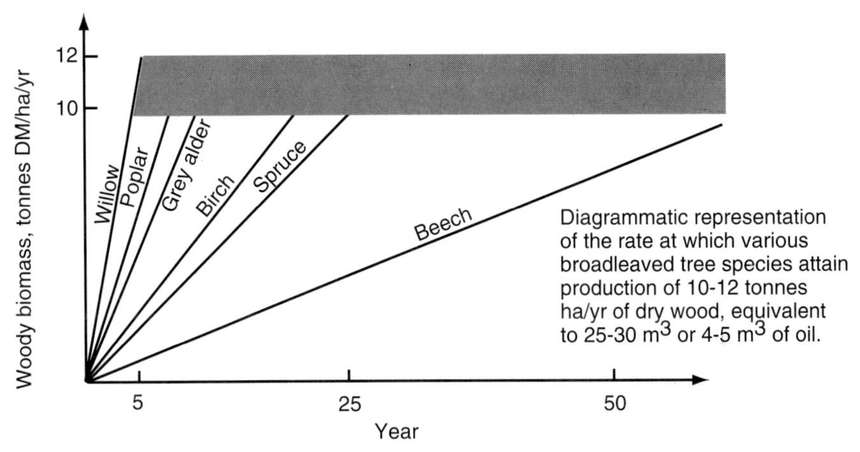

FIGURE 24-3
Comparative rates at which different broadleaved species produce 10–12 tonnes/ha of woody biomass annually (from Christersson et al. 1993).

America showed that spacings of 1.0 x 1.0 or 1.5 x 1.5 ft (0.3 x 0.3 or 0.5 x 0.5 m) provided significantly more volume over a 4-year period than plots planted at 0.5 x 0.5 ft (0.2 x 0.2 m). Survival did not differ appreciably between alternatives, but production at the closest spacing dropped below the others after the third cycle. Rates dropped substantially by the fifth cycle, either due to dry conditions during that year or some other factor that affected shoot development. All factors considered, the 1.5- x 1.5-ft (0.5- x 0.5-m) interval appeared best due to the lower establishment costs and higher levels of production in the fourth and fifth years (White et al. 1993).

Some tests indicate that willow coppice production does not reach peak until the second or third cycle after establishment (McElroy and Dawson 1986; White et al. 1993). In some cases, first-year growth measured only one-sixth to one-seventh the amount harvested in the fifth year (White et al. 1993). With hybrid poplars, benefits have accrued from delaying coppicing until the second year. In one experiment, stools of 48 different clones not cut until after the second growing season produced 230% higher yields after the third year (second annual coppicing) than those cut annually beginning with the first year (Tolsted and Hansen 1992). Apparently, in the extra year the root systems develop better and support better shoot development. At least with some species, stool mortality increases with each harvest, and production drops appreciably by the sixth cycle (Fig. 24-4). This suggests a benefit from replacement at that time (Gullberg 1993; White et al. 1993).

Managers must also include soil nutrient management to insure high levels of fertility over repeated minirotation cuttings. Several factors contribute to these changes. First, complete biomass removal exposes the soil to direct insolation for at least a short period of time annually, and reduces transpiration losses early in a growing season. The warmer and moister conditions promote accelerated decomposition, increasing the potential for nutrient leaching from the site. Harvesting would remove other amounts of the nutrients incorporated in the living tissues through photosynthesis, further reducing the total available in the soil. Much larger amounts cycle back to the soil by leaching from the leaves, and by decomposition of exfoliated plant parts. In fact, for species like *Eucalyptus* and *Leucaena leucocephala*, the annual uptake (including the amount recycled on site) exceeds that harvested by 3.3 times for nitrogen, 2.7 times for potassium, and 2.0 times for calcium. Even so, harvesting on short cycles eventually reduces available nutrients below the levels required for continued vigorous growth. Productivity will then decline, unless managers supplement the losses by fertilization or lengthen the rotation (Jorgensen and Wells 1986). Long-term experience in Sweden has led to a program of N, P, and K fertilization early in the first season, followed by a second N application in July. Landowners later apply more N annually, and add P and K following each harvest. Growing cover crops of legumes and selecting clones for nutrient use efficiency also help to maintain the fertility. Irrigation also helps to maximize production if growing season moisture deficits develop (Ledin et al. 1992). In fact, the combination of irrigation and fertilization has doubled annual yields from willow bioenergy plantations subjected to a growing-season moisture stress (White et al. 1992).

All factors considered, short- and minirotation coppice crops appear to offer considerable potential for producing high volumes of wood fiber on a relatively limited base of land. They also have promise for a component of agroforestry systems in tropical areas. In methods called *alley-cropping*, the landowners leave sufficient space between the rows of trees to grow food crops like maize or yams. If they select a leguminous species, the trees help replenish nitrogen, and landowners can produce several consecutive food crops without a decline in productivity (Longman and Jenik 1987). They can also collect the leaves for mulch, or incorporate them into the soil when planting the food crop. As with other coppicing, landowners must replace the stools after several rotations, digging the stumps and planting new trees, or interplanting between the declining stools.

COPPICE METHODS BASED UPON ROOT SUCKERS

Suckers arise from shallow roots, and generally as single stems at multiple points of origin (loci) along the root system (Fig. 24-5). Compared with stump sprouts, they fill the area more evenly, come up singly, develop independent root systems, and do not develop decay from the parent tree (Broun 1912). Suckering potential does not decline over repeated rotations. In some species (e.g. aspen and American beech) more than one sucker may initiate at a single locus. Eventually, the weakest ones die, leaving a single new tree (Schier and Campbell 1978; Jones and Rayual 1986; Jones et al. 1989). New sprouts appear at widely dispersed locations due to an interweaving of root systems. The new age class usually has a greater stem density than the parent stand, and a spatial distribution resembling well-stocked seed-origin communities.

Although suckering capacity differs between clones (Maini 1972), the potential does not decline with tree or stand age, or after repeated rotations. At least in aspen

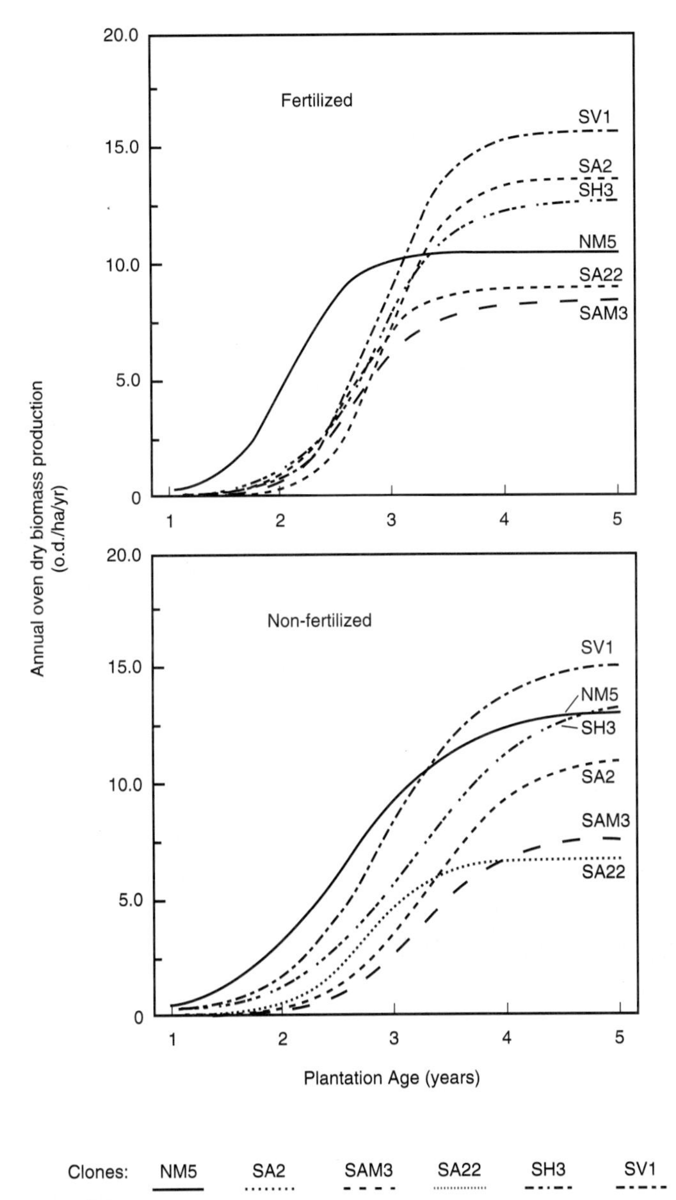

FIGURE 24-4
Yields from six willow clones declined sufficiently after 4–5 years of annual coppicing to warrant stool replacement in both fertilized and unfertilized bioenergy plantations in central New York (from Kopp et al. 1993).

(Kemperman 1978) and American beech (Jones and Rayual 1986; Jones et al. 1989), most new suckers initiate from relatively small-diameter lateral roots within 3 in. (7.62 cm) of the surface. Among aspen stands, removing thick accumulations of surface organic matter often increases the suckering capacity (Kemperman 1978; Perala 1987). Practices that sever or break apart the root systems reduce the suckering (Perala 1987), but small injuries (e.g., logging wounds) that break the bark and promote callus formation may enhance sprouting in some species (Jones and Rayual 1986). Generally practices like disking reduce sucker formation, but prescribed burning has proven advantageous at sites with a fairly thick organic layer. In fact, clearfelled and burned areas tend to produce suckers from deeper roots, perhaps due to added heat absorption by the blackened surface (Maini and Horton 1966; Zasada and Schier 1973; Schier and Campbell 1978). Among clones with a high capacity for suckering, burning may prove unnecessary, even at sites with well-developed humus and litter layers.

Stands of quaking aspen may produce four times as many suckers if cut in winter rather than in summer (Brinkman and Roe 1975). This poses an important problem for foresters who use coppicing to provide a steady year-round supply of wood to a mill. They cannot cut in only one season. At best they can identify the areas of poorest sprouting capacity (based on tree vigor, or clonal differences), and coppice those stands in the dormant season. Different clones have different capacities to sprout (Maini 1972), so stands with the highest suckering capacity have a better chance of regenerating sufficient shoots following growing season cutting.

SETTING ROTATION LENGTH IN COPPICE SYSTEMS

Because landowners use coppice systems primarily to produce fiber products, they will not thin the copse. Intermediate treatments might include a release cutting to eliminate undesirable tree or shrub species, and protection measures to insure stand health and safety. Landowners wait until the mean annual increment peaks, and then coppice the crop again by clearfelling. This applies to both sucker- and sprout-origin stands.

Figure 24-6 illustrates the appropriate rotation age for aspen on a good site in Saskatchewan, Canada. In this case the p.a.i. and m.a.i. converge at about age 35, signaling the time to reproduce a community. Repeated rotations of that length would optimize cordwood production. In some species like aspen, diseases become increasingly common as the stand matures, and this might recommend for shorter rotations in some areas (Anderson 1972; Perala 1977; Hinds and Wengert 1977). The actual length varies with soil and other site characteristics. In general, the better the site quality, the sooner m.a.i. peaks, and the shorter the rotation.

Landowners could extend a coppice rotation to produce large-diameter products, using m.a.i. for sawtimber production to set the optimum rotation. They might also thin to increase individual tree growth, and use financial maturity to time another coppicing. In cases of firewood, fence posts, and similar products foresters follow the same principle. They compile production data using whatever units of measure best describe the particular product they want to grow, and set rotation length at the time when the mean annual production peaks for unthinned stands. With total tree harvesting for fuelwood, they might use total tons or woody material per unit of area.

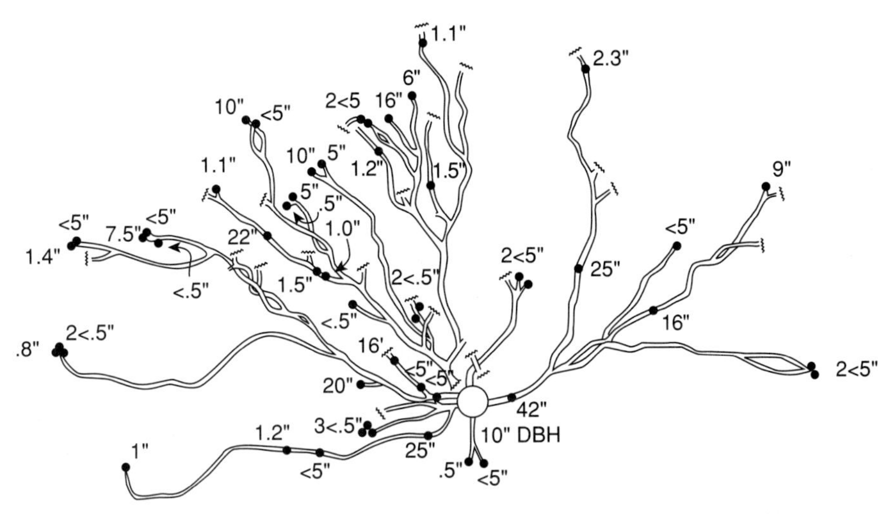

FIGURE 24-5
Root suckers originate from multiple points (loci) along the root system, giving a more uniformly distributed stand of new trees and a greater stem density (after Kormanik and Brown 1967).

ADVANTAGES AND DISADVANTAGES OF SIMPLE-COPPICE SYSTEMS

Coppice systems differ with the species, rotation length, spacing, and product goals. They may range from agriculturelike crops with short-rotation wood-grass systems, to more traditional forest stands to produce sawtimber. Nevertheless, coppicing offers some general advantages over high-forest methods (Hawley 1929; Cheyney 1942; Daniel et al. 1979; Smith 1986; Matthews 1989), including that

1 it involves relatively simple methods (clearfelling), and provides prompt and certain regeneration;

2 it effectively regenerates areas of any size and shape without concern for advance regeneration or seed source;

3 it produces abundant coppice shoots that grow rapidly in height and diameter, with high annual production per unit of area;

4 it allows landowners to produce high volumes of fiber crops, fuelwood, and forage over relatively short rotations;

5 it minimizes the health and disease problems associated with long rotations and old trees;

6 it produces stands of great uniformity and well suited to mechanized harvesting; and

7 it provides the potential for maintaining a variety of developmental stages between adjacent coupes, thereby diversifying the habitats for wildlife and herbs.

Similarly, short rotation biomass plantations have advantages (Matthews 1989) that include

1 their simplicity, and their adaptability to both large and small areas of land;

2 the short rotations, which allow considerable flexibility in adopting cultural innovations, introducing genetically improved stock, or changing the rotation to produce materials of different sizes; and

3 the quick return on the investment of establishment and management provided by short rotations.

The simplicity reduces costs of maintaining and reproducing a coppice forest, and the short rotations provide a relatively quick payoff on the investment of management.

Besides their use for growing small-diameter fiber products, coppice methods have proven valuable in some wildlife management programs. These include schemes that require frequent (short rotation) community renewal to maintain a component of dense stands with small-diameter and short trees across the landscape. For example, in North America, wildlife managers often use coppicing in aspen communities to enhance habitat conditions for species like the ruffed grouse (Gullion 1970, 1972, 1990), but these treatments have mixed effects on other species (Yahner 1990). They intersperse cutting areas across time and space to create an intermixing of stands at different stages of development. This maintains a diversity of cover and food conditions in close proximity (Hunter 1990). Landowners can also use coppice methods to maintain willow communities as filterstrips and buffers along streams, rivers, and other wet areas. They provide dense wildlife cover along or adjacent to the bank, and reduce the erosion potential from soil in proximity to the water. Some landowners use coppice methods to produce succulent shoots for cattle feed, fuel, and other subsistence products. In addition, coppicing has become an accepted method for short-rotation sawlog production in many tropical regions (Laarman and Sedjo 1992).

Low-forest methods have some distinct shortcomings as well. These, too, derive from the character of a copse and the materials produced. Important ones (Hawley 1929; Cheyney 1942; Daniel et al. 1979; Smith 1986; Matthews 1989) include the following:

1 Financial success depends upon access to markets for small-diameter pieces and wood chips rather than sawlogs.

2 Coppice systems serve a limited set of management goals, and landowners have traditionally used them with only a few species.

3 Frequent entry for harvesting requires extra caution to minimize soil disturbance on slopes, and may increase the loss of soil nutrients after repeated rotations.

4 Succulent coppice shoots suffer damage from early or late growing season freezing temperatures, and by browsing of wild animals.

5 For sprout stands, landowners must eventually replace the old stools with new rootstocks or seedling-origin trees to maintain the sprouting capacity.

6 Replacing existing coppice stands with new trees often proves difficult and costly.

7 Coppice stands have limited amenity and other nonmarket values due to the small trees, and the uniformity of size classes in a single stand.

Short-rotation coppice schemes have shortcomings (Matthews 1989; Jorgensen and Wells 1986) that include the following:

1 They require guaranteed markets to justify the initial investments, and low production costs to keep production competitive with more traditional fuels.
2 The systems require fertile soils with an abundant moisture (or irrigation), as well as fertilization to maintain critical nutrients.
3 The plantations may require protection if accessible to cattle or wild animals, and intensive programs to protect the succulent shoots from damage by insects and diseases.
4 The mechanized systems needed for efficient harvesting require fairly level sites, with uniform surfaces and highly trafficable soils.

Landowners interested in commercial-scale production must limit coppicing systems to sites they can access via good roads, and operate with specialized agricultural equipment. For small-scale production in regions with inexpensive labor, soil fertility and moisture supply determine the yields over repeated short rotations.

COPPICE-WITH-STANDARDS SYSTEMS

Some stands contain both seed- and sprout-origin trees, due to the species mixture or past cutting practices. Landowners may coppice these communities to establish a low-forest system. In other cases, they may keep some seed-origin trees at wide spacings for long periods, and produce coppice crops between them. Such a scheme may enhance recreational uses, maintain a more favorable habitat for some wild creatures, or have some other special value to a landowner.

Foresters call the system of combining short-rotation coppice crops with older-aged trees a *coppice-with-standards system* (after Ford-Robertson 1971; Soc. Am. For. 1989):

- a coppice system with selected trees left at wide spacing to grow some multiple of the coppice rotation
- a method of coppicing that leaves selected trees to grow to large sizes with a stand of coppice growth beneath them

Figure 24-7 illustrates coppice with standards, and its two-aged character. The reserved trees become established vegetatively, from seed, or by planting. *Seedling coppice* has both seedling and coppice trees, forming a *composite stand* or *composite forest*. The standards serve as parents for seedlings to renew the coppice growth as sprouting vigor declines, and to produce large-diameter sawtimber (Broun 1912).

Foresters may also use a system called *compound coppice* (Smith 1986). For this they harvest a portion of the old standards, and leave the remainder to grow longer. In addition, they select some new standards from the mature coppice trees (or those of seedling origin) to leave for continued growth as well. This creates a multiple-aged stand with (1) one age class for the new coppice regeneration; (2) one age class of standards from the coppice crop just harvested; and (3) one age class of

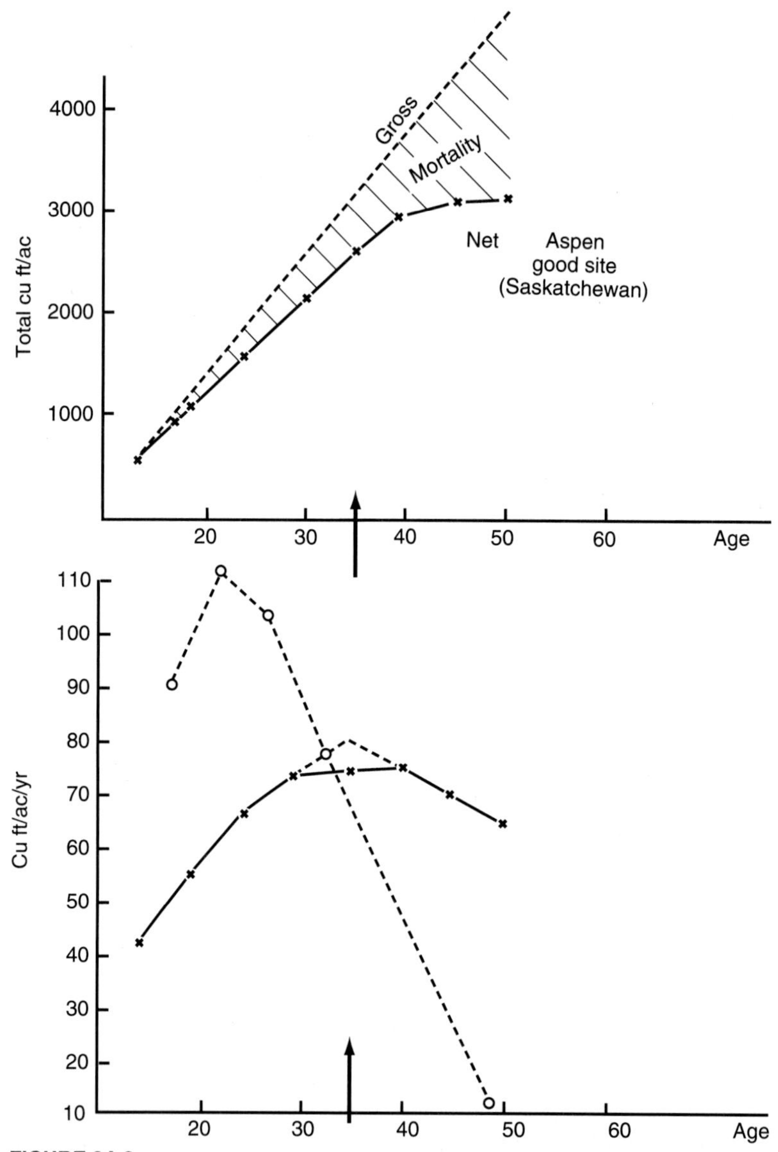

FIGURE 24-6
A typical production function for the development of even-aged aspen stands in Saskatchewan, Canada.

even older standards left as reserve trees from previous cuttings. Figure 24-7c illustrates this condition. The scheme gradually converts a coppice stand to a community suitable for sawlog production.

A coppice-with-standards system offers landowners some distinct advantages compared with simple coppice (Cheyney 1942; Smith 1986; Matthews 1989): They include the following:

1 It yields materials of several different sizes, with some large-diameter trees of high value.
2 It provides regular returns at short intervals, and insures periodic revenues from each stand.
3 It retains only a relatively low residual value per unit of area, to the benefit of landowners requiring high compound rates of return and who also wish to grow large and high value trees.
4 The standards grow rapidly, and increase in volume and value at above-average rates.
5 The standards eventually produce viable seed, allowing landowners to establish a seed-origin stand, or maintain both vegetative and seed-origin growing stock.
6 The continuous partial cover of tall trees and dense understory of coppice growth protect the soil better than simple coppice systems.
7 The dense coppice between the standards prevents site occupancy by undesirable trees and other woody growth.
8 The standards enhance the appearance of a stand, both immediately following a felling and during the interim between successive entries.
9 Landowners can maintain a more diverse array of species and age classes, and provide habitat for a broader community of wildlife.

Foresters can use coppice-with-standards systems to grow mixed-species communities, and maintain species that do not reproduce vegetatively. Most landowners use a coppice-with-standards system where markets take both small- and large-diameter products.

A coppice-with-standards system has some important limitations (Cheyney 1942; Smith 1986; Matthews 1989). These include the following:

1 Its complexity makes the system difficult to apply, particularly in maintaining an appropriate balance of growing space between the coppice growth and standards.
2 The dense growth of coppice often obscures crowns of the taller trees, making reserve tree selection difficult during marking.
3 Foresters must develop regular markets for large volumes of small-diameter pieces from the coppice growth, as well as sawlogs.
4 Exposure increases the likelihood of epicormic branching and sunscald on the standards, potentially degrading the main trunk.
5 Shading by the standards may oppress the coppice growth beneath and reduce its development and yield, particularly in stands with multiple age classes.
6 With compound coppice systems, shading may inhibit most shade-intolerant species in the coppice growth.
7 Use of large machines and high-production logging systems that offer the greatest cost effectiveness for harvesting fiber crops may injure the standards.
8 Standards freed by heavy cutting may suffer wind and snow damage at exposed sites, and blow down in areas with shallow soils.

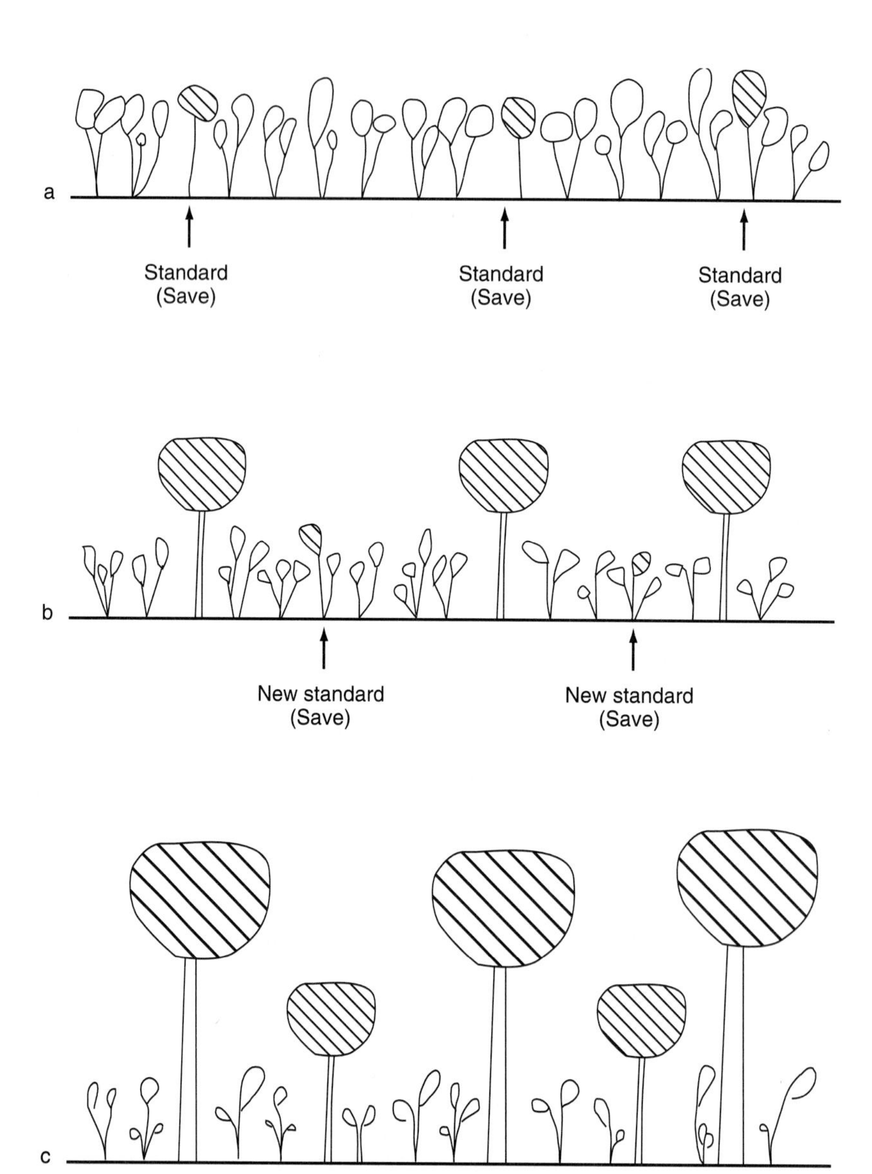

FIGURE 24-7
Coppice-with-standards system (a) leads to a two-aged stand with a scattered overstory of older trees, and a dense understory of coppice growth (b). By leaving some standards for multiple coppice rotations, the stand become multi-aged (c), known as *compound coppice.*

9 When harvested, the standards may not sprout, necessitating replacement by planting or regeneration from seed.

10 Young trees intended as future standards require early release (cleaning) to insure adequate rates of development.

11 Landowners may need to prune the standards to insure high-quality material along the lower bole.

12 Standards often develop poor form and heavy branching, making them more susceptible to ice and snow loading, breakage, and blowdown.

For stands at risk of wind damage, foresters can arrange the coupes as long and fairly narrow stands oriented at right angles to the prevailing winds. They would cut in progressive coupes (like progressive strip clearcutting) into the wind, leaving areas of tall growth on the upwind side for protection (Broun 1912). Most commonly, foresters use coppice-with-standards and compound-coppice systems to convert from coppice to high-forest stands, eventually applying a method much like selection system. Yet in the case of compound coppice, the different cohorts of standards must have a discontinuous canopy cover to insure sufficient understory brightness for the coppice growth over repeated cycles.

CONVERSION FROM COPPICE TO HIGH-FOREST SYSTEMS

Conversion from coppice to high-forest systems usually takes a long time. During the interim, landowners may realize limited production, particularly where they lack access to diverse markets for small-diameter pieces. This transition (after Hawley 1929; Cheyney 1942; Smith 1986; Matthews 1989) will include

1 holding the coppice growth to an advanced age, to weaken its sprouting capacity and reduce its crown density;

2 increasing the numbers of standards and growing them to seed-bearing ages to serve as parents for seed-origin trees, and to increasingly shade the coppice undergrowth;

3 thinning the coppice growth to maintain its vigor to an extended age; and

4 gradually removing the sprout-origin trees to make space for seedlings to develop.

For sprout-origin communities of long-lived species, foresters can simply let the sprout clumps grow until the trees begin producing seed and intertree competition reduces the stem density. They then make a regeneration cut much like that with two-aged silviculture. After the new cohort becomes established and develops for a period of time equivalent to a cutting cycle, they make another cutting to establish a third age class. Thereafter, they can apply uneven-aged silviculture. For stands of short-lived species they must gradually increase the number of long-lived standards to oppress the coppice growth. This may take a long time. Eventually, they can begin the conversion process described. For stands with multiple-sprout clumps, they might thin out all but a single stem per clump as one means to promote more rapid development of new standards.

Where a well-distributed cohort of seedlings forms following the first entry, foresters can remove the remaining coppice and the standards, leaving a young even-aged community. They could also keep selected standards to maintain a two-aged community, perhaps similar to the reserve shelterwood method. The rate at which new seedlings form beneath the coppice growth determines whether this conversion occurs relatively promptly (leading to an even-aged stand), or more gradually. In the latter case, the new community will have trees of several ages. Where landowners want a more rapid conversion, they might use site preparation and planting to establish long-lived species under the standards. This may also prove necessary in communities of short-lived, early-successional species like quaking aspen or willow. In none of these cases will foresters have an easy job converting coppice communities to stands of seedling origin, unless they employ herbicides and intensive site preparation to control sprouting and enhance seedling development (Smith 1986).

25

ADJUSTING TO ADMINISTRATIVE DEMANDS

SILVICULTURE IN PERSPECTIVE

The Society of American Foresters (Ford-Robertson 1971) describes *silviculture* as the bringing together of biologic and economic knowledge to devise practical treatments to *satisfy the objectives* of a landowner, and to make forest stands *more useful.* In managing any single stand, foresters should account for conditions and opportunities across the forest, and even on a greater landscape scale, to insure that the management provides the desired conditions (ecologically and economically) on an appropriate scale of time and space (Fig. 25-1). They also look for ways to maintain or enhance the basic productivity of a site, and prevent irreversible ecologic changes. Within this constraint, foresters let the economic and administrative requirements determine which of the biologically appropriate options to use. Further, they deal with more than just financial requirements, but respond to institutional and social interests as well. These include personal interests of a nonfinancial nature, operational requirements of an enterprise, and any other opportunity or constraint that reflects preferences of an owner about what to produce and how to do it.

A PROBLEM-SOLVING PROCEDURE FOR SILVICULTURE

The practice of silviculture involves making choices, and then implementing them through various kinds of stand treatments. In most cases, foresters can trigger more than a single kind of ecologically acceptable response in a stand. They simply select different combinations of treatments, arrange them in a different sequence, and apply them with different intensities over time and space. To make the proper choices, foresters work with a landowner or a firm's administrative representative in review-

FIGURE 25-1
By carefully scheduling silvicultural treatments to create a diversity of age classes and vegetal conditions, foresters can maintain different kinds of ecologic values at a landscape scale, and keep many options open for the future as well (courtesy of U.S. Forest Service).

ing the options and selecting the combinations to use. This involves asking and seeking answers to several key questions:

- What is the situation?
- What do I and others know about it?
- What must I still learn?
- How will I obtain this information?
- What will I hope to learn from the facts?
- What ecologically viable options can I foresee?
- What would happen if I implemented each one?
- What single option seems to best serve the interests of the landowner?

In silviculture, the *problem* is to satisfy interests of a landowner by appropriately manipulating the available tree resources to provide a continuing supply of the goods and values required. Foresters must do this in an ecologically sound manner that takes cognizance of long-term forest health and vitality.

The procedure that follows incorporates terminology and analytic techniques common in forestry, and works well in combining biophysical and economic considerations to reach silvicultural decisions. The process can help foresters to identify suitable components for a silvicultural system, or in selecting appropriate practices to apply at the moment. It begins by determining the exact objectives for the management, and ends by evaluating the degree to which the silviculture served those purposes. Steps along the way include these:

1 Determine the landowner's objectives
—determine what an owner seeks as an outcome from the management
—identify the economic constraints and financial requirements that will ultimately control the management
—recognize the time frame that the objectives impose
—ascertain what fiscal, institutional, and human resources a landowner will dedicate to the task

2 Evaluate existing stand conditions
—reconnoiter the stand to assess its suitability relative to the objectives
—inventory stand conditions, if the reconnaissance proves favorable
—analyze the data to describe physical attributes of the stand and site

3 Identify the options
—use the reconnaissance and inventory information to identify an array of potential options
—develop a prescription to describe each option
—compute the costs from implementing these treatments

4 Quantify likely outcomes of each
—describe likely post-treatment effects
—determine the probable conditions and values expected from each treatment
—project future effects and results
—ascertain if each option serves a landowner's interests, and would prove economically feasible

5 Drop unsuitable options
—compare the assessment of results with landowner objectives and constraints
—discard any ecologically viable option that clashes with the objectives or violates the financial constraints
—evaluate whether the remaining options fulfill the purposes
—seek clarification or adjustment of the objectives if all recognized options fail, and reinitiate the planning sequence

6 Describe the viable alternatives for the landowner
—review and enhance the prescriptions for each viable option
—determine if the assessment suggested any additional possibilities, and assess these as above
—assign preference rankings based upon the landowner objectives, particularly if the options differ appreciably

7 Aid the landowner in making a decision
—discuss the options
—render advice
—answer questions
—make a recommendation
—review the choice
—secure agreement to proceed

8 Implement the prescription
—review the prescription for completeness and requirements
—refine as needed to facilitate implementation
—arrange for the work
—implement the treatment, including any requisite marking
—supervise the work to insure compliance with the prescription

9 Evaluate the results
—assess post-treatment conditions by reconnaissance or reinventory
—determine the degree of success
—identify any requisite follow-up

Particularly at steps 2, 5, and 6 foresters may recognize that conditions within a stand or limitations of the site prevent a landowner from realizing the benefits at the permissible intensity of management. In other cases, financial limitations imposed by a landowner may encourage a forester to recommend postponing any stand treatment until tree sizes or stocking levels increase to make a treatment feasible by a commercial sale. In either of these situations, the forester and landowner will realign the objectives or consider an adjustment of administrative mandates that govern the management.

SILVICULTURE FROM AN ADMINISTRATIVE PERSPECTIVE

Silviculture must address administrative requirements of an ownership or enterprise, and silviculturists must mold the forest into an economically attractive resource suited to a landowner's interests. This applies to ownerships that put high demand products to recovery for commercial sale (e.g., marketable sawlogs, huntable animals, or salable water), or amenity values of different kinds (e.g., enhancing a recreational setting, providing habitat for songbirds, or managing to enhance biological diversity). In all cases, landowners must clearly describe the sought-after output of management in tangible terms, including a basis for measuring success (Fig. 25-2). They can often quantify traditional forest products fairly easily, but have considerable difficulty setting standards for measuring outputs of nonmarket values like recreational and aesthetic experiences. These may require indirect assessments like measuring the abundance of essential habitat for wildlife, the population levels of songbirds seen or heard during a census, and the numbers of trails and recreational use options afforded throughout a stand or forest. Foresters and landowners must agree on ways to account for the accomplishments from management. Other-

FIGURE 25-2
Foresters must understand the administrative requirements before recommending any specific silviculture, since treatments like this clearcutting might prove acceptable in some cases but not fit the objectives of other landowners.

wise, they cannot realistically determine how to modify the forest vegetation to make it more useful or if investments in silviculture paid the dividends required by a landowner.

Generally, these and other kinds of administrative demands fall into two categories, depending upon *when* they affect silvicultural plans and programs. They include the following:

1 demands that influence *initial* decisions when formulating an original set of silvicultural plans
—convey the objectives of an ownership, define demands that will influence future decisions, or set the economic/institutional/social framework for the forestry operations
—generally delineate the types of products and values sought, and set the intensity of management to follow in providing them
—represent the projected prevailing business interests of an enterprise, the financial constraints for silvicultural practice, and the required rate of return from investments in stand treatments

2 demands that alter the conduct of *existing* silvicultural programs
—bring change in programs previously planned and currently under implementation
—redefine the objectives, and set a new economic/institutional/social context for the silviculture and management
—force unscheduled replanning to move ownership, management, and use in unexpected new directions

This separation implies that foresters operate differently when undertaking initial planning and when unexpectedly forced to adjust ongoing silviculture. They periodically reassess existing plans as a routine and going process, and usually at intervals identified in the initial plans. The second set of demands reflects an unexpected change of status within the organization or the markets that it depends upon. These do not necessarily follow a crisis or disaster, but may reflect an economic improvement that permits a more intensified approach or a consideration of new options for management.

ADMINISTRATIVE CONSIDERATIONS FOR INITIAL PLANNING

Administrative demands that bear on initial planning, or scheduled replanning as undertaken routinely, have three effects:

1 They determine the intensity of effort foresters may expend on any silvicultural operation, plus the choice of treatments to include in a silvicultural system.
2 They control the emphasis of silvicultural activity, and delineate the end-point of management by defining what an owner expects as output in goods and values.
3 They specify how much an owner will invest to get them.

These point the direction for silvicultural undertakings, and fix the intensity of activity (level of expenditures) an owner will permit to achieve the desired results.

In essence, administrative concerns that guide initial planning and periodic routine assessment of silviculture reflect *economic* interests of an ownership, and delimit the biologically-ecologically viable options to pursue. Specifically, landowners will ask foresters to take account of factors like the following:

- market opportunities, including internal needs of an enterprise, that determine the kinds and amounts of products a landowner can sell or use, and the relative prices expected from each alternative product
- accessibility of resources relative to available markets, recognizing that if prices rise sufficiently a landowner can afford to recover even previously inaccessible resources
- taxes levied against both the land and real estate values, and upon income or profit from the operations
- overhead and operating costs, including those incurred as a result of complying with a host of government rules and regulations
- interest levels, as reflected in payments on borrowed money and the required alternate rate of return from unpaid costs of applying different kinds of silvicultural practices

Additionally, each landowner or enterprise will have certain personal and social interests. These often prove difficult to articulate in readily quantified terms.

Requirements like these emphasize that forestry is a business, and foresters must continually take cognizance of the financial, institutional, and social factors that control forestry operations. The relative importance of financial concerns compared with institutional and social concerns depends upon who owns the land and why. Yet fairly universally, landowners expect that the values realized from an investment in forestry must exceed the cost of the operations, compounded at the prevailing alternate rate of return for the firm or individual. Also, they must usually realize some minimum profit as an incentive to continue operating into the future, whether a landowner measures this profit in clearly tangible or less easily quantified terms.

Deliberately planned silviculture can satisfy almost any set of interests, as long as a landowner and forester can quantify the values sought, and settle upon the kind of vegetation conditions that will provide those benefits. The emphasis of silviculture will differ from one landowner to another, and reflect objectives at the time of initial planning. To illustrate, a corporate owner will normally require regular cash revenues and a profit based upon some minimum rate of return. Foresters usually operate each of these commercial forests as a profit center, and limit stand treatments to those that meet fairly narrow financial criteria. By contrast, foresters who manage public lands generally respond to a broader array of commodity and social interests, and give greater attention to nonmarket benefits. They must wisely allocate available funds to realize the greatest package of diverse benefits, rather than some minimum rate of return. Yet under most circumstances, public foresters also must generate sufficient revenues to help pay costs of their operations.

For nonindustrial private landowners, the interests and requirements of management may range even more widely. In fact, foresters working with these owners often find the purposes less clearly defined, and an individual's knowledge of forestry potentials quite limited. Some have a strong business orientation and strong commodity production goals, and these people operate their lands essentially as a small business. That has been a historic pattern with farm and ranch owners who view their woodlands as units of production and require regular yields as part of their entire agricultural enterprise. Some farm owners and many other nonindustrial owners espouse multiple-use objectives, and hope to integrate the sale of commodities with other uses of the land. They look to the timber to help pay costs of management for recreational and aesthetic values, and view various silvicultural treatments as the means to manipulate their stands in favor of the noncommodity values they seek (Fig. 25-3). Still other nonindustrial private landowners more narrowly focus on enhancing nonmarket benefits independent of timber harvesting. They willingly invest in practices that bring these interests to fruition, but may prohibit logging as a constraint of management. Silviculture has a role in all of these cases.

FACTORS THAT FORCE UNSCHEDULED CHANGES IN SILVICULTURAL PRACTICE

Despite their diverse purposes, most landowners want to operate as efficiently as possible, while still maintaining a healthy and viable forest. That requires cognizance of many forces that affect any business endeavor, as enumerated above (Davis 1954). Foresters watch for shifts in competition and supply, and how these affect market prices and the demand for goods and services ready for sale or use. Prices may improve and demand increase in times of prosperity, particularly for nontimber uses of forests. This may open opportunities to intensify management. At other times, demand declines, and costs of doing business increase. This forces a reduction in anticipated sales, and in the margin between income and expenditures. It precipitates a reduction in intensity of management. Whatever the change, foresters must usually adjust their management and reevaluate the options. The change may prove positive (stimulating) or negative (dampening), but tends to have two kinds of outcomes: (1) to totally abandon silvicultural practice, and begin liquidating assets; or (2) to substantially alter the intensity and nature of silviculture without abandoning interests in long-term sustained-yield management. In the former case, foresters may shift roles from silviculturist to protector of the environment, and work primarily to insure that nothing destroys the long-term ecological character of the land, nor limits its overall productive potential. In the second case, they restructure their silviculture programs, usually by lessening the intensity or altering the levels of timber harvesting and other resource use.

In general, foresters and landowners alter the intensity of any silvicultural program in three ways (Smith 1986):

1 increasing or decreasing the amounts invested in silvicultural operations within immature stands and age classes

2 changing the frequency and intensity of harvest cuttings, including scheduling removals sooner or delaying them

3 taking revenues from a stand immediately, or delaying the harvest to some future time

These may reflect shifts in market conditions, accessibility of the resources relative to prices, the productive potential of a site, and changes in the objectives and nature

FIGURE 25-3
In many cases, foresters can help landowners to derive many different values from a deliberate program of management, such as using their logging roads as hiking trails during the interim between entries to a stand.

of an ownership. The appropriate intensity may differ between stands to optimize the net effect for the enterprise as a whole. Factors that force change within the private sector include new taxes, new product opportunities, loss of market share, fiscal crisis, natural disaster, availability of new technologies for management or use, new government rules and regulations, and a change of ownership with the sale of a property. These either stimulate or dampen silvicultural activity. The new economic reality may even force a landowner to abandon deliberate management, or bring opportunities for an entirely new endeavor previously considered inappropriate.

EXAMPLES OF CHANGING ADMINISTRATIVE DEMANDS

Foresters commonly adjust silvicultural practice after a change in product values. These may reflect a shift in quality criteria used in valuing roundwood products, the development of new technologies that introduce new marketing opportunities or dampen the demand for traditional products, or a shift of preference by consumers that alters existing markets or opens new ones for previously low-value species and roundwood pieces. These changes force a reassessment of the growing stock value, of potentials for generating continuing revenues, or of the changes for turning a profit. Then foresters must alter their silvicultural plans, because the changes

1 act like a windfall to open opportunities for silvicultural treatments previously judged unprofitable
—as when a rise of pulpwood prices or an expansion of pulpwood markets increases the demand for small-diameter and low-quality pieces, making thinning and other intermediate treatments more commercially attractive
—as when some new harvesting technology permits a break-even operation with less volume per unit area, opening opportunities to treat stands previously proving unattractive for a commercial sale

2 give rise to sudden demand for a particular species and increase its value appreciably
—as when a rise in stumpage price encourages rescheduling to operate stands with abundant stocking of the newly demanded species
—as when an increase of tree value motivates landowners to shift from conservative to exploitative cuttings to liquidate the higher-priced assets

3 reduce demand, reducing prices, slowing sales, and forcing a termination of programmed silviculture
—as when a market crash suddenly leaves owners with an oversupply of worthless or low-value products
—as when previously usable species no longer have marketplace value

Changes of the first kind stimulate silviculture and encourage more intensive management (Fig. 25-4). Those of the second group promote rescheduling in the sequence and nature of ongoing silviculture to capitalize on the new opportunities for product sales. Changes of the third kind usually force new planning, and more than a simple rescheduling or restructuring of the silviculture. They force more extensive management.

Sometimes an enterprise or individual may face a financial crisis. Then the forestry staff may find its operating funds withdrawn, including the money to meet payrolls or make investments in silviculture. Suddenly, a landowner must find assets to liquidate. Common fiscal crises and some likely consequences include the following:

1 Unexpected increases in the alternate rate of return or interest on borrowed money tend to dampen silvicultural activity and intensify the rates of timber harvest
—as when higher guiding rates of return make previously acceptable practices suddenly unattractive, and a firm abandons forestry in favor of exploitive treatments

FIGURE 25-4
An improvement in prices or demand for timber can encourage management by providing added revenues to pay the costs of more intensive silviculture, and offering greater long-term revenues to a landowner.

—as when higher guiding rates of return change the conditions for financial maturity, encouraging the removal of younger age classes in uneven-aged stands, and shortening the rotations for even-aged ones

2 A financial collapse usually triggers accelerated cutting to provide essential capital to keep the enterprise alive
—as when payments on debt exceed planned revenues, forcing an enterprise to abandon forestry in favor of exploitation
—as when a natural disaster destroys or damages the growing stock, forcing unscheduled salvage operations, and leaving little opportunity for future sales at a level that will sustain the enterprise

The first group represents relatively minor disasters that mainly alter the intensity of an existing forestry program, and represent an economic belt-tightening. They force landowners to develop new plans for the future, and adjust their practice of silviculture across the ownership. The second group forces foresters to suddenly become timber sale agents, and serve as protectors of the environment during the rapidly moving harvesting and sales program. In severe cases, a firm or landowner may use uncontrolled cutting to save on costs, and dismiss its forestry staff. This usually ends controlled forestry operations. If the firm succeeds and decides to continue its forestry program, the staff must often look for ways to rehabilitate the heavily cutover stands, and begin new planning for the future.

TAKING AN ECOSYSTEM APPROACH

Even landowners who face a financial crisis or see some unique chance to capitalize on unexpected developments should not abandon their concern about maintaining healthy and resilient forests. Foresters have the professional and ethical responsibility to remind their employers of that obligation, but the primary opportunity for taking an ecosystem approach to forest resource management and use comes with initial planning, or scheduled replanning. Then landowners usually face fewer and less overwhelming financial constraints and feel less pressure to respond exclusively to their short-term interests.

Silviculturists base their work on the information and experience that forestry has accumulated over long periods through research and practice. This provides a perspective on how to solve the silvicultural challenges, and how to uniquely address different situations and needs. Taking such a historic perspective gives them an idea about what to expect from alternative methods, and what kinds and magnitudes of responses commonly follow different treatments. This helps them to identify the options, and to forecast the likely outcomes. Even so, foresters do not dictate what values a landowner should hold at any given time, or what relative (economic) importance to assign to each option. In the best of circumstances, foresters develop a close working relationship with the landowners that they serve, and use their continuing dialogue to suggest new opportunities for management and use as these unfold over time.

Prior to about 1960–1965, most forest landowners in eastern North America held largely commodity objectives. Forest products companies, government agencies, and farmers or ranchers owned most of the forests. But as active agriculture declined across many regions of the continent, and as society realized new prosperity and came to enjoy greater amounts of leisure time, many people purchased inactive farms and ranch lands to use for rural home sites or for recreational pursuits. Construction of new roads, improvement of others, and access to affordable cars and low-cost fuel triggered a boom in outdoor recreation across the entire continent. This even further encouraged forest land acquisition by a growing number of nonindustrial, nonfarm owners. The change also put new demands on public lands to serve a broad array of nonmarket interests. In response, forestry embraced the philosophy of multiple-use management, and worked to diversify the values that people could derive from a forest.

Among industrial ownerships and some public lands, foresters could adapt the former language and concepts of timber production to reflect changing management objectives and implement new approaches to realizing them. They introduced some new strategies to enhance nonmarket values along with the commodities, but continued to use traditional decision making and post-treatment evaluations. In many cases, the new nonindustrial private landowners showed no interest in the traditional aspects of forestry, or rejected timber cutting as a viable alternative for their properties. Those people seemed to hold the land for intangible reasons, and many had no need to engage in active timber management. They used their lands as sanctuaries, or regarded timber harvesting and tree cutting as incongruous with their primary interests. Others recognized that silviculture could enhance many nonmarket values, and organized active forest management to bring their dreams to fruition. Many owners could not articulate their interests in terms that foresters could translate into readily applied silviculture. In other cases, foresters seemed unable to communicate with people who never before had owned forests, who had no experience in thinking about management options on large spatial scales, and who had difficulty envisioning how communities would change and develop over long periods of time. A similar gap developed for public lands throughout North America, as citizens increasingly assigned greater value to the noncommodity benefits from recreation and other nonconsumptive uses of forests.

Perhaps most puzzling about these new perspectives of forest ownership and use, foresters had to formulate prescriptions and implement the practices without any real certainty about how to quantify the returns of intangible values or help landowners to account for the payoff from investments in management. This has profound implications for silviculture. Above all, the inability of landowners to describe their interests in quantifiable terms prevented foresters from delineating clear objectives for management. They could continue to apply some kind of treatment to a stand, but had no exact way of determining if the program would succeed relative to a landowner's interests. Also, because landowners could not articulate their interests, foresters had difficulty in deciding what specific kind of stand conditions to create and maintain through silvicultural practice. Even when landowners could define their purposes clearly, the lack of experience in dealing with similar kinds of demands often left foresters uncertain about what to do.

By the 1990s, shifts of attitudes and operating conditions like these began to trigger another important transition for forestry within North America. Often expressed as the onset of a *new perspectives management* (or *sustainable development,* and later *ecosystem management*), it reflected a response to the shift of societal interests from a dominant historic view of forests as sites for commodity production, to an acceptance of forests as dynamic ecosystems that can potentially provide a broad array of benefits to people as an outcome of their structure and function. This new approach seeks a balance between protecting natural systems and using them for economic benefit. It has the goal of meeting the needs of people, improving the well-being of communities and nations, and maintaining a balance and harmony between people and the land. It shifts the focus of management from sustaining the yields of outputs and balancing competing uses, to sustaining ecosystems. It recognizes both the ecological and utilitarian values of forests, and gives high priority to maintaining their health, diversity, and productivity (Kessler et al. 1992; Salwasser and Pfister 1994). Yet these emerging concepts do not necessarily change the practice of silviculture per se. Rather they change the way silviculturists evaluate the impacts in both time and space (beyond the stand) and upon all the components of an ecosystem. In this context, silviculturists have the role of continuing to meet people's needs, while also ensuring healthy and sustainable forests in perpetuity (Smith 1994).These concepts probably reflect a maturing of the multiple-use concept that dominated forestry thinking in North America from the 1960s through the 1980s. Its basic tenets include an emphasis on sustaining essential ecological conditions (after Brooks and Grant 1992a) by

1 accounting for the changing objectives and values that interest landowners and society as a whole;
2 insuring a minimum necessary level of integrity, resiliency, and diversity across the stands and forests under management;
3 accounting for effects of management and use at a broad landscape level, and for critical ecological balances on that spatial scale;
4 considering effects on a long-term scale, not just in relation to immediate payoffs;
5 recognizing if the planned activities will constrain future options for management and use, or cause an unacceptable or irreversible ecologic change; and
6 assessing the financial and general economic viability of different potential treatments, including any necessary ameliorative practices to insure fulfillment of the silvicultural and ecosystem management plans.

To accomplish these goals, foresters and landowners must take a holistic approach that links administration and management with silviculture. Further, they must integrate planning and decision making with eventual implementation of in-forest practices to insure an ecologically and socially-economically responsive outcome.

From this perspective, the concept of ecosystem management involves a carefully coordinated program of planning, administration, silviculture, and use. Key components (after Kaufmann et al. 1994) include

1 understanding landowner, public, and policy demands, and setting appropriate goals and objectives;
2 planning a program of management responsive to these concerns, and conceived on an appropriate spatial scale and from an ecologic perspective;
3 securing and allocating human, fiscal, and logistical resources to accomplish the task in an appropriate and timely fashion;
4 applying the treatments stand by stand in a logical pattern and sequence that create and maintain desirable ecological conditions across the forest and landscape;
5 doing the work in an ecologically sensitive manner; and
6 monitoring conditions on a spatial scale to insure success in fulfilling the goals and providing a basis for anticipating necessary changes.

FIGURE 25-5
By taking cognizance of the effects from a treatment in one stand upon the balance of conditions across a broader landscape, foresters can maintain many diverse values in perpetuity from a forest.

To make this possible, foresters first identify the critical ecosystem values (needs and opportunities) that they must deal with on different geographic scales, and for the present and the future. They must also articulate the desired forest conditions that will keep these critical factors at safe levels, and the reasons for doing so. They must plan a strategy for creating, maintaining, and sustaining those conditions on a forestwide scale, and beyond (Fig. 25-5).

Generally, to develop an appropriate program of ecosystem management, foresters will need to focus their planning in three areas of concern. One includes all aspects of *administration,* directed at

- understanding client/landowner/societal demands;
- planning how to address the goals and objectives;
- arranging the logistical support and allocating human, fiscal, and other resources to bring the program to fruition; and
- monitoring developments to keep track of past accomplishments and future needs.

A second area of planning deals with *ecological considerations,* including

- judging the minimum acceptable and desirable threshold levels and admixture of biologic resources to encourage and maintain across a forest and a landscape;
- determining the essential supporting physical environment conditions to maintain these communities and populations;
- insuring an appropriate diversity of plants, animals, and supporting environmental conditions for long-term resilience and stability of the desired ecological conditions; and
- monitoring the status of critical plants and animals to measure success of the program.

The third, focuses on *economic factors,* like

- providing opportunities to derive human value from biotic and physical elements of the forest and landscape;
- determining appropriate levels of use to insure an adequate payoff, while still insuring ecosystem health and resilience;
- sustaining availability of desired values that people derive from conditions in the forest and in the surrounding landscape; and
- accomplishing the task in a financially efficient manner.

The latter involves deriving revenues from harvest of commodities, charging fees for recreational use, and finding other ways to generate sufficient income to pay costs of owning and managing the land, and to provide a reasonable profit when people engage in forestry as a business (Fig. 25-6).

Ultimately, foresters must engage in two levels of planning and practice. First, they must determine what to do across time and space, and on an appropriate ecological scale. This includes planning an appropriate sequence for the treatments, and deciding what stands to work in as a part of each successive year's management. Second, they must prescribe an appropriate silviculture to apply stand by stand, and the proper time for each treatment. This also involves decisions about the intensity for management necessary to fulfill the ecologic-biologic objectives, and the possibilities a forester can consider based upon financial and other economic policies of the landowner. From this perspective, *use* of resources becomes an expected aftermath or byproduct of having created and maintained an appropriate set and admixture of desired conditions across a forest and its surrounding landscape. Also, *use* becomes a mechanism for creating and maintaining the desired conditions necessary to long-term ecologic and economic success.

SILVICULTURE IN RETROSPECT

Overall, administrative policies dictate the direction of silviculture, and the intensity of work planned within a stand and across a forest. Among ongoing programs, changes of administrative demands can have three basic effects on silviculture. These include (1) stimulating more intensive management and greater investment in stand treatments; (2) requiring a shift toward more extensive operations; and (3)

FIGURE 25-6
Realistic management must maintain critical ecological values while also providing economic benefits to people, including the sale of products to maintain private forestry as a viable business endeavor.

forcing a withdrawal from deliberate management in favor of liquidating assets over the short run. In all of these cases, foresters hope to exert some positive influence over the available resources, if simply to safeguard the long-term productivity of the land during exploitation to serve short-range interests.

Clearly, foresters cannot perfectly forecast the future. Rather, they look to the past and present for indicators of probable change, and then attempt to organize their programs to capitalize on those deemed most likely to develop. They must continuously monitor events and remain flexible when changes occur. They must approach change as an opportunity, and creatively plan ways to adjust to events that emerge.

Foresters have a diverse toolkit of techniques and methods to draw upon in addressing a broad spectrum of landowner interests and concerns. They must exercise good judgment and thoughtfulness in assessing each situation, and strive to find the best alternative that benefits a landowner and proves ecologically acceptable. When opportunities arise, they draw upon the organized body of theory that underlies silviculture, and use it as a basis for decision making. They succeed by identifying the widest possible array of ecologically and biologically viable alternatives, and recommending those that make the most economic sense. Above all, they act decisively to implement their programs in a timely fashion.

REFERENCES

Abbott, H.G. 1974. Direct seeding in the United States. In J.H. Cayford (Ed.), *Direct Seeding Symposium,* pp. 1–10. Sept. 11–13, 1973, Timmins, Ont. Dept. Environ., Can. For. Serv., Publ. No. 1339.

Abbott, J.E. 1982. Operational planting of container slash pine seedlings on problem sites. In R.W. Guldin and J.P. Barnett (Eds.), *Proceedings of the Southern Containerized Forest Tree Conference,* pp. 115–117. U.S. For. Serv. Gen. Tech. Rpt. SO-37.

Abell, J. 1954. The "wait and see" method of thinning. In S.O. Heiberg (Ed.), *Thinning Problems and Practices in Denmark,* pp. 51–58. SUNY Coll. For. at Syracuse. World For. Ser. Bull., No. 1.

Adams, D.A., and A.R. Ek. 1974. Optimizing the management of uneven-aged forest stands. *Can. J. For. Res.* 4:274–287.

Adams, D.L. 1980. The northern Rocky Mountain region. In J.W. Barrett (Ed.), *Regional Silviculture of the United States,* 2d ed., pp. 341–390. New York: Wiley.

Adams, D.L. 1995. The northern Rocky Mountain region. In J.W. Barrett (Ed.), *Regional Silviculture of the United States,* 3d ed., Ch. 9, pp. 387–440. New York: Wiley.

Agee, J.K. 1990. The historical role of fires in Pacific Northwest forests. In J.D. Walstad, S.R. Radosevich, and D.V. Sandbert (Eds.), *Natural and Prescribed Fire in Pacific Northwest Forests,* Ch. 3, pp. 25–38. Corvallis: Ore. State Univ. Press.

Agee, J.K. 1993. *Fire Ecology of Pacific Northwest Forests.* Washington, DC: Island Press.

Aim, A.A., V.M. Vaughn, and H.M. Rauscher. 1991. Bareroot nursery production and practices for white spruce: A literature review. U.S. For. Serv. Gen. Tech. Rpt. NC-142.

Alexander, R.R. 1971. Initial partial cutting in old-growth spruce-fir: A field guide. U.S. For. Serv. Res. Pap. RM-76A.

Alexander, R.R. 1974a. Silviculture of subalpine forests in the central and southern Rocky Mountains. The status of our knowledge. U.S. For. Serv. Res. Pap. RM-121.

Alexander, R.R. 1974b. Silviculture of central and southern Rocky Mountain forests: A summary of the status of our knowledge. U.S. For. Serv. Res. Pap. RM-120.

Alexander, R.R. 1977. Cutting methods in relation to resource use in central Rocky Mountain spruce-fir forests. *J. For.* 75(7):395–400.

Alexander, R.R. 1984. Natural regeneration of Engelmann spruce after clearcutting in the central Rocky Mountains in relation to environmental factors. U.S. For. Serv. Res. Pap. RM-254.

Alexander, R.R. 1986a. Silvicultural systems and cutting methods for old-growth spruce-fir forests in the central and southern Rocky Mountains. U.S. For. Serv. Gen. Tech. Rpt. RM-126.

Alexander, R.R. 1986b. Silvicultural systems and cutting methods for old-growth lodgepole pine forests in the central Rocky Mountains. U.S. For. Serv. Gen. Tech. Rpt. RM-127.

Alexander, R.R. 1986c. Silvicultural systems and cutting methods for ponderosa pine forests in the front range of the central Rocky Mountains. U.S. For. Serv. Gen. Tech. Rpt. RM-128.

Alexander, R.R. 1986d. Engelmann spruce seed production and dispersal, and seedling establishment in the central Rocky Mountains. U.S. For. Serv. Gen. Tech. Pap. RM-134.

Alexander, R.R. 1987a. Silvicultural systems, cutting methods, and cultural practices for Black Hills ponderosa pine. U.S. For. Serv. Gen. Tech. Rpt. RM-139.

Alexander, R.R. 1987b. Ecology, silviculture, and management of the Engelmann spruce-subalpine fir type in the central and southern Rocky Mountains. U.S. For. Serv., Agric. Handbk. No. 659.

Alexander, R.R., and C.B. Edminster. 1977a. Regulation and control of cut under uneven-aged management. U.S. For. Serv. Res. Pap. RM-182.

Alexander, R.R., and C.B. Edminster. 1977b. Uneven-aged management of old growth spruce-fir forests: Cutting methods and stand structure goals for initial entry. U.S. For. Serv. Res. Pap. RM-186.

Alexander, R.R., and C.B. Edminster. 1983. Engelmann spruce seed dispersal in the central Rocky Mountains. U.S. For. Serv. Res. Note RM-424.

Alexander, R.R., and W.D. Shepperd. 1990. *Pices engelmannii* Parry ex Engelm. Engelmann spruce. In R.M. Burns and B.H. Honkala (Eds.), *Silvics of North America. Volume 1. Conifers*, pp. 187–203. U.S. For. Serv., Agric. Handbk. 654.

Alexander, R.R., W.D. Shepperd, and C.B. Edminster. 1975. Yield tables for managed even-aged stands of spruce-fir in the central Rocky Mountains. U.S. For. Serv., Res. Pap. RM-134.

Allen, D.C. 1987. Insects, declines and general health of northern hardwoods: Issues relevant to good forest management. In R.D. Nyland (Ed.), *Managing Northern Hardwoods,* pp. 252–285. SUNY Coll. Environ., Sci. and For., Fac. For. Misc. Publ. No. 13 (ESF 87-200), Soc. Am. For. Publ. 87-03.

Allen, D.C. 1988. Insect problems of hardwood forests. In *Healthy Forests, Healthy World,* pp. 106–113. Proc. 1988 Soc. Am. For. Nat. Conv., Rochester, NY, Oct. 16-19, 1988. Soc. Am. For., Washington, DC. SAF Publ. 88-01.

Altsuler, S. 1985. Plug seedling success in the high country. In T.D. Landis (Ed.), *Proceedings: Western Forest Nursery Council—Intermountain Nurseryman's Association Combined Meeting*, pp. 1–3. U.S. For. Serv. Gen. Tech. Rpt. INT-185.

Amman, G.D. 1978. Biology, ecology, and causes of outbreaks of the mountain pine beetle in lodgepole pine forests. In E.A. Berryman, G.D. Amman, and R.W. Stark (Eds.), *Theory and Practice of Mountain Pine Management in Lodgepole Pine Forests*, pp. 39–53. Proc. Symp., April 25-27, 1978, Wash. St. Univ., Pullman, WA. For., Wildl, and Range Exp. Stn., Univ. Idaho, Moscow, and U.S. For Serv., For. Insect and Disease Res., Washington, DC, and U.S. For. Serv. Intermt. For. and Range Exp. Stn., Odgen, UT.

Amman, G.D., M.D. McGregor, D.B. Cahill, and W.H. Klein. 1977. Guidelines for reducing losses of lodgepole pine to the mountain pine beetle in unmanaged stands in the Rocky Mountains. U.S. For. Serv. Gen. Tech. Rpt. INT-36.

Anderson, G.W. 1972. Diseases. In *Aspen Symposium Proceedings* pp. 74–82. U.S. For. Serv. Gen. Tech. Rpt. NC-1.

Anderson, H.W. 1994. Some implications of logging damage in the tolerant hardwood forests of Ontario. In J.A. Rice (Ed.), *Logging Damage: The Problems and Practical Solutions,* pp. 3–27. Ont. Min. Nat. Resourc., Ont. For Res. Inst., For Res. Rpt. No. 117.

Anderson, H.W., B.D. Batchelor, C.M. Corbett, A.S. Corlett, D.T. Deugo, C.F. Husk, and W.R. Wilson. 1990. *A Silvicultural Guide for the Tolerant Hardwoods Working Group in Ontario*. Ont. Min. Nat. Resourc., For. Res. Gp., Sci. and Technol. Ser., Vol. 7, Toronto.

Arbogast, C., Jr. 1950. How much does felling large cull trees damage the understory? U.S. For. Serv., Lake States For. Exp. Stn., Tech. Note 337.

Arbogast, C., Jr. 1957. Marking guides for northern hardwoods under the selection system. U.S. For. Serv., Lake States For. Exp. Stn., Sta. Pap 56.

Armitage, F.B. 1974. Statement for Canada on tree planting and direct seeding. In *Symposium Stand Establishment*, pp. 410–421. Intl. Union For. Res. Organ., IUFRO Joint Meet., Div. 1 & 3, Oct 15–19, 1974. Wagenigen, The Netherlands.

Armson, K.A. 1977. *Forest Soils: Properties and Processes*. Univ. Toronto Press, Toronto.

Armson, K.A., and V. Sadreika. 1974. *Forest Tree Nursery Soil Management and Related Practices*. Ont. Min. Nat. Resourc., Div. For., For. Manage. Br., Toronto.

Arvidsson, A. 1986. Pruning for quality. In *Small Scale Forestry,* pp. 1–7. Vol. 1/86, Newsletter of the Dept. Operat. Efficiency, Swedish Univ. Agr., Garpenberg.

Assmann, E. 1970. *The Principles of Forest Yield Study*. Transl. by S.H. Gardiner. Oxford: Pergamon Press Ltd.

Aubertin, G.M. 1971. Nature and extent of macropores in forest soils and their influence on subsurface water movement. U.S. For. Serv. Res. Pap. NE-192.

Aubertin, G.M., and J.H. Patric. 1972. Quality water from clearcut forest land? *N. Logger and Timber Process.* 20(8):14–15, 22, 23.

Aust, W.M., T.W. Reisinger, B.J. Stokes, and J.A. Burger. 1993. Tire performance as a function of width and number of passes on soil bulk density and porosity in a minor stream bottom. In J.C. Brissette (Ed.), *Proceedings of the 7th Biennial Southern Silvicultural Research Conference*, pp. 137–141. U.S. For. Serv. Gen. Tech. Rpt. SO-93.

Avery, T.E. 1975. *Natural Resource Measurements*, 2d ed. New York: McGraw-Hill.

Baathe, P. 1957. *Thinnings in Even-aged Stands. A Summary of European Literature*. Fac. For., Univ. New Brunswick, Fredericton, NB.

Bacon, C.G., and S.N. Zedaker. 1985. First year growth response of young loblolly pine (*Pinus tadea L.*) to competition control in the Virginia piedmont. In E. Shoulders (Ed.), *Proceedings of the Third Biennial Southern Silvicultural Research Conference*, pp. 309–314. U.S. For. Serv. Gen. Tech. Rpt. SO-54.

Bain, J. 1979. Forest monocultures—How safe are they? An entomologist's view. *N.Z. J. For. Sci.* 9:37–42.

Baker, F.S. 1934. *Theory and Practice of Forestry*. New York: McGraw-Hill.

Baker, F.S. 1950. *Principles of Silviculture.* New York: McGraw-Hill.

Baker, J.B. 1991a. Alternative silvicultural systems—South. In *Silvicultural Challenges and Opportunities*. pp. 51–60. Proc. 1989 Natl. Silvi. Workshop, July 10-13, 1989, Petersburg, AK. U.S. For. Serv., Washington, DC.

Baker, J.B. 1991b. Uneven-aged management of loblolly-shortleaf pine. In H. Chung, Y. Chen, and Y.S. Haung (Eds.), *Forestry Administration and Forestry Technology*, pp.

93–98. Proc. 1990 Nov. 7–8, Taipei, Taiwan, R.O.C., Taiwan For. Inst. and Counc. Agric., Taipei, R.O.C.

Baker, J.B. 1992. Natural regeneration of shortleaf pine. In J.C. Brissette and J.P. Barnett (Eds.), *Proceedings of the Shortleaf Pine Regeneration Workshop*, pp. 102–112. U.S. For. Serv. Gen. Tech. Rpt. SO-90.

Baker, J.B., M.D. Cain, J.M. Guldin, P.A. Murphy, and M.G. Shelton. 1996. Uneven-aged Silviculture for the Loblolly and Shortleaf Pine Cover Types. U.S. For. Serv. Gen. Tech. Rpt. SO-118.

Baldwin, V.C., Jr., and D.P. Feduccia. 1987. Loblolly pine growth and yield prediction for managed West Gulf plantations. U.S. For. Serv. Res. Pap. SO-236.

Baldwin, V.C., Jr., D.P. Feduccia, and J.D. Haywood. 1988. Postthinning growth and yield of row-thinned and selectively thinned loblolly and slash pine plantations. *Can. J. For. Res.* 19:247–256.

Balmer, W.E., and N.G. Little. 1978. Site preparation methods. In T. Tippin (Ed.), *Proceedings: A Symposium on Principles of Maintaining Productivity on Prepared Sites,* pp. 61–64. Proc. Symp., Mar. 21–22, 1978. Miss. State Univ., U.S. For. Serv., S. For. Exp. Stn., New Orleans.

Balmer, W.E., and H.L. Williston. 1973. The need for precommercial thinning. For. Manage. Bull., SE Area State and Priv. For., U.S. For. Serv., Atlanta.

Barlowe, R. 1958. *Land Resource Economics. The Political Economy of Rural and Urban Land Resource Use*. Englewood Cliffs, NJ: Prentice-Hall.

Barnes, B.V. 1982. Ecosystem classification—Number 1 priority. In G.D. Mroz and J.F. Berner (Eds.), *Artificial Regeneration of Conifers in the Upper Great Lakes Region*, pp. 8–30. Mich. Technol. Univ., Houghton, MI.

Barnett, J.P. 1989. Seed, cultural practices, and seedling uniformity. In J. D. Mason, J. D. Deans, and S. Thompson (Eds.), *Producing Uniform Conifer Planting Stock,* pp. 95–105. Oxford: Oxford University Press.

Barnett, J.P., and J.B. Baker. 1991. Regeneration methods. In M.L. Duryea and P.M. Dougherty (Eds.), *Forest Regeneration Manual*, Ch. 3, pp. 3–7, 35–50. Boston: Kluwer Academic Press.

Barnett, J.P., and J.M. McGilvray. 1981. Container planting systems for the South. U.S. For. Serv. Res. Pap. SO-167.

Barnett, J.P., and J.M. McGilvray. 1992. Carry-over of loblolly pine seeds on cutover sites. *Tree Plant. Notes* 42(4):17–18.

Barnett, J.P., D.K. Lauer, and J.C. Brissette. 1990. Regenerating longleaf pine with artificial methods. In *Proceedings of the Symposium on the Management of Longleaf Pine*, pp. 72–93. U.S. For. Serv. Gen. Tech. Rpt. SO-75.

Barrett, J.W. 1972. Large-crowned planted ponderosa pine respond well to thinning. U.S. For. Serv. Res. Note PNW-179.

Barrett, J. W. 1979. Silviculture of ponderosa pine in the Pacific Northwest: The state of our knowledge. U. S. For. Serv. Gen. Tech. Rpt. PNW-97.

Barrett, J.W. 1980. The Northeastern region. In J.W. Barrett (Ed.), *Regional Silviculture of the United States,* 2d ed., pp. 25–66. New York: Wiley.

Barrett, J.W. 1983. Growth of ponderosa pine poles thinned to different stocking levels in central Oregon. U.S. For. Serv. Res. Pap. PNW-311.

Bartuska, A.M. 1994. Ecosystem management in the Forest Service. In L.H. Foley (Ed.), *Silviculture: From the Cradle of Forestry to Ecosystem Management,* pp. 12–15. Proc. Natl. Silvi. Workshop, Hendersonville, NC, Nov. 1–4, 1993. U.S. For. Serv. SE For. Exp. Stn., Asheville, NC.

Baskerville, G.L. 1975. Spruce budworm: Super silviculturist. *For. Chron.* 51:138–140.

Bassmann, J.H., and L. Fins. 1988. Genetic considerations for culture of immature stands. In W.C. Schmidt (Ed.), *Proceedings—Future Forests of the Mountain West: A Stand Culture Symposium,* pp. 146–152. U.S. For. Serv. Gen. Tech. Rpt. INT-243.

Bates, C.G. 1924. Forest types in the central Rocky Mountains as affected by climate and soil. U.S. Dept. Agric. Bull. 1233.

Bates, C.G., H.C. Hilton, and T. Kruger. 1929. Experiments in the silvicultural control of natural reproduction of lodgepole pine in the central Rocky Mountains. *J. Agric. Res.* 38(4):229–243.

Baughman, M.J. 1991. Windbreaks are bulwarks of the prairie. In M.E. Dix and M. Harrell (Eds.), *Insects of Windbreaks and Related Plantings: Distribution, Importance, and Management,* pp. 1–4. U.S. For. Serv. Gen. Tech. Rpt. RM-204.

Baumgartner, D.M., and C.A. Hyldahl. 1991. Using price data to consider risk in the evaluating of forest management investments. U.S. For. Serv. Gen. Tech. Rpt. NC-144.

Baumgartner, D.M., R.G. Krebill, J.T. Arnott, and G.F. Weetman (Eds.). 1985. *Lodgepole Pine: The Species and Its Management.* Symp. Proc., May 8–10, Spokane, WA and May 14–16, Vancouver, BC. Wash. State Univ., Coop. Ext., Pulmann, WA.

Beck, D.E. 1970. Effect of competition on survival and height growth of red oak seedlings. U.S. For. Serv. Pap. SE-56.

Beck, D.E. 1977. Growth and development of thinned versus unthinned yellow-poplar sprout clumps. U.S. For. Serv. Res. Pap. SE-173.

Beck, D.E. 1983. Thinning increases forage production in southern Appalachian cover hardwoods. *S. J. App. For.* 7(1):53–57.

Beck, D.E. 1987. Management options for southern Appalachian hardwoods: The two-aged stand. In D.R. Phillips (Ed.), *Proceedings of the Fourth Biennial Southern Silviculture Conference*, pp. 451–454. U.S. For. Serv. Gen. Tech. Rpt. SE-42.

Beck, D.E. 1989. Selection thinning in southern Appalachian hardwoods. In J.H. Miller (Ed.), *Proceedings of the Fifth Biennial Southern Silviculture Research Conference*, pp. 291–294. U.S. For. Serv. Gen. Tech. Rpt. SO-74.

Beers, T.W. 1962. Components of forest growth. *J. For.* 60(4):245–248.

Belanger, R.P. 1979. Stump management increased coppice yield of sycamore. *S. J. Appl. For.* 3(3):101–103.

Belanger, R.P. 1981. Silvicultural guidelines for reducing losses to the southern pine beetle. In R.C. Thatcher, J.L. Searcy, J.E. Coster, and G.D. Hertel (Eds.), *The Southern Pine Beetle*, Ch. 9, pp. 165–177. U.S. For. Serv., Sci. and Ed. Admin. Tech. Bull. 1631.

Belanger, R.P., and R.F. Malac. 1980. Silviculture can reduce losses from the southern pine beetle. Southern Pine Beetle Handbook. U.S. For. Serv., Agric. Handbk. No. 576.

Belcher, E.W. 1982. Basic concepts for obtaining better seed. In G.D. Mroz and J.F. Berner (Eds.), *Artificial Regeneration of Conifers in the Upper Great Lakes Region*, pp. 246–253. Mich. Technol. Univ., Houghton, MI.

Bella, I.E., and J.P. DeFranceschi. 1977. Young lodgepole pine responds to strip thinning, but . . . Can. For. Serv., N. For. Res. Ctr. Info. Rpt. NOR-X-192.

Bennett, F.A. 1956. Growth of slash pine plantations on the George Walton Experimental Forest. U.S. For. Serv., SE For. Exp. Stn., Stn. Pap. No. 66.

Bennett, F.A. 1975. A technique for comparing thinning schedules for natural slash pine stands. U.S. For. Serv. Res. Pap. SE-138.

Benson, D.A. 1976. Nursery operations—What the nursery can provide. In D.M. Baumgartner and R.J. Boyd (Eds.), *Tree Planting in the Inland Northwest*, pp. 82–103. Wash. State Univ. Ext. Serv., Wash. State Univ., Pullman, WA.

Benson, M.K. 1982. Hand or machine planting. In G.D. Mroz and J.F. Berner (Eds.) *Artifi-*

cial Regeneration of Conifers in the Upper Great Lakes Region, pp. 332–336. Mich. Technol. Univ., Houghton, MI.

Benzie, J.W. 1977. Manager's handbook for red pine in the North Central states. U.S. For. Serv. Gen. Tech. Rep. NC-32.

Benzie, J.W. 1982. Direct seeding upland conifers. In G.D. Mroz and J.F. Berner (Eds.), *Artificial Regeneration of Conifers in the Upper Great Lakes Region*, pp. 292–298. Mich. Techol. Univ., Houghton, MI.

Berglund, J.V. 1975. *Silvics*. SUNY Coll. Environ. Sci. and For., Syracuse, NY.

Berndt, H.W., and R.D. Gibbons. 1958. Root distribution of some native trees and understory plants growing on three sites within ponderosa pine watershed in Colorado. U.S. For. Serv., Rocky Mt. For. and Range Exp. Stn., Stn. Pap. 37.

Berry, A.B. 1964a. A time study in pruning plantation white spruce and red pine. *For. Chron.* 40(1):122–128.

Berry, A.B. 1964b. Effect of strip width on proportion of daily light reaching the ground. *For. Chron.* 40(1):130–131.

Berry, A.B. 1981. Study of single-tree selection for tolerant hardwoods. Can. For. Serv., Petawawa Natl. For. Inst., Info. Rpt. PI-X-8.

Berry, F.H. 1982. Reducing decay losses in high value hardwoods—A guide for woodland owners and managers. U.S. For. Serv., Agric. Handbk. No. 595.

Betters, D.R., and R.F. Woods. 1981. Uneven-aged stand structure and growth of Rocky Mountain aspen. *J. For.* 79(10):673–676.

Bigg, W.L., and J.W. Schalau. 1990. Mineral nutrition and the target seedling. In R. Rose, S.J. Campbell, and T.D. Landis (Eds.), *Target Seedling Symposium: Proceedings, Combined Meeting of the Western Forest Nursery Association,* Ch. 10, pp. 139–160. U.S. For. Serv. Gen. Tech. Rpt. RM-200.

Biolley, H. 1934. Nombres d'arbres et régimes de futaies. *J. For. Suisse* 4–8, 27–33. (Original not seen).

Bjorkbom, J.C. 1967. Seedbed-preparation methods for paper birch. U.S. For. Serv. Res. Pap. NE-79.

Bjorkbom, J.C. 1971. Production and germination of paper birch seed and its dispersal into a forest opening. U.S. For. Serv. Res. Pap. NE-209.

Blackburn, W.H., J.C. Wood, H.A. Pearson, and R.W. Knight. 1987. Storm flow and sediment loss from intensively managed forest watersheds in East Texas. U.S. For. Serv. Gen. Tech. Rpt. SO-68.

Blair, R.M., and H.G. Enghardt. 1976. Deer forage and overstory dynamics in a loblolly pine plantation. *J. Range Manage.* 29(2):104–108.

Blais, J.R. 1985. The ecology of the eastern spruce budworm: A review and discussion. In C.J. Sanders, R.W. Stark, E.J. Mullins, and J. Murphy (Eds.), *Recent Advances in Spruce Budworm Research*, pp. 49–59. Proc. CANUSA Spruce Budworm Res. Symp., Sept. 16-20, 1984. Bangor, ME, Can For. Serv., Cat. No. Fo18-5/1984.

Blauer, A.C., E.D. McArthur, R. Stevens, and S.D. Nelson. 1993. Evaluation of roadside stabilization and beautification plantings in south-central Utah. U.S. For. Serv. Res. Pap. INT-462.

Blum, B.M. 1961, Age-size relationships in all-aged northern hardwoods. U.S. For. Serv., NE For. Exp. Stn., For. Res. Note No. 125.

Blum. B.M., and S.M. Filip. 1963. A demonstration of four intensities of management in northern hardwoods. U.S. For. Serv. Res. Pap. NE-4.

Blum, B.M., and D.S. Solomon. 1980. Growth trends in pruned red spruce trees. U.S. For. Serv. Res. Note. NE-294.

Boerker, R.H. 1916. A historical study of forest ecology; its development in the fields of botany and forestry. *Quart. J. For.* Sept. 1916.

Boldt, C.E. 1974. Silviculture of ponderosa pine in the Black Hills: The status of our knowledge. U.S. For. Serv. Res. Pap. RM-124.

Bonner, F.T. 1974. Seed testing. In C.S. Schopmeyer (Ed.), *Seeds of Woody Plants in the United States*, pp. 136–152. U.S. Dept. Agric., For. Serv., Agric. Handbk. No. 450.

Bonner, F.T. 1991. Seed management. In M.L. Duryea and P.M. Dougherty, (Eds.), *Forest Regeneration Manual*, Ch. 4, pp. 51–73. Dordrecht, The Netherlands: Kluwer Academ. Publ.

Bonner, F.T., B.F. McLemore, and J.P. Barnett. 1974. In C.S. Schopmeyer (Ed.), *Seeds of Woody Plants in the United States*, pp. 126–135. U.S. Dept. Agric., For. Serv., Agric. Handbk. No. 450.

Bormann, D.B., and G.E. Likens. 1979. *Pattern and Process in a Forested Ecosystem*. New York: Springer-Verlag.

Botkin, D.B., J.F. Janak, and J.R. Wallis. 1972a. Rationale, limitations, and assumptions of a northeastern forest growth simulator. *IBM J. Res. Develop.* 16(2):101–116.

Botkin, D.B., J.F. Janak, and J.R. Wallis. 1972b. Some ecological consequences of a computer model for forest growth. *J. Ecol.* 60:948–972.

Botti, W. 1982. Disc-trencher. In *Proceedings Artificial Regeneration of Conifers in the Upper Great Lakes Region*, pp. 224–226. Mich. Technol. Univ., Houghton, MI.

Boyce, J.S. 1948. *Forest Pathology*, 2d ed., New York: McGraw-Hill.

Boyce, S.G., and R.W. Carpenter. 1968. Provisional grade specifications for hardwood growing-stock trees. U.S. For. Serv. Res. Pap. NE-97.

Boyd, R.J. 1976. The biology of planting. In D.M. Baumgartner and R.J. Boyd (Eds.), *Tree Planting in the Inland Northwest*, pp. 10–17. Proc. Conf. at Wash. State Univ., Feb. 17–19, 1976. Wash. State Univ., Coop. Ext. Serv., Pullman, WA.

Boyd, R.J. 1986. Conifer performance following weed control site preparation treatments in the Inland Northwest. In D.M. Baumgartner, R.J. Boyd, D.W. Breuer, and D.L. Miller (Eds.), *Weed Control for Forest Productivity in the Interior West*, pp. 95–104. Proc. Symp., Feb. 5–7, 1989. Spokane, Wash., Wash. State Univ., Coop. Ext. Serv., Pullman, WA.

Boyer, W.D. 1985. Timing of longleaf pine seedling release from overtopping hardwoods: A look 30 years later. *S. J. Appl. For.* 9(2):114–116.

Boyer, W.D. 1989. Response of planted longleaf pine bare-root and container stock to site preparation and release: Fifth-year results. In J.H. Miller (Ed.), *Proceedings of the Fifth Biennial Southern Silviculture Research Conference*, pp. 165–168. Memphis, TN., Nov. 1–3, 1989. U.S. For. Serv. Gen. Tech. Rpt. SO-74.

Boyer, W.D., and J.B. White. 1990. Natural regeneration of longleaf pine. In R.M. Farrar, Jr. (Ed.), *Proceedings of the Symposium on the Management of Longleaf Pine*, pp. 94–113. U.S. For. Serv. Gen. Tech. Rpt. SO-75.

Braastad, H. 1979. Vekst of stabilitet i et forbandsforsok med gran. *Norsk Inst. Skogf.* 34(7):169–215. (Summary only).

Brandle, J.R., and D.L. Hintz. 1987. An ill wind meets a windbreak. *Sci. of Food and Agric.* 5(4):8–12.

Bray, J.R. 1956. Gap phase replacement in a maple-basswood forest. *Ecol.* 37(3):598–600.

Brender, E.V. 1973. Silviculture of loblolly pine in the Georgia Piedmont. Ga. For. Res. Counc. Rpt. No. 33.

Brennan, L.A., and S.M. Hermann. 1994. Prescribed fire and forest pests: Solutions for today and tomorrow. *J. For.* 92(11):34–37.

Briggs, R.D., and R.C. Lemin, Jr. 1991. Early response of balsam fir (*Abies balsamea*

(L.)Mill.) to precommercial thinning in the context of site quality. In C.M. Simpson (Ed.), *Proceedings of the Conference on Natural Regeneration Management*, pp. 201–211. March 27–28, 1990., Fredericton, NB For. Can., Maritimes Reg., NB For. Res. Adv. Com., Fredericton, Cat. No. Fo42-162/991E.

Brinkman, K.A. 1955. Epicormic branching on oaks in sprout stands. U.S. Dept. Agric., Tech. Pap. 146.

Brinkman, K.A., and E.I. Roe. 1975. Quaking Aspen: Silvics and management in the Lake States. U.S. For. Serv., Agric. Handbk. No. 486.

Brissette, J.C. 1991. Development and function of the root systems of southern pine nursery stock. In *Proc. S. For. Nursery Assn.*, pp. 67–81. July 23–26, 1990, Biloxi, MS, Miss. State For. Comm. Proc. Reprint., U.S. For. Serv., S. For. Exp. Stn., New Orleans.

Brissette, J.C. (Ed.). 1993. *Proceedings of the 7th Biennial Southern Silvicultural Research Conference*. U.S. For. Serv. Gen. Tech. Rpt. SO-93.

Brissette, J.C., J.P. Barnett, and T.D. Landis. 1991. Container seedlings. In M.L. Duryea and P.M. Dougherty (Eds.), *Forest Regeneration Manual*, Ch. 7, pp. 117–141. Dordrecht, The Netherlands: Kluwer Academ. Publ.

Brissette, J.C., M. Elliott, and J. Barnett. 1990. Producing container longleaf pine seedlings. In R.M. Farrar, Jr. (Ed.), *Proceedings of the Symposium on the Management of Longleaf Pine*, pp. 52–71. U.S. For. Serv. Gen. Tech. Rpt. SO-75.

Broadfoot, W.M., B.G. Blackmon, and J.B. Baker. 1971. Soil management for hardwood production. In *Proceedings of the Symposium On Southeastern Hardwoods*, pp. 17–29. U.S. For. Serv. SE Area State and Priv. For., Atlanta, GA.

Brodie, J.D., and Tedder, P. L. 1982. Regeneration delay: Economic cost and harvest loss. *J. For.* 80(1):26–28.

Brodie, J.D., P.J. Kuch, and C. Row. 1987. Economic analysis of the silvicultural effects of vegetation management at the stand and forest levels. In J.D. Walstad and P.J. Kuch (Eds.), *Forest Vegetation Management for Conifer Production*, Ch. 12, pp. 365–395. New York: Wiley.

Brooks, D.J., and G.E. Grant. 1992a. New approaches to forest management, Part one. *J. For.* 90(1):25–28.

Brooks, D.J., and G.E. Grant. 1992b. New approaches to forest management, Part two of two. *J. For.* 90(2):21–24.

Broun, A.F. 1912. *Silviculture in the Tropics.* London: MacMillan and Co, Ltd.

Brown, G. 1974. Direct seeding in Ontario. In J.H. Cayford (Ed.), *Direct Seeding Symposium*, pp 119–124. Sept. 11–13, 1973. Timmins, Ont. Dept. Environ., Can. For. Serv., Publ. No. 1339.

Brown, G.W. 1973. The impact of timber harvest of soil and water resources. Ore. State Univ. Ext. Serv., Ext. Bull. 827.

Buckman, R.E., and R.F. Wambach. 1966. Physical response and economic implications of thinning methods in red pine. In *Proceedings of the Annual Meeting of the Society of American Foresters,* pp. 185–189. Oct. 24–28, 1965, Detroit, MI, Soc. Am. For., Washington, DC.

Bull, E.L., and A.D. Partridge. 1986. Methods of killing trees for use by cavity nesters. *Wild. Soc. Bull.* 14:142–146.

Buongiorno, R.G., and B.R. Michie. 1980. A matrix model of uneven-aged forest management. *For. Sci.* 26:609–625.

Burdett, A.N., and D.G. Simpson. 1984. Lifting, grading, packaging, and storage. In M.L. Duryea and T.D. Landis (Eds.), *Forest Nursery Manual: Production of Bareroot Seedlings.* Ch. 21, pp. 227–234. Martinus Nijhoff/Dr W. Junk Publ. The Hague/Boston/Lancaster. For. Res. Lab., Ore. State Univ., Corvallis.

Burger, J.A., K.J. Wimme, W.B. Stuart, and T.A. Walbridge, Jr. 1989. Site disturbance and machine performance from tree-length skidding with a rubber-tired skidder. In J.H. Miller (Ed.), *Proceedings of the Fifth Biennial Southern Silviculture Research Conference,* pp. 521–536. U.S. For. Serv. Gen. Tech. Rpt. SO-74.

Burgess, R.L., and D.M. Sharpe (Eds.), 1981. *Forest Island Dynamics in Man-dominated Landscapes.* Ecol. Stud. No. 1, New York: Springer-Verlag.

Burkhart, H.E., and B.V. Bredenkamp. 1989. Product-class proportions for thinned and unthinned loblolly pine plantations. *S. J. Appl. For.* 13(4):192–195.

Burkhart, H.E., G.R. Glover, and P.T. Sprinz. 1987. Loblolly pine growth and yield response to vegetation management. In J.D. Walstad and P.J. Kuch (Eds.), *Forest Vegetation Management for Conifer Production,* pp. 243–271. New York: Wiley.

Burns, R.M. 1965. Machine seeding—Row and broadcast. In *Proceedings Direct Seeding Workshop,* pp. 18–20. Alexandria, LA, Oct. 5–6, 1965, and Tallahassee, FL, Oct. 20–21, 1965. U.S. For. Serv. GSA Atlanta, GA. 66-4542.

Burns, R.M., and B.H. Honkala (Eds.). 1990a. *Silvics of North America. Volume 1, Conifers.* U.S. For. Serv., Agric. Handbk. 654.

Burns, R.M., and B.H. Honkala (Eds.). 1990b. *Silvics of North America. Volume 2, Hardwoods.* U.S. For. Serv., Agric. Handbk. 654.

Burr, K.E. 1990. The target seedling concept: Bud dormancy and cold-hardiness. In *Target Seedling Symposium: Proceedings, Combined Meeting of the Western Forest Nursery Association,* Ch. 7, pp. 79–90. U.S. For. Serv. Gen. Tech. Rpt. RM-200.

Burroughs, E.R., Jr., and J.G. King. 1989. Reduction of soil erosion on forest roads. U.S. For. Serv. Gen. Tech. Rpt. INT-264.

Busby, R.L. 1989. Economic returns using sulfometuron methyl (oust) for herbaceous weed control in southern pine plantations. In J.H. Miller (Ed.), *Proceedings of the Fifth Biennial Southern Silviculture Research Conference,* pp. 359–364. U.S. For. Serv. Gen. Tech. Rpt. SO-74.

Cain, M.D. 1983. Precommercial thinning for the private, nonindustrial landowner: A methodology report. In E.P. Jones (Ed.), *Proceedings of the Second Biennial Southern Silviculture Research Conference,* pp. 200–205. U.S. For. Serv. Gen. Tech. Rpt. SE-24.

Cain, M.D. 1989. A simple competition assessment system associated with intensive competition control in natural loblolly-shortleaf pine seedling stands. *S. J. Appl. For.* 13(1):8–12.

Cain, M.D. 1991a. The influence of woody and herbaceous competition on early growth of naturally regenerated loblolly and shortleaf pines. *S. J. Appl. For.* 15(4):179–185.

Cain, M.D. 1991b. Hardwoods on pine sites: Competition or antagonistic symbiosis. *For. Ecol. and Manage.* 44(1991):147–160.

Cain, M.D. 1993. Ten-year results from precommercial strip-thinning: Paradigm lost or reinforced? *S. J. Appl. For.* 17(1):16–21.

Cain, M.D., and D.A. Yaussy. 1983. Reinvasion of hardwoods following eradication in an uneven-aged pine stand. U.S. For. Serv. Res. Pap. SO-188.

Campbell, R.A. 1951. Tree grades, yields, and values for some Appalachian hardwoods. U.S. For. Serv. SE For. Exp. Stn., Stn. Pap. No. 9.

Campbell, R.L. 1965. Mechanical row and broadcast seeding techniques on Georgia Kraft company's lands. In *Proceedings Direct Seeding Workshops,* pp. 88–91. Oct. 5–6, 1965, Alexandria, LA, and Oct. 20–21, Tallahassee, FL. U.S. For. Serv. GSA Atlanta GA 66-4542.

Campbell, T.E. 1983. Effects of initial seedling density on spot-seeded loblolly and slash pines at 15 years. In *Proceedings of the Second Biennial Southern Silvicultural Research Conference,* pp. 118–127. U.S. For. Serv. Gen. Tech. Rpt. SE-24.

Campbell, T.E., and W.F. Mann, Jr. 1973. Regenerating loblolly pine by natural seeding, and planting. U.S. For. Serv. Res. Pap. SO-84.

Canham, H.O. 1986. Comparable valuation of timber and recreation in forest planning. *J. Environ. Manage.* 23:335–339.

Cannell, M.G.R. 1980. Productivity of closely-spaced young poplar on agricultural soils in Britain. *For.* 53:1–21.

Carlson, C.E., and W.C. Schmidt. 1989. Influence of overstory removal and western spruce budworm defoliation on growth of advance conifer regeneration in Montana. U.S. For. Serv. Res. Pap. INT-409.

Carlson, L.W. 1979. Guidelines for rearing containerized conifer seedlings in the Prairie Provinces. Environ. Can., For. Serv., N. For. Res. Ctr. Info. Rpt. NOR-X-214.

Carlson, W.C. 1991. Lifting, storing, and transporting southern pine seedlings. In M.L. Duryea and P.M. Dougherty (Eds.), *Forest Regeneration Manual*, Ch. 16, pp. 219–320. Dordrecht, The Netherlands: Kluwer Academ. Publ.

Carmean, W.H. 1982. Soil-site evaluation for conifers in the upper Great Lakes region. In G.D. Mroz and J.F. Berner (Eds.), *Artificial Regeneration of Conifers in the Upper Great Lakes Region*, pp. 31–52. Mich. Technol. Univ., Houghton, MI.

Carmean, W.H., and S.G. Boyce. 1973. Hardwood log quality in relation to site quality, U.S. For. Serv. Res. Pap. NC-103.

Carter, M.C., and E.H. White. 1971. The necessity for intensive cultural treatment in cottonwood plantations. Auburn Univ. Agric. Exp. Stn. Circ. 189.

Carvell, K.L. 1979. Factors affecting the abundance, vigor, and growth responses of understory oak seedlings. In H.A. Holt and B.C. Fischer (Eds.), *Regenerating Oaks in Upland Hardwood Forests*, pp. 23–26. 1979 J.S. Wright For. Conf., West Lafayette, Ind., Purdue Univ. Res. Found.

Cassel, J.F., and J.M. Wiehe. 1980. Uses of shelterbelts by birds. In R.M. DeGraaf and N.G. Tilghman (Eds.), *Management of Western Forests and Grasslands for Nongame Birds,* pp. 78–87. U.S. For. Serv. Gen. Tech. Rpt. INT-86.

Cayford, J.H. 1972. Container planting systems in Canada. *For. Chron.* 48:235–239.

Chamberlain, D. 1991. Prediction of advanced regeneration: A Quebec experience. In C.M. Simpson (Ed.), *Proceedings of the Conference on Natural Regeneration Management*, pp. 55–61. For. Can., Maritimes Reg., Fredericton, NB.

Chamberlin, T.W. 1982. Influence of forest and rangeland management in Anadromous fish habitat in western North America: Timber harvest. U.S. For. Serv. Gen. Tech. Rpt. PNW-136.

Champion, H.G., and G. Trevor. 1938. *Manual of Indian Silviculture.* Oxford: Oxford Univ. Press.

Chappell, D.E. 1962. Value growth of pine pulpwood on the George Walton Experimental Forest. U.S. For. Serv., S. For. Exp. Stn., Stn. Pap. 140.

Chavasse, C.G.R. 1974. A review of land clearing for site preparation for intensive plantation forestry. In *Symposium Stand Establishment,* pp. 109–128. Proc. IUFRO Joint Meet., Div. 1 and 3, Oct. 15–19, 1974. Wageningen, Netherlands. Intl. Union For. Res. Organ.

Cheyney, E.G. 1942. *American Silvics and Silviculture*. Minneapolis, MN: The Univ. Minn. Press.

Childs, T.W. 1949. Shade trees for the North Pacific Area. In *Trees, the Yearbook of Agriculture 1949*, pp. 82–85. 81st Congr., 1st Sess., House Doc. No. 29. U.S. Gov. Print. Off., Washington, DC.

Chisman, H.H., and F.X. Schumacher. 1940. On the tree-area ratio and certain of its applications. *J. For.* 38:311–317.

Chou, C.K. 1979. Monoculture, species diversification and disease hazard in forestry. *N. Z. J. For. Res.* 9:21–35.

Christersson, L., L. Sennerby-Forsse, and L. Zsuffa. 1993. The role and significance of woody biomass plantations in Swedish agriculture. *For. Chron.* 69(6):687–693.

Chroust, L. 1980. Erziehung von Fichtenbestaenden in durch Schnee und Wind gefaehrdeten Lagen. In H. Kramer (Ed.), *Biologische, technische und wirtschaftliche Aspekte der Jungbestandspflege*, pp. 206–213. SchrReihe Forstl. Fak. Univ. Goettingen, Bd. 67, (Transl. D. Clerc, 1985).

Clark, F.B., and R.F. Watt. 1971. Silvicultural methods for regenerating oaks. In *Oak Symposium Proceedings*, pp. 37–43. U.S. For. Serv., NE For. Exp. Stn., Upper Darby, PA.

Clark, R.G., and E.E. Starkey. 1990. Use of prescribed fire in rangeland ecosystems. In J.D. Walstad, S.R. Radosevich, and D.V. Sandberg (Eds.), *Natural and Prescribed Fire in Pacific Northwest Forests*, pp. 81–91. Corvallis: Ore. State Univ. Press.

Clatterbuck, W.K. 1993. Are overtopped white oak good candidates for management? In J. Brissette (Ed.), *Proceedings of the Seventh Biennial Southern Silvicultural Research Conference*, pp. 497–499. U.S. For. Serv. Gen. Tech. Rpt. SO-93.

Clayton, J.L. 1990. Soil disturbance resulting from skidding logs on granitic soils in central Idaho. U.S. For. Serv. Res. Pap. INT-436.

Cleary, B.D. 1978. Vegetation management and its importance in reforestation. Ore. St. Univ. For. Res. Lab. Res. Note 60.

Cleary, B., and R. Greaves. 1976. Determining planting stock needs. In D.M. Baumgartner and R.J. Boyd (Eds.), *Tree Planting in the Inland Northwest*, pp. 60–81. Wash. State Univ. Ext. Serv., Wash. State Univ., Pullman, WA.

Cleary, B.D., R.D. Greaves, and P.W. Owston. 1978. Seedlings. In R.D. Greaves, R.K. Hermann (Eds.), *Regenerating Oregon's Forests*, Ch. 6, pp. 63–97. Ore. State Univ., Sch. For., Corvallis.

Cleary, E.J. 1967. *The ORSANCO Story: Water Quality Management in the Ohio Valley Under an Interstate Company.* Baltimore, MD: The Johns Hopkins Press.

Cline, A.C., and E.D. Fletcher. 1928. *Pruning for Profit as Applied to Eastern White Pine.* A joint study by The Harvard For. and the Mass. For. Assoc., Boston.

Cochran, P.H., and W.W. Oliver 1988. Growth rates for managed stands of white fir. In W.C. Schmidt (Ed.), *Proceedings—Future Forests of the Mountain West: A Stand Culture Symposium*, pp. 197–200. U.S. For. Serv. Gen. Tech. Rpt. INT-243.

Coffman, M.S. 1982. Regeneration prescriptions: The need to be holistic. In G.D. Mroz and J.F. Berner (Eds.), *Artificial Regeneration of Conifers in the Upper Great Lakes Region,* pp. 109–208. Mich. Technol. Univ., Houghton, MI.

Cole, D.M. 1984. Crop-tree thinning a 5-year-old western larch stand: 25-year results. U.S. For. Serv. Res. Pap. INT-328.

Cole, D.M., and W.C. Schmidt. 1986. Site treatments influence development of a young mixed-species western larch stand. U.S. For. Serv. Res. Pap. INT-364.

Coleman, A.E. 1953. *Vegetation and Watershed Management.* New York: Ronald Press Co.

Coleman, S.S., and D.G. Neary (Ed.), 1991. *Proceedings of the 6th Biennual Southern Silviculture Research Conference.* U.S. For. Serv. Gen. Tech. Rpt. SE-70.

Connell, J.H., and R.D. Slatyer. 1977. Mechanisms of succession in natural communities and their role in community stability and organization. *Am. Nat.* 111:1119–1144.

Conway, S. 1982. *Logging Practices: Principles of Timber Harvesting Systems,* 2d ed. San Francisco: Miller Freeman Publ.

Cooper, A.W. 1981. Above-ground biomass accumulation and net primary production during the first 70 years of succession in *Populus grandidentata* stands on poor sites in northern lower Michigan. In D.C. West, H.H. Shugart, and D.B. Botkin (Eds.), *Forest*

Succession Concepts and Applications, Ch. 21, pp. 339–360. New York: Springer-Verlag.

Cordell, C.E., G.E. Hatchell, and D.H. Marx. 1990. Nursery culture of bare-root longleaf pine seedlings. In *Proceedings on the Symposium on the Management of Longleaf Pine*, pp. 38–51. U.S. For. Serv. Gen. Tech. Rpt. SO-75.

Coulson, R.N. 1981. Evolution of concepts of integrated pest management in forests. *J. Ga. Entomol. Soc.* 16(2):301–316.

Cox, S.K., and D.H. Van Lear. 1985. Biomass and nutrient accretion on Piedmont sites following clearcutting and natural regeneration of loblolly pine. In E. Shoulders (Ed.), *Proceedings of the Third Biennial Southern Silvicultural Research Conference,* pp. 501–506. U.S. For. Serv. Gen. Tech. Rpt. SO-54.

Crawford, H.S., and R.M. Frank. 1987. Wildlife habitat responses to silvicultural practices in spruce-fir forests. In 52nd Trans. N.A. Wildl. Nat. Resourc. Conf., pp. 92–100. Wildl. Manage. Inst., Washington, DC.

Creighton, J.L., B.R. Zutter, G.R. Glover, and D.H. Gerstad. 1987. Planted pine growth and survival responses to herbaceous vegetation control, treatment duration, and herbicide application technique. *S. J. Appl. For* 4(1):223–227.

Cremer, K.W., and E.M. Meredith. 1976. Growth of radiata pine after row thinning compared to selective thinning. *Aust. For.* 39(3):193–200.

Crocker, T.C., Jr., and W.D. Boyer. 1975. Regenerating longleaf pine naturally. U.S. For. Serv. Res. Pap. SO-105.

Croft, A.R. 1950. A water cost of runoff control. *J. Soil and Water Conserv.* 5:13–15.

Croft, B.A. 1985. Integrated pest management: The agricultural-environmental rationale. In R.E. Frisbie and P.L. Adkisson (Eds.), *Integrated Pest Management on Major Agricultural Systems*, pp. 712–728. Texas Agric. Exp. Stn. MP-1616. Texas A&M Univ., College Stn., TX.

Crouch, G.L. 1986. Effects of thinning pole-sized lodgepole pine on understory vegetation and large herbivore activity in central Colorado. U.S. For. Serv. Res. Pap. RM-268.

Crow, A.B., and C.L. Shilling. 1980. Use of prescribed burning to enhance southern pine timber production. *S. J. Appl. For.* 4(1):15–18.

Crow, T.R., and F.T. Metzger. 1987. Regeneration under selection cutting. In R.D. Nyland (Ed.), *Managing Northern Hardwoods*, pp. 81–94. SUNY Coll. Environ. Sci. and For., Syracuse, NY, SUNY-ESF Fac. For. Misc. Publ. No. 13 (ESF 87-002) (Soc. Am. For. Publ. No. 87-03).

Crow, T.R., R.D. Jacobs, R.R. Oberg, C.H. Tubbs. 1981. Stocking and structure for maximum growth in sugar maple selection stands. U.S. For. Serv. Res. Pap. NC-199.

Cubbage, F.W., W.C. Siegel, and T.K. Haines. 1987. Water quality laws affecting forestry in eastern United States. In S.J. Nix and P.E. Black (Eds.), *Monitoring, Modeling, and Mediating Water Quality*, pp. 597–609. Proc. Symp., Am. Water Resourc. Assoc., Bethesda, MD.

Cunningham, J.B., R.P. Balda, and W.S. Grant. 1980. Selection and use of snags by secondary cavity-nesting birds of the ponderosa pine forest. U.S. For. Serv. Res. Pap. RM-222.

Curtis, R.O. 1970. Stand density measures: An interpretation. *For. Sci.* 16:403–414.

Curtis, R.O., G.W. Clendenen, D.L. Reukema, and D.J. DeMars. 1982. Yield tables for managed stands of coast Douglas-fir. U.S. For. Serv. Gen. Tech. Rpt. PNW-135.

Dale, M.E. 1968. Growth response from thinning young even-aged white oak stands. U.S. For. Serv. Res. Rpt. NE-112.

Dale, M.E. 1972. Growth and yield predictions for upland oak stands, 10 years after initial thinning. U.S. For Serv. Res. Pap. NE-241.

Daniel, T.W. 1980. The middle and southern Rocky Mountain region. In J.W. Barrett (Ed.), *Regional Silviculture of the United States*, 2d ed. pp. 277–340. New York: Wiley.

Daniel T.W., J.A. Helms, and F.S. Baker. 1979. *Principles of Silviculture,* 2d ed. New York: McGraw-Hill.

Davey, C.B., and H.H. Kraus. 1980. Functions and maintenance of organic matter in forest nursery soils. In L.P. Abrahamson and D.H. Bickelhaupt (Eds.), *Proceedings North American Forest Tree Nursery Soils Workshop,* pp. 130–165. July 28–Aug. 1, 1980, SUNY Coll. Environ. Sci. and For., Syracuse, NY.

Davidson, H., R. Mecklenburg, and C. Peterson. 1994. *Nursery Management. Administration and Culture,* 3d ed. Englewood Cliffs, NJ: Prentice Hall Career and Technol.

Davis, C.J., and R.D. Nyland. 1991. Designated skid trails: Results after 18 years. Am. Soc. Agric. Eng., ASAE Meet. Presentation Pap. No. 917522.

Davis, K.P. 1954. *American Forest Management.* New York: McGraw-Hill.

Davis, K.P. 1966. *Forest Management: Regulation and Valuation.* New York: McGraw-Hill.

Davis, K.P., G.M. Byram, and W.R. Krumm. 1959. *Forest Fire Control and Use.* New York: McGraw-Hill.

Davis, L.S., and K.N. Johnson. 1987. *Forest Management*, 3d ed. New York: McGraw-Hill.

Davis, R.M. 1976. Great Plains windbreak history: An overview. In R.W. Tinus (Ed.), *Shelterbelts on the Great Plains.* Great Plains Agric. Counc. Publ No. 78.

Davis, W.C. 1991. The role of advance growth of red spruce and balsam fir in east central Maine. In C.M. Simpson (Ed.), *Proceedings of the Conference on Natural Regeneration Management,* pp. 157–168. For. Can., Maritimes Reg., Fredericton, NB.

Day, M.W., and V.J. Rudolph. 1972. Thinning plantation red pine. Mich. State Univ., Agric. Exp. Stn. Nat. Resourc. Rpt. 151.

Day, R.J. 1984. Water management. In M.L. Duryea and T.D. Landis (Eds.), *Forest Nursery Manual: Production of Bareroot Seedlings.* Ch. 11, pp. 93–106. Martinus Nijhoff/Dr W. Junk Publ. The Hague/Boston/Lancaster. For. Res. Lab., Ore. State Univ., Corvallis.

de Liocourt, F. 1898. De l'amènagement des sapinières. *Bull. de la Société forestière de Franche-Comte et du Territoire de Belfort* 4:396–409, 645–647.

Dean, T.J., and E.J. Jokela. 1992. A density-management diagram for slash pine plantations in the Lower Coastal Plain. *S. J. Appl. For.* 16(4):178–185.

DeBell, D.S., and L.T. Alford. 1972. Sprouting characteristics and cutting practices evaluated for cottonwood. *Tree Plant. Notes* 23:1–3.

DeGraaf, R.M. 1987. Managing northern hardwoods for breeding birds. In R.D. Nyland (Ed.), *Managing Northern Hardwoods*, pp. 348–362. SUNY Coll. Environ. Sci. and For., Syracuse, NY, SUNY-ESF Fac. For. Misc. Publ. 13 (ESF 87-002) (Soc. Am. For. Publ. 87-03).

DeGraaf, R.M., and A.L. Shigo. 1985. Managing cavity trees for wildlife in the northeast. U.S. For. Serv. Gen. Tech. Rpt. NE-101.

Deisch, M.S., and R.W. Sage, Jr. 1989. Summer songbird abundance and diversity on Huntington Wildlife Forest: 1983–1989. SUNY Coll. Environ. Sci. and For., Adk. Ecol. Ctr., Newcomb, NY, ALTEMP Proj. No. 2 (Breeding bird survey), Unpubl. report.

Deisch, M.S. and R.W. Sage, Jr. 1990. Songbird diversity following shelterwood cutting in two northern hardwood stands in the central Adirondacks. *Abstr. Northeast Fish and Wildl. Conf.,* Nashua, NH, April 1990.

Della-Bianca, L. 1983. Effect of intensive cleaning on natural pruning of cove hardwoods in the southern Appalachians. *For. Sci.* 29(1):27–32.

Dengler, A. 1944. *Waldbau auf okologischer Grundlage.* Vienne: Springer-Verlag.

Derr, H.J., and W.F. Mann, Jr. 1971. Direct-seeding pines in the south. U.S. For. Serv., Agric. Handbk. No. 391.

Derr, H.J., and W.F. Mann, Jr. 1977. Bedding poorly drained sites for planting loblolly and slash pines in southwest Louisiana. U.S. For. Serv. Res. Pap. SO-134.

Dicke, S.G., and J.R. Toliver. 1988. Effects of crown thinning on baldcypress height, diameter, and volume growth. *S. J. Appl. For.* 12(4):252–256.

Dierauf, T.A. 1989a. Loblolly pine release study report number 13. Va. Dept. For. Occ. Rpt. 84.

Dierauf, T.A. 1989b. Loblolly pine release study report number 14. Va. Dept. For. Occ. Rpt. 85.

Dierauf, T.A. 1989c. Loblolly pine release study report number 15. Va. Dept. For. Occ. Rpt. 86.

Dierauf, T.A. 1990a. Loblolly pine release study report number 16. Va. Dept. For. Occ. Rpt. 87.

Dierauf, T.A. 1990b. Loblolly pine release study report number 17. Va. Dept. For. Occ. Rpt. 88.

Dierauf, T.A. 1990c. Loblolly pine release study report number 18. Va. Dept. For. Occ. Rpt. 89.

Dierauf, T.A. 1990d. Loblolly pine release study report number 19. Va. Dept. For. Occ. Rpt. 90.

Dierauf, T.A. 1990e. Loblolly pine release study report number 20. Va. Dept. For. Occ. Rpt. 91.

Dierauf, T.A. 1990f. Loblolly pine release study report number 21. Va. Dept. For. Occ. Rpt. 92.

Dimock, E.J., II. 1981. Herbicide and conifer options for reforesting upper slopes in the Cascade Range. U.S. For. Serv. Res. Pap. PNW-292.

Dingle, R.W. 1976. Principles of site preparation. In D.M. Baumgartner and R.J. Boyd (Eds.), *Tree Planting in the Inland Northwest*, pp. 107–116. Wash. State Univ. Ext. Serv., Wash State Univ., Pullman, WA.

Doucet, R. 1991. The influence of stocking of regeneration on the yield of naturally regenerated jack pine and black spruce. In C.M. Simpson (Ed.), *Proceedings of the Conference on Natural Regeneration Management*, pp. 181–192. Fredericton, NB For. Can., Maritimes Reg., Fredericton. Cat. No. Fo42-162/991E.

Dougherty, P.M., and M.L. Duryea. 1991. Regeneration: An overview of past trends and basic steps needed to ensure future success. In M.L. Duryea and P.M. Dougherty (Eds.), *Forest Regeneration Manual*, Ch. 1, pp. 3–7. Dordrecht, The Netherlands: Kluwer Academ. Publ.

Drew, T.J., and J.W. Flewelling. 1979. Stand density management: An alternate approach and its application to Douglas-fir plantations. *For. Sci.* 25(3):518–532.

Drew, A.P., J.V. Berglund, and D.J. Robison. 1985. Responses from thinning a 50-year-old Norway spruce plantation in central New York. SUNY Coll. Environ, Sci. and For., Sch. For. Tech. Publ. No. 1 (ESF 85-002).

Dronogen, S.I. 1984. Windbreaks in the Great Plains. *N. J. Appl. For.* 1(3):55–59.

Dryness, C.T. 1965. Soil surface condition following tractor and high-lead logging in the Oregon Cascades. *J. For.* 63(4):272–275.

Duddles, R.E., and P.W. Owston. 1990. Performance of conifer stocktypes on National Forests in the Oregon and Washington coastal ranges. In R. Rose, S.J. Campbell, and T.D. Landis (Eds.), *Target Seedlings Symposium: Proceedings, Combined Meeting of the Western Forest Nursery Association*, pp. 263–268. U.S. For. Serv. Gen. Tech. Rpt. RM-200.

Duerr, W.A. 1960. *Fundamentals of Forestry Economics.* New York: McGraw-Hill.

Duerr, W.A. 1988. *Forestry Economics As Problem Solving.* W.A. Duerr, Blacksburg, VA, Distrib. Orange Student Book Store, Syracuse, NY.

Duerr, W.A., J.F. Fedkiw, and S. Guttenberg. 1956. Financial maturity: A guide to profitable timber growing. U.S. Dept. Agric. Tech. Bull. No. 1146.

Dunning, D. 1928. A tree classification for forests of the Sierra Nevada. *J. Agric. Res.* 36:755–771.

Dunsworth, B.G. 1988. Impact of lift date and storage on field performance for Douglas-fir and western hemlock. In T.D. Landis (Ed.), *Proceedings, Combined Meeting of the Western Forest Nursery Association*, pp. 199–206. U.S. For. Serv. Gen. Tech. Rpt. RM-167.

Duryea, M.L. 1984. Nursery cultural practices: Impacts on seedling quality. In M.L. Duryea and T.D. Landis (Eds.), *Forest Nursery Manual: Production of Bareroot Seedlings.* Ch. 15, pp. 143–164. Martinus Nijhoff/Dr W. Junk Publ. The Hague/Boston/Lancaster. For. Res. Lab., Ore. State Univ., Corvallis.

Duryea, M.L., and T.D. Landis (Eds.). *Forest Nursery Manual: Production of Bareroot Seedlings.* Martius Nijhoff/Dr W. Junk Publ. The Hague/Boston/Lancaster. For. Res. Lab., Ore. State Univ., Corvallis.

Dutrow, G.F. 1971. Economic implications of silage sycamore. U.S. For. Serv. Res. Pap. SO-66.

Dutrow, G.F., and J.R. Saucier. 1976. Economics of short-rotation sycamore. U.S. For. Serv. Res. Pap. SO-114.

Edgren, J.W. 1984. Nursery storage to planting hole: A seedling's hazardous journey. In M.L. Duryea and T.D. Landis (Eds.), *Forest Nursery Manual: Production of Bareroot Seedlings.* Ch. 22, pp. 235–242. Martinus. Nijhoff/Dr W. Junk Publ. The Hague/Boston/Lancaster. For. Res. Lab., Ore. State Univ., Corvallis.

Einspahr, D.W. 1982. Selection of tree species for planting. In G.D. Mroz and J.F. Berner (Eds.), *Artificial Regeneration of Conifers in the Upper Lake States Region,* pp. 129–133. Mich. Technol. Univ., Houghton, MI.

Ek, A.R. 1982. Closing the knowledge gaps in conifer regeneration research and practice in the Lake States. In G.D. Mroz and J.F. Berner (Eds.), *Artificial Regeneration of Conifers in the Upper Great Lakes Region,* pp. 420–428. Mich. Technol. Univ., Houghton, MI.

Engle, L.G. 1947. Skidding damage by tractors and horses in selective logging of northern hardwoods. U.S. For. Serv., Lake States For. Exp. Stn. Tech. Note 277.

Erdmann, G.G. 1967. Chemical weed control increases survival and growth in hardwood plantings. U.S. For. Serv. Res. Note NC-34.

Erdmann, G.G. 1987. Methods of commercial thinning in even-aged northern hardwood stands. In R.D. Nyland (Ed.), *Managing Northern Hardwoods,* pp. 191–210. SUNY Coll. Environ. Sci. and For., Fac. For. Misc. Publ. No. 13 (ESF 87-002) (Soc. Am. For. Publ. No. 87-03).

Erdmann, G.G. 1990. *Betula alleghaniensis* Britton. Yellow birch. In R.M. Burns and B.H. Honkala (Eds.), *Silvics of North America. Volume 2. Hardwoods,* pp. 133–147. U.S. For. Serv., Agric. Handbk. 654.

Erdmann, G.G., R.M. Peterson, Jr., and R.M. Godman. 1981. Cleaning yellow birch seedling stands to increase survival, growth, and crown development. *Can. J. For. Res.* 11:62–68.

Ernst, R.L., and W.H. Knapp. 1985. Forest stand density and stocking: Concepts, terms, and the use of stocking guides. U.S. For. Serv. Gen. Tech. Rpt. WO-44.

Evans, J. 1982. *Plantation Forestry in the Tropics.* Oxford: Clarendon Press.

Eyre, F.H., and P. Zehngraff. 1948. Red pine management in Minnesota. U.S. Dept. Agric. Circ. No. 778.

Eyre, F.H., and W.H. Zillgitt. 1953. Partial cuttings in northern hardwoods of the Lake States. U.S. Dept. Agric. Tech. Bull. 1076.

Ezell, A.W., L.R. Nelson, S.H. Schoenholz, and J.G. Williams. 1993. Controlling undesirable stems using low intensity injections of imazapyr. In J.C. Brissette (Ed.), *Proceedings of the Seventh Biennial Southern Silvicultural Research Conference,* pp. 455–459. U.S. For. Serv. Gen. Tech. Rpt. SO-93.

Farnsworth, E.C., and J.W. Barrett. 1966. Responses to silvicultural treatment: Northern hardwoods type five years after logging. In Vol. 2. Proc. Sixth World For. Congr., pp. 2338–2343. Madrid. Tech. Comm. I. Afforestation Techniques and Tree Improvement.

Farrar, R.M. Jr. 1985. Volume growth predictions for thinned even-aged natural longleaf pine stands in the east Gulf area. U.S. For. Serv. Res. Pap. SO-220.

Farrar, R.M., Jr., and W.D. Boyer. 1990. Managing longleaf pine under selection system—Promises and problems. In S.S. Coleman and D.G. Neary (Eds.), Vol. 1. *Proceedings of the 6th Biennial Southern Silviculture Research Conference,* pp. 357–368. U.S. For. Serv. Gen. Tech. Rpt. SE-70.

Farrar, R.M., Jr., T.J. Straka, and C.E. Burkhardt. 1989. A quarter century of selection management on Mississippi State farm forestry forties. Miss. Agric. and For. Ext. Stn. Tech. Bull. 164.

Feduccia, D.P., T.R. Dell, W.F. Mann, Jr., T.E. Campbell, and B.H. Polmer. 1979. Yields of unthinned loblolly pine plantations on cutover sites in the West Gulf Region. U.S. For. Serv. Res. Pap. SO-148.

Ferguson, D.E. 1984. Needed: Guidelines for defining acceptable advance regeneration. U.S. For. Serv. Res. Note INT-341.

Ferguson, D.E., and D.L. Adams. 1979. Guidelines for releasing advance grand fir from overstory competition. Univ. Idaho Wildl. and Range Exp. Stn. Note. 35.

Ferrell, W.K. 1953. Effect of environmental conditions on survival and growth of forest tree seedlings under field conditions in the Piedmont region of North Carolina. *Ecol.* 34:667–688.

Ffolliott, P.F. 1990. Opportunities for fire management in the future. In J.S. Krammes, Tech. Coord. *Effects of Fire Management of Southwestern Natural Resources,* pp. 152–167. U.S. For. Serv. Gen. Tech. Rpt. RM-191.

Ffolliott, P.F., and G.J. Gottfried. 1991. Mixed conifer and aspen regeneration in small clearcuts within a partially harvested Arizona mixed conifer forest. U.S. For. Serv. Res. Pap. RM-294.

Ffolliott, P.F., E.A. Hamson, and A.D. Sander. 1965. Snow in natural openings and adjacent ponderosa pine stands on the Beaver Creek Watersheds. U.S. For. Serv. Res. Note RM-53.

Fight, R.D., T.D. Fahey, and S. Johnson. 1992. Timber quality and pruning: An analysis of management regimes for the Siuslaw National Forest. In D. Murphey (Ed.), *Getting to the Future Through Silviculture—Workshop Proceedings,* pp. 93–95. U.S. For. Serv. Gen. Tech. Rpt. INT-291.

Filip, S.M. 1967. Harvesting costs and returns under four cutting methods in mature beech-birch-maple stands in New England. U.S. For. Serv. Res. Pap. NE-87.

Florence, R.G. 1977. The silvicultural decision. *For. Ecol. and Manage.* 1(4):293–306.

Foiles, M.W. 1956. Effects of thinning a 55-year-old western white pine stand. *J. For.* 54(2):130–132.

Foiles, M.W. 1972. Responses in a western white pine stand to commercial thinning methods. U.S. For. Serv. Res. Note INT-159.

Ford-Robertson, F.C. (Ed.). 1971. *Terminology of Forest Science, Technology, Practice and Products.* Multilingual For. Terminol. Ser. No. 1. Soc. Am. For., Washington, DC.

Forward, P.W. 1982. Stock production specifications—Bareroot stock "state of the art." In G.D. Mroz and J.F. Berner (Eds.), *Artificial Regeneration of Conifers in the Upper Great Lakes Region,* pp. 260–272. Mich. Technol. Univ., Houghton, MI.

Fowells, H.A. (Ed.). 1965. *Silvics of Forest Trees of the United States.* U.S. For. Serv., Agric. Handbk. 271.

Frank, B., and A. Netboy. 1950. *Water, Land and People.* New York: Knopf.

Frank, R.M. 1977. Indications of silvicultural potential from long-term experiments in spruce-fir types. In *Symposium on the Intensive Culture of Northern Forest Types,* pp. 159–177. U.S. For. Serv. Gen. Tech. Rpt. NE-29.

Frank, R.M., and J.C. Bjorkbom. 1973. A silvicultural guide for spruce-fir in the Northeast. U.S. For. Serv. Gen. Tech. Rpt. NE-6.

Frank, R.M., and B.M. Blum. 1978. The selection system of silviculture in spruce-fir stands—Procedures, early results, and comparisons with unmanaged stands. U.S. For. Serv. Res. Pap. NE-425.

Franklin, J. 1989. Toward a new forestry. *Am. For.* 95(11 & 12):37–44.

Franklin, J.F. 1990. *Abies procera* Rehd. Noble fir. In R.M. Burns and B.H. Honkala (Eds.), *Silvics of North America. Volume 1. Conifers,* pp. 80–87. U.S. For. Serv., Agric. Handbk. 654.

Fraser, J.W. 1974. Seed treatments (including repellants). In J.G. Cayford (Ed.), *Direct Seeding Symposium,* pp. 77–90. Timmins, Ont., Sept. 11, 12, 13 1973. Can. For. Serv., Dept. For. Publ. No. 1339.

Fraser, J.W. 1981. Operational direct seeding trials with black spruce on upland cutovers. Dept. Environ., Can. For. Serv., Great Lakes For. Res. Ctr. Rpt. 0-X-321.

Froehlich, H.A., and D.H. McNabb. 1984. Minimizing soil compaction in Pacific Northwest forests. In E.L. Stone (Ed.), *Forest Soils and Treatment Impacts,* pp. 159–192. Proc. 6th N. Amer. For. Soils Conf., Univ. Tenn., Knoxville.

Funk, D.T. 1961. Pruning white pine. A literature review. U.S. For. Serv. Cent. States For. Exp. Stn. Tech. Pap. 185.

Gagnon, R., H. Morin, and H. St-Pierre. 1991. Natural seed regeneration of black spruce (*Picea mariana*) stands in the Quebec boreal forest. In C.M. Simpson (Ed.), *Proceedings of the Conference on Natural Regeneration Management,* pp. 103–113. For. Can., Maritimes Reg., Fredericton, NB.

Gates, F.C., and G.E. Nichols. 1930. Relation between age and diameter in trees of the primeval northern hardwood forest. *J. For.* 28(3):395.

Gholz, H.L., and L.R. Boring. 1991. Characterizing the site: Environment, associated vegetation, and site potential. In M.L. Duryea and P.M. Dougherty (Eds.), *Forest Regeneration Manual,* Ch. 9, pp.163–182. Dordrecht, The Netherlands: Kluwer Academ. Publ.

Gibbs, C.B. 1963. Tree diameter a poor indicator of age in West Virginia hardwoods. U.S. For. Serv. Res. Note NE-11.

Gibson, I.A.S., and T. Jones. 1977. Monoculture as the origin of major forest pests and diseases. In J.M. Cherrett and G.R. Sagar (Eds.), *Origin of Pest, Parasite, Disease and Weed Problems,* pp. 139–161. Oxford: Blackwell Sci. Publ.

Gibson, H.G., and P.E. Pope. 1985. Design parameters for a biomass harvester for short-rotation hardwood stands. In E. Shoulders (Ed.), *Proceedings of the Third Biennial Southern Silvicultural Research Conference,* pp. 146–153. U.S. For. Serv. Gen. Tech. Rpt. SO-54.

Gilbert, A.M. 1954. What is this thing called growth? U.S. For. Serv., NE. For. Exp Stn., Stn. Pap. No. 71.

Gilbert, A.M. 1965. Stand differentiation ability in northern hardwoods. U.S. For. Serv. Res. Pap. NE-37.

Gilbert, A.M., and V.S. Jensen. 1958. A management guide for northern hardwoods in New England. U.S. For. Serv., NE For. Exp. Stn., Stn. Pap. No. 112.

Gill, L.S. 1949. Shade trees for the Rockies. In *Trees, the Yearbook of Agriculture 1949,* pp. 72–76. 81st Congr., 1st Sess., House Doc. No. 29. U.S. Gov. Print. Off., Washington, DC.

Gingrich, S.F. 1967. Measuring and evaluating stocking and stand density in upland hardwood forests of the central states. *For. Sci.* 13:38–53.

Gingrich, S.F. 1971. Management of young and intermediate stands of upland hardwoods. U.S. For. Serv. Res. Pap. NE-195.

Gjerstad, D.H., and B.L. Barber. 1987. Forest vegetation problems in the South. In J.D. Walstad and P.J. Kuch (Eds.), *Forest Vegetation Management for Conifer Production,* Ch. 3, pp. 55–76. New York: Wiley.

Godman, R.M., H.W. Yawney, and C.H. Tubbs. 1990. *Acer saccharum* Marsh. Sugar maple. In R.M. Burns and B.H. Honkala (Eds.), *Silvics of North America. Volume 2. Hardwoods,* pp. 78–91. U.S. For. Serv., Agric. Handbk. 654.

Goelz, J.C.G. 1990. Generation of a new type of stocking guide that reflects stand growth. In S.S. Coleman and D.G. Neary (Eds.), *Proceedings of the Sixth Biennial Southern Silviculture Research Conference,* Vol. 1, pp. 240–247. U.S. For. Serv. Gen. Tech. Rpt. SO-70.

Goff, F.G., and D. West. 1975. Canopy-understory interaction effects on forest populations. *For. Sci.* 21:98–108.

Gordon, D.T. 1973. Released advance reproduction of white and red fir: Growth, damage, mortality. U.S. For. Serv. Res. Pap. PNW-95.

Gordon, G.T. 1979. Successful natural regeneration cuttings in California true firs. U.S. For. Serv. Res. Pap. PSW-140.

Gottschalk, K.W. 1983. Management strategies for successful regeneration: Oak-hickory. In J. Finley, R.S. Cochran, and J.R. Grace (Eds.), *Regenerating Hardwood Stands,* pp. 190–213. 1983. Proc. Penn. State For. Issues Conf., March 15–16, 1983, Penn. State Univ., Univ. Park, PA.

Graber, R.E. 1974. Ground application methods in northeastern United States. In J.H. Cayford (Ed.), *Direct Seeding Symposium,* pp. 113–118. Sept. 11–13, 1973. Timmins, Ont. Dept. Environ., Can. For. Serv., Publ. No. 1339.

Graber, R.E., and D.F. Thompson. 1969. A furrow-seeder for the Northeast. U.S. For. Serv. Res. Pap. NE-150.

Graham, J.N., E.W. Murray, and D. Minore. 1982. Environment, vegetation, and regeneration after timber harvest in the Hungry-Pickett Area of southwest Oregon. U.S. For. Serv. Res. Note PNW-400.

Graham, R.T. 1988. Influences of stand density on development of western white pine, redcedar, hemlock, and grand fir in the northern Rocky Mountains. In W.C. Schmidt (Ed.), *Proceedings—Future Forests of the Mountain West: A Stand Culture Symposium,* pp. 175–184. U.S. For. Serv. Gen. Tech. Rpt. INT-243.

Graham, R.T., and R.A. Smith. 1983. Techniques for implementing the individual tree selection method in the grand fir-cedar-hemlock ecosystems of northern Idaho. U.S. For. Serv. Res. Note INT-332.

Graham, S.A. 1926. The biology and control of white pine weevil, *Pissodes strobi* Peck. Cornell Univ. Agric. Exp. Stn. Bull. 449:3–32.

Graham, S.A. 1952. *Forest Entomology,* 3d ed., New York: McGraw-Hill.

Graham, S.A. 1956. Forest insects and the law of natural compensation. *Can. Ent.* 88:45–55.

Gratkowski, H. 1975. Silvicultural use of herbicides in Pacific Northwest forests. U.S. For. Serv. Gen. Tech. Rpt. PNW-37.

Gratkowski, H. 1977. Site preparation and conifer release in Pacific Northwest forests. In *27th Annual Weed Conference Proceedings,* pp. 29–32. U.S. For. Serv. GPO 999-712.

Greaves, R.D. 1978. Planting and seeding. In B.D. Cleary, R.D. Greaves, and R.K. Hermann (Eds.), *Regenerating Oregon's Forests,* Ch. 8, pp. 131–161. Ore. St. Univ., Coop. Ext. Serv. Corvallis.

Gregory, G.R. 1987. *Resource Economics for Foresters.* New York: Wiley.

Griffin, R.H., and B.W. Carr. 1974. Aerial seeding of spruce in Maine. In J.H. Cayford (Ed.), *Direct Seeding Symposium,* pp. 131–138. Sept. 11–13, 1973. Timmins, Ont. Dept. Environ., Can. For. Serv., Publ. No. 1339.

Grisez, T.J. 1954. Hurricane damage on Penobscot Experimental Station. U.S. For. Serv., NE For. Exp. Stn. Res. Note 39.

Grisez, T.J. 1960. Slash helps protect seedlings from deer browsing. *J. For.* 58:385–387.

Grisez, T.J. 1978. Pruning black cherry in understocked stands. U.S. For. Serv. Res. Pap. NE-395.

Grisez, T.J., and M.R. Peace. 1973. Requirements for advance reproduction in Allegheny hardwoods—An interim guide. U.S. For. Serv. Res. Note NE-180.

Groome, J.P. 1988. Mutual support of trees. *Scottish For.* 42:12–14.

Guldin, J.M. 1991. Uneven-aged BDq regulation of Sierra Nevada mixed conifers. *W. J. App. For.* 6(2):27–32.

Guldin, R.W. 1981. An innovative step in plantation management: Planting container-grown seedlings. In *Proceedings Southern Timber Supply Constraints on NIPF Lands,* pp. 103–121. S. For. Econ. Workshop., March 17–19, 1981, Mt. Vies, Ark., U.S. For. Serv., S. For. Exp. Stn., Proc. Reprint.

Guldin, R.W. 1982. What does it cost to grow seedlings in containers? *Tree Plant. Notes* 33(1):34–37.

Guldin, R.W. 1984. Economic returns from spraying to release loblolly pine. In *Biotechnology and Weed Science,* pp. 248–254. Proc. S. Weed Sci. Soc., 37th Ann. Meet., Jan. 17–19, 1984., Hot Sp., AK. Weed Sci. Soc. Am., Champaign, IL.

Guldin, R.W. 1985. Loblolly and shortleaf pine differ in their response when released using herbicide sprays. In E. Shoulders (Ed.), *Proceedings of the Third Biennial Southern Silvicultural Research Conference,* pp. 287–291. U.S. For. Serv. Gen. Tech. Rpt. SO-54.

Gullberg, U. 1993. Towards making willows pilot species for coppicing production. *For. Chron.* 69(6):721–726.

Gullion, G.W. 1970. Factors influencing ruffed grouse populations. *Trans. N. Amer. Wildl. and Nat. Resourc. Conf.* 35:93–105.

Gullion, G.W. 1972. *Improving Your Forested Lands for Ruffed Grouse.* The Ruffed Grouse Soc., Coraopolis, PA.

Gullion, G.W. 1990. Management of aspen for ruffed grouse and other wildlife—An update. In R.D. Adams (Ed.), *Aspen Symposium '89,* pp. 133–145. U.S. For. Serv. Gen. Tech. Rpt. NC-140.

Guo, Y., and B.J. Karr. 1989. Influence of trafficking and soil moisture on bulk density and porosity on Smithdale sandy loam in north-central Mississippi. In J.H. Miller (Ed.), *Proceedings of the Fifth Biennial Southern Silviculture Research Conference,* pp. 533–538. U.S. For. Serv. Gen. Tech. Rpt. SO-74.

Gutzwiler, J.R. 1976. Mechanical site preparation for tree planting in the Inland Northwest. In D.M. Baumgartner and R.J. Boyd (Eds.), *Tree Planting in the Inland Northwest,* pp. 117–133. Proc. Conf., Feb. 17–19, 1976. Wash. State Univ., Coop. Ext. Serv., Pullman, WA.

Hafley, W.L., W.D. Smith, and M.A. Buford. 1982. A new yield prediction model for unthinned loblolly pine plantations. Sch. For. Resourc., NC State Univ., S. For. Res. Ctr. Tech. Rpt. No. 1.

Hague, R.A. 1976. SCS technical assistance in windbreak forestry. In R.W. Tinus (Ed.), *Shelterbelts on the Great Plains.* Great Plains Agric. Counc., Lincoln, NB, Publ. No. 78.

Hahn, P.F. 1982. A historical overview of the use of containerized seedlings for operational reforestation—How did we get where we are today? In R.W. Guldin and J.P. Barnett (Eds.), *Proceedings of the Southern Containerized Forest Tree Seedling Conference.* pp. 7–12. U.S. For. Serv. Gen. Tech. Rpt. SO-37.

Haig, I.T. 1936. Factors controlling initial establishment of western white pine and associated species. Yale Univ. Sch. For. Bull. 41.

Haig, I.T., K.P. Davis, and R.H. Weidman. 1941. Natural regeneration in the western white pine type. U.S. Dept. Agric. Tech. Bull. No. 767.

Haight, R.G., J.D. Brodie, and D.M. Adams. 1985. Optimizing the sequence of diameter distributions and selection harvests for uneven-aged stand management. *For. Sci.* 31:451–462.

Haila, Y. 1986. North European land birds in forest fragments: Evidence for area effects? In J. Verner, M.L. Morrison, and C.J. Ralph (Eds.), *Wildlife 2000. Modeling Habitat Relationships of Terrestrial Vertebrates,* pp. 315–319. Madison: The Univ. Wis. Press.

Hall, O. F. 1955. Where does thinning fit into the management of Lake States pulpwood stands. Proc. Soc. Am. For. Meet., Oct. 24–27, 1954, Milwaukee, WS. Soc. Am. For., Washington, DC.

Hallette, R.D., and T.S. Murray. 1980. Recent developments and current practices in forestation in Canada. Environ. Can., Can. For. Serv., Maritimes For. Res. Ctr. Info. Rpt. M-X-116.

Hallin, W.E. 1968. Soil surface temperatures on cutovers in southwest Oregon. U.S. For. Serv. Res. Note PNW-78.

Hanks, L.F. 1971. Interim hardwood tree grades for factory lumber. U.S. For. Serv. Res. Pap. NE-199.

Hann, D.W., and B.B. Bare. 1979. Uneven-aged forest management: State of the art (or science?). U.S. For. Serv. Gen. Tech. Rpt. INT-50.

Hannah, P.R. 1987. Regeneration methods for oaks. *N. J. Appl. For.* 4(2):97–101.

Hannah, P.R. 1988. The shelterwood method in northeastern forest types: A literature review. *N. J. Appl. For.* 5(1):70–77.

Hansen, E.A., and D.A. Netzer. 1985. Weed control using herbicides in short-rotation intensively cultured poplar plantations. U.S. For. Serv. Res. Pap. NC-260.

Hansen, E.A., D.A. Netzer, and D.N. Tolsted. 1993. Guidelines for establishing poplar plantations in the north-central U.S. U.S. For. Serv. Res. Note NC-363.

Hansen, G.D. 1987. Choosing diameter distributions to aid in marking uneven-aged northern hardwood stands. In R.D. Nyland (Ed.), *Managing Northern Hardwoods,* pp. 95–112. SUNY Coll. Environ. Sci. and For., Fac. For. Misc. Publ. No. 13 (ESF 87-002) (Soc. Am. For. Publ. No. 87-03).

Hansen, G.D., and R.D. Nyland. 1987. Effects of diameter distribution on the growth of simulated uneven-aged sugar maple stands. *Can. J. For. Res.* 17(1):1–8.

Hansen, G.D., and R.D. Nyland. nd. Target residual basal area by size class for selection system in northern hardwoods using different length cutting cycles. SUNY Coll. Environ. Sci. and For., Fac. For., Syracuse, NY 13210. Unpubl. manuscr.

Harlow, W.H., E.S. Harrar, and F.M. White. 1978. *Textbook of Dendrology,* 6th ed., New York: McGraw-Hill.

Harper, J. L. 1977. *Population Biology of Plants.* New York: Academic Press.

Harrington, M.G., and R.G. Kelsey. 1979. Influence of some environmental factors on initial establishment and growth of ponderosa pine seedlings. U.S. For. Serv. Res. Pap. INT-200.

Harrington, M.G., and S.S. Sackett. 1990. Using fire as a management tool in southwestern ponderosa pine. In J.S. Krammes, Tech. Coord. *Effects of Fire Management of Southwestern Natural Resources,* pp. 122–133. U.S. For. Serv. Gen. Tech. Rpt. RM-191.

Harris, A.S. 1990. *Picea sitchensis* (Bong.) Carr. Sitka spruce. In R.M. Burns and B.H. Honkala (Eds.), *Silvics of North America. Volume 1. Conifers,* pp. 260–267. U.S. For. Serv., Agric. Handbk. 654.

Harris, R.W. 1992. *Arboriculture: Integrated Management of Landscape Trees, Shrubs, and Vines,* 2d ed., Englewood Cliffs, NJ: Prentice Hall.

Hart, G. 1961. Humus depth under cut and uncut northern hardwood forests. U.S. For. Serv. NE For. Exp. Stn., Res. Note 113.

Hartung, R.E., and J. M. Kress. 1977. Woodlands of the Northeast: Erosion and sediment control guides. U.S. Soil Conserv. Serv., NE Tech. Serv. Ctr., Broomall, PA, and U.S. For. Serv., NE Area State and Priv. For., Upper Darby, PA.

Haussman, R.F., and E.W. Pruett. 1973. Permanent logging roads for better woodlot management. U.S. For. Serv., NE Area State and Priv. For., Broomall, PA. 383.1.

Hawley, R.C. 1921. *The Practice of Silviculture.* New York: Wiley.

Hawley, R.C. 1929. *The Practice of Silviculture,* 2d ed., New York: Wiley.

Hawley, R.C., and D.M. Smith. 1954. *The Practice of Silviculture,* 6th ed., New York: Wiley.

Hawley, R.C., and P.W. Stickel. 1948. *Forest Protection,* 2d ed., New York: Wiley.

Haywood, J.D. 1986. Response of planted *Pinus taeda* L. to brush control in northern Louisiana. *For. Ecol. and Manage.* 15:129–134.

Haywood, J.D., A.E. Tiarks, and E. Shoulders. 1990. Loblolly and slash pine height and diameter are related to soil drainage in winter on poorly drained silt loams. *New For.* 4:81–96.

Healy, W.M. 1987. Habitat characteristics of uneven-aged stands. In R.D. Nyland (Ed.), *Managing Northern Hardwoods,* pp. 338–347. SUNY Coll. Environ. Sci. and For., Fac. For. Misc. Publ No. 13 (ESF 87-002) (Soc. Am. For. Publ. No. 87-03).

Healy, W.M., R.T. Brooks, and R.M. DeGraaf. 1989. Cavity trees in sawtimber-sized oak stands in central Massachusetts. *N. J. Appl. For.* 6(2):61–65.

Hedden, R.L., S.J. Barnes, and J.E. Coster (Eds.). 1981. Hazard rating systems in forest insect pest management: Symposium proceedings. U.S. For. Serv. Gen. Tech. Rpt. WO-27.

Heede, B.H. 1991. Response of a stream in disequilibrium to timber harvest. *Environ. Manage.* 15(2):251–255.

Heiberg, S.O. 1942. Cutting based upon an economic increment. *J. For.* 40(8):645–651.

Heiberg, S.O. 1954. Introduction. In S.O. Heiberg (Ed.), *Thinning Problems and Practices in Denmark,* pp. 1–4. SUNY Coll. For. at Syracuse, World For. Ser. Bull. No. 1, Tech. Publ. No. 76.

Heilgmann, R.B., and J.S. Ward. 1993. Hardwood regeneration twenty years after three distinct diameter-limit cuts in upland central hardwoods. In A.R. Gillespie, G.R. Parker, and P.E. Pope (Eds.), *Proceedings of the 9th Central Hardwood Forest Conference,* pp. 261–270. U.S. For. Serv. Gen. Tech. Rpt. NC-161.

Heinselman, M.L. 1971. The natural role of fire in conifer forests. In *Fire in the Natural Environment—A Symposium,* pp. 61–72. U.S. For. Serv. Pac. NW For. and Range Exp. Stn., Portland, OR.

Heitzman, E., and R.D. Nyland. 1991. Cleaning and early crop-tree release in northern hardwoods: A review. *N. J. Appl. For.* 8(3):111–115.

Heitzman, E., and R.D. Nyland. 1994. Influences of pin cherry (*Prunus pennsylvanica* L. f.) on growth and development of young even-aged northern hardwoods. *For. Ecol. and Manage.* 67:39–48.

Hellum, A.K. 1974. Direct seeding in western Canada. In J.H. Cayford (Ed.), *Direct Seeding Symposium,* pp 103–118. Sept. 11–13, 1973. Timmins, Ont. Dept. Environ., Can. For. Serv., Publ. No. 1339.

Helms, J.A. 1995. The California region. In J.W. Barrett (Ed.), *Regional Silviculture of the United States,* 3d ed., Ch. 10, pp. 441–497. New York: Wiley.

Herrick, A.M., and C.L. Brown. 1967. A new concept in cellulose production: Silage sycamore. *Agric. Aci. Rev.* 5(4):7–12.

Hertel, G.D., and D.M. Benjamin. 1977. Intensity of site preparation influences on pine webworm and tip moth infestations of pine seedlings in north-central Florida. *Environ. Ento.* 6(1):118–122.

Hesterberg, G.A. 1957. Deterioration of sugar maple following logging damage. U.S. For. Serv. Lake States For. Exp. Stn. Stn. Pap. 51.

Hett, J.M., and O.L. Loucks, 1976. Age structure of balsam fir and eastern hemlock. *J. Ecol.* 64:1029–1044.

Hibbs, D.E. 1982. Gap dynamics in an hemlock-hardwood forest. *Can. J. For. Res.* 12:522–527.

Hibbs, D.E., and A.A. Ager. 1989. Red alder: Guidelines for seed collection, handling, and storage. Ore. State Univ., Coll. For., For. Res. Lab. Special Pub. 18.

Hilt, D.E., and M.E. Dale. 1982. Effects of repeated precommercial thinnings in central hardwood sapling stands. *S. J. Appl. For.* 6(1):53–58.

Hinds, T.E., and E.M. Wengert. 1977. Growth and decay losses in Colorado aspen. U.S. For. Serv. Res. Pap. RM-193.

Hite, W.A. 1976. Selecting an optimum planting microsite. In D.M. Baumgartner and R.J. Boyd (Eds.), *Tree Planting in the Inland Northwest,* pp. 177–184. Proc. Conf. at Wash. State Univ., Feb 17–19, 1976. Wash. State Univ., Coop. Ext. Serv., Pullman, WA.

Hocker, H.W., Jr. 1979. *Introduction to Forest Biology.* New York: Wiley.

Hocking, D., and R.D. Nyland. 1971. Cold storage of coniferous seedlings: A review. SUNY Coll. Environ Sci. and For., Appl. For. Res. Inst., AFRI Res. Rpt. No. 6.

Hodges, J.D. 1995. The southern bottomland hardwood region and brown loam bluffs subregion. In J.W. Barrett (Ed.), *Regional Silviculture of the United States,* 3d ed., Ch. 6, pp. 227–269. New York: Wiley.

Hoff, R.J. 1984. Role of genetics in pest management. In D.M. Baumgartner and R. Mitchell (Eds.), *Silvicultural Management Strategies for Pests of the Interior Douglas-fir and Grand Fir Forest Types,* pp. 121–128. Proc. Symp., Feb. 14–16, 1984. Spokane, WA., Wash. St. Univ, Coop Ext. Serv., Pullman, WA.

Holdridge, L.R., W.C. Grenke, W.H. Hatheway, T. Lainbg, and J.A. Tosi, Jr. 1971. *Forest Environments in Tropical Life Zones—A Pilot Study.* New York: Pergamon Press.

Holsoe, T. 1947. The relation of tree development to the timing of the first thinning in even-aged hardwood stands. Harvard For. Pap. Vol. I, No. 2.

Holsoe, T. 1951. Yellow-poplar. Reaction to crown release and other factors influencing growth. W. Va. Univ. Agric. Exp. Stn. Bull. 344T.

Horn, D.J. 1988. *Ecological Approach to Pest Management.* New York: Guilford Press.

Hornbeck, J.W. 1970. The radiant energy budget of clearcut and forested sites in West Virginia. *For. Sci.* 16:139–145.

Hornbeck, J.W., and W.B. Leak. 1992. Ecology and management of northern hardwoods in New England. U.S. For. Serv. Gen. Tech. Rpt. NE-159.

Hornbeck, J.W., C.W. Martin, R.S. Pierce, F.H. Bormann, G.E. Likens, and J.S. Eaton. 1987. The northern hardwood forest ecosystem: Ten years of recovery from clearcutting. U.S. For. Serv., NE For. Exp. Stn. NE-RP-596.

Horsley, S.B. 1994. Regeneration success and plant species diversity of Allegheny hardwood stands after Roundup application and shelterwood cutting. *N.J. Appl. For.* 11(4):109–116.

Horsley, S.B., and D.A. Marquis. 1983. Interference by weeds and deer with Allegheny hardwood reproduction. *Can. J. For. Res.* 13:61–69.

Horton, K.W. 1962. Regenerating white pine with seed trees and ground scarification. Can. Dept. For., For. Res. Br., Tech. Note No. 118.

Horton, K.W. 1966. Profitability of pruning white pine. *For. Chron.* 42(2):294–305.

Horton, K.W., and G.H.D. Bedell. 1960. White and red pine ecology, silviculture, and management. Can. Dept. N. Affairs and Nat. Resourc., For. Br., Bull. 124.

Horton, K.W., and B.S.P. Wang. 1969. Experimental seeding of conifers in scarified strips. *For. Chron.* 45(1):1–8.

Horton, R.E. 1919. Rainfall interception. *Month. Weather Rev.* 47:603–623.

Hotvedt, J.E., Y.F. Abernathy, and R.M. Farrar, Jr. 1989. Optimum management regimes for uneven-aged loblolly-shortleaf pine stands managed under selection system in the West Gulf Coast. *S. J. Appl. For.* 13(3):117–122.

Hough, Stansbury + Associates Ltd. 1973. Design Guidelines for Forest Management. Ontario Min. Nat. Resourc., Toronto.

Hoyer, G.E., and J.D. Swanzy. 1986. Growth and yield of western hemlock in the Pacific Northwest following thinning near the time of initial crown closure. U.S. For. Serv. Res. Pap. PNW-365.

Hughes, E.L. 1967. Studies in stand and seedbed treatment to obtain spruce and fir reproduction on the mixedwood slope type of northwestern Ontario. Can. Dept. For. and Rural Develop., For. Br., Dept. Publ. No. 1189.

Hummel, F.C. 1948. The Bowmont Norway spruce sample plots. *For.* 21:30–42.

Hungerford, R.D., and R.E. Babbitt. 1987. Overstory removal and residue treatments affect surface, air, and soil temperature: Implications for seedling survival. U.S. For. Serv. Res. Pap. INT-377.

Hungerford, R.D., R.E. Williams, and M.A. Marsden. 1981. Thinning and pruning western white pine for reducing mortality due to blister rust. U.S. For. Serv. Res. Note INT-322.

Hunter, J.L., Jr. 1990. *Wildlife, Forests, and Forestry. Principles of Managing Forests for Biological Diversity.* Englewood Cliffs, NJ: Prentice Hall.

Hurst, G.A., J.J. Campo, and M.B. Brooks. 1982. Effects of precommercial thinning and fertilizing on deer forage in a loblolly pine plantation. *S. J. Appl. For.* 6(3):140–144.

Husch, B. 1963. *Forest Mensuration and Statistics.* 1963. New York: Ronald Press.

Husch, B., C.I. Miller, and T.W. Beers. 1982. *Forest Mensuration,* 3d ed., New York: Wiley.

Hutnik, R.J. 1958. Three diameter-limit cuttings in West Virginia hardwoods a 5-year report. U.S. For., Serv. NE For. Exp. Stn., Stn. Pap. No. 106.

Isaac, L.A. 1940. Vegetative succession following logging in the Douglas-fir region with special reference to fire. *J. For.* 38:716–721.

Isaac, L.A. 1963. Fire—A tool not a blanket rule in Douglas-fir ecology. *Tall Timbers Fire Ecol. Conf. Proc:* 2:1–17.

Isaacson, J.A. 1984. Seed collection and handling. In D.M. Baumgartner and R.J. Boyd

(Eds.), *Tree Planting in the Inland Northwest,* pp. 49–59. Proc. Conf. at Wash. State Univ., Feb 17–19, 1976. Wash. State Univ., Coop. Ext. Serv., Pullman, WA.

Jacobs, R. D. 1966. A silvicultural evaluation of four methods of marking second-growth northern hardwood stands. U.S. For. Serv. Res. Note NC-1.

Jacobs, R.D. 1974. Damage to northern hardwood removal of shelterwood overstory. *J. For.* 72(10):654–656.

Jagd, T. 1954. The "active" method of thinning In S.O. Helberg (Ed.), *Thinning Problems and Practices in Denmark,* pp. 59–73. SUNY Coll. For. at Syracuse, World For. Ser. Bull, No. 1.

Jemison, G.M. 1934. The significance of the effect of stand density upon weather beneath the canopy. *J. For.* 32:446–451.

Jenkinson, J.L., and J.A. Nelson. 1985. Cold storage increases resistance to dehydration stress in Pacific Douglas-fir. In T.D. Landis (Ed.), *Proceedings: Western Forest Nursery Council-Intermountain Nurseryman's Association Combined Meeting,* Aug. 14–16, 1984. U.S. For. Serv. Gen. Tech. Rpt. INT-185.

Jensen, V.S. 1943. Suggestions for the management of northern hardwoods in the Northeast. *J. For.* 41:180–185.

Johann, K. 1980. Bestandesbehandlung und Schneebruchgefaehrdung. In H. Kramer (Ed.), *Biologische, technische und wirtschaftliche Aspekte der Jungbestandspflege,* pp. 296–286. SchrReihe Forstl. Fak. Univ. Goettingen, Bd. 67. (Transl. D. Clerc, 1985).

Johnson, H.J. 1968. Pre-scarification and strip clearcutting to obtain lodgepole pine regeneration. *For. Chron.* 44(6):1–4.

Johnson, J.A. 1984. Small-woodlot management by single-tree selection: 21-year results. *N. J. Appl. For.* 1(4):69–71.

Johnson, J.E. 1985. The Lake States region. In J.W. Barrett (Ed.), *Regional Silviculture of the United States,* 3d ed., Ch. 3, pp. 81–127. New York: Wiley.

Johnson, J.W. 1961. Thinning practices in short-rotation stands. In A.B. Crow (Ed.), *Advances in Management of Southern Pines,* pp. 50–60. 10th Ann. For. Sympos., La. State Univ. Press, Baton Rouge, LA.

Johnson, J.W. 1972. Silvicultural considerations in clearcutting. In R.D. Nyland (Ed.), *A Perspective on Clearcutting in a Changing World,* pp. 19–24. Proc. 1972 Winter Meet., NY Soc. Am. For., Feb. 23–25, 1972, Syracuse, NY. SUNY Coll. Environ. Sci. and For., Appl. For. Res. Inst., AFRI Misc. Rpt. No. 4.

Johnson, N.E. 1976. Biological opportunities and risks associated with fast-growing plantations in the tropics. *J. For.* 74(4):206–211.

Johnson, P.S. 1977. Predicting oak stump sprouting development in the Missouri Ozarks. U.S. For. Serv. Res. Pap. NC-149.

Johnson, R.S. 1984. Effect of small aspen clearcuts on water yield and water quality. U.S. For. Serv. Res. Pap. INT-333.

Johnson, V.L. 1984. Prescribed burning: Requiem or renaissance? *J. For.* 82(2):82–91.

Johnson, J.D., and M.L. Cline. 1991. Seedling quality for southern pines. In M.L. Duryea and P.M. Dougherty (Eds.), *Forest Regeneration Manual,* Ch. 8, pp. 143–162. The Netherlands: Klumer Academ. Publ.

Johnson, L.C., and R.P. Overton. 1984. Tree improvement plan for the northeastern area, state and private forestry. U.S. For. Serv. NE Area State and Priv. For., Broomall, PA.

Johnston, W.F. 1982. Direct seeding lowland conifers. In G.D. Mroz and J.F. Berner (Eds.), *Artificial Regeneration of Conifers in the Upper Great Lakes Region,* pp. 299–303. Mich. Technol. Univ., Houghton, MI.

Johnstone, W.D., and D.M. Cole. 1988. Thinning lodgepole pine: A research review. In W.C.

Schmidt (Ed.), *Proceedings—Future Forests of the Mountain West: A Stand Culture Symposium,* pp. 160–164. U.S. For. Serv. Gen. Tech. Rpt. INT-243.

Jokela, E.J., H.L. Allen, and W.W. McFee. 1991. Fertilization of southern pines at establishment. In M.L. Duryea and P.M. Dougherty (Eds.), *Forest Regeneration Manual,* pp. 263–277. The Netherlands: Kluwer Academ. Publ.

Jones, E.P., Jr. 1965. Spot seeding–A method of direct seeding. In *Proceedings Direct Seeding Workshops,* pp. 85–87. Oct. 5–6, 1965, Alexandria, LA, and Oct. 20–21, Tallahassee, FL. U.S. For. Serv., GSA Atlanta GA 66-4542.

Jones, G.T. 1993. A guide to logging aesthetics. Practical tips for loggers, foresters, and landowners. NE Reg. Agric. Eng. Serv., Coop. Ext., and Soc. Protect. NH For., NE For. Resourc. Ext. Counc. Ser. NRAES-60.

Jones, J.R. 1974. A spot seeding trial with southwestern white pine and blue spruce. U.S. For. Serv. Res. Note RM-265.

Jones, R.H., and D.J. Raynal. 1986. Spatial distribution and development of root sprouts in *Fagus grandifolia* (Fagaceae). *Am. J. Bot.* 73:1723–1731.

Jones, R.H., R.D. Nyland, and D.J. Raynal 1989. Response of American beech to selection cutting of northern hardwoods in New York. *N. J. Appl. For.* 6(1):34–36.

Jorgensen, J.R., and C.G. Wells. 1986. Tree nutrition and fast-growing plantations in developing countries. *Intl. Tree Crops J.* 3(1986):225–244.

Kaiser, H.F., P.J. Brown, and R.K. Davis. 1988. The need for values of amenity resources public natural resources management. In G.L. Peterson, B.L. Driver, and R. Gregory (Eds.), *Amenity Resource Valuation: Integrating Economics with Other Disciplines,* pp. 7–12. State College, PA: Venture Publ.

Kattelmann, R.C., N.H. Berg, and J. Rector. 1983. The potential for increasing streamflow from Sierra Nevada watersheds. *Water Resourc. Bull.* 19(3):395–402.

Kauffman, J.B. 1990. Ecological relationships of vegetation and fire in Pacific Northwest forests. In J.D. Walstad, S.R. Radosevich, and D.V. Sandberg (Eds.), *Natural and Prescribed Fire in Pacific Northwest Forests,* pp. 39–54. Corvallis: Ore. State Univ. Press.

Kauffman, M.R., R.T. Graham, D.A. Boyce, Jr., W.H. Moir, L. Perry, R.T. Reynolds, R.L. Bassett, P. Mehlhop, C.B. Edminster, W.M. Block, and P.S. Corn. 1994. An ecological basis for ecosystem management. U.S. For. Serv. Gen. Tech. Rpt. RM-246.

Keen, F.P. 1936. Relative susceptibility of ponderosa pine to bark-beetle attack. *J. For.* 34:919–927.

Keister, T.D. 1972. Thinning slash pine plantation results in little growth or economic gain after 40 years. La. St. Univ & Agric. and Market. Coll., LSU For. Note. 102.

Kelty, M.J. 1987. Shelterwood cutting as a reproduction method. In R.D. Nyland (Ed.), *Managing Northern Hardwoods,* pp. 128–142. SUNY Coll. Environ. Sci. and For., Fac. For. Misc. Publ. 13 (ESF 87-002) (Soc. Am. For. Publ. 87-03).

Kelty, M.J., and R.D. Nyland. 1981. Regenerating Adirondack northern hardwoods by shelterwood cutting and control of deer density. *J. For.* 79(1):22–26.

Kemperman, J.A. 1978. Sucker-root relationships in aspen. Ont. Min. Nat. Resource. For. Res. Note. No. 12.

Kennedy, H.E., Jr. 1975. Influence of cutting cycle and spacing on coppice sycamore yield. U.S. For. Serv. Res. Note. SO-193.

Kennedy, R.F. 1961. Thinning practices in sawlog-rotation stands. In A.B. Crow (Ed.), *Advances in Management of Southern Pines,* pp. 61–67. 10th Ann. For. Sympos., La. State Univ. Press, Baton Rouge, LA.

Kershaw, J.A., and B.C. Fischer. 1991. A stand density management diagram for sawtimber-sized mixed upland central hardwoods. In L.H. McCormick and K.W. Gottshalk (Eds.),

Proceedings of the 8th Central Hardwood Forest Conference, pp. 414–428. U.S. For. Serv. Gen. Tech. Rpt. NE-148.

Kessler, W.B. 1992. Science challenges of New Perspectives. Presented at Sustaining Ecological Systems, New Perspectives Nat. Conf., Dec. 3–5, 1991, Roanoke, VA. U.S. For. Serv.

Kessler, W.B., H. Salwasser, C.W. Cartwright, Jr., and J.A. Caplan. 1992. New perspectives for sustainable natural resources management. *Ecol. Appl.* 2(3):221–225.

Kimmins, J.P. 1987. *Forest Ecology.* New York: MacMillan Publ.

King, J.G. 1989. Streamflow responses to road building and harvesting. A comparison with the equivalent clearcut area procedure. U.S. For. Serv. Res. Pap. INT-401.

Kingsley, N.P., and P.S. DeBald. 1987. Hardwood lumber and stumpage prices in two eastern hardwood markets: The real story. U.S. For. Serv. NE-RP-601.

Kittredge, J. 1948. *Forest Influences.* New York: McGraw-Hill.

Klinka, K., and R.E. Carter. 1991. A stand-level guide to the selection of reproduction methods for regenerating forest stands in the Vancouver forest region. Contract rpt. to Min. For., Vancouver For. Reg., Burnaby, BC, Prov. BC Min. For., Burnaby.

Klinka, K., R.E. Carter, and M.C. Feller. 1990. Cutting old-growth forests in British Columbia: Ecological considerations for forest regeneration. *N. W. Environ. J.* 6:221–242.

Klock, G.O., and W. Lopushinsky. 1980. Soil water trends after clearcutting in the Blue Mountains of Oregon. U.S. For. Serv. Res. Note. PNW-361.

Knight, F.B., and H.J. Heikkenen. 1980. *Principles of Forest Entomology,* 5th ed. New York: McGraw-Hill.

Knuchel, H. 1953. *Planning and Control in the Managed Forest.* Transl. by M.L. Anderson. Edinburgh: Oliver and Boyd.

Koch, P., and D.W. McKenzie. 1977. Machine for row-mulching logging slash to enhance site—A concept. Trans. Amer. Soc. Agric. Eng. *ASAE* 20(1):13–17.

Kochenderfer, J.N. 1970. Erosion control on logging roads in the Appalachians. U.S. For. Serv. Res. Pap. NE-158.

Kochenderfer, J.N., and P.J. Edwards. 1991. Effectiveness of three streamside management practices in the central Appalachians. In S.S. Coleman and D.G. Neary (Eds.), *Proceedings of the Sixth Biennial Southern Silvicultural Research Conference, Volume 2,* pp. 688–700. U.S. For. Serv. Gen. Tech. Rpt. SE-70.

Kochenderfer, J.N., G.W. Wende, and H.C. Smith. 1984. Cost of and soil loss on "minimum standard" forest truck roads constructed in the central Appalachians. U.S. For. Serv. Res. Pap. NE-544.

Kohls, R.L., and J.N. Uhl. 1985. *Marketing of Agricultural Products,* 6th ed. New York: MacMillian Publ.

Koop, R. F., E. H. White, L. P. Abrahamson, C. A. Nowak, L. Zsuffa, and K. F. Burns. 1993. Willow biomass trials in central New York State. *Biomass and Bioenergy* 5(2): 179–187.

Kormanik, P.P., and C.L. Brown. 1967. Root buds and the development of root suckers in sweetgum. *For. Sci.* 13(4):338–345.

Kostler, J. 1956. *Silviculture.* Transl. by M.L. Anderson. Edinburgh and London: Oliver and Boyd.

Kotar, J., and M.S. Coffman. 1982. Application of habitat type concept to regeneration of conifers in the Upper Lake States. In G.D. Mroz and J.F. Berner (Eds.), *Artificial Regeneration of Conifers in the Upper Great Lakes Region,* pp. 53–60. Mich. Technol. Univ., Houghton, MI.

Kraft, G. 1884. *Beitrage z. Lehre v. d. Durchforstungen usw.* Hanover. (Original not seen).

Kramer, H. 1975. Erhoehung der Produktionssicherheit zur Foerderung einer nachhaltign Fichtenwirtschaft. *Forstarchiv* 46(1):9–13. (Trans. D. Clerc, 1985).

Kramer, H. 1980. Einfluss verschiedenartiger Durchforstungen auf Bestandessicherheit und Zuwachs in einem weitstaendigen begruendeten Fichtenbestand. In H. Kramer (Ed.), *Biologische, technische und wirtschaftliche Aspekte der Jungbestandspflege,* pp. 224–235. SchrReihe Forstl. Fak. Univ. Goettingen, Bd. 67. (Transl. D. Clerc, 1985).

Kramer, P.J., and T.T. Kozlowski. 1960. *Physiology of Trees.* New York: McGraw-Hill.

Krugman, S.I., W.I. Stein, and D.M. Schmitt. 1974. Seed biology. In C.S. Schopmeyer (Ed.), *Seeds of Woody Plants in the United States,* Ch. 1, pp. 5–40. U.S. For. Serv., Agric. Handbk. No. 450.

Krutilla, J.V., M.D. Bowes, and P. Sherman. 1983. Watershed management for joint production of water and timber: A provisional assessment. *Water Resource. Bull.* 19(3):403–414.

Laacke, R.J., and G.O. Fiddler. 1986. Overstory removal: Stand factors related to success and failure. U.S. For. Serv. Res. Pap. PSW-183.

Laarman, J.G., and R.A. Sedjo. 1992. *Global Forests. Issues for Six Billion People.* New York: McGraw-Hill.

LaBonte, G.A., and R.W. Nash. 1978. Cleaning and weeding paper birch—A 24-year case history. *J. For.* 76(4):223–225.

Laidly, P.R., and R.G. Barse. 1979. Spacing affects knot surface in red pine plantations. U.S. For. Serv. Res. Note NC-246.

Lamson, N.I., and G.W. Miller. 1983. Logging damage to dominant and codominant residual stems in thinned cherry-maple stands. In *Proceedings of the Fourth Annual Central Hardwoods Conference,* Nov. 9–10, 1982. Univ. Ky, Lexington, KY.

Lamson, N.I., and H.C. Smith. 1987. Precommercial treatments of 15- to 40-year old northern hardwood stands. In R.D. Nyland (Ed.), *Managing Northern Hardwoods.* SUNY Coll. Environ. Sci. and For., Fac. For. Misc. Publ. No. 13 (ESF 87-002) (Soc. Am. For. Publ. No. 87-03).

Lamson, N.I., and H.C. Smith. 1989. Crop-tree release increases growth of 12-year-old yellow-poplar and black cherry. U.S. For. Serv. Res. Pap. NE-622.

Lamson, N.I., and H.C. Smith. 1991. Stand development and yields of Appalachian hardwood stands managed with single-tree selection for the last 30 years. U.S. For. Serv. Res. Pap. NE-655.

Lamson, N.I., H.C. Smith, and G.W. Miller. 1984. Residual stocking not seriously reduced by logging damage from thinning of West Virginia cherry-maple stands. U.S. For. Serv. Res. Pap. NE-541.

Lamson, N.I., H.C. Smith, A.W. Perkey, and S.M. Brock. 1990. Crown release increases growth of crop trees. U.S. For. Serv. Res. Pap. NE-635.

Langdon, O.G., and F.A. Bennett. 1976. Management of natural stands of splash pine. U.S. For. Serv. Res. Pap. SE-147.

Langsaeter, A. 1941. Om tynning i enaldret gran- og furuskog. *Meddel. f. d. Norske Skogforseksvesen* 8:131–216.

Lantz, R.L. 1971. Guidelines for stream protection in logging operations. Ore. State Game Comm., Res. Div., Portland, OR.

Lareau, J.P. 1985. *Growth Among Second-growth Northern Hardwoods At Two Locations in New York State Following Thinnings to Various Levels of Residual Relative Density.* M.S. thesis. SUNY Coll. Environ. Sci. and For., Syracuse, NY.

Larsen, C.S. 1956. *Genetics in Silviculture.* Transl. by M.K. Anderson. Edinburgh: Oliver and Boyd.

Larson, J.A. 1922. Effect of removal of the virgin white pine stand on the physical factors of site. *Ecol.* 3(4):302–305.

Larson, J.A. 1924. Some factors affecting reproduction after logging in northern Idaho. *J. Agric. Res.* 28:1149–1157.

Larson, M.M., and G.H. Schubert 1969. Effects of osmotic water stress on germination and initial development of ponderosa pine seedlings. *For. Sci.* 15:30–36.

Lavalle, A., and M. Lortie. 1968. Relationship between external features and trunk rot in living yellow birch. *For. Chron.* 44(2):5–10.

Lavander, D.P. 1984. Plant physiology and nursery environment: Interactions affecting seedling growth. In Duryea, M.L. and T.D. Landis (Eds.), *Forest Nursery Manual: Production of Bareroot Seedlings.* Ch. 14, pp. 133–141. Martinus Nijhoff/Dr. W. Junk Publ. The Hague/Boston/Lancaster. For. Res. Lab., Ore. State Univ., Corvallis.

Law, J.R., and C.G. Lorimer. 1989. Managing uneven-aged stands. In F.B. Clark (Ed.), *Central Hardwood Notes,* pp. 60.08(1)–6.08(6). U.S. For. Serv., NC For. Exp. Stn, St. Paul, MN.

Leak, W.B. 1978. Stand structure. In *Uneven-aged silviculture and management in the United States,* pp. 104–114. U.S. For. Serv., Timber Manage. Res., Washington, DC.

Leak, W.B. 1985. Relationships of tree age to diameter in old-growth northern hardwoods and spruce-fir. U.S. For. Serv. Res. Note NE-329.

Leak, W.B., and S.M. Filip. 1977. Thirty-eight years of group selection in New England northern hardwoods. *J. For.* 75:641–643.

Leak, W.B., and R.E. Graber. 1976. Seedling input, death, and growth in uneven-aged northern hardwoods. *Can. J. For. Res.* 6:368–374.

Leak, W.B., and D.S. Solomon. 1975. Influence of residual stand density on regeneration of northern hardwoods. U.S. For. Serv. Res. Pap. NE-310.

Leak, W.B., and R. Wilson, Jr. 1958. Regeneration after cutting of old-growth northern hardwoods in New Hampshire. U.S. For. Serv., NE For. Exp. Stn., Pap. No. 103.

Leak, W.B., D.S. Solomon, and P.S. DeBald. 1987. Silvicultural guide for northern hardwood types in the Northeast. (revised). U.S. For. Serv. Res. Pap. NE-603.

Leak, W.B., D.S. Solomon, and S.M. Filip. 1969. A silvicultural guide for northern hardwoods in the Northeast. U.S. For. Serv. Res. Pap. NE-143.

Ledin, S. 1986. Planting. In L. Sennerby-Forsee (Ed.), *Handbook for Energy Forestry,* p. 15. Sect. Energy For., Dept. Ecol. and Environ. Res., Swedish Univ. Agric. Sci., Uppsala.

Ledin, S., L. Sennerby-Forsse, and H. Johansson. 1992. Implementation of energy forestry on private farmland in Sweden. In C.P. Mitchell, L. Sennerby-Fosse, and L. Zsuffa (Eds.), *Problems and Perspectives of Forest Biomass Energy,* pp. 44–50. Swedish Univ. Agric. Sci., Dept. Ecol. and Environ. Res., Sect. Short Rotation For. Rpt. 48.

Lees, J.C. 1963. Partial cutting with scarification in Alberta spruce-aspen stands. Can. Dept. For., For. Res. Br., Dept. For. Publ. No. 1001.

Lees, J.C. 1978. Hardwood silviculture and management: An interpretive literature review for the Canadian Maritime Provinces. Can. For. Serv., Maritimes For. Res. Ctr., Info. Rpt. M-X-93.

Lees, J.C. 1981. Three generations of red maple stump sprouts. Can. For. Serv. Info. Rpt. M-X-119.

Lees, J.C. 1987. Clearcutting as an even-aged reproduction method. In R.D. Nyland, (Ed.), *Managing Northern Hardwoods.* SUNY Coll. Environ. Sci. and For., Fac. For. Misc. Publ. No. 13 (ESF 87-002) (Soc. Am. For. Publ. No. 87-03).

Lees, J.S. 1970. Natural regeneration of white spruce under spruce-aspen shelterwood, B-18a Forest Section, Alberta. Can. Dept. Fish. and For. Publ. No. 1274.

Leopold, D.J., and G.R. Parker. 1985. Vegetation patterns on a Southern Appalachian watershed after successive clearcuts. *Castenea* 50:164–186.

Leopold, D.J., G.R. Parker, and W.T. Swank. 1985. Forest development after successive clearcuts in the Southern Appalachians. *For. Ecol. and Manage.* 13(1985):83–120.

Leuschner, W.A. 1990. *Forest Regulation, Harvest Scheduling, and Planning Techniques.* New York: Wiley.

Lexen, B. 1939. Space requirement of ponderosa pine by tree diameter. U.S. For. Serv. SW For. and Range Exp. Stn., Res. Note No. 63.

Li, T.L. 1926. Soil temperature as influenced by forest cover. Yale Univ., Sch. For. Bull. 18.

Liegel, L.H., and C.R. Venator. 1987. A technical guide for forest nursery management in the Caribbean and Latin America. U.S. For. Serv. Gen. Tech. Rpt. SO-67.

Lilieholm, R.J., D.E. Teeguarden, and D.T. Gordon. 1989. Thinning stagnated ponderosa and Jeffrey pine stands in Northeastern California: 30-year effects. U.S. For. Serv. Res. Note PSW-407.

Limstrom, G.A. 1962. Forest planting practice in the central states. U.S. For. Serv., Cent. For. Exp. Stn., Misc. Rel. 34.

Lindgren, R.M., R.P. True, and E.R. Toole. 1949. Shade trees for the Southeast. In *Trees, the Yearbook of Agriculture 1949,* pp. 60–65. 81st Congr., 1st Sess., House Doc. No. 29. U.S. Gov. Print. Off., Washington, DC.

Little, E.L., Jr. 1949. Fifty trees from foreign lands. In A. Stefferud (Ed.), *Trees: The Yearbook of Agriculture 1949,* pp. 815–832. 81st Congr., 1st Sess., House Doc. No. 29. U.S. Gov. Print. Off., Washington, DC.

Little, S., and J.J. Mohr, 1963. Five-year effects from row thinning in loblolly pine plantations of eastern Maryland. U.S. For. Serv. Res. Pap. NE-12.

Little, S., and H.A. Somes. 1959. Viability of loblolly pine seed stored in the forest floor. *J. For.* 57:848–849.

Lloyd, F.T., D.L. White, J.A. Abercrombie, Jr., and T.A. Waldrop. 1991. Releasing four-year-old pines in mixed shortleaf-hardwood stands. In S.S. Coleman and D.G. Neary (Eds.), *Proceedings of the Sixth Biennial Southern Silvicultural Research Conference,* Vol. 2. pp. 852–857. U.S. For. Serv. Gen. Tech. Rpt. SE-70.

Lockard, C.R., J.A. Putnam, and R.D. Carpenter. 1950. Log defects in southern hardwoods. U.S. For. Serv., Agric. Handbk. No. 4.

Lockard, C.R., J.A. Putnam, and R.D. Carpenter. 1963. Grade defects in hardwood timber and logs. U.S. For. Serv., Agric. Handbk. No. 244.

Loftis, D.L. 1985. Preharvest herbicide treatment improves regeneration in southern Appalachian hardwoods. *S. J. Appl. For.* 9(3):177–180.

Loftis, D.L. 1990a. Predicting post-harvest performance of advance red oak reproduction in the southern Appalachians, *For. Sci.* 36(4):908–916.

Loftis, D.L. 1990b. A shelterwood method for regenerating red oak in the southern Appalachians. *For. Sci.* 36(4):917–929.

Logan, K.T. 1965. Growth of tree seedlings as affected by light intensity. I. White birch, yellow birch, sugar maple, and silver maple. Can. Dept. For., Publ. No. 1121.

Logan, K.T. 1966a. Growth of tree seedlings as affected by light intensity. II. Red pine, white pine, jack pine and eastern larch. Can. Dept. For., Publ. No. 1160.

Logan, K.T. 1966b. Growth of seedlings as affected by light intensity. III. Basswood and white elm. Can. Dept. For., Publ. No. 1176.

Logan, K.T. 1969. Growth of seedlings as affected by light intensity. IV. Black spruce, white spruce, balsam fir, and eastern white cedar. Can. Dept. For., Publ. No. 1156.

Logan, K.T. 1973. Growth of tree seedlings as affected by light intensity. V. White ash, beech, eastern hemlock, and general conclusions. Can. For. Serv., Publ. No. 1323.

Lohrey, R.E. 1977. Growth responses of loblolly pine to precommercial thinning. *S. J. Appl. For.* 1(3):19–22.

Lohrey, R.E., and E.P. Jones, Jr. 1983. In E.L. Stone (Ed.), *The Managed Slash Pine Ecosystem,* pp. 183–193. Proc. Symp., June 9–11, 1983. Sch. For. Resourc. and Conserv., Univ. Fla., Gainsville, FL.

Long, A.J. 1991. Proper planting improves performance. In M.L. Duryea and P.M. Dougherty (Eds.), *Forest Regeneration Manual.* Dordrecht, The Netherlands: Kluwer Academ. Publ.

Long, J.N. 1988. Density management diagrams: Their construction and use in timber stand management. In W.C. Schmitt (Ed.), *Proceedings—Future Forests of the Mountain West: A Stand Culture Symposium,* pp. 80–86. U.S. For. Serv. Gen. Tech. Rpt. INT-243.

Long, J.N. 1995. The middle and southern Rocky Mountain region. In J.W. Barrett (Ed.), *Regional Silviculture of the United States,* 3d ed. New York: Wiley.

Longman, K.A., and J. Jenik. 1987. *Tropical Forest and Its Environment,* 2d ed., Singapore: Longman Singapore Publ (PTE) Ltd.

Lorimer, C.G., and L.E. Frelich. 1984. A simulation of equilibrium diameter distributions of sugar maple *(Acer saccharum). Bull. Torrey Bot. Club* 111(2):193–199.

Lotan, J.E. 1986. Silvicultural management of competing vegetation. In D.M. Baumgartner, R.J. Boyd, D.W. Breuer, and D.L. Miller. (Eds.), *Weed Control for Forest Productivity in the Interior West,* pp. 10–16. Proc. Symp. Feb. 5–7, 1985, Spokane, Wash., Coop. Ext., Wash. State Univ., Pullman, WA.

Lotan, J.E., C.E. Carlson, and J.D. Chew. 1988. Stand density and growth of interior Douglas-fir. In W.C. Schmitt (Ed.), *Proceedings—Future Forests of the Mountain West: A Stand Culture Symposium,* pp. 185–1914. U.S. For. Serv. Gen. Tech. Rpt. INT-243.

Loveless, R.V., and H.H. Page, Jr. 1991. Granular imazapyr and hexazinone rate study—efficacy of competition control and effects on pine growth. In S.S. Coleman and D.G. Neary (Eds.), *Proceedings of the Sixth Biennial Southern Silvicultural Research Conference,* Vol. 2. pp. 852–863. U.S. For. Serv. Gen. Tech. Rpt. SE-70.

Lowery, R.F., and D.H. Gjerstad. 1991. Chemical and mechanical site preparation. In M.L. Duryea and P.M. Dougherty (Eds.), *Forest Regeneration Manual,* pp. 251–261. The Netherlands: Klumer Academ. Publ.

Lull, H.W. 1965. Ecological and silvicultural aspects. In V.T. Chow (Ed.), *Handbook of Applied Hydrology.* New York: McGraw-Hill.

Lull, H.W., and K.G. Reinhart. 1967. Increasing water yield in the Northeast by management of forested watersheds. U.S. For. Serv. Res. Pap. NE-66.

Lull, H.W., and K.G. Reinhart. 1972. Forests and floods in the eastern United States. U.S. For. Serv. Res. Pap. NE-226.

Lutz, H.J. 1959. Forest ecology, the biological basis of silviculture. The H.R. MacMillan Lectureship in Forestry, 1959. Univ. Brit. Columbia, Vancouver, BC.

Lynch, J.A., and E.S. Corbett. 1991. Long-term implications of forest harvesting on nutrient cycling in central hardwood forests. In L.H. McCormick and K.W. Gottschalk (Eds.), *Proceedings: 8th Central Hardwood Forest Conference,* pp. 500–518. U.S. For. Serv. Gen. Tech. Rpt. NE-148.

Lyon, L.J. 1977. Elk use as related to characteristics of clearcuts in western Montana. In *Proceedings of the Elk-Logging-Roads Symposium, 1976,* pp. 69–72. Univ. Idaho, Moscow, ID.

MacArdle, R.E., W.H. Meyer, and D. Bruce. 1949. The yield of Douglas-fir in the Pacific Northwest. U.S. Dept. Agric. Tech. Bull 201 (Rev.).

MacArthur, R.H., and J.W. MacArthur. 1961. On bird species diversity. *Ecol.* 42(3):116–140.

MacLean, C.D. 1979. Relative density: The key to stocking assessment in regional analysis—A Forest Survey viewpoint. U.S. For. Serv. Gen. Tech. Rpt. PNW-87.

Mader, S.F., and R.D. Nyland. 1984. Six-year responses of northern hardwoods to selection system. *N. J. Appl. For.* 1:87–91.

Mahoney, R. 1986. Advantages and disadvantages of manual release vs. chemical release of conifers. In D.M. Baumgartner, R.J. Boyd, D.W. Breuer, and D.L. Miller (Eds.), *Weed Control for Forest Productivity in the Interior West,* pp. 61–62. Proc. Symp., Feb. 5–7, 1989. Spokane, Wash., Wash. State Univ., Coop. Ext. Serv., Pullman, WA.

Maini, J.S. 1972. Silvics and ecology in Canada. In *Aspen: Symposium Proceedings,* pp. 67–73. U.S. For. Serv. Gen. Tech. Rpt. NC-1.

Maini, J.S., and K.W. Horton. 1966. Influence of temperature on formation and initial growth of aspen suckers. *Can. J. Bot.* 44:1183–1189.

Majcen, Z., Y. Richard, M. Menard, and Y. Grenier. 1990. Choix des Tiges à Marquer pour le Jardinage d'Érablières Inequiennes. Gov. Quebec. Min. de l'Energie et des Ressources. (For.), Dir. Rec. du Dévelop., Ser. de la Rec. Appl., Mem. No. 96.

Manion, P.D. 1991. *Tree Disease Concepts,* 2d ed. Englewood Cliffs, NJ: Prentice Hall.

Manion, P.D. 1992. *Forest Decline Concepts.* St. Paul, MN: APS Press.

Mann, W.F., Jr. 1952. Response of loblolly pine to thinning. *J. For.* 50(6):443–446.

Mann, W.F., Jr. 1965a. Progress in direct-seeding the southern pines. In H.G. Abbott (Ed.), *Direct Seeding in the Northeast,* pp. 9–13. Proc. Symp., Aug. 25–27, 1964. Amherst, MA, Univ. Mass. Agric. Exp. Stn., Amherst, MA.

Mann, W.F., Jr. 1965b. Direct seeding the southern pines: Development and application. In *Proceedings Direct Seeding Workshops,* pp. 2–3. Oct. 5–6, 1965, Alexandria, LA, and Oct. 20–21, Tallahassee, FL. U.S. For. Serv., GSA Atlanta GA 66-4542.

Mann, W.F., Jr., T.E. Campbell, and T.W. Chappel. 1974. Status of aerial row seeding. *For. Farmer* 34(2):12–13.

Mar:Moller, C. 1945. Untersuchingen uber Laubmenge, Stoffverlust und Stoffproduktion des Waldes. Der forstl. Forsogsv. i Danmark, Bd. XVII. Kandrup und Wunsch, Copenhagen. (Cited in Baker 1950, Mar:Moller 1954).

Mar:Moller, C. 1954. The influence of thinning on volume increment. 1. Results of investigations. In *Thinning Problems and Practices in Denmark,* pp. 5–32. SUNY Coll. For. at Syracuse, World For. Ser. Bull. No. 1, Tech. Publ. No. 76.

Marks, P.L. 1974. The role of pin cherry *(Prunus pennsylvanica)* in the maintenance of stability in northern hardwood ecosystems. *Ecol. Monogr.* 44:73–88.

Marlega, R.R. 1981. Operational use of prescribed fire. In S.D. Honns, and O.T. Helgerson (Eds.), *Proceedings of the Workshop on Reforestation of Skeletal Soils,* pp. 71–72. Ore. State Univ., For. Res. Lab., Ore. State Univ., Corvallis, OR.

Marquis, D.A. 1965a. Regeneration of birch and associated hardwoods after patch cutting. U.S. For. Serv. Res. Pap. NE-32.

Marquis, D.A. 1965b. Controlling light in small clear-cuttings. U.S. For. Serv. Res. Rpt. NE-85.

Marquis, D.A. 1966. Germination and growth of paper birch and yellow birch in simulated strip cuttings. U.S. For. Serv. Res. Pap. NE-54.

Marquis, D.A. 1967. Clearcutting in northern hardwoods: Results after 30 years. U.S. For. Serv. Res. Pap. NE-85.

Marquis, D.A. 1972. Effect of forest clearcutting on ecological balances. In R.D. Nyland (Ed.), *A Perspective on Clearcutting in a Changing World,* pp. 47–60. Proc. 1972 Winter Meet., NY Soc. Am. For., Feb. 23–25, 1972, Syracuse, NY, SUNY Coll. Environ. Sci. and For., Appl. For. Res. Inst., AFRI Misc. Rpt. No. 4.

Marquis, D.A. 1981a. Removal or retention of unmerchantable saplings in Allegheny hardwoods: Effect on regeneration after clearcutting. *J. For.* 79(5):280–283.

Marquis, D.A. 1981b. Survival, growth, and quality of residual trees following clearcutting in Allegheny hardwood forests. U.S. For. Serv. Res. Pap. NE-277.

Marquis, D.A. 1986. Thinning Allegheny hardwood pole and small sawtimber stand. In H.C. Smith and M. Eyes (Ed.), *Guidelines for Managing Immature Appalachian Hardwood Stands,* pp. 68–84. W. Va. Univ., Morgantown, WV, Soc. Am. For. Publ. 86-02.

Marquis, D.A. 1987. Assessing the adequacy of regeneration and understanding early development patterns. In R.D. Nyland (Ed.), *Managing Northern Hardwoods,* pp. 143–159. SUNY Coll. Environ. Sci. and For., Fac. For. Misc. Publ. No. 13 (ESF 87-002) (Soc. Am. For. Publ. No. 87-03).

Marquis, D.A. 1990. *Prunus serotina* Ehrh. Black cherry. In R.M. Burns and B.H. Honkala (Eds), *Silvics of North America. Volume 2. Hardwoods,* pp. 594–604. U.S. For. Serv., Agric. Handbk. 654.

Marquis, D.A. 1991. Independent effects and interactions of stand diameter, tree diameter, crown class, and age on tree growth in mixed-species, even-aged hardwood stands. In L.H. McCormick and K.W. Gottshalk (Eds.), *Proceedings of the 8th Central Hardwood Forest Conference,* pp. 442–458. U.S. For. Serv. Gen. Tech. Rpt. NE-148.

Marquis, D.A., and T.W. Beers 1969. A further definition of some forest components. *J. For.* 67:493.

Marquis, D.A., and J.C. Bjorkbom. 1960. How much scarification from summer logging? U.S. For. Serv., NE For. Exp. Stn. For. Res. Note No. 110.

Marquis, D.A., J.C. Bjorkbom, and G. Yellenosky. 1964. Effect of seedbed condition and light exposure on paper birch regeneration. *J. For.* 62(12):876–881.

Marquis, D.A., R.L. Ernst, and S.L. Stout 1984. Prescribing silvicultural treatments in hardwood stands of the Alleghenies. U.S. For. Serv. Gen. Tech. Rpt. NE-96.

Marquis, D.A., R.L. Ernst, and S.L. Stout. 1992. Prescribing silvicultural treatments in hardwood stands of the Alleghenies (Revised). U.S. For. Serv. Gen. Tech. Rpt. NE-96.

Marquis, D.A., T.J. Grisez, J.C. Bjorkbom, and B.A. Roach. 1975. Interim guide to regeneration of Allegheny hardwoods. U.S. For. Serv. Gen. Tech. Rpt. NE-19.

Marshall, J.K. 1967. The effect of shelter on the productivity of grasslands and field crops. *Field Crop Abstr.* 20(1):1–14.

Martin, C.W., and J.W. Hornbeck. 1989. Revegetation after strip cutting and block clearcutting in northern hardwoods. A 10-year history. U.S. For. Serv. Res. Pap. NE-625.

Martin, C.W., and J.W. Hornbeck. 1990. Regeneration after strip clearcutting and block clearcutting in northern hardwoods. *N. J. Appl. For.* 7:65–68.

Martin, C.W., R.S. Pierce, G.E. Likens, and F.H. Bormann. 1986. Clearcutting affects stream chemistry in the White Mountains of New Hampshire. U.S. For. Serv. Res. Pap. NE-579.

Martin, G.L. 1982. Investment-efficient stocking guides for all-aged northern hardwood forests. Univ. Wis. (Madison), Dept. For., Res. Pap. R3129.

Martin, R.E. 1976. Prescribed burning for site preparation in the Inland Northwest. In D.M. Baumgartner and R.J. Boyd (Eds.), *Tree Planting in the Inland Northwest,* pp. 134–156. Proc. Conf., Feb. 17–19, 1976. Wash. State. Univ., Coop. Ext. Serv., Pullman, WA.

Martin, R.E. 1982. Shrub control by burning before timber harvest. In D. M. Baumgartner

(Ed.), *Site Preparation and Fuels Management on Steep Terrain,* pp. 35–40. Symp. Proc., Coop. Ext. Serv., Wash. State Univ., Pullman, WA.

Martin, R.E. 1990. Goals, methods, and elements of prescribed burning. In J.D. Walstad, S.R. Radosevich, and D.V. Sandberg (Eds.), *Natural and Prescribed Fire in Pacific Northwest Forest,* pp. 55–66. Ore. State Univ. Press., Corvallis, OR.

Martin, W.C., and R.S. Pierce. 1980. Clearcutting patterns affect nitrate and calcium in streams of New Hampshire. *J. For.* 78(5):268–272.

Maser, C. 1994. *Sustainable Forestry. Philosophy, Science, and Economics.* Delray Beach, FL: St. Lucie Press.

Matthews, D.M. 1935. *Management of American Forests.* New York: McGraw-Hill.

Matthews, J.D. 1964. Seed production and seed certification. *Unasylva* 18:2–3.

Matthews, J.D. 1989. *Silvicultural Systems.* Oxford: Oxford Univ. Press.

Maynard, C.A. 1994. Six-year field test results of micropropagated black cherry *(Prunus serotina). In Vitro Cellular Develop. Biol.* 22:231–233.

Maynard, C.A., R.P. Overton, and L.C. Johnson. 1987. The silviculturist's role in tree improvement in northern hardwoods. In R.D. Nyland (Ed.), *Managing Northern Hardwoods,* pp. 35–46. SUNY Coll. Environ. Sci, and For., Fac. For. Misc. Publ. No. 13 (ESF 87-002), Soc. Am. For. Publ. No. 87-03.

McAlpine, R.G., C.L. Brown, A.M. Herrick, and H.E. Ruark. 1966. "Silage" sycamore. *For. Farmer* XXVI:67-,16.

McArdle, R.E., W.H. Meyer, and D. Bruce. 1961. The yield of Douglas-fir in the Pacific Northwest. U.S. Dept. Agric., For. Serv., Tech. Bull. 201 (Rev.).

McCarter, J.B., and J.N. Long. 1986. A lodgepole pine density management diagram. *W. J. Appl. For.* 1(1):6–11.

McCauley, O.D., and G.R. Trimble. 1975. Site quality in Appalachian hardwoods. The biological and economic response under selection silviculture. U.S. For. Serv. Res. Pap. NE-312.

McClurkin, D.C., P.D. Duffey, and N.S. Nelson. 1987. Changes in forest floor and water quality following thinning and clearcutting of 20-year-old pine. *J. Environ. Qual.* 16(3):237–241.

McConnell, G. 1954. The Conservation movement—Past and present. *W. Political Quart.* 3:189–201.

McCormack, M.L., Jr. 1991. Herbicide technology for securing naturally regenerating stands. In C.M. Simpson (Ed.), *Proceedings of the Conference on Natural Regeneration Management,* pp. 193–200. For. Can., Maritimes Reg., Fo42-162/991E.

McDaniel, K.C., and L. WhiteTrifaro. 1987. Selective control of pinyon-juniper with herbicides. In R.E. Lverett (Ed.), *Proceedings—Pinyon-Juniper Conference,* pp. 448–455. U.S. For Serv. Gen. Tech. Rpt. INT-215.

McDonald, P.M. 1976. Shelterwood cutting in a young-growth, mixed-conifer stand in north central California. U.S. For. Serv. Res. Pap. PSW-117.

McDonald, P.M. 1980. Seed dissemination in small clearcuts in north-central California. U.S. For. Serv. Res. Pap. PSW-150.

McDonald, P.M. 1983. Clearcutting and natural regeneration . . . Management implications for the northern Sierra Nevada. U.S. For. Serv. Gen. Tech. Rpt. PSW-70.

McDonald, P.M., and G.O. Fiddler. 1986. Release of Douglas-fir seedlings: Growth and treatment costs. U.S. For. Serv. Res. Pap. PSW-182.

McDonald, P.M., and G.O. Fiddler. 1989. Competing vegetation in ponderosa pine plantations: Ecology and control. U.S. For. Serv. Gen. Tech. Rpt. PSW-113.

McDonald, P.M., and G.O. Fiddler. 1990. Ponderosa pine seedlings and competing vegetation: Ecology, growth, and cost. U.S. For. Serv. Res. Pap. PSW-199.

McDonald, P.M., and O.T. Helgerson. 1990. Mulches aid in regenerating California and Oregon forests: Past, present, and future. U.S. For. Serv. Gen. Tech. Rpt. PSW-123.

McDonald, P.M., and R.A. Tucker. 1989. Manual release in an "old" Douglas-fir plantation increases diameter growth. U.S. For. Serv. Res. Note PSW-405.

McDonald, S.E. 1984. Irrigation in forest-tree nurseries: Monitoring and effects on seedling growth. In M.L. Duryea and T.D. Landis (Eds.), *Forest Nursery Manual: Production of Bareroot Seedlings.* Ch. 12, pp. 107–121. Martinus Nijhoff/Dr W. Junk Publ. The Hague/Boston/Lancaster. For. Res. Lab., Ore. State Univ., Corvallis.

McElroy, G.H., and W.M. Dawson. 1986. Biomass from short-rotation coppice willows on marginal land. *Biomass* 10(1986):225–240.

McGee, C.E. 1989. Estimating tree ages in uneven-aged hardwood stands. S. J. *Appl For.* 13(1):40–42.

McGuire, W.S., and D.B. Hannaway. 1984. Cover and green manure crops for Northwest nurseries. In M.L. Duryea, and T.D. Landis (Eds.), *Forest Nursery Manual: Production of Bareroot Seedlings.* Ch. 10, pp. 87–92. Martinus Nijhoff/Dr W. Junk Publ. The Hague/Boston/Lancaster. For. Res. Lab., Ore. State Univ., Corvallis.

McKee, W.H., Jr., G.E. Hatchell, and A.E. Tiarks. 1985. A loblolly pine management guide. Managing site damage from logging. U.S. For. Serv. Gen. Tech. Rpt. SE-32.

McLemore, B.F. 1983. Recovery of understocked, uneven-aged pine stands and suppressed trees. In E.P. Jones, Jr. *Proceedings of the Second Biennial Southern Silviculture Research Conference,* pp. 226–229. U.S. For. Serv. Gen. Tech. Rpt. SE-24.

McLintock, T.F. 1952. Cost of pruning red spruce in natural stands. *J. For.* 50:485–486.

McLintock, T.F. 1954. Factors affecting wind damage in selectively cut stands of spruce and fir in Maine and northern New Hampshire. U.S. For. Serv. NE For. Exp. Stn., Stn. Pap. No. 70.

McMinn, J.W. 1965. Precommercial thinning in dense young slash pine in north Florida. *For. Farmer* 24(12):10–11.

McMinn, J.W. 1985. Coppice regeneration of upland hardwoods from intensive whole-tree harvesting. In E. Shoulders (Ed.), *Proceedings of the Third Biennial Southern Silvicultural Research Conference,* pp. 159–162. U.S. For. Serv. Gen. Tech. Rpt. SO-54.

McNab, W.H. 1977. An overcrowded loblolly pine stand thinned with fire. *S. J. Appl. For.* 1(1):24–26.

McNab, W.H. 1991. Factors affecting temporal and spatial soil moisture variation in and adjacent to group selection openings. In L.H. McCormick and K.E. Gottschalk (Eds.), *Proceedings of the Eighth Central Hardwood Forest Conference,* pp. 475–488. Penn. State Univ., March 3–5, 1991. U.S. For. Serv. Gen. Tech. Rpt. NE-148.

McQuilkin, R.A. 1975. Growth of four types of white oak reproduction after clearcutting in the Missouri Ozarks. U.S. For. Serv. Res. Pap. NC-116.

McQuilkin, W.E. 1965. Direct seeding in the Northeast—Status and needs. In H.E. Abbott (Ed.), *Direct Seeding in the Northeast—1964,* pp. 25–27. Proc. Symp. Aug. 25–27, 1964. Univ. Mass. Agric. Exp. Stn., Amherst, MA.

McRae, D.J. 1994. Prescribed fire converts spruce budworm-damaged forest. *J. For.* 92(11):38–40.

Medin, D.E. 1985. Breeding bird responses to diameter-cut logging in west-central Idaho. U.S. For. Serv. Res. Pap. INT-355.

Megahan, W.F. 1977. XIV. Reducing erosional impacts of roads. In *Guidelines for Watershed Management,* pp. 237–261. Food and Agric. Org., United Nat., FAO Conserv. Guide.

Metzger, F., and C.H. Tubbs. 1971. The influence of cutting method on regeneration of second growth northern hardwoods. *J. For.* 69(9):559–564.

Metzger, F.T. 1980. Strip clearcutting to regenerate northern hardwoods. U.S. For. Serv. Res. Pap. NC-186.

Mexal, J.G., and J.T. Fisher. 1987. Organic matter: Short-term benefits and long-term opportunities. In T.D. Landis (Ed.), *Meeting the Challenge of the Nineties: Proceedings, Intermountain Forest Nursery Association,* pp. 18–23. U.S. For. Serv. Gen. Tech. Rpt. RM-151.

Mexal, J.G., and D.B. South. 1991. Bareroot seedling culture. In M.L. Duryea and P.M. Dougherty (Eds.), *Forest Regeneration Manual,* Ch. 6, pp. 89–115. Dordrecht, The Netherlands: Kluwer Academ. Publ.

Meyer, H.A. 1952. Structure, growth, and drain in balanced uneven-aged forests. *J. For.* 50:85–92.

Meyer, H.A., and D.D. Stevenson. 1943. The structure and growth of virgin beech, birch, maple, hemlock forests in northern Pennsylvania. *J. Agric. Res.* 67:465–484.

Meyer, H.A., A.B. Recknagel, D.D. Stevenson, and R.A. Bartoo. 1961. *Forest Management,* 2d ed. New York: Ronald Press.

Meyer, W.H. 1961. Yield of even-aged stands of ponderosa pine. U.S. Dept. Agric. Tech. Bull. 630 (Rev).

Michael, J.L. 1985. Growth of loblolly pine treated with hexazinone, sulfometuron, methyl, and metsulfuron methyl for herbaceous weed control. *S. J. Appl. For.* 9(1):20–26.

Millar, R.H. 1965. Seed spotting—A method of direct seeding. In *Proceedings Direct Seeding Workshops,* pp. 11–13. Oct. 5–6, 1965, Alexandria, LA, and Oct. 20–21, Tallahassee, FL. U.S. For. Serv., GSA Atlanta GA 66-4542.

Miller, D.L. 1986a. Conifer release in the inland northwest—effects. In D.M. Baumgartner, R.J. Boyd, D.W. Breuer, and D.L. Miller. *Weed Control for Forest Productivity in the Interior West,* pp. 17–24. Proc. Symp. Feb. 5–7, 1985. Spokane, Wash., Coop. Ext., Wash State Univ., Pullman, WA.

Miller, D.L. 1986b. Manual and mechanical methods of vegetation control—What works and what doesn't. In D.M. Baumgartner, R.J. Boyd, D.W. Breuer, and D.L. Miller. (Eds.), *Weed Control for Forest Productivity in the Interior West,* pp. 55–60. Proc. Symp. Feb. 5–7, 1985, Spokane, Wash., Coop. Ext., Wash State Univ., Pullman, WA.

Miller, G.W. 1984. Releasing young hardwood crop trees—Use of a chain saw costs less than herbicides. U.S. For. Serv. Res. Pap. NE-500.

Miller, G.W. 1993. Financial aspects of partial cutting practices in central Appalachian hardwoods. U.S. For. Serv. Res. Pap. NE-673.

Miller, G.W., and H.C. Smith. 1991. Comparing partial cutting practices in central Appalachian hardwoods. In L.H. McCormick and K.W. Gottschalk (Eds.), *Proceedings of the 8th Central Hardwood Forest Conference,* pp. 105–119. U.S. For. Serv. Gen. Tech. Rpt. NE-148.

Miller, G.W., N.I. Lamson, and S.M. Brock. 1984. Logging damage associated with thinning central Appalachian hardwood stands with a wheeled skidder. In P.A. Peters and J. Luchok (Eds.), *Proceedings of the Mountain Logging Symposium,* June 5–7, 1984, W. Va. State Univ., Morgantown, WV.

Miller, J.H. 1991. Application methods for forest herbicide research. In J.G. Miller and G.R. Glover (Eds.), *Standard Methods For Forest Herbicide Research*, Ch. 6, pp. 45–60. S. Weed. Sci. Soc., Auburn Univ. Silvi. Herbicide Coop., and U.S. For. Serv. S. For. Exp. Stn., New Orleans, LA.

Miller, J.H., B.R. Zutter, S.M. Sedaker, M.B. Edwards, J.D. Haywood, and R.A. Newbold. 1991. A regional study on the influence of woody and herbaceous competition on early loblolly pine growth. *S. J. Appl. For.* 15:169–178.

Miller, R.E., R.E. Bigley, and S. Webster. 1993. Early development of matched planted and naturally regenerated Douglas-fir stands after slash burning in the Cascade Range. *W. J. Appl. For.* 8(1):5–10.

Miller, R.E., R.L. Williamson, and R.R. Silen. 1974. Regeneration and growth of coastal Douglas-fir. In O.P. Cramer (Ed.), *Environmental Effects of Forest Residues Management in the Pacific Northwest, a State-of-knowledge Compendium,* pp. J1–J41. U.S. For. Serv. Gen. Tech. Rpt. PNW-24.

Mills, W.L., Jr., and J.C. Callahan. 1981. Financial maturity: A guide to when trees should be harvested. Purdue Univ., Coop. Ext. Serv., For and Nat. Resourc. Bull. FNR 91.

Minckler, L.S., and A.C. Chapman. 1948. Tree planting in the Central, Piedmont, and Southern Appalachian regions. U.S. Dept. Agric., Farmers Bull. No. 1994.

Minckler, L.S., and J.D. Woerheide. 1965. Reproduction of hardwoods 10 years after cutting as affected by site and opening size. *J. For.* 63(2):103–107.

Minckler, L.S., J.D. Woerheide, and R.C. Schlesinger. 1973. Light, soil moisture, and tree reproduction in hardwood forest openings. U.S. For. Serv. Res. Pap. NC-89.

Minogue, P.J., R.K. Cantell, and H.C. Griswold. 1991. Vegetation management after plantation establishment. In M.L. Duryea and P.M. Dougherty (Eds.), *Forest Regeneration Manual,* Ch. 19, pp. 335–358. Dordrecht, The Netherlands: Kluwer Academ. Publ.

Minogue, P.J., B.R. Zutter, and D.H. Gjerstad. 1985. Comparison of liquid and solid hexazione formulations for pine release. In E. Shoulders (Ed.), *Proceedings of the Third Biennial Southern Silvicultural Research Conference,* pp. 292–299. U.S. For. Serv. Gen. Tech. Rpt. SO-54.

Minore, D. 1990. *Thuja plicata* Donn ex D. Don. Western redcedar. In R.M. Burns and B.H. Honkala (Eds.), *Silvics of North America. Volume 1. Conifers,* pp. 590–600. U.S. For. Serv., Agric. Handbk. 654.

Minore, D., and M.E. Dubrasich. 1981. Regeneration after clearcutting in subalpine stands near Windigo Pass, Oregon. *J. For.* 79(9):619–621.

Mitchell, C.P., and J.B. Ford-Robertson. 1992. Supply systems for short rotation energy forestry in the UK. In C.P. Mitchell, L. Sennerby-Fosse, and L. Zsuffa (Eds.), *Problems and Perspectives Of Forest Biomass Energy,* pp. 31–35. Swedish Univ. Agric. Sci., Dept. Ecol. and Environ. Res., Sect. Short Rotation For. Rpt. 48.

Mitchell, H.L. 1961. A concept of intrinsic wood quality, and nondestructive methods for determining quality in standing timber. U.S. For. Serv., For. Prod. Lab. Rpt. No. 2233.

Mitchell, J.A. 1930. Interception of rainfall by the forest. *J. For.* 28:101–102.

Mitchell, J.A., J.P. Dwyer, R.A. Musbach, H.E. Garrett, G.S. Cox, and W.B. Kurtz. 1988. Crop tree release of a scarlet-black oak stand. *N. J. Appl. For.* 5(2):96–99.

Mitscherlich, E.A. 1921. Das Wirkungsgesetz der Wachstumfaktoren. *Landwirtschaft Jahrhbuch Bog.* 11:15.

Miyata E.S., H.N. Steinhib, R.L. Sajdak, and M. Coffman. 1982. Roller chopping for site preparation in Wisconsin: A case study. U.S. For. Serv. Res. Pap. NC-223.

Mobley, H.E., R.S. Jackson, W.E. Balmer, W.E. Ruziska, and W.A. Hough. 1978. *A Guide for Prescribed Fire in Southern Forests.* U.S. For. Serv. SE Area State and Priv. For., Atlanta, GA.

Morash, R., and B. Freedman. 1983. Seedbanks in several recently clear-cut and mature hardwood forests in Nova Scotia. *Proc. N.S. Inst. Sci.* 33:85–94.

Morby, F.E. 1984. Nursery-site selection, layout, and development. In Duryea, M.L. and T.D. Landis (Eds.), *Forest Nursery Manual: Production of Bareroot Seedlings.* Ch. 2, pp. 9–16. Martinus Nijhoff/Dr W. Junk Publ. The Hague/Boston/Lancaster. For Res. Lab., Ore. State Univ., Corvallis.

Morgenstern, E.K., and L. Roche. 1969. Using concepts of selection to delimit seed zones. In Second World Consul. On For. Tree Breeding, pp. 203–215. FAO-FO-FTB-69—2/16. Vol. 1.

Moriarty, J.J., and W.C. McComb. 1983. The long-term effects of timber stand improvement on snag and cavity densities in the central Appalachians. In J.W. Davis, G.A. Goodwin, and R.A. Ockenfels (Eds.), *Snag Habitat Management,* pp. 40–44. U.S. For. Serv. Gen. Tech. Rpt. RM-99.

Morris, L.A., and R.G. Campbell. 1991. Soil and site potential. In M.L. Duryea and P.M. Dougherty (Eds.), *Forest Regeneration Manual,* Ch. 10, pp. 183–206. Dordrecht, The Netherlands: Kluwer Academ. Publ.

Morrow, R.R. 1978. Growth of European larch at five spacings. NYS Coll. Agric. and Life Sci., Cornell Univ., NY Food and Life Sci. Bull. No. 75.

Morrow, R.R., L.S. Hamilton, and F.E. Winch, Jr. 1981. Planting forest trees in rural areas. Cornell Univ., NYS Coll. Agric. and Life Sci., Plant Sci. Nat. Resourc. 13, Info. Bull. 174.

Moser, J.W. 1976. Specification of density for the inverse J-shaped diameter distribution. *For. Sci.* 22:177–180.

Moser, J.W., Jr., C.H. Tubbs, and R.D. Jacobs. 1979. Evaluation of a growth prediction system for uneven-aged northern hardwoods. *J. For.* 77(7):421–423.

Mracek, Z. 1980. Bestandessicherheit junger Fichten in Abhaengigkeit zu ihrem Wuchsraum. In H. Kramer (Ed.), *Biologische, technische und wirtschaftiche Aspekte der Jungbestandspflege,* pp. 247–259. SchrReihe Forstl. Fak. Univ. Goettingen, Bd. 67. (Transl. D. Clerc, 1985).

Muelder, D. 1966. Comparative analysis of costs and benefits to the various systems of silvicultural management. In *Proceedings of the Sixth World Forestry Congress,* Vol. II, pp. 2287–2294. Madrid.

Munzel, E. 1962. *Relationships Between Stand Density and Volume Increment.* M.F. thesis, NYS Coll. For. at Syracuse Univ., Syracuse, NY.

Murphy, P.A., and R.W. Guldin. 1987. Financial maturity of trees in selection stands revisited. U.S. For. Serv. Res. Pap. SO-242.

Murphy, P.A., and M.G. Shelton. 1991. Stand development five years after cutting different diameter limits in loblolly-shortleaf pine stands. In S.S. Coleman and D.G. Neary (Eds.), *Proceedings of the Sixth Biennial Southern Silvicultural Research Conference,* pp. 384–393. U.S. For. Serv. Gen. Tech. Rpt. SE-70.

Murphy, P.A., M.G. Shelton, and D.L. Graney. 1993. Group selection—Problems and possibilities for the more shade-intolerant species. In A.R. Gillespie, G.R. Parker, and P.E. Pope (Eds.), *Proceedings of the 9th Central Hardwood Forest Conference,* pp. 229–247. U.S. For. Serv. Gen. Tech. Rpt. NC-161.

Mutch, R.W. 1994. Fighting fire with prescribed fire. A return to ecosystem health. *J. For.* 92(11):31–33.

Myers, C.A. 1967. Growing stock levels in even-aged ponderosa pine. U.S. For. Serv. Res. Pap. RM-33.

Nair, P.K.R. 1991. State-of-the-art agroforestry systems. In P.G. Jarvis (Ed.), *Agroforestry: Principles and Practice,* pp. 5–29. Amsterdam: Elsevier.

Nat. Res. Counc. 1981. *Sowing Forests From the Air.* Rpt. Ad Hoc Panel, Adv. Comm. on Technol. Innovation, Bd. Sci. and Technol. for Internal. Develop., Comm. Internal. Rel., Nat. Res. Counc., Washington, DC: National Academy Press.

Naumann, J.R. 1994. The role of silviculturists in ecosystem management. In L.H. Foley (Ed.), *Silviculture: From the Cradle of Forestry to Ecosystem Management,* pp. 22–25. Proc. Natl. Silvi. Workshop, Hendersonville, NC, Nov. 1–4, 1993. U.S. For. Serv., SE For. Exp. Stn., Asheville, NC.

Nebeker, T.E., and J.D. Hodges. 1985. Thinning and harvesting practices to minimize site disturbances and susceptibility to bark bettle and disease attack. In S.J. Branham and R.C. Thatcher (Eds.), *Integrated Pest Management Research Symposium: The Proceedings,* pp. 263–271. U.S. For. Serv. Gen. Tech. Rpt. SO-56.

Nebeker, T.E., Mizell, R.F., III, and Bedwell, N.J. 1984. Management of bark beetle populations: Impact of manipulating predator cues and other control tactics. In W.Y. Garner and J. Harvey, Jr. (Eds.), *Chemical and Biological Controls in Forestry,* pp. 25–33. Am. Chem. Soc. Symp. Ser. 238. Am. Chem. Soc., Washington, DC.

Nebeker, T.E., J.D. Hodges, B.K. Karr, and D. M. Moehring. 1985. Thinning practices in southern pines—With pest management recommendations. U.S. For. Serv. Tech. Bull. 1703.

Needham, T., and S. Clements. 1991. Fill planting to achieve yield targets in naturally regenerated stands. In C.M. Simpson (Ed.), *Proceedings of the Conference on Natural Regeneration Management,* pp. 213–224. For. Can., Maritimes Reg., Fredericton, NB.

Neisess, J. 1984. Integrated pest management (IPM) in forestry. In W.Y. Garner and J. Harvey, Jr. (Eds.), *Chemical and Biological Controls in Forestry,* pp. 69–75. Am. Chem. Soc. Symp. Ser. 238. Am. Chem. Soc., Washington, DC.

Nelson, L.R., and J.H. Miller. 1991. Forest herbicide technology. *For. Farmer* 50(3):73–76.

Newton, M., E.C. Cole, D.E. White, and M.L. McCormack, Jr. 1992a. Young spruce-fir forests released by herbicides. I. Response of hardwoods and shrubs. *N. J. Appl. For.* 9(4):126–130.

Newton, M., E.C. Cole, M.L. McCormack, Jr., and D.E. White. 1992b. Young spruce-fir forests released by herbicides. II. Conifer response to residual hardwoods and overstocking. *N. J. Appl. For.* 9(4):130–135.

Newton, M., M.L. McCormack, Jr., R.L. Sajdak, and J.D. Walstad. 1987. Forest vegetation problems in the Northeast and Lake States/Provinces. In J.D. Walstad and P.J. Kuch (Eds.), *Forest Vegetation Management For Conifer Production,* Ch. 4, pp. 77–103. New York: Wiley.

Nicholls, T.H. 1989. Prevention, detection, and control of nursery tree diseases. U.S. For. Serv. Res. Note NC-348.

Nienstaedt, H., and E.B. Synder. 1974. Principles of genetic improvement of seed. In C.S. Schopmeyer, Tech. Coordinator. *Seeds of Woody Plants of the United States,* pp. 5–40. U.S. For. Serv., Agric. Handbk. No. 450.

Noble, D.L., and F. Ronco, Jr. 1978. Seedfall and establishment of Engelmann spruce and subalpine fir in clearcut openings in Colorado. U.S. For. Serv. Res. Pap. RM-200.

Noble, I.R., and R.O. Slatyer. 1977. Post-fire succession of plants in Mediterranean ecosystems. In *Proceedings of the Symposium on the Environmental Consequences of Fire and Fuel Management in Mediterranean Ecosystems,* pp. 27–36. U.S. For. Serv. Gen. Tech. Rpt. WO-3.

Nolley, J.W. 1991. Bulletin of hardwood market statistics: Fall 1990. U.S. For. Serv. Res. Note. NE-343.

Noon, B.R., and K.P. Able. 1978. A comparison of avian community structure in the northern and southern Appalachian Mountains. In D. DeGraaf (Ed.), *Management of Southern Forests for Non-game Birds,* pp. 98–117. U.S. For. Serv. Gen. Tech. Rpt. SE-14.

Noon, B.R., V.P. Bingman, and J.P. Noon. 1979. The effects of changes in habitat on northern hardwood forest bird communities. In R. DeGraaf (Ed.), *Management of Northcentral and Northeastern Forests for Nongame Birds,* pp. 33–48. U.S. For. Serv. Gen. Tech. Rpt. NC-51.

Norris, L.A. 1990. An overview and synthesis of knowledge concerning natural and prescribed fire in Pacific Northwest forests. In J.D. Walstad, S.R. Radosevich, and D.V. Sandberg (Eds.), *Natural and Prescribed Fire in Pacific Northwest Forests,* pp. 7–22. Ore. State Univ. Press, Corvallis.

Noss, R.F. and A.Y. Cooperrider. 1994. *Saving Nature's Legacy.* Washington, DC: Island Press.

NY Soc. Am. For. 1975. *Timber Harvesting Guidelines for New York.* NY Soc. Am. For. Pamph.

Nyland, R.D. 1972. Over-winter cold storage and its use in New York State. In *Proceedings,* pp. 20–29. Ont. Min. Nat. Resourc., Proc. Provincial Nurserymen's Meet., June 20–22, 1972, Midhurst, Ontario.

Nyland, R.D. 1974a. Cold storage delays flushing of conifers. SUNY Coll. Environ. Sci. and For., Appl. For. Res. Inst., AFRI Res. Note 10.

Nyland, R.D. 1974b. Fall lifting for over-winter cold storage of conifers. SUNY Coll. Environ. Sci. and For., Appl. For. Res. Inst., AFRI Res. Rpt. No. 22.

Nyland, R.D. 1986a. Important trends and regional differences in silvicultural practice for northern hardwoods. In G.D. Mroz and D.D. Reed (Eds.), *The Northern Hardwood Resource: Management and Potential.* Proc. Conf., Aug. 18–20, 1986, Mich. Technol. Univ., Houghton, MI.

Nyland, R.D. 1986b. Logging damage during thinning in even-aged hardwood stands. In H.C. Smith and M.C. Eye (Eds.), *Guidelines for Managing Immature Appalachian Hardwood Stands.* Proc. Symp. W. Va. Univ, Morgantown, WV, May 28–30, 1986. Soc. Am. For. Publ. 86-02. W. Va. Univ., Coll. Agric. and For., Morgantown, WV.

Nyland, R.D. 1987. Selection system and its application to uneven-aged northern hardwoods. In R.D. Nyland (Ed.), *Managing Northern Hardwoods.* SUNY Coll. Environ. Sci. and For., Fac. For. Misc. Publ. 13 (ESF 87-002) (Soc. Am. For. Publ. 87-03).

Nyland, R.D. 1988. Past and present silviculture and harvesting practices in central and northern hardwoods. In M.E. Demeritt (Ed.), *Proceedings,* pp. 21–39. 31st Northeast For. Tree Improve Conf. and 6th For. Tree Improve Assoc. Ann. Meet., Joint Meet., NE Tree Improve. Conf., c/o U.S. For. Serv., Berea, KY.

Nyland, R.D. 1989. Logging damage. In F.B. Clark and J.G. Hutchinson (Eds.), *Central Hardwood Notes.* U.S. For. Serv. N. Cent. For. Exp. Stn., St. Paul., MN.

Nyland, R.D. 1990a. Selection marking for uneven-aged northern hardwoods. In G. van der Kelen (Ed.), *Les fôrets feuillues . . . un potentiel à connaître,* pp. 21–28. Proc. 21 Semaine des Sci. Fôr. Fac. de fôr. et de géomatique, Univ. Laval. Ste-Foy, Que.

Nyland, R.D. 1990b. Logging damage during conventional harvesting in northern hardwood stands. In G. van der Kelen (Ed.), *Les fôrets feuillues . . . un potential à connaître,* pp. 53–58. Proc. Semaine des Sci. Fôr., Fac. de fôr. et de géomatique, Univ. Laval. Ste-Foy, Que.

Nyland, R.D. 1991. A perspective on northern hardwood silviculture. In Proc. Soc. Am. For. Nat. Conv., Aug. 4–7, 1991, San Francisco, CA. Soc. Am. For., Washington, DC.

Nyland, R.D. 1992. Exploitation and greed in eastern hardwood forests. *J. For.* 90(1)33–37.

Nyland, R.D. 1994a. Careful logging in northern hardwoods. In B. Batchelor (Ed.), *Proceedings of the Careful Logging Workshop,* pp. 28–45. March 18–19, 1991. Pinewood Park, North Bay, Ont. Ont. Min. Nat. Resourc., Cent. Reg. Sci. and Technol. Tech Rpt. #33.

Nyland, R.D. 1994b. Careful logging in northern hardwoods. In J.A. Rice (Ed.), *Logging Damage: The Problems and Practical Solutions,* pp. 29–51. Ont. Min. Nat. Resourc. For. Res. Rpt. No. 117.

Nyland, R.D., and W.J. Gabriel. 1971. Logging damage to partially cut hardwoods stands in New York State. SUNY Coll. Environ. Sci. and For., Appl. For. Res. Inst., AFRI Res. Rpt. No. 5.

Nyland, R.D., L.P. Abrahamson, and K.B. Adams. 1982. Use of prescribed fire for regenerating red and white oak in New York. In *America's Hardwood Forests—Opportunities Unlimited,* pp. 163–167. Proc. Soc. Am. For. Natl. Conv., Cincinnati, OH, Sept. 19–22, 1982. Soc. Am. For., Washington, DC.

Nyland, R.D., L.M. Alban, and R.L. Nissen, Jr. 1993. Greed or sustention: Silviculture or not. In R.D. Briggs and W.B. Krohn (Eds.), *Nurturing the Northeastern Forest,* pp. 37–52. Proc. Joint Meet. New Engl. Soc. Am. For. and Me. Wildl. Soc., March 3–5, 1993, Portland, ME, Me. Agric. and For. Exp. Stn., Misc. Rpt. 382, SAF Publ. No. 93-05.

Nyland, R.D. C.C. Larson, and H.L. Shirley. 1983. *Forestry and Its Career Opportunities,* 4th ed. New York: McGraw-Hill.

Nyland, R.D., A.L. Leaf, and D.H. Bickelhaupt. 1979. Litter removal impairs growth of direct-seeded Norway spruce. *For. Sci.* 25(1):244–246.

Nyland, R.D., D.M. Marquis, and D.W. Whittemore. 1981. Northern hardwoods. In *Choices in Silviculture for American Forests,* pp. 16–22. Soc. Am. For., Washington, DC.

Nyland, R.D., W.C. Zipperer, and D.B. Hill. 1986. The development of forest islands in exurban central New York State. *Landscape and Urban Plann.* 13(1986):111–123.

Nyland, R.D., D.F. Behrend, P.J. Craul, and H.E. Echelberger. 1977. Effects of logging in northern hardwood forests. *Tappi* 60(6):58–61.

Nyland, R.D., P.J. Craul, D.F. Behrend, H.E. Echelberger, W.J. Gabriel, R.L. Nissen, Jr., R. Uebler, and J. Zarnetske. 1976. Logging and its effects in northern hardwoods. SUNY Coll. Environ. Sci. and For., Appl. For. Res. Inst., AFRI Res. Rpt. No. 31.

NYS Dept. Environ. Conserv. 1981. *A Clean Harvest. Protecting Water Quality During Timber Harvesting.* NYS Dept. Environ. Conserv. LF-P171 (7/81)

NYS Dept. Environ. Conserv. 1992. *Timber Harvesting Guidelines. What Are They?* NYS Dept. Environ. Conserv. LF P-185 (7/92)—18A

O'Hara, K.L., and C.D. Oliver. 1992. Silviculture: Achieving new objectives through stand and landscape management. *W. Wildlands* 17(4):28–33.

Ohman, J.H. 1970. Value losses from skidding wounds in sugar maple and yellow birch. *J. For.* 68(5):226–230.

Oliver, C.D. 1981. Forest development in North America following disturbances. *For. Ecol. and Manage.* 3:153–168.

Oliver, C.D., and B.C. Larson. 1990. *Forest Stand Dynamics.* New York: McGraw-Hill.

Oliver, W.W. 1979. Early response of ponderosa pine to spacing and brush: Observations on a 12-year-old plantation. U.S. For. Serv. Res. Note PSW-341.

Oliver, W.W. 1986. Growth of California red fir advance regeneration after overstory removal and thinning. U.S. For. Serv. Res. Pap. PSW-180.

Oliver, W.W., and C.B. Edminster. 1988. Growth of ponderosa pine thinned to different stocking levels in the western United States. In W.C. Schmidt (Ed.), *Proceedings—Future Forests of the Mountain West: A Stand Culture Symposium,* pp. 153–159. U.S. For. Serv. Gen. Tech. Rpt. INT-243.

Olszewski, R.J. 1989. Forestland drainage and regulation in the southern coastal plain. In D.D. Hook and R. Lea (Eds.), *The Forested Wetlands of the Southern United States,* pp. 59–62. Proc. Symp., July 12–14, 1988. Orlando, FL. U.S. For. Serv. Gen. Tech. Rep. SE-50.

Ontario Ministry of Natural Resources. 1983. *Management of Tolerant Hardwoods in Algonquin Provincial Park.* Wildl. Br., Ont. Min. Nat. Resourc., Toronto.

Ontario Ministry of Natural Resources. 1989. *Timber Management Guidelines for the Protection of Tourism Values.* Ont. Min. Nat. Resourc. and Ont. Min. Tourism and Recreation. 3877 (lk P.R., 89 03 31) Rp.

Osmaston, F.C. 1968. *The Management of Forests.* New York: Gafner Publ.

Ostrander, M.D., and D.L. Brisbin. 1971. Sawlog grades for eastern white pine. U.S. For. Serv. Res. Pap. NE-205.

Owston, P.W. 1990. Target seedling specifications: Are stocktype designations useful? In R. Rose, S.J. Campbell, and T.D. Landis (Eds.), *Target Seedling Symposium: Proceedings, Combined Meeting of the Western Forest Nursery Association,* pp. 9–16. U.S. For. Serv. Gen. Tech. Rpt. RM-200.

Owston, P.W., and L.P. Abrahamson. 1984. Weed management in forest nurseries. In M.L. Duryea, and T.D. Landis (Eds.), *Forest Nursery Manual: Production of Bareroot Seedlings.* pp. 193–202. Martinus Nijhoff/Dr W. Junk Publ. The Hague/Boston/Lancaster. For. Res. Lab., Ore. State Univ., Corvallis.

Packer. P.E. 1967. Criteria for designing and locating logging roads to control sediment. *For. Sci.* 13(1):2–18.

Paddock, R.W. 1982. Site preparation using the KG blade. In *Artificial Regeneration of Conifers in the Upper Great Lakes Region,* pp. 215–219. Mich. Technol. Univ., Houghton, MI.

Pait, J.A., III, D.M. Flinchum, and C.W. Lantz. 1991. Species variation, allocation, and tree improvement. In M.L. Duryea and P.M. Dougherty (Eds.), *Forest Regeneration Manual,* Ch. 11, pp. 207–231. Dordrecht, The Netherlands: Kluwer Academ. Publ.

Palik, B.J., and K.S. Pregitzer. 1991. The relative influence of establishment time and height growth rates on species vertical stratification during secondary forest succession. *Can. J. For. Res.* 21(10):1481–1490.

Palmer, J.F., S. Shannon, M.A. Harrilchak, P.H. Gobster, and T. Kokx. 1995. Esthetics of clearcutting: Alternatives in the White Mountain National Forest. *J. For.* 93(5):37–42.

Parker, G.R., and C. Merritt. 1995. The Central Region. In J.W. Barrett (Ed.), *Regional Silviculture of the United States,* 3d ed., Ch. 4, pp. 129–172. New York: Wiley.

Parson, G.A. 1950. Management of ponderosa pine in the Southwest. U.S. Dept. Agric. Monogr. 6.

Patton, D.R. 1992. *Wildlife Habitat Relationships in Forested Ecosystems.* Portland, OR: Timber Press.

Pavel, C., and R. Kellison. 1993. The impacts of timber harvesting and soil disturbance on the vegetation in a blackwater swamp. In J.C. Brissette (Ed.), *Proceedings of the 7th Biennial Southern Silvicultural Research Conference,* pp. 147–150. U.S. For. Serv. Gen. Tech. Rpt. SO-93.

Pearson, G.A. 1923. Natural reproduction of western yellow pine in the Southwest. U.S. Dept. Agric. Bull. 1105.

Pearson, G.A. 1928. Measurement of physical factors in silviculture. *Ecol.* 9:404–411.

Peet, R.K. 1981. Changes in biomass and production during secondary forest succession. In D.C. West, H.H. Shugart, and D.B. Botkin (Eds.), *Forest Succession Concepts and Applications,* Ch. 20, pp. 324–338. New York: Springer-Verlag.

Peet, R.K., and N.L. Christensen. 1988. Changes in species diversity during secondary forest succession on the North Carolina Piedmont. In H.J. During, M.J.A. Werger, and J.H. Willems (Eds.), *Diversity And Pattern In Plant Communities,* pp. 233–245. The Hague, Netherlands: SPB Academic Publ.

Perala, D.A. 1977. Manager's handbook for aspen in the north-central states. U.S. For. Serv. Gen. Tech. Rpt. NC-36.

Perala, D.A. 1982. Early release—Current technology and conifer release. In G.D. Mroz and J.F. Berner (Eds.), *Artificial Regeneration of Conifers in the Great Lakes Region,* pp. 396–410. Proc. Conf., Oct. 26–28, 1982. Green Bay, WI, Mich. Technol. Univ., Houghton, MI.

Perala, D.A. 1987. Aspen sucker production and growth from outplanted cuttings. U.S. For. Serv. Res. Note NC-241.

Perison, D.M., R. Lea, and R. Kellison. 1993. The response of soil physical and chemical properties and water quality to timber harvest and soil disturbance: Preliminary results. In J.C. Brissette (Ed.), *Proceedings of the 7th Biennial Southern Silvicultural Research Conference,* pp. 143–146. U.S. For. Serv. Gen. Tech. Rpt. SO-93.

Perry, D.A., and J.E. Lotan. 1977. Regeneration and early growth on strip clearcuts in lodgepole pine/bitterbrush habitat type. U.S. For. Serv. Res. Note INT-238.

Peterman, R.M. 1978. The ecological role of mountain pine beetle in lodgepole pine forests. In E.A. Berryman, G.D. Amman, and R.W. Stark (Eds.), *Theory and Practice of Mountain Pine Management in Lodgepole Pine Forests,* pp. 16–26. Proc. Symp., April 25-27, 1978, Wash. St. Univ., Pullman, WA. For., Wildl, and Range Exp. Stn., Univ. Idaho, Moscow, ID, and U.S. For Serv., For. Insect and Disease Res., Washington, DC, and U.S. For. Serv. Intermt. For. and Range Exp. Stn., Odgen, UT.

Petersen, T.D. 1986. Ecological principles for managing forest weeds. In D.M. Baumgartner, R.J. Boyd, D.W. Breuer, and D.L. Miller (Eds.), *Weed Control for Forest Productivity in the Interior West,* pp. 3–7. Proc. of Symp., Feb. 5–7, 1985, Spokane, WA, Coop. Ext., Wash. State Univ., Pullman, WA.

Pfister, R.D. 1976. Choosing tree species for planting. In D.M. Baumgartner and R.J. Boyd, (Eds.), *Tree Planting in the Inland Northwest,* pp. 192–204. Wash. State Univ. Ext. Serv., Wash. State Univ., Pullman, WA.

Philbrook, J.S., J.P. Barrett, and W.B. Leak. 1973. A stocking guide for eastern white pine, U.S. For. Serv. Res. Note NE-168.

Pickett, S.T.A., and P.S. White, ed. 1985. *The Ecology of Natural Disturbance and Patch Dynamics.* New York: Academic Press.

Pieper, R.D., and R.D. Wittie. 1990. Fire effects in southwestern chaparral and pinyon-juniper vegetation. In J.S. Krammes, Tech. Coord. *Effects of Fire Management of Southwestern Natural Resources,* pp. 87–93. U.S. For. Serv. Gen. Tech. Rpt. RM-191.

Pierce, R.S. 1971. Clearcutting and stream water. Soc. Protect. NH For. *For. Notes* 106:9, 13.

Pierce, R.S., J.W. Hornbeck, C.W. Martin, L.M. Tritton, C.T. Smith, C.A. Federer, and H.W. Yawney. 1993. Whole-tree clearcutting in New England: Guide to impacts on soils, streams, and regeneration. U.S. For. Serv. Gen. Tech. Rpt. NE-172.

Pierce, R.S., C.W. Martin, C.C. Reeves, G.E. Likens, and F.H. Bormann. 1972. Nutrient loss from clearcutting in New Hampshire. In S.C. Csallany, T.G. McLaughlin, and W.D. Striffler (Eds.), *Watersheds in Transition,* pp. 285–295. Proc. Symp., June 12–22, Ft. Collins, CO, Colo. State Univ., Fort Collins, CO.

Pinchot, G. 1947. *Breaking New Ground.* New York: Harcourt, Brace.

Pritchett, W.L. 1979. *Properties and Management of Forest Soils.* New York: Wiley.

Pritchett, W.L., and C.G. Wells. 1978. Harvesting and site preparation increase nutrient mobilization. In T. Tippin (Ed.), *Proceedings: A Symposium on Principles of Maintaining Productivity on Prepared Sites,* pp. 98–110. Proc. Symp., Mar. 21–22, 1978. Miss. State Univ., U.S. For. Serv., S. For. Exp. Stn., New Orleans, LA.

Prodan, M. 1948. *Normalisierung des Plenterwaldes?* Badische Forestliche Versuchsanstalt, 1949. (Original not seen).

Pruitt, L.R., and S.E. Pruitt. 1987. Effects of northern hardwood management on deer and other large mammal populations. In R.D. Nyland (Ed.), *Managing Northern Hardwoods,* pp. 363–379. SUNY Coll. Environ. Sci. and For., Fac. For. Misc. Publ. 13 (ESF 87-002) (Soc. Am. For. Publ. 87-03).

Rachal, J.M., and B.L. Karr. 1989. Effects of current harvesting practices on the physical properties of a loessial soil in west-central Mississippi. In J.H. Miller (Ed.), *Proceedings of the Fifth Biennial Southern Silviculture Research Conference,* pp. 527–532. U.S. For. Serv. Gen. Tech. Rpt. SO-74.

Radosevich, S.R., and K. Osteryoung. 1987. Principles governing plant-environment interactions. In J.D. Walstad and P.J. Kuch (Eds.), *Forest Vegetation Management for Conifer Production,* Ch. 5, pp. 105–156. New York: Wiley.

Ralston, R.A., and W. Lemmien. 1956. Pruning pine plantations in Michigan. Mich. State Univ., Agric. Exp. Stn. Circ. Bull. 221.

Randall, A. 1981. *Resource Economics.* New York: Wiley.

Randall, A., and G.L. Peterson. 1984. The valuation of wildland benefits: An overview. In G.L. Peterson and A. Randall (Eds.), *Valuation of Wildland Resource Benefits,* Ch. 1, pp. 1–52. Boulder, CO: Westview Press.

Randall, W.K. 1984. Advantages and disadvantages of bare-root and container seedlings and vegetatively propagated plant materials for progeny testing. In R. Miller (Ed.), *Progeny Testing,* pp. 263–274. Proc. Servicewide Genetics Workshop, Charleston, SC, Dec. 5–9, 1983. U.S. For. Serv., Timber Manage., Washington, DC.

Rast, E.D., J.A. Beaton, and D.L. Sonderman. 1989. Photographic guide of selected external defect indicators and associated internal defects in white oak. U.S. For. Serv. Res. Pap. NE-628.

Rast, E.D., D.L. Sonderman, and G.L. Gammon. 1973. A guide to hardwood log grading (Revised). U.S. For. Serv. Gen. Tech. Rpt. NE-1.

Rathfon, R.A., J.J. Johnson, and D.A. Groeschl. 1991. Response of ten-year-old yellow-poplar to release and fertilization. In S.S. Coleman and D.G. Neary (Eds.), *Proceedings of the Sixth Biennial Southern Silvicultural Research Conference,* Vol. 2, pp. 842–851. U.S. For. Serv. Gen. Tech. Rpt. SE-70.

Reaves, J.L., M.A. Palmer, and E.E. Nelson. 1989. Effects of root diseases on the health of western forests. In *Healthy Forests, Healthy World,* pp. 97–101. Proc. 1988 Soc. Am. For. Nat. Conv., Rochester, NY, Oct. 16–19, 1988. Soc Am. For., Washington, DC. SAF Publ. 88-1.

Reese, C.D. 1979. Evolution of pest management. In R.D. Gale (Ed.), *Proceedings for Integrated Pest Management Colloquium,* pp. 10–12. U.S. For. Serv. Gen. Tech. Rpt. WO-14.

Reese, K.H. 1982. Producing containerized seedlings for the northern forest. In G. Mroz and J.F. Berner (Eds.), *Artificial Regeneration of Conifers in the Upper Great Lakes Region,* pp. 273–282. Mich. Tech. Univ., Houghton, MI.

Rehfeldt, G.E., and R.J. Hoff. 1976. In D.M. Baumgartner and R.J. Boyd (Eds.), *Tree Planting in the Inland Northwest,* pp. 43–48. Proc. Conf. at Wash. State Univ., Feb 17–19, 1976. Wash. State Univ., Coop. Ext. Serv., Pullman, WA.

Reineke, L.H. 1933. Perfecting a stand-density index for even-aged forests. *J. Agric. Res.* 46:627–638.

Reinhart, K.G. 1972. Effects of clearcutting upon soil/water relations. In R.D. Nyland (Ed.), *A Perspective on Clearcutting in a Changing World,* pp. 67–74. Proc. 1972 Winter Meeting, NY Sect. Soc. Am. For., Feb. 23–24, 1972, Syracuse, NY, SUNY Coll. Environ. Sci. and For., Appl. For. Res. Inst., AFRI Misc. Rpt. No. 4.

Reisinger, T.W., D.B. Powell, Jr., A.M. Aust, and R.G. Oderwald. 1994. A postharvest evaluation of a mechanized thinning operation in natural loblolly pine. *S. J. Appl. For.* 18(1):24–28.

Reukema, D.L. 1961. Response of individual Douglas-fir trees to release. U.S. For. Serv. Res. Note PNW-210.
Reukema, D.L. 1972. Twenty-one-year development of Douglas-fir stands repeated thinned at varying intervals. U.S. For. Serv. Res. Pap. PNW-141.
Reukema, D.L. 1979. Fifty-year development of Douglas-fir stands planted at various spacings. U.S. For. Serv. Res. PNW-253.
Reukema, D.L., and D. Bruce 1977. Effects of thinning on yield of Douglas-fir: Concepts and some estimates obtained by simulation. U.S. For. Serv. Gen. Tech. Rpt. PNW-58.
Reynolds, R.R., J.B. Baker, T.T. Ku. 1984. Four decades of selection management on the Crossett farm forestry forties. Univ. Ark., Ark. Agric. Exp. Stn. Bull. 872.
Richards, N.A. 1974. Forestry in an urbanizing society. *J. For.* 72(8):458–461.
Richards, N.A. 1975. Greenspace silviculture. In pp. 80–84. Proc. 1974 Nat. Conv. Soc. Am. For., Sept. 22-26, 1974, NY City. Soc. Am. For., Bethesda, MD.
Richards, N.A. 1992. Optimum stocking of urban trees. *J. Arboriculture* 18(2):64–68.
Richards, N.A. u.d. *Greenspace Silviculture.* SUNY Coll. Environ. Sci. and For., Syracuse, NY. Unpubl. rpt.
Richardson, J. 1982. Release of young spruce and pine from alder competition. Environ. Can., Can. For. Serv., Newfoundland For. Res. Ctr., Info. Rpt. N-X-221.
Rimando, E.F., and M.V. Dalmacio. 1978. Direct seeding of Ipil-Ipil *(Leucaena leucocephala). Sylvatrop. Phillip. For. Res. J.* 3:171–175.
Ritchie, G.A. 1984. Assessing seedling quality. In E.L. Duryea and T.D. Landis (Eds.), *Forest Nursery Manual: Production of Bareroot Seedlings.* Ch. 23, pp. 243–259. Martinus Nijhoff/Dr W. Junk Publ. The Hague/Boston/Lancaster. For. Res. Lab., Ore. State Univ., Corvallis.
Roach, B.A. 1974a. Scheduling timber cutting for sustained yield of wood products and wildlife. U.S. For. Serv. Gen. Tech. Rpt. NE-14.
Roach, B.A. 1974b. Selection cutting and group selection. SUNY Coll. Environ. Sci. and For., Appl. For. Res. Inst., AFRI Misc. Publ. No. 5.
Roach, B.A. 1977. A stocking guide for Allegheny hardwoods and its use in controlling intermediate cuttings. U.S. For. Serv. Res. Rpt. NE-373.
Roach, B.A., and S.F. Gingrich. 1968. Even-aged silviculture for upland central hardwoods. U.S. Dept. Agric., Agric. Handbk. No. 355.
Roberge, M.R. 1977. Influence of cutting method and artificial regeneration of yellow birch in Quebec northern hardwoods. *Can. J. For. Res.* 7:175–182.
Roberts, M.R. 1992. Stand development and overstory-understory interactions in an aspen-northern hardwoods stand. *For. Ecol. And Manage.* 54:157–174.
Roberts, M.R., and H. Dong. 1991. Effects of forest floor disturbance on soil seed banks, germination and early survival after clearcutting a northern hardwood stand in central New Brunswick. In C.M. Simpson (Ed.), *Proceedings of the Conference on Natural Regeneration Management,* pp. 67–84. For. Can., Maritimes Reg., Fredericton, NB.
Robinson, W.L., and E.G. Bolen. 1989. *Wildlife Ecology and Management,* 2d ed. New York: Macmillan.
Robison, D., L.P. Abrahamson, and E.H. White. 1994. Silviculture of wood biomass crops as an industrial energy feedstock. *NY For. Owner* Sept/Oct 1994:4–5.
Roe, A.L., and G.M. DeJarnette. 1965. Results of regeneration cutting in a spruce-subalpine fir stand. U.S. For. Serv. Res. Pap. INT-17.
Roe, A.L., Alexander, R.R., and M.D. Andrews. 1970. Engelmann spruce regeneration practices in the Rocky Mountains. U.S. For. Serv. Prod. Res. Rpt. 115.

Roe, E.I. 1963. Direct seeding of conifers in the Lake States: A review of past trials. U.S. For. Serv. Res. Pap. LS-3.

Rogers, L.L. 1987. Effects of food supply and kinship on social behavior, movements, and population growth of black bears in northwestern Minnesota. *Wildl. Monogr.* No. 97.

Rogers, R. 1983. Guides for thinning shortleaf pine. In E.P. Jones, Jr. (Ed.), *Proceedings of the Seventh Biennial Southern Silvicultural Research Conference,* pp. 217–225. U.S. For. Serv. Gen. Tech. Rpt. SO-93.

Ronoco, F. Jr., C.B. Edminster, and D.P. Trujillo. 1985. Growth of ponderosa pine thinned to different stocking levels in northern Arizona. U.S. For. Serv. Res. Pap. RM-262.

Ronoco, F., Jr., G.J. Godfried, and R.R. Alexander. 1984. Silviculture of mixed conifer forests in the southwest. U.S. For. Serv. Technol. Transfer Pap. RM-TT-6.

Rose, D.W., and D.S. DeBell. 1978. Economic assessment of intensive culture of short-rotation hardwood crops. *J. For.* 76(11):706–711.

Rose, D.W., K. Ferrguson, D.C. Lothner, and J. Zavitkovski. 1981. An economic and energy analysis of poplar intensive cultures in the Lake States. U.S. For. Serv. Res. Pap. 196.

Rose, R., W.C. Carlson, and P. Morgan. 1990. The target seedling concept. In R. Rose, S.J. Campbell, and T.D. Landis (Eds.), *Target Seedling Symposium: Proceedings, Combined Meeting of the Western Forest Nursery Association,* pp. 1–8. U.S. For. Serv. Gen. Tech. Rpt. RM-200.

Roth, E.R. 1956. Decay following thinning of sprout oak clumps. *J. For.* 54:26–30.

Roth, E.R., and G.H. Hepting. 1943. Origin and development of oak stump sprouts as affecting their likelihood to decay. *J. For.* 41:27–36.

Rouse, G.D. 1984. Spain 1981. *Quart. J. For.* 78:104–111.

Row, C. 1987. Using costs and values in forest vegetation management analyses. In J.D. Walstad and P.J. Kuch (Eds.), *Forest Vegetation Management for Conifer Production,* Ch. 11, pp. 327–364. New York: Wiley.

Rudolf, P.O. 1974. Tree-seed marketing controls. In C.S. Schopmeyer (Ed.), *Seeds of Woody Plants in the United States,* pp. 153–166. U.S. For. Serv., Agric. Handbk. No. 450.

Rudolph, J.T. 1973. Direct seeding versus other regeneration techniques: Silvicultural aspects. In J.H. Cayford (Ed.), *Direct Seeding Symposium,* pp. 29–48. Sept. 11–13, 1973. Timmins, Ont. Dept. Environ., Can. For. Serv., Publ. No. 1339.

Ruel, J.C. 1991. Advance growth abundance and regeneration patterns after clearcutting in Quebec. In C.M. Simpson (Ed.), *Proceedings of the Conference on Natural Regeneration Management,* pp. 115–131. For. Can., Maritimes Reg., Fredericton, NB.

Runkle, J.R. 1981. Gap replacement in some old-growth mesic forests of Eastern North America. *Ecol.* 62(4):1041–1051.

Runkle, J.R. 1982. Patterns of disturbance in some old-growth mesic forests of Eastern North America. *Ecol.* 63(5):1522–1546.

Runkle, J.R. 1990. Gap dynamics in an Ohio Acer-Fagus forest and speculations on the geography of disturbance. *Can. J. For. Res.* 20:632–641.

Runkle, J.R. 1992. Guidelines as sampling protocol for sampling forest gaps. U.S. For. Serv. PNW-GTR-283.

Russell, T.E. 1975. Broadcast and spot-seeded pines grow equally well in central Tennessee. *Tree Plant. Notes* 25(3):20–22.

Ruth, R.H. 1976. Harvest cuttings and regeneration in young-growth western hemlock. In *Managing Young Forests in the Douglas-fir Region,* pp. 41–73. Proc., Vol. 5., Sch. For. Ore. State Univ., Corvallis.

Ruth, R.H., and A.S. Harris. 1979. Management of western hemlock-Sitka spruce forests for timber production. U.S. For. Serv. Gen. Tech. Rpt. PNW-88.

Rutherford, W., and E. Shafer, 1969. Selection cuts increased natural beauty in two Adirondack forest stands. *J. For.* 67(6):415–419.

Safford, L.O., and S.M. Filip. 1974. Biomass and nutrient content of 4-year-old fertilized and unfertilized northern hardwood stands. *Can. J. For. Res.* 4:549–554.

Stafford, L.O., J.C. Bjorkbom, and J.C. Zasada. 1990. *Betula papyrifera* Marsh. Paper birch. In R.M. Burns and B.H. Honkala (Eds.), *Silvics Of North America. Volume 2. Hardwoods,* pp. 158–171. U.S. For. Serv., Agric. Handbk. 654.

Safranyik, L., D.M. Shrimpton, and H.S. Whitney. 1974. Management of lodgepole pine to reduce losses from the mountain pine beetle. Environ. Can., For. Serv., For. Tech. Rpt. 1.

Sage, R.W., Jr. 1987. Unwanted vegetation and its effects on regeneration success. In R.D. Nyland (Ed.), *Managing Northern Hardwoods,* pp. 298–316. SUNY Coll. Environ. Sci. and For., Fac. For. Misc. Publ. 13 (ESF 87-002) (Soc. Am. For. Publ. 87-03).

Sajdak, R.L. 1982. Site preparation in the upper Great Lakes Region. In G.D. Mroz and J.F. Berner (Eds.), *Artificial Regeneration of Conifers in the Upper Great Lakes Region,* pp. 209–214. Proc. Conf. Oct. 26–28, 1982. Green Bay, WI, Mich. Technol. Univ., Houghton, MI.

Sall, H-O. 1986. Harvesting In L. Sennerby-Forsse (Ed.), *Handbook for Energy Forestry,* pp. 25–26. Sect. Energy For., Dept. Ecol. and Environ. Res., Swedish Univ. Agric. Sci., Uppsala.

Salwasser, H., and R.D. Pfister. 1994. Ecosystem management: From theory to practice. In W.W. Covington and L.F. DeBano (Eds.), *Sustainable Ecological Systems: Implementing an Ecological Approach to Land Management,* pp. 150–162. U.S. For. Serv. Gen. Tech. Rpt. RM-247.

Sammi, J.C. 1961. de Liocourt's method, modified. *J. For.* 59(4):294–295.

Sander, I.L. 1972. Size of oak advance reproduction: Key to growth following harvest cutting. U.S. For. Serv. Res. Pap. NC-79.

Sander, I.L. 1979. Regenerating oaks with the shelterwood system. In H.A. Holt and B.C. Fischer (Eds.), *Regenerating Oaks In Upland Hardwood Forests,* pp. 54–60. Proc. 1979 J.S. Wright For. Conf., Feb. 22–23, 1979, Purdue Univ., West Lafayette, IN.

Sander, I.L., and F.B. Clark. 1971. Reproduction of the upland hardwood forests in the Central States. U.S. For. Serv., Agric. Handbk. 405.

Sander, I.L., and D.L. Graney. 1993. Regenerating oaks in the central states. In D.L. Loftis and C.E. McGee (Eds.), *Oak Regeneration: Serious Problems, Practical Recommendations,* pp. 174–183. U.S. For. Serv. Gen. Tech. Rpt. SE-84.

Sander, I.L., and M.J. Williamson. 1957. Response of a mixed hardwood stand in eastern Kentucky to a harvest cutting. *J. For.* 55:291–293.

Sander, I.L., P.S. Johnson, and R. Rogers. 1984. Evaluating oak advance regeneration in the Missouri Ozarks. U.S. For. Serv. Res. Pap. NC-251.

Sander, I.L., P.S. Johnson, and R.F. Watt. 1976. A guide for evaluating the adequacy of oak advance reproduction. U.S. For. Serv. Gen. Tech. Rpt. NC-23.

Sander, I.L., C. Merritt, and E.H. Tryon. 1981. Oak-hickory. In D.H. Van Lear (Ed.), *Choices in American Silviculture For American Forests.* Soc. Am. For., Washington, DC.

Sander, I.L., C.E. McGee, K.G. Day, and R.E. Willard. 1983. Oak-hickory. In R.M. Burns (Ed.), *Silvicultural Systems for the Major Forest Types of the United States,* pp. 116–120. U.S. For. Serv., Agric. Handbk. No. 445.

Sandvik, M. 1974. Biological aspects of planting and direct seeding in forestry. In *Sympo-*

sium Stand Establishment, pp. 184–198. Proc. IUFRO Joint Meet., Div. 1 and 3, Internal. Union For. Res. Organ., Oct. 15–18, 1974. Wegeningen, The Netherlands.

Santillo, D.J. 1987. Response of small mammals and breeding birds to herbicide-induced habitat changes on clearcuts in Maine. MS thesis. Univ. Maine, Orono, ME.

Sassaman, R.W., J.A. Barrett, and A.D. Twombly. 1977. Financial precommercial thinning guides for northwest ponderosa pine stands. U.S. For. Serv. Res. Pap. PNW-226.

Saucier, J.R. 1981. Effects of fast growth rate on wood quality and product yield. TAPPI Ann. Meet., Mar. 2–5, 1981. Proc:435–437.

Savill, P.S., and J. Evans. 1986. *Plantation Silviculture in Temperate Regions with Special Reference to the British Isles.* Oxford: Clarendon Press.

Schadelin, W. 1942. *Die Auslesedurchforstung,* 3d ed. Bern, Haupt. Original not seen.

Schier, G.A., and R.B. Campbell. 1978. Aspen sucker regeneration following burning and clearcutting on two sites in the Rocky Mountains. *For. Sci.* 24(2):303–308.

Schmeckpeper, E.J., and R.P. Belanger. 1985. Coppice production following the harvest of a pole-size sycamore plantation. In E. Shoulders (Ed.), *Proceedings of the Third Biennial Southern Silvicultural Research Conference,* pp. 132–136. U.S. For. Serv. Gen. Tech. Rpt. SO-54.

Schmidt, F.L., and D.S. DeBell. 1973. Wood production and kraft pulping of short rotation hardwoods in the Pacific Northwest. In *Working Party on the Mensuration of Forest Biomass,* pp. 509–516. S401 Mensuration, Growth, and Yield. IUFRO Biomass Studies. Coll. Life Sci. and Agric., Univ. Me., Orono, ME.

Schmidt, W.C., and K.W. Seidel. 1988. Western larch and space: Thinning to optimize growth. In W.C. Schmitt (Ed.), *Proceedings—Future Forests of the Mountain West: A Stand Culture Symposium,* pp. 165–174. U.S. For. Serv. Gen. Tech. Rpt. INT-243.

Schmidt, W.C., R.C. Shearer, and A.L. Roe. 1976. Ecology and silviculture of western larch forests. U.S. Dept. Agric. For. Serv. Tech. Bull. No. 1520.

Schoenholts, S.H., and B.L. Barber. 1989. The impact of herbicides on loblolly pine plantation establishment in east Texas. In J.H. Miller (Ed.), *Proceedings of the Fifth Biennial Southern Silviculture Research Conference,* pp. 341–348. Memphis, TN., Nov. 1–3, 1989. U.S. For. Serv. Gen. Tech. Rpt. SO-74.

Schopmeyer, C.S. (Ed.). 1974. *Seeds of Woody Plants in the United States.* U.S. Dept. Agric., For. Serv., Agric. Handbk. No. 450.

Schowalter, T.D., and P. Turchin. 1993. Southern pine beetle infestation development: Interaction between pine and hardwood basal areas. *For. Sci.* 39(2):201–209.

Schowalter, T.D., R.N. Coulson, and D.A. Crossley, Jr. 1981. Role of southern pine beetle and fire management of structure and function of the southeastern coniferous forest. *Environ. Entomol.* 10(6):821–825.

Schubert, G.H. 1974. Silviculture of southwestern ponderosa pine: The status of our knowledge. U.S. For. Serv. Res. Pap. RM-123.

Schubert, G.H. 1976. Planting methods for the Inland Northwest. In D.M. Baumgartner and R.J. Boyd, (Eds.), *Tree Planting in the Inland Northwest,* pp. 239–261. Proc. Conf. at Wash. State Univ., Feb 17–19, 1976. Wash. State Univ., Coop. Ext. Serv., Pullman, WA.

Schubert, G.H. and H.A. Fowels. 1964. Sowing rates for reforestation by the seed-spotting method. U.S. For. Serv., Pacific SW For. and Range Exp. Stn. Res. Note. 44.

Schubert, G.H., L.J. Heidmann, and M.M. Larson. 1970. Artificial reforestation practices for the southwest. U.S. For. Serv., Agric. Handbk. No. 370.

Schweitzer, D.L., and E.G. Schuster. 1976. Economics and tree planting in the Inland Northwest. In D.M. Baumgartner and R.J. Boyd (Eds.), *Tree Planting in the Inland Northwest,* pp. 18–28. Wash. State Univ. Ext. Serv., Wash. State Univ., Pullman, WA.

Scott, V.E., and J.L. Oldemeyer. 1983. Cavity-nesting requirements and response to snag cutting in ponderosa pine. In J.W. Davis, G.A. Goodwin, and R.A. Ockenfels (Eds.), *Snag Habitat Management,* pp. 19–23. U.S. For. Serv. Gen. Tech. Rpt. RM-99.

Seidel, K.H., and P.H. Cochran. 1981. Silviculture of mixed conifer forests in eastern Oregon and Washington. U.S. For. Serv. Gen. Tech. Rpt. PNW-121.

Seidel, K.W. 1977. Levels-of-growing-stock study in thinned western larch pole stands in eastern Oregon. U.S. For. Serv. Res. Rap. PNW-22.

Seidel, K.W. 1979a. Regeneration in mixed conifer clearcuts in the Cascade Range and the Blue Mountains of Eastern Oregon. U.S. For. Serv. Res. Pap. PNW-248.

Seidel, K.W. 1979b. Natural regeneration after shelterwood cutting in a grand fir-Shasta red fir stand in central Oregon. U.S. For. Serv. Res. Pap. PNW-259.

Seidel, K.W. 1979c. Regeneration in mixed conifer shelterwood cuttings in the Cascade Range of eastern Oregon. U.S. For. Serv. PNW-264.

Seidel, K.W. 1980. Diameter and height growth of suppressed grand fir saplings after overstory removal. U.S. For. Serv. Res. Pap. PNW-275.

Seidel, K.W. 1982. Growth and yield of western larch: 15-year results of a levels-of-growing-stock study. U.S. For. Serv. Res. Note PNW-366.

Seidel, K.W. 1983a. Growth of suppressed grand fir and Shasta red fir in central Oregon—10-year results. U.S. For. Serv. Res. Note PNW-404.

Seidel, K.W. 1983b. Regeneration in mixed conifer and Douglas-fir shelterwood cuttings in the Cascade Range of Washington. U.S. For. Serv. Res. Pap. PNW-314.

Seidel, K.W. 1986. Tolerance of seedlings of ponderosa pine, Douglas-fir, grand fir, and Engelmann spruce for high temperatures. *N. W. Sci.* 60:1–7.

Sennerby-Forsse, L. 1986. Principles for energy forestry. In L. Sennerby-Forsse (Ed.), *Handbook for Energy Forestry,* p. 8. Sect. Energy For., Dept. Ecol. and Environ. Res., Swedish Univ. Agric. Sci., Uppsala.

Seymour, R.E. 1992. The red spruce-balsam fir forest of Maine: Evolution of silvicultural practice in response to stand development patterns and disturbances. In M.J. Kelty, B.C. Larson, and C.D. Oliver (Eds.), *The Ecology and Silviculture of Mixed-species Forests,* pp. 217–244. Norwell, MA: Kluwer Academ. Publ.

Seymour, R.S. 1995. The Northeast region. In J.W. Barrett (Ed.), *Regional Silviculture of the United States,* 3d ed., Ch. 2, pp. 31–79. New York: Wiley.

Seymour, R.S., R.A. Ebeling, and C.J. Gadzik. 1984. Operational density control in spruce-fir sapling stands—Production of a mechanical swath cutter and brush-saw workers. Univ. Maine, Coll. For. Resourc., Coop. For. Res. Unit., CFRU Res. Note. 14.

Sharpe, D.M., F.W. Stearns, R.L. Burgess, and W.C. Johnson. 1981. Spatio-temporal patterns of forest ecosystems in man-dominated landscapes. In *Proceedings of the International Congress,* pp. 109–115. Neth. Soc. Landscape Ecol., Veldhoven, 1981. Produc. Wegeningen.

Shaw, E.W., and G.R. Staebler. 1950. Financial aspects of pruning. U.S. For. Serv., Pacific NW For. and Range Exp. Stn., Portland, OR.

Shea, K.R. 1985. Integrated forest pest management in the south. In S.J. Branham and R.C. Thatcher (Eds.), *Integrated Pest Management Research Symposium: The Proceedings,* pp. 3–4. U.S. For. Serv. Gen. Tech. Rpt. SO-56.

Shelton, M.G., and J.B. Baker. 1992. Uneven-aged management of pine and pine-hardwood mixtures. In J.P. Barnett and J.C. Brissette (Eds.), *Proceedings Of The Shortleaf Pine Re-*

generation Workshop, pp. 217–224. Oct. 29–31, 1991. Little Rock, AR. U.S. For. Serv. Gen. Tech. Rpt. SO-90.

Shelton, M.G., and P.A. Murphy. 1990. Age and size structures of a shortleaf pine-oak structure in the Ouachita Mountains—Implications for uneven aged management. In *Proceedings of the Seventh Biennial Southern Silvicultural Research Conference,* Vol. 2, pp. 616–629. U.S. For. Serv. Gen. Tech. Rpt. SO-93.

Shelton, M.G., and R.F. Wittwer. 1992. Effects of seedbed condition on natural shortleaf pine regeneration. In J.C. Brissette and J.P. Barnett (Eds.), *Proceedings of the Shortleaf Pine Regeneration Workshop,* pp. 124–139. U.S. For. Serv. Gen. Tech. Rpt. SO-90.

Shigo, A.L. 1966. Decay and discoloration following logging wounds in northern hardwoods. U.S. For. Serv. Res. Pap. NE-47.

Shigo, A.L. 1985. Wounded forests, starving trees. *J. For.* 83(11):668–673.

Shigo, A.L. 1987. *New Tree Health: New Information on How To Keep Your Trees Healthy.* Durham, NH: Shigo And Trees, Assoc.

Shigo, A.L. 1989. *Tree Pruning: A Worldwide Photo Guide.* Durham, NH: Shigo And Trees, Assoc.

Shigo, A.L., and E. vH. Larson. 1969. A photo guide to the patterns of discoloration and decay in living northern hardwood trees. U.S. For. Serv. Res. Pap. NE-127.

Shigo, A.L., K. Vollbrecht, and N. Hvass. 1987. *Tree Biology and Tree Care: A Photo Guide.* SITAS—Skovvej 56. Denmark.

Shigo, A.L., E.A. McGinnes, Jr., D.T. Funk, and N. Rogers. 1979. Internal defects associated with pruned and non-pruned branch stubs in black walnut. U.S. For. Serv. Res. Pap. NE-440.

Shortle, W.C. 1987. Defect, discoloration, cull, and injuries. In R.D. Nyland (Ed.), *Managing Northern Hardwoods.* SUNY Coll. Environ. Sci. and For., Fac. For. Misc. Publ. 13 (ESF 87-002) (Soc. Am. For. Publ. 87-03).

Shoulders, E. 1990. Identifying longleaf pine sites. In R.M. Farrar, Jr. *Proceedings of the Symposium on the Management of Longleaf Pine,* pp. 23–37. U.S. For. Serv. Gen. Tech. Rpt. SO-75.

Shugart, H.H., D.C. West, and E.R. Emanuel. 1981. Patterns and dynamics of forests: An application of simulation models. In D.C. West, H.H. Shugart, and D.B. Botkin (Eds.), *Forest Succession Concepts and Applications,* Ch. 7, pp. 74–94. New York: Springer-Verlag.

Siegel, W.C. 1989. State water quality laws and programs to control nonpoint source pollution from forest land in eastern United States. In *Non-point Water Quality Concerns—Legal and Regulatory Aspects,* pp. 131–140. Proc. 1989. Ann. Symp., Am. Soc. Agric. Eng., Dec. 11–12, 1989, New Orleans. Am. Soc. Agric. Eng., St. Joseph, MI.

Siegel, W.C., and F.W. Cubbage. 1990. The impact of federal environmental law on forest resource management in the United States. In *Proceedings of XIXth IUFRO World Congress,* pp. 408–419. Aug. 5–11, 1990. Montreal, Can. Intl. Union For. Res. Organ. Vol. 4, Montreal.

Siegel, W.C., and T.K. Haines. 1990. State wetland protection legislation affecting forestry in northeastern United States. *For. Ecol. and Manage.* 33/34:239–252.

Silverborg, S.B. 1954. Northern hardwoods cull manual. SUNY Coll. For. at Syracuse Univ., Bull. 31.

Silverborg, S.B. 1959. Rate of decay in northern hardwoods following artificial inoculation with some common heart rot fungi. *For. Sci.* 5(3):223–228.

Simpson, D.G. 1988. Fixing the Edsel—Can bareroot stock quality be improved? In T.D. Landis (Ed.), *Proceedings, Combined Meeting of the Western Forest Nursery Associa-*

tions, pp. 24–30: Western Forest Nursery Council, Forest Nursery Association of British Columbia, and Intermountain Nursery Association. August 8–11, 1988. U.S. For. Serv. Gen. Tech. Rpt. INT-167.

Slagsvold, T. 1977. Bird population changes after clearance of deciduous shrub. *Biol. Conserv.* 12:229–243.

Sloan, J.P., L.H. Jump, and R.A. Ryker. 1987. Container-grown ponderosa pine seedlings outperform bareroot seedlings on harsh sites in southern Utah. U.S. For. Serv. Res. Pap. INT-384.

Smith, D.M. 1964. *The Practice of Silviculture,* 7th ed. New York: Wiley.

Smith, D.M. 1986. *The Practice of Silviculture,* 8th ed. New York: Wiley.

Smith, D.M. 1991. Natural regeneration from sprouts and advanced growth. In C.M. Simpson (Ed.), *Proceedings of the Conference on Natural Regeneration Management,* pp. 63–66. For. Can., Maritimes Reg., Fredericton, NB.

Smith, D.M., and P.M.S. Ashton. 1993. Early dominance of pioneer hardwood after clearcutting and removal of advanced regeneration. *N. J. Appl. For.* 10(1):14–19.

Smith, D.M., and R.S. Seymour. 1986. Relationship between pruning and thinning. In D.T. Funk (Ed.), *Eastern White Pine: Today And Tomorrow,* pp. 62–66. U.S. For. Serv. Gen. Tech. Rpt. WO-51.

Smith, D.W. 1994. The role of silviculture in the modern context. In L.H. Foley (Ed.), *Silviculture: From the Cradle of Forestry to Ecosystem Management,* pp. 16–21. Proc. Natl. Silvi. Workshop, Hendersonville, NC, Nov. 1–4, 1993. U.S. For. Serv., SE For. Exp. Stn., Asheville, NC.

Smith, D.W. 1995. The southern Appalachian hardwood region. In J.W. Barrett (Ed.), *Regional Silviculture of the United States,* 3d ed., Ch. 5, pp. 173–225. New York: Wiley.

Smith, H.C. 1977. Changes in tree density do not influence epicormic branching of yellow-poplar. U.S. For. Serv. Res. Note NE-239.

Smith, H.C. 1980. An evaluation of four uneven-aged cutting practices in Central Appalachian hardwoods. *S. J. Appl. For.* 4:193–200.

Smith, H.C. 1981. Diameters of clearcut openings influence central Appalachian hardwood stem development—A 10-year study. U.S. For. Serv. Res. Pap. NE-476.

Smith, H.C. 1988. Possible alternatives to clearcutting and selection harvesting practices. In H.S. Smith, A.W. Perkey, and W.E. Kidd, Jr. (Eds.), *Guidelines for Regenerating Appalachian Hardwood Stands,* pp. 276–289. Workshop Proc., Univ. W. Va., Morgantown, WV. Soc. Am. For. Publ. 88-03.

Smith, H.C., and N.I. Lamson. 1977. Stand development 25 years after a 9.0-inch diameter-limit first cutting in Appalachian hardwoods. U.S. For. Serv. Res. Pap. NE-379.

Smith, H.C., and N.I. Lamson. 1978. Response to crop-tree release: Sugar maple, red oak, black cherry, and yellow-poplar saplings in a 9-year-old stand. U.S. For. Serv. Res. Pap. NE-394.

Smith, H.C., and N.I. Lamson. 1986. Cultural practices in Appalachian hardwood sapling stands—If done, how to do them. In H.C. Smith and M.C. Eye (Eds.), *Guidelines for Managing Immature Appalachian Hardwood Stands,* pp. 46–61. Div. For., Coll. Agric. and For., W. Va. Univ., Morgantown, WV. SAF Publ. 86-02.

Smith, H.C., and N. Lamson. 1982. Number of residual trees: A guide for selection cutting. U.S. For. Serv. Gen. Tech. Rpt. NE-80.

Smith, H.C., and G.W. Miller. 1991. Releasing 75- to 80-year-old Appalachian hardwood sawtimber trees: 5-year d.b.h. response. In L.H. McCormick and K.W. Gottshalk (Eds.), *Proceedings of the 8th Central Hardwoods Forest Conference,* pp. 402–412. U.S. For. Serv. Gen. Tech. Rpt. NE-148.

Smith, H.C., N.I. Lamson, and G.W. Miller. 1989. An esthetic alternative to clearcutting? Deferment cutting in eastern hardwoods. *J. For.* 87(3):14–18.

Smith, W.D., and M.R. Strub. 1991. Initial spacing: How many trees to plant. In M.L. Duryea and P.M. Dougherty (Eds.), *Forest Regeneration Manual,* Ch. 15, pp. 281–289. Dordrecht, The Netherlands: Kluwer Academ. Publ.

Soc. Am. For. 1950. *Forestry Terminology.* Soc. Am. For., Washington, DC.

Soc. Am. For. 1961. Tree planting in the Allegheny Section. Silvi. Comm., Allegheny Sec., Soc. Am. For., U.S. For. Serv., NE For. Exp. Stn., Stn. Pap. 158.

Soc. Am. For. 1981. *Choices in Silviculture for American Forests.* Soc. Am. For., Washington, DC.

Soc. Am. For. 1989. *Recommended Changes in Silvicultural Terminology.* Silvi. Instructor's Subgp., Silvi. Work. Gp. (D2), Soc. Am. For., Washington, DC.

Solomon, D.S. 1977. The influence of stand density and structure on growth of northern hardwoods in New England. U.S. For. Serv. Res. Pap. NE-143.

Solomon, D.S., and B.M. Blum. 1967. Stump sprouting of four northern hardwoods. U.S. For. Serv. Res. Rpt. NE-59.

Solomon, D.S., and W.B. Leak. 1985. Simulated yields for managed northern hardwood stands in New England. U.S. For. Serv. Res. Pap. NE-578.

Sonderman, D.L. 1979. Guide to the measurement of tree characteristics important to the quality classification system for young hardwood trees. U.S. For. Serv. Gen. Tech. Rpt. NE-54.

Sonderman, D.L. 1986. Changes in stem quality on young thinned hardwoods. U.S. For. Serv. Res. Pap. NE-567.

Sonderman, D.L. 1987. Stem-quality changes on young, mixed upland hardwoods after crop tree release. U.S. For. Serv. Res. Pap. NE-RP-597.

Southerland, J.R. 1984. Pest management in Northwest bareroot nurseries. In Duryea, M.L. and T.D. Landis (Eds.), *Forest Nursery Manual: Production of Bareroot Seedlings.* Ch. 19, pp. 203–210. Martinus Nijhoff/Dr W. Junk Publ. The Hague/Boston/Lancaster. For. Res. Lab., Ore. State Univ., Corvallis.

Sparks, R.C., N.E. Linnartz, and H.E. Harris. 1980. Long-term effects of early pruning and thinning treatments on growth of natural longleaf pine. *S. J. Appl. For.* 4(2):77–79.

Speight, M.R., and D. Wainhouse. 1989. *Ecology and Management of Forest Insects.* Oxford Sci. Publ., Oxford: Clarendon Press.

Spurr, S.H., and B.V. Barnes. 1980. *Forest Ecology,* 3d ed. New York: Wiley.

Staebler, G.R. 1956. Effect of controlled release on growth of individual Douglas-fir trees. *J. For.* 54(9):567–568.

Stark, R.W. 1977. Integrated pest management in forest practice. *J. For.* 75(5):251–254.

Stark, R.W. 1984. Integrated forest protection: Implication for forest management. In D.M. Baumgartner and R. Mitchell (Eds.), *Silvicultural Management Strategies for Pests of the Interior Douglas-fir and Grand Fir Forest Types,* pp. 117–119. Proc. Symp., Feb. 14–16, 1984. Spokane, WA. Wash. State Univ. Coop. Ext., Wash. State Univ., Pullman, WA.

Starr, J.W. 1965. The role of herbicides in direct seeding. In *Proceedings Direct Seeding Workshops,* pp. 4–7. Oct. 5–6, 1965, Alexandria, LA, and Oct. 20–21, Tallahassee, FL. U.S. For. Serv., GSA Atlanta GA 66-4542.

Stein, W.I. 1976. Prospects for container-grown nursery stock. In D.M. Baumgartner and R.J. Boyd (Eds.), *Tree Planting in the Inland Northwest,* pp. 89–103. Wash. State Univ., Coop. Ext. Serv., Wash State Univ., Pullman, WA.

Stein, W.I. 1988. Nursery practice, seedling size, and field performance. In T.D. Landis

(Ed.), *Proceedings, Combined Meeting of the Western Forest Nursery Associations,* pp. 15–18. W. For. Nursery Counc., For. Nursery Assoc. of BC, and Intermountain Nursery Assoc. August 8–11, 1988. U.S. For. Serv. Gen. Tech. Rpt. INT-167.

Stein, W.I., P.E. Slabaugh, and A.P. Plummer. 1974. In C.S. Schopmeyer (Ed.), *Seeds of Woody Plants in the United States,* pp. 98–125. U.S. For. Serv., Agric. Handbk. No. 450.

Steinbeck, K, J.T. May, and R.G. McAlpine. 1968. Silage cellulose—A new concept. In *Forest Engineering Conference Proceedings,* pp. 104–105. Sept. 1968. East Lansing MI, Am. Soc. Agric. Eng., St. Joseph, MI.

Steneker, G.A. 1964. Ten-year results of thinning 14-, 19-, and 23-year-old aspen to different spacings. Can. Dept. For. Publ. No. 1038.

Steneker, G.A., and J.M. Jarvis. 1966. Thinning in trembling aspen stands in Manitoba and Saskatchewan. Can. Dept. For. Publ. No. 1140.

Stenzel, G., T.A. Walbridge, Jr., and J.K. Pearce. 1985. *Logging and Pulpwood Production,* 2d ed. New York: Wiley.

Stephens, E.P., and S.H. Spurr. 1948. The immediate response of red pine to thinning and pruning. In *Proc. Soc. Am. For. Meet.*, pp. 353–369. Dec. 17–20, 1947., Minn., MN, Soc. Am. For., Washington, DC.

Stewart, R.E. 1976. Chemical site preparation in the Inland Empire. In D.M. Baumgartner and R.J. Boyd (Ed.), *Tree Planting in the Inland Northwest,* pp. 158–173. Proc. Conf., Feb. 17–19, 1976. Wash. State Univ., Coop. Ext. Serv., Pullman, WA.

Stewart, R.E. 1978. Site preparation. In B.D. Cleary, R.D. Greaves, and R.K. Hermann (Eds.), *Regenerating Oregon's Forests,* pp. 100–129. Sch. For., Ore. State. Univ., Corvallis.

Stewart, R.E. 1987. Seeing the forest for the weeds: A synthesis of forest vegetation management. In J.D. Walstad and P.J. Kuch (Eds.), *Forest Vegetation Management for Conifer Production,* Ch. 14, pp. 431–480. New York: Wiley.

Stoeckeler, J.H., and C.F. Arbogast. 1947. Thinning and pruning in young second-growth hardwoods in northeastern Wisconsin. In pp. 328–346. *Proc. Soc. Am. For. Meet.,* Dec. 17–20, 1947, Minneapolis, MN. Soc. Am. For., Washington, DC.

Stoeckeler, J.H., and G.W. Jones. 1957. Forest nursery practice in the Lake States. U.S. For. Serv., Agric. Handbk. No. 110.

Stoeckeler, J.H., and R.A. Williams. 1949. Windbreaks and shelterbelts. In *Trees, the Yearbook of Agriculture 1949,* pp. 191–199. 81st Congr., 1st Sess., House Doc. No. 29. U.S. Gov. Print. Off., Washington, DC.

Stone, E.L. 1975. Soil and man's use of forest land. In B. Bernier and C.H. Winget (Eds.), *Forest Soils and Forest Land Management,* pp. 1–9. Laval Univ. Press, St. Foy, Quebec.

Stone, E.L., R. Feuer, and H.W. Wilson. 1970. Judging land for forest plantations in New York. Cornell Univ., NYS Coll. Agric., Ext. Bull 1075.

Stoszek, K. 1976. Protection concerns in plantation establishment, In D.M. Baumgartner and R.J. Boyd (Eds.), *Tree Planting in the Inland Empire,* pp. 291–311. Wash. State Univ., Coop. Ext. Serv., Wash. State Univ., Pullman, WA.

Stott, K.C., C. McElroy, W. Abernathy, and D.P. Davis. 1980. Coppice willow for biomass in the U.K. In W. Palz, P. Chartier, and D.O. Hall (Eds.), *Energy From Biomass,* pp. 198–209. London: Elsevier Appl. Sci. Publ., Ltd.

Stout, S.L. 1987. Planning the right residual: Relative density, stand structure, and species composition. In R.D. Nyland (Ed.), *Managing Northern Hardwoods.* SUNY Coll. Environ. Sci. and For., Fac. For. Misc. Publ. 13 (ESF 87-002) (Soc. Am. For. 87-03).

Stout, S.L, and B.C. Larson. 1988. Relative stand density: Why do we need to know. In W.C. Schmitt (Ed.), *Proceedings—Future Forests of the Mountain West: A Stand Culture Symposium,* pp. 73–79. U.S. For. Serv. Gen. Tech. Rpt. INT-243.

Stout, S.L., and R.D. Nyland. 1986. Role of species composition in relative density measurement in Allegheny hardwoods. *Can. J. For. Res.* 16:574–579.

Stroempl, G. 1984. Thinning clumps of northern hardwood stump sprouts to produce high quality timber. Ont. Min. Nat. Resourc., For. Res. Info. Pap. No. 104.

Strong, T. 1989. Rotation length and repeated harvesting influence *Populus* coppice production. U.S. For. Serv. Res. Note NC-350.

Strothman, R.O., and D.F. Roy 1984. Regeneration of Douglas-fir in the Klamath Mountains region, California and Oregon. U.S. For. Serv. Gen. Tech. Rpt. PSW-81.

Stuart, W.B., and J.L. Carr. 1991. Harvesting impacts on steep slopes in Virginia. In L.H. McCormick and K.W. Gottschalk (Eds.), *Proceedings: 8th Central Hardwood Forest Conference,* pp. 67–81. U.S. For. Serv. Gen. Tech. Rpt. NE-148.

Sturges, D.L. 1983. Long-term effects of big sagebrush control on vegetation and soil water. *J. Range Manage.* 36(6):760–765.

Sutton, R.F. 1982. Plantation establishment with bareroot stock: Some critical factors. In G.D. Mroz and J.F. Berner, *Artificial Regeneration of Conifers in the Upper Great Lakes Region,* pp. 304–331. Mich. Technol. Univ., Houghton, MI.

Swindel, B.F., L.F. Conde, and J.E. Smith. 1983. Plant cover and biomass response to clearcutting, site preparation, and planting in *Pinus elliotti* flatwoods. *Sci.* 219:1421–1422.

Swindel, B.F., J.E. Smith, and K.W. Outcalt. 1989. Long-term response of competing vegetation to mechanical site preparation in pine plantations. In J.H. Miller (Ed.), *Proceedings of the Fifth Biennial Southern Silviculture Research Conference,* pp. 123–128. U.S. For. Serv. Gen. Tech. Rpt. SO-74.

Tanaka, Y. 1984. Assuring seed quality for seedling production: Cone collection and seedling processing, testing, storage, and stratification. In M.L. Duryea, and T.D. Landis (Eds.), *Forest Nursery Manual: Production of Bareroot Seedlings.* Ch. 4, pp. 27–39. Martinus Nijhoff/Dr W. Junk Publ. The Hague/Boston/Lancaster. For. Res. Lab., Ore. State Univ., Corvallis.

Tappeiner, II, J.C., and R.G. Wagner. 1987. Principles of silviculture in prescriptions for vegetation management. In J.D. Walstad and P.J. Kuch (Eds.), *Forest Vegetation Management for Conifer Production,* Ch. 13, pp. 399–429. New York: Wiley.

Temple, S.A., and B.A. Wilcox. 1986. Introduction: Predicting effects of habitat patchiness and fragmentation. In J. Verner, M.L. Morrison, and C.J. Ralph (Eds.), *Wildlife 2000. Modeling Habitat Relationships of Terrestrial Vertebrates,* pp. 261–262. Madison, WI: The Univ. Wis. Press.

Tesch, S.D. 1995. The Pacific Northwest region. In J.W. Barrett (Ed.), *Regional Silviculture of the United States,* 3d ed., Ch. 11, pp. 499–558. New York: Wiley.

Tesch, S.D., and J.W. Mann. 1991. Clearcut and shelterwood reproduction methods for regenerating southwest Oregon forests. Ore. State Univ., For. Res. Lab. Res. Bull. 72.

Thatcher, R.C., G.N. Mason, and D.D. Hertel. 1986. Integrated pest management in southern pine forests. Integrated Pest Management Handbook. U.S. For. Serv. and Coop. State Res. Serv., Agric. Handbk. 650.

Thomas, J.W., C. Maser, and J.E. Rodiek. 1979. Riparian zones. In J.W. Thomas (Ed.), *Wildlife Habitats in Managed Forests: The Blue Mountains of Oregon and Washington,* pp. 40–47. U.S. For. Serv., Agric. Handbk. No. 553.

Thomas, J.W., G.L. Crouch, R.S. Bumstead, and L.D. Bryant. 1975. Silvicultural options and habitat values in conifer forests. In *Management of Forest and Range Habitats for Non-game Birds,* pp. 272–287. U.S. For Serv. Gen. Tech. Rpt. WO-1.

Thompson, B.E. 1984. Establishing a vigorous nursery crop: Bed preparation, seed sowing, and early seedling growth. In M.L. Duryea, and T.D. Landis (Eds.), *Forest Nursery Manual: Production of Bareroot Seedlings.* Ch. 5, pp. 41–49. Martinus Nijhoff/Dr W. Junk Publ. The Hague/Boston/Lancaster. For. Res. Lab., Ore. State Univ., Corvallis.

Thompson, E.F. 1966. Traditional forest regulation model: An economic critique. *J. For.* 64(11):750–752.

Throusdell, K.B., and M.D. Hoover. 1955. A change in ground-water level after clearcutting of loblolly pine in the coastal plain. *J. For.* 53(7):493–498.

Tiarks, A.E. 1987. Effect of site preparation and fertilization on second rotation slash pine. In *Proceedings of the Second Rotation Plantation Establishment Workshop,* pp. 56–62. Feb. 17–19, 1987, Georgetown, SC. Baruch For. Sci. Inst., Georgetown.

Tinus, R.W., and S.E. McDonald. 1979. How to grow tree seedlings in containers in greenhouses. U.S. For. Serv. Gen. Tech. Rpt. RM-60.

Tobalske, B.W., R.C. Shearer, and R.L. Hutto. 1991. Bird populations in logged and unlogged western larch/Douglas-fir forests in Northwestern Montana. U.S. For. Serv. Res. Pap. INT-442.

Tolsted, D., and D. Hansen. 1992. Age of hybrid poplar stools at first cut influences third-year cutting production. U.S. For. Serv. Res. Note NC-357.

Toumey, J.W., and C.F. Korstian. 1931. *Seeding and Planting in the Practice of Forestry,* 2d ed. New York: Wiley.

Toumey, J.W., and C.F. Korstian. 1947. *Foundations of Silviculture upon an Ecological Basis,* 2d ed. New York: Wiley.

Toumey, J.W., and E.J. Neethling. 1924. Insolation, a factor in the natural regeneration of certain conifers. Yale Univ., Sch. For. Bull. 11.

Trettin, C.C. 1986. Application of water management techniques to hardwood silviculture. In G.D. Mroz and D.D. Reed (Eds.), *The Northern Hardwood Resource: Management And Potential,* pp. 183–192. Proc. of Conf., Aug. 18–20, 1986. Houghton, MI. Sch. For. and Wood Prod., Mich. Technol. Univ., Houghton, MI.

Tricoli, D.M., C.A. Maynard, and A.P. Drew. 1985. Tissue culture of propagation of mature trees of *Prunus serotina* Ehrh. I. Establishment, multiplication, and rooting in vitro. *For. Sci.* 31:201–208.

Trimble, G.R., Jr. 1965. Species composition changes under individual tree selection cutting in cove hardwoods. U.S. For. Serv. Res. Note. NE-30.

Trimble, G.R., Jr. 1970. Twenty years of intensive uneven-aged management: Effect on growth, yield, and species composition in two hardwood stands in West Virginia. U.S. For. Serv. Res. Pap. NE-154.

Trimble, G.R. 1971a. Early crop-tree release in even-aged stands of Appalachian hardwoods. U.S. For. Serv. Res. Pap. NE-203.

Trimble, R.G. 1971b. Diameter-limit cutting in Appalachian hardwoods: Boon or bane. U.S. For. Serv. Res. Pap. NE-208.

Trimble, G.R. 1973a. Response to crop-tree release by 7-year-old stems of yellow-poplar and black cherry. U.S. For. Serv. Res. Pap. NE-253.

Trimble, G.R., Jr. 1973b. The regeneration of central Appalachian hardwoods with emphasis on the effects of site quality and harvesting practice. U.S. For. Serv. Res. Pap. NE-282.

Trimble, G.R. 1974. Response to crop-tree release by 7-year-old stems of red maple stump sprouts and northern red oak advance reproduction. U.S. For. Serv. Res. Pap. NE-303.

Trimble, G.R., and E.H. Tryon. 1966. Crown encroachment into openings cut in Appalachian hardwood stands. *J. For.* 64(2):104–108.

Trimble, G.R., Jr., and B.D. Fridley. 1963. Thirteen years of forestry research in West Virginia. U.S. For. Serv. Res. Pap. NE-5.

Trimble, G.R., Jr., and G. Hart. 1961. An appraisal of early reproduction after cutting in northern Appalachian hardwood stands in West Virginia. U.S. For. Serv., NE. For. Exp. Stn., Stn. Pap. NE-154.

Trimble, G.R., Jr., and H.C. Smith. 1976. Stand structure and stocking in Appalachian mixed hardwoods. U.S. For. Serv. Res. Pap. NE-340.

Trimble, G.R., Jr., J.J. Mendel, and R.A. Kennell. 1974. A procedure for selection marking in hardwoods: Combining silvicultural considerations and economic guidelines. U.S. For. Serv. Res. Pap. NE-292.

Trimble, G.R., Jr., E.H. Tryon, H.C. Smith, and J.D. Hillier. 1986. Age and stem origin of Appalachian hardwood reproduction following a clearcut and herbicide treatment. U.S. For. Serv. Res. Pap. NE-589.

Troendle, C.A. 1982. The effects of small clearcuts in water yield from Deadhorse Watershed, Fraser, Colorado. In *Proceedings of the 50th Annual Western Snow Conference.* April 19–23, 1982. Reno, NV, Colo. State Univ., Ft. Collins, CO.

Troendle, C.A. 1983. The potential for water yield augmentation from forest management in the Rocky Mountain Region. *Water Resourc. Bull.* 19(3):359–373.

Troendle, C.A., and R.M. King. 1985. The effect of timber harvest on the Fool Creek Watershed, 30 years later. *Water Resourc. Res.* 21(12):1915–1922.

Troendle, C.A., and C.F. Leaf. 1980. Effects of timber harvest in the snow zone on volume and timing of water yield. In *Interior West Watershed Management,* pp. 231–243. Proc. of Symp., Apr. 8–10, 1980, Spokane, WA, Wash. State Univ., Wash. Coop. Ext., Wash. State Univ., Pullman, WA.

Troendle, C.A., and J.R. Meiman. 1984. Options for harvesting timber to control snowpack accumulation. In *Proceedings of the Western Snow Conference,* pp. 86–97. Apr. 17–19, 1984. Sun Valley, ID. U.S. For. Serv., Ft. Collins, CO.

Troup, R.S. 1921. *The Silviculture of Indian Trees. Vol. 1. Dilleniaceae to Leguminosiae (Papilionaceae).* Publ. under authority His Majesty's Sec. State for India in Counc. Oxford: Clarendon Press.

Troup, R.S. 1928. *Silvicultural Systems.* Oxford Manuals For. London: Oxford Univ. Press.

Tubbs, C.H. 1977a. Manager's guide for northern hardwoods in the North Central States. U.S. For. Serv. Gen. Tech. Rpt. NC-39.

Tubbs, C.H. 1977b. Natural regeneration of northern hardwoods in the northern Great Lakes Region. U.S. For. Serv. Res. Pap. NC-150.

Tuttle, W.A. 1982. Frontiers of applied technology. In G.D. Mroz and J.F. Berner, *Artificial Regeneration of Conifers in the Upper Great Lakes Region,* pp. 416–419. Mich. Technol. Univ., Houghton, MI.

U.S. Corps Eng. 1988. Clearcutting: Effects on the forest environment. Tombigbee and Black Warrior River Basins, Alabama and Mississippi. Rpt. by Tenn. Mt. Manage. Contractor. U.S. Dept. Army, Corps Eng., Mobile Ala., Contract No: DACW 01-87-C-0140.

U.S. Environ. Protect. Agency. 1973. Processes, procedures, and methods to control pollution resulting from silvicultural activities. U.S. Environ. Protect. Agency, Off. Air and Water Prog., EPA 430/9-73-010.

U.S. For. Serv. 1935. Effect of shading in reducing soil surface temperature. U.S. For. Serv., Lake States For. Exp. Stn., Tech. Note 100.

U.S. For. Serv. 1978. *Uneven-aged Silviculture and Management in the United States.* Proc. Inservice Workshop, July 15–17, 1975 and Oct. 19–21, 1976. U.S. For. Serv., Timber Manage. Res., Washington, DC.

U.S. For. Serv. 1989. A guide to the care and planting of southern pine seedlings. U.S. For. Serv., S. Reg., Manage. Bull. R8-MB39.

Vallentine, J.F. 1983. The application and use of herbicides for range plant control. In S.B. Monsen and N. Shaw (Eds.), *Managing Intermountain Rangelands—Improvement of Range and Wildlife Habitats,* pp. 39–48. U.S. For. Serv. Gen. Tech. Rpt. INT-157.

van den Driessche, R. 1984. Soil fertility in forest nurseries. In M.L. Duryea and P. M. Dougherty (Eds.), *Forest Regeneration Manual,* Ch. 7, pp. 63–74. The Netherlands: Klumer Academ. Publ.

Van Lear, D.H. 1981. Forest types and treatments. In *Choices in Silviculture for American Forests,* Ch. 1, pp. 1–9. Soc. Am. For., Washington, DC.

Van Lear, D.H., and T.A. Waldrop. 1991. Prescribed burning for regeneration. In M.L. Duryea and P.M. Dougherty (Eds.), *Forest Regeneration Manual,* Ch. 12, pp. 235–250. The Netherlands: Kluwer Academ. Press.

Van Lear, D.H., J.R. Saucier, and J.G. Williams, Jr. 1977. Growth and wood properties for longleaf pine following silvicultural treatments. *Soil Sci. Soc. Am. J.* 41(5):989–992.

Vaughn, C.L., A.C. Wollin, K.A. McDonald, and E.H. Bulgrin. 1966. Hardwood log grades for standard lumber. U.S. For. Serv. Res. Pap. FPL-63.

von Althen, F.W. 1979. Preliminary guide to site preparation and weed control in hardwood plantations in southern Ontario. Can. Dept. Environ., Can. For. Serv., Sault Ste. Marie, Ont. Info. Rpt. O-X-288.

von Althen, F.W. 1981. Site preparation and post-planting weed control in hardwood afforestation: White ash, black walnut, basswood, silver maple, hybrid poplar. Can. For. Serv., Dept. Environ., Great Lakes For. Res. Ctr. Rpt. O-X-325.

Voorhis, N.G. 1990. Precommercial crop-tree thinning in a mixed northern hardwood stand. U.S. For. Serv. Res. Pap. NE-640.

Vyas, A.D., and S-Y. Shen. 1982. Analysis of short-rotation forests using the Argonne model for selecting economic strategy (MOSES). U.S. Dept. Energy and Environ. Sys. Div., Argonne Natl. Lab. ANL/CNSV-36.

Wackerman, A.E., W.D. Hagenstein, and A.S. Michell. 1966. *Harvesting Timber Crops,* 2d ed. New York: McGraw-Hill.

Wade, D.D., and L. Lundsford. 1990. Fire as a forest management tool: Prescribed burning in the southern United States. *Unasylva* 162 (41):28–38.

Wagener, W.W. 1949. Shade trees for California. In *Trees, the Yearbook of Agriculture 1949,* pp. 77–82. 81st Congr., 1st Sess., House Doc. No. 29. U.S. Gov. Print. Off., Washington, DC.

Wagner, R.G. 1982. A system to assess weed severity in young forest plantations for making vegetation management decisions. In *Forest Vegetation Management Notebook.* Coll. For. Ore. State Univ., Corvallis, OR.

Wahlenberg, W.G. 1965. A guide to loblolly and slash pine plantation management in southeastern USA. Ga. For. Res. Counc. Rpt. No. 14.

Wakeley, P.C. 1954. Planting the southern pines. U.S. For. Serv. Monogr. No. 18.

Waldron, R.M. 1974. Direct seeding in Canada 1900–1972. In J.H. Cayford (Ed.), *Direct Seeding Symposium,* pp. 11–29. Sept. 11–13, 1973. Timmins, Ont. Dept. Environ., Can. For. Serv., Publ. No. 1339.

Walker, L.C. 1995. The southern pine region. In J.W. Barrett (Ed.), *Regional Silviculture of the United States,* 3d ed., Ch. 7, pp. 271–333. New York: Wiley.

Wall, R.E. 1978. Disease control in forest nurseries. Can. For. Serv., Maritimes For, Res. Ctr., Info. Rpt. M-X-94.

Wall, R.E. 1982. Secondary succession on recently cut-over forest land in Nova Scotia. Can. Dept. Environ, For. Serv., Maritimes For. Res. Ctr. Info. Rpt. M-X-133.

Walstad, J.D., and P.J. Kuch. 1987. Introduction to forest vegetation management. In J.D. Walstad and P.J. Kuch (Eds.), *Forest Vegetation Management for Conifer Production,* Ch. 2, pp. 3–14. New York: Wiley.

Walstad, J.D., and K.W. Seidel. 1990. Use and benefits of prescribed fire in reforestation. In J.D. Walstad, S.R. Radosevich, and D.V. Sandberg (Eds.), *Natural and Prescribed Fire in Pacific Northwest Forests,* pp. 67–79. Ore. State Univ. Press., Corvallis.

Walstad, J.D., M. Newton, and R.J. Boyd, Jr. 1987a. Forest vegetation problems in the Northwest. In J.D. Walstad and P.J. Kuch (Eds.), *Forest Vegetation Management for Conifer Production,* Ch. 3, pp. 15–53. New York: Wiley.

Walstad, J.D., M. Newton, and D.H. Gjerstad. 1987b. Overview of vegetation management alternatives. In J.D. Walstad and P.J. Kuch (Eds.), *Forest Vegetation Management for Conifer Production,* Ch. 6, pp. 157–200. New York: Wiley.

Walters, E.W. 1974. Systems approach to managing pine bark beetles. In T.L. Payne, R.N. Coulson, and R.C. Thatcher (Eds.), *Southern Pine Beetle Symposium,* pp. 9–22. Proc. Symp., Texas Agric. Exp. Stn., College Station, TX.

Walters, R.S., and R.D. Nyland. 1989. Clearcutting central New York northern hardwood stands. *N. J. Appl. For.* 6(2):75–79.

Wang, Z., and R.D. Nyland. 1993. Tree species richness increased by clearcutting of northern hardwoods in central New York. *For. Ecol. and Manage.* 57(1993):71–84.

Waring, R.H., and W.H. Schlesinger. 1985. *Forest Ecosystems—Concepts and Management.* New York: Academic Press.

Waterman, A.M., R.U. Swingle, and C.S. Moses. 1949. Shade trees of the Northeast. In *Trees, the Yearbook of Agriculture 1949,* pp. 48–60. 81st Congr., 1st Sess., House Doc. No. 29. U.S. Gov. Print. Off., Washington, DC.

Waters, W.E., and E.B. Cowling. 1976. Integrated forest pest management: A silvicultural necessity. In J.L. Apple and R.F. Smith (Eds.), *Integrated Pest Management,* pp. 149–177. New York: Plenum Press.

Watt, A. 1947. Pattern and process in the plant community. *J. Ecol.* 35(1&2):1–22.

Watt, R., K.A. Brinkman, and B.A. Roach. 1979. Oak-hickory. In *Silvicultural Systems for the Major Forest Types of the United States.* U.S. Dept. Agric., Agric. Handbk. No. 445.

Way, M. 1977. Integrated control—Practical realities. *Outl. Agric.* 9:127–135.

Webb, W.L., and D.F. Behrend. 1970. Effect of logging on songbird populations in a northern hardwood forest: A preliminary analysis. The Wildl. Soc., Northeast. Sect., *Trans.* 211–217.

Webb, W.L., D.F. Behrend, and B. Saisorn. 1977. Effect of logging on songbird populations in a northern hardwood forest. *Wildl. Mongr.* 55:6–36.

Weber, R. 1964. Early pruning reduces blister rust mortality in white pine plantations. U.S. For. Serv. Res. Note LS-38.

Weitzman, S., and C.J. Holcomb. 1952. How many trees are destroyed in logging? U.S. For. Serv., NE For. Exp. Stn., For. Res. Note 17.

Wells, C.G. 1983. Impact of management practices on nutrient cycling and losses. In *Maintaining Forest Site Productivity,* pp. 40–51. Proc. Conf., Myrtle Beach, SC, 1983. SAF 83-09. Soc. Am. For., Bethesda, MD.

Welsh, C.J.E., W.M. Healy, and R.M. DeGraaf. 1992. Cavity-nesting bird abundance in thinned vs. unthinned Massachusetts oak stands. *N. J. Appl. For.* 9(1):6–9.

Wendel, G.W., and J.N. Kochenderfer. 1984. Aerial release of Norway spruce with Roundup in the central Appalachians. *N. J. Appl. For.* 1(2):29–32.
Wendel, G.W., J.N. Kochenderfer. 1988. Release of 7-year-old underplanted white pine using hexazinone applied with a spot gun. U.S. For. Serv. Res. Pap. NE-614.
Wendel, G.W., and N.I. Lamson. 1987. Effects of herbicide release on the growth of 8- to 12-year-old hardwood crop trees. U.S. For. Serv. NE-RP-598.
Wenger, K.F. 1984. *Forestry Handbook,* 2d ed. New York: Wiley.
Wenger, K.F. 1955. Growth and prospective development of hardwoods and loblolly pine seedlings on clearcut areas. U.S. For. Serv., S. For. Exp. Stn., Stn. Pap. No. 55.
Wenger, K.F., and K.B. Trousdell. 1958. Natural regeneration of loblolly pine in the south Atlantic coastal plain. U.S. Dept. Agric. Prod. Res. Rpt. No. 13.
West, D.C., H.H. Shugart, and D.B. Botkin. 1981. *Forest Succession Concepts and Applications.* New York: Springer-Verlag.
White, E.H., and D.D. Hook. 1975. Establishment and regeneration of silage plantings. *Iowa St. J. Res.* 49(3)Pt.2:287–296.
White, E.H., and R.L. Gambles. 1988. Experiences with willow as a wood biomass species. *Bio-joule* May 1988:4–7.
White, E.H., L.P. Abrahamson, R.F. Kopp, and C.A. Nowak. 1993. Willow bioenergy plantation research in the northeast. In *Energy, Environment, Agriculture, And Industry,* pp. 199–213. Proc. First Biomass Conf. of the Americas. Aug. 30–Sept 2, 1993, Burlington, VT. Natl. Renewable Energy Lab., NREL/CP-200-5768, DE93010050, UC Cat.: 241.
White, E.H., L.P. Abrahamson, R.F. Kopp, C.A. Nowak, and J. Sah. 1992. Integrated woody biomass systems in eastern North America. In C.P. Mitchell, L. Sennerby-Fosse, and L. Zsuffa, (Eds.), *Problems and Perspectives of Forest Biomass Energy,* pp. 68–75. Swedish Univ. Agric. Sci., Dept. Ecol. and Environ. Res., Sect. Short Rotation For. Rpt. 48.
Wiedemann, E. 1950. *Ertragskundliche und waldbauliche Grundlagen der Forstwirtschaft.* Teil I.J.D. Suaerlander's Verlag, Frankfort a. M. (Original not seen).
Wilde, S.A. 1958. *Forest Soils: Their Properties and Relation to Silviculture.* New York: Ronald Press.
Wiley, S., and B. Zeide. 1992. Development of a loblolly pine plantation thinned to different density levels in southern Arkansas. Univ. Ark., Ark. Agric. Exp. Stn., Div. Agric. Rpt. Ser. 322.
Wilhite, L.P., and E.P. Jones, Jr. 1981. Bedding effects in maturing slash pine stands. *S. J. Appl. For.* 5(1):24–27.
Willebrand, E., and T. Verwijst. 1993. Population dynamics of willow coppice systems and their implications for management of short-rotation forests. *For. Chron.* 69(6):699–704.
Willebrand, E., S. Ledin, and T. Verwust. 1993. Willow coppice systems in short rotation forestry: Effects of plant spacing, rotation length and clonal composition on biomass production, *Biomass and Bioenergy* 4(5):323–331.
Willett, R.L., and J.B. Baker. 1990. Low cost forest management alternatives for loblolly-shortleaf pine. In C.A. Hickman (Ed.), *Proceedings of Southern Forest Economics Workshop on Evaluating Even and All-aged Timber Management Options for Southern Forest Lands,* pp. 65–68. U.S. For. Serv. Gen. Tech. Rpt. SO-79.
Williams, R.D., and S.H. Hanks. 1976. Hardwood nurseryman's guide. U.S. For. Serv., Agric. Handbk. No. 473.
Williams, T.M. 1989. Site preparation on forested wetlands of the southeastern coastal plain. In D.D. Hook and R. Lea (Eds.), *The Forested Wetlands of the Southern United States,* pp. 67–71. Proc. Symp., July 12–14, 1988. Orlando, FL. U.S. For. Serv. Gen. Tech. Rep. SE-50.
Williamson, R.L. 1973. Results of shelterwood harvesting of Douglas-fir in the Cascades of western Oregon. U.S. For. Serv. Res. Pap. PNW-161.

Williston, H.L., W.E. Balmer, and L.P. Abrahamson. 1976a. Chemical control of vegetation in southern forests. U.S. For. Serv., S. Area State and Priv. For., For. Manage. Bull.

Williston, H.L., W.E. Balmer, and L.P. Abrahamson. 1976b. Chemical control of vegetation in southern forests. *For. Farmer* 50(3):73–76.

Wilson, L.F., and I. Millers. 1983. Pine root collar weevil—Its ecology and management. U.S. Dept. Agric. Tech. Bull. No. 1675.

Wilson, R.W. 1953. How second-growth northern hardwoods develop after thinning. U.S. For. Serv., NE For. Exp. Stn., Stn. Pap. 62.

Wilson, R.W., and V.S. Jensen. 1954. Regeneration after clear-cutting second growth northern hardwoods. U.S. For. Serv. Res. Note 27.

Wilusz, W., and M. Giertych. 1974. Effects of classical silviculture on the genetic quality of the progeny. *Silvae Genet.* 23(4):127–130.

Winters, R.K. 1977. *Terminology of Forest Science, Technology, Practice and Products.* Addendum Number One. Soc. Am. For., Washington, DC.

Wood, D.L. 1988. Forest Health Management for the future: Insect problems of coniferous forests. In *Healthy Forests, Healthy World,* pp. 114–120. Proc. 1988 Soc. Am. For. Nat. Conv., Rochester, NY, Oct. 16–19, 1988. Soc. Am. For., Washington, DC. SAF Publ. 88-01.

Wood, J.S., W.H. Blackburn, H.A. Pearson, T.K. Hunter, and R.W. Knight. 1987. Assessment of silvicultural and grazing treatment impacts on infiltration and runoff water quality of longleaf-slash pine forest, Kisatchie National Forest, Louisiana. U.S. For. Serv. Gen. Tech Rpt. SO-68.

Woods, K.D. and R.H. Whittaker. 1981. Canopy-understory interaction and the internal dynamics of mature hardwood and hemlock-hardwood forests. In D.C. West, H.H. Shugart, and D.B. Botkin (Eds.), *Forest Succession Concepts and Applications,* Ch. 19, pp. 305–323. New York: Springer-Verlag.

Worgan, D. 1974. Aerial seeding by fixed-wing aircraft. In J.H. Cayford (Ed.), *Direct Seeding Symposium,* pp. 125–130. Sept. 11–13, 1973. Timmins, Ont. Dept. Environ., Can. For. Serv., Publ. No. 1339.

Worthington, N.P. 1966. Response to thinning 60-year-old Douglas-fir. U.S. For. Serv. Res. Note PNW-35.

Wright, E., and T.W. Bretz. 1949. Shade trees for the Plains. In *Trees, the Yearbook of Agriculture 1949,* pp. 65–72. 81st Congr., 1st Sess., House Doc. No. 29. U.S. Gov. Print. Off., Washington, DC.

Wright, H. A., and A. W. Bailey. 1982. *Fire Ecology—United States and Southern Canada.* New York: Wiley.

Wright, J.P. 1976. First thinning options: Row thinning *v.* selection thinning. *N.Z. J. For. Sci.* 6(2):308–317.

Wright, J.W. 1962. *Genetics of Forest Tree Improvement.* FAO For. and For. Prod. Stud. No. 16. Food and Agric. Organ. of the U.N., Rome.

Wright, J.W. 1976. *Introduction to Forest Genetics.* New York: Academic Press.

Yahner, R.H. 1990. Nongame response to ruffed grouse habitat management. In R.D. Adams (Ed.), *Aspen Symposium '89,* pp. 145–153. U.S. For. Serv. Gen. Tech. Rpt. NC-140.

Young, J.A., and C.G. Young. 1992. *Seeds of Woody Plants in North America.* Portland, OR: Dioscorides Press.

Young, M.J., R.C. Kellison, and D.J. Kass. 1993. Effects of residuals on succeeding stand development following clearcutting. In J. Brissette (Ed.), *Proceedings of the Seventh Biennial Southern Silvicultural Research Conference,* pp. 361–356. U.S. For. Serv. Gen. Tech. Rpt. SO-93.

Youngberg, C.T. 1984. Soil and tissue analysis: Tools for managing soil fertility. In M.L. Duryea and P.M. Dougherty (Eds.), *Forest Regeneration Manual,* Ch. 8, pp. 75–80. The Netherlands: Klumer Academ. Press.

Zadoks, J.C., and R.D. Schein. 1979. *Epidemiology and Plant Disease Management.* New York: Oxford Univ. Press.

Zahner, R. 1970. Site quality and wood quality in upland hardwoods: Theoretical considerations of wood density. In C.T. Youngberg and C.B. Davey (Eds.), *Tree Growth and Forest Soils,* pp. 477–497. Ore. State Univ. Press, Corvallis.

Zarnoch, S.J., D.P. Feduccia, V.C. Baldwin, Jr., and T.R. Dell. 1991. Growth and yield predictions for thinned and unthinned slash pine plantations on cutover sites in the West Gulf region. U.S. For. Serv. Res. Pap. SO-264.

Zasada, J.C., and D.F. Grigal. 1978. The effects of silvicultural system and seedbed preparation on natural regeneration of white spruce and associated species in interior Alaska. In C.A. Hollis and A.E. Squillace (Eds.), *Proceedings: Fifth North American Forest Biology Workshop 1978,* pp. 213–220. Sch. For. Resourc. and Conserv., Univ. Fla., Gainsville, FL.

Zasada, J.C., and E.C. Packee. 1995. The Alaska region. In J.W. Barrett (Ed.), *Regional Silviculture of the United States,* 3d ed., Ch. 12, pp. 559–605. New York: Wiley.

Zasada, J.C., and G.A. Schier. 1973. Aspen root suckering in Alaska: Effect of clone collection date and temperature. *Northwest Sci.* 47:100–104.

Zedaker, S.M. 1983. The competition-release enigma: Adding apples and oranges and coming up with lemons. In C.P. Jones (Ed.), *Proceedings of the Second Biennial Southern Silviculture Research Conference,* pp. 357–364. Atlanta, GA., Nov. 4–5, 1982. U.S. For. Serv. Gen. Tech. Rpt. SE-24.

Zillgitt, W.M. 1945. Growth responses of sugar maple following light selective cutting. U.S. For. Serv., Lake States For. Exp. Stn., Tech. Note 229.

Zillgitt, W.M. 1947. Stocking in northern hardwoods under the selection system. In Proc. Soc. Am. For. Meet., pp. 320–327. Dec. 17–20, 1947. Minneapolis, MN, Soc. Am. For., Washington, DC.

Zipperer, W.C., R.D. Nyland, and R.L. Burgess. 1988. Interaction of land use and forest island dynamics in central New York. In *Healthy Forests, Healthy World,* pp. 137–140. Proc. 1988 Soc. Am. For. Nat. Conv., Rochester, NY., Oct. 16–19, 1988. Soc. Am. For. Washington, DC., SAF Publ. 88-01.

Zipperer, W.C., R.L. Burgess, and R.D. Nyland. 1990. Patterns of deforestation and reforestation in different landscape types in central New York, U.S.A. *For. Ecol. and Manage.* 36(1990):103–117.

Zobel, B.J., and J. Talbert 1984. *Applied Forest Tree Improvement.* New York: Wiley.

Zobel, B.J., G. van Wyk, and P. Stahl. 1987. *Growing Exotic Forests.* New York: Wiley.

Zutter, B.R., G.R. Grover, and D.F. Dickens. 1985. Competing vegetation assessment systems in young southern pine plantations. In E. Shoulders (Ed.), *Proceedings of the Third Biennial Southern Silvicultural Research Conference,* pp. 279–286. U.S. For. Serv. Gen. Tech. Rpt. SO-54.

INDEX

Page number set *italic* indicates figure; page number followed by "t" indicates table.

Acceptable growing stock (AGS), 219, 372
Accretion, determining growth and, 185, 188
Acquisition expenditures, 42
Active site preparation, 83–84
  methods used, 83
  reasons for, 83
Active surface, 223
Administrative demands, adjusting to, 533–549
  considerations for initial planning, 538–540
    accessibility of resources relative to available markets, 539
    deriving different values from deliberate program of management, 540, *541*
    interest levels, 539
    market opportunities, 539
    overhead and operating costs, 539
    taxes levied against both land and real estate values, 539
  examples of changing, 541–544
    altering silvicultural plans, 542
    fiscal crises and consequences, 542–543
  factors that force unscheduled changes in silviculture practice, 540–541
    altering intensity of silvicultural program, 540
  problem-solving procedures for silviculture, 533–536
  silviculture from administrative perspective, 536–538
  silviculture in perspective, 533
  silviculture in retrospect, 548–549
  taking an ecosystem approach, 544–548
Advance regeneration, 65, 318
Aerial application of herbicides, 94
  weeding by, 331
Aerial seeding, 169
Afforestation, 60
  definition of, 59
Age class
  control of, 17
  definition of, 28
  distribution, 203
  single, 29, *32*
  uneven-aged stands, 29, *30–31*
Aggradation phase, 190, 321
  release treatments and, 329
Agri-silvicultural systems, 119
Agroforestry in tree-poor regions, 118–120
  agri-silvicultural systems, 119
  agro-sylvo-pastoral systems, 119
  intercropping, 120
  multiple-purpose systems, 119
  sequential cropping, 120
  sylvo-pastoral systems, 119
Agro-sylvo-pastoral systems, 119
AGS. *See* Acceptable growing stock (AGS)
Ailanthus, 117
Airborne, various seeds, 66–67
Alder, dissemination and seed structure, 67
A-level stocking, 367, 372
Alley-cropping, 522
Alnus, 519
Alternate-patch method, 277

Alternate rate of return (ARR), 38
Alternate-strip method, 277
American beech, 64
  seed dispersal distances, 270t
American elm, seed dispersal distances, 270t
Animal browsing and feeding, 62
Animals
  clearcutting and protection of threatened and endangered, 289
  domesticated and wild, habitat, 23
  even-aged stands and, 37–38
Annosus root rot (*Heterobasidion annosum*), 454, 468
Aphids, 142
Arbogast guide, 215–219
Area control, 183, 184
Area-volume control, 184, 249
ARR. *See* Alternate rate of return (ARR)
Artificial crosses between species, 115
Artificial regeneration, 17, 60
  advantages, 70
  choosing between natural and, 71–72
  clearcutting with, 286–288
    regeneration control and flexibility, 288
  direct seeding, 159–160
  disadvantages, 71
  factors influencing success of program, *108*
  natural and, 63
  planning for, 106–132
    arrangement and operational considerations, 129–130
    bare-root and container stock comparison, 124
    examples of successful exotic species, 116–118
    factors relating to seedling quality, 125–126
    field assessment of site conditions, 120–121
    importance of soil conditions, 121–122
    kinds of planting stock, 122–124
    planning tree-planting project, *108*, 108–109
    planting plan, 130–132
    purposes and advantages over natural methods, 109–112
    requirement for successful, 112–113
    selecting a spacing, 126–129
    species selection, 113–115
    tree planting, 106–108
    use of exotic species, 115–116
    windbreaks and agroforestry in tree-poor regions, 118–120
  reasons for choosing, 177
Asexual reproduction, 63
Ash
  dissemination and seed structure, 66
  white, 191
Aspens
  dissemination and seed structure, 66
  even-aged stands regenerating by sprouting and suckering, 60
  regeneration by vegetative methods, 64
  seed dispersal distances, 270t
  thinned stands in Canada, 396
Atrazine, 94
Average physical product, 187

Balanced forests, 205
  stands, 203–205, 209–213
Bales, 145
Balsam fir, 174
  seed dispersal distances, 270t
Band treatment. *See* Spot treatment
Bank slopes and bottoms, 52
  minimizing disturbance of, 52–53
  protecting, 54
Bare-root seedling quality, condition, uniformity factors, 138
Bare-root stock, 124
  advantages of, 124
  comparison between container and, 124
  handling, 143–145, 147–148
  ideal time to plant, 147
  packaging techniques, 145
Basal sprays method, 338
Basswood, dissemination and seed structure, 66
Bedding, 85, *86*
Beech
  dissemination and seed structure, 67
    cached or buried by animals, 67
  European beech to northeast and southeast North America, 117
Beech bark disease, 473
Bell-shaped distributions, 352
Best management practices (BMP), 51, 53
  relevant measures, 52
Bimodal (or multimodal) distributions, 352
Biocides, 140
Biological potential (viability) of seed, 138
Biologic factors, unhealthy tree and, 461
  allelopathy of one plant to another, 461
  animal activity, 461
  disease, 461
  insect feeding, 461
Biomass, 519
Birch
  dissemination and seed structure, 67
  European white, 117
  paper
    regeneration, 82, 89
    seed dissemination by wind, 267
  seed dispersal distances, 270t
  yellow, 82, 89, 191
Black spruce, seed dispersal distances, 270t
Black stain root disease (*Leptographion wageneri*), 468
Black tupelo, seed dispersal distances, 270t
Blading, 85
Blister rust, 120, 473
Blue gum, seed dispersal distances, 270t
Blue spruce, seed dispersal distances, 270t

BMP. *See* Best management practices (BMP)
Borggreve method, 384
Broadcast chemical methods, 334, 336, 338
Broadcast seeding, 167, 169, 170
  formula for rates of seeding, 167–168
Broadcast spraying, 92
  disadvantages of ground applications, 92–93
Broadleafed species, 64, 65
Buckeye, 67
Bucking, definition of, 45
Buffer strips, uncut or lightly cut
  protecting water resources, 53
Bulk density, 48
Butternut seed
  cached or buried by animals, 67
  dispersal distances, 270t
Byproduct of management, 377

California red fir, seed dispersal distances, 270t
Camphor-tree, 117
Canopy gaps, 162
  development of trees underneath, 162, *162*
Caribbean pine grown in Africa and South America, 116
Cavities, culls, dead trees as habitat essentials, 482–484
  cavity dependent wildlife species, 482, *483*
Cedar
  Deodar, 117
  Incense, dissemination and seed structure, 66
  red, regeneration of, 82
  seed dispersal distances for N. white, 270t
Cellulose batting, 145
Center fire, 102
Certified orchard seed, 136
Chaining, 88, *88*, 89
Checkerboard pattern, 12, *13*
Check method, 205, 231
  assessing stand structural change under selection system, 249
  unmanaged old-growth hardwood forests of North America, high density European stands managed under, 206, *207*
Chemical applications, 83
  See also Herbicides
Cherry
  black, 82
    seed dispersal distances of, 270t
  pin, 191
  seeds ingested as food, 67
Chevron fire, 102
Chinese elm, 117
Chopping, 87, *87*
Cleaning in release treatment, 332–340, 347
  all chemical methods, 337
  basal sprays and stem injection, 338, *339*
  broadcast chemical methods, 336, 337, 338
  common methods of, 333–334
  during sapling stage, 329
  experimental strip treatments, 336
  ground methods directing herbicides between rows of crop trees, 335
  ground-spread granular, 338
  mechanical methods, 335, 337
  operational strip treatment, 336, 337
  releases saplings of desirable characteristics by removing overtopping trees, 333, *334*
  removing undesirable trees from sapling stands, 332, *332*
  spot treatments, 337, 338
  tree cutting, 338
  woody vines and shrubs removal, 329
Clearcutting
  accommodating species tolerances, 275–276
    sheltering seedlings, 276
  altering configuration of, 276–282
    choices between strip, patch, or large block, 281–282
    creating even-aged community by alternate or progressive strip or patch method, 277, *277*
    effects of patch size, shape, light pattern orientation, 279, *280*
    effect of strip width on proportion of daily summertime illumination, 279, 281t
    progressive cutting on upwind side, 281–282
    progressive strip cutting, *278*, 279
  with artificial regeneration, 286–287
  definition of, 263
  difference between shelterwood and, 292–293
  differentiating clearcutting method, clearcutting system, and, 263
  effect of environmental conditions near ground, 271–275
    effects of aspect and slope steepness on success of regeneration, 274, *274*
  effects on other ecosystem attributes, 282–285
    strip clearcutting as useful technique for snow management, 283, *283*
  importance of seed source and advance regeneration, 267–271
    dispersal distances for seed blown downwind, 269, 270t
    seed dispersal patterns, 267, *268*
  operational considerations in, 288–289
  as production method, 263–366
    attributes of, 266–267
    removing overstory to establish new cohort, 263, *264*
    values and limitations, 289–291
  site preparation and other ancillary treatments, 285–286
Clearcutting method
  definition of, 263
  differentiating clearcutting, clearcutting system, and, 263
Clearcutting system, 31, 178, 181
  definition of, 263
  differentiating clearcutting, clearcutting method, and, 263

Clearfelling, 517
Climate-controlled greenhouses, 123, *123*
Clonal forestry, 64
Codominant crown class, *355*, 356
  definition of, 356
Codominant trees, *383*
  in selection thinning, 384
Cohort, 28, 176
  establishing new, 263, *264*
Colorado blue spruce to northeastern North America, 117
Commercial clearcutting, 265, 503
Commercial thinning, 399
Commodity production, 176, *177*
Community composition, determinants to, 317–318
Compaction, 48
Competition
  definition of, 323
  interspecific, 323
  intraspecific, 323
  involving single or multiple resources, 323
  weeding to reduce, by herbaceous plants, 330, *331*
Composite forest, 527
Composite stand, 527
Composition
  controlling establishment, growth, and, 17–18
  definition of, 322
  determinants to community, 317–318
  effect of residual density on species, *226*
Compound coppice, 527
Concave slopes and drainage, 121–122
Conifer(s)
  replaced by northern hardwoods, 286
  scarification and regeneration of several, 90
  shelterwood method for natural regeneration and, 311
  Sierra Nevada mixed, 211
Conifer forests
  canopy cover of western North America, 303
Conservation
  deliberate silviculture and, 491–492
  maintaining ecologic standard, 492–494
  selective cutting and, 511–512
  silvicultural systems and, 25–28
Consumer costs, 42
Container(s), seedling definition of, 149
  molded blocks, 149
  reusable molds for plug seedlings, 149
  rigid-wall tubes, 149
  and rooting material requirements, 149
  size and design of, 149, *150*
  types of, 149
Container-grown seedlings, 123, *123*, 148–152
  advantages of, 148–149
  biological and operational advantages, 124
  comparing bare-root and, 124
  disadvantages of, 151
  forestation methods, 148
  planting process care requirements, 151–152
  production estimates for, 151
Containerized planting, 149
Container stock. *See* Container-grown seedlings
Continuous stimulation of final crop trees, 416
Contour stripping, 84
Control, ecologic continuity, 17
  Controlled burning, 99–100
  See also Prescribed burning
Control method
  assessing stand structural change under selection system, 249
Convex slopes and drainage, 121
Coppice
  conversion from coppice to high-forest systems, 531–532
  definition of, 514
Coppice method
  based on root suckers, 522–524
    originating from multiple points (loci) along root system, 522, *525*
  based on stump sprouting, 515–518
    broadleafed species and, 515
    clearfelling and, 517
    close cutting and decay resistance, 516–517
    definition of, 514
    desirable stool sprouts from root collar, 516
    effect of tree size on sprouting capacity of red and sugar maple, 516, *516*
    sprout-origin trees from stump versus sprouts of tops or upper parts, 517
Coppice shoot, definition of, 514
Coppice silviculture, 513
  advantages and disadvantages of simple-coppice systems, 525–527
  conversion from coppice to high-forest systems, 531–532
  coppice-with-standards systems, 527–531
  methods based on root suckers, 522–524
  methods based on stump sprouting, 515–518
  role in forest stand management, 513–515
  setting rotation length in coppice systems, 524
  short-rotation coppice systems, 518–522
Coppice systems, 63
  advantages and disadvantages of simple, 525–527
  definitions of, 514
  setting rotation length in, 524
  short-rotation, 518–522
Coppice-with-standards systems, 527–531
  advantages compared with simple coppice, 529
  limitations of, 529, 531
  two-aged stands with scattered overstory of older trees, 527, *530*
Coppicing, definition of, 514
Copse, definition of, 514
Cottonwoods, 94
  dissemination and seed structure, 66
  seed dispersal distances, 270t
  See also Poplars; *Populus*
Coupe, definition of, 514
Crop tree release, 391, *392*
  See also Free thinning

Crop trees
continuous stimulation of final, 416
crown thinning and, 381
fillers and, 381
response of, to release, 340–341
trainers and, 383
Cross-pollination, 63
Crown canopy, 320
closing shading ground, 314, *315*
closure, release treatments and, 329
seed cutting creating permanent openings in main, 293, *294*
Crown canopy area, *357*
Crown classes, 352–358
codominant, *355*, 356
definition of, 356
complex interacting biophysical factors determining growth rate and longevity, 352, *354*
dominant, 354, *355*
definition of, 354
indicating relative past rates of development, 357, *357*, 358
intermediate, *355*, 356
definition of, 356
overtopped, *355*, 356
definition of, 356
oppressed trees, 356
suppressed trees, 356
Crown cover removal, 46
Crown fires, 98
Crown length ratio, 202
Crown position, 217
Crown ratio over time for slash pine, 126, 127, *128*
Crown size, branching habit, and stem form, 217
Crown thinning, 380–384
effects of thinning on growth rates, 383, *383*
maintaining uniform spacing between residual tree, 381, *382*
Crown touching crop tree release, 393
Crown touching method, 341
Culled forest, 503
Culled stand, 503
Culls, cavities, dead trees as habitat essentials, 482–484
Culmination of mean annual increment (m.a.i.), 189, 194
Cutting cycle, 29, 185, 194, 200, 223, 242
integrating residual density, structure, and, 252–256
changing selection system stand conditions, *255*
relationship between stand density and growth, *254*
Cuttings, 63, 183
Cutting series, 199
Cypress
dissemination and seed structure
bald, 67
true, 67
Damage, discoloration, decay in living trees, 467–469
injuries in, 468, *469*
physical defects caused by logging and natural agents, 467
Damping-off fungi, 141–142
Dead shade, 276
Dead standing trees (snags), 482
Dead trees, culls, cavities as habitat essentials, 482–484
Decay, damage, discoloration in living trees, 467–469
injuries in, 468, *469*
physical defects caused by logging and natural agents, 467
Decline-causing defects, 430
Decline phase of growth, 358
Decomposition of organic matter, 140
Defects in standing trees
decline-causing defects, 430
that degrade trees, 430–431, *432–433*
external indicators, 431, *432*
internal defects, 431
value-affecting defects, 430
Degrees of thinning, 378
See also Thinning from below
Delayed shelterwood method, 304
Delimbing, definition of, 45
Density of seed to spread, 137, 138
Deodar cedar, 117
Diameter class distribution, 203
Diameter distribution, 218
of uneven-aged upland hardwood stands resembles rotated-S or reverse-J structure, 208, *209*, 210
Diameter-limit cuttings, 229
in even- and uneven-aged communities, 504
expected changes in uneven-aged stand diameter distribution following repeated, 246, *247*, 248
Dibble, 152, *154*
Dicamba, 92
Direct seeding, 159–160
above- and below-ground ecologic space management, 160
advantages and shortcomings of, 165
in Canada, 174
conditions causing failure, 164–165
difficulties with, 174
ecologic concepts, 160–161
for forest regeneration, 163–164, *164*, 165–166
insuring physical site resources, 160
in North America, 173
reasons for decline, 173
objections to, 165–166
preparation for, 166–169
in the tropics, 174
Discoloration, decay, damage in living trees, 467–469
injuries in, 468, *469*

physical defects caused by logging and natural agents, 467
Dissemination, 66
common kinds of seeds and mechanisms, 66–67
Disturbance, 183
of bank slopes and bottoms, 52–53
by machines compacting, 47–48
major (human or natural), altering community, 162–163
Ditching, 87, 88
Dogwood, seeds ingested as food, 67
Domesticated forest, 35
Dominant crown class, 354, *355*
definition of, 354
Dominant trees, in selection thinning, 384
Douglas-fir
as Christmas trees, 76, 116
and clearcutting, 319
communities in northwestern North America, 316
dissemination and seed structure, 66
financial maturity in, growing, *40*
to Great Britain, 76, 116
from Pacific to east, 76, 117
in plains states, 117
and prescribed burning, 319
scarification and, 90
seed dispersal distances, 270t
stocking guide to control intensity and frequency of thinning, *420*, 421
storage temperatures, 145
thinned stand in Pacific northwest North America, 396
Drainage
raised beds and, 140
soil conditions, 121
timber harvesting and, 45
Drilled seeding, 167
formula for rates of seeding, 168
Drought, as external agent, 62
Dry pruning, 438

Early stand development
even-aged communities at mesic sites, 316–321
even-aged community establishment and formation, 314–316
intermediate treatments for even-aged stands, 321–323
definition of, 321
judging established seedling or stand of seedlings, 323–325
vegetative competition stand development, 324, *325*
judging regeneration success guidelines, 327
vegetation and regeneration surveys, 325–326
Early wood, 436
Eastern black walnut, 117
Eastern hemlock, seed dispersal distances, 270t
Eastern white pine, seed dispersal distances, 270t
Ecological fragmentation, 11
Ecologic continuity
control and, 17
facilitate harvesting, management, use, 17
protecting site, 17
protecting trees, 17
salvaging, 17
Ecologic space, 263
Ecologic standard, maintaining an, 492–494
essential minimum requirements, 493
Economic clearcutting, 265
Ecosystem approach, taking, 544–548
developing appropriate program of management, 547–548
ecological considerations, 548
economic factors, 548
planning strategy for creating, maintaining, sustaining conditions on forestwide scale, 547, *547*
Ecosystem management, 15, 545
Ecosystems
government and corporate owned forests and, 10–11
landscape diversity and parcelization, 14–15
private landowners and, 11
protecting, 23
Edaphic features, 120, 121
Edges created by fragmentation, 11
Elm
Chinese elm to northeastern North America, 117
dissemination and seed structure, 67
Siberian, 117
Empirical yield table, 399
Engelmann spruce, 82, 90
seed dispersal distances, 270t
seed dispersal patterns, 267, *286*
Enrichment plantings, 70
Environment
effects of timber harvesting, 46–49
associated with logging, 46
physical, 2, 45
Environmental pollution
preventing, 49–53, *54–55*
Equal area per age class (balanced), 205
Erosion
circumventing potential, 51–52
common best management practices for minimizing soil erosion and protecting water quality during timber harvesting, 51t
controlling potential after harvesting, 54–56
designing trail and road features, 55
inhibiting, 23
matching equipment and practices to needs and conditions, 56
preventing, 50
scheduling operations, 54–55
siting and maintaining landings, 55–56
tree cutting and, 46, 48
using skidding, forwarding, yarding strategies, 55
watersheds against, 4

Ethics of resource use, 25–28
  flow resources, 26–27
Eucalyptus, 64, 115, 522
Eucalyptus globulus, 517
European linden, 117
European white birch, 117
Evapotranspiration, 46, 69–70
Even-aged communities at mesic sites, 316–321
  conditions species survive and develop, 320
  determinants to community composition, 317–318
  herbs and shrubs beneficial to ecologic functions, 318
  patterns changes influenced by physical site conditions and biotic factors, 320–321
  peaks and decreases in biomass, 320
Even-aged community establishment and formation, 314–316
  clearcutting, seed-tree, shelterwood methods physical environment characteristics, 315–316
  crown canopy closing shading ground, 314, *315*
Even-aged methods
  characteristics of, 179, 181
  definition of, 178
  differences between uneven and, 178–179, *180*
  timber harvesting and, 179
Even-aged stand development
  production function as model of, 361–362
Even-aged stands, 28–29
  aspens regenerating by sprouting and suckering, 60
  comparison of stand under silviculture and casual cutting, 508–511
  comparison of uneven-aged and, 28–32
  dynamics among intermediate-aged, 349–352
  intermediate treatments for, 321–323
  lodgepole pine, 60
  northern hardwoods, 60
  phases of development, 190–192
    change in numbers of trees over time, 191, *191*
    patterns of development among unmanaged natural hardwood stand, 192, *192*
  production function depicting development, 187, *188*
  single age class, 29, *32*
  well-managed features of, 508–509
Even-flow sustained yield, 24, *25*
Exhaustible resources, 26–27
Exotic species
  dealing with potential problems, 117–118
  definition of, 75
  problems with, 117
  trees from one geographic area moved considerable distances, 115–116
  use of, 75–78, 115–116
    benefits using, 76–77
    biological advantages, 77
    common problems, 78
    examples of successful, 116–118
    landscaping, from one region to another, 117
    windbreaks and agroforestry in tree-poor regions, 118–119, *119*, 120
Experimental strip treatment, 336
Exploited forests, 33
Extended shelterwood method, 304
Extensive edge, 12
External agents, 62
Extraction, definition of, 45

Factory sawlogs, 429, 429t
Felling, definition of, 45
Fertilization, 87
Fertilizers, use of, 140
Fiber production, 414
Field assessment of site conditions, 120–121
  importance of soil conditions, 121–122
Field storage, 147–148
  heeling-in, 148
Final cutting, 293
Financially mature, 196
Financial maturity, 39–41, 193
  and different product options, 193–194, *194*, 195–196
    culmination of mean annual increment (m.a.i.), 194
    cutting cycle, 194
    financially mature, 196
    mean annual value (m.a.v.), 195
    mean annual yield (m.a.y.), 195
    periodic annual value (p.a.v.), 195
    periodic annual yield (p.a.y.), 195
    production function for multi-aged communities under selection system, *194*
  in Douglas-fir growing, *40*
Financial yield, 39
Fir
  Balsam, 174, 270t
  California red, 270t
  Douglas-fir, 40, *40*, 76, 116, 270t, 396
  Fraser, 270t
  Grand, 211, 270t
  Noble, 82, 270t
  Pacific Silver fir, 270t
  Subalpine, 267, *268*, 270t
  White, 117, 270t
Fire
  as external agent, 62
  its effects in forests, 97–99
  kinds of, 97–98
  use of mechanical, chemical, fire, and biological methods for site preparation, 105t
Fixed-wing aircraft aerial applications, 93–94
Flank fire, 102
Flooding, as external agent, 62
Flow resources, 26, 27
Forestation, 60
  definition of, 59
Forest capital
  definition of, 39

Forest ecosystem maintenance and renewal, 15
Forest fragmentation, 12
  agriculture areas and, 14
Forest management, 8
  regulating stand stability and growth, 248–249
Forest regeneration. *See* Regeneration
Forest regulation, 182
  silviculture and, 181–184
Forestry
  agriculture and, 1–2
  changing context of, 3–6
  defined, 1
Forests
  definition of
    ecologic sense, 13
    ownership meaning of, 13–14
    silvicultural/management perspective, 14
  ownership of, 10
  stands, 2
Forest stand protection
  factors protecting site, 450
  and health management, 450–472
    See also Health management of stand
  health management programs for protecting trees, 451
  reducing accelerated erosion following harvesting, 450–451
Forest stands, 366
  community occupying a specified area, 265
  control of, 17–18
  definition of, 2
  depicting development patterns in, 186–190
    average physical product, 187
    classic function model for input-output relationships, 187, *187*
    culmination of mean annual increment, 189
    marginal physical product, 187
    mean annual increment (m.a.i.), 189
    periodic annual increment (p.a.i.), 188
    physiological maturity, 189, *189*
    production function depicting even-aged stand development, 187, *188*
    standwide balance between accretion, mortality, ingrowth, 188
  determining quality of, 426–428
  essential characteristics, 265
  factors influencing trees and, 460–461
  foresters work with, 3–4
  silviculture and treatments to create and maintain, 183
  spatial arrangement to distinguish from adjacent communities, 265
  sufficient uniformity, 265
  untreated, 6–7
Forest tree improvement, 72–73, *73*, 74–75
Forwarding, 46, 47–48
  definition of, 45
  preventing environmental pollution, 49
Fosamine, 92
Fragmentation, 11–12
  See also Forest fragmentation
Fraser fir, seed dispersal distances, 270t
Free thinning, 391–393
  crop tree release, 391, *392*
French method, 381
Frilling, 92, *93*
Fuelwood, 428, 429t
Full vigor phase of growth, 358
Fully stocked stand, definition of, 366
Fund resources, 26, 27
Fungi, 23
  beneficial, 140
  pathogenic, killing with high temperature, 140
Fungicides, 142
Fusiform rust, 116

Gap-phase replacement, 162, *163*, 222–223
Genetic and species selection
  preventive measure in health management of stand, 463–467
    considering intrinsic site conditions, 465–466
    large-scale monocultures, 466
    risk assessment of management decisions, choices of species, 464, *465*
    wildfire serves as natural means for restoration and renewal in unmanaged forest, *464*
Genotype, 72–74
Geometric thinning, 387
German thinning, 378
Germinated seed, 63, 65
Germination and development, 68–70
Germination percent tests, 141
Germination tests, 138
Ginkgo to northeastern North America, 117
Glyphosate, 92
Gmelina arborea, 64
  grown in Africa and South America, 116
Government control of forests, 10
  silviculture programs in, 10–11
Grades of thinning, 378–379
Grafts, 63
Grand fir, 211
  seed dispersal distances, 270t
Green ash, seed dispersal distances, 270t
Greenhouses, climate-controlled, 123, *123*
Green pruning, 438
Greenspace, 42
Ground fire, 98
Ground methods of cleaning in release treatment, 335
  mechanical methods, 335
  operational strip treatments, 336
Group selection cutting, 222
Group selection method, 222, 226–229
  characteristic conditions, 227–228
  diameter-limit cuttings, 229
  family group regeneration, 226
  limitations with, 228
  shade-intolerant species and, 226–227

Group selection system, 31–32, 222
Group-shelterwood method, preference for, 308
Group-shelterwood systems, 306
Growing stock, 377
Growth, determining, 185–186
  accretion, 185
  assessing, 185
  ingrowth, 185
  mortality, 185
  regulation methods based on growth require data, 185, *186*
Growth form, definition of, 384
Growth-limiting conditions, 120
Gum, seeds ingested as food, 67
Gypsy moth, 473

Hackberry, seed, cached or buried by animals, 67
Hand planting, methods of, 152–156
  survival after, 152, *153*
  survival rate, 155–156
  tools used, 154
  use of dibble, 152, *154*
  use of power auger, 154, *155*
Hardening off, 151
Hardwood stands, patterns of development among unmanaged natural, 192, *192*
Harvest cuts, 502
Harvest cutting
  definition of, 45
Harvesting. *See* Timber harvesting
Harvest scheduling, 182
Health management of stand
  damage, discoloration, decay in living trees, 467–469
  differences between selection system and thinning, 470–471
  effects of noncatastrophic agents, 462–463
    effects of injurious agents, 462
    protective steps, 463
  factors that influence forest trees and stands, 460–461
  integrating, and silviculture, 454–458
  measures for containing logging damage, 471–472
  silviculture and, 450–454
  species and genetic selection as preventive measure, 463–467
  and stand protection, 450–472
  through silviculture, 459–460
    appropriate site preparation, 459
    control of species composition and age class distribution, 459
    intermediate treatments, 459
    managing vegetation community and environmental conditions, 459
    protection measures strategy, 460
    replacement of age class, 460
  trees becoming unhealthy, 461–462
Heeling-in, 148

Hemlock
  dissemination and seed structure, 66
  western, 211
    seed dispersal distances, 270t
Herbicides, 51, 142, 345
  broadcast chemical applications, 330
  chemical treatment shortcomings, 96–97
  conditions for success with, 95–97
  effective control formula, 95
  preemergence, 140
  in site preparation, 90–95
    advantages of, 92
    broadcast spraying, 92–94
    common uses, 94–95
    compounds used, 92
    granular, 94
    reasons for, 91
    soil-active chemicals, 94
    spot and band treatment, 92, 94
    techniques used, 92, *93*
    types of, 91
    use of mechanical, chemical, fire, and biological methods for, 105t
Hexazinone
  granular, 94
  ground-spread granular, 338
Hickory seed
  cached or buried by animals, 67
  dispersal distances, 270t
High forest systems, 63
  conversion from coppice to, 531–532
High thinning, 381
Himalayan pine, 117
Honey locust, seed dispersal distances, 270t
Hophornbeam, dissemination and seed structure, 66
Hornbeam, dissemination and seed structure, 66
Hybrid uneven-aged methods, 229–231
  combination of group selection and single-tree selection, 229
  patch-selection method, 229–230
    formula determining number of trees, 230–231
    implementing, 230
  single-tree selection, 229
Hybrid vigor, 115

IHM. *See* Integrated health management (IHM)
Imazapyr, ground-spread granular, 337
Improvement cuttings, 473–490
  assessing opportunities for, 475–480
  cavities, culls, dead trees as habitat essentials, 482–484
  definitions of, 475
  economic considerations in salvage and, 487–490
  integrating improvement into other treatments, 480–482
  other aspects of, 490
  removal of damaged trees by natural agents, 475, *476*
  role of improvement, 474–475

silvicultural responses to effects of injurious agents, 473–474
Incense cedar, dissemination and seed structure, 66
Incitants (harmful agents), 457
Increment, 182, 183
Induced edges, 11, 12, *13*, 14
Ingrowth, determining growth and, 185, 188
Inherent edges, 11, 14
Insect(s)
attack, as external agent, 62
harmful, 142
maintaining indigenous, 23
pupae, larvae, killing with high temperature, 140
Insecticides, 142
Integrated health management (IHM), 451, 452
activities incorporated in, 452
dealing with unique factors, 453
important benefits of, 453
and silviculture, 454–458
avoiding direct pest control measures, 454
pathologists' hypotheses on decline/dieback of old/large trees, 457
pine plantations of southern North America and, 454
role of predisposing, inciting, contributing agents to tree decline and mortality, 457–458, *458*
using silviculture to reduce susceptibility to harmful agents, 453
Integrated pest management (IPM), 451
important ideas in, 452
See also Integrated health management (IHM)
Intensity of thinning, 377, 397, 409
Intercropping, 120
Interference
definition of, 322
release treatments and, 322
results of, 324
on tree community survival and development, 324, *325*
weeding to reduce, by herbaceous plants, 330, *331*
Intermediate-aged even-aged stands, 349–352
common alternate diameter distributions, 352, *353*
differential rates of height growth within and between species, 349, *350*
three different types of diameter distributions, 351–352
unequal growth of survivor trees increases spread of heights and diameters, 349, *351*
Intermediate crown class, *355*, 356
definition of, 356
Intermediate cutting, 321, 362, 415
Intermediate phase of growth, 358
Intermediate stand treatments, 362
Intermediate treatments, 475
for even-aged stands, 321–323
aggradation phase, 321
composition, definition of, 322
criteria evaluating success of reproduction method, 321
definition of, 321
implemented cultural measures, 322
reducing interference and competition, 322–323
release cutting isolation decisions, 321–322
to temper stand development, 362–364
developing larger crowns and increasing radial growth results in, 363
Internal fragmentation, 11
Interspecific competition, 323
Intraspecific competition, 323
Intrinsic site conditions, 465–466
Intrinsic site factors, *61*, 62
IPM. *See* Integrated pest management (IPM)
Irregular distribution of age classes, 223
Irregular shelterwood method, 304
Isolation
definition of, 322
release cutting isolation decisions, 321–322

Jack pine, 66, 173, 174, 332
seed dispersal distances, 270t
Jack pine budworm, 455
Jeffrey pine, seed dispersal distances, 270t
Jelly-roll bundles, 145, *146*
Juniper, seeds ingested as food, 67
Juvenile phase of growth, 358

Ladder fuels, 98
Landscape, silvicultural systems and, 183
Landscape fragmentation, 11
Larch
dissemination and seed structure, 66
imported species to North America, 116
western, seed dispersal patterns, 267
Latewood, 436
Law of Minimum, 69, 161
Layering, definition of, 514
Leave-tree system, 32, 178, 181
Leguminosae, 67
Leucaena leucocephala, 522
Liberation cutting, 341–345, 347, 480, *481*
definition of, 341
techniques used in, 341
Liebig's Law, 69
Lifting, bare-root stock handling and, 143–145
Light-sensitive herbs and shrubs, 191
Linden, European, 117
Live crown ratio, 202, 356
Loblolly pine, 126, *128*, 173, 211
herbaceous plants and, 317
seed dispersal distances, 270t
seed dissemination by wind, 267
Locust, black, 67
Lodgepole pine, 66, 174, 211, 405, 455
scarification and, 90

Lodgepole pine *(cont.)*
seed dispersal distances, 270t
seed stored in cones, 267
stands, 60
to Sweden, 76
Loessial soil, 48
Logger education and training, 56
Logging, 44
as passive site preparation measure, 82
complying with contract terms, 44
effects of timber harvesting, 46–49
containing extent and severity of disturbance, 48
installing waterbars and grading trails after, 304
protecting natural drainage integrity, 44
protecting soil stability, 44
safe condition of roads and trails, 44
silvicultural requirements of, 45
Logging damage, measures for containing, 471–472
preventing measures, 471–472
Logging debris, 46, 82
Logging slash, 82, 83, 84
Lombardy poplar, 117
Longleaf pine, seed dispersal distances, 270t
Low-density stand, 29
Low forest method, 513
shortcomings of, 526
Low forest systems, 63
Low thinning, 378–380
intensities of thinning from below, 378, *380*
thinning from below, 378, 379t

Machine planting, 156–157, *157*
production rates for, 157
Macroenvironment, 68–69
Macroporosity, 48
Magnolia, 67
M.a.i. *See* Mean annual increment (m.a.i.)
Management
intensity of, 32–33, *34*, 35
silvicultural system as plan for, 19–21
components and character of, *21*
Managing quality in forest stand
determining quality in trees and stands, 426–428
factors promoting quality among trees and stands, 434–436
forest products and factors in grading trees and logs, 429t
marketing and quality factors, 423–426
pruning, 437–440
judging profitability of, 444–446
landscape trees, 446–447
managing pruning as investment, 442–444
and nonmarket values, 447–449
tools and equipment, 440–442
recognizing product quality in standing trees, 429–434
requirements for different wood products, 428–429
Maple
dissemination and seed structure, 66
Striped, 66
Sugar maple
history of prices for, 36, *37*
site preparation for, 82
Marginal physical product, 187
Marketable values, 8
Marketing factors in silvicultural planning and operations, 423–426
aggressive marketing components, 425
conditions affecting forestry valuation decisions, 424
influencing supply and price, 425–426
M.a.v. *See* Mean annual value (m.a.v.)
M.a.y. *See* Mean annual yield (m.a.y.)
Mean annual increment (m.a.i.), 189, 362
Mean annual value (m.a.v.), 195
Mean annual yield (m.a.y.), 195
Mechanical and physical agents, unhealthy trees and, 461–462
fire, 462
flooding, 462
mechanical breakage, 462
pollution, 462
Mechanical methods in release treatment cleaning, 335
Mechanical site preparation, 83, 84–90
with artificial reforestation, 85
bedding, 85, *86*
reasons for, 84–85
undesirable side effects of, 90
use of mechanical, chemical, fire, and biological methods for site preparation, 105t
Mechanical thinning, 387–390
row thinning and, 387, *389*, 390
spacing thinning, 387, *388*
Methods of thinning. *See* Thinning, methods of
Methyl bromide, 140
Microenvironment, 68–69
Microorganisms, 23
Mimosa, 117
Mineral sedimentation, 50
Mineral soil
exposing, disturbing, 46, 48, 49, 80–82, 85, 87–88, 90
Minimum stocking line, 477, *478*
Mistblowing, 92
Mistblowing herbicides prior to seed cutting, 311
Mixed regeneration
advantages, 70
disadvantages, 71
Monocultures, 466
definition of, 466
avoiding, 466
Monterey pine, 76
exporting from California, 116
Mortality, determining growth and, 185, 188, 189
Mortality-caused openings (canopy gaps), 161–162, *162*, 163

Mountain pine beetle, 455
Mowing, in release treatment cleaning, 334
Multiproduct yields in managed natural slash pine stands, 406t
Multiple-use management, 4, 544
Mycorrhizal organisms, 120, 140

NAC. *See* Number of age classes (NAC)
Natural branch shedding, 437
Natural pruning, 437
Natural reforestation, 60
Natural regeneration
  advantages, 70
  choosing between artificial and, 71–72
  disadvantages, 71
  reasons for choosing, 178
Needle casts, 142
New perspectives management, 545
Noble fir, 82
  seed dispersal distances, 270t
Noncommodity aspects of shelterwood and seed-tree methods, 312–313
Noncommodity objectives, 4–5, 10, 33–35, 41–42, 181, 253, 256–261, 282–285, 312–313, 482–484, 544–545
  considerations for serving goals, 262
  intermixing of trees having different ages and heights, *258–259*
  setting maximum diameter for, 260–261
Nonmarket values, 41–42
  nontimber benefits, 23
Nonpoint source, 50
Nonsystems harvest cuttings, 502–503
  qualifying as selective cutting, 503
  varying nature of selective cutting, 503
    primary concerns of, 503
  within the context of selective cutting, 502
Nonsystems timber harvesting, 502
Normal stand, definition of, 366
Normal yield tables, 205
Northern hardwoods
  alternative residual structures to produce large sawtimber with different length cutting cycles for stands, 210, 211t
  clearcutting in, 269
  conifers replaced by, 286
  diameter distribution of uneven-aged upland hardwood stands resembles rotated-S or reverse-J structure, 208, *209*, 210
  effect of individual tree development and stand production within uneven-aged, 244, *245*
  evaluating stand structure and developing marking guide for, 214t, 215
  even-aged stands, 60
  examples of uneven-aged silviculture, 234–235
  fifteen-year development of uneven-aged stand treated by selection cutting, 250, *251*
  prescription-making process based on tree quality and diameter, 218, 219t
  residual diameter distribution
    comparison of precut stand structure with, 215, *216*
    for uneven-aged, under selection system, 209, 210t
    for uneven-aged northern hardwoods, ponderosa pine, Norway spruce based on empirical studies of managed stands, 210, *212*
  unmanaged old-growth hardwood forests of North America, high density European stands managed under, 206, *207*
  using shelterwood method in communities lacking abundant regeneration, 303
Norway maple to northeastern North America, 117
Norway spruce, 76, 210, *212*
  aerial application of herbicide over, 331
  to northeastern North America, 116, 117
  soil conditions and, 121
Number of age classes (NAC), 200
Nursery and tree planting
  bare-root stock handling, 143–145, 147–148
  container stock, 148–152
  cultural practices, 142–143
  field storage, 147–148
  hand planting methods, 152–156
  health management of seedlings, 141–142
  lifting, 143–144
  machine planting, 156–157
  lifting, 143–144
  planting time, 145–147
  protecting seedlings, 141–142
  seed certification, 136
  seed processing and storage, 136–138
    biological potential, 138
    physical characteristics, 138
  seed production, 138
  seed selection and handling, 133–135
    provenance, 133
    seed zones, 134–135, *135*
  soil management, 138–140
  sowing, 140–141
  special planting considerations, 157–158
Nursery lifting, 143
  bare-root stock handling, 143–145

Oak
  northern red, 117
  red, 36, *37*
  seed dispersal distances, 270t
  Shumard, 117
  southern red, 117
Oak communities, 60
  selection system in, 239–240
  three-cut shelterwood method and, 295
  upland, 238–239
Oak seed, cached or buried by animals, 67
Old growth forests/timber, 12
One-cut overstory removal, 295

One-cut shelterwood method, 271, 295
100% relative density, 367
One-step overstory removal, 271
Operational strip treatments, 336
Opportunistic forest, 33
Ordinary thinning, 378
Organic materials, use of, 139–140
Overstocked, definition of, 366
Overtopped crown class, *355*, 356
  definition of, 356
Overwood, 293
  and shelterwood method, 304
Ownership of forested lands, 10–13
  government control 10
    concessions for private corporations, 10
  silvicultural systems and, 183
  timber harvesting, 43–44
    requirements of, 43
Ownership wants, 181–182
  planning management program
    allocation of treatments over time and geographic areas, 182
    matters that affect silvicultural choices,
Oxyfluorofen, 94

Pacific Silver fir, seed dispersal distances, 270t
P.a.i. *See* Periodic annual increment (p.a.i.)
Paper birch, 267
  seed dispersal distances, 270t
Parcelization
  ecosystem and, 14
  private lands, 11
Partial cuttings, 491–512
  comparison of stands under silviculture and casual cutting, 508–511
  deliberate silviculture and conservation, 491–492
  diameter-limit cutting in even- and uneven-aged communities, 504
  exploitation versus silviculture, 511–512
  maintaining an ecologic standard, 492–494
  nonsystems harvest cuttings, 502–503
  selective cuts resembling deliberate silviculture, 504–508
  two-aged silviculture, 494–500
    advantages and disadvantages of, 500–501
  two-aged treatments as pathway to selection silviculture, 501–502
Passes in skidding
  single versus multiple, 49
Passive site preparation, 82–83
Passive site preparation measure, 82
Past sapling stage, 475
Patch and strip cutting methods, 231–233
  ecologic, visual, commodity, and other concerns, 283
  formula to determine area to regenerate, 232
  hybrid methods benefits, 233
  location of patch openings, 231
  patch cutting system requirements, 232
  strip-selection method, 232
Patch-selection method, 229
Patch-selection system, 230
Patch-thinning treatments, 231
P.a.v. *See* Periodic annual value (p.a.v.)
P.a.y. *See* Periodic annual yield (p.a.y.)
Peat moss, 145
Perimeter fire, 102, *103*
Periodic annual increment (p.a.i.), 188–189, 190, 362
Periodic annual value (p.a.v.), 195
Periodic annual yield (p.a.y.), 195
Period of stand initiation, 190, 321
Period of stem exclusion, 190, 319
Pest, definition of, 451
Pesticides, 51
Pest management, 451
Phacidopycnis pseudotsuga, 438
Phenotype, 63, 72, 73
Phenoxy, 92
Physical and mechanical agents, unhealthy trees and, 461-462
Physical characteristics of seed, 138
Physical environment (site), 2, 45
  climatic features, 120
  edaphic features, 120
  topographic features, 120
Physiological maturity, 189, *189*, 455, 457
Picloram, 92
Piling logging slash, 84
Pin cherry, 66
Pine(s)
  blister rust and western white, 120
  Caribbean, 116
  Himalayan, 117
  Jack, 66, 173, 174, 332
  with large seeds, small wings, cached or buried by animals, 67
  loblolly, 126, *128*, 173, 211
  lodgepole, 60, 66–67, 76, 90, 174, 211, 267, 405, 455
  Monterey, 76, 116
  most, dissemination and seed structure, 66
  pitch, 270t
  ponderosa, 117, 210–211, *212*, 237–238, 405
    seed dispersal distances, 270t
    seed dissemination by wind, 267
  red, 121, 173
  scarification and red and white, 90
  Scotch, 76, 116, 117
  seed dispersal distances, 270t
  shortleaf, 211
    seed dissemination by wind, 267
  slash, 46, 126, 127, *128*, 173, 316
    seed dissemination by wind, 267
  southern, 239–240
  white, 173
Pine webworm, 84
Pioneer species, 345

Pitch pine, seed dispersal distances, 270t
Plant(s)
  clearcutting and protection of threatened and endangered, 289
  community attributes, 5–6
  even-aged stands and, 37–38
  improving native, 23
Planting
  special considerations, 157–158
    coarse-texture soils, 158
    moisture content of soil, 157–158
    optimum day for, 157
    threats to plantation, 158
  time of, 145–147
    bare-root stock, 147
    temporate zones, 147
    tropical and subtropical areas, 147
Planting plan, 112
  major factors to consider, 112
Planting stock, 122
Plantlets, clonally propagated, 63, 64
Plant parasitic nematodes, 454
Plant succession
  different processes causing change, 162
    major (human or natural) disturbance altering community, 162–163
    new trees developing underneath canopy gap, 162
  and forest regeneration, 161–163
  and gap phase replacement, 162, *163*
  Law of Minimum and, 69, 161
  replacement of one seral stage by another, 162
  series of changes after death of large tree, 162, *163*
Platanus, 519
Plus trees, 73
Point source, 50
  truck roads, 53
Poles and pilings, 429, 429t
Pollution
  addition of solid materials, 50
  changes in water temperature, 50
  classification of, 50
  debris, 50
  foreign chemicals, 50
  mineral sedimentation, 50
  preventing, 49–53, *54–55*
Ponderosa pine, 117, 210, 211, *212*, 237–238, 405
  seed dispersal distances, 270t
Poplars, 94
  dissemination and seed structure, 66
  hybrid, 64
  seed dispersal distances, 270t
  Yellow, 66
Populus, 115, 517, 519
  See also Aspens
Porosity, 48
  texture soil, 46
Postcutting germination, 318
Postlogging slash burns, 311
Postlogging stabilization, 53
Posts, 429, 429t
Power auger, 154, *155*
Precipitation, 46
Preemergence herbicides, 140
Preparatory cutting, definition of, 293
Presalvage cutting, 485
Prescribed burning, 99–100, 311
  firing methods for, 102–104
  maintaining control, 100
  methods for controlling effects of, 100–102
  reasons for, 99
  shortcomings of, 102, 104
Prescribed fire, 83
Prevention silviculture strategies used in pine forests, 456
Production function
  as model of even-aged stand development, 361–362
    ecologic and economic factors, 361
  for multi-aged communities under selection system, *194*
  for thinned stands, 364–365
    comparison with unmanaged stands, 364
    increasing yield of usable sawtimber volume, 364
    showing effects of stocking levels and patterns, 364, *365*
Product quality in standing trees, recognizing, 429–434
  decline-causing defects, 430
  defects that degrade trees, 430–431, *432–433*
  external indicators, 431, *432*
  high-value trees, 431
  internal defects, 431
  lumber for construction, 431–432
  value-affecting defects, 430
Progeny tests, 134
Progressive-patch method, 277, *277*
Progressive strip backfires, 102
Progressive strip headfire method, 102, *103*
Progressive-strip method, 277, *277*, *278*, 279
Provenance, 133–134
Pruning, 437–440
  appropriate technique, 438, *439*
  autumn pruning in Sweden and increase of *Phacidopycnis pseudotsuga,* 438
  definition of, 438
  dry, 438
  judging profitability of, 444–446
    example of compound costs from pruning 10-inch conifer, growing another 30 years to merchantable size, 446, 446t
  landscape trees, 446–447, *448*
  managing pruning as investment, 442–444
    better sites, fully stocked stands, 443, *444*
    factors for choosing crop trees, 442
    number crop trees formula, 443
  natural, 437–438
  and nonmarket values, 447–449

Pruning *(cont.)*
tools and equipment, 440–442
long-handle sectional pruning saws, *441*
Pulpwood, 428, 429t
Pythium, 142

QSD. *See* Quadratic stand diameter (QSD)
Q structures, 205–208, 210–211, 253–257, *258–259*, 260–265
Quadratic stand diameter (QSD), 367, 388, 399, 405
Quality factors in forest stands
conditions affecting forestry valuation decisions, 424
promoting, 434–437
factors determining quality at reproduction method cutting time, 436–437
hardwood and conifer species lumber quality factors, 434
indirect measures to promote crop-tree quality, 437
patterns of wood formation in growing trees, 434–435, *435*
in silvicultural planning and operations, 423–426
Quality in forest stand, managing. *See* Managing quality in forest stand
Quality in trees and stands, determining, 426–428
from buyer's perspective, 427
from technical or engineering perspective, 427
offering highest market value to landowner, 427
selectivity, 427
using standard grading and measurement system, 428
Quality requirements for different wood products, 428
end-point market requirements, 428
forest products and factors in grading trees/logs for those uses, 428, 429t

Recreation, even-aged stands and, 37–38
Recruitment, 176
Red alder, seed dispersal distances, 270t
Red-brown butt rot (*Phaeolus schweinitzzi*), 468
Redcedar, western, 211
seed dispersal distances, 270t
Red gum, dissemination and seed structure, 67
Red ironbark, 117
Red oak
historic price data and, 36, *37*
Red pine, 121, 173
seed dispersal distances, 270t
Red spruce, seed dispersal distances, 270t
Redwood
dissemination and seed structure, 67
seed dispersal distances, 270t
Reforestation, 60
definition of, 59
ecological and economical factors, 175
Regeneration, 17
advantages and disadvantages of different methods, 70–71
artificial methods, 17
as process, 60–62
chain of events, *61*
seed regeneration requirements, 62
choosing between natural and artificial, 71–72
concepts of, 58–78
defining term, 176–177
different processes causing change, 162
major (human or natural) disturbance altering community, 162–163
new trees developing underneath canopy gap, 162
direct seeding for, 163–164, *164*, 165–166
from direct seeding, 159–174
from seed, 65–68, 159–174
germination and development, 68–70
guidelines for judging success, 327
impediments to success following seed dispersal, 62–63
implications of silvicultural practice to forest tree improvement, 72–75
insuring timely and effective, 58–60
natural and artificial, 63
objectives of forest, 58
plant succession and, 161–163
seed source selecting, 78–79
using exotic species, 75–78
and vegetation surveys, 325–327
by vegetative methods, 64–65
Regeneration area, 177
Regeneration cutting, 176
Regeneration method, 79
Regeneration openings, 222
Regeneration period, 177
Regeneration phase, 321
Regulated forests, 35
Regulatory aspects of silviculture, 182
Reinforcement plantings, 70
Relative density
controlling through thinning, 372–375
to control thinning, 419, 421
definition of, 366
gross growth constant with, 372, *373*
in unthinned shands, 365–372
Release cutting, 321
broad economic considerations, 345–348
early treatment, 328–329
cleaning, 329, 332–340
weeding, 329, 330–332
liberation cutting, 341–345
response of crop trees to, 340–341
Release treatments, 328
cleaning, 329, 332–340
all chemical methods, 337
basal sprays and stem injection, 338
broadcast chemical methods, 336, 338
common methods of, 333–334

during sapling stage, 329
experimental strip treatments, 336
ground methods directing herbicides between rows of crop trees, 335
ground-spread granular, 338
mechanical methods, 335, 337
operational strip treatment, 336, 337
releases saplings of desirable characteristics by removing overtopping trees, 333
removing undesirable trees from sapling stands, 332, *333*
spot treatments, 337, 338
tree cutting, 338
woody vines and shrubs removal, 329
objections of, 329
weeding during seedling stage, 329, 330–332
reducing competition and interference by herbaceous plants, 330, *331*
Removal cutting
definition of, 293
and seedling damage, 308–310
containing regeneration losses, 309
harvesting plan provisions, 310
logging damage, 308
skidding or falling tree damage, 308–309
Reorganization phase, 190, 321
release treatments and, 329
Reproduction, defining, 176
Reproduction method cutting, quality determining factors at, 436–437
Reproduction methods, 79–80, *81*, 175–197
characteristics of different, 179, *180*, 181
depicting development patterns in forest stands, 186–190
determining the growth, 185–186
difference between even- and uneven-aged methods, 178–179
ecological and economical factors, 175
financial maturity and different product options, 193–196
patterns of uneven-aged stand development, 192–193
phases of even-aged stand development, 190–192
physical site conditions required, 176
reasons for choosing artificial methods, 177
reasons for choosing natural methods, 178
regulatory considerations, 181–184
relationship to regeneration strategy, 196–197
and their role in silviculture, 176–178
values and limitations of clearcutting as, 289–291
advantages of, 289–290
limitations of, 290–291
See also Uneven-aged reproduction methods
Reserve cutting, 299
Reserve shelterwood method, 304–305
Reserve shelterwood system, 32
operational and financial disadvantages, 306
Reserve trees, 299
criteria in choosing, 306
Residual density, 242, 243
integrating cutting cycle length, structure, and, 252–256
changing selection system stand conditions, *255*
relationship between stand density and growth, *254*
on species composition, effect of, *226*
Residual diameter distribution, 211
comparison of precut stand structure with, 215, *216*
for uneven-aged northern hardwoods under selection system, 209, 210t
for uneven-aged northern hardwoods, ponderosa pine, Norway spruce based on empirical studies of managed stands, 210, *212*
Residual overstory, 293
Residual shelter, 293
Residual trees, 293
determining appropriate level of residual stocking, 301–304
formula determining average number of seed trees to leave, 302
key characteristics, 301–302
seed cutting factors, 302
Revegetation, 46
Reverse-J curves, 203–204, *204*
Reverse-J distributions, 352
Rhizoctonia, 142
Ring fire, 102, *103*
Road surfaces, 53
effect of different kinds of surface obstructions, 53, *54*
effect of different surface treatments, 53, *55*
Root collar, 142
resprouting from, 309
Root disease, 142, 468
Annosus root rot (*Heterobasidion annosum*), 468
black stain (*Leptographion wageneri*), 468
red-brown butt rot (*Phaeolus schweinitzzi*), 468
Rooting depth, 121
Root pruning, 143
Root suckers, coppice method based on, 522–524
Root wrenching, 143
Rotation, 28
Roundwood, 429
Row mulching, 84
Row seeding, 169, 170
Row thinning, 387–388, *389*, *390*
Rubus, 191
early development of even-aged communities, 316, 319
Russian mulberry, 117
Russian-olive, 117
Rusts, 142
Rutting, 48

Salvage cuttings, 473–474, 484–486
economic considerations in improvement and, 487–490

Salvage cuttings *(cont.)*
other aspects of, 490
recovering timber value in trees harmed by injurious agent, 482, *484*
regeneration after, 486–487
silvicultural responses to effects of injurious agents, 473–474
Sanitation cuttings, 473–474, 484–486
aspects of, 490
definitions of, 485
regeneration after, 486–487
silvicultural responses to effects of injurious agents, 473–474
Sawtimber
accelerating diameter increment without permanently breaking canopy, 415
increasing yield of usable, 364
management schemes for, 414–417
purpose of thinning, 415
Sawtimber rotation, 365, *365*
Scarification, 81–82
and regeneration of several conifers, 90
of seed, 137
using heavy barrellike devices, 88, *89*
Scotch pine, 76, 117
as Christmas trees, 116
seed dispersal distances, 270t
Seed banks, 65
character and dynamics of, *66*
Seedbed density, effects of, 143
Seedbed preparation, 80–81, 85, 140, 163, 166
See also Scarification
Seed certification, 136
Seed cutting
creating permanent openings in main crown canopy, 293, *294*
definition of, 293
oak regeneration and seedling density encouraged by, 311–312
Seed dispersal distances, 267, 269
for bulk of seed blown downwind, 270t
Seed dispersal patterns, 267
Engelmann spruce, 267, *268*
subalpine fir, 267, *268*
western larch, 267
Seed dissemination, 66–67
airborne, 66–67
animal carried, 67
buried by, 67
ingested, 67
floating on water, 66
wind, 67
loblolly, shortleaf, slash pines, longleaf pines, ponderosa pine, paper birch, 267
Seeding cutting, definition of, 293
Seedling(s), 79, *80*
common diseases of older, 142
conditions/quality, 122
container-grown, 123
damage, removal cutting and, 308–310
dry winds and, 142
health management, 141–142
high temperatures and, 142
judging successfully established, 323–324
quality, factors related to, 125, *125*, 126
packaging, 145
protection, 141–142
regulating, 123–124
storage temperature, 145
time of planting, 145–147
weeding during seedling stage, 329
Seedling production, 138
Seedling transplants, 122
Seed lot, 136
Seed orchards, 73
Seed-origin regeneration, securing, 266–267
Seed processing and storage, 136–138
Seed production areas, 73
Seed rain (falling), 67, 267
Seed regeneration, 65–68
requirements, 62
sexual reproduction, 63
Seeds
angled, with aerodynamic shape, 67
with large terminal wings that whirl, 66
small, with cottonlike tufts, 66
small, with large marginal wings that glide, 67
small, with small marginal wings that glide, 67
with small terminal wings that do not whirl, 66
with specialized bracts, 66
Seed selecting and handling, 133–135
Seed source and advance regeneration, 267–271
Seed source selecting, 78–79
Seed-spot method, 170
Seed tree(s)
choosing, 300
definition of, 297
individual characteristics affecting sturdiness, 300–301
phenotypic characteristics of species, 300
physical site characteristics influencing anchorage, 300
physiographic features affecting wind velocity, 300
seed cutting leaving widely scattered trees of excellent phenotypes, 304, *305*
for shelterwood and seed-tree cutting, 299–301
windfirmness and, 300
withstanding blowdown and exposure after seed cutting, 300
Seed-tree method
as means to insure adequate seed supply, 296, *298*
characteristics of, 295–299
definition of, 297
determining appropriate level of residual stocking, 301–304
insuring adequate seed supply across regeneration area, 296, *298*
noncommodity aspects of, 312–313
removal cutting and seedling damage, 308–310

selecting seed trees for seed-tree cutting, 299–301
and shelterwood method, rationale for alternate even-aged reproduction methods, 292
site preparation with, 310–312
Seed-tree system, 31, 181
definition of, 297
northern hardwoods, 60
See also Even-aged methods
Seed zones, 135, *135*
Selected seed, 136
Selection cutting, 199, 222
contrasts between selective and, 246–248
expected changes in uneven-aged stand diameter distribution following repeated diameter-limit cutting, *247*
evaluating stand structure and developing marking guide for, 214t, 215
individual tree development and stand production within uneven-aged northern hardwood, effect of, 244, *245*
Selection method, 178, 179, 200, 221
See also Uneven-aged methods
Selection system, 198–220, 222, 229
applying, 213–214
balanced structures and other guides, 203–205, 209–213
alternate residual structures to produce large sawtimber with different length cutting cycles for northern hardwood stands, 210, 211t
diameter distribution of uneven-aged upland hardwood stands resembling rotated-S or reverse-J structure, *209*, 210
residual diameter distributions for uneven-aged northern hardwoods, ponderosa pine, Norway spruce, 209, 210, 210t, 211, *212*
characterizing conditions among stands, 201-205
age class distribution, 203
crown length ratio, 202
diameter class distribution, 203
live crown ratio, 202
production function, composite of multiple age class development from concurrent regeneration and tending treatments, 201, *202*
reverse-J curves, 203–204, *204*
structural features of uneven-aged stand containing four distinct age classes, 202, *203*
character of, 198–201
creating uneven-aged community, 200
cutting series, 199
maintaining uneven-aged community, 200
number of age classes (NAC), 200
selection cutting, 199
use of timber harvesting to regenerate new age class, 198, *199*
working circle, 199
defining residual structure, 205–20
check method, 205
equal area per age class (balanced), 205
normal yield tables, 205
Q structures, 205–208, 210–211, 253–257, *258–259*, 260–265,
unmanaged old-growth hardwood forests of North America, high density European stands managed under check method, 206, *207*
differences between thinning and, 470–471
enhancing quality and value, 217–220
acceptable growing stock (AGS), 219
crown position, 217
marking for timber management goals, 217–218
maturity, 217
prescription-making process based on tree quality and diameter, 218, 219t
risk, 217
soundness, 217
species, 217
stem form, crown size, branching habit, 217
truncating residual structure by removing high-risk trees, 219, *220*
unacceptable growing stock (UGS), 218–219
vigor, 217
an example of, 250–252
fifteen-year development of uneven-aged northern hardwood stand treated by selection cutting, 250, *251*
preparing marking guide, 214t, 215–217
Arbogast guide, 215–219
comparison of precut stand structure with residual diameter distribution, 215, *215*
procedures for evaluating stand structure and developing marking guide for selection cutting, 214t, 215
Selection system stands growth and development, 241–262
advantages of selection silviculture, 241–242
assessing stand structural change, 249–250
contrasts between selective and selection cutting, 246–248
expected changes in uneven-aged stand diameter distribution following repeated diameter-limit cutting, 246, *247*
vertical structural diversity, 248
disadvantages of selection silviculture, 242–243
effects of stand structure control on tree growth rates, 245–246
well-planned network of permanent skidding corridors, 243, *244*
example of, 250, *251*, 252
features of, 509
integrating cutting cycle length, residual density, structure, 252–256
maximum diameter for noncommodity objectives, 260–261
noncommodity objectives, 256–260
intermixing trees having different ages and heights, 257, *258–259*
other considerations for serving noncommodity goals, 262

Selection system stands growth and development (*cont.*)
regulating stand stability and growth, 248–249
requirements for success in, 243–245
effects of selection cutting on individual tree development and stand production, 243, *245*
Selection thinning, 384–386
and diameter-limit cutting, 384, *385*
removing trees larger than specified diameter, 505, *505*
growth form, 384
Selective cutting, 200, 229, 502–508
contrasts between selection and, 246–248, 506, *507*
expected changes in uneven-aged stand diameter distribution following repeated diameter-limit cutting, *247*
nonsystems harvest cuttings, 502–503
primary concerns of, 503
qualifying as, 503
varying nature of, 503
resembling deliberate silviculture, 504–511
features common to reproduction method cutting, 505
uneven-aged stands given diameter-limit or sawtimber-only cuts develop distorted diameter distributions, 506, *507*
within the context of, 502
Selectively cut stands, 510
common features of, 510
irregular distribution of residual trees, 510, *511*
Selectivity, differentiating between trees/logs of high/low end-use value, 427
Senescence phase of growth, 358
Sequential cropping, 120
Sexual reproduction
from germinated seed, 63, 65
Shearing, 87
Shelterbelts, 118, *119*
Shelter trees, 293
Shelterwood cutting, definition of, 293
Shelterwood method
characteristics of, 292–295
as desirable species develop across site, removal cutting takes away mature age class, 295, *296–297*
seed cutting and permanent openings in main crown canopy, 293, *294*
component treatments in, 293
definition of, 293
difference between clearcutting and, 292–293
noncommodity aspects of, 312–313
removal cutting and seedling damage, 308–310
reserve, group, and strip, 304–308
and seed-tree method, rationale for alternate even-aged reproduction methods, 292
selecting seed trees for, 299–301
site preparation with, 310–312
Shelterwood system, 31, 178
definition of, 293
shelterwood method one part of, 293
See also Even-aged methods
Shingle toe, 145
Shoot-root radio, 143
Shoots, 63
Shortleaf pine, 211, 267
seed dispersal distances, 270t
Short-rotation biomass plantation advantages, 526
Short-rotation coppice systems, 518–522
bioenergy plantations of shrub willows, 519, *520*
shortcomings of schemes, 527
yield from six willow cloves, 520, *523*
See also Coppice; Coppice method; Coppice silviculture; Coppice system
Shumard oak to northeastern North America, 117
Siberian elm, 117
Sierra Nevada mixed conifers, 211
Silage cellulose, 519
Silage sycamore, 519
Silvics, definition of, 7
Silvicultural planning and operations
marketing and quality factors in, 423–426
Silvicultural practice, factors that force unscheduled changes in, 540–541
Silvicultural system, 19–42
adjusting management intensity, 32–33, *34*, 35
alternate views of, 21, *22*
character of, 21–23
conservation and, 25–28
even- and uneven-aged systems, comparison of, 28–29, *30–31*, 31–32, *32*
financial maturity, 36, *37*, 37–41
management of stands across ownership or landscape, 183
management plan, 19–21
components and character of, *21*
nonmarket values, 41–42
objectives of, 21–23
resources use ethics, 25–28
sustaining yield of multiple values, 23–25, *25*
thinning regimes in even-aged, 395–422
Silviculture
defined, art and science, 7–8
basic science foundation, contributing to forest resources management, *9*
definition of, 1, 8, 16, 533
deliberate silviculture and conservation, 491–492
ecologic perspective, 13–15
ecosystems, 10–13
exploitation versus, 511–512
forestry and, 1–2
from administrative perspective, 536–538
demands that alter conduct of existing silvicultural programs, 538
demands that influence initial decisions, 538
understanding requirements, 536, *537*
future challenges, 13
harvesting and, 43–57
health management through, 459–460
integrating health management and, 454–458

new philosophy process of, 15
no-cutting recommendation, 6–7
objectives and costs, 8, 10–13
intangible values, 10
management objectives, 8
tangible values, 10
ownerships, 10–13
in perspective, 533
scheduling silvicultural treatments to create diversity of age class and vegetal conditions, 533, *534*
philosophy of, 2–3
administrative requirements, 2–3, 7
biologic factors, 2
ecologic factors, 2–3, 7
economic requirements, 2–3, 7
problem-solving procedure for, 533–536
steps involved in, 535–536
regulatory considerations in, 181–184
reproduction methods and their role in, 176–178
in retrospect, 548–549
silviculturist role, 2, 4–5, 16–17
and stand health, 450–454
tree planting in, 106–112
technology of, 2
Silviculturist role, 2, 4–5
controlling establishment, composition, growth, 17–18
from ecologic perspective, 16
from economic perspective, 16
primary functions, 17
stand or age class development treatments, 18
Simazine, 94
Simulated shelterwood cutting, 295
Single-tree method, 222
Single-tree selection method, 224
alternative approaches to, 224–225
effect of residual density on species composition, *226*
Single-tree system, 31
Site(s) (physical environment), 2
field assessment, 120–121
protecting, 23
Site preparation, 79–105
active, 83– 84
clearcutting and other ancillary treatments, 285–286
defining, 79
fire and its effects in forests, 97–99
mechanical, 84–90
passive, 82–83
postharvesting weed control, 285
prescribed burning, 99–100
controlling effects of, 100–102
firing methods for, 102–104
reasons foresters use, 80–81
role and scope of, 79–82
selecting, 104–105
assessment of factors, 104
with shelterwood and seed-tree methods, 310–312
timing and juxtaposition of, 80, *81*
use of mechanical, chemical, fire, and biological methods for, 105t
using herbicides in, 90–95
conditions for success with, 95–97
Sitka spruce, 82
exported to Great Britain, 76, 116
seed dispersal distances, 270t
Skidding, 46, 47, 48–49, 82
clearcutting and, 289
definition of, 45
preventing environmental pollution, 49
Skid trail, 243, *244*, 289
Slash, 46
See also Logging slash
Slash pine
multiproduct yields in managed stands, 406t
reasons for decline of direct seeding, 173
relationship between planting spacing and
crown ratio over time for, 126, 127, *128*
quadratic median DBH over time for loblolly and, 126, *128*
seed dispersal distances, 270t
seed dissemination by wind, 267
Snags (dead standing trees), 482
Snow, early season melt, 46
Society of American Foresters
silviculture definition, 16
Soil
conditions, importance of, 121–122
drainage, 46
protecting stability of, 44
goals to insure, 45
protection and clearcutting, 289
protective cover for, 38
removing organic surface layer, 46
safeguarding stability, 54
saturation after tree removal, 46
site disturbance by machines compacting, 47–48
stabilizing, 23
texture, porosity, 46
Soil-active chemicals, 94
Soil erosion. *See* Erosion
Soil management, 138–140
best location requirements, 139
seedbed preparation, 140
site condition factors, 138
use of fertilizers, 140
use of organic materials, 139–140
Soil surface layers and site preparation, 81–82
Soil texture, 121
Source-identified seed, 136
Southern pine, 239–240
Southern pine beetle, 455
Sowing, 140–141, 169–171
aerial seeding, 169
broadcast, 169
germination percent tests, 141
ground methods, 169
in North America, 140–141

Sowing *(cont.)*
seed-spot method, 170
sowing guides, 170, *171*
spot seeding, hand or mechanized, 171, *172*
Sowing depth, optimal, 141
Sowing rate, 136–137, 138
Spacing, selecting, 126–129
arrangement and operational considerations, 129–130
cost factors, 128
determining factors, 126
minimum number of trees factors, 127
relationship between planting spacing and
crown ratio over time for slash pine, 126, 127, *128*
quadratic median DBH over time for loblolly and slash pine, 126, *128*
relationship between stocking and Girard form class and stem taper for loblolly pine, *128*
Spacing thinning, 387, *388*
Species selection, 113–115
planted trees/seedlings versus seed distribution, 113–114
preventive measure in health management of stand, 463–467
Sphagnum moss, 145
Spot fire technique, 102, *103*
Spot seeding, 167, 169
formula for rates of seeding, 168–169
Spot (or band) treatment, 83, 84, 92, 94
advantages and disadvantages with, 84
mechanical and chemical methods, 337
Spot weeding, 330, *331*
Sprayers
portable, 84
tractor-mounted, 84
Spring wood, 436
Sprout method, definition of, 514
Sprouts, 63, 64
Spruce
black, 173, 332
Colorado blue, 117
dissemination and seed structure, 66
Engelmann, 82, 90, 267, *268*
scarification and red and white, 90
seed dispersal distances, 270t
Sitka, 76, 82, 116
white, 66
Spruce budworm, 455
Spruce-fir
eastern, 211, 236–237
western, 211, 235–236
Stand density, definition of, 366
Stand development, thinning effects on, 349–375
Stand protection and health management, 450–472
See also Health management of stand
Stands. *See* Forest stands
Standscape, 260, 262
Stand stability and growth, regulating, 248–249
Stand structural change under selection system, assessing, 249–250
Stand structure control on tree growth rates, effect of, 245–246
Stand volume production, effects of thinning on, 410–413
comparisons of row thinning and thinning from below, 413
long-term potential of thinning program, 410, *410*
sources of added volume realized, 411, *412*
Steady-state phase, 190
Stem cankers, 142
Stem injection method, 338
Stocking
definitions of, 365
guides showing patterns of stand development, 367, *368–369*, 370
and relative density among unmanaged stands, 365–372
A-level curves representing 100% relative density for stands, 370, *371*
change of standing volume and cumulative gross production in, 367, *370*
Stool, definition of, 514
Stool shoot, 515
definition of, 514
Storms, heavy, 46
Stratification
of seed, 137
of species by life zones, 120
Stratify seed, 137
Strip backfire technique, 102
Strip burning, 102
Strip clearcutting
alternate-strip method, 277
choices between patch, large block, or, 281–282
and patch methods, 231–233, 283
as useful technique for snow management, 283, *283*
Striped maple, 66
Strip-selection method, 232
Strip-shelterwood method
narrow strip cuts and, 306, *307*
reserve, group, and, 304–308
Strip-shelterwood systems, 306
Strip thinning, 390
Stumpage value, 41
Stump sprouting, 64
coppice methods based on, 515–518
Subalpine fir
seed dispersal distances, 270t
seed dispersal patterns, 267, *268*
Subsoiling, 87–88
Succession, definition of, 161
See also Plant succession
Suckers, 63, 64
definition of, 514
Sugar maple, 82
history of prices for, 36, *37*

seed dispersal distances, 270t
See also Maple, Sugar
Sugar pine, seed dispersal distances, 270t
Sulfometuron, 94
Summer wood, 436
Surface fires, 97–98
Surface/ground, effect of environmental conditions near, 271–275
clearcutting and slash to, 272
evapotranspiration and clearcutting, 273
humidity and, 272
macro- and microenvironment changes, 272
moisture in upper soil, 273, 275
organic materials and, 272, 273
temperature changes, 271–172, 275
slope and aspect amplify, 273–274, *274*
Sustainable development, 545
Sustained yield, 24, *25*, 182–183
continuous production, balance between increment and cutting, 182
enhancing intangible values, 23
essential ecosystem conditions, 24
forest producing regularly and continuously, 182
providing commodities crops, 23
Sweetgum, 117
seed dispersal distances, 270t
Sylvo-pastoral systems, 119

Tamerack, seed dispersal distances, 270t
Target seedling, 122
Temperatures, high
damage to seedlings, 142
killing weed seeds, pathogenic fungi, insect pupae, larvae, 140
Terracing, 84
Thinned stands, 395–397
Aspen stands in Canada, 396
Douglas-fir stand in Pacific Northwest North America, 396
to temper development of stand, 395
to temper successional trends, 395
Thinning
differences between selection system and, 470–471
effects on stand development, 349–375
controlling relative density through, 372–375
crown classes and silviculture importance, 352–358
intermediate-aged even-aged stands, and, 349–352
intermediate treatments to temper stand development, 362–364
for older even-aged stands, 363
potential to produce high-quality trees, 363
production function as model, 361–362
production function for thinned stands, 364–365
stocking and relative density among unmanaged stands, 365–372
tree aging implications, 358–361
without permanently breaking crown canopy, 363
methods of, 376–394
comparing intensity and, 376–378
controlling spacing and serving diverse landowner values, 394
crown thinning, 380–384
free thinning, 391–393
low thinning, 378–380
mechanical thinning, 387–390
selecting a method, 393–394
selection thinning, 384–387
setting thinning interval, 420–422
stocking guide to control intensity and frequency of thinning, *420*, 421
timing first thinning, 417–420
Thinning from above, 380
Thinning from below, 378, 379t, *380*
Thinning gaps, 223
Thinning of dominants, 384
Thinning regimes in even-aged silvicultural system, 395–422
characteristic responses among thinned stands, 395–397
definition of, 416
effects of thinning on diameter of tree of mean basal area, *407*
effects of thinning on stand volume production, 410–413
influencing yield by thinning, 397–410
long-term development of cubic volume in upland oak stand, *404*
management schemes for fiber production, 414
management schemes for sawtimber, 414–417
multiproduct yields in managed slash pine stands, 406t
setting a thinning interval, 420–422
timing a first thinning, 417–420
for upland oak stands, 402t–403t
Thinning schedule, 416
Three-cut shelterwood method, 295
oak communities and, 303
Tie logs, 429, 429t
Timber harvesting, 502
best management practices, preventing environmental pollution, 49–53, *54–55*
controlling erosion potential, 54–56
effects of, 46–49
from silviculturist perspective, 44
nature of, 44–45
protecting special habitats, 56–57
protecting vegetation, 56
to regenerate new age class, 198, *199*
silvicultural requirements of logging, 45
and silviculture, 43–44
Timber management scheduling, 182
Timbers for construction, 429, 429t
Timber stand improvement, 415
Tissue culture techniques, 64, *65*

Tools and equipment for pruning, 440–442
long-handle sectional pruning saws, *441*
Tools used for hand planting, 154
Topography, 46
Topping, definition of, 45
Tractor-drawn toothed disks, 84
Tractor-mounted sprayers, 84
Transition phase, 190
Transplants, 122
Tree aging, implications of, 358–361
growth patterns between individual trees, 358, *359*, 360
Tree-area ratios, 367
Tree community
survival and development, interference on, 324, *325*
and vegetation development, 318
Tree planting, 106–108
common circumstances for, 106–107
factors influencing artificial regeneration program, *108*
nursery and, 133–158
protection and health management, 141–142
plan, 130
check list of factors to consider, 131t
maps needed, 130
written documents, 132
planning project, 108–109, 130
critical factors to consider, 108–109
primary use for, 106
purposes and advantages over natural methods, 109–112
justifying costs, 111–112
noncommodity reasons, 109–110
recreation, parks, greenbelts, 110
requirements, 112–113
species selection, 113–115
using exotic species, 115–116
examples of, 116–118
Trees
becoming unhealthy, 461–462
biologic factors, 461
mechanical and physical agents, 461–462
determining quality of, 426–428
genetically engineered, 115
living, damage, discoloration, decay in, 467–469
Tree seedlings, 79, *80*
Tree trucks for posts, poles, pilings, 249, 249t
Tricopyr, 92
Truck roads, design and construction of, 53
Tulip-poplar, 117
See also Yellow-poplar
Tulip-tree, seed dispersal distances, 270t
seed dispersal distances, 270t
Two-aged community
features of, 509–510
with two distinct canopy layers, 304, *305*
Two-aged method, 178–179
definition of, 178
timber harvesting and, 179
Two-aged silviculture system, 29, 31–32
advantages and disadvantages of, 500
two-storied or two-aged stand advantages, 500–501
two-storied or two-aged stand disadvantages, 501
creating a two-aged community resembling shelterwood seed cutting, 498
deliberate control of residual tree density and site preparation measures, 498
diameter-limit cutting, 498
leaving low density of older reserve trees, 492, *496*
lowering potential for epicormic branching, 499
reserve shelterwood method, 494, *495*
primary characteristics in selecting trees, 494
reserve shelterwood stands with two-layered structure, 497
stratified mixed-species communities, 497
treatments as pathway to selection silviculture, 501
converting even-aged communities to multiaged arrangement, 501–502
Two-cut shelterwood method, 295

UGS. *See* Unacceptable growing stock (UGS)
Unacceptable growing stock (UGS), 218, 372
Unbalanced, irregular distribution of age classes, 223
Unclassified seed, 136
Uneven-aged methods
characteristics of, 179, 181
definition of, 178
differences between even and, 178–179, *180*
timber harvesting and, 179
Uneven-aged reproduction methods, 221–240
group selection method, 226–229
hybrid uneven-aged methods, 229–231
patch and strip cutting methods, 231–233
in selection system, 221–222
group selection method, 222
single-tree selection method, 222
single-tree selection method, 222–224
alternative approaches to, 224–226
uneven–aged silviculture examples in systems context, 234–240
eastern spruce-fir, 236–237
northern hardwoods, 234–235
ponderosa pine, 237–238
southern pines, 239–240
upland oak communities, 238–239
western spruce-fir, 235–236
Uneven-aged stands, 28
comparison of even-aged and, 28–32
creating community, 200
maintaining community, 200
patterns of development, 192–193
northern hardwoods, 60
structural features containing four distinct age classes, 202, *203*

User fees, 41

Value added in manufacturing, 42
Value-affecting defects, 430
Vegetation
programs to control interfering, 288
protecting, 56–57
and regeneration surveys, 325–327
and tree community development, 318
Vegetative reproduction, 63
regeneration by, 64–65
Veneer logs, 429, 429t
Vertical structural diversity, 223–224, *225*, 248
Volume control, 183–184
Volume-volume control, 184

Walnut, eastern black, 117
Walnut seed
cached or buried by animals, 67
dispersal distances, 270t
Water
buffer strips to protect water resources, 53
common best management practices for minimizing soil erosion and protecting water quality during timber harvesting, 51t
enhancing quality and yield, 23, 38
Federal Clean Water Act, 51
Federal Water Pollution Control Act, 51
natural supplies of, 26–27
preventing unnatural changes in, 51
protecting purity of, 50
resources protection and clearcutting, 289
soil conditions and, 121
Water-absorbing packing material, 145
Watersheds, 4, 11
strip and patch cuts to trap snow, 284
tree cutting and, 46, *47*
Weed(s), 140
species in seedbeds, 142
Weeding, 330–332, 347
aerial application of herbicide, 331
broadcast methods, 330
during seedling stage, 329
reducing competition and interference by herbaceous plants, 330, *331*
spot weeding, 330
total vegetative control, 331
vegetation management in western North America, 332
Weed seeds, killing with high temperatures, 140
Western larch, seed dispersal distances, 270t
Western white pine, seed dispersal distances, 270t
West Indies mahogany, 117
White fir, 117
seed dispersal distances, 270t
White spruce, 66
seed dispersal distances, 270t
Willow(s), 64, 94
coppice production, 520, *520*, 521
dissemination and seed structure, 66
Windbreaks in tree-poor regions, 118–119, *119*, 120
and high latitudes, 118
purpose for, 118
Wind-disseminated seed, 67
Windrows, 85
Winds, dry, damage or kill seedings, 142
Wood products
forest products and grading trees and logs, 428–429, 429t
quality requirements for different, 428
use-classification groups, 428
Wood products firm
timber harvesting and, 43–44
requirements of, 43
Working circle, 199

Yarding, 46, 47, 82
definition of, 45
preventing environmental pollution, 49
Yellow birch, seed dispersal distances, 270t
Yellow-poplar, dissemination and seed structure, 66
See also Tulip-poplar
Yew, seeds ingested as food, 67
Yield tables, 399

Zone of rational action, 361